Molecular Biology

A SERIES OF BOOKS IN BIOLOGY

CONSULTING EDITOR: Cedric I. Davern

Molecular Biology

A Comprehensive Introduction to Prokaryotes and Eukaryotes

DAVID FREIFELDER

University of California, San Diego
University of Alabama
Formerly of Brandeis University

Jones and Bartlett Publishers, Inc.
BOSTON   PORTOLA VALLEY

Editorial offices: Jones and Bartlett Publishers, Inc., 30 Granada Court, Portola Valley, CA 94025.

Sales and customer service offices: Jones and Bartlett Publishers, Inc., 20 Park Plaza, Boston, MA 02116.

Library of Congress Cataloging in Publication Data

Freifelder, David Michael, 1935–
   Molecular biology.

   Includes bibliographies and index.
   1. Molecular biology   I. Title.
QH506.F73 1983   574.8′8   82–17006
ISBN 0–86720–012–X

ISBN 0-86720-012-X

Publisher   Arthur C. Bartlett

Book and Cover Design   Hal Lockwood

Illustrator   Donna Salmon, Assisted by Cyndie Clark-Huegel, Evanell Towne, John and Judy Waller, Dorothy Beebe, Kelly Solis-Navarro, and Brenda Booth

Manuscript Editor   Kirk Sargent

Production   Bookman Productions

Composition   Typothetae

Printer and Binder   Halliday Litho

Printed in the United States of America
Printing Number (last digit)   10   9   8   7   6   5   4

To my parents,
MORRIS *and* FLORENCE,

and to my children,
RACHEL *and* JOSHUA

Preface

In 1975 I was introduced to the pleasures and agonies of writing a textbook. When I finished that book, I vowed never to undertake another one; but somehow five other books were conceived and then brought to fruition in the next five years. In 1980 I acceded to the insistence of my undergraduate students, who for several years had been asking me why I did not write an undergraduate molecular biology text, and decided to do just that.

Molecular Biology, A Comprehensive Introduction to Prokaryotes and Eukaryotes is the result of this most recent effort. A multipurpose college textbook in molecular biology, it is based, in part, on a course that I taught for many years at Brandeis University. Its primary use will be in undergraduate courses emphasizing basic molecular processes (such as the synthesis of DNA, RNA, and protein) and genetic phenomena in both prokaryotic and eukaryotic cells. Students who wish to make full use of the text should have knowledge of basic biology and general chemistry and have completed or be enrolled in a course in organic chemistry.

Because of the variety of subjects treated and the detailed presentation in some chapters, the book can also be used in other courses. For example, fairly complete coverage of the structure of macromolecules enables the book to serve as a text for a short course in structural biology. (An advanced course in this subject would require an additional book on physical biochemistry.) The material contained in the rather lengthy chapters on the chemical and physical properties of DNA and on nucleic acid synthesis has been used in a one-semester course in nucleic acids that I taught several years ago. Classical genetics, microbial genetics, and general features of genetic recombination are likewise treated in detail; my experience has been that when supplemented with sets of problems these topics can serve as a core for a molecular genetics course. (A separate problems book is available to make this application possible.) I expect that the book will be useful in advanced genetics courses, also. Finally, reviewers of the text who teach in medical schools have pointed out to me that the text can be used as a component of medical biochemistry and cell biology courses, in keeping with the current trend of emphasizing molecular processes in such courses. Before writing my text, I made a survey to determine the college year in which molecular biology and molecular genetics are usually taught. This was found to range from the sophomore year to the first year of graduate school. Thus, in an effort to accommodate various curricula, I have included in each chapter of my book basic information (sometimes in the form of a review or an overview), advanced topics, and discussion of many specific systems.

Molecular Biology differs in organization from other molecular biology texts: it follows my philosophy that molecular biology must emphasize both molecules and biology, and that to be molecular, it must also be chemical and physical—quite a tall order for a single book. With this goal in mind, I chose to present a biological introduction to the subject first (Chapter 1) before getting into the complexity of molecular detail, so that the student will know the phenomena that molecular biologists would at present like to explain and will be familiar with the biological systems used in research. A second introductory discussion, Chapter 2, reviews those aspects of organic chemistry needed by the reader and provides some basic biochemical concepts. The first major discussion following the introductory material is concerned with physical biochemistry; hence, there is extensive early coverage of the properties of macromolecules; this is a notable feature of the organization of this book. Early presentation of this material is based on my teaching experience, in which I have concluded that students have an easier time understanding the great variety of phenomena in which macromolecular interactions play a role if they have first acquired a thorough understanding of macromolecular structure. For example, the student should have a clear understanding of the significance of molecular shape and the interactions that give rise to the shape of a particular molecule; how, otherwise, can the concepts of templates, active sites, specificity of binding, and mutation be understood?

Following the treatment of macromolecules, the text turns to what is traditionally thought of as molecular biology—namely, DNA, RNA, and protein synthesis. In each case, both prokaryotic and eukaryotic systems are described in the same chapter; the prokaryotic systems are given in greater detail, as more is understood about these systems, and the similarities and differences between prokaryotes and eukaryotes are pointed out.

The next major unit of the book is concerned with how the basic synthetic processes are integrated to enable a prokaryotic cell or particle to function efficiently and to respond to a changing environment; here, the text examines in some detail metabolic regulation in bacteria and phages, stressing the notion of efficiency.

Features of genetic exchange—natural systems such as plasmids, transposons, and homologous recombination, and the artificial process of genetic engineering—form the next unit; parts of each chapter are presented as a survey, for a beginning course, but discussions of particular models are provided for use in courses of greater depth.

The final unit of the book is concerned exclusively with eukaryotic systems. It is divided into two parts, the first on animal and plant viruses—their basic biology and how they are regulated—the second on regulation in numerous eukaryotic systems, in both unicells and animal tissue.

A problems book (*Problems for Molecular Biology*, Jones and Bartlett Publishers, Inc., 1983), containing more than 800 problems with complete and detailed solutions, is also available as a supplement to this text and is matched to the text chapter by chapter. Drill questions to test a student's mastery of facts are given for each major topic and many problems of a more substantive and sometimes difficult nature are also included. These problems have all been used in prior courses and have been extensively checked for ambiguities and correctness of solutions by undergraduate students at Brandeis University and graduate students and faculty at the University of California, Berkeley.

Inasmuch as *Molecular Biology* can be used in its entirety in either a condensed one-semester course, such as I have taught, or a more relaxed full-year course, instructors will have to be selective of the material for one-semester courses. To aid in this selection, summaries of the chapters—and especially of the points of view taken in the chapters—are presented in the following paragraphs.

Since the *modus operandi* of molecular biologists requires knowledge of phage biology, bacteriology, cell biology, and classical and microbial genetics, these topics are presented in Chapter 1 in a phenomenological way. Commonly used genetic tools such as recombinational mapping and complementation are reviewed. A brief discussion is also given of the logical approach taken in molecular biology that distinguishes it from the biology of the earlier part of the century. The basic life cycles of particular biological systems used by molecular biologists—namely, *E. coli*, phage, viruses, yeasts, and animal cells—

and the means for handling these systems—liquid culture, growth on agar, colony and plaque counting, and eukaryotic cell culture—are described in some detail. Chapter 2 provides a parallel review of elementary organic chemistry (e.g., functional groups) and introduces the molecules commonly found in biological systems. Since in many schools the study of molecular biology precedes that of biochemistry and cell biology, such biochemical topics as carbohydrate metabolism, the role of ATP in biochemical reactions, and the concept of the metabolic pathway are also included. A discussion of growth media for various organisms is also provided. To some students, little of the introductory material will be new, but my experience as a teacher has demonstrated to me that most college students have forgotten a great deal of what they learned in previous courses and profit from such a review. Instructors for advanced courses may choose to omit Chapters 1 and 2, but I advise having the student read these chapters quickly and reflect on possibly forgotten topics.

The text then turns to a study of macromolecules. Chapter 3 presents the general properties of macromolecules and some of the more common techniques used in their study. Chapters 4 and 5 treat nucleic acids and proteins, respectively, in considerable detail, with respect to both their chemistry and physical properties; structure is related to function whenever possible. A goal in the treatment of nucleic acids is to provide the student with the information needed to understand the mechanisms of replication and transcription, which are examined in later chapters. Furthermore, the reader will examine in some detail the properties of nucleic acids that researchers utilize when designing experiments investigating fundamental biochemical mechanisms. Chapter 5 has three goals—to give the student an appreciation of the variety of possible protein structures, to relate protein structure to biological function, and to indicate how a single amino acid change in a mutant can have a profound effect on protein function. In recent years knowledge of macromolecular structure has reached the point where examination of complex systems of macromolecules—for example, antibodies, chromatin, membranes, nuclear bodies, cell walls, and so forth—is profitable; these and other topics are included in Chapter 6. Some of these topics may be omitted from a short course, but the sections on immunoglobulins and chromatin should be included, for knowledge of these structures will be required in later chapters of the book.

To keep their science courses contemporary, most instructors cannot afford to spend much time on the story of how molecular biology evolved as a science; indeed, I believe many students find the topic less interesting nowadays than instructors do. (An instructor's interest is often derived from acquaintance with the personages involved, but increasingly the material must be considered history.) Accordingly, very little history is given in this book; however, I have included a brief presentation of the discovery of the genetic nature of DNA, for it still seems worth while to me to give beginning students some idea of the foundation of the field of study on which they are embarking. This historical interlude is presented in Chapter 7 but is given as part of a more general discussion of those properties that enable nucleic acids to serve as the repository of genetic information.

The treatment of what is traditionally thought of as molecular biology begins in Chapter 8 with DNA replication—an uncommon starting point. I have chosen this organization for five reasons: (1) the essential concepts of a template and of chemical polarity are easily introduced with this approach; (2) DNA replication examplifies, in a simple way, the kinds of geometric problems faced by many biochemical systems; (3) this topic shows how complicated a "simple" process can be; (4) the concepts of both variability of genetic material (for example, mutational changes) and protection of existing information (for example, repair processes) are more obvious here than in other topics; (5) the notion of regulation (that is, when DNA synthesis begins in a cell) can be introduced. The chapter on DNA replication is very lengthy because I have chosen to present examples of almost every known mode of replication in both prokaryotes and eukaryotes. The instructor will probably have to be selective and may choose to delay certain topics—for example, phage DNA replication—to later chapters.

Chapter 9, the subject of which is DNA repair, and Chapter 10, which is concerned with mutations, follow logically from a treatment of DNA replication, for in these chapters I have discussed the imperfections of biological systems and how cells deal with the problem of damage to their DNA by many environmental agents and reagents. Most of the material in these chapters consists of well-understood subjects; the main exception is SOS repair, a subject still actively being investigated.

By the time the reader has completed Chapter 10, he or she will have become familiar with the structure of the important intracellular molecules and will have examined systems for synthesis of one type of macromolecule. In so doing, the foundation will have been laid for understanding how a cell manages to do what it has to do at the appropriate time and how it responds to environmental factors. Thus, the reader is ready to be presented with the mechanisms of RNA synthesis and protein synthesis in prokaryotes and eukaryotes, which is done in Chapters 11–13. Chapter 11 deals with transcription; as in Chapter 8, the basic phenomenon is described first; then this is elaborated on in considerable detail. Special attention has been paid to essential features (such as recognition sites) of the macromolecules required for transcription, and the value of understanding the complete nucleotide sequence is made evident. Chapters 12 and 13 are concerned with protein synthesis. This immense topic has been divided into two parts—namely, the information problem, which is presented in Chapter 12, and the chemistry, which is given in Chapter 13. The structures of the relevant molecules (tRNA and the aminoacyl synthetases) and particles (ribosomes) are described twice—first in a simple way and then in considerable detail. The chemistry of protein synthesis is presented fairly completely and, in some cases, as somewhat advanced material, but chapter sections have been arranged in such a way that more detailed sections may be omitted, if desired. For example, a student can learn about protein synthesis by reading an overview consisting of a few pages, but if detail is wanted, a sequence of sections consisting of nearly twenty pages can be used.

Having completed all units on macromolecular synthesis, the reader is then ready to study how such synthesis is regulated. Chapter 14 is concerned with regulation in prokaryotes and presents the operon concept. (Regulation in eukaryotic cells is delayed to Chapter 22 because of its complexity, but it may be covered after Chapter 14.) The *lac* operon serves as the basis for the chapter because most of the features and concepts of regulation were developed by investigating this system. Discussion of the *lac* operon is rather lengthy, though, and some instructors may wish to omit certain subsections. In the early days of molecular biological research, when the *lac* operon was first being studied, it appeared that a general mechanism of regulation—namely, via operons—had been uncovered and that the *lac* system was the prototype. However, work from numerous laboratories has shown that though operons are indeed a general phenomenon in prokaryotes, a great many different regulatory programs exist for different metabolic systems; furthermore, the *lac* system has been found to be considerably more complex than was earlier thought. In Chapter 14 the reader is exposed to a wide variety of systems—the negatively regulated, the positively regulated, those with one promoter or with more than one promoter, degradative systems and biosynthetic systems, and so forth. The chapter is rather long and I assume that instructors will pick and choose among the systems. (Although a large number of operons are presented, selectivity has nonetheless been necessary and no doubt some instructors will find that a favorite system has been omitted.)

Chapters 15 and 16 treat the bacteriophages, elegant systems that exemplify the principles of regulation and of macromolecular synthesis. The study of phage biology is not simple because, though the life cycles of most phages follow the same basic pattern, details of the modes of nucleic acid replication and regulation often differ from one phage species to the next. Also, the reader will be introduced in this chapter to the first exception to the rule that DNA is the genetic material—namely, the RNA-containing phages. Many phage species are examined in these chapters—each one selected to demonstrate some feature of phage biology that can be easily observed with that phage. For example, T4 illustrates the mechanisms of taking over a host cell and of cleaving DNA prior to packaging, λ is more useful for examining the regulation of transcription and for studying a system with alternate life styles, ϕX174 illustrates the problems faced by a single-stranded DNA molecule, and the RNA-containing phages show that a genetic system can function without utilizing DNA, though special problems are encountered. Chapter 15 deals exclusively with the lytic life cycle; in Chapter 16, lysogeny, primarily of phage λ, and the related phenomenon, specialized transduction, are examined.

At this point in the book, the student will have seen how prokaryotic cells grow and maintain themselves in a changing and often hostile environment, and how continuous existence from one generation to the next is achieved. However, one of the essential features of life is that living organisms change in time. Organisms generally evolve slowly by accumulating mutations, but discontinuities in genetic composition can also occur by exchange of genetic material. Such exchange, called genetic recombination, occurs in a variety of ways, which are treated in some detail in Chapters 17–19. These three chapters are concerned respectively with plasmids and their transmissibility, the mechanism of homologous recombination, and transposition. Again, basic phenomena are described simply early in each chapter and details are given in later sections. Inclusion of all of these chapters may be impossible in a short course; however, if genetic engineering is to be covered later in the course, Chapter 17, on plasmids, should be included. The relatively new subject of transposition—treated in Chapter 19—is also important if regulation in eukaryotic systems is to be part of the course; this chapter can, however, be covered briefly by omitting the models for transposition.

Genetic engineering, the basis of the new biotechnology, is presented in Chapter 20. I have chosen to place this important topic late in the book because it is difficult to appreciate the applications and problems of this technique before learning about macromolecular synthesis and regulation. However, many instructors will prefer to discuss this topic early in their courses, so the use of probes can be included in the study of transcription and regulation. If this is to be the case, I would recommend that this chapter be preceded by Chapter 17 (plasmids) and parts of Chapter 11 (transcription).

In all earlier chapters except for Chapter 14 (prokaryotic regulation), both prokaryotic and eukaryotic systems are described and compared. The remainder of the book deals exclusively with eukaryotes. The animal viruses, which are given a lengthy treatment in Chapter 21, exhibit great variety in structure and in life cycles. Here the reader can see that just as the study of phage has given important insight into macromolecular processes and regulation in bacteria, study of viruses gives information about regulation in eukaryotes. Of course, animal viruses possess a great deal of intrinsic interest also, both as highly regulated systems and as agents of disease. A great many viruses are examined in Chapter 21; each one exemplifies a particular mode of transcription, replication, or production of viral proteins, the mode varying with the particular genetic material possessed by that virus. Two points emphasized in this chapter are that different virus types use rather different strategies for generating mRNA and that viruses use a variety of mechanisms to deal with the problem of making a sufficient number of protein types from a rather limited amount of nucleic acid. In a full-year course students will profit by studying this chapter completely; in a short course there surely will not be enough time, but I urge the instructor to find a place at least for poliovirus or vesicular stomatitis virus in order to demonstrate the role of polyproteins in eukaryotic systems.

The final chapter in this book—Chapter 22—is concerned with regulation in eukaryotes. Both unicellular organisms, such as yeast, and differentiated animal cells are included. The unusual organization of eukaryotic DNA, and gene families, as well as various modes of DNA deletion and amplification and recombination are all important topics. The role of hormones as regulators is examined in some detail. Several topics—such as synthesis of antibodies and of vitellogenin—are given rather lengthy treatments. This chapter will necessarily be frustrating to some readers because more questions are raised than answered. Because the field of eukaryotic regulation is advancing so rapidly, the instructor will find it necessary to supplement the text with the latest information, for no textbook can be up to date in this rapidly progressing field. I have chosen to present systems that examplify general principles and are reasonably well understood. This chapter should certainly be included in any course in molecular biology except one devoted exclusively to prokaryotes, because this subject is the molecular biology of the future.

An introductory textbook must necessarily contain a large number of terms not previously encountered by the reader. To aid in recognizing these terms and in finding the definitions at a later time, new terms are printed in **boldface** type where they first appear in the text.

Molecular Biology is the result of the effort of a large number of people. Many scientists served as consultants, reviewers, and contributors of illustrations. My consultants examined a proposed seven-page table of contents, answered a questionnaire concerned with the content of various courses and the background of students taking these courses, and let me know what they felt should be in a molecular biology book. My reviewers read preliminary and revised drafts of every chapter, corrected my mistakes, and gave me the latest unpublished information from their and other laboratories. These reviewers are listed following this preface. Many people contributed photographs and electron micrographs. I owe special thanks to Robley Williams (Professor of Molecular

Biology at the University of California, Berkeley) for allowing me to look through his immense collection of electron micrographs and use whatever I needed. The contributions of three people were indispensable in completing the manuscript. Mary Jane Voll (Professor of Microbiology at the University of Maryland) and Peter Gray (Professor of Biochemistry at the University of Oklahoma) read the entire second draft, then still in need of a great deal of revision, and gave me valuable scientific and pedagogical advice; Cedric Davern (Vice-President of the University of Utah) read the entire final draft before manuscript editing was begun. I was exceedingly fortunate to have the help of Peter Geiduschek (Professor of Biology at the University of California, San Diego), who read the page proofs of the entire book and examined all of the completed figures; he found errors I had missed and provided last-minute updating.

Arthur Bartlett, my friend and publisher of this and my earlier books, supported me more than amply in all phases of the work. Hal Lockwood (of Bookman Productions in San Rafael, California) designed and was responsible for the production of the book. I was most fortunate to have a great scientific illustrator, Donna Salmon, already known for her magnificent drawings in Lubert Stryer's *Biochemistry,* consent to illustrate the book. Not only is Donna a skilled illustrator, designer, and technician, but, through her unusual understanding of the scientific message of each figure, frequently she was able to add something for which each student and reader must be grateful. I was also fortunate to have the advice and editorial criticism of Kirk Sargent, the manuscript editor of this and two of my earlier books, who established a standard of clarity of writing that I could not have achieved myself and who made many valuable suggestions about increasing the information content of the illustrations. My long-time secretary and friend, Mildred Kravitz, typed most of the rough draft, and Elizabeth Lindheim, Judith Day, and Thomas Armstrong entered the final copy into my word processor. My daughter Rachel helped me in proofreading; it was my computer-expert son Joshua who helped me to master the capabilities of the word processor. Indexing was accomplished by "Indexor," a new computer program (Compress, Inc., Wentworth, New Hampshire) written by Jeremy Sagan and provided to me before it was commercially available. To all of these people I owe my thanks.

Now, after two and a half years of the most difficult work my students have ever suggested that I undertake, this book is ready for them and for other readers. I hope that the effort invested in it will have, as its result, that instructors and students of biology everywhere will find the book a new and valuable tool. If so, my great love of teaching will be satisfied and I will be content.

February 1983   David Freifelder
San Diego, California

Prepublication Reviewers of *Molecular Biology*

Contents

Chapter 2   The Basics of Biochemistry   53

Chapter 3   Macromolecules   76

Chapter 9   DNA Repair   316

Chapter 10   Mutagenesis, Mutations, and Mutants   339

Chapter 13   Translation: II. The Machinery and Chemical Nature of Protein Synthesis   491

Chapter 14   Regulation of the Activity of Genes and of Gene Products in Prokaryotes   540

Chapter 15   Bacteriophages: I. Lytic Phages   596

Chapter 16   Bacteriophages: II. The Lysogenic Life Cycle   665

Chapter 17   Mechanisms for Rearrangement and Exchange of Genetic Material: I. Plasmids   701

Chapter 21   Eukaryotic Viruses   839

Chapter 22   Regulation in Eukaryotes   922

Index   971

1 Systems and Methods of Molecular Biology

For a century or so many biologists believed that living cells possessed some vital force that would bc lost once the cell was broken. Thus, the vitalists thought that little, if anything, could be learned about life by studying anything other than the intact cell. They were wrong, as events were to prove. When it was first decided to break open a living cell and study its inner workings, molecular biology was born.

A BRIEF OVERVIEW OF MOLECULAR BIOLOGY

The term molecular biology was first used in 1945 by William Astbury, who was referring to the study of the chemical and physical structure of biological macromolecules. By that time, biochemists had discovered many fundamental intracellular chemical reactions, and the importance of specific reactions and of protein structure in defining the numerous properties of cells was appreciated. However, the development of molecular biology had to await the realization that the most productive advances would be made by studying "simple" systems such as bacteria and bacteriophages (bacterial viruses), which yield information about basic biological processes more readily than animal cells. Although bacteria and bacteriophages are themselves quite complicated biologically, they enabled scientists to identify DNA as the molecule that contains most, if not all, of the genetic information of a cell. Following this discovery, the new field of molecular genetics moved ahead rapidly in the late 1950's and early 1960's and provided new concepts at a rate that can be matched only by the development of quantum mechanics in the 1920's. The initial success and the accumulation of an enormous body of information enabled researchers to apply the techniques and powerful logical methods of molecular genetics to the subjects of muscle and nerve function, membrane structure, the mode of action of antibiotics, cellular differentiation and development, immunology, and so forth. Faith in the basic uniformity of life processes was an important factor in this rapid growth. That is, it was believed that the fundamental biological principles that govern the activity of simple organisms, such as bacteria and viruses, must apply to more complex cells; only the details should vary. This faith has been amply justified by experimental results.

HOW MOLECULAR BIOLOGY WILL BE PRESENTED IN THIS BOOK

Molecular biology is an inherently logical discipline, once the ground rules are understood. However, it is enormously complex and crosses traditional boundaries between genetics, biochemistry, cell biology, physics, organic chemistry, and biophysical chemistry, so that molecular biology can no longer be thought of as an indivisible subject. Unfortunately, molecular biology also leads the student through a circular process. That is, to understand how a cell replicates, one needs to know about DNA and RNA, yet a clear understanding of DNA and RNA functions requires that one refer to cells. This naturally poses the problem of where to begin.

The approach taken in this book is to explain the systems that must continually be referred to in order to understand basic phenomena. In our initial development of molecular biology we shall look at bacteria and phages; later we shall turn to more complicated systems such as animal viruses and animal cells. These cells and how they are handled will be described in this chapter. Genetic techniques provide one of the most powerful experimental tools in molecular biology, especially in the study of microorganisms, and for this reason these techniques will be reviewed extensively; special attention will be given to the use of mutants to analyze a system, as this is a common approach.

Biochemical techniques and knowledge of the chemical conversions that occur in living cells have played an enormous role in all biological research. Thus, we devote Chapter 2 to this topic.

Biological systems contain many types of gigantic molecules (**macromolecules**)—for example, proteins, nucleic acids, and polysaccharides. Discussion of detailed biochemical mechanisms is made easier if the structure of biological macromolecules is understood at the outset. Hence this subject is developed in Chapters 3 through 6.

Armed with the tools of molecular biological research and knowledge of both the structure of macromolecules and the chemical processes occurring in cells, we will be able to proceed to a detailed study of the fundamental activities that determine the biological properties of each cell—namely, DNA replication, RNA synthesis, and protein synthesis—the topics of Chapters 7 through 13. Then we can ask the profound question—since all possible chemical reactions that can occur in a cell do not occur at the same time, what determines the particular intracellular events that transpire at a particular time? Such intracellular regulation is examined in Chapters 14 through 16, in which we shall look at bacteria and bacterial viruses, and in Chapters 21 and 22, which are concerned with animal viruses and animal cells.

In Chapters 15 and 16 we will study bacteriophages, for their intrinsic interest and as a means of seeing how the synthesis of DNA, RNA, and protein are regulated and integrated to yield a complex functioning biological system.

As we approach the end of the book, we will examine the means of creating organisms having new properties. We begin in Chapters 17 through 19 with natural processes for scrambling DNA, such as genetic recombination, and then in Chapter 20 we turn to the new biotechnology—the technology of recombinant DNA or genetic engineering—in which knowledge of fundamental biochemical processes is used to design new organisms having features that are valuable for research, medicine, consumers, and industry.

Recombinant DNA technology also provides new techniques for laboratory science and with these methods the study of animal viruses can be approached (Chapter 21). Perhaps the greatest achievement of recombinant DNA technology in research to date is that it enables one to approach the problem of regulation in eukaryotes. This fascinating subject is reserved for the final chapter in the book, Chapter 22—a chapter that will almost surely require updating even by the time this book appears in print, inasmuch as an often repeated statement, "this is not yet understood," will often no longer apply.

The vocabulary of molecular biology is very large because it is drawn from so many disciplines. This often makes it difficult for students to familiarize themselves with the subject. Thus, as a study aid, throughout this book, essential terms will be printed in boldface type where each is first explained.

THE PHYSICAL APPROACH TO PROBLEMS IN MOLECULAR BIOLOGY

Throughout this book, physical and physicochemical techniques will be applied in order to understand the properties of molecules. There are two reasons to approach molecular biology in this way. One is the dictum that structure and function are intimately related; the other is that many biological questions can be answered by the data obtained from measurements of physical and physicochemical properties.

The structure-function relation appears repeatedly. For example, collagen, the protein of which tendon is made, is triple-stranded for strength, aggregates side-to-side for additional strength, and has special features that enable long, tough fibers to form. DNA, the single most important molecule in every cell, is double-stranded so that two copies of its precious genetic information will exist. Cell membranes are rich with nonpolar fatty acids, so that polar molecules cannot freely pass in and out of the cell; however, other molecules and special proteins provide passageways through the membranes for transit of specific substances. Hundreds of such examples are known and most of the information about them has come from precise physical measurements that enabled molecular structures to be elucidated. Several of these examples will be described in Chapters 3 through 6.

Physical measurements also provide other types of information. For example, by ultracentrifugation one sees that incompletely synthesized protein molecules always sediment together with intracellular particles called ribosomes. This observation was one of many that showed that the ribosome is the structure on which proteins are synthesized. By electron microscopy, DNA molecules are seen to be linear when isolated from a phage but circular when isolated from a phage-infected bacterium. Further studies have shown the reason for this: in an early stage in the phage life cycle, phage DNA becomes circularized. Electrophoretic studies indicate that many presumably pure enzymes have two components, each able to move in an electric field at a distinct rate; these enzymes have multiple forms that are important in metabolic regulation. The electrophoretic behavior of the hemoglobin isolated from a person with sickle-cell anemia differs from that of a normal person; in the sickle-cell hemoglobin, a particular amino acid has been replaced by another amino acid. Infrared absorption spectroscopy of thymine, one of the components of DNA, indicates that both a carbonyl and a hydroxyl group are present; this shows that both keto and enol forms are in equilibrium, which is important in understanding the mechanism of mutation. Hundreds of examples of this type can be taken from the history of numerous subfields of molecular biology.

In Chapter 3 several important physical techniques will be described in detail.

THE *IN VITRO* APPROACH TO UNDERSTANDING BIOLOGICAL REACTIONS

At some point in the study of living cells it is almost always necessary to examine individual cellular reactions separately. This approach usually has the advantage of some simplification of a system but ignores the fact that these reactions often interact with other cellular systems. Nonetheless, a great deal of information can be obtained in this way. Thus, one does not usually study directly a phenomenon as complex as overall cell growth but looks separately at the synthesis of DNA, proteins, and other components.

In many cases, it is not possible to study certain component reactions in a living cell. For example, the polymerization of phosphate-containing DNA monomers to form DNA molecules cannot be studied by adding these substances to a growth medium, because organic phosphorus-containing compounds are unable to pass through the cell membranes to enter the cell. Likewise, one cannot determine how the activity of an enzyme complex that freely associates and dissociates is regulated when there are 3000 other proteins present. Thus, it has become repeatedly necessary to break open cells and to purify the components of the reactions. Any investigation carried out with broken cells or pure components is called an *in vitro* study; this contrasts with an *in vivo* study, which is done with intact cells.

There are two different types of *in vitro* studies—those that use a **crude extract** and those that use a purified or a reconstituted system. A crude extract is obtained when cells suspended in a buffer are broken up and nothing is removed other than unbroken cells and, possibly, fragments of the cell wall and cell membrane. A crude extract is often used to assay for the presence of particular enzymes. For instance, the enzyme β-galactosidase, which breaks down the sugar lactose to glucose and galactose, is easily detected in a crude extract and can be measured quantitatively. Crude extracts are often used in early stages of analysis to determine whether a particular enzymatic activity is present. However, their use usually requires great cleverness on the part of the experimenter because so many substances are present in the extract. Conclusions obtained with crude extracts must always be accepted with caution, for often several reactions will occur simultaneously that compete with the reaction one thinks is being studied or that even antagonize the reaction. An example of the latter is that, without special care, the synthesis of nucleic acids is unobservable because of the ubiquity of enzymes that depolymerize nucleic acids (**nucleases**).

A more useful and less problematical approach to understanding an intracellular process is to purify the components of the subsystem, determine the chemical and physical properties of each component, and then reconstruct the system. Experimental conditions are then sought in which a process simulates as closely as possible that which occurs, or is believed to occur, within the cell. This is the ultimate *in vitro* experimentation and is often the only way to understand precisely how a system works. The *in vitro* approach requires extraordinary skill both in experimentation and interpretation, as will be seen throughout this book, because so many misleading possibilities can arise. For instance, one might be misled by choosing conditions that force an enzyme to catalyze a reaction that never occurs within a cell; examples are known of polymerases that have been used to drive depolymerization reactions and of depolymerases used to carry out synthetic reactions. Also, essential factors can be lost during purification. For example, in early studies of the synthesis of RNA, excessive purification of the polymerization system caused a protein to be lost that is needed for recognition of specific start sites on a DNA molecule and this loss prevented essential features of the reaction from being observed. Changes in the physical properties of the molecules are also common. For example, for 96 years DNA molecules were always fragmented during isolation until 1959, when unbroken molecules were first isolated, and it was not until then that it was realized that DNA molecules are huge and that a chromosome contains a single DNA molecule. Also, it is not at all uncommon for protein molecules to be isolated that are either damaged by protein-digesting enzymes or by the physicochemical procedures used in isolation and thus are devoid of biological activity.

The *in vitro* approach is nonetheless a powerful technique for understanding biological mechanisms. In many cases there is no other way to understand a system fully. We shall see numerous examples of this approach throughout the book.

THE LOGIC OF MOLECULAR BIOLOGY

Three kinds of logical reasoning appear repeatedly in molecular biology—arguments based on efficiency, examination of models, and strong inference. These are described in the following sections.

The Efficiency Argument

Living cells have had hundreds of millions of years to evolve. During this time the rigors of competition and survival selected for efficiency (though this may have been compromised through accommodation of potential efficiency to other demands on the system). Little energy and material are wasted, inasmuch as wastefulness is disadvantageous in the long run. That is, if a less wasteful mutant arises among a population of more wasteful cells (i.e., is more efficient), this mutant should grow slightly faster than the rest of the population. Ultimately, since population numbers are virtually unlimited, the population will consist entirely of this mutant. Thus, when proposing a mechanism, one usually tries to think of ways in which a system avoids both waste and errors. Experience has shown that this is often a productive approach and it has been an important factor in the early development of new notions of molecular genetics. For instance, the molecular geneticist is continually attuned to the presence of start signals, stop signals, and regulatory sites in DNA and RNA molecules, and similar ways to avoid waste. One should not take the efficiency notion unquestionably though, because it fails, at the present at least, to explain some very complex aspects of the properties of eukaryotic cells. This may, however, be a reflection of our lack of understanding.

Examination of Models

A model is a tentative explanation of the way a system works. It proposes the components, interactions, and sequences of events. An important function of a model is to suggest additional experiments. Models enable experimenters to make predictions and, to investigate a model, the predictions must be tested experimentally. If predictions do not agree with experimental results, the model must be considered incorrect *as it stands*. Such a contradiction calls for a change in the model. It is important to realize that a model cannot be proved to be correct merely by showing that it makes a correct prediction. However, if it makes *many* correct predictions, it is probably nearly, if not completely, correct. Whereas a rigorous proof of the reality of a model is logically impossible, some models are very persuasive in terms of their elegance and power.

Strong Inference

A great deal of biological reasoning is inductive. That is, something may be observed so many times that one infers that it is a real phenomenon, or a theory may explain so many things that one believes that the theory is correct. However, strong inference is a logical process based on exclusion of alternative possibilities. It is one of the most powerful logical approaches in molecular biology. In the method of strong inference one first states all possible explanations for a particular phenomenon and then experimentally eliminates the possibilities one by one. When only one possibility remains, that one is inferred to be correct. This is true as long as the possibilities are exclusive and all alternatives have been specified. Actually, it is not common in molecular biology (or probably in any branch of science) to consider all possibilities. Usually, the molecular biologist uses intuition and generalizations from other biological systems and considers only those alternatives that seem reasonable. For this reason, readers of papers in scientific journals will often encounter the phrase "it is likely that" when experimental data are being interpreted.

A FEW PROBLEMS STUDIED IN MOLECULAR BIOLOGY

It would be foolhardy to try to list all biological phenomena for which a molecular explanation is desired. It is useful, however, to list some of the important problems that have been and are being attacked by the methods of molecular biology and to describe the problems of the immediate future.

In the early days of what might be called modern molecular biology, the principal effort went toward understanding genetic phenomena. An account of this effort forms a major part of this book. The two earliest major problems were the identification of the genetic material and the mechanism of protein synthesis. Once DNA was shown to be the genetic substance, two questions immediately arose: how does DNA replicate (the *detailed* answer for which still remains to be discovered), and how is information obtained from a DNA molecule? The latter question became tied to the problem of protein synthesis as soon as it was learned that the amino acid sequence of every protein molecule is determined by the base sequence of its DNA. Basic questions about mechanisms have been answered more or less completely, though there are still many details to be learned.

It also was apparent that the protein-synthesizing mechanisms in bacteria are not identical to those used in animal cells, though they are related. Now, after 25 years of an intensive investigation of phage and bacterial systems, the technology of molecular biology has reached a stage at which more complex systems, such as the animal cell, can be profitably studied. Study of animal cells has been done primarily with cells grown in culture. Progress has been slow for many reasons, not the least of which is the 24-hour doubling time of most cells in culture compared to 25 minutes for many bacteria. In the mid-1970's it was realized that just as the most rapid progress in understanding basic phenomena had been made through study of simpler systems such as phages and bacteria, knowledge of the molecular biology of animal cells might come faster through concentration of study on a simple, rapidly growing cell. The baker's yeast, *Saccharomyces cerevisiae,* a unicellular organism, whose doubling time is 70 minutes, has turned out to be a valuable experimental object because its basic mechanisms of macromolecular synthesis and its regulatory mechanisms are nearly the same as those of the more complex cells.

Once basic synthetic mechanisms had become better understood, the question of metabolic regulation arose. Cells rarely make substances they do not need, yet when a particular substance is required, its synthesis usually starts up very soon after the "needed" signal has been received. The nature of the signal and, in general, the identity of all start and stop signals (which are present in many systems) have been and remain important questions. Some complex systems, such as the life cycle of phages, are regulated in an extraordinary fashion; in these systems, the timing, sequence, and amount of synthesis of all components are predetermined and everything is carried out with great efficiency. A few phage systems (particularly *E. coli* phages λ and T7) are very well characterized, and probably all of the start and stop signals and regulatory elements are known.

The internal concentrations of many cellular substances are remarkably constant. Furthermore, cells are exceedingly selective with respect to the substances allowed to enter the cell. Selectivity is determined by the chemical and physical properties of membranes and by various carrier molecules and pumping systems. The identity of these transport systems, how they work, and the structures of different membranes are exciting questions whose answers are being actively sought.

In the specialized animal cells, such as muscle and nerve cells, there are many mysteries. How does contractility occur? What causes the transmission of an electrical impulse along nerve cells and what maintains the electrical charge across the membranes of resting nerve cells? Why does one type of cell make hemoglobin and another make insulin? Some of these questions are answered in Chapter 22.

One of the most precise and profitable branches of molecular biology is structural biology, actually the birthplace of modern molecular biology. The driving force for the great activity in this field is the doctrine that no structure exists without a function, and its corollary, that functions are carried out by neatly designed, efficient systems whose activities are usually determined by specific macromolecular structures. For example, the analysis of the structure of DNA led to an understanding of mutation and suggested how replication and genetic recombination might occur. A look at the various forms of RNA opened up the field of protein synthesis, a problem that is still not completely solved. However, the machinery of protein synthesis—that is, the ribosome—has been well characterized, and the sites on the ribosome at which various stages of protein synthesis occur are known. Also, elucidation of the three-dimensional structure of the transfer RNA molecule has explained how the information in DNA is translated into the amino acid sequence of proteins.

At present, structural biology has two subfields—the detailed arrangement of atoms in individual macromolecules and the organization of macromolecules in large aggregates. An example of the former is the structure of DNA, which is already known. Current research focuses on the structure of proteins—especially enzymes. The analysis of enzyme structure shows where on the surface of the enzyme the chemical reaction catalyzed by the enzyme occurs. Biochemists interested in the mechanism of enzyme action anxiously await this information for each enzyme being studied. Unfortunately, definition of structural details at the atomic scale is obtained only by the tedious and difficult technique of x-ray diffraction analysis, and analyzing a structure usually takes a very long time. As for the second subfield—the structures of large aggregates—some structures, such as tendon, hair, and wool are well known. However, many questions remain about myosin (the principal muscle protein) and about actin and tubulin (two proteins that are responsible for the shapes of different cell types), and these proteins are being actively studied. Both technique and our understanding of structural rules have advanced to the point that one can attempt to solve the structure of aggregates containing molecules of different types. Thus, a great deal of research effort is directed at present toward the structure of chromosomes, which contain DNA and several proteins, and of membranes, which contain several types of proteins and lipoproteins. Of special interest with respect to membranes is how the structure enables a membrane to be selectively permeable—that is, permeable to some, but not all, molecules. Differences in the structure of the cell membranes of normal and cancer cells are especially tantalizing.

Chapter 6 is devoted specifically to structural biology.

WHAT ISN'T MOLECULAR BIOLOGY?

By now the reader may have the idea that molecular biology includes all biology. This is true in the sense that the ultimate goal of molecular biology is to understand the molecular basis of all biological phenomena. However, certain topics, which are not included in this book, are not usually thought of as molecular biology by those who call themselves molecular biologists. Yet some aspects of even these topics interest molecular biologists. For example, the chemical reactions of metabolism are considered to be biochemistry. If these reactions are regulated simply by the effects of reactant and product concentrations, as in a standard chemical reaction, the study of this regulation remains biochemistry; however, if an enzyme-catalyzed reaction is regulated by alteration of the structure of the enzyme, a molecular biologist would probably lay claim to the topic (though a biochemist might not agree). Similarly, study of the arrangement and structure of the inner components of a cell is usually considered to be cell biology. However, as time passes, this topic is becoming amenable to molecular analysis also. Behavioral biology seems clearly at first not to be molecular biology; however, some people have isolated behavioral mutants of insects and certain protozoa and these too are being subjected to molecular biological analysis. Therefore, the border between molecular biology and other branches of biology is thin and hazy and is apt to be adjusted as molecular biology is extended to fresh territory. Perhaps someday, subdivisions of biological research will no longer be important and the unifying term biology will again be meaningful.

PROKARYOTES AND EUKARYOTES

Living organisms can be classed in two categories—the prokaryotes and the eukaryotes. A **prokaryote** is a cell whose genetic material is present throughout the cell. In a **eukaryote** the genetic material is organized in a well-defined compartment, the nucleus, and the DNA is organized transiently into compact structures called chromosomes, prior to and in preparation for cell division. Another important distinction between prokaryotes and eukaryotes is in the way that the DNA molecule itself is organized. For instance, the DNA of eukaryotes is tightly associated with specific proteins to form nucleoproteins (which are reorganized to form chromosomes) whereas in the prokaryotes the DNA is free of such "structural" proteins. The distinction between prokaryotes and eukaryotes is exceedingly important because, for unknown reasons, there are profound biochemical differences between prokaryotes and eukaryotes. For example, the assembly of amino acids into proteins occurs by the same mechanism in human cells and in unicellular eukaryotes, such as yeast and algae; in contrast, protein synthesis in bacteria, which are prokaryotes, differs in many ways from that in eukaryotes, yet the mechanism is basically the same in all bacteria.

Some organisms, such as phages and viruses, which are subcellular in organization but dependent on cells as an environment for their reproduction, are neither prokaryotes nor eukaryotes. The bacteriophages grow only in bacteria and thus adopt a prokaryotic strategy for reproduction while the animal and plant viruses are obligated to manufacture their component macromolecules by using eukaryotic rules.

Much more is known about prokaryotes than eukaryotes because most of the prokaryotic systems are simpler.

In this book prokaryotes and eukaryotes will be discussed separately and compared and contrasted. Usually prokaryotes will be discussed first because they are simpler. In keeping with this, we begin by describing the properties of bacteria that are important in molecular biological research.

BACTERIA

Bacteria are free-living unicellular organisms. They have a single chromosome, which is not enclosed in a nucleus (they are prokaryotes), and compared to eukaryotes they are simple in their physical organization. For all practical purposes, a bacterium can be thought of as a solution of several thousand chemicals and a few organized particles enclosed in a rigid cell wall.

Bacteria have many features that make them suitable objects for the study of fundamental biological processes. For example, they are grown easily and rapidly and, compared to cells in multicellular organisms, they are relatively simple in their needs. The bacterium that has served the field of molecular biology best is *Escherichia coli* (usually referred to as *E. coli*), which divides every 20 minutes at 37°C under optimal conditions. Thus, a single bacterium becomes 10^9 bacteria in about 20 hours. How one grows bacteria and counts their number is described in the following section. These procedures also apply to the yeasts, which are discussed later in this chapter.

The terms used in this section are also used in discussing the growth of all microorganisms and frequently of animal cells.

Growth of Bacteria

Bacteria can be grown in a **liquid growth medium** or on a solid surface. A population growing in a liquid medium is called a bacterial **culture.** If the liquid is a complex extract of biological material, it is called a **broth.** Some examples are tryptone broth, which is the milk protein casein hydrolyzed by the digestive enzyme trypsin to yield a mixture of amino acids and small peptides, and nutrient broth, which is an extract of beef. If the growth medium is a simple mixture containing no organic compounds other than a carbon source such as a sugar, it is called a **minimal medium.** A typical minimal medium contains each of the ions Na^+, K^+, Mg^{2+}, Ca^{2+}, NH_4^+, Cl^-, HPO_4^{2-}, and SO_4^{2-}, and a source of carbon such as glucose, glycerol, or lactate. If a bacterium can grow in a minimal medium—that is, if it can synthesize *all* necessary organic

Table 1-1
Standard substances used in microbial genetics, which are encountered in this book, and their abbreviations

Substance	Genotype symbol[1]	Substance	Genotype symbol
Amino acids		*DNA and RNA bases*[2]	
Alanine	*ala*	Purine	*pur*
Arginine	*arg*	Pyrimidine	*pyr*
Asparagine	*asn*	Adenine	*ade*
Aspartic acid	*asp*	Cytosine	*cyt*
Cysteine	*cys*	Guanine	*gua*
Glutamic acid	*glu*	Thymine	*thy*
Glutamine	*gln*	Uracil	*ura*
Glycine	*gly*	*Vitamin*	
Histidine	*his*	Biotin	*bio*
Isoleucine	*ile*	*Sugars*	
Leucine	*leu*	Arabinose	*ara*
Lysine	*lys*	Galactose	*gal*
Methionine	*met*	Glucose	*glu*
Phenylalanine	*phe*	Lactose	*lac*
Proline	*pro*	*Antibiotics*	
Serine	*ser*	Ampicillin	*amp*
Threonine	*thr*	Chloramphenicol	*cam*
Tryptophan	*trp*	Kanamycin	*kan*
Tyrosine	*tyr*	Penicillin	*pon*
Valine	*val*	Streptomycin	*str*
		Tetracycline	*tet*

1. When stating a substance or a phenotype rather than a genotype, the same three-letter symbol is used, but it is capitalized and not italicized—for example, His for histidine.

2. When stating base sequences, a purine or a pyrimidine is usually abbreviated Pu or Py respectively, if the identity of the base is unimportant. Particular bases will be denoted by A (adenine), G (guanine), T (thymine), U (uracil), and C (cytosine).

substances such as amino acids, vitamins, and lipids—the bacterium is said to be a **prototroph.** If any organic substances other than a carbon source must be added for growth to occur, the bacterium is termed an **auxotroph.** For example, if the amino acid leucine is required in the growth medium, the bacterium is a leucine auxotroph; the genetic symbol for such a bacterium is Leu$^-$. A prototroph would be indicated Leu$^+$. Table 1-1 lists standard abbreviations for various nutrients and genetic characteristics encountered in microbiological studies.

Bacteria are frequently grown on solid surfaces. The earliest surface used for growing bacteria was a slice of raw potato. This was later replaced by media solidified by gelatin. Because many bacteria excrete enzymes that digest gelatin, an inert gelling agent was sought. **Agar,** which is a gelling agent obtained from a variety of seaweed and used extensively as a thickening agent in Japanese cuisine, is resistant to bacterial enzymes and has been universally used. A solid growth

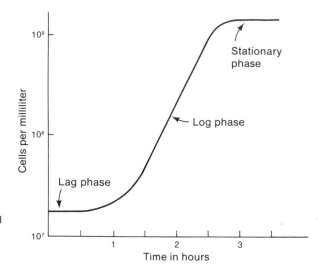

Figure 1-1
A typical growth curve for a bacterial culture. The *y* axis is logarithmic, so that the curve is a straight line in log phase.

medium is called a nutrient agar, if the liquid medium is a broth, or a minimal agar, if a minimal medium is gelled. Solid media are typically placed in a **petri dish.** In lab jargon a petri dish containing a solid medium is called a **plate** and the act of depositing bacteria on the agar surface is called **plating.**

When bacteria are placed in a liquid medium, they slowly start to grow and divide. After an initial period of slow growth called the **lag phase,** they begin a period of rapid growth in which they divide at a fixed time interval called the **doubling time.** The number of cells per milliliter, the **cell density,** doubles repeatedly, giving rise to a logarithmic increase in cell number; this stage of growth of the bacterial culture is called the **log phase.** This stage of growth continues until a cell density (for *E. coli*) of about 10^9 cells/ml is reached, at which point nutrients and O_2 become limiting and the growth rate decreases. (This cell density is often five- to tenfold lower for many other bacterial species.) Ultimately, at a cell density of 2 to 3×10^9 cells/ml, no further growth is possible and the cell number becomes constant. This stage is the **stationary phase.** A typical growth curve for a bacterial culture is shown in Figure 1-1.

Counting Bacteria

A bacterium growing on an agar surface also divides. Since most bacteria are not very motile on a solid surface, the progeny bacteria remain very near the location of the original bacterium. The number of progeny increases so much that a visible cluster of bacteria appears.

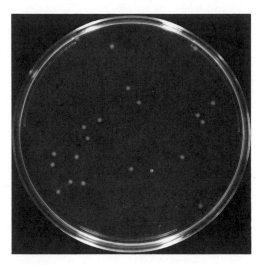

Figure 1-2
A petri dish containing colonies of *E. coli*
grown on agar. (Courtesy of Robert Haynes.)

This cluster is called a **bacterial colony** (Figure 1-2). Colony formation allows one to determine the number of bacteria in a culture. For instance, if 100 cells are plated, 100 colonies will appear the next day. If 0.1 ml of a 10^6-fold dilution of a bacterial culture is plated and 200 colonies appear, the cell density in the original culture is $(200/0.1)(10^6)$ = 2×10^9 cells/ml.

Identification of Nutritional Requirements of Bacteria

Plating is a convenient way to determine if a bacterium is an auxotroph. This is done in the following way. Minimal agar and nutrient agar plates are prepared. Several hundred bacteria are plated on each plate and the plates are incubated overnight in an oven. Several hundred colonies are subsequently found on the nutrient agar because it contains so many substances that it can satisfy the requirements of nearly any bacterium. If colonies are also found on the minimal agar, the bacterium is a prototroph; if no colonies are found, it is an auxotroph and some required substance is not present in the minimal agar. Minimal plates are then prepared with various supplements. If the bacterium is a leucine auxotroph, the addition of leucine alone will enable a colony to form. If both leucine and histidine must be added, the bacterium is auxotrophic for both of these substances.

Table 1-2 shows the result of a plating experiment in which the nutritional requirements of a bacterium are deduced. The fact that there is no growth with supplement 1 shows that something other than histidine, leucine, and thymine is needed. Growth in the absence of

Table 1-2
Plating data enabling the determination of the nutritional requirements of a bacterium

Medium supplement	Growth	Conclusion
1. His, Leu, Thy	−	Needs some nutrient
2. Leu, Ala, His	+	Thy not needed
3. Ala, Thy, His	+	Leu not needed
		Ala needed (cf. plate 1)
4. Leu, Ala, Thy	−	His needed (cf. plates 1 and 3)

Note: Abbreviations of names of amino acids are given in Table 1-1, note 1.

thymine (supplement 2) and of leucine (supplement 3) indicates that neither thymine nor leucine is required. Note, however, that the presence of alanine in supplement 3 overcomes the deficiency of supplement 1, so alanine must be required. Finally, the lack of growth with supplement 4, in which alanine is present, shows that there is still another requirement. Since leucine is not needed, the only significant difference between supplements 3 and 4 is the presence of histidine in supplement 3. Therefore, histidine is also required and the bacterium is auxotrophic for both alanine and histidine.

These techniques apply to all microorganisms.

The Physical Organization of a Bacterium

The intracellular organization of bacteria is poorly understood, but some points are clear (Figure 1-3). The bacterial cell does not have a defined nucleus and hence is a prokaryote. The intracellular material is enclosed in a rigid multilayered cell wall that gives it a defined shape—spherical, rodlike, crescent-shaped, and so forth (Figure 1-4). If bacteria are treated with the enzyme **lysozyme,** which is found in egg white and in tears, some of the cell wall components are removed and the rigidity of the wall is lost. All bacteria treated in this way become spherical and ultimately burst. The spherical form, which can be stabilized in suspending media having a high osmotic strength (e.g., 20 percent sucrose), is called a **spheroplast** if some cell wall remains and a **protoplast** if the cell wall is completely removed. Beneath the cell wall is a very thin layer called the **cell membrane** (see Chapter 6). This layer provides a permeability barrier that determines which substances can pass in and out of the cell. Some substances are transported by proteins called **transport proteins** or **permeases**; these will be discussed in Chapter 14.

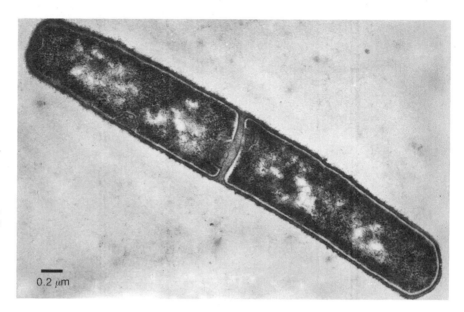

Figure 1-3
A cross-section of a dividing bacterium. The layers of the cell wall can be seen. The light areas inside the cell are the DNA; note that the DNA is distributed throughout the cell. The fine dark particles are ribosomes, the units on which proteins are synthesized. (Courtesy of J. C. Benichou and A. Ryter.)

0.2 μm

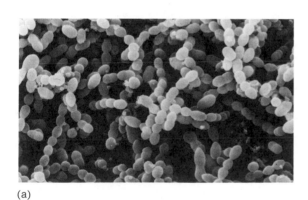

(a)

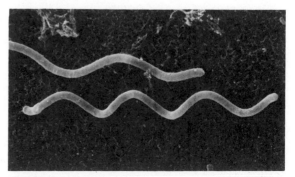

(b)

Figure 1-4
Three forms of bacteria. (a) Cocci (spherical chains). (b) Bacilli (rods). (c) Spirillae (helices). (From G. Shih and R. Kessel. 1982. *Living Images: Biological Microstructures Revealed by Scanning Electron Microscopy.* Science Books International.)

(c)

Metabolic Regulation in Bacteria

Bacteria are well-regulated and highly efficient organisms. For example, they rarely synthesize substances that are not needed. Thus the enzymatic system for synthesizing the amino acid tryptophan is not formed if tryptophan is present in the growth medium, but when the tryptophan in the medium is used up, the enzymatic system will be rapidly formed. The systems responsible for utilization of various energy sources are also efficiently regulated. A well-studied example is the metabolism of the sugar lactose as an alternate carbon source to glucose.

Glucose is metabolized by all living cells by a series of chemical conversions in which the molecule is progressively broken down. Glucose is a fundamental carbon source in the sense that other sugars must either be converted to glucose or to a glucose-degradation product in order to be metabolized. Thus bacteria use glucose more efficiently than lactose. The first step in the metabolism of lactose is its cleavage to form two sugars, glucose and galactose. However, the enzyme needed to catalyze this cleavage is not present in cells in significant quantities unless lactose is in the growth medium. If lactose is provided to a cell, through a complex multistep process the enzyme can then be formed, and glucose produced from lactose is broken down and used as a source of energy. The galactose that is also produced is converted to a component that enters the glucose utilization pathway and is thereby also metabolized. That the system is regulated can be seen by the fact that if a cell is supplied with both glucose and lactose in the growth medium, there is no reason for the cell to synthesize the lactose-cleaving enzyme and, in fact, a simple system prevents the cell from wasting its energy synthesizing this enzyme until all the glucose is utilized. This system is discussed in detail in Chapter 14.

The control of both tryptophan synthesis and lactose degradation are two examples of **metabolic regulation.** This very general phenomenon will be explored extensively throughout the book. Both simple and complex regulatory systems will be seen, all of which act to determine how much of a particular compound is utilized and how much of each intracellular compound is synthesized at different times and in different circumstances. This will demonstrate the length to which the so-called simple cells have gone to utilize limited resources efficiently and to optimize their metabolic pathways for efficient growth.

Metabolic regulation occurs in all living organisms.

BACTERIOPHAGE

Bacteria are subject to attack by smaller organisms called **bacterio-phage** or simply **phage.** These are small particles, part of the general

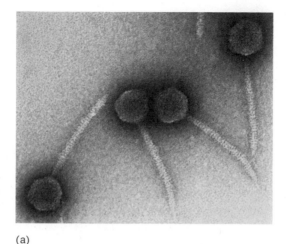
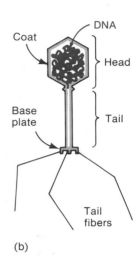

Figure 1-5
(a) *E. coli* phage λ. (Courtesy of M. Wurtz.)
(b) Schematic drawing of a tailed phage showing the major components.

(a)

(b)

class of particles called viruses, and they are capable of growing only inside bacteria. Phage* have been the object of choice for a great many types of experiments because they are much simpler than bacteria in their structures (usually having between two and ten components) as well as their life cycles and yet possess the most essential, if minimal, attributes of life. A typical life cycle is outlined briefly in the following sections.

Phage Structures

A typical phage contains only a few different types of molecules—usually several hundred protein molecules of one to ten types (depending on the complexity of the phage) and one nucleic acid molecule. The protein molecules are organized in one of three ways. In the most common mode the protein molecules form a protein shell called the **coat** or **phage head,** to which a **tail** is generally attached; the nucleic acid molecule is contained in the head (Figure 1-5). Another form of a phage is a tailless head. The least common form is a filament in which the protein molecules form a tubular structure in which the nucleic acid is embedded. Phages are known that contain double-stranded DNA (the most common variety), single-stranded DNA, single-stranded RNA, and double-stranded RNA (least common). A few phage species contain more than one nucleic acid molecule, but this is quite rare. Double-stranded DNA-containing phages typically contain 50 percent of their weight as DNA, so they are a useful source of DNA for physical studies.

*The plural word phages refers to different species; the word phage is both singular and plural and in the plural sense refers to particles of the same type. Thus, T4 and T7 are both phages, but a test tube might contain either 1 T7 phage or 100 T7 phage.

An Outline of the Life Cycle of a Phage

Phage are parasites and cannot multiply except within a host bacterium. Thus, a phage must be able to enter a bacterium, multiply, and then escape. There are many ways by which this can be accomplished. However, a basic life cycle is outlined below and depicted in Figure 1-6.

The life cycle of a phage begins when a phage particle adsorbs to the surface of a susceptible bacterium. The phage nucleic acid then leaves the phage particle through the phage tail (if the phage has a tail) and enters the bacterium through the bacterial cell wall. In a complicated but understandable way the phage converts the bacterium to a phage-synthesizing factory. Within about an hour, the time varying with the phage species, the infected bacterium bursts or **lyses** and several hundred progeny phage are released. The suspension of newly synthesized phage is called a **phage lysate.**

Phage Multiply Very Rapidly

Phage multiply much faster than bacteria. A typical bacterium doubles in about half an hour, while a single phage particle gives rise to more than 100 progeny in the same time period. Each of these phage can then infect more bacteria, and those released in this second cycle of infection can infect even more. Thus, in two hours there are four cycles of infection for both a bacterium and a phage yet a single bacterium has become $2^4 = 16$ bacteria and a single phage becomes $100^4 = 10^8$ phage particles.

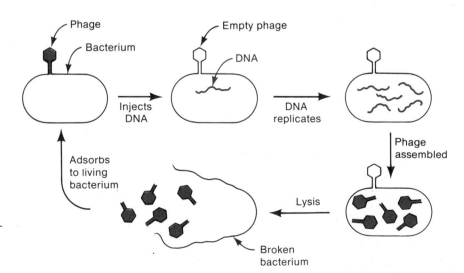

Figure 1-6
A schematic diagram of the life cycle of a typical bacteriophage. The number of phage released usually ranges from 20 to 500.

Counting Phage

Phage are easily counted by a technique known as the **plaque assay**; this technique is performed in the following way.

In a previous section it was explained that 100 bacteria on agar yield 100 colonies. Likewise, if 10^8 bacteria are plated on agar, the 10^8 colonies that result can be made to appear as a confluent, turbid layer of bacteria called a **lawn.** During plating, to achieve maximal uniformity of the turbidity of the lawn that will form, the bacteria are mixed into a small volume of warm liquid agar which is poured onto the surface of the solid medium. The liquid, known as **top agar** or **soft agar,** rapidly hardens, providing a very smooth surface, and the bacteria grow in this thin agar layer with much uniformity (Figure 1-7). If a phage is present in the hardened top agar, it can adsorb to a bacterium in the agar; shortly afterward, the infected bacterium lyses and releases about 100 phage, each of which will adsorb to nearby bacteria; these bacteria in turn will release a burst of phage which then can infect other bacteria in the vicinity. These multiple cycles of infection continue and after several hours, the phage will have destroyed all of the bacteria at a single localized area in the agar, giving rise to a clear, transparent hole in the turbid, confluent layer. This hole is called a **plaque.** Since one phage forms one plaque, the individual phage particles put on the plate can be counted.

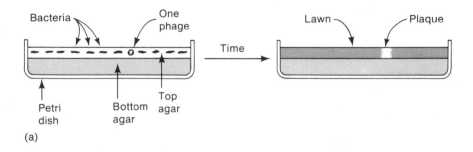

Figure 1-7
(a) Schematic drawing of plaque formation. Bacteria grow and form a translucent lawn. There are no bacteria in the vicinity of the plaque, which remains transparent. (b) Plaques of *E. coli* phage T4. Two types of plaques are present. The smaller plaques are made by wild-type phage; the larger plaques are those of an *rII* mutant. Note the halo around the larger plaques—it is a result of a large amount of lysozyme diffusing outward and lysing uninfected cells. (Courtesy of A. H. Doermann.)

Regulation of the Phage Life Cycle

The life cycle of a phage is highly regulated, though in a slightly different way from the metabolic regulation of a bacterium. Phages are totally dependent on the metabolism of their host bacteria, so that usually the regulatory systems of the hosts control the basic metabolic processes such as energy generation and synthesis of the precursors of DNA, RNA, and proteins. The job of an infecting phage is to reproduce itself by synthesizing its own nucleic acid and structural proteins, and finally to cause the bacterial cell wall to break so that progeny phage can escape. This requires a strict temporal regulation. For instance, if the system that causes lysis were to act immediately after infection, the infected bacterium would be destroyed before any new phage had been produced; clearly, the lytic system must act last. A nonproductive infection would also occur were the phage coat to be synthesized so early as to envelop the incoming nucleic acid molecule before phage reproduction could occur. Such failures are avoided by the existence of a phage-specified, finely tuned, temporal regulatory system. The basic program is the same for all phage species, though the particular regulatory mechanisms differ from one species to the next. This temporal regulation will be examined in detail in Chapter 15; in that chapter we shall see how several different phage species manage to determine how much of each phage protein is made at different times after infection.

Infection of Bacteria with Purified Phage DNA

Infection of a bacterium can be initiated in another way that has been exceedingly useful. Bacteria can be treated in several ways that break down the permeability barriers of the cell membrane. One of these methods is treatment with lysozyme to form a spheroplast, as explained earlier in the chapter. Another useful procedure is to incubate the bacteria in iced 0.1 M $CaCl_2$. Cells treated in either of these ways gain the ability to take in DNA present in a solution. Thus, following such a treatment, a bacterium can be infected with phage DNA that has been isolated from a suspension of phage. An infection with free DNA is called **transfection.** It is an important procedure in molecular biology, having two main uses. First, it allows the experimenter to alter the DNA molecule either chemically or physically and then study the effect of the alteration by infecting a cell with the altered DNA. Second, it enables one to identify as phage DNA a DNA sample isolated from an infected cell before it lyses. For example, there are many experiments in which the DNA from a phage-infected cell is isolated, separated into different fractions, and transfected to other bacteria; the fraction containing the phage DNA can be identified because it will cause phage production in the transfected bacteria.

YEASTS

Yeasts are unicellular organisms that have been used for millennia for producing wine and beer (Figure 1-8 (a,b,c)). A great deal of early biochemical research was carried out with yeasts rather than bacteria,

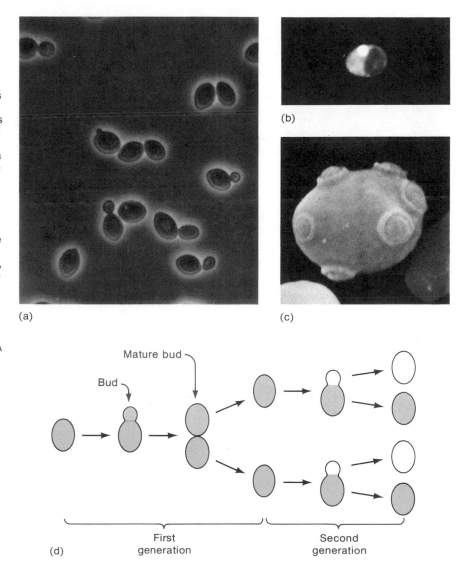

Figure 1-8
(a) A phase-contrast light micrograph of several cells of the yeast *Saccharomyces cerevisiae*. Many cells are budding. (Courtesy of Breck Byers.) (b) A fluorescence micrograph of a single stationary cell. The fluorescent dye, acridine orange, which binds to DNA (giving green fluorescence) and to RNA (giving orange fluorescence), has been added to the culture medium and the cells are viewed by ultraviolet light, which excites the fluorescence. The bright spot is the nucleus, which has a green fluorescence. The dark region is the vacuole, a liquid-filled sac that is free of nucleic acids. (c) A scanning electron micrograph of an old cell that has budded many times and has numerous bud scars. (From G. Shih and R. Kessel. 1982. *Living Images: Biological Microstructures Revealed by Scanning Electron Microscopy.* Science Books International.) (d) Drawing of budding, showing the mother-daughter relation. The first generation daughter, which is also the second generation mother, is shown in red.

work stimulated mainly by interest in understanding and improving fermentation. Researchers in beer-producing countries, such as Denmark, have also studied yeast genetics intensively in an effort to produce organisms that can make a better beer.

Yeast cells are propagated in the laboratory and counted in much the same way as bacteria. They grow in liquid suspensions in either chemically defined media or in complex broths. They also grow on a solid surface to form colonies. The multiplication mechanism of all but the fission yeasts differs from the simple splitting of a mature bacterium in that yeast cells do not divide but multiply by budding (Figure 1-8(d)). In contrast with bacteria which divide to yield twin progeny cells, with yeast there is a clear mother-daughter relation in the sense that the mother retains a scar on the cell wall at the site of budding.

Whereas the molecular biology of prokaryotes has undergone enormous advances in the past twenty-five years, at the present time much of the research activity is directed toward studying eukaryotes, which have a more complex organization. There are thus many advantages to beginning the study of eukaryotes with a unicellular organism that can be handled as easily as bacteria. For this reason, the yeast *Saccharomyces cerevisiae,* which is a eukaryote, has become an important experimental organism in recent years. An initial step in its study is to see if mechanisms worked out for *E. coli* and which apply to many other bacteria as well are also valid for a simple eukaryote. Furthermore, one can use the yeast as a model to study some of the important features of higher cells.

The properties of yeasts that are especially useful (and which are quite different from the properties of bacteria) are the following:

1. Yeasts are eukaryotes. That is, genetic material is located in a well-defined nucleus enclosed in a nuclear membrane (Figure 1-8(a,b,c)). Thus, one can use yeast to examine mechanisms for transfer of material across the nuclear membrane; this transfer is apparently very important in eukaryotes, and very little information has been gained from the study of animal cells.

2. *Saccharomyces* has seventeen chromosomes in its haploid set, i.e., a set containing one copy of each chromosome. (Bacteria have one chromosome.) Thus, one can determine if special regulatory mechanisms are required in organisms possessing several chromosomes.

3. Yeasts have both haploid and diploid (two copies of each chromosome) phases. There are two haploid mating types, a and α, which can mate to form a stable $a\alpha$ diploid. Under certain conditions a diploid cell can undergo meiosis to form four haploid cells—called **spores**—encased in a sac called an **ascus.** Two of the spores are type a and two are type α. (These mating

types will be discussed again in Chapter 22.) The existence of the mating types means that genetic experiments can be done that frequently complement physicochemical studies. Furthermore, this mating system affords an opportunity to study both the mechanisms of meiosis and the chromosomal interactions responsible for genetic recombination.

Other unicellular eukaryotes, namely the alga *Chlamydomonas* and the protozoan *Tetrahymena,* also possess many of the attributes of yeast and are being used more frequently in eukaryotic molecular biology.

ANIMAL CELLS

As more information about fundamental mechanisms in bacteria has been gained, increasing efforts have been made to study animal cells. Research with animal cells is exciting because we are beginning to understand such complex processes as hormonal regulation and the development of an egg into an adult organism, and to gain some insight into the differences between normal cells and cancer cells. However, research with animal cells proceeds much more slowly than work with bacteria. There are two reasons for this. First, animal cells divide every 24 to 48 hours, whereas many bacteria divide every 25 to 50 minutes, so experiments with animal cells often take much longer than experiments with bacteria. Second, bacteria growing in culture are not significantly different from bacteria in nature but experiments with animal cells require that the cells be removed from the animal and often separated from one another. Cells treated in this way have lost the normal route for receiving nutrients and are definitely in an unnatural state. Many growth media that enable cells to grow in culture have been developed. They have been designed to keep the cells alive as long as possible but, as shown in the next section, they do not always maintain a normal state for the cell.

The Unnatural State of Animal Cells in Culture

In contrast with the behavior of microorganisms, the behavior of most animal cells in culture is far different from that in the animal. There are two main kinds of difference—in culture, animal cells grow as individuals and the cells grow continually.

In nature, most animal cells grow in organized tissue units—that is, they are in contact with one another. Except for special circumstances, in culture animal cells require a substitute surface on which to grow

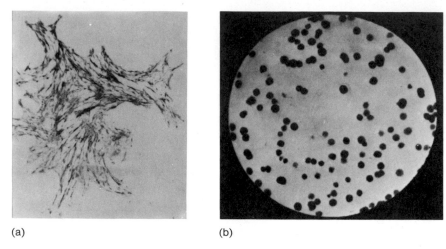

(a)   (b)

Figure 1-9
(a) A microcolony of Chinese hamster fibroblasts which have been growing for a few
days on a glass surface. The form of the cluster is determined by the spindle shape of
each cell. (b) A petri dish on which are numerous colonies of human HeLa cells. After
growth of the cells for about two weeks the plate was filled with a dye solution to stain
the individual clones to make them visible. Each clone consists of the progeny of a
single cell. (Courtesy of Theodore Puck.)

(Figure 1-9). The usual surface that is provided in the laboratory is the
bottom of a plastic petri dish that is filled with a liquid growth medium.
No agar is used—cells placed in the dish slowly adsorb to the plastic
surface. The cells grow and divide until they come in contact with one
another and form a confluent monolayer of cells on the bottom of the
dish. At this stage growth and cell division stop, as in organized tissue.
Continued growth and division are obtained only by removing the
confluent layer, separating the cells, and placing a small number of
dispersed cells in another plastic dish. This handling is certainly an
unnatural treatment and, in the course of many generations of growth,
results in the accumulation of abnormal (mutant) cells that can tolerate
these manipulations and the physiologically alien environment.

A related problem is that indefinite cell division of animal cells is
itself exceptional. In an animal, cells grow and divide only until
adulthood is reached, a state that is attained in no more than 50 or 60
generations in a human. In adulthood, cell division is rare, except in
specialized cells such as those in the skin, gut, mucous membranes,
and bone marrow, and those that divide during the healing of wounds.
In fact, cell division in animals can be considered quite rare except in a
developing embryo. Inasmuch as cells obtained from many tissues
other than an embryo have previously received signals that limit growth
or terminate growth, their propagation in a culture medium is a novel
cellular circumstance.

The result of forcing cells—which in their natural state grow only in contact with other cells and then only for a limited number of generations—to divide for an extended time in cell cultures, is that they become quite different from the cells originally taken from the animal tissue, and are often tumorlike. The number of chromosomes in progeny cells is almost always increased beyond the normal diploid number and is usually not even the same in all cells of one culture. The longer these cells are grown, the more heterogeneous a culture becomes.

Primary Cell Cultures and Established Cell Lines

Two kinds of animal cell cultures are in common use—the primary culture and the established cell line. The distinction between these two types can be understood by examining how a cell culture is established and the characteristics of the culture in further growth. A culture is established by dispersing fresh tissue, done by treating the tissue with a **proteolytic** (protein-digesting) enzyme and placing the individual cells (and sometimes some of the undispersed tissue) in a plastic petri dish containing growth medium. Cells then adsorb to the plastic surface, begin to grow and divide after a few days, continue growth for one week to three months (the time depends on the cell type), and then gradually begin to disintegrate and die. During the final few weeks the culture consists mostly of a sick and dying population. A culture showing these characteristics is called a **primary culture** because each cell presumably has the same characteristics of the cells in the original tissue and is following the program of limited cell division set up when the tissue formed.

If a primary culture that has lost its ability to reproduce is maintained for many months by removing some cells and transferring them to a fresh dish each time confluent growth is achieved, frequently cells arise that in an unknown way have gained the ability to grow and divide indefinitely (like a bacterium). A population derived from such a cell is called an **established cell line.** Studying primary cultures has a particular advantage even though most of the cells eventually die— namely, that the cells obtained from a particular tissue retain, for at least a few generations, the properties of "normal" cells. There is, of course, a disadvantage too—namely, that the primary population is short-lived, so that in any experiment that takes more than a few weeks, one does not know whether the results obtained apply to living or dying cells. An established cell line has the advantage of continued growth; however, the cells also are quite variable and often have the capacity to form tumors. Nevertheless, a great many experiments are done with established cell lines because of their immortality. Those cell lines that are used frequently in molecular biological experiments are HeLa (a human tumor cell), 3T3 (a mouse embryo cell), CHO (an apparently normal cell from Chinese hamster ovary), L (a mouse tumor), BHK (Syrian hamster kidney), Balb-c (a mouse tumor), and African green monkey cells.

Useful Primary Cell Cultures

Certain animal cell types are easy to maintain in culture for short periods of time. For instance, **reticulocytes,** which are hemoglobin-forming cells, are easily obtained from bone marrow as a nearly pure paste of cells, which are easily dispersed. These cells are programmed to make hemoglobin almost exclusively, and have been widely studied as a means of understanding protein synthesis in eukaryotes. Some cell types can also be easily studied while still in the parent tissue. For example, the **oviduct** of the chicken consists almost exclusively of cells whose function is to make egg white proteins. Synthesis of these proteins occurs in response to various sex hormones. Macerated tissue freshly obtained from a chicken oviduct and suspended in standard growth media is one of the most important systems for studying how cells respond to specific hormones.

Eggs and Oocytes

In one type of animal cell many of the problems just mentioned are not present—namely, in the developing eggs of amphibians and echino-derms, which normally develop in water. In particular, the eggs of the frog, toad, and sea urchin have been especially useful in studies of the molecular biology of differentiation. Eggs have the additional advantage that they are resting cells that can be switched on merely by adding sperm. The egg precursor (the **oocyte**) can also be obtained in large quantities from amphibian ovaries; the oocyte of the South African clawed toad, *Xenopus laevis,* is a favorite object for studying cell differentiation, and will be studied in Chapter 22.

Counting Animal Cells

In certain conditions animal cells form colonies on the surface of a plastic petri dish and this method is used occasionally to count the cells (Figure 1-9(b)). However, it is more common to remove the cells from the surface by treatment with trypsin and then either perform a count of a known microvolume with a microscope or use an electronic cell counter. The latter two procedures are always used when cells are grown in liquid suspension.

ANIMAL VIRUSES

Animal viruses, which superficially resemble bacteriophages, consist of a nucleic acid molecule encased in a protein shell; they adsorb to host cells, reproduce within the cells, and often are released in huge numbers by a variety of means. They are usually grown by infecting cultured cells, though occasionally a living animal is used. (Plant viruses are at this time propagated only in living plants.) Counting viruses is quite tedious and inaccurate, as will be seen in Chapter 21.

Viruses are very useful tools for studying basic processes in eukaryotes, such as DNA replication and RNA synthesis, for they give one the opportunity to examine eukaryotic systems acting in a way that is controlled by the experimenter—namely, the systems are turned on when the viruses are added. Viruses, whose external shells consist of only one or two types of protein molecules, have also been especially valuable for studying the structure and assembly of macromolecular aggregates. Viruses are, of course, interesting objects of study in their own right, too.

Most viruses are able to infect only a single cell type (as is the case in nature) and sometimes will grow only in a primary culture. Of special interest, though, are those viruses that reproduce in one cell type yet in other cell types induce a transformation of the host cell to a cancer cell. A well-studied example is the polyoma virus, which produces progeny viruses in embryonic mouse cells yet converts hamster cells to tumor cells. We will have more to say about animal viruses and the viral transformation process in Chapter 21.

GENERA, SPECIES, AND STRAINS

In the Linnean scheme of classification, genus and species names are assigned to all organisms. For example, the bacterium *Bacillus subtilis* is a member of the genus *Bacillus* and is of the species *subtilis*. Two organisms in the same genus are usually related evolutionarily and exhibit many similar characteristics. Thus, the bacterium *Bacillus brevis* has many charactertistics in common with *Bacillus subtilis*. The binary combination of a generic name and a specific epithet describes an organism but sometimes it is necessary to make further divisions. If differences are substantial the term **variety** is used but if they are small, the term **strain** is used. In molecular biology the term strain is encountered repeatedly. For example, thousands of times in various parts of the world E. coli has been isolated from nature—it grows in the intestines of large animals and has been sampled from individuals and from fecal matter in sewage. By the standard taxonomic criteria these isolates are classified as E. coli yet they may show subtle differences, such as differences in colony morphology, differences in susceptibility to various phages, different motility, and so forth. These differences define strains. Strains are designated by adding an unitalicized capital letter or number after the species name. That is, strains A, B, etc., of *E. coli* are written *E. coli* A, *E. coli* B, and so forth. Often variation is even found within a strain; a number is then added after the strain letter; for example, K1, K2, etc. The three strains of *E. coli* used most commonly in molecular biology are *E. coli* B (the host for the so-called T phages), *E. coli* C (the host for the phage ϕX174), and *E. coli* K12 (the most widely used strain).

Variants of animal and plant viruses distinguished by immunological criteria are extremely common. The viruses that cause influenza in humans are usually designated by stating a basic immunological type (A, B, C, . . .) and the geographical location where they were first isolated, as with the influenza virus known as type B: Hong Kong that was epidemic in the late 1960's.

Genetic Analysis as an Approach to Solving Biological Problems

Genetics is an abstract and logical system by which the genetic factors determining certain properties of living cells can be located with respect to one another and by which changes in these properties can be expressed in quantitative terms. Genetics alone cannot prove anything about molecular mechanisms because the conclusions it leads to are independent of the molecular basis of biological phenomena. However, genetic analyses have been the major source of intuition about molecular mechanisms and have forced scientists to consider phenomena that might otherwise be ignored. Genetics-based arguments can be so suggestive of a particular mechanism, and the suggested mechanism has so often turned out to be the correct one, that molecular biologists often view alternative and conflicting ideas as not even worthy of consideration. The following sections are an introduction to genetic analysis of organisms as well as to the genetic phenomena most commonly used in such analyses.

Genetic Notation, Conventions, and Terminology

In order to discuss genetics one must first distinguish the phenotype of a cell from its genotype and develop a consistent notation for describing the genetic properties of an organism. **Phenotype** refers to an observable property of an organism. Thus a cell that can synthesize the amino acid leucine is denoted Leu$^+$; if it cannot do so, it is denoted Leu$^-$. Note that the symbol has three letters, is capitalized, and is not italicized.

Genotype refers to the genetic composition of an organism. The Leu⁻ cell certainly must have a defective gene that keeps it from synthesizing leucine; this would be denoted *leu⁻* (or in some books, *leu,* without the minus sign). Note that the genotype is written in lowercase letters and in italics. Several genes may be required to synthesize leucine. These would usually be denoted *leuA, leuB,* . . . , with all letters italicized. A Leu⁺ (leucine-synthesizing) cell must have one functional copy of every requisite gene; thus, its genotype must be *leuA⁺ leuB⁺,* . . . , which would normally be summarized by writing *leu⁺* unless it is important for some reason to state the genotype of each gene. A Leu⁺ cell might also be diploid and have a defective gene (e.g., *leuA*) in one chromosomal set. The two haploid sets are separated in their notation by a diagonal line; the genotype would therefore be *leu⁺/leuA⁻*.

Some genes are responsible for resistance and sensitivity to certain extracellular products such as antibiotics. The genotypes of a penicillin-resistant cell and a penicillin-sensitive cell are written *pen-r* (or *pen⁺*) and *pen-s* (or *pen⁻*), respectively; the phenotypes are correspondingly Pen-r (Pen⁺) and Pen-s (Pen⁻). Many plasmids contain drug-resistance genes and these plasmids must be distinguished from another plasmid that lacks the gene—that is, from a plasmid that does not contain a mutant form of the gene nor have a copy of the gene at all. In this case, the genotype of a plasmid-containing cell would always be written *pen⁺* (because the gene is invariably *pen-r*), and if there is no plasmid, the notation *pen⁻* is used; however, the phenotypes can be expressed either as Pen-r versus Pen-s or Pen⁺ versus Pen⁻.

In summary, for bacteria and other eukaryotes, we use the following conventions:

1. Abbreviations of phenotypes contain three roman letters (the first capitalized) with a superscript + or − to denote presence or absence of the designated character and "s" and "r" for sensitivity and resistance, respectively.
2. The genotype designation is always lowercase and all components of the symbol are italicized.

This convention has unfortunately not been adopted for bacteriophages, for which one- and two-letter symbols are widely used.

We shall also have occasion to designate particular mutants of a gene. Mutants are usually numbered in the order in which they are isolated. Thus the leucine mutants numbered 58 and 79 would be written *leu58* and *leu79*. If one wanted to denote mutants of a particular gene, one would write *leuA58* and *leuB79,* if the mutations were in the *leuA* and *leuB* genes, respectively.

Finally, it is often necessary to denote the protein products of a particular gene. There is not a standard notation, but in this book the following conventions are used. An enzyme, integrase, for which the genetic symbol is *int,* will be written out as integrase, or it will be abbreviated and termed the *int* product or the Int protein, or it will simply be designated Int. All three designations are convenient and will be used in keeping with the three common modes seen in scientific literature.

Mutants

We have seen that a gene can be either functional or nonfunctional and that these states are denoted by a superscript + or − after the gene abbreviation. The functional form of a gene is sometimes called **wild type** because presumably this is the form found in nature. This term is ambiguous though, because often the − form is the one that is prevalent; for example, many bacterial species isolated from nature carry the genes for metabolism of lactose, yet in many strains the gene is defective—that is, *lac⁻*. In this book we shall minimize the use of the term. The precise genetic term **allele** is used to indicate that there are alternative forms of a gene; sometimes the + and − forms are called the + allele and the − allele, respectively. It is very common to call the defective − form of a gene a mutant form. Strictly speaking, a mutant is an organism whose genotype differs from that found in nature. However, it is more convenient (and definitely it is common jargon) to equate the terms mutant gene and defective gene, and we shall use the word mutant in that sense in this book.

Mutants can be classified in several ways. One classification is based on the conditions in which the mutant character is expressed. An **absolute defective mutant** displays the mutant phenotype under all conditions; that is, if a bacterium requires leucine for growth in all culture media and at all temperatures, it is an absolute defective. A **conditional mutant** does not always show the mutant phenotype; its behavior depends on physical conditions and sometimes on the presence of other mutations. An important example of a conditional mutant is a **temperature-sensitive (Ts) mutant,** which behaves normally below 34°C and as a mutant above 39°C; an intermediate state is usually observed between 34°C and 39°C. Note that the gene does not mutate above 39°C; rather, the product of the gene is inactive above 39°C. Temperature-sensitive mutants have been of great use in the laboratory because they enable one to turn off the activity of a gene product simply by raising the temperature. In many cases the temperature-sensitive defect is reversible, so that the activity of the gene product can be regained by lowering the temperature.

The most widely encountered conditional mutants are the **suppressor-sensitive mutants;** these exhibit the mutant phenotype in some strains but not in others. The difference is a consequence of the presence of particular gene products, called **suppressors.** These suppressor gene products either compensate for the defect in the mutant or, in a variety of ways, enable an altered gene to produce a functional gene product. In the jargon of molecular biology, one says that the phenotype of a suppressor-sensitive mutant "depends on the genetic background." A mutation is designated as a suppressor-sensitive one by adding the roman letters Am (refers to amber) or Oc (for ochre) in parentheses.* The symbols are roman and capitalized because they represent a phenotype; if the mutant has a number, this is also added. Thus, if mutation 35 in the *leu* gene is a suppressor-sensitive Am type, it would be written *leu35*(Am).

The process of formation of a mutant organism is called **mutagenesis.** In nature and in the laboratory, mutants sometimes arise spontaneously, without any help from the experiment. This is called **spontaneous mutagenesis.** Mutagenesis can also be induced by the addition of chemicals called mutagens or by exposure to radiation, both of which result in chemical alterations of the genetic material.

The Number of Mutations in a Mutant

Another classification of mutants is based on the number of changes that have occurred in the genetic material (that is, the number of DNA base pairs that have changed). If only one change has occurred, the mutant is called a **single** or **point mutant.** If two changes are present, it is a **double mutant.** Sometimes the mutation occurs by means of removal of all or part of a gene; in that case, the mutation is a **deletion.** Occasionally, other material replaces the deleted gene or gene segment. Such a mutation is called a **deletion-substitution.** These types of mutations are shown schematically in Figure 1-10.

Revertants and Reversion

A mutant organism sometimes regains its original character. This occurs by means of chemical changes in the mutant genetic material, which restore the genetic material to a functional state. The process of regaining the original phenotype is called **reversion,** and the organism that has reverted is called a **revertant.**

Reversion can be spontaneous or be induced by mutagens. The **reversion frequency**—that is, the fraction of cells in a population of mutants that after many generations of growth have regained the functional phenotype—is a useful criterion for identifying a point

*Amber and ochre are terms that first arose out of a private laboratory joke.

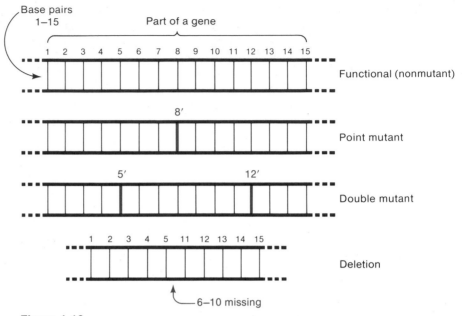

Figure 1-10
Schematic diagram of a normal DNA molecule and various kinds of mutants. In the point mutant, base pair 8′ replaces the normal base pair 8. The double mutant has two base pairs (5′ and 12′) not present in the nonmutant organism. In the deletion, five base pairs (6–10), present in the nonmutant, are absent.

mutant, because a small alteration in the nucleic acid is sufficient to revert a point mutant, and such a change can occur at some measurable, albeit low, frequency. With a deletion, however, the probability that the missing DNA will be replaced with material having an equivalent sequence, thereby restoring a functional gene, is virtually zero. Thus *reversions are not observed with deletions*.

A revertant bacterium might be detected in the following way. If 100 cells of a *leu⁻* bacterium are placed on agar lacking leucine, no colonies will form. If 10^7 *leu⁻* bacteria are plated, about 50 colonies arise; these colonies must consist of Leu⁺ bacteria and are spontaneous revertants. The reversion frequency in this case is $50/10^7 = 5 \times 10^{-6}$. Such a frequency value is characteristic of the reversion of point mutants. The production of a spontaneous revertant is a random process so the reversion of a double mutant would require two independent events. If each were to occur at a frequency of 5×10^{-6}, then the frequency of reversion of a double mutant would be $(5 \times 10^{-6})^2 = 2.5 \times 10^{-11}$; that is, 25 colonies would be found if 10^{12} double mutants were plated. Usually, no more than 10^{10} bacteria can be plated in a single petri dish, so that, experimentally, one would need 100 plates to score one revertant. Reversion will be discussed in detail in Chapters 10 and 12.

Uses of Mutants: Some Examples

Some of the most significant advances in molecular biology have come about through the use of mutants. In the following the kinds of approaches that have been taken are described.

1. *A mutant defines a function.* For example, the intake of Fe^{3+} ions by bacteria might be by passive diffusion through the cell membrane, or some system might be responsible for the process. Wild-type *E. coli* can take in the Fe^{3+} ion from a 10^{-5} *M* solution but mutants have been found that cannot do so unless the ion concentration is very high. This finding indicates that a genetically determined system for Fe^{3+} intake exists, though the observation does not tell what this system is. Temperature-sensitive mutants are especially useful in defining functions. For example, temperature-sensitive mutants of *E. coli* have been isolated that fail to synthesize DNA. These mutants fall into at least ten distinct classes, suggesting that there may be at least ten different proteins required for DNA synthesis. One of these mutants lacks a particular protein that is normally located in the cell membrane; the interpretation of such an experimental result is not unambiguous but does suggest that there is some connection between DNA synthesis and the cell membrane.

2. *Mutants can introduce biochemical blocks that aid in the elucidation of metabolic pathways.* The metabolism of the sugar galactose, for example, requires the activity of three distinct genes called *galK, galT,* and *galE.* If radioactive galactose ([^{14}C]Gal) is added to a culture of *gal*$^+$ cells, many different radioactive compounds can be found as the galactose is metabolized. At very early times after addition of [^{14}C]Gal, three related compounds are detectable: [^{14}C]galactose 1-phosphate (Gal-1-P), [^{14}C]uridine diphosphogalactose (UDP-Gal), and [^{14}C]uridine diphosphoglucose (UDP-Glu). Different mutant genes will block different steps of the metabolic pathway. If the cell is a *galK*$^-$ mutant, the [^{14}C]Gal label is found only in galactose. Thus the *galK* gene is known to be responsible for the first metabolic step. If the mutant *galT*$^-$ is used, Gal-1-P accumulates. Thus, the first step in the reaction sequence is found to be the conversion of galactose to Gal-1-P by the *galK* gene product (namely, the enzyme galactokinase). If a *galE*$^-$ mutant is used, some Gal-1-P is found but the principal radiochemical is UDP-Gal. Thus the biochemical pathway must be

$$\text{Gal} \xrightarrow[\substack{galK \\ product}]{} \text{Gal-1-P} \xrightarrow[\substack{galT \\ product}]{} \text{UDP-Gal} \xrightarrow[\substack{galE \\ product}]{} \text{X}$$

The identity of X cannot be determined from these genetic experiments, but one might guess that it is UDP-Glu (which it is).

3. *Mutants enable one to learn about metabolic regulation.* Many mutants have been isolated that alter the amount of a particular protein that is synthesized or the way the amount synthesized responds to external signals. These mutants define regulatory systems. For example, the enzymes corresponding to the *galK, galT,* and *galE* genes are normally not present in bacteria but appear only after galactose is added to the growth medium. However, mutants have been isolated in which these enzymes are always present, whether or not galactose is also present. This indicates that some gene is responsible for turning the system of enzyme production on and off and this regulatory gene must be responsive to the presence and absence of galactose.

4. *Mutants enable a biochemical entity to be matched with a biological function or an intracellular protein.* For many years an E. *coli* enzyme called DNA polymerase I was studied in great detail. Purified polymerase I is capable of synthesizing DNA *in vitro,* so it was believed that this enzyme was also responsible for *in vivo* bacterial DNA synthesis. However, an E. *coli* mutant (*polA⁻*) was isolated in which the activity of polymerase I was reduced 50-fold yet the mutant bacterium grew and synthesized DNA normally. This observation suggested strongly that polymerase I could not be the only enzyme that synthesizes intracellular DNA. Indeed, biochemical analysis of cell extracts of the *polA⁻* mutant showed the existence of two other enzymes, polymerase II and polymerase III, which could, when purified, also synthesize DNA. In further study, a temperature-sensitive mutation in a gene called *dnaE* was found to be unable to synthesize DNA at 42°C, although synthesis was normal at 30°C. The three enzymes, polymerases I, II, and III, were isolated from cultures of the *dnaE⁻*(Ts) mutant and each enzyme was assayed. Although polymerases I and II were active at both 30°C and 42°C, polymerase III was active at 30°C but not at 42°C, so that polymerase III was determined to be the product of the *dnaE* gene and the enzyme responsible for intracellular DNA synthesis.

5. *Mutants locate the site of action of external agents.* The antibiotic rifampicin prevents synthesis of RNA. When first discovered, it was not known whether rifampicin might act by preventing synthesis of precursor molecules (by binding to DNA and thereby preventing the DNA from being transcribed into RNA) or by binding to RNA polymerase, the enzyme responsible for synthesizing RNA. Mutants were isolated that were resistant to rifampicin. These mutants were of two types—those in which the bacterial cell wall was altered so that rifampicin could not enter the cell (an uninformative type of mutant) and those in which the RNA polymerase was slightly altered. The finding of the latter mutants proved that the antibiotic acts by binding to RNA polymerase.

6. *Mutants can indicate relations between apparently unrelated systems.* Bacteriophage λ, which normally adsorbs to and grows in *E. coli,* fails to adsorb to a bacterial mutant unable to metabolize the sugar maltose. Such failure is not associated with mutants incapable of metabolizing other sugars or with any other phages, and this knowledge implicated some product or agent of maltose metabolism in the adsorption of λ. Similarly, *E. coli* mutants unable to adsorb the phage φ80 require exceedingly high concentrations of the Fe^{3+} ion in the growth medium if the bacterium is to grow. This is because the adsorption site of φ80 and the protein responsible for transport of the Fe^{3+} ion are structurally related.

7. *Mutants can indicate that two proteins interact.* How this occurs is best shown by a hypothetical example. Suppose that mutants in two genes *a* and *b,* which are responsible for synthesizing the proteins A and B, fail to carry out some process P. Thus, the products of both genes are necessary for this process to occur. These products may act consecutively or interact to form a single functional unit consisting of both products. Interaction as a single unit is often indicated by reversion studies. When revertants of an *a⁻* mutant are sought, it is sometimes found (by additional genetic analysis) that reversion is a result of production of a mutation in gene B. When reversion of this type occurs, it is usually observed that reversion of other *b⁻* mutants (those

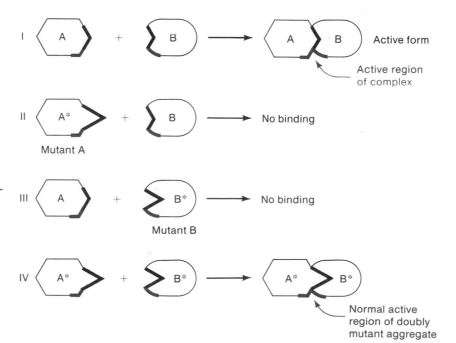

Figure 1-11
Schematic diagram showing how two separately inactive mutants can combine to make a functioning protein complex. Sites of interaction of proteins A and B are denoted by heavy lines. Components of the active site of the A-B complex are shown in red. Only complexes I and IV have active binding sites.

not formed by reversion of an a^- mutant) is often the result of (different) mutations in gene a. The interpretation of these results is the following. Proteins A and B interact to form a protein aggregate (Figure 1-11), which is designated an active structure. Some alterations in either A or B prevent aggregation inasmuch as the protein will have been changed at a binding site. A compensating alteration in the other protein can then enable the interaction to occur again. Such an interpretation has frequently been found to be correct.

GENETIC ANALYSIS OF MUTANTS

A great deal of information about molecular mechanisms can be obtained by combining separately isolated mutations in various combinations. This is done by **genetic recombination.** Another important use of genetic recombination is the construction of an array that indicates the positions of genes on a chromosome with respect to one another. When this is done by genetic techniques, the array is called a **genetic map. Physical maps** have also been constructed by using various physical techniques, and some elegant experiments have shown that the gene positions in the two maps are identical. Another genetic technique is **complementation.** By this procedure it is possible to determine the number of genes responsible for a particular phenotype and to distinguish genes from regulatory sites. In the following sections both genetic recombination and complementation will be described.

Genetic Recombination

Genetic recombination is the process of combining two genetic loci, initially on two different chromosomes, onto a single chromosome (hence, historically, recombination was termed **crossing over**). The molecular mechanism, which is very complex and, at present, not well understood, is discussed in Chapter 18. For the present purposes it is sufficient to refer to the "scissors-and-tape" mechanism. In this, one assumes that two chromosomes align with one another, that a cut is made in both chromosomes at random but matching points, and that the four fragments are then joined together to form two new combinations of genes. This is a naive and incorrect model; it enables one to account for only the simpler features of gene exchange, but these features are in fact the only ones that are of concern at this time. The process is depicted in Figure 1-12. Two parental chromosomes having the genotypes a^+b^- and a^-b^+ pair, are cut, and are then joined to form two recombinant chromosomes whose genotypes are a^+b^+ (wild type) and a^-b^- (double mutant).

Genetic recombination frequently occurs when one bacterium is

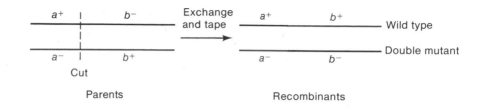

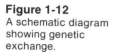

Figure 1-12
A schematic diagram
showing genetic
exchange.

simultaneously infected with two mutant phage. That is, if the parental phage have the genotypes a^+b^- and a^-b^+, as in Figure 1-12, of the hundred or so progeny phage released when the infected bacterium lyses, there are a few a^+b^+ and a^-b^- recombinant phage. The ratio

$$\text{Number of recombinant phage/Number of total phage}$$

is called the **recombination frequency.**

Genetic Mapping

In genetic mapping, the distance along the chromosome between two recombining genetic loci (or mutations) determines the recombination frequency. As long as the two loci are not too near one another, the recombination frequency is proportional to distance, because chromosomal cuts are made at random. Thus, in the following crosses between chromosomes, the genotypes of which are $a^+b^-c^-$ and $a^-b^+c^+$, and the genes of which are in alphabetical order and equally spaced,

$$
\begin{array}{ccc}
a^+ & b^- & c^- \\
\hline
 & \times & \\
\hline
a^- & b^+ & c^+
\end{array}
$$

there will be twice as many a^+c^+ recombinants as a^+b^+ recombinants, because loci a and c are twice as far apart as loci a and b.

Because the recombination frequency is proportional to distance, recombination frequency can be used to determine the arrangement of genes on the chromosome. This can be seen in a simple example (Figure 1-13). Consider three genes a, b, and c, whose arrangement is unknown. Using the notation $p \times q = m$ percent to denote a recombination frequency of m percent between genes p and q, we assume it has been shown that $a \times b = 1$ percent and that $b \times c = 2$ percent. The two arrangements shown in the figure are consistent with these values and can be distinguished by determining the recombination frequency between a and c. Let us assume that this is 1 percent. If that is the case,

Figure 1-13
Two arrangements of three genes for which $a \times b = 1$ percent and $b \times c = 2$ percent.
See text for discussion.

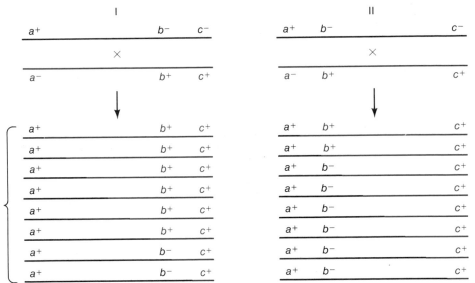

Figure 1-14
A three-factor cross that shows linkage. Eight possible a^+c^+ progeny arising from
equally spaced exchange points are shown for each arrangement. In I, 6 of 8 a^+c^+ are
$a^+b^+c^+$; in II, 6 of 8 a^+c^+ are $a^+b^-c^+$.

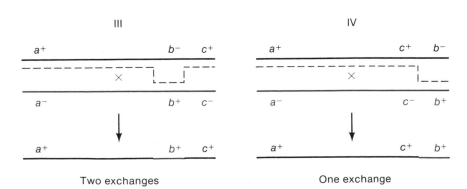

Figure 1-15
Determination of gene order by a three-factor cross. The frequency of appearance of
$a^+b^+c^+$ is much higher for order IV because only one exchange is required.

only arrangement II is possible. The order c a b for these genes and the relative separation constitute a genetic map.

Any number of genes can be mapped in this way. For instance, consider a fourth gene, d, in the preceding example. If $d \times b = 0.5$ percent, d must be located 0.5 unit either to the left or to the right of b. If $a \times d = 1.5$ percent, d is clearly to the right of b and the gene order is c a b d. If $a \times d = 0.5$ percent, the gene order would be c a d b.

Linkage and Multifactor Crosses

Mapping can be carried out with a fewer number of crosses if the cross is done with three loci simultaneously (a **three-factor cross**). The procedure utilizes an effect known as **linkage** and is occasionally called the "method of unselected markers." Consider the cross shown in Figure 1-14 between two parents whose genotypes are $a^+b^-c^-$ and $a^-b^+c^+$. Instead of measuring recombination frequencies, let us just select all recombinants that are a^+c^+ and then in a second test ask whether they are also b^+ or b^-. If the genes were arranged as in panel I, the recombinants would mostly be b^+, because the distance from a to b is greater than that from b to c. On the contrary, if the genes were located as in panel II, most of the a^+c^+ recombinants would be b . This simple analysis locates b with respect to a and c. For arrangements I and II, one says that b and c or a and b, respectively, are linked.

The procedure just described does not give the gene order. That is, if b and c are linked, the gene order could be a b c or a c b and these two orders would not be distinguished by the analysis just given. Gene order can in theory always be determined by the mapping procedures that we have used to distinguish the two arrays shown in Figure 1-13; in the case of two closely linked loci, however, determination of gene order is not feasible from data derived from two-factor crosses, because the data are usually not sufficiently accurate. For example, suppose $a \times b = 7$ percent and $b \times c = 0.2$ percent. If $a \times c = 6.8$ percent, the order is a c b, and if it is 7.2 percent, the order is a b c. However, experimentally observed values might be $a \times b = 7.0 \pm 0.3$ percent and $a \times c = 7.1 \pm 0.3$ percent, so that the order would not be established with certainty in this way. Figure 1-15 shows how a three-factor cross clearly gives the order. A cross is performed between parents having genotypes $a^+b^-c^+$ and $a^-b^+c^-$. Two types of data can be obtained. In the first, the linkage method is used; that is, b^+c^+ recombinants are selected and these are tested to determine whether they are a^+ or a^-. They are usually a^- for order III and a^+ for order IV. In the second and more rapid method, the frequency of $a^+b^+c^+$ recombinants is measured. These recombinants arise by two exchanges in case III but only one exchange in case IV. The frequency for a double exchange is the product of the frequencies of each simple exchange. Thus, for case III, the recombination frequency is $(.07)(.002) = 0.00014 = 0.014$ percent but for case IV, in which only a single exchange is needed, the frequency is 0.2 percent (the same as that for forming $b^+ c^+$ recombinants). Hence one three-factor cross yields the gene order unambiguously.

Further discussion of mapping techniques and the analysis of recombination frequencies can be found in the genetics texts listed at the end of this chapter.

Complementation

As we have seen earlier, a particular phenotype is frequently the result of the activity of many genes. In the study of any genetic system it is always important to know the number of genes and regulatory elements that constitute the system. The genetic test used to evaluate this number is called **complementation.**

Complementation is best explained by example. The test requires that two copies of the genetic unit to be tested be present in the same cell. In bacteria this can be done by constructing a **partial diploid**—that is, a cell containing one complete set of genes and duplicates of some of these genes. How cells of this type are constructed is described in a later section. A partial diploid is described by writing the genotype of each set of genes on either side of a diagonal line. As an example of this, $b^+ c^+ d^+ / a^+ b^- c^- d^- e^+ \cdots z^+$ indicates that a chromosomal segment containing genes b, c, and d is present in a cell whose single chromosome contains all of the genes a, b, c, . . . , z. Usually, only the duplicated genes are indicated, so that this partial diploid would be designated $b^+ c^+ d^+ / b^- c^- d^-$. We now consider a hypothetical bacterium that synthesizes a green pigment from the combined action of genes a, b, and c. The genes encode the enzymes A, B, C, which we assume to act sequentially to form the pigment. If there is a mutation in any of these genes, no pigment is made. Pigment is made by the partial diploid $a^- b^+ c^+ / a^+ b^- c^+$, however, because the cell contains a set of genes that produce functional proteins A, B, and C; B will be made from the $a^- b^+ c^+$ chromosome and A from the $a^+ b^- c^+$ chromosome. In a partial diploid $a_1^- b^+ c^+ / a_2^- b^+ c^+$, in which a_1^- and a_2^- are two different mutations in gene a, no pigment can be made because the bacterium will not contain a functional A protein. The two mutations a^- and b^- of the diploid are said to complement one another because the phenotype of the partial diploid containing them is $A^+ B^+$; the mutations a_1^- and a_2^- do not complement one another because the phenotype of the partial diploid containing these mutations is A^-.

Suppose that now a mutation x^- has been isolated but the gene in which the mutation has occurred has not been ascertained. By constructing a set of partial diploids, this gene can be identified by a complementation test. As a start, we might test the genes a, b, and c with the partial diploids a^-/x^- (I), b^-/x^- (II), and c^-/x^- (III). If diploids I and II make pigment, the mutation cannot be in genes a or b; if no pigment is made by diploid III, the mutation must be in gene c. If pigment were made in all three diploids, then the important conclusion that mutation x^- is in none of the genes a, b, or c could be drawn; furthermore, since we have assumed that a, b, and c are each pigment genes, the fact that x is not in any of these genes and yet prevent pigment formation would be evidence that pigment formation requires at least four genes ("at least" because more genes might be discovered).

A common approach to the initial characterization of a genetic system is to isolate about fifty mutants and perform complementation tests between them, as will be shown in the example of the next section. The analysis, which may be tedious, is nonetheless a straightforward one. The basic rule is the following:

Rule I. If x_1^- and x_2^- complement, they are in different genes. The converse statement—that is, if two mutations do not complement, they are in the same gene—does not always follow.

There are three explanations for the lack of complementation of two mutations; these are stated in the following rule:

Rule II. If mutations x_1^- and x_2^- fail to complement, then one of the following is true:

(a) They are in the same gene, or
(b) At least one of the mutations is in a regulatory site for the other gene, or
(c) At least one of the mutations yields an inhibitory gene product.

Complementation Analysis: An Example

Use of rules I and II allows every mutation to be placed in a mutually exclusive complementation group. Let us first examine these rules for a set of mutations, 1 through 8, in which rules I and II(a) account for all of the data. After studying this simple example, we will add first one mutation that requires that rule II(b) be invoked and then a second mutation that can be explained by II(c). The data are presented in Table 1-3. To analyze the first example, use rows 1 through 8 of the table. Each + or − entry designates a single pairing of the mutation numbers shown at the top and side of the table. Any − entry denotes that the pair does not complement; any + entry denotes complementation. In particular, notice any − entries that can be aligned in a single direction (i.e., that align down one column or across one row). These are noncomplementing mutations within a single gene. For example, in column 1, mutations 1 and 5 fail to complement; other noncomplementing (aligning) mutations can be found in columns 2, 3, 4, and 6, and in rows 4, 5, 6, 7, and 8. Three genes can be identified in this way: the first

Table 1-3
An example of complementation data

	1	2	3	4	5	6	7	8	9	10
				Mutation number						

Mutation number

	1	2	3	4	5	6	7	8	9	10
1	−									
2	+	−								
3	+	+	−							
4	+	−	+	−						
5	−	+	+	+	−					
6	+	+	−	+	+	−				
7	+	+	−	+	+	−	−			
8	+	−	+	−	+	+	+	−		
9	−	−	−	−	−	−	−	−	−	
10	±	−	±	−	±	±	±	−	−	−

Note: + = complementation; − = no complementation; ± = weak complementation.
An entry at the intersection of a horizontal row and a vertical column represents the result of one complementation test between two mutations. For example, the + entry at the intersection of row 3 and column 2 indicates that mutations 2 and 3 complement; the corresponding entry for row 2 and column 3 is not given since it is the same as the one just stated—that is, the table is symmetric about the diagonal. The analysis first discussed in the text uses only mutations 1–8, which is the reason for the dashed line between rows 8 and 9.

contains mutations 1 and 5; the second contains mutations 2, 4, and 8; and the third contains mutations 3, 6, and 7. (Inasmuch as the table is symmetrical, a row or column need be counted just once.) Single mutations (a row or column) that intersect at a + sign are usually complementing. The three genes (or groups of mutations) found in the table are therefore complementing. We can give each complementing gene an identifying letter for convenience:

Group	Mutation
A	1, 5
B	2, 4, 8
C	3, 6, 7

The simplest explanation for the data is that the phenotype being studied consists of at least three genes.

We now examine the effect of a mutation in a regulatory site rather than in a gene (rule II(b)). Suppose a ninth mutant had been isolated having the complementing property shown below:

	1	2	3	4	5	6	7	8	9
9	−	−	−	−	−	−	−	−	−

Alone these data might suggest that all of the mutations are in the same gene, but the data just analyzed show that this is not the case, for clearly, when mutation 9 is present in a particular chromosome, that chromosome will not yield the products of genes A, B, and C. Mutation 9 is called a **cis-dominant mutation,** because such a mutation prevents expressions of related genes residing on the same chromosome in which the mutation is located. Such mutations invariably prove to be regulatory site mutations; that is, they are in some site on the chromosome that must be active if gene products in the *same* chromosome and in the particular gene system are to be made. For instance, the mutation might be in a site that signals start of synthesis of the gene products.

This interpretation of mutation 9 has not been done fairly, because mutations 1 through 8 were classified first and then 9 was introduced. Let us consider how all nine mutations would have been classified if they were examined simultaneously. Once again, − entries would be observed and clustered. However, mutation 9 would immediately be anomalous because it would appear in more than one complementation group—in fact, in all groups. Thus, whenever a mutation appears in more than one group, the need for rule II(b) should be suspected and that mutation should temporarily be ignored in making the basic classification. It is, of course, not always the case that a mutation in a regulatory site will fail to complement all other mutations being examined, because the site might be required for activity of only one, or a few, genes in the set. That is, the regulatory mutation would only fail to complement mutations in the genes regulated.

Many examples of *cis*-dominant mutations will be seen later in Chapter 14.

Let us now consider the rare case in which rule II(c) would be applied. Inasmuch as mutation 10 fails to complement mutations 2, 4, and 8, it might appear to be a separate, additional mutation in the group B gene. The + response is weak with mutants 1, 3, 5, 6, and 7, however, and mutant 10 cannot be considered complementary to groups A and C because, tested against the mutants of those groups, 10 would also yield a −. The correct interpretation is that 10 is not only a B-group mutation but that the mutant gene product also inhibits the activity of a good copy of B. Thus, in each cell in which one would expect B to be fully active, the + response is weak because the total activity of the inhibited B is not adequate for a normal + response. In this case, mutation 10 is said to be anticomplementing.

Complementation in Phage-Infected Bacteria

So far we have discussed complementation only in bacteria. The concept is also applicable to phages and is a common way to assign phage mutations to particular genes. The equivalent of the partial diploid test is the infection of a bacterium with two mutant phages, neither of which can grow alone; one then notes whether phage are produced by the doubly infected cells. This test is usually done with conditional mutants in the following way. Consider two temperature-sensitive mutations a^-(Ts) and b^-(Ts) in genes a and b of two respective phage; neither gene product is active at 42°C, so the mutants cannot grow at 42°C and phage production depends upon the complementation of separate Ts mutations. The phage are allowed to adsorb at 42°C to a host bacterium and then the infected cell is incubated at that temperature. If only a^-(Ts) or only b^-(Ts) phage have adsorbed, no phage will be produced. If both have adsorbed, however, and the Ts$^-$ mutations are in different genes, the infected cell will contain a good copy of product A and a good copy of product B; complementation can then occur and progeny phage will be produced. Note that in addition to the parental types of a^-(Ts) and b^-(Ts) phage that are present in the resulting lysate, there are also some Ts$^+$ and a^-(Ts)b^-(Ts) recombinants.

GENETIC SYSTEMS IN BACTERIA

The bacterium E. *coli* possesses two mating types: donors or **males,** and recipients or **females.** The determinant of maleness is a small circular piece of extrachromosomal DNA denoted by **F** and called the **sex plasmid;** as donor, a male cell is designated F^+. A female cell lacks the F plasmid and is designated F^-. When a culture of males is mixed with a culture of females, male-female pairs form (this is called **conjugation**), each pair being joined by a conjugation bridge. In a way that is not understood, pairing signals the replication of F, and one copy of F is transferred to the female in about one minute (Figure 1-16). In contrast with other sexual systems, the female is converted to a male inasmuch as, after the mating, the recipient cell contains the F plasmid.

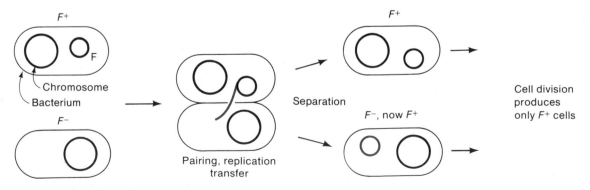

Figure 1-16
Diagram showing the events in an $F^+ \times F^-$ mating.

Hfr Cells

Another property possessed by the F plasmid is its ability to become integrated into the bacterial chromosome. (The mechanism for this integration will be discussed in Chapter 18.) When this occurs, the chromosome remains a single, circular DNA molecule and F behaves as if it were part of this chromosome, increasing the size of the chromosome. The integration process happens rarely but it is possible to purify a culture of cells that are the progeny of the cell in which integration occurred. The cells in such a culture are called **Hfr males.** (Hfr is an acronym for *high frequency of recombination.*) When a culture of Hfr cells is subsequently mixed with an F^- culture, conjugation also occurs as described earlier, though the material transferred is slightly different from that in an $F^+ \times F^-$ mating (Figure 1-17). This time, under the influence of F, DNA replication begins in the Hfr cell and a replica is transferred to the F^- cell; however, the direction of replication is such that a small portion of F is transferred first and the major portion is transferred last. Moreover, because bacteria are very small and in constant motion by being bombarded by solvent molecules (Brownian motion), and because it takes 100 minutes to transfer an entire chromosome, the mating pair usually breaks apart before transfer is completed. Thus, the female receives both a large fragment of the male chromosome, which may contain hundreds or thousands of genes, and a small functionless fragment of F. Because of this, the exconjugant female remains a female in an Hfr $\times$ F^- mating.

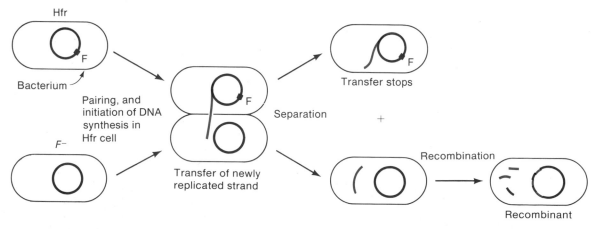

Figure 1-17

A diagram showing mating between an Hfr male bacterium and an F^- female bacterium. The two bacteria form a pair. Then, under the influence of a unit called F, DNA replication begins in the male, adjacent to F, and a replica of F is transferred to the female. By random motion the mating cells break apart. At a later time a portion of the transferred DNA exchanges with the corresponding piece in the female. The intact female chromosome then replicates; the fragments do not replicate and are ultimately lost in the course of cell division.

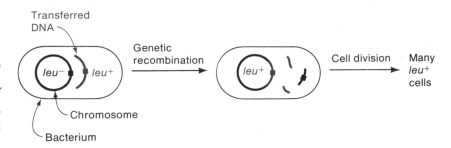

Figure 1-18
Conversion of a *leu⁻* cell to a *leu⁺* cell by incorporation of a segment of the transferred DNA. The fragments remaining after recombination do not replicate and are gradually diluted as the recombinant cell divides.

The presence of the new chromosomal fragment in the female sets in motion a recombination system that causes genetic exchanges to occur, so that a recombinant F^- cell often results. Thus, in a mating between an Hfr *leu⁺* culture and an F^- *leu⁻* culture, F^- *leu⁺* cells arise (Figure 1-18). In order to recognize these cells, some means is needed to distinguish a male from a female cell. This is done by using cells that have genetic differences that can be recognized by growth of a colony on agar. A common method is to use antibiotic resistance. For instance, consider an Hfr that is not only Leu⁺ but also sensitive to streptomycin (Str-s) and a female that is both Leu⁻ and resistant to streptomycin (Str-r). If the streptomycin locus is so far from the origin of transfer that the mating pairs, which inevitably break apart, have separated before the *str* locus has been transferred, then plating the mated cell mixture onto agar containing streptomycin but lacking leucine (1) will selectively kill the Hfr Str-s cells and (2) will not lead to growth of the F^- cells unless these also possess the *leu⁺* locus. Thus, any cell that survives will have the genotype *leu⁺str-r* and will be a recombinant female. Only these Leu⁺ Str-r recombinants can form a colony. When a mating is done in this way, the transferred locus that is selected by means of the agar conditions (*leu⁺* in this case) is called a **selected marker,** and the locus used to prevent growth of the male (*str-r* in this case) is called the **counterselective marker.**

Genetic Mapping by an Hfr × F⁻ Mating

An important feature of an Hfr × F^- mating is that the transfer of the Hfr chromosome proceeds at a constant rate from a fixed point determined by the site at which F has been inserted in the Hfr chromosome. This means that the times at which particular genetic loci enter a female are proportional to the positions of these loci on the chromosome. Thus a genetic map can be obtained from the time of entry of each gene. This is done in the following way. An Hfr whose genotype is $a^+b^+c^+d^+e^+$ *str-s* is mated with an $a^-b^-c^-d^-e^-$ *str-r* female. At various times after mixing the cells, a sample is taken and agitated violently in order to break apart all of the pairs simultaneously. The

Figure 1-19
Construction of the circular genetic map of *E. coli.*
(a) Time-of-entry curves for a particular Hfr strain. (b), (c), (d), (e) Four linear maps obtained from four different Hfr types. (f) The circular map derived from the data in panels (a)—(e).

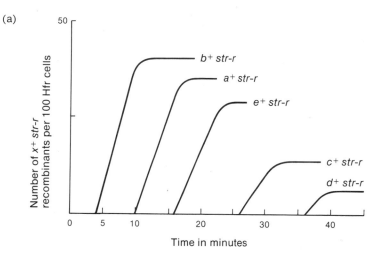

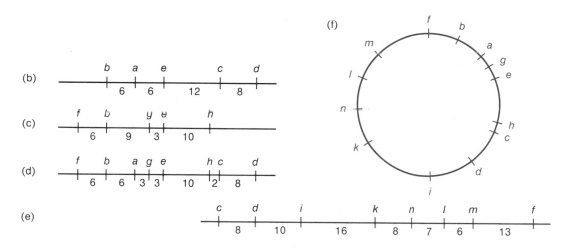

sample is then plated on five different agars containing streptomycin, each of which also contains a different combination of four of the five substances A through E. Thus colonies that grow on agar lacking A are a^+str-r, those growing without B are b^+str-r, and so forth. All of these data can be plotted on a single graph to give a set of time-of-entry curves, as shown in Figure 1-19(a). Extrapolation of each curve to the time axis yields the time of entry of each locus $a^+, b^+, \ldots, e^+$. These times can be placed on a map, as shown in part (b) of the figure. Use of a second female that is $b^- e^- f^- g^- h^- str$-r can then provide the relative positions of three additional genes. These might give a map such as that in Figure 1-19(c). Since genes b and e are common to both maps, the two maps can be combined to form a more complete map, as shown in Figure 1-19(d).

The Circular Genetic Map of *E. coli*

The F plasmid can integrate at numerous sites in the chromosome to generate Hfr cells that have different origins of transfer. Each of these Hfr strains can also be used to obtain maps. It has been found that when separate maps are combined, a circle is eventually obtained. For instance, the map obtained from another Hfr might be that shown in Figure 1-19(e), which, when combined with that in panel (d), would yield the circular map shown in panel (f). This mapping technique has been used repeatedly with hundreds of *E. coli* genes to generate an extraordinarily useful genetic map—one with great significance in the development of molecular genetics. The discovery in the early 1950's that the genetic map of *E. coli* is circular provided the first suggestion that the *E. coli* chromosome might be a circular DNA molecule, as is now known to be the case.*

We shall have many occasions in this book to refer to isolated segments of the *E. coli* map. From a practical point of view, it has provided information needed to construct bacterial strains having particular properties. Also, the map shows that genes whose functions are related are usually clustered. For example, the three genes responsible for the metabolism of the sugar galactose are adjacent to one another. The reason for this clustering has now been explained as one of the necessary aspects of the now-classical theory for the regulation of gene expression in bacteria.

E. coli Recombination Genes

Genetic recombination in the female cell is catalyzed by a set of enzymes and proteins that constitute the **Rec system.** Many mutants in this system are known; the best studied are those in the genes *recA*, *recB*, and *recC*. A mutation in the Rec system does not prevent transfer to the female, but the segment of the transferred male DNA cannot be recombined or recombines very infrequently with the DNA of the female. Since the transferred fragment is unable to replicate, it is usually diluted out in subsequent cell division and recombinants are rarely found.

F′ Plasmids

An Hfr cell is produced when F integrates stably into the chromosome, as we have already stated. At a low frequency, F can excise itself. When

---

*A circular map can result even if the chromosome is not circular—for example, if crossovers usually occur in pairs. Nonetheless, the circular map was highly suggestive of a circular DNA molecule.

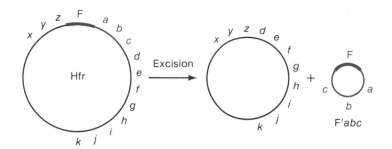

Figure 1-20
Diagram showing excision of an F′ plasmid from an Hfr chromosome.

this happens, the excised circular DNA is sometimes found to contain genes that were adjacent to F in the chromosome (Figure 1-20). A plasmid containing both F genes and chromosomal genes is called an F′ plasmid. It is usual to describe an F′ plasmid by stating the genes it is known to possess—for example, F′*lac pro* contains the genes for lactose utilization and proline synthesis. F′ plasmids can also be transferred from an F′ male to a female. This occurs sufficiently rapidly that the entire F′ is usually transferred before the mating pair breaks apart. Thus the female recipient is converted to an F′ male in an $F′ \times F^-$ mating.

F′ plasmids have been useful in the production of partial diploid bacteria. For example, if an F′*lac⁺*/*str-s* male is mated with a *lac⁻ str-r* female and a Lac⁺ Str-r colony is isolated, the cells in the colony will carry two copies of the *lac* gene—the *lac⁺* version brought to the female in the F′ and the *lac⁻* version already present in the female chromosome. To denote this, the genotype of the cell is written F′*lac⁺*/*lac⁻ str-r,* as is usual with partial diploid cells. By convention, genes carried on the F′ plasmid are written at the left of the diagonal line.

In addition to F and F′ plasmids there are numerous other plasmids; their properties and their utility in the laboratory are discussed in detail in Chapters 17 and 20.

REFERENCES

Davis, B. D., R. Dulbecco, H. N. Eisen, H. S. Ginsberg, and W. B. Wood. 1973. *Microbiology.* Harper & Row.

Gardner, E. J. 1975. *Principles of Genetics.* 5th Ed. Wiley.

Hayes, W. 1968. *The Genetics of Bacteria and Their Viruses.* Wiley.

Keeton, W. 1980. *Biological Science.* W. W. Norton.

Moat, A. G. 1979. *Microbial Physiology.* Wiley.

Puck, T. T. 1972. *The Mammalian Cell as a Microorganism.* Holden-Day.

Sheeler, P., and D. E. Bianchi. 1980. *Cell Biology.* Wiley.

Srb, A., R. Owen, and R. Edgar. 1965. *General Genetics.* W. H. Freeman and Co.

Stent, G. S., and R. Calendar. 1978. *Molecular Genetics*. W. H. Freeman and Co.

Stryer, L. 1981. *Biochemistry*. 2nd Ed. W. H. Freeman and Co.

Suzuki, D. T., and A. J. F. Griffiths. 1976. *Introduction to Genetic Analysis*. W. H. Freeman and Co.

Swanson, C. P. 1977. *The Cell*. 4th Ed. Prentice-Hall.

Tooze, J. (ed.) 1980. *The Molecular Biology of Tumour Viruses*. Cold Spring Harbor Laboratory.

2 The Basics of Biochemistry

A relatively simple bacterium such as E. coli contains about 2000 genes, each of the great majority of which is responsible for the synthesis of a particular protein molecule. Roughly half of these proteins are enzymes, which are responsible for catalyzing chemical reactions in which a variety of substances participate. Thus, a bacterium can be thought of as a tiny sac containing thousands of distinct chemicals that are continually being interconverted. The sac does not contain a homogeneous solution but is somewhat subdivided into small domains by complex structures consisting mainly of gigantic molecules, some of which form membranes. The intracellular concentrations of all of these thousands of molecules are delicately regulated by small molecules and macromolecules, as well as by the needs of the cell as determined by environmental factors. This brief description should suffice to convince the reader that an understanding of biological phenomena requires knowledge of biochemistry. Fortunately, however, the essentials (not the details) of molecular biology can be understood and appreciated with only a minimal amount of chemistry. In fact, little is required other than familiarity with some simple chemical groups and types of compounds. A review of essential information is provided in the following section. It is not necessary at this stage to know about the chemistry of the reactions in which intracellular biochemicals participate. Later we will discuss the chemical needs of a cell and the concept of a metabolic pathway. Macromolecules will be defined briefly in this chapter, but a detailed presentation is reserved for Chapters 3–6.

This chapter is meant to be only an introduction to biochemistry for the reader who has not yet studied the subject. For details, the reader should consult the many excellent texts listed in the references at the end of this chapter.

THE MOLECULES OF BIOLOGICAL SYSTEMS

Many types of organic compounds are found in living cells but only a few of these are of concern in this book. The student should, however, be thoroughly familiar with the functional groups and bond types, the major classes of organic compounds found in cells, and the properties and formulas of the components of proteins and nucleic acids. These are reviewed in the following sections.

Functional Groups and Bond Types

The functional groups and bond types with which the reader should be familiar are the following:

Amino group, or $-NH_2$. This behaves as a base in aqueous solution; at neutral pH it usually attracts a proton and becomes positively charged. When in a charged state, this group is usually written $-NH_3^+$.

Hydroxyl group, or $-OH$. This is the defining group for the alcohols. It is important in biological systems because it is somewhat polar and increases the solubility in water of compounds containing this group. The $-OH$ group readily forms a hydrogen bond with an amino group.

Carboxyl group, or $-COOH$. This group is acidic in aqueous solution. At neutral pH it is usually ionized and negatively charged, having the structure $-COO^-$. Because of its strong charge, it easily forms an ionic bond.

Keto or **carbonyl group,** or

$$\overset{|}{\underset{|}{C}}=O$$

This is often a component of hydrogen bonds. If a hydrogen atom is attached to the carbon atom, the group is called an **aldehyde.**

Sulfhydryl group, or $-SH$. This group, which is found primarily in proteins, has two important properties: (1) it can bind certain cations, such as Zn^{2+}, very tightly; and (2) two sulfhydryl groups can be joined to form a —S—S— or **disulfide bond.** The disulfide bond is a major factor in determining the structure of protein molecules.

Amide group, or

$$-\overset{\overset{\displaystyle O}{\|}}{C}-\overset{\overset{\displaystyle H}{|}}{N}-$$

This group is found in many small molecules but is most important for connecting amino acid monomers in protein molecules, in which case the C—N bond is called a **peptide bond.** The C—N bond has a partial double-bond character and thus lacks the ability to rotate. Both the O and H atoms frequently engage in hydrogen bonds. The peptide bond will be discussed in detail when the strucure of protein molecules is presented.

Phosphate ester, or

$$-\overset{|}{\underset{|}{C}}-O-\overset{\overset{\displaystyle O}{\|}}{\underset{\underset{\displaystyle O^-}{|}}{P}}-O^-$$

This important group, which is always negatively charged, is used both to join two sugars in the backbone of a nucleic acid, in which case the unit is called a **phosphodiester group,** and as part of various molecules that store energy. Phosphate esters will be encountered in the discussions of nucleic acid structure and of energy metabolism.

Pyrophosphate group, or

$$-\overset{\overset{\displaystyle O}{\|}}{\underset{\underset{\displaystyle O^-}{|}}{P}}-O-\overset{\overset{\displaystyle O}{\|}}{\underset{\underset{\displaystyle O^-}{|}}{P}}-O^-$$

Phosphate groups frequently exist in sequence; one important example is the **triphosphate group** having the structure

$$-\overset{\overset{\displaystyle O}{\|}}{\underset{\underset{\displaystyle O^-}{|}}{P}}-O-\overset{\overset{\displaystyle O}{\|}}{\underset{\underset{\displaystyle O^-}{|}}{P}}-O-\overset{\overset{\displaystyle O}{\|}}{\underset{\underset{\displaystyle O^-}{|}}{P}}-O^-$$

The segment indicated by each of the two brackets is the pyrophosphate linkage. The role of this linkage will be discussed in detail in Chapter 8.

Major Classes of Organic Compounds in Biochemistry

The following types of compounds will be encountered frequently.

Alcohols. The alcohols found in cells are usually simple hydrocarbon chains containing one or more OH groups. The simplest intracellular alcohol is ethanol, CH_3CH_2OH. A common polyhydroxy alcohol is glycerol, $HOCH_2(CHOH)CH_2OH$.

Sugars. A sugar is a polyhydroxy molecule containing a C=O group either as an aldehyde or a ketone. Sugars are classified by the number of their carbon atoms. A sugar is a triose, tetrose, pentose, or hexose if it contains 3, 4, 5, or 6 carbon atoms, respectively. The pentoses and hexoses may form ring structures. These monomeric sugars will be encountered often in this book. One of these is **glucose,** a hexose, shown below in linear and ring forms.

Glucose (linear)   Glucose (ring)

The two others are pentoses, **ribose** and **deoxyribose,** shown below.

Ribose   2-Deoxyribose

The numbers assigned to the carbon atoms above for ribose and 2-deoxyribose are by convention given primes when the sugar is in the ring form.

Disaccharides. The sugars just described are called monosaccharides. Two monosaccharides can be joined by elimination of water to form a disaccharide. An important example, shown below, is **lactose,** which consists of one glucose and one galactose component:

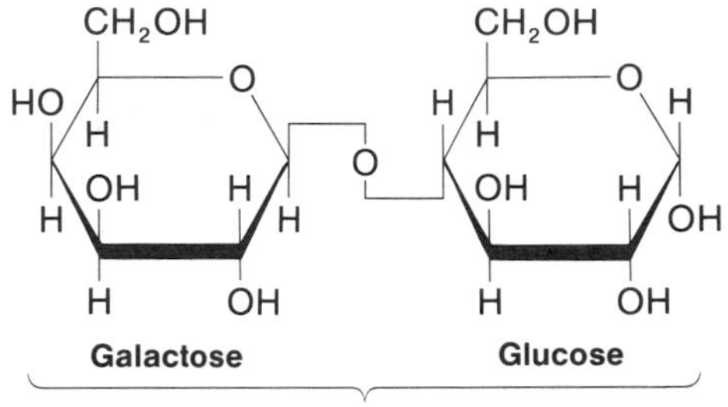

In Chapter 14 we will discuss in detail a system that cleaves lactose to produce these two monosaccharides.

Carboxylic acids. The common simple carboxylic acids found in living cells consist of an aliphatic chain (a chain containing only carbon and hydrogen atoms) terminated by a COOH group. Two examples are acetic acid, CH_3COOH, and propionic acid, CH_3CH_2COOH. More complex carboxylic acids having more than one carboxyl group are often found in living material. For example, citric acid, which is found in citrus fruits and which is also an important intermediate in one of the major energy-generating systems, is a tricarboxylic acid.

Hydroxy acids. These carboxylic acids have an OH group on the α-carbon atom (not the carboxyl carbon) of the hydrocarbon chain. An important example is lactic acid,

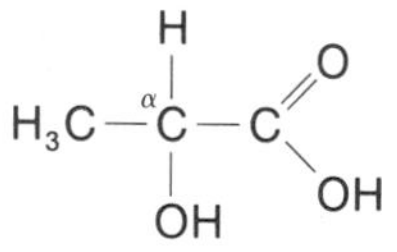

Lactic acid

which is a major intermediate in the energy-generating system of all living cells.

Keto acids. These carboxylic acids have a keto group in the hydrocarbon chain. Pyruvic acid is the most important example of this group:

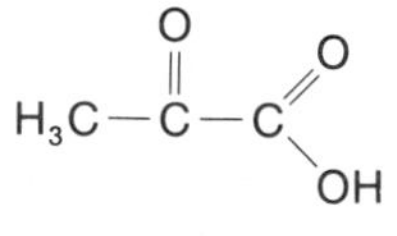

Pyruvic acid

Components of Proteins

The principal components of protein molecules are the amino acids, whose properties are described in this section.

Amino acids. These are carboxylic acids to which an amino group and a side chain are attached. In naturally occurring biological amino acids the amino group is bonded to the carbon atom adjacent to the carboxyl group.

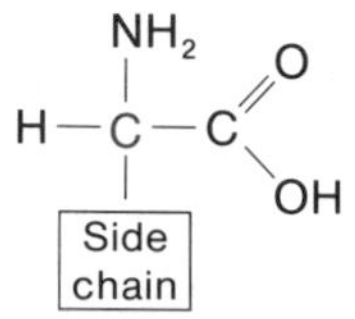

This atom is called the α carbon (shown in red). The α carbon is also the site of attachment of the side chain that distinguishes one amino acid from another. There are many naturally occurring amino acids. The most important ones and their common abbreviations are listed below; their chemical structures are given in Figure 3-2 in Chapter 3.

Alanine (Ala)	Leucine (Leu)
Arginine (Arg)	Lysine (Lys)
Asparagine (Asn)	Methionine (Met)
Aspartic acid (Asp)	Phenylalanine (Phe)
Cysteine (Cys)	Proline (Pro)
Glutamic acid (Glu)	Serine (Ser)
Glutamine (Gln)	Threonine (Thr)
Glycine (Gly)	Tryptophan (Trp)
Histidine (His)	Tyrosine (Tyr)
Isoleucine (Ile)	Valine (Val)

Pyrimidines and Purines—Components of Nucleic Acids

Each monomeric unit of the polymers that constitute DNA and RNA consists of a sugar linked to a phosphate group and to an exchangeable organic base, which is either a substituted pyrimidine (abbreviated Py) or a substituted purine (abbreviated Pu).

Pyrimidine. The compound pyrimidine has the following structure:

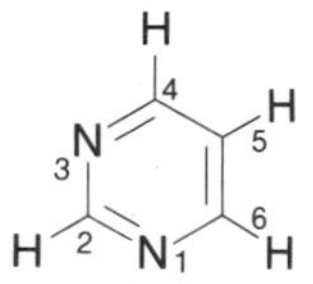

Pyrimidine

The pyrimidines present in nucleic acids are cytosine (C), thymine (T), and uracil (U), whose chemical structures are given in Figure 3-6 in Chapter 3.

Purine. The compound purine has the structure shown below.

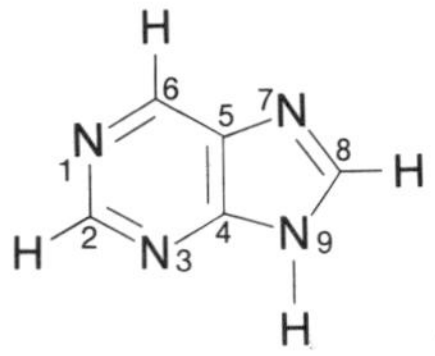

Purine

The purines present in nucleic acids are adenine (A) and guanine (G), whose chemical structures are also given in Figure 3-6 in Chapter 3.

Macromolecules

The molecular weights of biological molecules range from 18 (water) to more than 10^{11} (eukaryotic DNA molecules). It is usually convenient to divide all biological molecules into two size classes—**small molecules,** whose molecular weights are generally less than 500 and which are simple monomeric substances, and **macromolecules,** whose molecular weights are greater than 2000 (and usually much greater) and which are always polymers.

There are three major classes of biological macromolecules—proteins, polysaccharides, and nucleic acids. The composition of these macromolecules is the following.

Proteins are linear polymers of amino acids in which the amino group of one amino acid is linked to the carboxyl group of the adjacent amino acid.

Polysaccharides are polymers of sugars. Many polysaccharides are branched molecules.

Nucleic acids are of two types—**ribonucleic acid (RNA)** and **deoxyribonucleic acid (DNA).** Both of these are polymers of nucleo-

Figure 2-1
Structure of the
nucleotides.

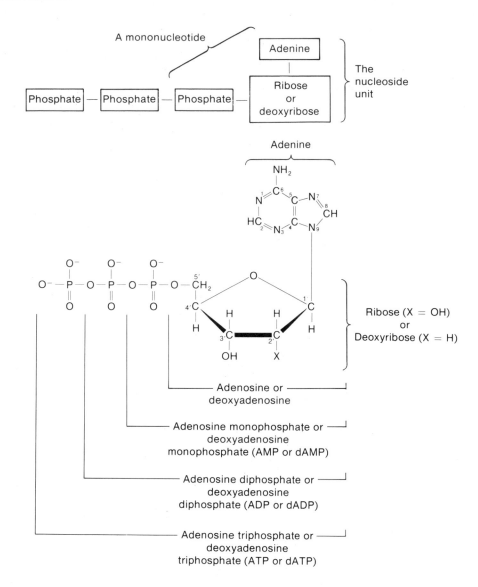

tides. A **nucleotide** consists of a sugar (ribose for RNA, deoxyribose for DNA) to which is linked one purine or one pyrimidine, and one phosphate group (Figure 2-1). The purine-sugar and pyrimidine-sugar units are called **nucleosides.** A nucleotide is therefore a nucleoside phosphate. Separate nucleotides can be joined together by the bonding of the phosphate group of one to the sugar of another. This generates a chain in which sugars and phosphates alternate; the purines or pyrimidines extend outward from the chain. The structures of nucleotides and of DNA and RNA are described in Chapter 3.

NUTRITIONAL REQUIREMENTS OF CELLS

Small molecules are formed both by synthesis from more elementary constituents and by interconversion from other molecules. In both kinds of formation, raw materials must be provided to a cell from an outside source—namely, the growth medium, or, in the case of eukaryotic cells organized into tissues, from extracellular fluids such as blood and lymph. The smallest class of nutrients that must be supplied to a cell if it is to grow is called its **nutritional requirement.** In this section, we describe the nutritional requirements for bacteria, algae, and animal cells.

Nutritional Requirements of Bacteria

More than 30 elements are essential for growth of bacteria but of these only 9 are required in substantial quantities; the others are called **trace elements.** The major elements are added to all chemically defined (minimal) growth media. The others are needed in such small quantities that they are obtained as trace contaminants of the major components. The major elements, their uses, and their sources are listed below.

Carbon. Carbon is found in every intracellular compound except water. Cells usually obtain carbon by degradation of sugars. Other sources are carboxylic acids and amino acids.

Oxygen. Oxygen is widely distributed in intracellular compounds. Molecular oxygen is required as an oxidizing agent in the final stage of respiration but most of the oxygen utilized in chemical compounds is derived from water.

Hydrogen. This is found in virtually every compound. A cell obtains hydrogen from water and from any compound used as a carbon source, since that compound invariably contains hydrogen.

Nitrogen. Nitrogen is found in all nucleic acids and proteins, as well as in other compounds essential for growth. While a few bacterial species (the nitrogen-fixing bacteria) can utilize atmospheric nitrogen, most bacteria derive nitrogen from the ammonium ion, and occasionally from nitrate and nitrite ions. Animal cells obtain the bulk of their nitrogen from organic sources, such as from breakdown of amino acids, nucleotides, and more complex compounds. Algae and plants also can utilize the NH_4^+ and NO_3^- ions. In nutrient broths, nitrogen is provided as amino acids. In minimal media, the nitrogen source is almost always the NH_4^+ ion, or when algae are grown in minimal media, the nitrogen source is the NO_3^- ion.

Phosphorus. Phosphorus is found in all nucleic acids, in some proteins, and in many essential small molecules. Phosphate ester compounds are the major repositories of energy stored in living cells. Phosphorus is provided in growth media exclusively as inorganic phosphate, because phosphorylated organic molecules cannot pass through the outer cell membrane of a bacterium. Within the cell, the PO_4^{3-} ion is released by hydrolysis of many intracellular compounds. However, because intracellular pH is near neutrality, the ion does not remain in this form but is converted either to HPO_4^{2-} ion or an $H_2PO_4^-$ ion; a mixture of these ions then exists within the cell.

Sulfur. Sulfur is needed in smaller amounts than the elements just mentioned; it is found, for example, in only two amino acids (cysteine and methionine), in a few RNA molecules, and in two important metabolic intermediates (coenzyme A and *S*-adenosylmethionine). Sulfur is usually added to growth media as the SO_4^{2-} ion.

Mg^{2+} and Ca^{2+} ions. These ions are required for the activity of many enzymes. The free ions can be supplied in growth media.

K^+ ion. The K^+ ion is the principal intracellular positive ion. It neutralizes negative charges and its intracellular concentration is adjusted to the correct osmotic and ionic strengths.

Cl^- and HCO_3^- ions. The Cl^- and HCO_3^- ions are the major intracellular monovalent anions. The Cl^- ion is added to growth media as NaCl or KCl. The HCO_3^- ion is not added to growth media but is formed as a result of carbon metabolism and from atmospheric CO_2 dissolving in water; its principal function is to maintain intracellular pH. Although NaCl is present in almost all growth media, the Na^+ ion is kept at a very low intracellular concentration by specific ion pumps present in the cell membranes of most cells. The principal function of NaCl in growth media is to maintain the medium at the same osmotic pressure as the intracellular fluid.

Trace metal ions. Many metal ions are components of numerous enzymes. Among the more prevalent ones are the Zn^{2+}, Fe^{3+}, and Cu^{2+} ions. Except occasionally for the Fe^{3+} ion, trace metals are rarely added to growth media, because the required concentration of each is so low (10^{-6} to 10^{-10} M) that the ions are invariably present as contaminants in the other components of the medium.

Lack of Precipitation of Insoluble Ionic Compounds

The intracellular concentrations of the major ions (expressed as moles per volume of a cell) are sufficiently high that sparingly soluble salts such as $Ca_3(PO_4)_2$ ought to come out of solution (precipitate). However, this does not happen because only a small fraction of the ions are free; the majority are bound to the various intracellular macromolecules, so that the concentration of the free ions is not high enough for precipitation to occur.

Factors Affecting the Growth Rate of Bacteria

In a log phase culture the time required for the total number of cells to double is called the **doubling time.** In general, doubling times are determined by the availability of nutrients, the source of energy, and the temperature. As more nutrients are added to a growth medium, the doubling time decreases.

Intracellular nutrients—that is, those that a bacterium can synthesize—are called **endogenous nutrients.** Those that are present in the growth medium (and that may or may not be required) are called **exogenous nutrients.** (In Chapter 14 we shall describe the elegant mechanisms that bacteria have for saving energy when exogenous nutrients are present.)

Each organism usually has an optimal temperature for growth. For *E. coli* this is 37°C (the temperature of the human body) and there is about a 50 percent increase in doubling time at 34°C and 41°C. Optimal temperatures for other microorganisms are also related to the ambient temperatures at which the organisms usually grow in nature.

A major factor that affects the growth rate is the energy source or carbon source. *The best energy source is glucose.* Other compounds either yield less net energy per unit of energy consumed in the energy-yielding reaction, or their metabolism produces energy more slowly than metabolism of glucose. For example, if the succinate ion is used as a carbon source, cells grow at about half the rate occurring when glucose is present.

Let us examine a typical growth medium to see how it influences growth of a bacterium.

A particularly simple bacterial growth medium is the M9 medium (*M* for Jacques Monod, who developed this and many other growth media). Its composition is the following:

0.009 M NaCl	0.01 M $MgSO_4$
0.019 M NH_4Cl	0.0001 M $CaCl_2$
0.022 M KH_2PO_4	0.006 M glucose ($C_6H_{12}O_6$)
0.048 M Na_2HPO_4	

Each of the essential elements mentioned earlier is represented. Two different phosphate ions are present in order to form a buffer that can maintain the medium at near-neutral pH. Since no organic compound other than glucose is present, only a prototroph could grow in this medium, unless the medium is supplemented. If the bacterium to be grown is an auxotroph, the appropriate required substances (for example, an amino acid or a purine) must be added. At 37°C an *E. coli* prototroph divides every 50 minutes in this medium. By adjusting the composition of the medium, the doubling time can be decreased, which is often desirable, inasmuch as many experiments require large populations of cells within as short a time as possible. For example, since glucose is the only organic compound provided, all intracellular organic compounds must be synthesized. If all twenty amino acids are added (for example, by adding a complete set of amino acids obtained by hydrolyzing milk protein or casein, a mixture known as casamino acids), the bacteria do not waste energy in synthesizing these substances and will divide every 35 minutes at 37°. If many nutrients (for example, nucleic acid bases, vitamins, fatty acids, etc.) are added, growth can be accelerated until a 22-minute doubling time is achieved.

Nutritional Requirements of Algae, Yeasts, and Protozoa

The nutritional requirements for all organisms are not the same. For instance, algae are able to dispense with an organic carbon source because they can manufacture glucose from atmospheric CO_2 by photosynthesis. In fact, algae are often found growing in the laboratory in simple inorganic solutions in which there is a source of nitrogen; often they obtain their phosphorus from residues of phosphate-containing compounds used to clean glassware. A typical growth medium for algae contains the Na^+, K^+, Cl^-, Mg^{2+}, Ca^{2+}, NO_3^-, and HPO_4^{2-} ions; light must, of course, be provided. Many of the protozoa, on the other hand, grow very poorly in synthetic media and may even require bacteria as a food source. Yeast grow in media similar to bacterial growth media.

Nutritional Requirements of Animal Cells

Animal cells are very fastidious. They are unable to synthesize many of the amino acids, vitamins, and fatty acids from simple inorganic compounds and sugars. Some amino acids, the so-called essential amino acids, cannot even be obtained by chemical transformation of other amino acids and must be added to the growth medium. (Humans, for example, must obtain at least 12 amino acids directly from the diet.) Thus, growth media for animal cells are very complex, usually containing all of the amino acids and vitamins, and several fatty acids. In addition to salts and trace elements, animal cells in culture need fetal blood serum, which is a source of various growth hormones required by individual cell types (for example, fibroblast growth factor, epithelial growth factor, and nerve growth factor).

METABOLISM AND BIOCHEMICAL PATHWAYS

All intracellular organic molecules are formed ultimately by chemical conversion of the constituents of the growth medium. Since thousands of compounds must be synthesized, many thousands of chemical reactions must occur. These reactions have two features in common: (1) they are almost always catalyzed by specialized proteins called enzymes and (2) the reactions invariably occur in several steps. These features can now be examined.

Enzymes

Most organic reactions proceed far too slowly to be useful to living cells and hence must be catalyzed. The catalysts used by cells are specialized proteins called **enzymes,** and they can increase the reaction rate by a factor of 10^4 to 10^{15}. Virtually every intracellular chemical reaction, in which covalent bonds are made or are broken, utilizes enzymes as catalysts.

Enzymes are highly specific in the sense that an enzyme typically catalyzes only a single type of chemical reaction, such as ester hydrolysis or phosphate transfer. Moreover, they often act on only a single kind of molecule. For example, the enzyme that catalyzes the hydrolysis of maltose, a disaccharide consisting of two glucose units, cannot act on lactose, a glucose-galactose disaccharide; a distinct enzyme is needed for each of these reactions.

The properties of enzymes will be discussed in detail in Chapter 5.

Degradative and Biosynthetic Reactions

In thinking about biochemical reactions it is worth while to separate them into two categories. One of these is the **degradative** or **catabolic reaction** in which molecules are broken down to smaller molecules. An example is the breakdown of glucose. Catabolic reactions serve two functions: (1) they transfer the bond energy released during degradation to a form of energy that can be utilized in other reactions (usually as phosphorus-containing energy storage compounds), and (2) they provide the small units that are used to build up other essential molecules.

The second class of reactions is the **biosynthetic** or **anabolic reaction,** in which small molecules are built up into compounds that are successively altered to form all essential intracellular·constituents. For example, the carbon atoms of glucose are used in every organic compound if glucose is the sole source of carbon in a growth medium.

The energy for this synthesis is obtained from energy-storage systems, such as the pyrophosphate linkages in adenosine triphosphate (ATP), which will be discused shortly.

All of the degradative and biosynthetic reactions of a cell are collectively called the metabolism of the cell.

Multistep Pathways

Degradative and biosynthetic reactions usually consist of a series of reactions. A group of connected reactions is called a **metabolic pathway.** For biosynthesis these pathways are necessary because it is rarely possible for an organism to create a needed compound from a source compound in a single chemical step. In the case of degradation of nutrients, a multistep pathway allows the cell to transfer the bond energy of the initial compound to a large number of energy storage molecules from which energy can be obtained at a later time when it is needed.

In a series of reactions that constitute a metabolic pathway, each step requires a particular enzyme as a catalyst.

Many metabolic pathways are linear; that is, the series of steps may be written

$$A \xrightarrow{E_1} B \xrightarrow{E_2} C \xrightarrow{E_3} D$$

in which A is a starting material, D is the **product,** B and C are pathway **intermediates** and E_1, E_2, and E_3 are enzymes that catalyze the successive steps. Substances A, B, and C are called **precursors** of D, because providing any one of them to a cell will result in their conversion to D.

Biosynthetic pathways can also branch. This means that an intermediate can be converted into two or more diffferent compounds. An example of a branched pathway is

$$A \longrightarrow B \longrightarrow C \begin{cases} \nearrow D \longrightarrow E \\ \searrow F \longrightarrow G \end{cases}$$

In Chapter 14 several ways are described by which a cell is able to turn off a biosynthetic pathway when the product is supplied exogenously or accumulates in the cell.

Energetics of Degradative and Biosynthetic Pathways

The direction of all chemical reactions can be described in terms of a thermodynamic quantity called the **free energy change,** denoted ΔG. A reaction that proceeds spontaneously in a particular direction is said to have a negative value of ΔG for the reaction as written (from left to right)—that is, the reaction loses free energy as it progresses. A reaction that does not proceed spontaneously but requires an input of energy has a positive value of ΔG. Degradative and biosynthetic reaction sequences can be distinguished by their energetics—that is, by the sum of individual ΔG values. In a degradative pathway, a clear decrease in free energy is measurable ($\Delta G < 0$), so degradation is a spontaneous process (one whose rate is merely accelerated by enzymes). In a biosynthetic process, small molecules are successively built up into larger molecules. If the reaction is considered simply as an assembly of components, it almost always will have a positive value of ΔG and thus cannot occur unless coupled with a "driving" system that loses at least the same amount of free energy that the biosynthetic process requires.

One of the most important features of degradative pathways is that, associated with the degradation, often a series of reactions culminates in the synthesis of compounds containing several phosphate esters. These important compounds can yield a large negative free energy through hydrolysis; this energy can be utilized by other reactions for which $\Delta G > 0$. This has the effect that many biosynthetic reactions, which do not occur alone, can be driven in the direction of synthesis when the phosphoester bonds are hydrolyzed. The most important substance of this type is adenosine triphosphate, which will be discussed in the next section.

Energy Metabolism and Storage: ATP

All living cells require a continual supply of free energy for two main purposes: for biosynthesis of both small molecules and macromolcules, and for the active transport of ions and molecules across the cell membrane. Other than chlorophyll-containing cells, which derive energy from light, all cells obtain this free energy ultimately from the oxidation of exogenous nutrients. The energy of such an oxidation is not used directly but, by means of a complex and long sequence of reactions, is channelled into a special energy-storing molecule, **adenosine triphosphate (ATP).** This molecule contains several phosphoester linkages and is able to release energy when needed by hydrolysis of these phosphoester bonds. The structure of ATP is shown in Figure 2-1. ATP is often said to contain high-energy-phosphate bonds and the notation $\sim$ is used to denote the particular bonds that possess the high energy. This is actually a misnomer because the bond energy (i.e., the energy required to break the bond) is not significantly higher than that of chemical bonds in other molecules; rather, what constitutes high energy is the amount of energy released by a reaction with water in which the bond is hydrolyzed and a water molecule is split. Thus, *ATP has a high free energy of hydrolysis.*

The principal feature of the structure of ATP that gives rise to the large negative free energy of hydrolysis is the physical arrangement of the four negatively charged oxygen atoms, as shown in the figure. These charges repel one another very strongly because they are very close. This electrostatic repulsion is reduced when ATP is hydrolyzed; therefore, hydrolysis releases rather than consumes free energy. This released energy is utilized by means of phosphate-transfer reactions, as will be seen shortly.

ATP can be hydrolyzed in two ways: to form adenosine diphosphate (ADP) and inorganic phosphate (P_i) or to yield adenosine monophosphate (AMP) and pyrophosphate (PP_i). The reactions are

$$\text{I.} \qquad ATP + H_2O \rightarrow ADP + P_i + H^+$$

or

$$\text{II.} \qquad ATP + H_2O \rightarrow AMP + PP_i + H^+$$
$$PP_i \rightarrow 2P_i$$

Each of these reactions is enzymatically catalyzed.

Synthesis of ATP

ATP is synthesized primarily from ADP. A cell expends a great deal of work in synthesizing ATP, in order that an adequate supply of free energy will always be available. For example, about 25 distinct enzymatic reactions are coupled to degrade glucose to CO_2 and H_2O, and to utilize as much of the bond energy of glucose as possible. The reaction sequence is a clear example showing why cells have evolved with multistep pathways. That is, the free energy change for complete oxidation of glucose is 688 kilocalories (kcal)* per mole, while the free energy required for synthesis of ATP is about 7 kcal per mole. Hence, one glucose molecule can provide enough free energy to make 100 ATP molecules. However, it is not possible to design a single chemical reaction in which one molecule of glucose can simultaneously react with oxygen, 100 P_i, and 100 ADP molecules, water, and CO_2. On the other hand, by carrying out the oxidative degradation stepwise, single ATP molecules can be generated in particular steps; the result is the formation of 38 ATP molecules per glucose molecule, or 38 percent efficiency of energy conversion, which thermodynamically is very efficient.

*A calorie is a measure of energy, defined as the energy required to raise the temperature of 1 gram of water from 15° to 16°C; it is also equal to 4.184 joule.

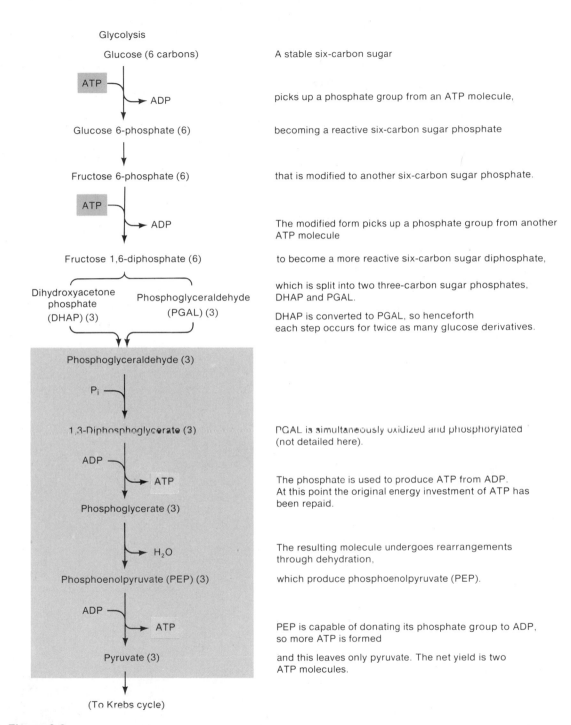

Glycolysis

Glucose (6 carbons) — A stable six-carbon sugar

ATP → ADP — picks up a phosphate group from an ATP molecule,

Glucose 6-phosphate (6) — becoming a reactive six-carbon sugar phosphate

Fructose 6-phosphate (6) — that is modified to another six-carbon sugar phosphate.

ATP → ADP — The modified form picks up a phosphate group from another ATP molecule

Fructose 1,6-diphosphate (6) — to become a more reactive six-carbon sugar diphosphate,

Dihydroxyacetone phosphate (DHAP) (3)   Phosphoglyceraldehyde (PGAL) (3) — which is split into two three-carbon sugar phosphates, DHAP and PGAL.

DHAP is converted to PGAL, so henceforth each step occurs for twice as many glucose derivatives.

Phosphoglyceraldehyde (3)

P$_i$

1,3-Diphosphoglycerate (3) — PGAL is simultaneously oxidized and phosphorylated (not detailed here).

ADP → ATP — The phosphate is used to produce ATP from ADP. At this point the original energy investment of ATP has been repaid.

Phosphoglycerate (3)

→ H$_2$O — The resulting molecule undergoes rearrangements through dehydration,

Phosphoenolpyruvate (PEP) (3) — which produce phosphoenolpyruvate (PEP).

ADP → ATP — PEP is capable of donating its phosphate group to ADP, so more ATP is formed

Pyruvate (3) — and this leaves only pyruvate. The net yield is two ATP molecules.

(To Krebs cycle)

Figure 2-2
An outline of the glycolysis pathway (left) and an explanation (right). Each step is catalyzed by a separate enzyme. All steps in the shaded area occur twice for each glucose molecule—that is, for each of the three-carbon molecules, DHAP and PGAL. Two ATP molecules shaded in red are used and those shaded in black are produced in the pathway. The number of carbon atoms in each molecule is shown in the parentheses. (After C. Starr and R. Taggart. 1981. *Biology.* Wadsworth.)

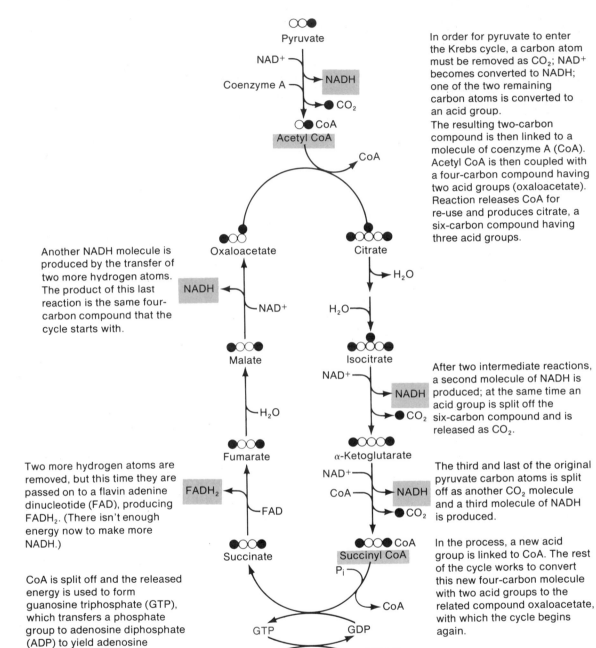

In order for pyruvate to enter the Krebs cycle, a carbon atom must be removed as CO_2; NAD^+ becomes converted to NADH; one of the two remaining carbon atoms is converted to an acid group.

The resulting two-carbon compound is then linked to a molecule of coenzyme A (CoA). Acetyl CoA is then coupled with a four-carbon compound having two acid groups (oxaloacetate). Reaction releases CoA for re-use and produces citrate, a six-carbon compound having three acid groups.

Another NADH molecule is produced by the transfer of two more hydrogen atoms. The product of this last reaction is the same four-carbon compound that the cycle starts with.

After two intermediate reactions, a second molecule of NADH is produced; at the same time an acid group is split off the six-carbon compound and is released as CO_2.

Two more hydrogen atoms are removed, but this time they are passed on to a flavin adenine dinucleotide (FAD), producing $FADH_2$. (There isn't enough energy now to make more NADH.)

The third and last of the original pyruvate carbon atoms is split off as another CO_2 molecule and a third molecule of NADH is produced.

CoA is split off and the released energy is used to form guanosine triphosphate (GTP), which transfers a phosphate group to adenosine diphosphate (ADP) to yield adenosine triphosphate (ATP).

In the process, a new acid group is linked to CoA. The rest of the cycle works to convert this new four-carbon molecule with two acid groups to the related compound oxaloacetate, with which the cycle begins again.

Figure 2-3

The Krebs cycle. Each open circle represents a carbon atom in the molecule and each solid circle represents an acid group (COOH) or a carbon dioxide molecule produced when an acid group is split off. Energy-rich molecules are solid red. Molecules whose oxidation yields ATP are shaded red. The NADH molecules are ultimately oxidized, producing most of the utilizable energy of the cell. Remember that for each glucose molecule being metabolized, two pyruvate molecules have been formed. Thus, two turns of the Krebs cycle must occur for each molecule of glucose oxidized. (After C. Starr and R. Taggart. 1981. *Biology*. Wadsworth.)

Most ATP molecules are not actually direct products of the glucose degradation pathway. Glucose is degraded in two stages. The first stage, **glycolysis,** is a multistep sequence in which two ATP molecules are generated directly in separate steps (Figure 2-2). The remaining product of glycolysis is pyruvate, which is oxidized (though not by molecular O_2) in a pathway known as the **Krebs cycle,** the second stage of glucose degradation (Figure 2-3). At no step in the Krebs cycle is an ATP molecule formed directly. Instead, two molecules of a related compound, guanosine triphosphate (GTP), are formed: each GTP can then generate ATP by the reaction

$$GTP + ADP \rightleftharpoons GDP + ATP$$

Figure 2-3 also shows that the substances NADH (reduced nicotine adenine dinucleotide) and $FADH_2$ (reduced flavin adenine dinucleotide) are produced. These molecules contain the unused free energy of glucose remaining after the two pyruvates (formed as a result of glycolysis) have passed through the Krebs cycle. In order that their unused free energy may be extracted, the NADH and $FADH_2$ molecules are oxidized in a complex and incompletely understood series of reactions, collectively called **oxidative phosphorylation,** yielding 32 additional ATP molecules. Thus, the complete conversion of glucose to CO_2 and H_2O, which is summarized in Figure 2-4, yields 38 ATP molecules.

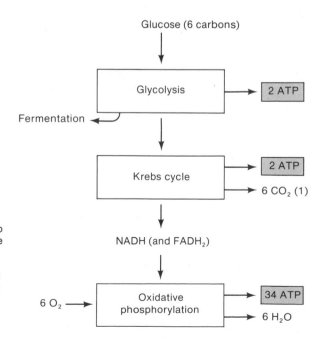

Figure 2-4
Summary of glucose metabolism. The overall oxidative reaction yields 38 ATP molecules. In the absence of oxygen, fermentation occurs, yielding either lactic acid or ethanol, and no ATP molecules form other than the two produced by glycolysis. Note that in the oxidative sequence, 34 of the 38 ATP molecules are produced in the oxidative phosphorylation step so that glycolysis and the Krebs cycle may be thought of as a means of converting glucose to NADH, rather than as a direct source of ATP.

Utilization of the Energy of ATP

The mechanism by which ATP makes its energy available to a cell is the **phosphate-transfer reaction,** which is a special case of a general process known as the **group-transfer reaction.** A group-transfer reaction is one in which molecules exchange functional groups; for example,

$$AX + BY \rightleftarrows AY + BX$$

The rationale for using a group-transfer reaction can be seen in the following example.

Consider the hypothetical synthetic reaction $M + N \rightarrow Q$, in which M and N are reactants and Q is a product. Let us assume also that at the concentrations of M and N in the cell where the reaction proceeds, the value of ΔG for this reaction is 3 kcal per mole. (For reasons that can be found in a physical chemistry textbook, ΔG for a chemical reaction becomes greater if it is negative, or smaller if it is positive, as the concentration of the reactants increases.) Thus we know that the reaction will not occur spontaneously. Let us now consider coupling the reaction to the hydrolysis of ATP, for which ΔG is about -13 kcal per mole at the ATP concentration that exists within a cell. If these two reactions could be joined together in some way, the combined value of ΔG would be $+3 - 13 = -10$ kcal per mole, which is less than zero (i.e., there would be excess free energy); then the reaction would occur spontaneously. Obviously, simply cleaving ATP without channeling the free energy into the synthetic reaction is an insufficient beginning because the energy released would appear as the kinetic energy of the products (ADP, P_i, and H^+) and would merely be dissipated as heat. However, another reaction,

$$M + ATP \rightarrow M\text{—}P + ADP$$

might be possible for which $\Delta G = -5$ kcal per mole. This reaction would happen spontaneously since $\Delta G < 0$. It may be the case then, that M—P can react with N to yield the product Q in the reaction

$$M\text{—}P + N \rightarrow Q + P_i$$

for which $\Delta G = -3$ kcal per mole. Since again $\Delta G < 0$, this reaction would also occur spontaneously. Note that it is the synthesis of M—P, which occurs by transfer of phosphate from ATP to M, that allows the reaction sequence, for which ΔG is $-5 - 3 = -8$ kcal per mole, to proceed. An intermediate substance of this sort (that is, M—P) is called an **activated intermediate.**

This type of mechanism, in which a phosphate or AMP or ADP is transferred to a reactant to form an activated intermediate, occurs repeatedly in biosynthetic reactions such as protein synthesis and in processes such as muscle contraction, in which mechanical work is done. We shall see several examples of this in this book. For example, in the first step of catabolism of galactose, an intermediate, galactose 1-phosphate, is synthesized by transfer of a phosphate from ATP. In protein synthesis, AMP rather than a single phosphate group is transferred to the carboxyl end of one amino acid prior to joining it to the amino group of an adjacent amino acid:

$$\underset{\text{Glycine}}{H_3N^+ - \overset{\overset{H}{|}}{\underset{\underset{H}{|}}{C}} - C\overset{\diagup O}{\diagdown O^-}} + ATP \longrightarrow \underset{\text{"Activated" glycine}}{H_3N^+ - \overset{\overset{H}{|}}{\underset{\underset{H}{|}}{C}} - C\overset{\diagup O}{\diagdown O \sim AMP}} + PP_i$$

$$\underset{\text{"Activated" glycine}}{H_3N^+ - \overset{\overset{H}{|}}{\underset{\underset{H}{|}}{C}} - C\overset{\diagup O}{\diagdown O \sim AMP}} + \underset{\text{Alanine}}{H_3N^+ - \overset{\overset{H}{|}}{\underset{\underset{CH_3}{|}}{C}} - C\overset{\diagup O}{\diagdown O^-}} \longrightarrow \underset{\text{Glycylalanine}}{H_3N^+ - \overset{\overset{H}{|}}{\underset{\underset{H}{|}}{C}} - \overset{\overset{O}{\|}}{C} - N - \overset{\overset{H}{|}}{\underset{\underset{CH_3}{|}}{C}} - C\overset{\diagup O}{\diagdown O^-}} + AMP + H_2O$$

In the synthesis of DNA and of RNA, each of the four mononucleotide precursors is converted to a nucleoside triphosphate by two separate phosphate transfers from two ATP molecules. The nucleoside triphosphate, which is the activated intermediate, can then react with the growing end of the nucleic acid chain, as will be seen in Chapter 8.

Elucidation of Metabolic Pathways

One of the goals of biochemistry is to identify the steps in metabolic pathways and the enzymes that catalyze these steps. This can be done in several ways. Three of the more important procedures are tracer analysis, blockage of particular steps by metabolic inhibitors, and blockage of steps by mutation. These procedures, which will be encountered several times in this book, can be examined by referring to a linear pathway in which a molecule A is converted to E in four reactions catalyzed by the enzymes 1, 2, 3, and 4.

$$A \overset{1}{\longrightarrow} B \overset{2}{\longrightarrow} C \overset{3}{\longrightarrow} D \overset{4}{\longrightarrow} E$$

In tracer analysis, a radioactive substance is added to the growth medium or to a cell extract and the appearance of other radioactive substances is monitored. For example, if A* (the asterisk indicates that a substance is a radiochemical) is added and the reaction is terminated at various time intervals, the substances B*, C*, D*, and E* will be detected successively in time. Alternatively, if the reaction is allowed to reach equilibrium, it will usually be found that [E*] > [D*] > [C*] > [B*] > [A*], the [ ] denoting concentration. If instead of A*, C* is added, then D* and E* will be found and since many enzymatic reactions are reversible, small amounts of A* and B* may also be detected. In this case, it will typically be found that [E*] > [D*] > [C*] > [B*] > [A*].

A metabolic inhibitor I of the synthesis of E would be used in the following way. Suppose I inhibits step 3 but that this is not known by the researcher. Although addition of A* or B* would lead to accumulation of C*, no D* would be formed; thus one would know that products D and E come after C in the pathway. Since C* accumulates, then C must be further along than A or B. Furthermore, if addition of D* yields E* but no A*, B*, or C*, then D → E must be the last step and C → D (the inhibited step) must be the next-to-last step.

The mutant-block procedure is in principle the same as the metabolic inhibitor procedures, except that a particular enzyme activity is eliminated by a mutation in the gene specifying the enzyme. The mutant-block procedure is a standard technique of molecular biology and will be encountered repeatedly in this book.

A common problem in studying any metabolic reaction by the addition of radioactive substances is that the added substance becomes diluted by the intracellular material. In the reaction sequence just described, some A, B, C, D, and E are present at all times. These intracellular substances are called the **metabolic pool.** If the internal concentration of B is very high and only a very small fraction of B is converted to C, then, if B* is added, the radioactivity is immediately diluted by the pool of B, so that only a fraction of B* might be converted to C*. If the amount of C* is too small to be detected, the B → C reaction might be unnoticed. This is a major problem when studying animal cells, in which the metabolic pools are sometimes very large. When this occurs, it is necessary to increase the number of cells used in the experiment or the amount of added radioactive substance, in order to obtain a measurable amount of the radioactive product.

REFERENCES

Davis, B. D., R. Dulbecco, H. N. Eisen, H. S. Ginsberg, and W. B. Wood. 1973. *Microbiology.* Harper & Row.

Frieden, E. 1972. "The chemical elements of life." *Scient. Amer.* 227 (July), 52–64.

Hinkle, P. C., and R. E. McCarty. 1978. "How cells make ATP." *Scient. Amer.* 238 (March), 104–123.

Keeton, W. 1980. *Biological Science.* W. W. Norton.

Lehninger, A. L. 1971. *Bioenergetics.* 2nd Ed. Benjamin-Cummings.

Lehninger, A. L. 1975. *Biochemistry.* Worth.

Moat, A. G. 1979. *Microbial Physiology.* Wiley.

Puck, T. T. 1972. *The Mammalian Cell as a Microorganism.* Holden-Day.

Sheeler, P., and D. E. Bianchi. 1980. *Cell Biology.* Wiley.

Speakman, J. C. 1966. *Molecules.* McGraw-Hill.

Stryer, L. 1981. *Biochemistry.* 2nd Ed. W. H. Freeman and Co.

Tooze, J. (ed.) 1973. *The Molecular Biology of Tumour Viruses.* Cold Spring Harbor Laboratory. (This book has an excellent discussion of growth media and cell culture.)

Watson, J. D. 1976. *Molecular Biology of the Gene.* Benjamin-Cummings.

3 Macromolecules

A typical cell contains 10^4–10^5 different kinds of molecules. Roughly half of these are small molecules—namely, inorganic ions and organic compounds whose molecular weights usually do not exceed several hundred. The others are polymers that are so massive (molecular weights from 10^4 to 10^{12}) that they are called **macromolecules.** These molecules are of three major classes: proteins, nucleic acids, and polysaccharides are polymers of amino acids, nucleotides, and sugars, respectively. There are also subclasses of macromolecules, inasmuch as macromolecules can be modified in various ways—for example, glycoproteins (proteins carrying sugar groups), lipoproteins (proteins carrying lipids or fats), lipopolysaccharides (polysaccharides bearing lipids), and glucosylated nucleic acids (nucleic acids bearing sugar groups on the bases).

Macromolecules perform a variety of functions. For example, nucleic acids store and carry information, polysaccharides provide an energy reserve and also constitute the external cell wall of plants and many microorganisms, and lipoproteins enclose animal cells and are responsible for determining the substances that can pass in and out of a cell. The most versatile macromolecules—the proteins—catalyze chemical reactions, regulate the flow of information, and are major components of many structural units.

Knowledge of the properties of macromolecules is essential for understanding living processes. This and the next two chapters will emphasize nucleic acids and proteins. Information about other types of macromolecules will be given as needed in other chapters, but for details, the references at the end of this chapter should be consulted.

CHEMICAL STRUCTURES OF THE MAJOR CLASSES OF MACROMOLECULES

In this unit we examine the chemical structure of proteins, nucleic acids, and polysaccharides—in particular, the monomers and the chemical linkages by which they are joined. Physical properties of these macromolecules and the interactions that exist between various parts of a single macromolecule are described in later units.

Proteins

A protein is a polymer consisting of several amino acids (a **polypeptide**). Each amino acid can be thought of as a single carbon atom (the α carbon) to which there is attached one carboxyl group, one amino group, and a side chain denoted R (Figure 3-1). The side chains are generally carbon chains or rings to which various functional groups may be attached. The simplest side chains are those of glycine (a hydrogen atom) and alanine (a methyl group). The complete chemical structures of each amino acid are given in Figure 3-2. Note that proline differs from the basic structure shown in Figure 3-1 in that the N atom is included in a ring. Properly speaking, proline is an *imino* acid. Because of its structure, proline introduces a kink into a polypeptide chain with a consequent dramatic effect on the three-dimensional structure of a protein, in the vicinity of a proline.

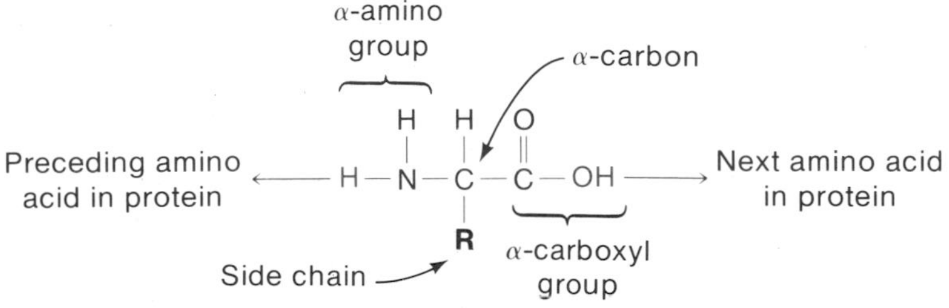

Figure 3-1
Basic structure of an α-amino acid. The NH_2 and COOH groups are used to connect amino acids to one another. The red OH of one amino acid and the red H of the next amino acid are removed when two amino acids are linked together (see Figure 3-3).

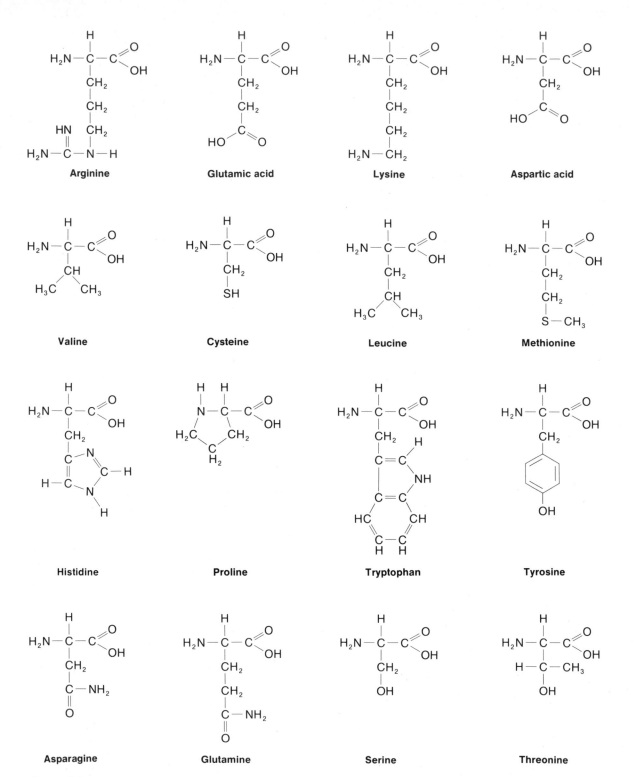

Figure 3-2
Chemical structures of the amino acids.

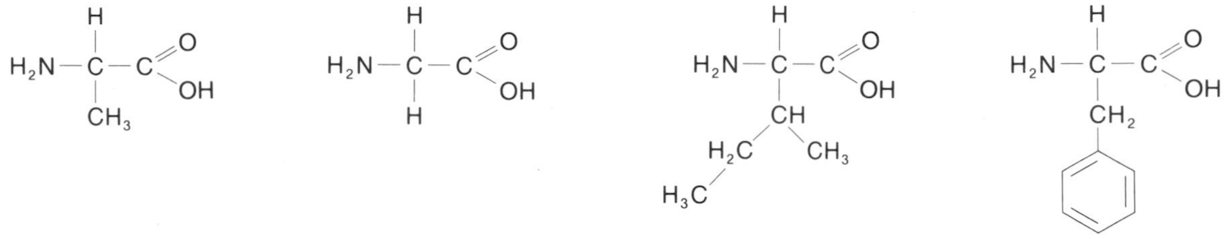

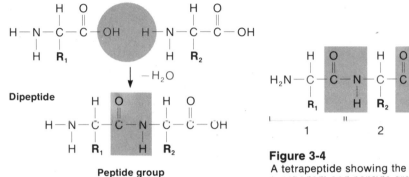

Alanine   **Glycine**   **Isoleucine**   **Phenylalanine**

Figure 3-2 (continued)

Dipeptide

Peptide group

Figure 3-3
Formation of a dipeptide from two amino acids by elimination of water (shaded circle) to form a peptide group (shaded rectangle).

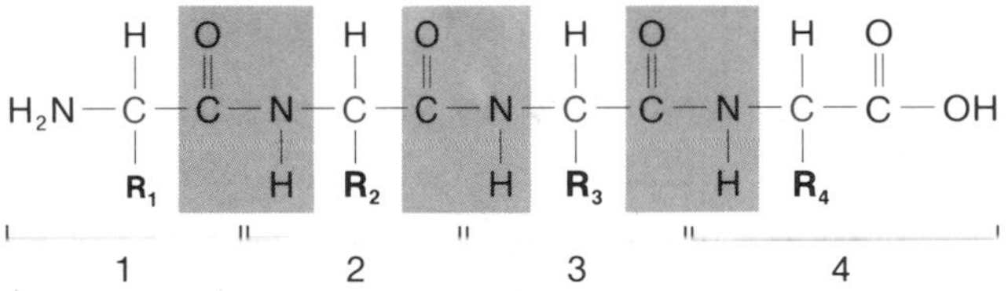

Figure 3-4
A tetrapeptide showing the alternation of α-carbon atoms (unshaded) and peptide groups (shaded). The four amino acids are numbered below.

To form a protein the amino group of one amino acid reacts with the carboxyl group of another; the resulting chemical bond is called a **peptide bond** (Figure 3-3). Amino acids are joined together in succession to form a linear **polypeptide chain** (Figure 3-4). When the number of peptide bonds exceeds about 15 (the number is arbitrary), the polypeptide is called a **protein.** (A protein may also be an aggregate of several polypeptide chains, as will be seen later.) Thus, a protein is a polymer in which α-carbon atoms alternate with peptide units to form a linear chain having an ordered array of side chains. This linear chain is called the **backbone** of the molecule.

The two ends of every protein molecule are distinct. One end has a free —NH_2 group and is called the **amino terminus**; the other end has a free —COOH group and is the **carboxyl terminus.** The ends are also called the N (or NH_2) and C termini, respectively.

The amino-acid side chains usually do not engage in covalent bond formation; an exception is the —SH group of cysteine, which often forms a disulfide (S—S) bond by reaction with a second cysteine in the same or different polypeptide chain. This is important in determining the three-dimensional structure of a protein, as will be examined in Chapter 5.

Many proteins contain metal ions engaged in coordination complexes with groups in the side chains. Common metal ions are Zn^{2+}, Cu^{2+}, and Fe^{3+}, which are often bound to histidine and glutamic acid.

Nucleic Acids

A nucleic acid is a polynucleotide—that is, a polymer consisting of nucleotides. Each nucleotide has the three following components (Figure 3-5):

1. A cyclic five-carbon sugar. This is ribose, in the case of ribonucleic acid (RNA), and deoxyribose, in deoxyribonucleic acid (DNA). The difference in the structure of ribose and 2'-deoxyribose is shown in Figure 3-5. Note that they differ only in the absence of a 2'-OH group in deoxyribose, a difference that makes DNA chemically more stable than RNA, as will be seen later.
2. A purine or pyrimidine base attached to the 1'-carbon atom of the sugar by an N-glycosylic bond. The bases, which are shown in Figure 3-6, are the purines, adenine (A) and guanine (G), and the pyrimidines, cytosine (C), thymine (T), and uracil (U). DNA and RNA contain A, G, and C; however, T is found only in DNA and U is found only in RNA. There are rare exceptions to this rule—namely, T is present in some tRNA molecules and there are a few phages whose DNA exclusively contains U rather than T.
3. A phosphate attached to the 5' carbon of the sugar by a phosphoester linkage. This phosphate is responsible for the strong negative charge of both nucleotides and nucleic acids.

A base linked to a sugar is called a **nucleoside**; thus *a nucleotide is a nucleoside phosphate*. The terminology used to describe nucleic acid components is listed in Table 3-1.

The nucleotides in nucleic acids are covalently linked by a second phosphoester bond that joins the 5'-phosphate of one nucleotide and the 3'-OH group of the adjacent nucleotide (Figure 3-7). Thus the phosphate is esterified to both the 3'- and 5'-carbon atoms; this unit is often called a **phosphodiester group.**

The purine and pyrimidine bases are not engaged in any covalent bonds to each other. Thus, a polynucleotide consists of an alternating sugar-phosphate backbone having one 3'-OH terminus and one 5'-phosphate (5'-P) terminus. In the laboratory, polynucleotides can be prepared that have 3'-P and 5'-OH termini, but such molecules do not occur naturally.

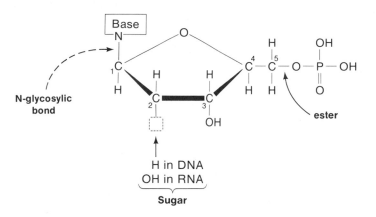

Figure 3-5
Structure of a mononucleotide. The carbon atoms in the sugar ring are numbered.

N-glycosylic bond

ester

H in DNA
OH in RNA

Sugar

Adenine

Guanine

Figure 3-6
The bases found in nucleic acids. The weakly charged groups are shown in red.

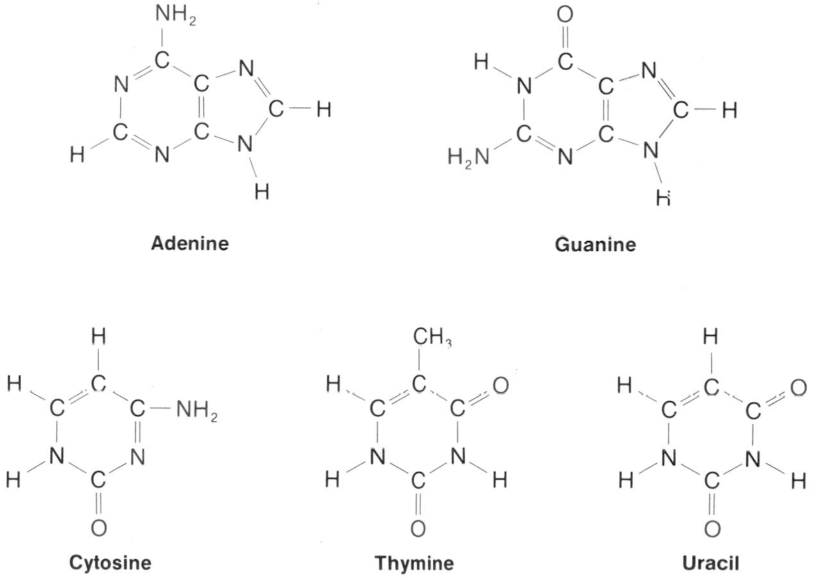

Cytosine

Thymine

Uracil

Figure 3-7
The structure of a dinucleotide. The vertical arrows show the bonds in the phosphodiester group about which there is free rotation. The horizontal arrows indicate the N-glycosylic bond about which the base can freely rotate. A polynucleotide would consist of many nucleotides linked together by phosphodiester bonds.

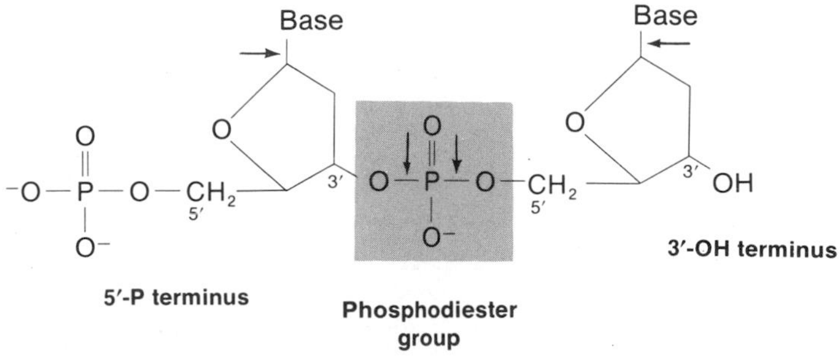

Base

Base

5'-P terminus

Phosphodiester group

3'-OH terminus

A typical RNA molecule is the single-stranded polyribonucleotide that has just been described. However, except in unusual cases, DNA contains two polydeoxynucleotide strands wrapped around one another to form a double-stranded helix. This remarkable structure will be described in the following chapter.

Table 3-1
Nucleic acid nomenclature

Base	Nucleoside[1]	Nucleotide[2]
Purines (Pu)		
Adenine (A)	Adenosine (rA)	Adenylic acid, or adenosine monophosphate (AMP)
	Deoxyadenosine (dA)	Deoxyadenylic acid, or deoxyadenosine monophosphate (dAMP)
Guanine (G)	Guanosine[3] (rG)	Guanylic acid, or guanosine monophosphate (GMP)
	Deoxyguanosine (dG)	Deoxyguanylic acid, or deoxyguanosine monophosphate (dGMP)
Pyrimidines (Py)		
Cytosine (C)	Cytidine (rC)	Cytidylic acid, or cytidine monophosphate (CMP)
	Deoxycytidine (dC)	Deoxycytidylic acid, or deoxycytidine monophosphate (dCMP)
Thymine (T)	Thymidine[4] (dT)	Thymidylic acid, or thymidine monophosphate (TMP)
Uracil (U)	Uridine[5] (rU)	Uridylic acid, or uridine monophosphate (UMP)

[1]Note that the names of purine nucleosides end in -osine and the names of pyrimidine nucleosides end in -idine.

[2]Note that each nucleotide has two names for the same substance.

[3]Guanosine should not be confused with guanidine, which is not a nucleic acid base.

[4]Thymidine is the deoxy- form. The ribo- form, ribosylthymine, is not generally found in nucleic acids.

[5]Uridine is the ribo- form. Deoxyuridine is not commonly found, although deoxyuridylic acid is on the pathway for synthesis of thymidylic acid—i.e., deoxyuridylic acid is methylated to yield thymidylic acid.

Polysaccharides

Polysaccharides are polymers of sugars (most often glucose) or sugar derivatives. These are very complex molecules because sometimes covalent bonds occur between many pairs of carbon atoms. This has the effect that one sugar unit can be joined to more than two other sugars, which results in the formation of highly branched macromolecules. These branched structures are sometimes so enormous that they are almost macroscopic. For example, the cell walls of many bacteria and plant cells are single gigantic polysaccharide molecules (as is described in Chapter 6).

NONCOVALENT INTERACTIONS THAT DETERMINE THE THREE-DIMENSIONAL STRUCTURES OF PROTEINS AND NUCLEIC ACIDS

The biological properties of macromolecules are mainly determined by noncovalent interactions that result in each molecule acquiring a unique three-dimensional structure. In this section the noncovalent interactions that are important and the chemical groups responsible for them are described.

The Random Coil

Linear polypeptide and polynucleotide chains contain several bonds about which there is free rotation. In the absence of any intrastrand interactions* each monomer would be free to rotate with respect to its adjacent monomers; this is limited only by the fact that atoms cannot occupy the same space. The three-dimensional configuration of such a chain is called a **random coil**; it is a somewhat compact and globular structure that *changes shape continually* owing to constant bombardment by solvent molecules.

Few, if any, nucleic acid or protein molecules exist in nature as random coils because there are many interactions between elements of the chains. These interactions are hydrogen bonds, hydrophobic interactions, ionic bonds, and van der Waals interactions. For example, the bases of nucleic acids attract one another by means of both hydrogen bonds and hydrophobic interactions; and amino-acid side chains have attractive interactions of all four types and can repel one another if they have charges of the same sign.

*An intrastrand interaction is an interaction between two regions of the same strand. An interstrand interaction is between different strands. It is important to remember the distinction between these similar words.

The properties and consequences of these interactions are described in the following sections.

Before discussing these interactions and the configurations they produce, it is essential to realize that the interactions and their consequences take place in an aqueous environment, unless otherwise noted.

Hydrogen-bonding

The most common hydrogen bonds found in biological systems are shown in Figure 3-8. In nucleic acids, hydrogen-bonding causes intrastrand pairing between nucleotide bases, which, if it were random, would cause a single polynucleotide strand to be more compact than a random coil. In DNA, interstrand hydrogen bonds are responsible for the double-stranded helical structure. In proteins, intrastrand hydrogen-bonding occurs between a hydrogen atom on a nitrogen adjacent to one peptide bond and an oxygen atom adjacent to a different peptide bond. This interaction gives rise to several particular polypeptide chain configurations, which will be discussed shortly.

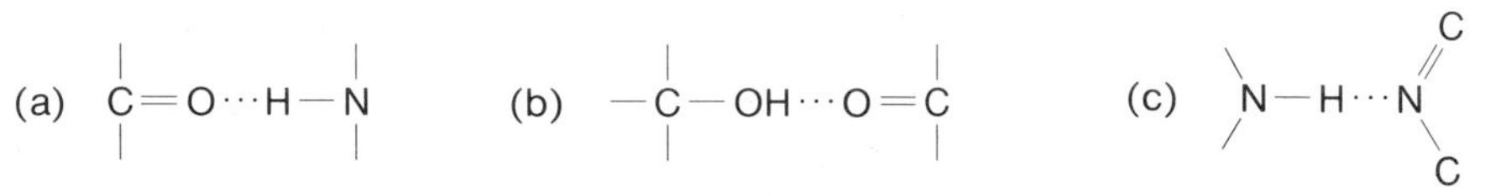

(a) $C=O \cdots H-N$ (b) $-C-OH \cdots O=C$ (c) $N-H \cdots N$

Figure 3-8
Structures of three types of hydrogen bonds (indicated by three dots). (a) A type found in proteins and nucleic acids. (b) A weak bond found in proteins. (c) A type found in DNA and RNA.

The Hydrophobic Interaction

A hydrophobic interaction is an interaction between two molecules (or portions of molecules) that are somewhat insoluble in water. The phenomenon is simply that *two molecules (which may be different) that are poorly soluble in water tend to associate*. The explanation is the following. Water molecules form hydrogen bonds with one another and a molecule is soluble if it can hydrogen-bond with water. An insoluble molecule in water causes (for complicated and poorly understood reasons) a shell of hydrogen-bonded water molecules to form around it. This ordering of water is equivalent thermodynamically to a decrease in entropy*, which is not a probable situation, as is explained by the

---

*The entropy of a system or of a process is a thermodynamic quantity whose value increases as the degree of *disorder* increases. The entropy of the universe *increases* in all processes.

second law of thermodynamics. Two separate insoluble molecules would form two such shells. If the two molecules were in contact, a single shell would surround the pair and, for strictly geometric reasons, the shell would have fewer than twice the number of ordered water molecules that would surround a single molecule of the solute. In general, the number of ordered water molecules per poorly soluble solute molecule is always smaller if the solute molecules are in contact; hence, *clustering of weakly soluble molecules is thermodynamically favored.* This tendency to cluster is called a **hydrophobic interaction.** Note that no bonds are formed, it is only that a cluster is the more probable arrangement, or, in other words, the disorder (entropy) of *all* molecules, both solvent and solute, is greater.

Many components in nucleic acids and proteins have the hydrophobic property just described. For example, the bases of nucleic acids are planar organic rings carrying localized weak charges (Figure 3-6). The localized charges are sufficient to maintain solubility, but the large, poorly soluble organic ring portion causes the ordering of water just described, so that the bases tend to cluster. The most efficient kind of clustering is one in which the faces of the rings are in contact, an array known as **base-stacking** (Figure 3-9). Note that since the bases are adjacent in the chain, stacking gives some rigidity to single polynucleotide strands and tends to extend them rather than allow a random coil to form. In the following chapter we will see how stacking is of major importance in determining nucleic acid structure. Many amino-acid side chains are very poorly soluble, and this causes (1) the benzene ring of phenylalanine to stack and (2) the hydrocarbon chains of alanine, leucine, isoleucine, and valine to form tight unstacked clusters. Since these amino acids are not necessarily adjacent, *hydrophobic interactions tend to bring distant hydrophobic parts of a polypeptide chain together.*

Figure 3-9
The meaning of stacking.
(a) Three stacked bases.
(b) Structures of stacked and unstacked polynucleotides. The stacked polynucleotide is more extended because the stacking tends to decrease the flexibility of the molecule.

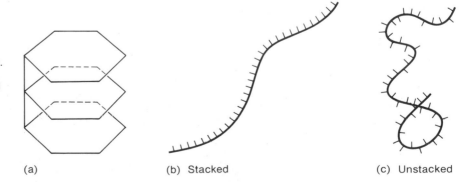

(a)   (b) Stacked   (c) Unstacked

Ionic Bonds

An ionic bond is the result of attraction between unlike charges. Several amino-acid side chains have ionized, negatively charged carboxyl groups (aspartic and glutamic acids) and positively charged amino groups (lysine, histidine, and arginine) that can form such bonds. This tends to bring together distant parts of the chain. Ionic interactions can also be repulsive—namely, between two like charges; thus, it would be unlikely for a polypeptide chain to fold in such a way that two aspartic acids are very near one another. Ionic bonds are the strongest of the noncovalent interactions. However, they are destroyed by extremes of pH, which can change the charge of the groups, and by high concentrations of salt, the ions of which shield the charged groups from one another.

Van der Waals Attraction

Van der Waals forces exist between all molecules and are a result of both permanent dipoles and the circulation of electrons. The dependence of the attractive force between two atoms is proportional to $1/r^6$, in which r is the distance between their nuclei. Thus, the attraction is a very weak force and is significant only if two atoms are very near one another (1 to 2 Å apart). A powerful repulsive force also comes into play if the two outer electron shells overlap. The **van der Waals radius** is defined as the distance at which the attractive and repulsive forces balance precisely. The van der Waals radii of atoms differ from one another; some representative values are shown in Table 3-2.

The shape of a molecule is in essence the surface formed by the van der Waals spheres of each atom. Figure 3-10 shows the shapes of two

Table 3-2
Van der Waals radii of several atoms found in biological molecules

Atom	Radius, Å
H	1.2
O	1.4
N	1.5
S	1.85
P	1.9
C	2.0

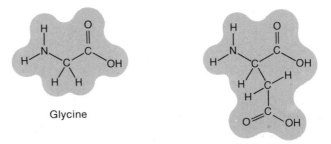

Glycine

Aspartic acid

Figure 3-10
The shapes of two amino acids, glycine and aspartic acid, determined from van der Waals radii. This is only a schematic drawing because the molecules are three-dimensional and all bonds are not in the plane of the paper.

molecules when defined in this way. The interaction energy of two atoms separated by the sum of their van der Waals radii is about 1 kcal/mole. Since the average energy of thermal motion at room temperature is about 0.6 kcal/mole and, according to the Boltzmann distribution (a mathematical expression that relates the mean energy of a system of particles and the fraction of the total number of particles having each energy value), the energy of many molecules will exceed this value (and even exceed 1 kcal/mole), the van der Waals interaction between two atoms is not sufficient to maintain these atoms in proximity. However, if the interaction of *several* pairs of atoms are combined, the cumulative attractive force can be great enough to withstand being disrupted by thermal motion. Thus, two molecules can attract one another if several of their component atoms can mutually interact. However, because of the $1/r^6$-dependence, the intermolecular fit must be nearly perfect. What this means is that *two molecules can bind to one another if their shapes are complementary*. This is true also of two separate regions of a polymer—that is, the regions can hold together if their shapes match. Sometimes the van der Waals attraction between two regions is not large enough to effect this; however, it can significantly strengthen other weak interactions such as the hydrophobic interaction, if the fit is good.

Summary of the Effects of Noncovalent Interactions

The effect of noncovalent interactions is to constrain a linear chain to fold in such a way that different regions of the chain, which may be quite distant in a chain (if the chain were linear), are brought together. An example of a molecule showing the four kinds of noncovalent folding that have been described is shown in Figure 3-11.

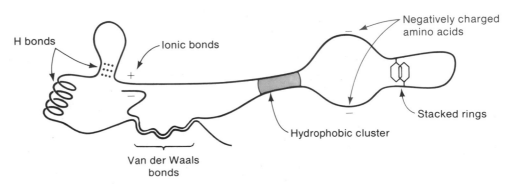

Figure 3-11
A hypothetical polypeptide chain showing attractive (black) and repulsive (red) interactions.

In Chapters 4, 5, and 6 we will discuss the specific structures brought about by these interactions and the physical properties of nucleic acids and proteins. In order to appreciate the kinds of experiments that have provided this information, several methods for studying macromolecules must be understood. These are described in the following section.

METHODS USED TO STUDY MACROMOLECULES

Several methods of studying macromolecules are used repeatedly in molecular biology. Some of these—e.g., spectrophotometry and chromatography—are usually covered in elementary chemistry courses and will not be discussed here. Three other techniques, ultracentrifugation, electrophoresis, and electron microscopy, will be described briefly in this section so that the experiments presented in this book can be understood. For additional details, the reader should consult the references at the end of this chapter.

Velocity Sedimentation

Several important properties of macromolecules can be determined from sedimentation data. These studies cannot be done with an ordinary laboratory centrifuge because this instrument cannot produce a great enough centrifugal force to sediment molecules sufficiently rapidly that their positions are not randomized by diffusion. Modern ultracentrifuges can, however, generate forces as great as 700,000 times gravity, which is more than adequate to cause macromolecules to sediment through a solution.

The velocity with which a molecule moves is mainly a function of two properties of the molecule:

1. *Its molecular weight* M. As M increases, the velocity increases.
2. *Its shape.* The motion of any particle through a fluid is impeded by friction. If a ball and a stick having the same masses are moving through a liquid, the ball, being more compact, will encounter less frictional resistance and, hence, will move faster. This is true of macromolecules also: for a given value of M, the less extended a polymer chain is, the more rapidly it will move.

The ratio of molecular velocity to centrifugal force is called the **sedimentation coefficient, s.** That is,

$$s = \text{velocity/centrifugal force}.$$

The value of s for a particular molecule is often the same in many different solutions, so that an s value is frequently considered to be a constant that characterizes a molecule. Furthermore, since the value of s depends on molecular weight and shape, changes in the s value, as experimental conditions are varied, can be used to monitor changes in molecular weight (such as aggregation or degradation) and in shape (such as unfolding an extended molecule to form a random coil).

For most macromolecules the value of s is between 1×10^{-13} and 100×10^{-13} sec. Thus in honor of The Svedberg, who invented the ultracentrifuge, 10^{-13} seconds is called one **svedberg** or one S. It is very common to refer to a molecule whose s value is 30 svedbergs as a 30S molecule.

The most common type of sedimentation in use today is called **zonal centrifugation.** In this procedure (Figure 3-12), a centrifuge tube is filled with a sucrose solution (occasionally, glycerol or other solutes are

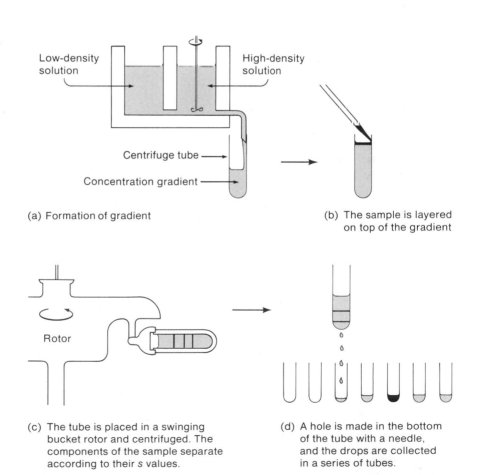

Figure 3-12
Operations in zonal centrifugation. (From D. Freifelder. 1982. *Physical Biochemistry.* 2nd Ed. W. H. Freeman and Co.)

(a) Formation of gradient

(b) The sample is layered on top of the gradient

(c) The tube is placed in a swinging bucket rotor and centrifuged. The components of the sample separate according to their s values.

(d) A hole is made in the bottom of the tube with a needle, and the drops are collected in a series of tubes.

used) whose concentration decreases continuously from the bottom of the centrifuge tube to the top of the tube. Since the density of the solution increases with increasing concentration, the tube contains a solution having a density gradient; the highest density is of course at the bottom of the tube. The density of the solution of molecules to be sedimented is adjusted to be lower than the density of the sucrose solution at the top of the tube, so that the sample can be layered on the surface of the sucrose solution, thus forming a band or zone. Because of the density gradient, this procedure is often called **sucrose gradient centrifugation.** After layering the sample, the tube is centrifuged in a swinging bucket rotor for a particular time. After centrifugation is completed, a tiny hole is punched in the bottom of the the tube and drops of the solution are collected. These drops represent successive layers of the tube. The drops are assayed for a particular macromolecule to obtain the distribution of the concentration of this macromolecule along the tube. One common assay uses a spectrophotometer to determine macromolecular concentration, as long as the macromolecule of interest absorbs visible or ultraviolet light and is present in a detectable amount. Another common assay requires using a radioactive macromolecule; radioactivity, which is proportional to the concentration of the macromolecule, is then measured. Usually, before centrifuging, a molecule whose s value is known is added to the sample as a standard for comparison. The ratio of the distances moved by this standard molecule and the macromolecule of interest is then also the ratio of the s values.

Equilibrium Centrifugation

Another centrifugation technique, **equilibrium centrifugation in a density gradient,** is also widely used. In this procedure, the macromolecules (usually nucleic acids or virus particles) are dissolved in a solution of CsCl whose concentration is chosen so that its density, ρ, is approximately equal to that of the macromolecules. Initially, no sedimentation can occur because the buoyant force of the CsCl solution equals the centrifugal force.*

Under the influence of a powerful centrifugal force, the Cs^+ and the Cl^- ions themselves sediment somewhat, but they do not accumulate on the bottom of the centrifuge tube because the centrifugal force is not great enough to counteract the tendency for diffusion to maintain a uniform distribution of the ions. The result is that after a period of

*To understand this, one should realize that if a centrifuge tube were to contain particles of wood floating on water, no amount of centrifugal force would make the wood reach the tube bottom, because the value of the force on the water, which is denser than the wood, is greater than the value of the force on the wood.

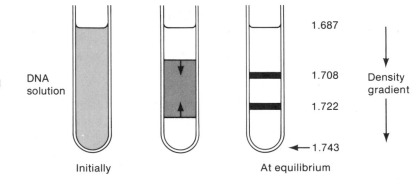

Figure 3-13
Formation of bands in a CsCl density gradient. The DNA solution contains a mixture of two DNA molecules that differ only in that one contains the normal isotope ^{14}N and the other contains the heavy isotope ^{15}N.

several hours the ions achieve an equilibrium concentration distribution in which there is a nearly linear concentration gradient, and hence a nearly linear density gradient, in the centrifuge tube. The density is maximal at the bottom of the tube. Once a density gradient has formed, macromolecules can begin to sediment. Those in the upper reaches of the tube move toward the bottom, stopping at the position at which their density equals the solution density. Similarly, macromolecules in the lower part of the tube move upward to the same position. In this way the macromolecules form a narrow band in the tube. (For this reason, the technique is often called CsCl "banding.") If the solution contains macromolecules having different densities, each macromolecule forms a band at the position in the gradient that matches its own density, and thus the macromolecules can be separated. The resolution of the technique is extraordinary. For example, DNA molecules with a density of 1.708 g/cm³ can be separated from other DNA molecules in which the naturally occurring ^{14}N atoms have been substituted by ^{15}N atoms and which therefore have a density of 1.722 g/cm³, as shown in Figure 3-13. In this book several examples will be given of separations based on isotopic substitution.

Electrophoresis

Most biological macromolecules carry an electrical charge and thus can move in an electric field. For example, if the terminals of a battery are connected to the opposite ends of a horizontal tube containing a solution of positively charged protein molecules, the molecules will move from the positive end of the tube to the negative end. The direction of motion obviously depends on the sign of the charge but the rate of movement depends on the magnitude of the charge and, as in sedimentation, on the shape of the molecule (that is, its frictional resistance). The mass of the molecule plays no direct role in the rate of migration (in contrast with sedimentation) and influences the rate only indirectly when the surface area of the molecule, which affects its frictional coefficient, is a function of its mass.

The most common type of electrophoresis used in molecular biology is zonal electrophoresis through a gel, or **gel electrophoresis.** This procedure can be performed so that the rate of movement depends only on the molecular weight of the molecule, as will be seen below.

An experimental arrangement for gel electrophoresis of DNA is shown in Figure 3-14. A thin slab of an agarose or polyacrylamide gel is prepared containing small slots ("wells") into which samples are placed. An electric field is applied and the negatively charged DNA molecules penetrate and move through the agarose. A gel is a complex network of molecules, and migrating macromolecules must squeeze through narrow, tortuous passages. The result is that smaller molecules pass through more easily and thus the migration rate increases as the molecular weight, M, decreases. For unknown reasons the distance moved, D, depends logarithmically on M, obeying the equation

$$D = a - b \log M,$$

in which a and b are empirically determined constants that depend on the buffer, the gel concentration, and the temperature. Figure 3-15 shows the result of electrophoresis of a collection of DNA molecules.

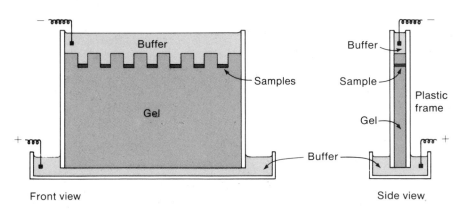

Front view   Side view

Figure 3-14
Apparatus for slab-gel electrophoresis capable of running seven samples simultaneously. The liquid gel is allowed to harden in place. An appropriately shaped mold is placed on top of the gel during the hardening in order to make wells for the samples (red). After electrophoresis, the slab is stained by removing the plastic frame and immersing the gel in a solution of the stain. Excess stain is removed by washing. The components of the sample appear as bands, which may be either visibly colored or fluorescent when irradiated with ultraviolet light. The region of the gel in which the components of one sample can move is called a *lane.* Thus, the gel shown has seven lanes. (After D. Freifelder. 1982. *Physical Biochemistry.* 2nd Ed. W. H. Freeman and Co.)

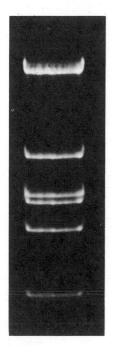

Figure 3-15
A gel electrophoregram of six fragments of *E. coli* phage λ DNA. The direction of movement is from top to bottom. The DNA is made visible by the fluorescence of bound ethidium bromide. (Courtesy of Arthur Landy and Wilma Ross.)

Gel electrophoresis of proteins is in principle carried out the same way. However, proteins can be either positively or negatively charged, and a sample containing several different proteins must be placed in a centrally located well in order that migration can occur in both directions. The charge per unit mass, which is very small because most amino acids are uncharged, varies from one protein to the next (in contrast with DNA or RNA, which have one negative charge per nucleotide) and proteins come in a variety of shapes, so there is no simple way to predict the migration rate. In one important technique the anionic detergent sodium dodecyl sulfate (SDS)

$$H_3C-(CH_2)_{10}-CH_2OSO_3^-Na^+$$

and the disulfide bond-breaking agent mercaptoethanol

$$HO-\overset{\overset{\displaystyle H}{|}}{\underset{\underset{\displaystyle H}{|}}{C}}-\overset{\overset{\displaystyle H}{|}}{\underset{\underset{\displaystyle H}{|}}{C}}-SH$$

are added to the protein solution. One molecule of SDS binds per amino acid, giving each protein molecule the same charge per amino acid residue. Furthermore, *when SDS is bound and there are no disulfide bonds, all proteins have nearly the same shape—namely, the near-random coil.* The net effect is that, as in the case of DNA, the migration rate increases as M decreases and the dependence is logarithmic and described by the equation given above. Because a *polyacrylamide* gel is used for protein electrophoresis, the technique is called the **SDS-PAGE technique;** it is discussed in more detail in a later section.

Electron Microscopy

Macromolecules can be viewed directly by electron microscopy. Two of the many techniques of sample preparation will be described here. Others will be presented, as needed, elsewhere in the book.

The usual method for observing nucleic acids is the **Kleinschmidt spreading technique,** illustrated in Figure 3-16. The nucleic acid molecules, which are considerably larger than any protein molecule, are embedded in a protein film, as shown, and are coated with protein molecules. The protein molecules serve two functions:

1. They cause the nucleic acid molecule to be extended, for the following reason. If a very small volume of a protein solution is layered

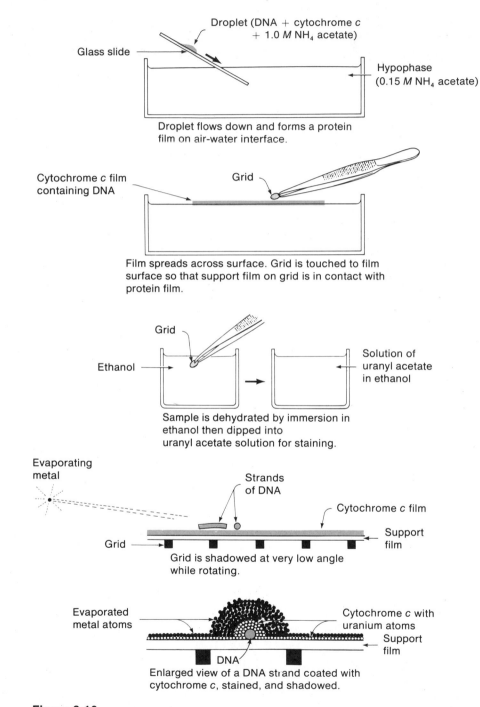

Droplet (DNA + cytochrome *c* + 1.0 *M* NH₄ acetate)

Glass slide

Hypophase (0.15 *M* NH₄ acetate)

Droplet flows down and forms a protein film on air-water interface.

Cytochrome *c* film containing DNA

Grid

Film spreads across surface. Grid is touched to film surface so that support film on grid is in contact with protein film.

Grid

Ethanol

Solution of uranyl acetate in ethanol

Sample is dehydrated by immersion in ethanol then dipped into uranyl acetate solution for staining.

Evaporating metal

Strands of DNA

Cytochrome *c* film

Grid

Support film

Grid is shadowed at very low angle while rotating.

Evaporated metal atoms

Cytochrome *c* with uranium atoms

Support film

DNA

Enlarged view of a DNA strand coated with cytochrome *c*, stained, and shadowed.

Figure 3-16
The Kleinschmidt method for preparing DNA for electron microscopy. (From D. Freifelder. 1982. *Physical Chemistry for Students of Biology and Chemistry.* Science Books International.)

onto the surface of an ionic solution, the protein molecules will spread out to form a monolayer film of protein. If nucleic acid molecules are in the protein solution, they will also spread out as the protein film expands; this spreading enables entire nucleic acid molecules to be seen without the tangling and intramolecular aggregation that is observed if spreading does not occur.

2. Proteins form a thick coating on the nucleic acid molecule and this coating makes the molecule easier to see. Something else is also needed for good observation because molecules containing only small atoms are ordinarily nearly transparent to electrons. Thus, to make a nucleic acid molecule visible, the protein-coated molecule is additionally coated with a thick layer of a heavy metal. This is accomplished either by evaporating platinum or uranium atoms onto the molecule (**shadowing**) or, in the case of uranium, chemically depositing uranium atoms onto the protein coating (**staining**). These heavy atoms are good absorbers of electrons, whereas the background support absorbs relatively few electrons. Thus, the metal ions provide visual contrast, and both shadowing and staining are very useful techniques for observing very long molecules when one is primarily interested in their length and linear topology. An example of an electron micrograph of DNA molecules is shown in Figure 3-17.

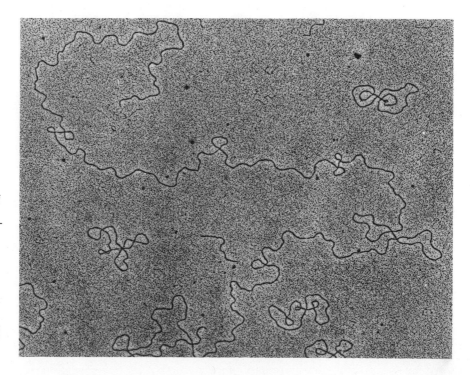

Figure 3-17
Electron micrograph containing a linear *E. coli* phage λ DNA molecule and several smaller circular molecules of *E. coli* phage φX174 RFI DNA. The molecular weight of the φX174 DNA is accurately known, so that the molecular weight of the λ DNA molecule can be measured from the relative lengths of λ DNA and φX174 DNA. (Courtesy of Manuel Valenzuela.)

The **negative contrast technique** provides an alternative way to visualize molecules that cannot readily be seen by the protein mono-layer technique just described and when one is interested in the shape and surface details of the molecule being examined. For example, a protein molecule would not be seen if it were embedded in a protein film. It would also not suffice to coat a single protein molecule with a thick layer of metal atoms because the metal layer would probably be comparable in size to or larger than the protein molecule, and obscure its shape as well. Even with larger particles, such as viruses, whose surface structures are of great interest, a thick metal layer will obscure all surface detail. An alternative is to use a very thin coating of metal atoms. These atoms are sufficiently small that, if the layer is only one or two atoms thick, the atoms will follow the contours of the particle. This can be done by staining with a very small amount of uranium atoms but is more conveniently done by the negative contrast technique. In this technique, the protein molecules or virus particles that are to be viewed are mixed with a solution of phosphotungstic acid (PTA) or uranium salts (usually uranyl acetate or uranyl formate), both of which contain atoms that absorb electrons strongly. The solution is then deposited as microdroplets on a supporting film. The droplets containing the mac-romolecules or particles of interest dry on the support film. The thickness of metal ions in the droplet is less where the macromolecule is present than elsewhere in the droplet. As shown in Figure 3-18, when the sample is placed in a beam of electrons, more electrons pass through the droplet in the region of a macromolecule; thus a negative image of a protein or a small virus particle can be obtained. Electron micrographs of a virus particle and a protein molecule viewed in this way are shown in Figure 3-19.

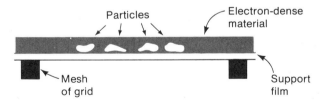

Figure 3-18
The negative-contrast method of visualizing particles by electron microscopy. Four particles are embedded in a substance that absorbs electrons strongly. As the beam passes through the sample, the fraction of the electrons in the beam that is absorbed depends on the total thickness of the substance; therefore, more electrons will pass through the regions containing each particle and the particle will appear bright against a dark background.

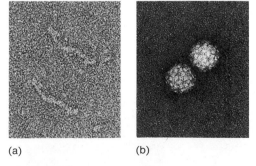

(a)   (b)

Figure 3-19
Electron micrographs obtained by the negative contrast procedure. (a) Molecules of the blood-clotting protein, fibrin. (b) Tomato bushy stunt virus particles—note the surface detail, which shows the individual protein molecules of which each particle is composed. (Courtesy of Robley Williams.)

DETERMINATION OF THE MOLECULAR WEIGHT OF A MACROMOLECULE

The large size of macromolecules eliminates the possibility of determining molecular weight M by procedures used for small molecules. For example, if the freezing-point-depression method, which is a standard procedure for small inorganic and organic molecules, were used for a typical DNA molecule, it would be necessary to detect a depression of about $10^{-6}\,°C$. Thus, a variety of specialized techniques have been developed for nucleic acids and proteins. A few of the more common ones in current use will be described.

Determination of M for DNA Molecules

The means used to determine the molecular weight of a DNA molecule depend on the size of the molecule. That is, there are distinct procedures applicable to small, midsize, and large molecules. These will be discussed separately.

Small DNA molecules ($M < 5 \times 10^6$)

For molecules in this size range, gel electrophoresis is best. Molecules for which M is accurately known are added to the sample and the relative distances migrated are measured. The data are plotted as shown in Figure 3-20.

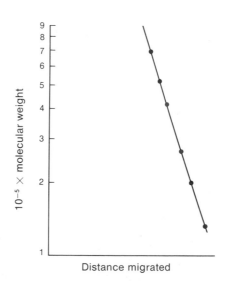

Figure 3-20
Determination of the molecular weights of DNA molecules (red points) by gel electrophoresis with DNA molecules whose molecular weights are known (black points). The graph is drawn of the black points; the molecular weights of the molecules being studied are then determined from the positions of the red points.

Midsize DNA molecules ($M = 5$ to 100×10^6)

For molecules in this size range, electron microscopy is most convenient. A measurement of the length of the molecule yields the value of M, since the mass per unit length of double-stranded DNA is 2×10^6 molecular weight units per micrometer. Often errors arise in determining the value of the magnification of the molecules. A common technique that avoids the necessity of knowing this value is to add a molecule of precisely known molecular weight to the sample being studied and to determine the ratio of the lengths of the standard molecule and the molecule of interest. The most commonly used standard molecule is the circular double-stranded replicating form of *E. coli* phage ϕX174, which contains exactly 5386 nucleotide pairs. The smaller molecules in Figure 3-17 are ϕX174 DNA molecules.

Large DNA molecules ($M > 100 \times 10^6$)

For very large molecules (or for midsize molecules when only an estimate of M is needed) the value of M can be determined by measuring the s value. However, the s value is dependent on the DNA concentration for large molecules. To obtain a standard s value that can be related directly to M, the s value is measured at various concentrations and a value called s^0 is obtained by extrapolation of the measured values of s to zero concentration. Then the equation

$$s^0 = 2.8 + 0.00834\, M^{0.497},$$

which applies to double-stranded DNA in 1 M NaCl, can be used to evaluate M. Other equations exist for single-stranded DNA, for RNA, and for other ionic conditions. A simpler method is to use a radioactive label for the DNA molecules of interest (so that they can be detected at a very low concentration) and to mix these DNA molecules with other

Figure 3-21
Determination of the molecular weight of a DNA molecule (red) by zonal centrifugation with two standards. (a) The concentration distributions in the centrifuge tube. (b) A plot of the molecular weight of the standards (50×10^6 and 176×10^6) versus the distance sedimented gives a standard curve from which the molecular weight (157×10^6, red point), is determined.

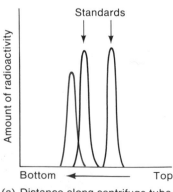

(a) Distance along centrifuge tube

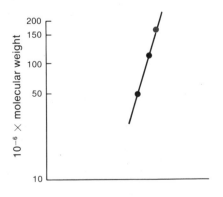

(b) Log (distance sedimented)

radioactive DNA molecules whose molecular weights are accurately known, and to sediment the mixture through a sucrose gradient. The logarithm of the distance d sedimented is proportional to log M for a variety of ionic conditions, so that a plot of log d versus log M yields the value of M. An example of the type of data obtained is shown in Figure 3-21.

Measurement of the Molecular Weight of Proteins

For accurate values of M for proteins, various centrifugation procedures (sedimentation equilibrium, Archibald method, Yphantis method), which are not described in this chapter, are used. These methods require several additional precise measurements. Information concerning these procedures can be found in references given at the end of this chapter.

The most convenient procedure for determining the molecular weights of proteins is SDS-PAGE electrophoresis (described earlier). This is a fairly accurate method. It is performed by adding a series of molecules for which M is known to the sample containing molecules of unknown M and plotting the relative distances migrated against log M, as shown in Figure 3-22. There is a complication in this procedure—namely, the subunit effect. As mentioned earlier, SDS binds to polypeptide chains in a way that produces a fixed ratio of charge to mass and, furthermore, breaks all noncovalent interactions, so that the molecule assumes a random coil configuration. However, many proteins consist of several identical polypeptide chains called subunits, joined together by noncovalent interactions. (This will be discussed in

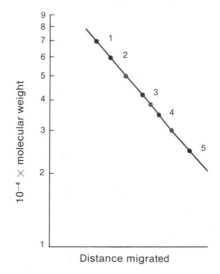

Figure 3-22
Typical semilogarithmic plot of M versus distance migrated for determining molecular weight by SDS-polyacrylamide-gel electrophoresis. Proteins were treated with a reducing agent to eliminate disulfide bonds. The known proteins are: (1) bovine serum albumin; (2) catalase; (3) ovalbumin; (4) carboxypeptidase A; (5) chymotrypsinogen. The unknown proteins are represented by red points.

detail in Chapter 5.) Because SDS breaks noncovalent interactions, it dissociates these protein aggregates. Thus, the value of M that is measured is that of a single subunit. To determine M for the entire protein, the number of subunits must be known. This number can be determined in a variety of ways that will not be described here. The problem is slightly more complex if the subunits are not identical, in which case several different components are seen in the gel. Again, the number of each type of subunit can be determined and then the molecular weight of the protein can be calculated from the molecular weight of each subunit.

REFERENCES

Brewer, J. M., A. J. Pesce, and R. B. Ashworth. 1974. *Experimental Techniques in Biochemistry*. Prentice-Hall.

Cantor, C. R., and P. R. Schimmel. 1980. *Biophysical Chemistry. Part I. The Conformation of Biological Macromolecules. Part II. Techniques for the Study of Biological Structure and Function*. W. H. Freeman and Co.

Dickerson, R. E., and I. Geis. 1969. *The Structure and Action of Proteins*. Harper & Row.

Freifelder, D. 1982. *Physical Biochemistry*. 2nd Ed. W. H. Freeman and Co.

Freifelder, D. 1982. *Physical Chemistry for Students of Biology and Chemistry*. Science Books International.

Kornberg, A. 1980. *DNA Replication*. W. H. Freeman and Co.

Lehninger, A. L. 1982. *Biochemistry*. Worth.

Meselson, M., F. W. Stahl, and J. Vinograd. 1957. "Equilibrium sedimentation of macromolecules in density gradients." *Proc. Nat. Acad. Sci.*, 43, 581–588.

Stryer, L. 1981. *Biochemistry*. 2nd Ed. W. H. Freeman and Co.

Tanford, C. 1961. *Physical Chemistry of Macromolecules*. Wiley. (This is for advanced students.)

Watson, J. D. 1976. *Molecular Biology of the Gene*. Benjamin-Cummings.

Wold, F. 1971. *Macromolecules: Structure and Function*. Prentice-Hall.

Younghusband, H. B., and R. B. Inman. 1974. "The electron microscopy of DNA." *Ann. Rev. Biochem.*, 43, 605–619.

Zwan, J. 1967. "Estimation of molecular weights by polyacrylamide gel electrophoresis." *Analyt. Biochem.*, 21, 155–168.

4 Nucleic Acids

Deoxyribonucleic acid (DNA) is the single most important molecule in living cells and contains all of the information that specifies the cell. In this chapter its remarkable structure and many of its physical and chemical properties are described. Some properties of ribonucleic acid (RNA), and how nucleic acids are studied, will also be presented.

PHYSICAL AND CHEMICAL STRUCTURE OF DNA

In the previous chapter the structure of a nucleotide and how nucleotides are joined to form a polynucleotide, or a nucleic acid, were described. In this section we show how nucleotides interact with one another to produce a double helix from two polynucleotides having complementary base sequences, and we present the Watson-Crick model for the DNA double helix.

Base-pairing and the DNA Double Helix

In early physical studies of DNA a variety of experiments indicated that the molecule is an extended chain having a highly ordered structure. The most important technique was x-ray diffraction analysis, by which information was obtained about the arrangement and dimensions of various parts of the molecule. The most significant observations were that the molecule is helical and that the bases of the nucleotides are stacked with their planes separated by a spacing of 3.4 Å.

Chemical analysis of the molar content of the bases (generally called the **base composition**) adenine, thymine, guanine, and cytosine in DNA molecules isolated from many organisms provided the important fact that [A] = [T] and [G] = [C], in which [ ] denotes molar concentration, from which followed the corollary

$$[A+G] = [T+C] \qquad \text{or} \qquad [\text{purines}] = [\text{pyrimidines}].$$

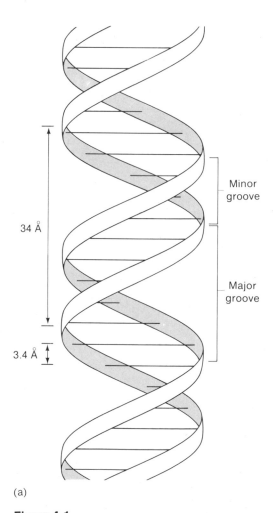

34 Å

3.4 Å

Minor groove

Major groove

(a)

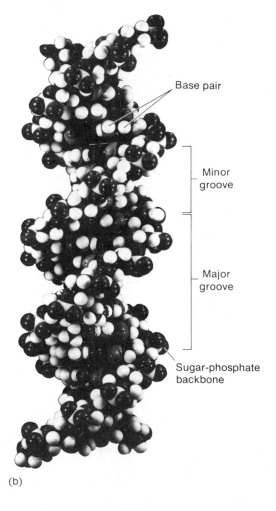

Base pair

Minor groove

Major groove

Sugar-phosphate backbone

(b)

Figure 4-1
(a) Diagramatic model of the DNA double helix in the common B form. (b) Space-filling model of the DNA double helix, again in the B form. (Courtesy of Sung-Hou Kim.)

James Watson and Francis Crick combined chemical and physical data for DNA with a feature of the x-ray diffraction diagram that suggested that two helical strands are present in DNA and showed that the two strands are coiled about one another to form a double-stranded helix (Figure 4-1). In this model the sugar-phosphate backbones follow a helical path at the outer edge of the molecule and the bases are in a helical array in the central core. The bases of one strand are hydrogen-bonded to those of the other strand to form the purine-to-pyrimidine base pairs A · T and G · C (Figure 4-2). Because each pair contains one two-ringed purine (A or G) and one single-ringed pyrimidine (T or C, respectively), the length of each pair (in the sugar-to-sugar direction) is about the same and the helix could hypothetically fit into a smooth cylinder.

The two bases in each base pair lie in the same plane and the plane of each pair is perpendicular to the helix axis. The base pairs are rotated 36° with respect to each adjacent pair, so that there are ten pairs per helical turn. The diameter of the double helix is 20 Å and the molecular weight per unit length of the helix is approximately 2×10^6 per micrometer. Since the molecular weight of a typical bacterial DNA molecule is about 2×10^9, such a molecule is 1 millimeter or 10^6 Å long and is very long and thin indeed.

The helix has two external helical grooves, a deep wide one (the **major groove**) and a shallow narrow one (the **minor groove**); both of

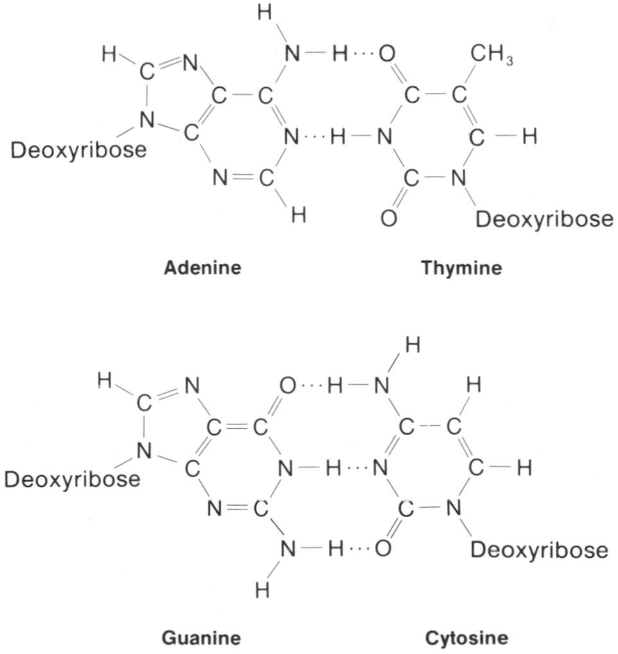

Figure 4-2
The two common base pairs of DNA. Note that hydrogen bonds (dotted lines) are between the weakly charged groups noted in Figure 3-6.

```
ATGGTCAACTG
| | | | | | | | | | |
TACCAGTTGAC
```

Figure 4-3
A diagram of a DNA molecule showing complementary base-pairing of the individual strands. The light lines represent hydrogen bonds.

these grooves are large enough to allow protein molecules to come in contact with the bases.

Base-pairing is one of the most important features of the DNA structure because it means that the base sequences of the two strands are complementary (Figure 4-3); that is, if one strand has the base sequence AATGCT, the other strand has the sequence TTACGA, reading in the same direction. This has deep implications for the mechanism of DNA replication because, in this way, the replica of each strand is given the base sequence of its complementary strand.

Antiparallel Orientation of the Two Strands of DNA

The two polynucleotide strands of the DNA double helix are antiparallel—that is, the 3'-OH terminus of one strand is adjacent to the 5'-P (5'-phosphate) terminus of the other (Figure 4-4). The significance of this is twofold. First, in a linear double helix there is one 3'-OH and one 5'-P terminus at each end of the helix. The second and generally more significant point is that the orientations of the two strands are different. That is, if two nucleotides are paired, the sugar portion of one nucleotide lies upward along the chain whereas the sugar of the other nucleotide lies downward. We will see in Chapter 8 that this structural feature poses an interesting constraint on the mechanism of DNA replication.

Figure 4-4
A stylized drawing of a segment of a DNA duplex showing the antiparallel orientation of the complementary chains. The arrows indicate the 5' to 3' direction of each strand. The phosphates (P) join the 3' carbon of one deoxyribose (horizontal line) to the 5' carbon of the adjacent deoxyribose.

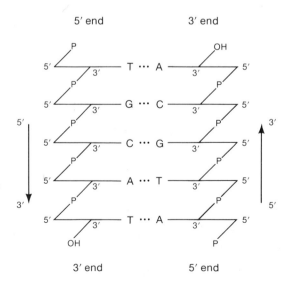

Because the strands are antiparallel, a convention is needed for stating the sequence of bases of a single chain. The convention is to write a sequence with the 5'-P terminus at the left; for example, ATC denotes the trinucleotide P-5'—ATC—3'-OH. This is also often written as pApTpC, again using the conventions that the left side of each base is the 5' terminus of the nucleotide and that a phosphodiester group is represented by a p between two capital letters.

Variation of Base Composition of the DNA of Different Organisms

Although it is always true that [A] = [T] and [G] = [C], there is no rule governing the ratio of total concentrations of guanine and cytosine to adenine and thymine in a DNA molecule—that is, ([G] + [C])/([A] + [T]). In fact, there is enormous variation in this ratio, ranging from 0.37 to 3.16 for different species of bacteria. The usual way that the base composition of the DNA molecule from a particular organism is expressed is not by the ratio just given but by the fraction of all bases that are G · C pairs; that is, ([G] + [C])/[all bases]; this is termed the G+C content, or percent G+C. The base composition of hundreds of organisms has been determined. Generally speaking, the value of the G+C content is near 0.50 for the higher organisms and has a very small range from one species to the next (0.49–0.51 for the primates); for the lower organisms the value of the G+C content varies widely from one genus to another. For example, for bacteria the extremes are 0.27 for the genus *Clostridium* and 0.76 for the genus *Sarcina*; *E. coli* DNA has the value 0.50. A convenient way to measure the base composition is by centrifuging the DNA to equilibrium in CsCl (see Chapter 3), because the density of DNA in CsCl solutions is linearly related to the G+C content, as shown in Figure 4-5.

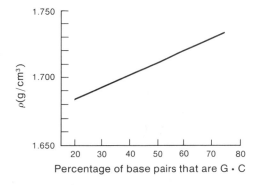

Figure 4-5
Density (ρ) of DNA in CsCl as a function of guanine-plus-cytosine content. The equation that describes the curve is ρ = [percent (G+C) $\times$ 0.00098] + 1.660 g/cm³.

Forms of the DNA Helix

The double helix shown in Figure 4-1 is a right-handed helix. This means that when an observer looks down the axis of the helix in either direction, each strand follows a clockwise path as it moves away from the observer. If the path were counterclockwise, the helix would be left-handed.

Naturally occurring DNA molecules are generally right-handed helices.

In a later section we will examine a synthetic DNA molecule, which in certain conditions forms a left-handed helix, and discuss the recent observation that some naturally occurring DNA molecules apparently have regions that are left-handed helices.

The helical structure of DNA was worked out from x-ray diffraction data, as has been stated. This required a sample in which the DNA molecules were oriented; this was accomplished by using a gellike fiber drawn from a very viscous solution of DNA. In order to prevent the fiber from drying and becoming disoriented, it was maintained in an atmosphere of high humidity. Two forms of DNA were observed: the A form at 66 percent relative humidity and the B form at 92 percent relative humidity. In the A form there are 11 base pairs per turn of the helix and the planes of the base pairs are tilted 20° away from the perpendicular to the helix axis. In the B form, which is the one shown in Figure 4-1, there are ten base pairs per turn and the base pairs are perpendicular to the helix axis. The B form is the "classic" Watson-Crick structure. Since DNA is normally studied in solution and since within a cell there is a high water content, it is important to know which of these forms is present in these conditions. A great deal of work has shown that the B form is present in solution; presumably this is also the intracellular structure but, as will be seen, the molecule is often tightly bound to proteins, which certainly alters the structure of the DNA in the region of the binding.

The Size of DNA Molecules

In viruses and prokaryotes, the total genome* specifying the virus or cell is usually encompassed in a single DNA molecule (the RNA viruses are an exception to this rule). In eukaryotes, including the unicells such as algae, yeast, and protozoa, the DNA is partitioned into a number of chromosomes, each of which is believed to contain a single gigantic

---

*The **genome** is the total DNA content of a haploid organism or, for a diploid organism, the DNA that contains one complete set of genes.

Table 4-1
Sizes of various DNA molecules

Organism or particle	Molecular weight, M	Organism or particle	Molecular weight, M
15 plasmid*	1.4×10^6	Vaccinia virus	121×10^6
Polyoma virus	3.2×10^6	Fowlpox virus	178×10^6
Phage 186*	18×10^6	F'450 dimer plasmid*	210×10^6
Phage T7*	25×10^6	Mycoplasma homina	5.3×10^8
Phage λ*	32×10^6	Most bacteria	$2.0–2.6 \times 10^9$
F plasmid*	62×10^6	Yeast	6×10^8
F'lac plasmid*	95×10^6	Drosophila (fruit fiy)	7.9×10^{10}
Phage T4*	106×10^6	Human	8×10^{10}

Note: Phages and plasmids marked with an asterisk have *E. coli* as a host. *Mycoplasma homina* is the smallest known bacterium. For yeast, *Drosophila*, and humans the molecular weight of the largest DNA molecule in the organisms is given.

DNA molecule. Table 4-1 shows the molecular weights, M, of the individual DNA molecules of various organisms. The values for many phage, viral, and bacterial DNA molecules are very accurately known (within 5 percent and, in some cases, exactly); the uncertainty in the values for eukaryotic DNA molecules is 50 percent in some cases. Note that the range of size is very great among the viruses and phages but much less so for bacteria.

The length of these molecules can be obtained from this relation: $1 \, \mu m = 2 \times 10^6$ molecular weight units. Thus, the molecules listed in the table range from 0.7 μm to 40,000 μm (4 cm!). The width of a DNA molecule is 20 Å $= 0.002 \, \mu m$.

In general, the genomes of the more complex organisms require much more DNA than for simpler organisms (though the cells of both the toad and the South American lungfish have considerably more DNA than human cells).

Fragility of DNA Molecules

The great length of DNA molecules makes them extremely susceptible to breakage by the hydrodynamic shear forces resulting from such ordinary operations as pipetting, pouring, and mixing. Unbroken DNA molecules for which $M < 2 \times 10^8$ can usually be isolated with ease from phages and viruses. However, when DNA is isolated from bacteria and higher organisms, unless great care is taken, the DNA molecules are almost always broken. In fact, the mean value of M for a sample is usually about 25×10^6, so that bacterial DNA, for instance, is frequently fragmented into about a hundred pieces. For plant and animal cells the yield of unbroken molecules in the least broken samples is usually a small fraction of one percent. Later in this chapter and elsewhere in the book we will see several instances in which the fact that the DNA of bacteria and of higher organisms is invariably fragmented by manipulation has important experimental consequences.

THE FACTORS THAT DETERMINE THE STRUCTURE OF DNA

The helical structure of nucleic acids is determined by stacking between adjacent bases in the same strand, and the double-stranded helical structure of DNA is maintained by hydrogen-bonding between the bases in the base pairs. This conclusion, as well as features of the structure that have not yet been described, has come from studies of denaturation. This analysis is presented in this section.

Denaturation and Melting Curves

The free energies of the weaker noncovalent interactions described in Chapter 3 are not much greater than the energy of thermal motion at room temperature, so that, at elevated temperatures, the three-dimensional structures of both proteins and nucleic acids are disrupted. A macromolecule in a disrupted state, in which the molecules are in a nearly random coil configuration, is said to be **denatured**; the ordered state, which is presumably that originally present in nature, is called **native.** A transition from the native to the denatured state is called **denaturation.** When double-stranded DNA or native DNA is heated, the bonding forces between the strands are disrupted and the two strands separate; thus, *denatured DNA is single-stranded.*

A great deal of information about structure and stabilizing interactions has been obtained by studying nucleic acid denaturation. This is done by measuring some property of the molecule that changes as denaturation proceeds—for example, the absorption of ultraviolet light. Originally, denaturation was always accomplished by heating a DNA solution, so that a graph of a varying property as a function of temperature is called a **melting curve.** Reagents are known that either break hydrogen bonds or weaken hydrophobic interactions. These are powerful denaturants. Thus, denaturation is also studied by varying the concentration of a denaturant at constant temperature. In the case of DNA, the simplest way to detect denaturation is to monitor the ability of DNA in a solution to absorb ultraviolet light whose wavelength is 260 nanometers (nm). The bases of nucleic acids absorb 260-nm light strongly. A convenient measure of the absorption is the absorbance (A) of a solution—that is, $-\log_{10}$[(intensity of light transmitted by a solution 1 cm thick)/(intensity of incident light)]. The absorbance at 260 nm, A_{260}, is proportional to concentration; a useful relation is

A DNA concentration of 50 micrograms per milliliter has a value of A_{260} equal to 1.

More important, the amount of light absorbed by nucleic acids is dependent on the structure of the molecule. The more ordered the structure, the less light is absorbed. Therefore, free nucleotides absorb more light than a single-stranded polymer of DNA or RNA and these in turn absorb more light than a double-stranded DNA molecule. For example, three solutions of double-stranded DNA, single-stranded DNA, and free bases, each at 50 μg/ml, have the following A_{260} values:

Double-stranded DNA $A_{260} = 1.00$
Single-stranded DNA $A_{260} = 1.37$
Free bases $A_{260} = 1.60$

This relation is often described by stating either that double-stranded DNA is **hypochromic** or that the bases are **hyperchromic.**

If a DNA solution is slowly heated and the A_{260} is measured at various temperatures, a melting curve such as that shown in Figure 4-6 is obtained. The following features of this curve should be noted:

1. The A_{260} remains constant up to temperatures well above those encountered by living cells in nature.
2. The rise in A_{260} occurs over a range of 6–8°C.
3. The maximum A_{260} is about 37 percent higher than the starting value.

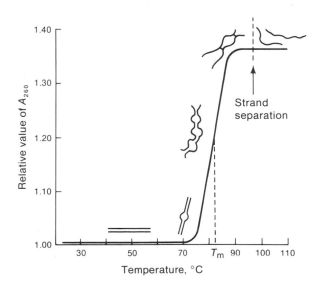

Figure 4-6
A melting curve of DNA showing T_m and possible
molecular configurations for various degrees of melting.

The state of a DNA molecule in different regions of the melting curve is also shown in the figure. Before the rise begins, the molecule is fully double-stranded. In the rise region, base pairs in various segments of the molecule are broken; the number of broken base pairs increases with temperature. In the initial part of the upper plateau a few base pairs remain to hold the two strands together until a critical temperature is reached at which the last base pair breaks and the strands separate completely.

A convenient parameter to characterize a melting transition is the temperature at which the rise in A_{260} is half complete. This temperature is called the **melting temperature**; it is designated T_m.

In the next section we will see how melting curves are used to understand the interactions that stabilize DNA.

Evidence for Hydrogen Bonds in DNA

A great deal of information is obtained by observing the variation of T_m with base composition and experimental conditions. For example, DNA can be isolated from various bacterial species in which the base compositions vary from 20 percent G + C to 80 percent G + C. The values of T_m from many such DNA molecules are plotted versus percent G + C in Figure 4-7. T_m increases with increasing percent G + C. This fact is interpreted in terms of the relative number of hydrogen bonds in a G · C pair (three) and an A · T pair (two). That is, a higher temperature is required to disrupt a G · C pair than an A · T pair, because the double-stranded structure is stabilized, at least in part, by hydrogen

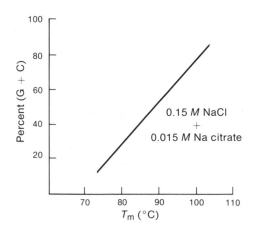

Figure 4-7
Plot of T_m versus G + C content of various DNA molecules.

bonds. This conclusion has been substantiated by measuring the value of T_m in the presence of reagents such as urea and formamide,

$$
\begin{array}{cc}
\text{O} & \text{O} \\
\| & \| \\
\text{H}_2\text{N}-\text{C}-\text{NH}_2 & \text{H}_2\text{N}-\text{C}-\text{H} \\
\textbf{Urea} & \textbf{Formamide}
\end{array}
$$

These two substances are capable of hydrogen-bonding with the DNA bases. Broken and unbroken base pairs are in thermal equilibrium and these reagents compete with one base for pairing to another base; thus, they can maintain the unpaired state at a temperature at which a broken base pair might normally pair again. Hence, permanent melting of a section of paired bases requires less input of thermal energy and T_m is reduced. A reduction of T_m is a characteristic of all known reagents that can form hydrogen bonds with nucleotides.

Evidence for Hydrophobic Interactions in DNA

Any reagent that would either enhance the interaction of weakly soluble substances with water or would disrupt the water shell about the substance should weaken hydrophobic interactions. An example of the former type of substance is methanol, which increases the solubility of the bases. The salt sodium trifluoracetate is an example of the second type. The addition of both of these reagents reduces T_m enormously, which suggests that hydrophobic interactions are also important in stabilizing the DNA structure. (Figure 4-8 shows data for the salt.) This conclusion follows from another experiment. As was discussed in Chapter 3, low solubility of substances favors hydrophobic clustering. The solubility of adenine in solutions containing a wide variety of reagents, such as alcohols, amines, and other soluble organic compounds, was measured. Many of the added substances increased the solubility of adenine and at the same time decreased the T_m of DNA suspended in a solution having the same composition. In fact, the extent to which a reagent lowered T_m correlated nicely with the increased solubility of the base. To confirm that this is an effect on the bases only, the solubility of sodium pyrophosphate, a highly charged molecule that presumably has the solubility properties of the sugar-phosphate backbone, was also measured. It was found that reagents that increase the solubility of adenine and decrease T_m either have no effect on the solubility of sodium pyrophosphate or decrease it slightly. These results suggest that a hydrophobic interaction is present in DNA. Furthermore, the results imply that a polynucleotide would tend to have a three-dimensional structure that maximizes the contact of the highly soluble

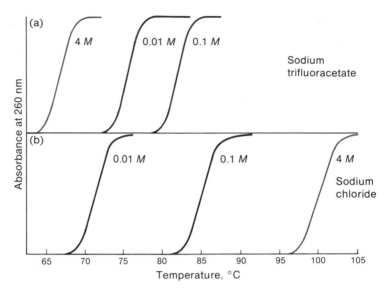

Figure 4-8
A comparison of the effect of sodium trifluoracetate and NaCl on the melting behavior of DNA. With NaCl the melting temperature increases with concentration because the high ionic strength neutralizes the negatively charged phosphates and stabilizes hydrogen bonds. With sodium trifluoracetate the melting temperature rises initially as the concentration increases but then decreases as binding of water by the trifluoracetate ion weakens the hydrophobic interaction.

phosphate group with water and minimizes contact between bases and water. This explains why, in DNA, the sugar-phosphate chain is on the outside and the bases are on the inside.

Base-stacking

The experiments just described implicate hydrophobic interactions as a stabilizing effect but do not indicate that the bases are stacked. That they do stack became clear when the techniques of optical rotatory dispersion (ORD) and circular dichroism (CD) were applied to single-stranded DNA and to RNA; these experiments showed that a large fraction of the bases in single-stranded polynucleotides are arranged in a helix. A parallel array of bases was even detected in dinucleotides, which suggests that base-stacking is present even in these small molecules. In double-stranded polynucleotides (both DNA and synthetic model compounds) stacking was also found to be present. Furthermore, in all cases it was found that

1. Base-stacking is eliminated by the addition of reagents that weaken hydrophobic interactions.
2. If a DNA sample is heated, base-stacking is reduced during the same transition in which the A_{260} increases.
3. Reagents that break down hydrogen-bonding have no effect on base-stacking of single-stranded polynucleotides but reduce base-stacking of double-stranded DNA to the degree found in denatured DNA.

These results lead to the conclusion that the bases of double-stranded DNA are *more* stacked than those in single-stranded DNA. The increased stacking in double-stranded DNA can be accounted for by considering the effect of hydrogen bonds between the two strands on the structure of each of the individual strands. Both hydrogen bonds and hydrophobic interactions are weak and easily disrupted by thermal motion. Maximum hydrogen-bonding is achieved when all bases are pointing in the right direction. Similarly, stacking is enhanced if the bases are unable to tilt or swing out from a stacked array. Clearly, stacked bases are more easily hydrogen-bonded and correspondingly, hydrogen-bonded bases, which are oriented by the bonding, stack more easily. Thus, the two interactions act cooperatively to form a very stable structure. If one of the interactions is eliminated, the other is weakened; this explains why T_m drops so markedly after the addition of a reagent that destroys either type of interaction.

Cooperativity of Base-stacking

Base-stacking itself is also a cooperative interaction. That is, in a sequence of stacked bases, ABCDEFGHIJ, it would be very unlikely for base E to swing out of the stacked array because the plane of the base tends to be parallel to the planes of both D and F. However, the tendency to conform to an orderly stacked array is not so great at the ends of the molecule. That is, the orientation of base A is stabilized by only a single base, namely, B; thus a rapidly moving solvent molecule might crash into A and cause it to rotate out of the stack more easily than collision with E would cause disorientation of E. Since A has a lower probability of being stacked than B, then of course B must also be more easily disoriented than is C. This slight tendency toward instability is also present in a double-stranded molecule. This is most noticeable in the value of T_m for double-stranded polynucleotides containing fewer than twenty base pairs (oligonucleotides). For example, if a molecule having 10^5 base pairs is broken down to fragments having 10^3 base pairs, there is no detectable change in T_m. However, with conditions for

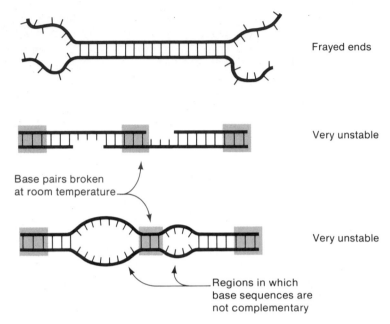

Frayed ends

Very unstable

Base pairs broken
at room temperature

Very unstable

Regions in which
base sequences are
not complementary

Figure 4-9
Several effects of cooperativity of base-stacking. Each shaded area indicates base pairs that would be broken at room temperature; if these tracts
contained more than fifteen base pairs, they would be stable.

which the T_m for *E. coli* DNA is 90°C, the value of T_m for a double-
stranded hexanucleotide (six nucleotides per strand) can be as low as
30°C; the exact value depends on the base composition and sequence.
This effect has several consequences:

1. The ends of a linear double-stranded DNA molecule are usually
   not hydrogen-bonded, but are frayed (Figure 4-9), typically with
   seven base pairs broken. This has been detected in very large
   DNA molecules by using an antibody that reacts only with
   single-stranded DNA.*
2. Short double-stranded oligonucleotides, having fewer than 15
   base pairs per single strand, have particularly low values of T_m.
   A double-stranded trinucleotide (three bases per strand) is not
   stable at room temperature.
3. Molecules in which the paired regions are very short and are
   flanked by unpaired regions, such as those shown in the lower
   part of Figure 4-9, cannot maintain that configuration at physiological temperatures.

*Some base sequences stack better than others and are even stacked at an end of a molecule. An
example is the sequence CCA, which is found at the 3′ terminus of tRNA molecules.

Effect of the Ionic Strength of a Solution on the Structure of DNA

In addition to the cooperative attractive interactions between adjacent DNA bases and between the two strands, there is an interstrand electrostatic repulsion between the negatively charged phosphates. (There is also an intrastrand repulsion, which is probably not important.) This strong force would drive the two strands apart if the charges were not neutralized. By examining the variation of T_m as a function of the ionic concentration of the buffer solution, it is found that T_m decreases sharply as salt concentration decreases; indeed, in distilled water, DNA denatures at room temperature. The interpretation of the data is the following. In the absence of salt the strands repel one another. As salt is added, positively charged ions (e.g., Na^+) form "clouds" of charge around the negatively charged phosphates (as explained by the Debye-Hückel theory, which is presented in most physical chemistry texts) and effectively shield the phosphates from one another. Ultimately all of the phosphates are shielded and repulsion ceases; this occurs near the physiological salt concentration of about 0.2 M. However, T_m continues to rise as the [NaCl] is increased, because the solubility of the bases decreases at high [NaCl], which increases the hydrophobic interaction.

A second effect also occurs—namely, neutralization of the negative charges by binding of the Na^+ ion. It might be thought that since ionization of Na^+ salts is always complete, such binding would not occur. However, since the molar concentrations of most DNA solutions are rarely more than 10^{-4} and we are talking about solutions in which [Na^+] is in 10–10^4-fold excess, it should not be surprising that some Na^+ ions are bound to the DNA. Several lines of evidence indicate that positive ions are bound by DNA. For example, if the molecular weight of a particular DNA molecule is measured first in NaCl and then in CsCl, it is found that the values obtained are different, and that the ratio of these values is 0.75. This is explained by noting the molecular weights of a nucleotide (330 on the average), the Na^+ ion (23), and the Cs^+ ion (137). Thus, a nucleotide of NaDNA has a molecular weight of $23 + 330 = 353$ and CsDNA has a molecular weight of $137 + 330 = 467$. The ratio of these molecular weights, namely, $353/467 = 0.75$, is the measured ratio for the DNA molecules, implying that in NaCl and CsCl solutions, most of the phosphate groups have bound Na^+ and Cs^+ ions, respectively.

Fluctuation of the Structure of the DNA Molecule

An important structural feature of the DNA molecule becomes apparent when the DNA is examined in the presence of formaldehyde:

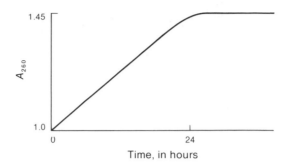

Figure 4-10
The increase in A_{260} of a DNA solution containing
4 percent formaldehyde, as a function of time
at 25°C.

Formaldehyde

Formaldehyde can react with the NH_2 group of the bases and thus eliminate their ability to hydrogen-bond. The addition of formaldehyde causes a slow and irreversible denaturation of DNA, as shown in Figure 4-10. Thus, the amino groups must be available to the formaldehyde to allow the reaction, which means that bases are continually being paired and unpaired (that is, hydrogen bonds are breaking and reforming). A related phenomenon is observed when DNA is dissolved in tritiated water (^3H_2O) — there is a rapid exchange between the hydrogen-bonded protons of the bases and the $^3H^+$ ions in the water. These two observations indicate that DNA is a dynamic structure in which double-stranded regions frequently open to become single-stranded bubbles. This important phenomenon, called **breathing,** is thought to enable specialized proteins to interact with the DNA molecule and to "read" its encoded information. Note that since a G • C pair has three hydrogen bonds and an A • T pair has only two, breathing occurs more often in regions rich in A • T pairs than in regions rich in G • C pairs.

Denaturation and Strand Separation

Certain changes in the physical properties of DNA solutions that accompany denaturation — for example, a decrease in viscosity and the ability to rotate polarized light — indicate that when hydrogen bonds and hydrophobic interactions are eliminated, the helical structure of DNA is disrupted and the molecule loses its rigidity. This collapse of the

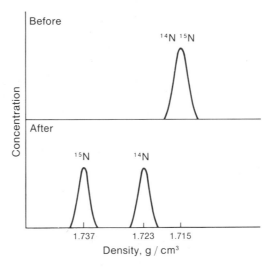

Figure 4-11
Demonstration of strand separation by equilibrium
centrifugation in CsCl. Before denaturation,
[$^{14}N^{15}N$]DNA gives a single band. After denaturation
two bands result; the density of the [^{14}N]DNA is
greater than that of [$^{14}N^{15}N$]DNA because the
density of single-stranded DNA is greater than that
of double-stranded DNA by 0.014 g/cm³, even if
the isotopic composition is the same.

ordered structure may be accompanied by complete disentanglement of
the two strands. At one time it was thought that DNA strands are so long
that complete unwinding of the helix and strand separation would be
impossible, but several lines of evidence clearly indicate that strand
separation does occur. One of these experiments will be described.

In this experiment DNA was prepared that had one ^{14}N-labeled
strand and one ^{15}N-labeled strand. Such a molecule (designated
[$^{14}N^{15}N$]DNA) has a unique density in CsCl equal to the average of the
densities of [$^{14}N^{14}N$] and [$^{15}N^{15}N$]DNA (i.e., molecules having the same
isotope in both strands). The DNA was heated to various temperatures
and A_{260} was measured to obtain a melting curve. Each sample was also
subjected to equilibrium centrifugation in a CsCl density gradient. It
was found that shortly after the maximum A_{260} was reached, two bands
appeared in the CsCl solution; these bands had the densities of
^{14}N-labeled and ^{15}N-labeled single-stranded DNA respectively and
contained the separated single strands (Figure 4-11).

Other types of experiments allowed the measurement of the time
required for strand separation. This rate was found to be 12×10^7
molecular weight units per second at 37°C. Since the rate of DNA
replication in bacteria is only 10^6 molecular weight units per second,

strand separation, which is known to be part of the DNA replication process (see Chapter 8), cannot limit the replication process.

In the course of studying strand separation, another important fact emerged. If a DNA solution is heated to a temperature at which most but not all hydrogen bonds are broken and then cooled to room temperature, A_{260} drops immediately to the initial undenatured value. Additional experiments show that the native structure is restored. Thus, if strand separation is not complete and denaturing conditions are removed, the helix rewinds. Thus, if two separated strands were to come in contact and form even a single base pair at the correct position in the molecule, the native DNA molecule should reform. We will encounter this phenomenon again when renaturation is described.

Denaturation of DNA by Helix-destabilizing Proteins

There are many proteins that can unwind a DNA helix; these are called **relaxation, helix-destabilizing,** or **melting proteins.** Of these the simplest to understand is the protein made by gene *32* of E. coli phage T4, commonly called the *32*-protein. This protein has two properties that enable it to denature DNA—(1) it binds tightly to the bases of single-stranded DNA and (2) the individual molecules of the *32*-protein prefer to line up adjacent to one another along a single strand (Figure 4-12).

Binding of the first molecule of the *32*-protein is made possible by the breathing of the DNA. Once one molecule has invaded the helix, it binds to several adjacent bases. The binding is very tight so that the protein stays in place. This not only renders the bases on the opposite strand available for binding, but also destabilizes adjacent base pairs, because paired bases adjacent to unpaired bases cease to be optimally stacked and their hydrogen bonds become less stable. Because of the preference for *32*-proteins to bind in close juxtaposition, another molecule of the *32*-protein will tend to bind next to the first; this breaks still other base pairs so that additional protein molecules bind. The process continues until the DNA is totally denatured.

Figure 4-12
Cooperative binding of gene-*32* protein to single- and double-stranded DNA. At low concentrations the protein molecules are located at random positions but at high concentrations they are adjacent to one another, forming numerous clusters.

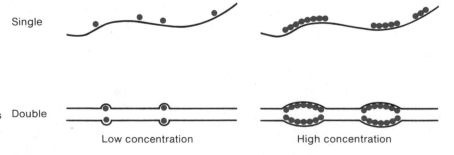

As will be seen later, proteins of this type are needed to unwind the helix during replication and to facilitate joining of single strands during genetic recombination.

Denaturation of DNA in Alkali

We have seen by now that the helical structure of polynucleotides is maintained by base-stacking and that two strands are held together by hydrogen-bonding. Most of the studies of denaturation have used heat as a means of disrupting the structure. This can, of course, be used whenever it is necessary to prepare denatured DNA, which is often an essential step in many experimental protocols. However, as will be seen in a later section, phosphodiester bonds may be broken at high temperatures, so the product of heat denaturation often is a collection of broken single strands. At high pH the charge of several of the groups engaged in hydrogen-bonding is changed and a base bearing such a group loses its ability to form these bonds. At a pH greater than 11.3 all hydrogen bonds are eliminated and DNA is completely denatured. Since DNA is quite resistant to alkaline hydrolysis, this procedure is the method of choice for denaturing DNA.

Structure of Denatured DNA

To obtain the data for the melting curves of the sort that have been shown, A_{260} is measured at various temperatures, which are plotted on the x axis. Denaturation is usually complete at a temperature above 90°C. In most experiments there is a total increase in A_{260} of 37 percent and the solution consists entirely of single strands whose bases are unstacked. However, if the solution is rapidly cooled to room temperature and the salt concentration is above 0.05 M, the value of A_{260} reached at the maximum temperature drops sigificantly but not totally (Figure 4-13), because in the absence of disrupting thermal motion, random intrastrand hydrogen bonds re-form between distant short tracts of bases whose sequences are sufficiently complementary. Typically, the value of A_{260} drops to 1.12 times the initial value for the native DNA, suggesting that, after cooling, about two-thirds of the bases are either hydrogen-bonded or in such close proximity that stacking is restored. The molecule will be very compact, as shown in Figure 4-13. The situation is quite different if the salt concentration is 0.01 M or less. In this case, the electrostatic repulsion due to unneutralized phosphate groups keeps the single strands sufficiently extended that the bases cannot approach one another. Thus, after cooling no hydrogen bonds are re-formed and base-stacking remains at a minimum.

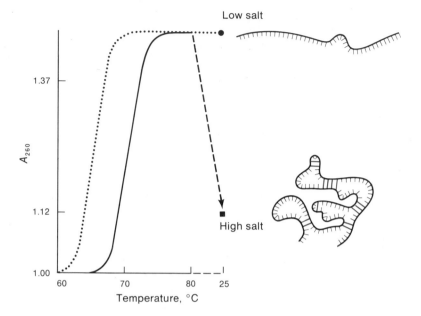

Figure 4-13
The effect of lowering the temperature to 25°C after strand separation has occurred. The types of molecules obtained when the DNA is in a solution having either a low or a high salt concentrations are shown. Base pairs are shown in red.

Denatured DNA can be maintained in a state in which there is no intrastrand base-pairing at a high salt concentration and at room temperature. One way is to add formaldehyde while the temperature is high, for as we have mentioned, formaldehyde reacts with the NH_2 groups of bases and thereby prevents hydrogen-bonding. Another method is to use high pH for in alkali hydrogen-bonding is not possible, as we discussed in an earlier section.

At a sufficiently high DNA concentration and in a high salt concentration, interstrand hydrogen-bonding should compete with the intrastrand bonding mentioned above. In the following section we will see how this effect can be used to re-form native DNA from denatured DNA.

RENATURATION

A solution of denatured DNA can be treated in such a way that native DNA re-forms. The process is called **renaturation** or **reannealing** and the re-formed DNA is called **renatured DNA.**

Renaturation has proved to be a valuable tool in molecular biology since it can be used to demonstrate genetic relatedness between different organisms, to detect particular species of RNA, to determine whether certain sequences occur more than once in the DNA of a particular organism, and to locate specific base sequences in a DNA molecule.

Requirements for Renaturation

Two requirements must be met for renaturation to occur:

1. The salt concentration must be high enough that electrostatic repulsion between the phosphates in the two strands is eliminated—usually 0.15 to 0.50 M NaCl is used.
2. The temperature must be high enough to disrupt the random, intrastrand hydrogen bonds described in the previous section. However, the temperature cannot be too high, or stable interstrand base-pairing will not occur and be maintained. The optimal temperature for renaturation is 20–25° below the value of T_m.

Renaturation is a slow process compared to denaturation. The rate-limiting step is not the actual rewinding of the helix (which occurs in roughly the same time as unwinding) but the precise collision between complementary strands such that base pairs are formed at the correct positions. Since this is a result only of random motion, it is a concentration-dependent process; at concentrations normally encountered in the laboratory this takes several hours. We shall have more to say about the concentration dependence in the section entitled C_0t curves.

Mechanism of Renaturation

The molecular details of renaturation can be understood by referring to the hypothetical molecule shown below, having a sequence (shown in red) that is repeated several times.

IA	IB	II	IC
$\cdots$ ATGA $\cdots$	ATGA $\cdots$	CCCC $\cdots$	ATGA $\cdots$
$\cdots$ TACT $\cdots$	TACT $\cdots$	GGGG $\cdots$	TACT $\cdots$
IA'	IB'	II'	IC'

Assume that each single strand contains 50,000 bases, and that the base sequences are complementary. However, any short sequence of bases (say, four to six bases long) will certainly appear many times in such a molecule and can provide sites for base-pairing. Random collision between noncomplementary sequences such as IA and II' will be ineffective but a collision between IA and IC' will result in base-pairing. This will be short-lived, though, because the bases surrounding these short complementary tracts are not able to pair and stacking stabilization will not occur. Thus, at the elevated temperatures that have brought about strand separation, these paired regions rapidly become disrupted. However, as soon as two sequences such as IB and IB′ pair, the adjacent bases will also rapidly pair and the entire double-stranded DNA molecule will "zip up" in a few seconds.

It is important to realize that each renatured native DNA molecule is not formed from its own original single strands: in a solution of denatured DNA, the single strands freely mix so that during renaturation strands join at random. This was shown in an experiment using two DNA samples isolated from *E. coli* grown, in one case, in a medium containing $^{14}NH_4Cl$ and in the other, $^{15}NH_4Cl$. The two DNA samples were mixed, denatured and renatured, and centrifuged in a CsCl density gradient. The result was a mixture containing three types of native DNA molecules—25 percent contained ^{14}N in both strands, 50 percent contained ^{14}N in one strand and ^{15}N in the other, and 25 percent contained ^{15}N in both strands, which indicates random mixing of the strands during renaturation. This type of molecular mixing is called **hybridization.** One example of its use is described in the next section.

Hybridization of DNA of Different Organisms

Hybridization has been used to obtain many kinds of information. For example, it is possible to show by this technique that two different species have common base sequences. How this is done is shown in Figure 4-14. DNA from two bacterial species, labeled with either ^{14}N or

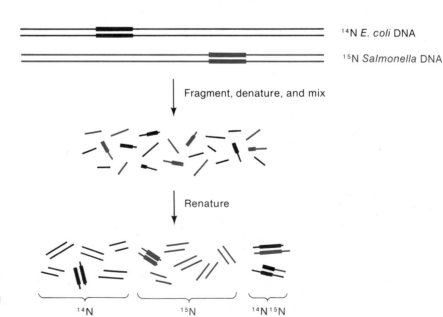

Figure 4-14
Demonstration that there are common sequences (heavy line) in *E. coli* and *Salmonella* DNA.

^{15}N, is isolated and broken into many small fragments. The DNA molecules are mixed, denatured, renatured, and centrifuged to equilibrium in a CsCl density gradient. All renatured molecules consisting of two single strands from the same bacterial species will contain either ^{14}N only or ^{15}N only. However, if there are common sequences in the two species, hybrid [^{14}N^{15}N] DNA will be detected; the fraction of the DNA of E. coli that is common to Salmonella is twice the fraction of the total renatured DNA that is a hybrid of [^{14}N]DNA and [^{15}N]DNA. This type of experiment has confirmed the basic expectations of evolutionary relatedness—that is, taxonomically related organisms have common base sequences and the degree of sequence overlap is connected to the relatedness determined by other criteria.

Filter-binding Assays for Renaturation

Very thin filters (**membrane filters**) made of nitrocellulose are commercially available. These filters bind single-stranded DNA very tightly but fail to bind either double-stranded DNA or RNA. They provide an important technique for measuring hybridization.

Consider a sample of denatured DNA that is being renatured. At any time after renaturation begins, the sample will consist of both double-stranded renatured molecules and single strands that have not yet renatured. If the sample is filtered through a nitrocellulose filter, the renatured DNA passes through and the single strands are retained on the filter, thus providing a measurement of the fraction of the DNA that is renatured.

Renaturation can also be carried out on a filter in the following way (Figure 4-15). A sample of denatured DNA is filtered. The single strands bind tightly to the filter along the sugar-phosphate backbone, but the bases remain free. The filter is then placed in a vial with a solution

Figure 4-15
Method of hybridizing DNA to nitrocellulose filters containing bound, single-stranded DNA. In the final step the filter is treated with a single-strand-specific DNase, an enzyme that depolymerizes single-stranded, but not double-stranded DNA.

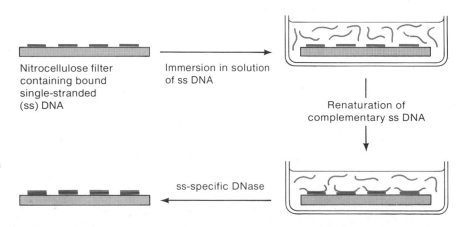

Nitrocellulose filter containing bound single-stranded (ss) DNA

Immersion in solution of ss DNA

Renaturation of complementary ss DNA

ss-specific DNase

containing a small amount of radioactive denatured DNA and a reagent that prevents additional binding of single-stranded DNA to the filter. After a period of renaturing the filter is washed. Radioactivity is found on the filter only if renaturation has occurred.

Filter binding is also another way to determine whether two organisms have common base sequences. For example, if excess E. coli DNA is on the filter and a very small amount of denatured ^{14}C-labeled Salmonella DNA is applied, the fraction of the applied ^{14}C that is retained on the filter is the fraction of Salmonella DNA that has a sequence with sufficient complementarity with that of E. coli to hybridize under the annealing conditions used and indicates the degree of evolutionary homology.

The most important use of filter hybridization is the detection of sequence homology between a single strand of DNA and an RNA molecule—this is called **DNA-RNA hybridization** and it is the method of choice to detect an RNA molecule that has been copied from a particular DNA molecule. In this procedure, a filter to which single strands of DNA have been bound, as above, is placed in a solution containing radioactive RNA. After renaturation the filter is washed and hybridization is detected by the presence of radioactive RNA on the filter. Examples of the use of this technique will be found in Chapters 11 and 14.

Concentration Dependence of Renaturation: C_0t Curves

The initial event in renaturation is a collision; thus, the renaturation rate obeys the law of mass action and increases with DNA concentration, as shown in Figure 4-16. This concentration dependence has been used to discover some surprising properties of the DNA of eukaryotes and prokaryotes.

Consider two DNA solutions having equal concentrations in terms of g/ml (not molar concentration). The DNA molecules are phage T7 DNA, whose molecular weight M is 2.5×10^7, and phage T4 DNA, for which $M = 1.1 \times 10^8$. The molar concentration of the T7 DNA is 4.4 times that of the T4 DNA but the A_{260} values of both solutions are the same. We further assume that there is no base sequence homology (that is, no common sequences) between the two DNA molecules. The two solutions are mixed, denatured, and renatured, and A_{260} is measured as a function of time. A curve such as that in Figure 4-17 is obtained. Two features of this curve should be noted. (1) The curve consists of two steps. Each accounts for half of the decrease in A_{260} because the initial DNA concentration (in g/ml) of each was the same. (2) The times at which the renaturation of each component is half-complete differ.

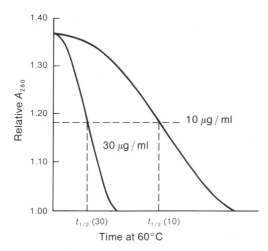

Figure 4-16
Dependence of renaturation time on the concentration of phage T7 DNA. The DNA is heated to 90°C to separate the strands and then cooled to 60°C. Renaturation is complete when the relative value of A_{260} reaches 1. The times required for renaturation to be half completed, $t_{1/2}$, are obtained by drawing the red horizontal line, which divides each curve equally in the vertical direction, and then extending the red vertical lines to the time axis.

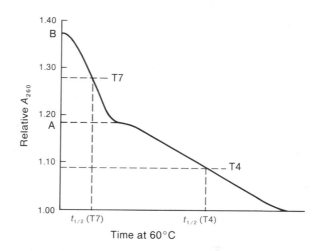

Figure 4-17
Renaturation of a mixture of T7 and T4 DNA molecules, each at 30 μg/ml. Extrapolation (black dashed line) of the inflection region at the junction between the upper and lower components of the curve to the vertical axis yields the early portion of the T4 curve; the ratio of the values of A_{260} at points A and B yields the fraction of the total DNA that is T4 DNA. The times required for renaturation of each type of DNA to be half completed, $t_{1/2}$, are calculated as in Figure 4-16 and shown in red.

Comparison with Figure 4-16 shows that the slower component is the T4 DNA whose molar concentration is lower.

A very important point about renaturation kinetics is the following. If the two kinds of DNA molecules described in Figure 4-17 were broken into thousands of fragments, the renaturation kinetics would differ from that shown in the figure, but two features—the A_{260} values of each step and the relative values of the time required for 50 percent renaturation—remain the same. The first is true because the ratio of the two components and hence the A_{260} values are unchanged. The latter follows from the fact that, whereas the renaturation rates are definitely less for the fragments, the molar concentrations of each base sequence are unchanged, so that the ratio of the molar concentrations of the T4 and T7 fragments is unaffected by the fragmentation.

Fragmentation of the DNA molecules allows a new feature of the base sequences to be seen. Consider a very long molecule having a repeating base sequence as shown in Figure 4-18(a). This sequence contains 3 percent of the total number of bases of the DNA and there are five copies of the repeated sequence in the DNA. If this large molecule is broken to fragments much smaller than 1 percent of the total length of

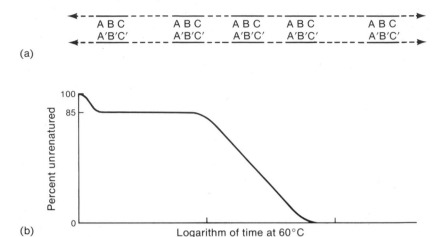

(a)

(b)

Figure 4-18
(a) A hypothetical DNA molecule having a base sequence that is 3 percent of the total length of the DNA and is repeated five times. The dashed lines represent the non-repetitive sequences; they account for 85 percent of the total length. (b) A renaturation curve for the DNA in part (a). Time is logarithmic in order to keep the curve on the page. The y axis—percent unrenatured DNA—is equivalent to the unit, relative value of A_{260}, used in previous figures with $A_{260} = 1.37$ equal to 100 percent.

the molecule, the molar concentration of this sequence becomes five times larger than that of the remainder but the weight fraction remains $(5 \times 3)/100 = 0.15$. Such a DNA would have a renaturation curve such as that shown in Figure 4-18(b). Thus, this curve, which is obtained with DNA from a single organism, contains two components. The more rapidly renaturing component accounts for 15 percent of A_{260}, from which one can conclude that 15 percent of the DNA (on a weight base) contains a repeating sequence. In the following we show that from the kinetics one can determine the number of bases in the repeating sequence (called its **complexity**), and thus the number of copies of this sequence and the number of bases in a single nonrepeating sequence.

The following equation describes the kinetics of renaturation in terms of the initial DNA concentrations C_0 (expressed in moles of bases per liter), the concentration C of unrenatured DNA at times t (in minutes), and k_2, a rate constant that depends on the temperature and the size of the DNA fragments:

$$C/C_0 = 1/(1 + k_2C_0t).$$

Experimentally, one chooses several values of C_0 and then measures C as a function of time; the data obtained are plotted as C/C_0 versus log

(a)

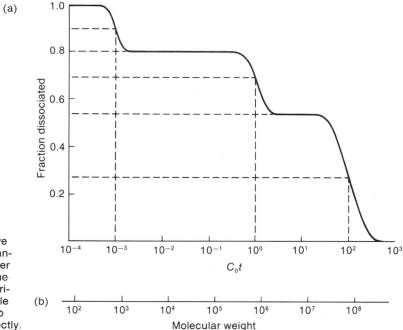

Figure 4-19
$C_0 t$ analysis. (a) The $C_0 t$ curve analyzed in the text. (b) A standard scale relating the number of nucleotide pairs in a unique sequence and $C_0 t_{1/2}$. The horizontal axis in (a) and the scale in (b) are aligned vertically so values can be compared directly.

$C_0 t$. When $C/C_0 = 1/2$, then $C_0 t_{1/2} = 1/k_2$. This is a significant expression because it can be shown that k_2 is inversely proportional to the number N of bases per repeating unit. Thus, the value of $C_0 t_{1/2}$ is directly proportional to N. Note that if there are no repeating sequences, so that the DNA molecule itself represents a unique sequence, then N is the number of base pairs in the complete DNA molecule of the organism. Figure 4-19(a) shows the kind of graph obtained. To analyze the sequence complexity of the DNA, one first notes that 53 percent of the sequences have a $C_0 t_{1/2}$ of 10^2, 27 percent have a $C_0 t_{1/2}$ of 1, and 20 percent have a $C_0 t_{1/2}$ of 10^{-3}. These data are then used to determine the number of copies and the sizes of each sequence. From analysis of pure samples of molecules having a unique sequence of known length, a size scale, such as that shown in Figure 4-19(b), can be obtained. However, molecular size cannot be read directly from the observed $C_0 t_{1/2}$ values, because the values of C_0 used in plotting the horizontal axis in panel (a) is the *total* DNA concentration in the renaturation mixture, rather than the concentration of each component. To utilize the size scale, one must calculate the concentration of each component; this is done by simply multiplying each observed $C_0 t_{1/2}$ value by the fraction of the total DNA that it represents. Thus, the "real" $C_0 t_{1/2}$ values are 0.53×10^2, 0.27×1, and 0.20×10^{-3}; the corresponding sizes are 4.2×10^7, 2.2×10^5, and

160 base pairs; respectively. The number of copies of each sequence is inversely proportional to $t_{1/2}$ and hence to the observed (uncorrected) $C_0 t_{1/2}$ value. Thus, if one assumes that the genome contains one copy of the longest sequence (possible only in a haploid organism), then the cell from which the DNA has been isolated would contain 1, 10^2, and 10^5 copies of sequences having 4.2×10^7, 2.2×10^5, and 160 base pairs, respectively. The total number of base pairs per genome would be

$$4.2 \times 10^7 + 100(2.2 \times 10^5) + 10^5(160) = 8 \times 10^7,$$

and the molecular weight of the total cellular DNA would be

$$(8 \times 10^7)(660) = 5.3 \times 10^{10}.$$

Note that if an independent measurement of the total molecular weight of the DNA were to yield the value $1 \times 10^{11} \cong 2(5.3 \times 10^{10})$, then there would be 2, 200, and 2×10^5 copies of the 4.2×10^7-, 2.2×10^5-, and 160-base-pair sequences, respectively.

Bacterial DNA contains a few repeating units such as the genes for transfer RNA but these account for such a small fraction (0.1 percent) of the DNA that they are not readily detected. However, $C_0 t$ analysis has shown that in eukaryotic DNA a significant fraction of the DNA consists of repeated sequences and in some cases there are a million or more copies of a particular repeated sequence. This will be discussed in detail in Chapter 22.

DNA Heteroduplexes

Renaturation has been combined with the technique of electron microscopy in a procedure that allows the localization of common, distinct, and deleted sequences in DNA. This procedure is called **heteroduplexing.** Consider the DNA molecules #1 and #2 shown in Figure 4-20(a). These molecules differ in sequence only in one region. If a mixture of the two molecules is denatured and renatured, hybrid molecules having unpaired single strands are produced, as shown in Figure 4-20(b). Figure 4-21 shows an actual electron micrograph of a heteroduplex. Measurement of the lengths of the single and double-stranded regions yields the endpoints of the regions of nonhomology. Note that the two single strands of the bubble always have the same length.

Consider now molecule #3, shown in Figure 4-20(a). In this molecule, region A is deleted. If a hybrid is made between this molecule and molecule #2, the result is a molecule with a single loop, as shown in Figure 4-20(b).

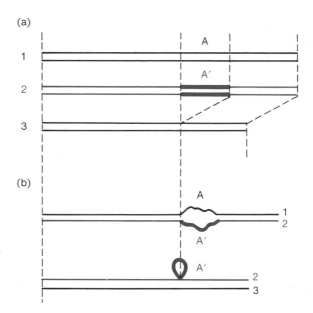

Figure 4-20
(a) Three DNA molecules to be hetero-duplexed. Sequences A and A′ of mole-cules 1 and 2 differ. Neither sequence is present in the deletion molecule 3. The dashed lines indicate reference points. (b) Heteroduplexes resulting from rena-turing molecules 1 and 2 or 2 and 3.

Figure 4-21
An electron micrograph of a heteroduplex be-tween λimmλ b2 DNA, which carries the b2 de-letion, and λimm434 b2+, in which the immλ seg-ment is replaced by the shorter, nonhomologous imm434 segment. (a) Two bubbles of nonhomology are seen. The identity of each single-stranded seg-ment is indicated. The arrow is explained in part (c). For technical reasons the circular λ DNA mole-cule is interrupted at the cohesive end region, denoted c.e. (b) An en-largement of the imm434–immλ segment of another molecule. (c) An interpretive draw-ing of panel (b). The arrow indicates a region of homology between the imm434 and immλ seg-ments. The same region is indicated by the arrow in panel (a). (Courtesy of Barbara Westmoreland and W. Szybalski.)

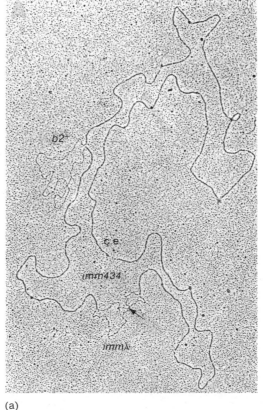

(a)

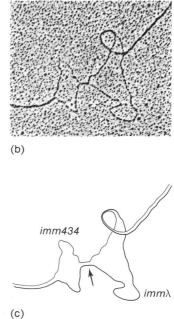

(b)

(c)

CIRCULAR AND SUPERHELICAL DNA

The intact DNA molecules of most prokaryotes and viruses are circular, though this was not noticed for many years because the DNA molecules usually broke during isolation, as mentioned in an earlier section. A circular molecule may be a **covalently closed circle,** which consists of two unbroken complementary single strands, or it may be a **nicked circle,** which has one or more interruptions **(nicks)** in one or both strands, as shown in Figure 4-22. With few exceptions, covalently closed circles are twisted, as shown in Figure 4-23. Such a circle is said to be a **superhelix** or a **supercoil.** Let us now examine what is meant by a circular DNA molecule being twisted.

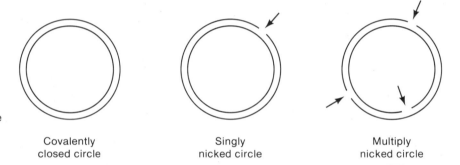

Figure 4-22
A covalently closed circle and two kinds of nicked circles. Arrows indicate the nicks.

Covalently
closed circle

Singly
nicked circle

Multiply
nicked circle

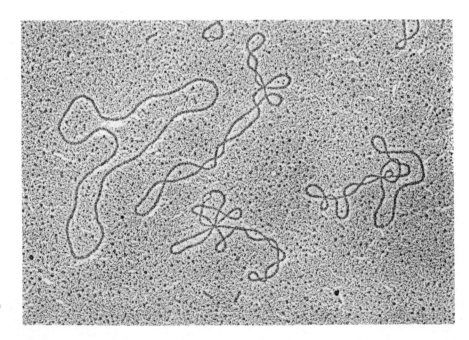

Figure 4-23
Nicked circular and supercoiled DNA of phage PM2. (Courtesy of K. G. Murti.)

Twisted Circles

The two ends of a linear DNA helix can be brought together and joined in such a way that each strand is continuous. If in so doing one of the ends is rotated 360° with respect to the other to produce some unwinding of the double helix, and then the ends are joined, the resulting covalent circle will, if the hydrogen bonds re-form, twist in the opposite sense to form a twisted circle, in order to relieve strain. Such a molecule will look like a figure 8 (that is, have one crossover point). If it is instead twisted 720° prior to joining, the resulting superhelical molecule will have two crossover points. The reason for the twisting is the following. In the case of a 720° unwinding of the helix, 20 base pairs must be broken (because the linear molecule has 10 base pairs per turn of the helix). However, a DNA molecule has such a propensity for maintaining a right-handed (positive) helical structure with 10 base pairs per turn that it will deform itself in such a way that the underwinding is rewound and compensated for by negative (left-handed) twisting of the circle. Similarly, the initial rotation might instead be in the sense of overwinding, in which case the joined circle will twist in the opposite sense; this is called a positive superhelix. *All naturally occurring superhelical DNA molecules are initially underwound and hence form negative superhelices.* Furthermore, the degree of twisting—that is, the **superhelix density**—is about the same for all molecules; namely, one negative twist is produced per 200 base pairs, or, 0.05 twists per turn of the helix.

Single-Stranded Regions in Superhelices

We have just pointed out that the strain of underwinding can be accommodated by negative supercoiling. It might seem that there are three other possible arrangements that would counteract the strain of underwinding: (1) The number of base pairs per turn of the helix could change. This does not happen, though, for thermodynamic reasons. (2) All of the underwinding could be taken up by having one or more large single-stranded regions (Figure 4-24). (3) The underwinding could be taken up in part by superhelicity and in part by bubbles. The real situation is alternative 3—that is, a state intermediate between forms (b) and (c) of Figure 4-24—because a DNA molecule is a dynamic structure that breathes. If a circular molecule were made that was not underwound, breathing (which is a transient unwinding that results from thermal motion) would introduce compensating transient negative twists. If the DNA is initially superhelical, the superhelix density will fluctuate as breathing occurs: the strain produced by the underwinding

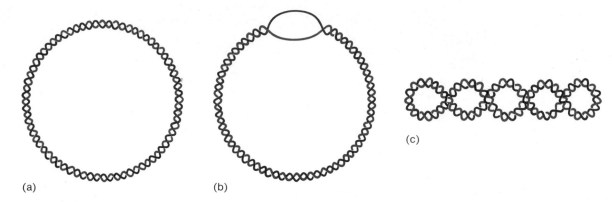

Figure 4-24
Different states of a covalent circle. (a) A nonsuper-coiled covalent circle having 36 turns of the helix. (b) An underwound covalent circle having only 32 turns of the helix. (c) The molecule in part (b), but with four super-helical turns to eliminate the underwinding. In solution, (b) and (c) would be in equilibrium; the equilibrium would shift toward (b) with increasing temperature.

is relieved in a superhelix both by the superhelicity and by an increase in the number and size of the bubbles and the duration of their existence. Thus, in a supercoil, the fraction of the molecule that is single-stranded at any moment is greater than in a nicked circle. Although it has not yet been demonstrated, it is likely that a sequence containing more than 90 percent A·T pairs may be permanently unpaired in a superhelical molecule. It is thought that this may play a role in such processes as initiating genetic recombination, the initiation of DNA replication, and the initiation of messenger RNA synthesis.

The Origin of Supercoiling

In bacteria, the underwinding of superhelical DNA is not a result of unwinding prior to end-joining but is introduced into preexisting circles by an enzyme called **DNA gyrase,** which will be described in Chapter 8, when replication is examined. In eukaryotes, the under-winding is a result of the structure of chromatin, a DNA-protein complex of which chromosomes are composed. In this complex, DNA is wound about specific protein molecules in a direction that introduces underwinding. Chromatin will be discussed in Chapter 6.

Experimental Detection of Covalently Closed Circles

In the life cycles of many organisms, the DNA molecules cycle through the various circular forms that have just been described. Several

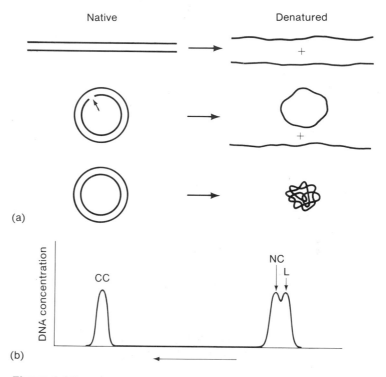

(a)

(b)

Figure 4-25
(a) Products of the denaturation of different forms of DNA. (b) Separation of covalently closed circles (CC), nicked circles (NC), and linear (L) molecules by sedimentation in alkali. The horizontal axis represents the length of a centrifuge tube. Sedimentation is from right to left.

techniques have been developed for distinguishing these forms. Covalently closed (but not necessarily superhelical) circles can be detected in two ways:

1. *Sedimentation at alkaline pH*. Above pH 11.3 all hydrogen bonds are broken and DNA molecules unwind. For a linear DNA molecule two single strands result whose s value is about 30 percent greater than that of the native DNA if the salt concentration is 0.3 M or greater. On the contrary, the two strands of a covalently closed circle cannot separate, so that the molecules collapse in a tight tangle whose s value is about three times as great as that of native DNA. If the circle has a single nick, one linear molecule and one single-stranded circle result, the s value of the latter being 14 percent greater than the former (Figure 4-25(a)). Figure 4-25(b) shows a sedimentation pattern for a mixture composed of linear molecules, nicked circles, and covalently closed circles in an akaline sucrose gradient.

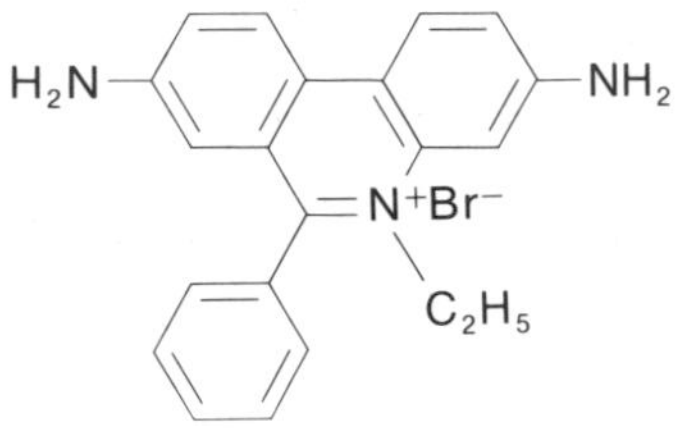

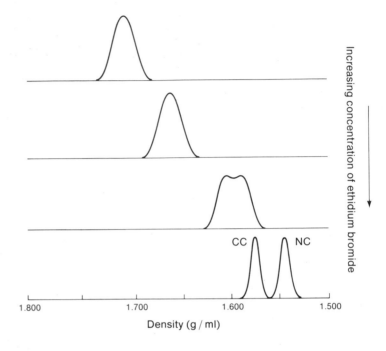

Figure 4-26
Chemical formula for ethidium bromide.

Figure 4-27
Effect of ethidium bromide on the density of DNA in a CsCl solution. A mixture of equal parts of nicked circles (NC) and covalently closed circles (CC) is centrifuged in CsCl containing various concentrations of ethidium bromide. The density of the DNA molecules decreases until, at saturation, the two components separate. The covalent circles bind less ethidium bromide and are therefore at a higher density.

2. *Equilibrium centrifugation in CsCl containing ethidium bromide.* **Ethidium bromide** (Figure 4-26) binds very tightly to DNA (in both dilute and concentrated salt solutions) and, in so doing, decreases the density of the DNA by approximately 0.15 g/cm^3. It binds by intercalating between the DNA base pairs and thereby causes the DNA molecule to unwind as more ethidium bromide is bound. A covalently closed, circular DNA molecule has no free ends so that, as it unwinds, the entire molecule twists in the opposite direction. For example, an O-shaped molecule that has bound enough ethidium bromide to produce one clockwise turn will twist in the counterclockwise direction to produce a molecule shaped like the figure 8. As more and more of the molecules intercalate, the 8-shaped molecule will become more twisted. Ultimately, the DNA molecule is unable to twist any more, so that no more ethidium bromide molecules can be bound. On the contrary, a linear DNA molecule or a nicked circle does not have the topological constraint of reverse twisting and can therefore bind more of the ethidium bromide molecules. Because the density of the DNA and ethidium bromide complex decreases as more ethidium bromide is bound and because more ethidium bromide can be bound to a linear molecule or an open circle than to a covalent circle, the covalent circle has a higher density at saturating concentrations of ethidium bromide. Therefore, covalent circles can be separated from the other forms in a density gradient, as shown in Figure 4-27.

Experimental Detection of Superhelicity

Superhelicity can be detected in three ways:

1. *Electron microscopy.* This has already been shown in Figure 4-23.

2. *Sedimentation at neutral pH.* A circle is more compact than a linear molecule so that its s value is higher. A nicked and a covalently closed circle have the same shape and thus the same s values. A superhelix is more compact than either form. The relative s values of a linear molecule, nonsuperhelical circle, and a superhelix are 1.00, 1.14, and 1.41, respectively. Sedimentation in the presence of varying concentrations of ethidium bromide is instructive because the intercalation of ethidium bromide introduces positive twists. Thus, as ethidium bromide is added, the DNA gradually loses its negative superhelicity until it is no longer superhelical; correspondingly, the s value decreases to the value for nonsuperhelical DNA. However, additional ethidium bromide introduces more positive twists and the s value rises again. This is shown in Figure 4-28. The concentration of ethidium bromide giving the minimum s value can be used to calculate the superhelix density.

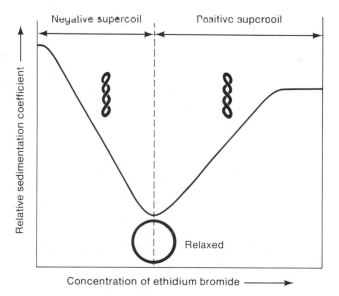

Figure 4-28
Intercalation of ethidium bromide into covalently closed circular DNA molecules. A negative supercoil, at a particular concentration of ethidium bromide, is relaxed by intercalation to yield a nonsupercoiled, slowly sedimenting form. More extensive intercalation introduces positive supercoiling, making the molecule more compact and rapidly sedimenting once again.

Figure 4-29
A gel electrophoregram showing the conversion of a covalent circle to a supercoil by DNA gyrase. (a) Maximally supercoiled molecules (intense band), used as a position standard. The faint band near the top contains contaminating nicked circles. (b) The species present early in the enzymatic reaction. Each band contains molecules having two more superhelical turns than the molecules in the band just above. (c) A later time—several bands containing molecules with additional twists are present. (d) Still later—by this time some molecules have become maximally supercoiled. Note the order of the lanes: a c b d. The direction of migration is from top to bottom. (Courtesy of James Wang and Tadaatsu Goto.)

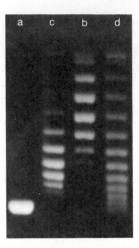

3. *Gel electrophoresis.* Because of the compactness of a superhelix, it has a very high mobility for its molecular weight. Furthermore, the resolution of gel electrophoresis is so great that it is possible to distinguish two molecules that differ by only a single twist. For example, each of the intermediates present in a reaction mixture in which the enzyme DNA gyrase is introducing twists can be separated, as shown in Figure 4-29.

THE STRUCTURE OF RNA

A typical cell contains about ten times as much RNA as DNA yet we have said little about RNA up to this point. With the exception of the RNA of one phage and a few viruses, RNA is a single-stranded polynucleotide. There are three primary types of RNA—ribosomal RNA (of which there are three or four forms), transfer RNA (of which there are about fifty different structures), and messenger RNA (of which there are almost as many different molecules as there are genes). These molecules superficially resemble single-stranded DNA in that single-stranded regions are interspersed with intramolecular double-stranded regions. Between 1/2 and 2/3 of the bases are paired. In single-stranded DNA the pairing is random and the paired regions tend to contain six or fewer pairs. Furthermore, if a sample of identical DNA molecules is denatured and intramolecular hydrogen bonds are allowed to form, the base-pairing pattern may differ from one molecule to the next. On the contrary, in RNA the double-stranded regions may contain up to twenty base pairs and a particular molecule has a definite base-pairing pattern.

The structures of the different classes of RNA molecules will be discussed in detail in Chapters 11 and 12.

HYDROLYSIS OF NUCLEIC ACIDS

Both DNA and RNA can be hydrolyzed to free nucleotides either chemically or enzymatically.

At very low pH (1 or less), phosphodiester hydrolysis of both DNA and RNA occurs. This is accompanied by breakage of the N-glycosylic bond between the base and the sugar so that free bases are produced. This procedure has been used in determining the base composition of nucleic acids. At a pH of about 4 the N-glycosylic bond by which purines are attached to deoxyribose in DNA is broken. Long-term exposure of DNA to this pH removes all of the purines, yielding a molecule known as **apurinic acid.**

Certain reagents add methyl groups to A, G, and C. This methylation makes the N-glycosylic bond especially labile at low pH and also when heated at neutral pH. In the DNA sequencing procedure to be described shortly, we shall make use of the ability of heat to remove methylated bases.

At high pH the behavior of DNA is strikingly different from that of RNA. DNA remains very resistant to hydrolysis at pH 13 (about 0.2 phosphodiester bonds broken per million bonds per minute at 37°C) whereas at pH 11 a typical RNA molecule is totally hydrolyzed to ribonucleotides in a few minutes at 37°. In one situation DNA is cleaved by alkali; if a base has been removed either by the pH or depurination reaction just described or by heating methylated DNA, the phosphodiester bond at the 5′ end of the deoxyribose lacking a base is broken by alkali as rapidly as RNA is cleaved.

A variety of enzymes hydrolyze nucleic acids. They are called **nucleases.** They usually show chemical specificity and are either deoxyribonucleases (DNase) or ribonucleases (RNase). Many DNases act on only single-stranded or only double-stranded DNA, although some act on either kind. Furthermore, some nucleases act only at the end of a nucleic acid, removing a single nucleotide at a time; these are called **exonucleases** and they may also be specific for the 3′ or 5′ end of the strand. Most nucleases act within the strand and are **endonucleases**; some of these are specific in that they cleave only between particular bases.

Nucleases serve a variety of biological functions and have also been useful in the laboratory for removing unwanted nucleic acids and as an early step in determining the base sequence of RNA molecules. Specific nucleases are described in Table 4-2; others will be described as needed in the text.

The **S1 endonuclease** isolated from *Aspergillus oryzae* has been of special use in recent years. This enzyme acts exclusively on single-stranded polynucleotides or on single-stranded regions of double-stranded molecules; it differs from other single-strand-specific enzymes

Table 4-2
Properties of selected nucleases

Nuclease	Substrate*	Site of cleavage	Product
Pancreatic ribonuclease	RNA	Endonuclease; adjacent to pyrimidines	Mono- or oligo-nucleotide terminating with pyrimidine nucleoside 3'-P
T1 RNase	RNA	Endonuclease; adjacent to guanosine	Mono- or oligo-nucleotide terminating with guanosine 3'-P
Pancreatic DNase I	DNA, RNA	Endonuclease	Oligonucleotides
Venom phospho-diesterase	RNA or DNA	Exonuclease at 3'-OH end	5'-P mononucleotides
Spleen phospho-diesterase	RNA or DNA	Exonuclease at 5'-OH end	3'-P mononucleotides
Micrococcal nuclease	SS DNA	Endonuclease	3'-P mononucleotides
E. coli endonuclease I	ds DNA	Many 3' → 5'-phosphoester sites	Oligonucleotides with with 5'-P end
Neurospora endonuclease	ss DNA or RNA	Many sites	5'-P mononucleotides
S1 nuclease	ss regions of DNA and RNA	Endonuclease; any ss site	5'-P mononucleotides
E. coli exonuclease I	ss DNA	Exonuclease at 5'-OH end	5'-P mononucleotides plus a terminal dinucleotide
E. coli exonuclease III	ds DNA	Exonuclease at 3'-OH end	5'-P mononucleotides
E. coli exonuclease V	DNA	Exonuclease at 3'-OH end	5'-P mononucleotides
E. coli exonuclease VII	ss DNA	Exonuclease at 3'-OH end	5'-P mononucleotides
Phage λ exonuclease	ds DNA	Exonuclease at 5'-P end	5'-P mononucleotides

*ds and ss are abbreviations for double-stranded and single-stranded, respectively.

in that the single-stranded region can be very small. We have noted earlier that supercoiled DNA contains somewhat long-lived single-stranded bubbles because of increased breathing. Supercoiled DNA can be broken by S1 nuclease because of these regions. In fact, this nuclease can be used to distinguish supercoiled from both nonsupercoiled covalent circles and nicked circular DNA, both of which are resistant to the enzyme. Interestingly, S1 nuclease makes a double-strand break because it acts on both single-stranded branches of the bubble. The single-stranded region can be as small as one base pair. For example, if two DNA molecules that differ by only one base pair are denatured and renatured, the heteroduplexes, which have one mismatched base pair, can be broken by S1 nuclease at the site of the mismatch. This technique has been used for physical mapping of mutations. That is, mutant and wild-type molecules are mixed, denatured and renatured, and then treated with S1 nuclease. A double-strand break occurs exactly at the mutant site. Electron microscopic or gel electrophoretic analysis of the length of the resulting DNA fragments can localize the mutation.

POLYNUCLEOTIDE KINASE, A USEFUL ENZYME

Polynucleotide kinase is an enzyme of unknown biological function which was first isolated from E. *coli* infected with phage T4 and which has since been found in a variety of mammalian cells. It catalyzes transfer of ^{32}P from γ-^{32}P-labeled-ATP to a 5'-OH end of RNA and DNA chains of all sizes, even to a mononucleotide. Naturally occurring nucleic acids possess 5'-P but not 5'-OH termini, but the 5'-P group is easily removed by the enzyme alkaline phosphatase to expose an OH terminus. Thus, the following reaction sequence can be used to label DNA with ^{32}P.

Some examples of the use of polynucleotide kinase are the following:

1. *Identification of the 5'-terminal nucleotide(s) of a nucleic acid.* Complete hydrolysis of the nucleic acid to 5'-P mononucleotides followed by electrophoretic separation of the products yields two ^{32}P-labeled mononucleotides. These are the terminal nucleotides. One is found in the case of a single-stranded polynucleotide.

2. *Measurement of the molecular weight of nucleic acids.* A DNA sample of known concentration (g/ml) is terminally labeled with

^{32}P by polynucleotide kinase. The amount of DNA-bound radioactivity in the sample is measured, which yields the number of ^{32}P atoms per sample. From the specific activity of the ATP (that is, the fraction of the ATP molecules containing a ^{32}P atom) one can calculate the total number of terminal phosphates per milliliter. If the DNA is double-stranded, the molecular weight in grams is

(g/ml of the sample)/(number of terminal phosphates × 1/2).

If the nucleic acid is single-stranded, the 1/2 is omitted.

Polynucleotide kinase is also used for terminal labeling in the base sequencing procedure described later in this chapter.

RADIOACTIVE LABELING OF NUCLEIC ACIDS

Many experiments require the use of radioactive nucleic acids. Several techniques are available for labeling biological molecules by growth of cells in radioactive media. The underlying principle is to add to the growth medium a precursor that only enters the molecule of interest, when this is possible. For DNA this is uncomplicated because thymine and thymidine are utilized only in DNA synthesis. Thus, if [^{3}H]thymidine or [^{14}C]thymidine is put in the growth medium, newly replicated DNA will contain ^{3}H and ^{14}C, respectively. This is by far the most common procedure for preparing radioactive DNA. Sometimes a greater amount of radioactivity is needed than can be attained with radioactive thymidine; then ^{32}P is used; it is added as the ^{32}PO$_4^{3-}$ ion. However, when this is done, both DNA and RNA become labeled as well as other phosphorus-containing compounds. If only radioactive DNA is needed, labeled cells can be treated with either RNase or dilute alkali to hydrolyze the RNA, as was explained in an earlier section. It then only becomes necessary to separate DNA molecules from the free RNA nucleotides, for which there are several simple procedures.

The only substance that is unique to RNA is ribose. However, if radioactive ribose (or any other sugar) is added to a growth medium, it would be metabolized and the label would appear in almost all compounds containing carbon. The best precursor is [^{3}H]uridine or [^{14}C]uridine, which is converted to uridine triphosphate (UTP) and then polymerized into RNA. However, some uridine is converted to dUTP, which is methylated to yield thymidine triphosphate and then incorporated into DNA. Thus the use of radioactive uridine yields both radioactive RNA and radioactive DNA. In order to have radioactive RNA only, it is necessary to treat the cell extract with a DNase to hydrolyze the DNA. RNA can also be labeled with ^{32}P; again, a final step must be removal of DNA with DNase.

SEQUENCING NUCLEIC ACIDS

Throughout this book, base sequences within DNA molecules will be described as a means of understanding the kinds of physical and chemical signals that are used by cells. This section describes how these sequences are determined.

There are two methods for determining the base sequence of a DNA molecule—the Maxam-Gilbert procedure and the Sanger procedure. Although both methods are equally effective, we shall describe only the Maxam-Gilbert procedure, as it is used more frequently at present. The other procedure is described in a reference listed at the end of this chapter.

In the Maxam-Gilbert procedure single-stranded DNA (derived from double-stranded DNA) is subjected to several treatments that cleave the DNA molecules and generate a family of short single-stranded molecules. The number of nucleotides in each fragment is determined by gel electrophoresis, which can separate molecules whose lengths differ by only a single nucleotide. Several cleavage protocols are used, which distinguish the four bases and provide the base sequence by the following rule:

> If a fragment is found that contains n nucleotides and is generated by a treatment active against a particular base, then that base is present in position n + 1 of the DNA strand, the position being counted from the 5' end.

Thus, if a 36-base fragment results from the protocol that identified guanine, then it is known that guanine is the 37th base in the original molecule.

We will now look more closely at the Maxam-Gilbert procedure, by which a single-stranded fragment containing a maximum of about 100 nucleotides can be sequenced. Such a fragment is obtained from a large DNA molecule by digestion with one or more restriction endonucleases, enzymes that cleave DNA at particular sites (Chapter 20). The fragment to be sequenced is processed in the following manner. First, the two 5'-P termini of the double-stranded molecule (which are at opposite ends of any fragment because the two strands are chemically antiparallel) are enzymatically removed by alkaline phosphatase and then replaced, by means of the polynucleotide-kinase procedure described earlier in this chapter, with ^{32}P, which is radioactive. The radioactive DNA is then denatured with alkali and, by using a special electrophoretic technique, the two single strands are then purified. Each of these strands is sequenced separately. (Because the base sequences are complementary, the information is redundant; however, not only do the two determinations serve as a check on one another, but, as will be seen shortly, it is necessary to sequence the other strand also in order to obtain the complete sequence.)

The sequencing procedure is the following (Figure 4-30 and Table 4-3). A sample containing the purified single strands is divided into two portions. To one sample (which we denote I) dimethyl sulfate is added. This methylates both adenine (A) and guanine (G), but mainly G. An essential feature of the protocol is that the reaction is not carried to completion, but only to the extent that about one purine per single strand is methylated. Since methylation occurs at random positions, the particular A or G that is methylated differs in each strand. The methylated DNA is next divided into two aliquots, which are called Ia and Ib. Sample Ia is heated; the effect of heating is to remove all methylated bases. The sample is then treated with alkali, which cleaves the sugar-phosphate chain at the site of the base that has been removed. This cleavage generates a set of fragments of varying size, and the differing number of nucleotides in each fragment is determined by the different positions of the methylated G or A. Sample Ia is said to contain the G-only fragments, because G was primarily methylated. Sample Ib is not heated but instead is treated with diluted acid, which removes mainly methylated A; then it is treated with alkali to cleave the sugar-phosphate chain at the site where an A was removed. Thus, in sample Ib, fragments are produced whose size is determined mainly by the position of the methylated A—these are called the A + G fragments. Note that every G-only fragment is also present in the (A + G) collection.

The two samples Ia (G-only) and Ib (A + G) are now electrophoresed in a 20 percent polyacrylamide gel containing 8 M urea, a denaturant that prevents hydrogen-bonding and hence keeps the fragments single-stranded. After electrophoresis, the electrophoretic bands are located by placing autoradiographic film (a film that responds to radioactivity) on the gel and exposing the film for several days. The single terminal ^{32}P atom, which was enzymatically added before the methylation reaction, is the source of the radioactivity. Note that when a 5′-^{32}P-labeled molecule is cleaved, only one of the two fragments produced contains ^{32}P and only that one is detected.

The positions of A and G in the single strand are determined by the following rules (which are special cases of the rule stated on page 141):

If a band containing n nucleotides is present in both the (A + G) lane and the G-only lane, then a G exists at position $n + 1$ in the original molecule,

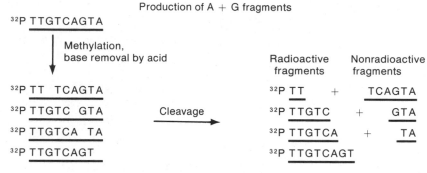

Production of A + G fragments

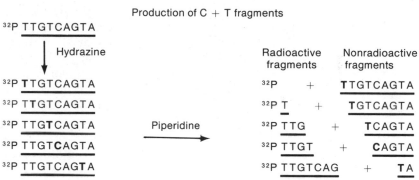

Figure 4-30
A diagram showing the production of the A+G and C+T fragments by the Maxam-Gilbert sequencing procedure. In the hydrazine-treated molecules the bold-faced letters indicate the affected bases. The generation of the G-only and C-only fragments is not shown but is explained in the text.

Production of C + T fragments

Table 4-3
Four treatments used in the Maxam-Gilbert sequencing procedure

Treatment	Base-modifying agent	Modification	Postmodification treatment	Base identified
Ia	Dimethyl sulfate	Purine methylation	Heating at pH7, then NaOH	G
Ib	Dimethyl sulfate	Purine methylation	Cold HCl, then hot NaOH	A + G
IIa	Hydrazine	Opening of pyrimidine ring	Heating in piperidine	T + C
IIb	Hydrazine + 2 M NaCl	Opening of cytosine ring	Heating in piperidine	C

Dimethyl sulfate   **Hydrazine**   **Piperidine**

Figure 4-31
A portion of a sequencing gel. The sequence is read from the bottom to the top. Each horizontal row represents a single base. Each vertical column represents a sample treated to indicate the position of G, A or G, T or C, and C, respectively. The sequence from a portion of the gel is shown.

and

> If a band containing n nucleotides is present only in the (A + G) lane, then an A exists at position $n + 1$ in the original molecule.

So far, the analysis has used only sample I, which was methylated and then treated in two different ways to generate the (A + G) and the G-only classes of fragments. Sample II is used to identify the positions of cytosine (C) and thymine (T). This sample is also divided into two portions IIa and IIb, which are reacted with hydrazine in either dilute buffer (sample IIa) or in 2 M NaCl (sample IIb). Hydrazine reacts with C and T only (and neither A nor G) but in 2 M NaCl the reaction is with C only (Table 4-2). Cleavage at the site of the hydrazine reaction is accomplished by treatment with piperidine, which breaks the sugar-phosphate backbone at the 5′ side of each base that has reacted with hydrazine. The sizes of the fragments are then determined by the positions of both C and T and by C only. Thus, after electrophoresis and autoradiography, the positions of C and T are determined by the following rules:

> If a fragment containing n nucleotides is present only in the (C + T) lane (sample II), there is a T at position $n + 1$ in the original molecule.

and

> If a fragment containing n nucleotides is present in both the (C + T) lane and the C-only lane, there is a C at position $n + 1$ in the original molecule.

All four samples Ia, Ib, IIa, and IIb are usually electrophoresed simultaneously so that all bands are seen in a single gel. This enables the sequence to be read directly from the gel. Figure 4-31 is a photograph of a sequencing gel. The shortest fragments are those that move the fastest and farthest. Each fragment contains the original 5′-^{32}P group so that the sequence can be read from the bottom to the top of the gel. Note that where a band in the figure is labeled with base X, this means that X was removed from the 3′ end of the molecule to generate the fragment.

If only one single strand of the original double-stranded DNA molecule were analyzed in the way just described, the complete base sequence would not be obtained. This is for two reasons. First, if the cleavage is at position $n + 1$ (counting from the ^{32}P-labeled terminus), the number of bases in the fragment is n. This means that no fragment identifies the 5′-terminal nucleotide. Second, the mononucleotide that would identify the penultimate base cannot for technical reasons be detected unambiguously. The identity of these bases can be obtained by sequencing the complementary single strand, in which case the two unidentified bases from the first strand are complementary to the 3′-terminal bases of the second strand. Sequencing both strands also has the advantage that the sequence determined from one strand confirms the sequence of the other strand.

Although the gels can resolve two fragments differing in length by one nucleotide, they are not capable of distinguishing one hundred different fragments simultaneously, unless the gel is exceedingly long. To avoid this problem, after all of the chemical treatments one usually divides all samples into three aliquots and electrophoreses each set of four samples for different times. For example, run 1 might give the sequence of bases 3 through 40, run 2 that of bases 35 through 75, and run 3 that of bases 65 through 100. By observing the common sequences (for example, bases 35–40 of runs 1 and 2), the complete sequence can be worked out.

This method, as well as other procedures, has been used to determine the base sequence of a DNA molecule having 40,000 base pairs (see Chapter 15).

LEFT-HANDED DNA HELICES

The Watson-Crick structure of DNA (the B form) and the related A form of DNA are both right-handed helices. Subtle changes in the structure of DNA occur if the salt concentration is very high and if divalent cations (Ca^{2+}, Mg^{2+}, and Mn^{2+}) are present but the helix remains right-handed. Several years ago it was observed that the double-stranded hexanucleotide

CGCGCG

GCGCGC

undergoes a drastic and reversible structural change when the NaCl concentration exceeds 2 M or if the $MgCl_2$ concentration is allowed to exceed 0.7 M. The most striking aspect of the change is that the optical activity of the solutions changes sign when the transition occurs. X-ray diffraction analysis indicates that in these high-ionic-strength solutions the DNA is a left-handed helix. This left-handed structure is called **Z-DNA** (because of its zig-zag structure). It has been found that this structure is not confined to the poly($dC \cdot dG$) molecule but extends to any polydeoxynucleotide in which purines and pyrimidines alternate—for example,

ACACACAC

TGTGTGTG

Furthermore, if an alternating purine-pyrimidine sequence is contained in a long tract of DNA, for example,

···TGATCCGCGCGCGAGTCTT···
···ACTAGGCGCGCGCTCAGAA···

the purine-pyrimidine alternating sequence can still assume the Z configuration at 2 M NaCl while the remainder of the DNA has the B configuration.

Z-DNA is not simply a mirror image of a right-handed B-type helix. First, the sugar-phosphate backbone in Z-DNA follows a zig-zag path, as shown in Figure 4-32. Second, if the Z-DNA region is extensive, the

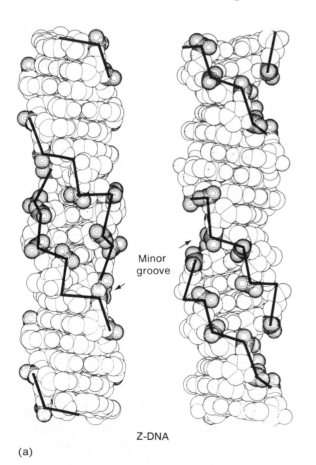

Minor groove

Z-DNA

(a)

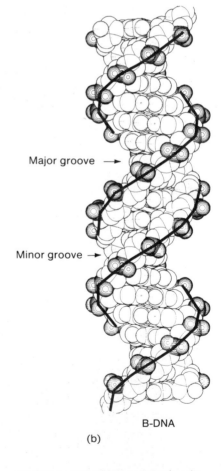

Major groove →

Minor groove →

B-DNA

(b)

Figure 4-32
Side views of (a) left-handed Z-DNA and (b) right-handed B-DNA. The heavy lines indicate the sugar-phosphate backbone. The zig-zag path of the backbone in Z-DNA is evident. Although it cannot be seen in panel (a), the minor groove of Z-DNA is quite deep, penetrating the helix axis. In contrast, the sugar-phosphate backbone of B-DNA is smooth and the grooves are shallow. (Courtesy of Andrew Wang.)

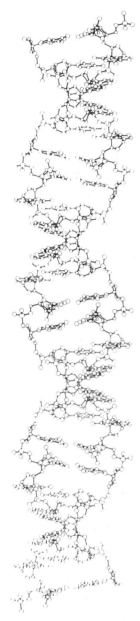

Figure 4-33
The structure of double-stranded poly(dGC)—an alternating copolymer of dG and dC—showing how the helix axis continuously changes its direction, producing a "wavy" molecule. (Courtesy of S. Arnott and R. Chandrasekaran.)

molecule becomes wavy, as shown in Figure 4-33. There are also other differences. Two important differences are the following: (1) In Z-DNA there are 12 base pairs per turn of the helix, in contrast with B-DNA, in which there are 10 turns. (2) There is only a single groove in the helix, rather than a major and a minor groove, as in B-DNA. The details of the differences between Z-DNA and B-DNA are not important at this point—the significant fact is that the structures are very different.

The immediate question that arises is whether the left-handed structure has biological significance. Certainly the intracellular salt concentration does not ever approach 2 M, so *in vivo* salt could not cause the transition. However, cells contain many highly charged DNA-binding proteins (which will be discussed in Chapters 6 and 8) and these could certainly produce a very high ionic strength in the vicinity of the DNA. For this reason and since the transition from B-DNA to Z-DNA occurs spontaneously at room temperature (that is, with little input of energy), it is believed that left-handed helical regions do exist in intracellular DNA.

The best evidence for the existence of Z-DNA in nature comes from recent studies with antibodies to Z-DNA. Bromination of the dC · dG alternating double-stranded molecule produces a stable Z form that maintains the Z structure at all concentrations. Antibodies to this molecule have been prepared by coupling the molecule to bovine serum albumin and injecting the coupled molecule into a rabbit. The antibody has been isolated from immunized rabbits and has been found to react with Z-DNA and not with B-DNA. Antibodies of another sort can be prepared by injecting a goat with rabbit immunoglobulin; this antibody, called goat anti-rabbit-immunoglobulin, reacts specifically with any rabbit immunoglobulin G. A highly fluorescent molecule, fluorescein, is then coupled to the goat antibody. Cells from the salivary gland of the fruit fly *Drosophila,* showing chromosomes, are then prepared for light microscopy. Anti-Z-DNA is added and then fluorescent goat anti-rabbit-immunoglobulin. The cells are then washed and observed by fluorescence microscopy. If there is Z-DNA in the chromosomes, anti-Z-DNA will be bound and fluorescent goat antibody will also be bound at the same point. Thus, Z-DNA, if present, can be detected by the fluorescence. Figure 4-34 shows the fluorescence of chromosomes treated by this protocol. Thus, Z-DNA must be present in the *Drosophila* chromosomes. The significance of this very recent observation is not yet known.

Throughout this book we will see examples of recognition of particular base sequences by regulatory proteins. It is usually assumed that the protein recognizes the sequence of bases. However, it is possible that DNA conformation is also important. Some circumstantial evidence supports the hypothesis that Z-DNA might be just such a region. For example, in one class of animal viruses (the rodent parvoviruses) there are two neighboring segments of DNA containing

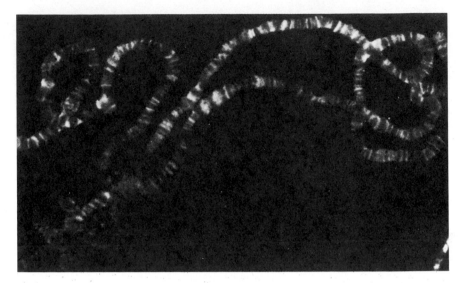

Figure 4-34
Chromosomes of the fruit fly, to which a fluorescent antibody to Z-DNA is bound. (From A. Nordheim, *et al*. 1981. *Nature*, 294: 417–422.

lengthy alternating dC • dG sequences; these are adjacent to the site at which DNA replication begins.

Z-DNA has only been known since 1980 and is being actively studied; new information about its possible role in gene regulation is accumulating rapidly.

SATELLITE DNA

The mean base composition of DNA varies widely throughout the microbiological world. The extremes are 23 percent G+C for the *Clostridium* genus of bacteria and 76 percent G+C for *Sarcina* and the *Micrococci. E. coli,* the bacterium that we will study most often in this book, is 50 percent G+C. Proceeding upward along the evolutionary scale, the variation becomes smaller. For instance, for most plants and animals, the extremes are 48 percent and 52 percent.

A convenient way to measure the mean base composition is by sedimenting the DNA to equilibrium in buoyant CsCl because the density of DNA is linearly related to base composition (Figure 4-35). When this was first done, it was observed that even when the DNA is broken randomly into about 1000 fragments, the bands that resulted are usually very narrow. This indicates that the mean base composition does not vary very much from fragment to fragment—that is, the DNA of an organism is not usually 25 percent G+C in one region and 75 percent G+C in another.

However, when the DNA of the crab was examined, two discrete bands were observed (Figure 4-35). The main band accounted for 70

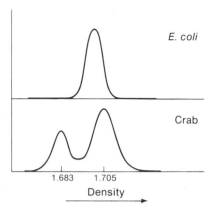

Figure 4-35
The concentration distribution of the DNA of the bacterium *E. coli* and the crab *Cancer borealis* after equilibrium centrifugation in CsCl. As is common with bacteria, the *E. coli* DNA has a narrow range of densities. The crab DNA consists of two discrete fractions, one of very low density. This minor band is called *satellite DNA*. (From N. Sueoka, 1961. *J. Mol. Biol.*, 3: 31–40. Copyright: Academic Press Inc. [London] Ltd.)

percent of the total DNA and was 52 percent G + C. The smaller band, which was 3 percent G + C, was termed **satellite DNA.** This phenomenon has been observed often in eukaryotes in which the satellite DNA may have a higher or lower G + C content than the main fraction. Some organisms have two satellites. C_0t analysis has shown that satellite DNA is usually highly repetitive, consisting sometimes of 10^6 copies of a single sequence.

Satellite DNA will be discussed in greater detail in Chapter 22.

REFERENCES

Bauer, W. R. 1978. "Structure and reactions of closed duplex DNA." *Ann. Rev. Biophys. Bioeng.*, 7, 287–313.

Bauer, W. R., F. H. C. Crick, and J. H. White. 1980. "Supercoiled DNA." *Scient. Amer.*, 243 (July), 118–133.

Bloomfield, V. A., D. M. Crothers, and I. Tinoco. 1974. *Physical Chemistry of the Nucleic Acids.* Harper & Row.

Cantor, C. R., and P. R. Schimmel. 1980. *Biophysical Chemistry. Parts I and III.* W. H. Freeman and Co.

Cozzarelli, N. 1980. "DNA gyrase and supercoiling of DNA." *Science,* 207, 953-960.

Cohn, W. (ed.) 1976–81. *Progress in Nucleic Acid Research and Molecular Biology.* Vols. 16–26. Academic Press.

Crick, F. H. C. 1976. "Linking numbers and nucleosomes." *Proc. Nat. Acad. Sci.,* 73, 2639–2643.

Davidson, J. N. 1977. *The Biochemistry of the Nucleic Acids.* Academic Press.

Duchesne, J. 1973. *Physico-Chemical Properties of Nucleic Acids.* Academic Press.

Fiddes, J. C. 1977. "The nucleotide sequence of a viral DNA." *Scient. Amer.,* 237 (December), 54–67.

Freifelder, D. (ed.) 1978. *The DNA Molecule: Structure and Properties.* W. H. Freeman and Co.

Freifelder, D. 1982. *Physical Biochemistry.* 2nd Ed. W. H. Freeman and Co.

Kornberg, A. 1980. *DNA Replication.* W. H. Freeman and Co.

Maxam, A. M., and W. Gilbert. 1977. "A new method for sequencing DNA." *Proc. Nat. Acad. Sci.,* 74, 560–564.

Nordheim, A., M. L. Pardue, E. M. Lafer, A. Möller, B. D. Stollar, and A. Rich. 1981. "Antibodies to left-handed Z-DNA bind to interband regions of *Drosophila* polytene chromosomes." *Nature,* 294, 417–422.

Sanger, F., S. Nicklen, and A. R. Coulson. 1977. "DNA sequencing with chain-terminating inhibitors." *Proc. Nat. Acad. Sci.,* 74, 5463–5467.

Tanford, C. 1978. "The hydrophobic effect and the organization of living matter." *Science,* 200, 1012–1018.

Ts'o, P. O. P. 1974. *Basic Principles in Nucleic Acid Chemistry.* Academic Press.

Watson, J. D., and F. H. C. Crick. 1953. "Molecular structure of nucleic acid. A structure for deoxyribose nucleic acid." *Nature,* 171, 737–738.

Watson, J. D., and F. H. C. Crick. 1953. "Genetic implications of the structure of desoxyribosenucleic acid." *Nature,* 171. 964–867.

Younghusband, H. B., and R. B. Inman. 1974. "The electron microscopy of DNA." *Ann. Rev. Biochem.,* 43, 605–617.

5 The Physical Structure of Protein Molecules

All proteins are polymers of amino acids, yet each species of protein molecule has a unique three-dimensional structure determined principally by the amino acid sequence of that protein, in contrast with DNA, which has a universal structure—the double helix. This makes the study of proteins very complex but, on the other hand, the diversity of protein structures enables these molecules to carry out the thousands of different processes required by a cell and makes proteins fascinating objects to study.

The study of the detailed structure of proteins is beyond the scope of this book, being heavily dependent on a variety of optical techniques, especially the mathematically complex technique of x-ray diffraction. For this reason the discussion of proteins in this chapter will be a survey rather than a detailed description. For further information the reader should consult the references at the end of this chapter.

SIZES OF PROTEIN MOLECULES

In the previous chapter we saw that nucleic acid molecules are very large, having molecular weights often as high as 10^{10}. Protein molecules are much smaller; in fact, the molecular weight of a typical protein molecule is comparable to that of the smallest nucleic acid molecules, the transfer RNA molecules. The molecular weights of hundreds of different proteins have been measured. Typical polypeptide chains have molecular weights ranging from 15,000 to 70,000. The average molecular weight of an amino acid is 110, which means that typical polypeptide chains contain some 135 to 635 amino acids. The sizes of some proteins are outside this range. For example, the polypeptide hormones found in higher animals contain 8 to 12 amino acids, and thus, are relatively small. The largest known polypeptide chain has a molecular weight of 165,000 and therefore contains about 1500 amino acids. It will be seen later in this chapter that many proteins consist of several polypeptide chains joined by various noncovalent interactions. The molecular weights of these multisubunit proteins typically range from 75,000 to 200,000; the largest known enzyme, polymerase III holoenzyme, has fourteen subunits and a molecular weight of 760,000.

The length of a typical polypeptide chain, if it were fully extended, would be 1000 to 5000 Å. A few of the longer fibrous proteins, such as myosin (from muscle) and tropocollagen (from tendon) are in this range—with lengths of 1600 Å and 2800 Å, respectively. However, most proteins are highly folded and their longest dimension usually ranges from 40 to 80 Å. This folding is described in the next section.

STRUCTURE OF A POLYPEPTIDE CHAIN

In Chapter 3 the basic chemical features of protein molecules were described. That is, a protein is a polymer of amino acids in which carbon atoms and peptide groups alternate to form a linear polypeptide chain and specific groups—the amino-acid side chains—project from the α-carbon atom (Figure 3-1). The term "linear" requires careful consideration, for, as will be seen in the following sections, a polypeptide chain is highly folded and can assume a variety of three-dimensional shapes; each shape in turn consists of several standard elementary three-dimensional configurations and other configurations which may be unique to that molecule.

The Folding of a Polypeptide Chain

A fully extended polypeptide chain, if it were to exist, would have the configuration shown in Figure 5-1. (The chain is not perfectly straight because the C—N and C—C bonds in which the α-carbon atom participates are not collinear.) Such an extended zig-zagged molecule could not exist without stabilizing interactions to maintain the extension. In fact, a single polypeptide is never completely extended but is folded in a complex way. If there were free rotation about every bond in the chain and no interaction between different parts of the chain, the

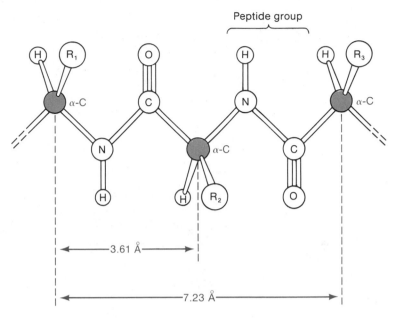

Peptide group

Figure 5-1
The configuration of a hypothetical fully extended polypeptide chain. The length of each amino acid residue is 3.61 Å; the repeat distance is 7.23 Å. The α-carbon atoms are shown in red. Side chains are denoted by R.

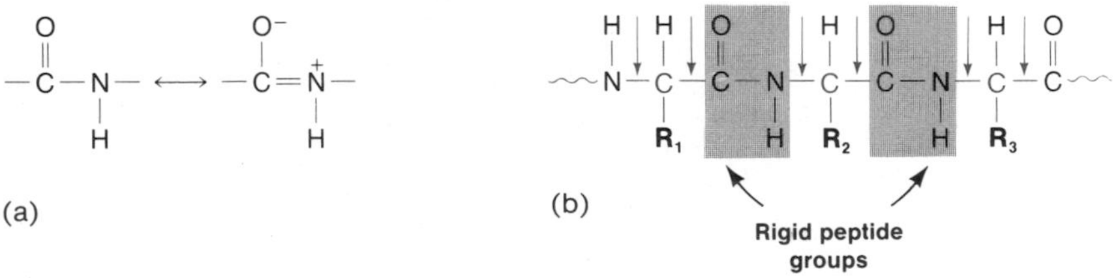

(a)

(b)

Rigid peptide
groups

Figure 5-2
The planarity of the peptide group. (a) Two forms of the peptide bond in equilibrium. In the form at the right the double bond creates a rigid unit. Owing to the equilibrium, the peptide group has a partial double-bond character and is rigid. (b) The peptide groups in a protein molecule. They are planar and rigid, but there is freedom of rotation about the bonds (red arrows) that join peptide groups to α-carbon atoms (red).

folding would be random. Instead, three rules govern the manner of folding:

1. The peptide bond has a partial double-bond character (Figure 5-2) and hence is constrained to be planar. Free rotation occurs only between the α carbon and the peptide unit. Thus, the

polypeptide chain is flexible but is not as flexible as would be the case if there were free rotation about all of the bonds.

2. The side chains of the amino acids cannot overlap. Thus, the folding can never be truly random because certain orientations are forbidden.

3. Two charged groups having the same sign will not be very near one another. Thus, like charges tend to cause extension of the chain.

However, in addition to these rules, folding behaves according to general tendencies, a few of which are the following:

1. Amino acids with polar side chains tend to be on the surface of the protein in contact with water.

2. Amino acids with nonpolar side chains tend to be internal. Very hydrophobic side chains tend to cluster.

(These two tendencies have been **described** by likening a protein molecule to an oil droplet with a polar **coat**.)

3. Hydrogen bonds tend to form **between** the carbonyl oxygen of one peptide bond and the **hydrogen** attached to a nitrogen atom in another peptide bond. This **hydrogen-bonding** gives rise to two fundamental polypeptide **structures** called the α helix and the β structure, which will be **described** in the following section.

4. The sulfhydryl group of the **amino** acid cysteine tends to react with an —SH group of a second **cysteine** to form a covalent S—S (**disulfide**) bond. Such bonds **pose** powerful constraints on the structure of a protein (Figure 5-3). **Most** proteins contain several cysteines and it is not **generally** possible to predict which cysteines will be paired.

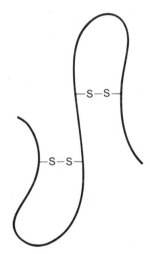

Figure 5-3
A polypeptide chain in which four cysteines are engaged in two disulfide bonds.

These tendencies should not **be** considered invariable because there are many exceptions. Nevertheless they indicate that the structure of a protein may be changed markedly by a single amino acid substitution—for example, a polar amino acid for a nonpolar one; similarly, the change might be minimal if one nonpolar amino acid replaces another nonpolar one. This notion will be encountered again in Chapter 10 when mutations are considered.

The three-dimensional shape of a polypeptide chain is a result of a balance between all of the rules and tendencies just described and can be very complex. However, in examining many polypeptide chains, it has become apparent that certain geometrically regular arrays of the chain are found repeatedly in different polypeptide chains and in different regions of the same chain. These are the arrays resulting from hydrogen-bonding between different peptide groups. These arrays are described in the next section.

Hydrogen Bonds and the α Helix

In the absence of any interactions between different parts of a polypeptide chain, a random coil would be the expected configuration. However, hydrogen bonds easily form between the H of the N—H group and the O of the carbonyl group. Figure 5-4 shows several types of hydrogen-bonding possibilities, each of which causes a folding of the polypeptide chain. Such bonding occurs when peptides can hydrogen-bond to one another more strongly than they can form hydrogen bonds with water, which is often the case.

Studies of the effect, on proteins, of urea,

$$H_2N-\overset{\displaystyle \overset{O}{\|}}{C}-NH_2$$

a molecule that forms hydrogen bonds with a variety of substances, indicates that hydrogen bonds are present in polypeptide chains. That is, addition of urea causes most proteins to undergo a structural transition from some definite shape to the random coil configuration. A large number of proteins contain regions having a repeat distance of 5.5 Å. The existence of such repetition implies that some order is present in this region. However, since the distance repeated is less than 7.2 Å (the value that would be present if the chain were fully extended), such a region must contain a polypeptide segment that is foreshortened in some way. Linus Pauling and Robert Corey showed that these facts were consistent with a helical structure, which they named the **α helix.** In the α helix the polypeptide chain follows a helical path that is stabilized by hydrogen-bonding between peptide groups. Each peptide group is hydrogen-bonded to two other peptide groups, one three units ahead and one three units behind in the chain direction (Figure 5-5). The

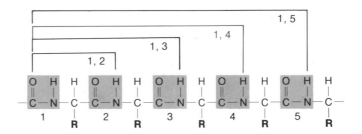

Figure 5-4
Possible hydrogen bonds between the C=O group in peptide unit 1 (lower numbers) and the N—H groups of other peptide units. The red line shows the hydrogen bond present in the α helix. The C=O and N—H bonds have been drawn in the same direction for clarity.

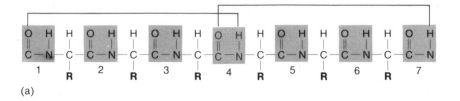

(a)

Figure 5-5

Properties of an α helix. (a) The two hydrogen bonds in which peptide group 4 (red) is engaged. The peptide groups are numbered below the chain. (b) An α helix drawn in three dimensions, showing how the hydrogen bonds stabilize the structure. The red dots represent the hydrogen bonds. The hydrogen atoms that are not in hydrogen bonds are omitted for the sake of clarity.

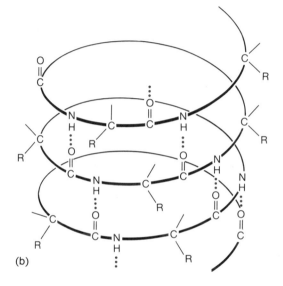

(b)

Figure 5-6

A schematic diagram of a cross-section of one turn of an α helix. The α-carbon atoms are red and the side chains are black. The carboxyl C is pink and the amino N is gray. The polypeptide is rotating clockwise and advancing toward the viewer. The van der Waals radii of the backbone atoms are larger than drawn and almost totally fill the core of the helix.

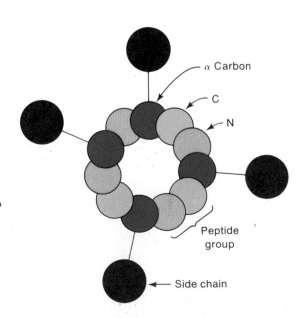

helix has a pitch of 5.4 Å, which is the repeat distance, a diameter of 2.3 Å, and contains 3.6 amino acids per turn. Thus, it is a much tighter helix than the DNA helix (Figure 5-6). The side chains, including those that are ionized, do not participate in forming the α helix.

In the absence of all interactions other than the hydrogen-bonding just described, the α helix is the preferred form of a polypeptide chain because, in this structure, all monomers are in identical orientation and each forms the same hydrogen bonds as any other monomer. Thus, polyglycine, which lacks side chains and hence cannot participate in any interactions other than those just described, is an α helix.

If all monomers are not identical or if there are secondary interactions that are not equivalent, the α helix is not necessarily the most stable structure. Then, not only is it true that the amino-acid side chains do not participate in forming the α helix, but also that the side chains are responsible for preventing α-helicity. A striking example of the disruptive effect of a side chain on an α helix is evident with the synthetic polypeptide polyglutamic acid, a polypeptide containing only glutamic acid. At a pH below about 5, the carboxyl group of the side chain is not ionized and the molecule is almost purely an α helix. However, above pH 6, when the side chains are ionized, electrostatic repulsion totally destroys the helical structure (Figure 5-7). With the synthetic polypeptide polylysine the pH dependence is reversed since the NH_2 group in the side chain of lysine (the ϵ-NH_2 group) is charged below pH 10.

If the amino acid composition of a real protein is such that the helical structure is extended a great distance along the polypeptide backbone, the protein will be somewhat rigid and fibrous (not all rigid fibrous proteins are α helical, though). This structure is common in many structural proteins, such as the α-keratin in hair.

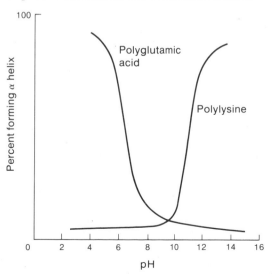

Figure 5-7
Dependence of α-helical content on pH, for two polyamino acids.

β Structures

Another common hydrogen-bonded configuration is the **β structure.** In this form, the molecule is almost completely extended (repeat distance = 7 Å) and hydrogen bonds form between peptide groups of polypeptide segments lying adjacent and parallel with one another (Figure 5-8(a)). The side chains lie alternately above and below the main chain.

Two segments of a polypeptide chain (or two chains) can form two types of β structure, which depend on the relative orientations of the segments. If both segments are aligned in the N-terminal-to-C-terminal direction or in the C-terminal-to-N-terminal direction, the β structure is **parallel.** If one segment is N-terminal to C-terminal and the other is C-terminal to N-terminal, the β structure is **antiparallel.** Figure 5-8(b) shows how both parallel and antiparallel β structures can occur within a single polypeptide chain.

When many polypeptides interact in the way just described, a pleated structure results called the **β-pleated sheet** (Figure 5-8(c)). These sheets can be stacked and held together in rather large arrays by van der Waals forces and are often found in fibrous structures such as silk.

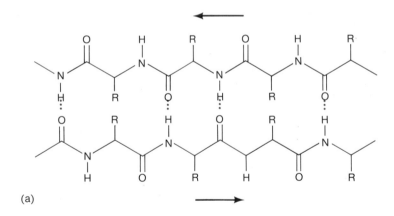

(a)

Figure 5-8
β structures. (a) Two regions of nearly extended chains are hydrogen-bonded (red dots) in an antiparallel array (arrows). The side chains (R) are alternately up and down. (b) Antiparallel and parallel β structures in a single molecule. (c) A large number of adjacent chains forming a β pleat.

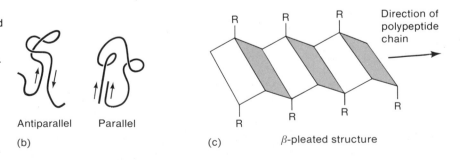

Antiparallel   Parallel

(b)

(c)   β-pleated structure

Direction of polypeptide chain

Fibrous and Globular Proteins

Few proteins are pure α helix or β structure; usually regions having each structure are found within a protein. Since these configurations are rigid, a protein in which most of the chain has one of these forms is usually long and thin and is called a **fibrous protein.** In contrast are the quasi-spherical proteins called the **globular proteins** in which α helices and β structures are short and interspersed with randomly coiled regions and compact structures.

The fibrous proteins are typically responsible for the structure of cells, tissues, and organisms. Some examples of structural proteins are collagen (the protein of tendon, cartilage, and bone), elastin (a skin protein), tubulin (a protein that maintains the shape of nerve cells), and actin (a ubiquitous protein that contributes to the shape of almost all animal cells). Some of the fibrous proteins are not soluble in water—examples are the proteins of hair and silk.

The catalytic and regulatory functions of cells are performed by proteins that have a well defined but deformable structure. These are the globular proteins, of which the catalytic proteins, or enzymes, are the most widely studied. Globular proteins are compact molecules having a generally spherical or ellipsoidal shape. Large segments of the polypeptide backbone of a typical globular protein are α-helical. However, the molecule is extensively bent and folded. Usually, the stiffer α-helical segments alternate with very flexible randomly coiled regions, which permit bending of the chain without excessive mechanical strain. Numerous segments of the chain, which might be quite distant along the backbone, form short parallel and antiparallel β structures; these also are responsible in part for the folding of the backbone (Figure 5-8(c)). The α helix and β structures are called the **secondary structure** of the molecule. The extensive folding of the backbone is usually called the **tertiary structure** or **tertiary folding.**

A very important distinction can be made between secondary and tertiary structure; namely,

> Secondary structure results from hydrogen-bonding between peptide groups whereas tertiary structure is formed from β structures and several different side-chain interactions.

The most prevalent interactions responsible for tertiary structure are the following:

1. Ionic bonds between oppositely charged groups in acidic and basic amino acids.
2. Hydrogen bonds between the hydroxyl group in tyrosine and a carboxyl group of aspartic or glutamic acids.

3. Hydrophobic clustering between the hydrocarbon side chains in phenylalanine, leucine, isoleucine, and valine.
4. Metal-ion coordination complexes between amino, hydroxyl, and carboxyl groups, ring nitrogens, and pairs of SH groups.

Hydrophobic clustering (item 3 above) is the most important stabilizing feature.

Figure 5-9 shows a schematic diagram of a hypothetical protein (in two dimensions) in which several of these interactions determine the structure. This figure should be examined carefully for it indicates the role of different features of a protein molecule in determining the overall conformation of the molecule. One can see the following:

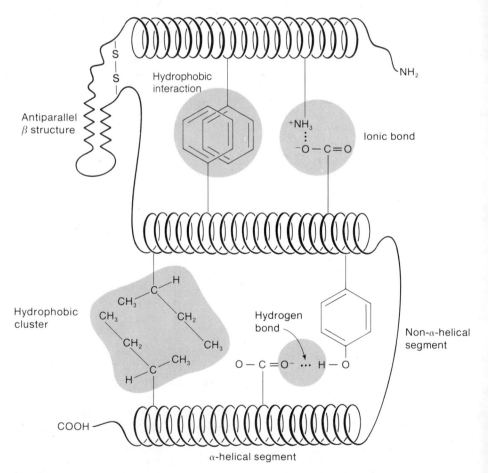

Figure 5-9
A hypothetical globular protein having several types of side-chain interactions.

1. Disulfide bonds bring distant amino acids together.
2. Hydrophobic interactions bring distant amino acids together.
3. Hydrogen bonds sometimes bring distant amino acids together, but usually a single hydrogen bond makes a more subtle change in position.
4. Electrostatic interactions bring amino acids together or keep them apart, depending on the signs of the charges.
5. A β structure brings distant segments of the polypeptide backbone together and creates rigidity.
6. An α helix makes adjacent regions of a polypeptide backbone stiff and linear.
7. Van der Waals forces produce specific interactions between clusters of amino acids which may or may not be nearby in the polypeptide chain.

Figure 5-10 is an idealized drawing of the three-dimensional structure of the enzyme carbonic anhydrase, showing a β structure, a hydrophobic cluster, and three amino acids in a coordination bond with a metal ion.

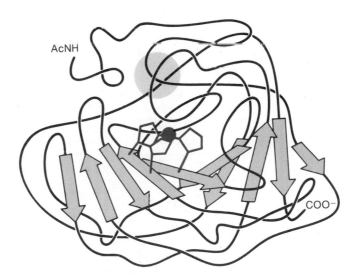

Figure 5-10
An idealized drawing of the tertiary structure of human carbonic anhydrase. Shown are the peptide chain and the three histidines (red) that coordinate to a zinc ion at the active site. Individual β-sheet strands are drawn as arrows from the amino to carboxyl ends. Note the twist of the β sheets. A hydrophobic cluster is shaded in red. (After K. K. Kannan, et al. 1971. *Cold Spring Harbor Symp. Quant. Biol.,* 36: 221).

BINDING SITES OF PROTEINS

Each protein in a cell carries out a particular function, which may be structural, enzymatic, or regulatory. The structural proteins are usually fibrous, whereas the other two functions are generally carried out by globular proteins utilizing highly specific interactions between the protein molecule and one or more intracellular molecules. These interactions occur on the surface of the protein, primarily with the amino-acid side chains.

Because of the folding of the polypeptide chain, the side chains produce a unique surface configuration of charged and hydrophobic groups. Usually a localized portion of the surface, having a size of 5–10 Å, is responsible for an interaction between the protein and other molecules. This is called a **binding site.** In an enzyme, the binding site is the catalytically active region, therefore called the **active site.** It is frequently the case that a binding site is formed by bringing together amino acids from distant parts of the polypeptide chain rather than from adjacent amino acids. When this occurs, most, if not all, of the intrachain interactions contribute to the creation of the binding site. (This enables one to understand how, in a mutant, substitution of one amino acid for another at some place other than at the binding site can markedly affect or even destroy a binding site by altering the manner of folding of other parts of the chain.)

The binding of a small molecule to a binding site can occur in many ways, a few of which are the following (Figure 5-11):

1. Two charged groups in the protein can be spaced in such a way that they make a precise match with two other charged groups on a small molecule.
2. A hydrophobic cluster of side chains in the protein can provide a nonpolar surface onto which a molecule having a large nonpolar group can adsorb.
3. Hydrogen-bonding groups in the protein can be arranged to facilitate hydrogen bonds to complementary groups on a small molecule.
4. An electronic configuration is formed on the protein surface in an arrangement that allows attractive van der Waals interactions between the protein and many portions of the molecule being bound. In this way the very weak van der Waals forces can sum to produce very tight binding if both molecules have precisely the right shapes. This is called **steric fit.**

Usually several of the modes of binding are simultaneously present. For example, it is almost always the case with enzymes that the shape of the active site nearly complements the shape of the molecule on which the enzyme acts. This allows a fairly strong van der Waals attraction to

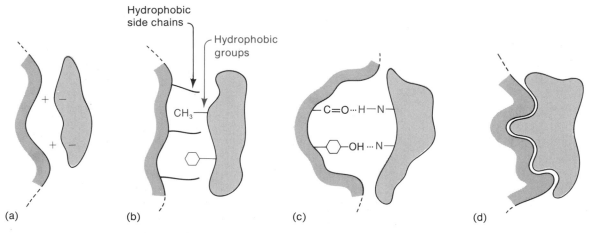

Figure 5-11
Four types of binding of small molecules (red) to the binding site of a protein (shaded). (a) Electrostatic. (b) Hydrophobic. (c) Hydrogen bonds. (d) Van der Waals bonds.

be present and has the effect that molecules that do not fit are excluded from the binding site. The binding is usually stabilized by one or possibly two of the attractive interactions just listed—namely, ionic bonds, hydrophobic interactions, or hydrogen bonds.

Binding a small molecule or a portion of a larger molecule to the binding site of a protein may cause the protein to undergo a small shape change because binding often alters the strength of the weak forces that determine the shape of a protein. Alteration in shape serves two important functions. First, the shape change of an enzyme can improve contact between the active site and the substrate molecule on which an enzyme acts. This property of enzymes is called **induced fit.** Second, through adjustments in shape the biological activity of molecules such as enzymes can be regulated. If an enzyme has two binding sites, a regulatory molecule can bind to one site and thereby affect the active site in a way that either facilitates or prevents binding to the active site. This phenomenon will be discussed in greater detail in Chapter 14.

It should not be thought that the molecules that bind to proteins are necessarily smaller or of different composition than proteins. The surface of one protein molecule can also bind another protein molecule. Some examples are protein-hydrolyzing enzymes (**proteases**) and proteins that aggregate to form complex structures such as membranes and virus shells (see Chapter 6). A particularly common event is self-aggregation to form multisubunit complexes. This phenomenon is described in the following section.

PROTEINS WITH SUBUNITS

A polypeptide chain usually folds such that nonpolar side chains are internal—that is, isolated from water. However, it is rarely possible for a polypeptide chain to fold in such a way that *all* nonpolar groups are internal. Thus, it is often the case that nonpolar amino acids on the surface form clusters in an effort to minimize contact with water. A protein molecule having a large hydrophobic patch can further reduce contact with water by pairing with a hydrophobic patch on another protein molecule. Similarly, if a molecule has several distantly located hydrophobic patches, a structure consisting of several protein molecules in contact effectively minimizes contact with water. The protein is then said to consist of identical **subunits**; this is in fact a very common phenomenon, with two, three, four, and six subunits occurring most frequently. (A multisubunit protein may also contain unlike subunits. This will be discussed shortly.)

One can take the point of view expressed above that multisubunit proteins exist because they dispose of hydrophobic groups in an efficient way. An alternate and probably more correct view is that many proteins have evolved to fold in such a way as to form hydrophobic patches in order that subunit assembly will occur. Such evolution has been valuable because multisubunit systems have the following advantages:

1. *Subunits are an economical way to utilize DNA.* To synthesize a protein containing 6000 amino acids requires an amount of DNA containing 18,000 base pairs (three pairs for each amino acid) or 0.04 percent of the coding capacity of a typical bacterium. If the protein is assembled from six identical subunits, only 3000 DNA base pairs are needed. One might wonder if the resulting proteins are comparable—if, for example, a multisubunit protein is as able to form binding sites as a single larger protein is. It is, because just as a binding site is usually formed by bringing together distant amino acids in a particular polypeptide chain, a binding site can be formed by conjunction of amino acids in separate subunits.

2. *Possession of subunits may reduce the effect of random errors in protein synthesis on protein activity.* If one time in 6000 an incorrect amino acid were inserted during protein synthesis, most good copies of a polypeptide chain consisting of 6000 amino acids would be nonfunctional.* If there are six subunits, each consisting of 1000 amino acids, one in six subunits would be defective. This arrangement preserves function in two ways. First, it is often possible that with five good

---

*Note that we are assuming that one incorrect amino acid is detrimental. This may not always be true but is an assumption in the example given.

subunits and one defective one, a protein may be functional to some degree; thus, a redundant system has a lower probability of forming nonfunctional proteins than a ɩonredundant system. Second, a defective subunit is likely to have a different surface structure and this will rarely be incorporated into the polymeric protein. Hence, usually only normal proteins are formed, thereby completely negating the effect of errors that would result if proteins were synthesized as individual polypeptide chains.

3. *The activity of multisubunit proteins is very efficiently and rapidly switched on and off.* Throughout this book we shall see examples of regulation of activity of enzymes. Often an enzyme must be active for one instant and inactive the next, as, for example, in cycles of muscular contraction and relaxation. In many cases, activity is prevented by dissociation of a multisubunit protein and restored by reassociation. One way to accomplish this cycle is the following. A signal or effector molecule binds to one subunit whose shape is changed very slightly. As a consequence of the shape change, the subunit falls away from the aggregate and the activity is lost. Once free from the aggregate, the subunit undergoes another shape change that disrupts its binding to the effector molecule, too. Having lost the effector, the subunit then regains its original shape and returns to the aggregate, thus re forming an active protein. This kind of switching is very rapid, achieves a reduction of activity to the zero level, and is accomplished with a very small expenditure of energy. Details of this modulation of activity are described in the section that follows.

The multisubunit proteins that have been described so far consist of several identical subunits. Different polypeptide chains can also aggregate and form proteins made up of nonidentical subunits and in fact this is quite common. For example, hemoglobin, the oxygen carrier of blood, consists of four subunits, two each of two different types; likewise, RNA polymerase, which catalyzes synthesis of RNA has five subunits of which four are different; and DNA polymerase III, which synthesizes DNA in *E. coli,* consists of ten different subunits.

It is quite rare for a very small protein to contain subunits. Since so many proteins do consist of subunits, however, there is an implication that these larger proteins are "better" than smaller proteins. This is probably too strong a generalization, though, in that sometimes the multiplicity of subunits may just have been an evolutionary accident. For example, a mutation could occur that allows a particular polypeptide chain to self-aggregate; this could in turn result in the formation of an active site that would catalyze a novel reaction. If this enzyme were of value to the cell in which the mutation occurred, selective pressure would retain the structure even though a simpler enzyme might have sufficed. There are two cases in which a large protein does have an advantage. First, in some processes the molecules to be bound are so large or so complicated that a small protein would find it difficult to produce an effective active site within itself. Second, when a binding reaction is regulated (and many are), the binding protein needs at least two binding sites—one for the molecule being bound and one for the regulator. It may not be possible for a small-molecular-weight molecule to have several binding sites. Furthermore, as will be seen in the following section, regulation of multisubunit proteins can be accomplished with elegance and fine attunement.

A FEW MEANS OF STUDYING PROTEIN STRUCTURE

Many of the methods used to study the structure of protein molecules are complex physicochemical techniques such as x-ray diffraction, circular dichroism, nuclear magnetic resonance, and fluorescence spectroscopy. These techniques are described in texts on physical biochemistry listed in the references at the end of this chapter. Other methods are simple; a few of these are described in this section.

Denaturation of Proteins

As in the case of nucleic acids, protein molecules can be denatured by exposure to reagents that eliminate hydrogen bonds and hydrophobic interactions and by disruptive agents such as high temperature. When a protein contains disulfide bonds, which is often the case, renaturation is often rapid and complete following removal of the denaturant. Thus, reagents such as mercaptoethanol ($HOCH_2CH_2SH$) that break disulfide bonds are usually incorporated into a denaturing mixture.

Denaturation studies are usually carried out to learn something about three-dimensional structure. Most denaturation methods and their interpretations are complex but the following examples illustrate the kind of information that can be obtained.

Example 1. A protein is known to have six tyrosines. If a tyrosine is on the surface of a protein, it can react with iodine. Suppose radioactive I_2 is used and the reaction is carried out twice, once on the native protein and again after denaturation. If the radioactivity of a sample iodinated prior to denaturation is one-third of that acquired if the sample is iodinated after denaturation, we may conclude that $(1/3) \times 6 = 2$ tyrosines are on the protein surface and $6 - 2 = 4$ are internal.

Example 2. The addition of a denaturant causes a fourfold reduction in molecular weight. This indicates that the protein contains four subunits.

Example 3. Denaturation causes an increase in the sedimentation coefficient of a particular protein without a change in molecular weight. Since the s value of a rod is less than that of a sphere and a denatured protein is nearly a random coil, the native protein is likely to be rodlike and to have an extensive α-helical region or β structure.

Renaturation of Proteins

Renaturation is another process studied in an effort to understand how a polypeptide chain folds. Since a single amino acid substitution often alters the three-dimensional structure of a protein, the idea was developed that the structure of a protein molecule should be entirely determined by its amino acid sequence. What this means is that a polypeptide chain, guided by the side-chain interactions, should spontaneously fold and re-fold until it achieves the configuration having minimum free energy. Furthermore, the appropriate disulfide bonds should form. With this view, it was expected that a denatured protein would re-fold to form the native molecule if appropriate conditions were created. Such results were in fact achieved with the enzyme ribonuclease A (RNase). This protein is a single chain containing 124 amino acids; eight of these are cysteines which form four disulfide bonds. If RNase is treated with a denaturant such as urea and then the urea is removed, renaturation occurs immediately, because the disulfide bonds provide reference points for proper folding of the chain. If mercaptoethanol is also added, the disulfide bonds are cleaved and RNase becomes a random coil. If the urea and the mercaptoethanol are slowly removed, perfect renaturation occurs, including formation of the four correct disulfide bonds (disulfides can form spontaneously by air oxidation). This latter finding is remarkable because there are 105 possible ways to form four disulfide bonds from eight cysteines. The interpretation of this observation is that the folding of this protein is determined exclusively by the amino acid sequence and that the proper disulfide bonds are formed because, during folding, the cysteines are correctly placed for pairing. Evidence that disulfide bond formation does not direct the folding comes from an experiment in which the mercaptoethanol was removed first and oxidation was allowed to occur prior to removal of the urea—that is, while the RNase was a random coil. With this protocol, the native molecule was not formed.

From the experiment described we can certainly conclude that, in ribonuclease A at least, the tertiary structure is determined by the amino acid sequence and is the one having the minimum free energy. Similar observations have been made for a few other proteins but, in general, it is not precisely the case. That is, perfect renaturation is not usually attained, though in many cases a structure results that is near that of the native molecule, which means that correct folding must be directed in some way. The following idea may be an explanation but it is not yet proved.

Protein molecules are synthesized in an N-terminus-to-C-terminus direction (Chapter 13). It seems likely that some folding begins before synthesis is complete, so certain parts of the molecule might form a stable configuration that is not altered. For instance, consider a protein that is arbitrarily divided into ten segments. If the protein is denatured and then allowed to renature, segments 2 and 9 might interact so rapidly and strongly that all subsequent folding is determined by this initial interaction. However, if folding occurs prior to completion of synthesis, segments 2 and 6, whose interaction is less strong than that between 2 and 8, might interact first and this interaction might be stabilized either by a disulfide bond or pairing of segments 1 and 4. When the molecule is complete, segments 3 and 8 might join. Thus, if this were the case, folding would be determined *stepwise* by the amino acid sequence and the native configuration would not be the one having the least free energy. Recent experiments suggest that this sequential process may be responsible for the proper folding of the enzyme lysozyme.

Hydrolysis of Proteins

Proteins can be hydrolyzed to free amino acids by 1 M HCl at high temperature.* There are also numerous hydrolytic enzymes, whose principal role is in digestion (for example, stomach pepsin and intestinal trypsin) or in destruction of bacteria and viruses by phagocytic cells. These enzymes are called **proteolytic enzymes** or **proteases.** Most proteases act only in particular regions of a protein. For example, carboxypeptidase, an exoprotease, removes amino acids one by one from the carboxyl terminus of a protein chain, and trypsin, an endo-peptidase, cleaves only on the carboxyl side of arginine and lysine. These enzymes not only have functional biological roles but are useful in the laboratory, as will be seen in the following section.

---

*A disadvantage of this procedure compared to the enzymatic methods is that tryptophan is destroyed by acid.

Determination of the Amino Acid Sequence of a Protein

The determination of the amino acid sequence is essential in understanding the detailed structure of a protein. This is accomplished easily by breaking the molecule into short fragments having at most fifteen amino acids, purifying each fragment, and determining the sequence of each one. The sequence of a fragment is obtained by a procedure known as the Edman degradation. Phenyl isothiocyanate reacts exclusively with the N terminus of a polypeptide. By appropriate choice of hydrolytic conditions the labeled N-terminal peptide can be removed and later identified. The next N-terminal amino acid is then reacted, removed, and identified. This process is repeated until the sequence of the peptide is obtained and then it is carried out for each peptide. The problem is then to determine the order of the peptides. This is done by cleaving the protein in several ways so that there is more than one set of polypeptides. For example, suppose two polypeptides obtained by trypsin hydrolysis have the sequences Ala-Gly-Trp-Gly-Lys and Leu-Asn-Val-Arg, and a fragment obtained by chymotrypsin hydrolysis is Val-Arg-Ala-Gly-Trp; these sequences can be matched as follows:

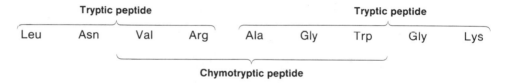

Usually, by obtaining three sets of peptides and sequencing each set, the puzzle can be solved. For determining the position of disulfide bonds and the sequences for proteins consisting of two polypeptide chains held together by disulfide bonds (for example, insulin), the reader should consult the references at the end of the chapter.

REGULATION OF THE ACTIVITY OF PROTEINS

The activity of many proteins varies with both intracellular and extracellular conditions. In this unit we examine several ways by which the activity of a protein is regulated.

End-Product Inhibition

If an enzyme catalyzes a reaction A → B and the product B is supplied exogenously, there is no reason for the enzyme to continue to carry out the reaction; in fact, if A has other uses, it would be efficient to turn off the activity of the enzyme. A simple way to accomplish this would be for the enzyme to have two binding sites—one for A and one for B. If B were bound, the enzyme could undergo a conformational change which would eliminate the binding site for A. In this way B would prevent the enzyme from carrying out the conversion. This does occur and is an example of **end-product inhibition.**

Regulation of Multisubunit Proteins—Allostery

Many enzymes contain several subunits for reason of economy of genetic information, as has been stated earlier. These enzymes are also regulated. However, a common arrangement is that the binding sites for the molecule that is acted on (the **substrate**) and the inhibitor (which may be the product) are located on different subunits. If binding of the inhibitor prevents binding of the substrate, the information that a site on one subunit is occupied must somehow be transmitted to the other subunit. This can be mediated by the subunit contact regions in the following way. Binding of the inhibitor molecule alters the shape of the subunit to which it is bound and this results in changes in the sites on this subunit with which it interacts with other subunits (Figure 5-12(a)). If the subunits remain in contact, all subunits adjoining the first will undergo a shape change at their respective subunit-interaction sites and this in turn alters the binding site of other subunits. Proteins capable of undergoing such modification are called **allosteric proteins.**

So far we have considered only the reduction of binding activity of a protein by an inhibitor. The reverse can also occur—that is, an inactive protein can be activated. In other words, a substrate-binding site on one

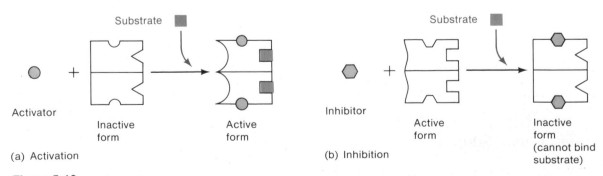

(a) Activation   (b) Inhibition

Figure 5-12
An example of allosteric (a) activation and (b) inhibition. The effector and the substrate can bind to the same or different subunits.

subunit may be not quite right for binding but may acquire the right configuration when an activator molecule is bound to an activation site on another subunit (Figure 5-12(b)). Thus, some proteins undergo allosteric activation, whereas others undergo allosteric inhibition.

The two cases just considered may be designated on→off (inhibition) and off→on (activation) processes. Many multisubunit proteins are subject to more subtle changes—that is, their binding activity is modulated by a regulator.

So far we have considered how a second molecule can affect the binding of a substrate molecule. However, it is fairly common that the binding activity of a protein is modulated by the substrate itself. For example, suppose an enzyme that carries out the conversion A→B is to be only weakly active when the concentration of A is low but must be very active when it is high. This would be difficult to accomplish with a single-subunit protein. But consider a multisubunit protein in which *each* subunit can bind A; the affinity of each subunit for A might usually be low but once A has bound to a subunit, the binding could cause a conformational change that increases the affinity of the binding sites on the other subunits. If there is little A, the initial binding will occur at a low rate and in general only one binding site will be occupied, so the conversion of A to B will occur infrequently. If the concentration of A is high and occupation of the first binding site increases the affinity of the other binding sites, then these binding sites will be rapidly filled. When one A is converted to B, the binding site for A will immediately be occupied again, so the conversion rate will be very high.

This kind of modulation of protein activity occurs commonly. An important example is hemoglobin, whose ability to bind O_2 increases with O_2 concentration.

Two Mechanisms for Allosteric Modulation

Allosteric changes can occur in several ways. Two important models have been presented to describe the mechanism by which the initial shape change occurs and how this results in modification of the ability of a multisubunit protein to bind a molecule A. These are called the **symmetry** or **concerted model** and the **sequential model.** In both models it is assumed that:

1. Each subunit is capable of assuming two forms, T and R, which bind A with low and high affinity respectively.
2. When A is bound to a subunit firmly, that subunit must be in the R form.

The models also differ in several ways (Figure 5-13):

1. In the concerted model it is assumed that both T and R forms are initially present in equilibrium, that only the R forms of the subunits make firm bonds with A (that is, only rarely and insubstantially do T forms bind to A), and that the binding to a single R-form subunit (high affinity) causes *all* other subunits in the same protein to change from T form to R form, thus shifting the equilibrium strongly in the R direction. In contrast, in the sequential model it is assumed that A binds first to the T form (low affinity) and that this binding somehow *induces* that T subunit (but not others) to transform into an R form (thus strengthening the binding). Note that in the concerted model binding of a substrate molecule to one subunit indirectly affects the form of the other subunits whereas, in the sequential model, each subunit is affected directly.

2. The number of possible forms that an allosteric protein whose binding sites are not all filled can have is not the same for the two models. Thus, in the concerted model symmetry in the arrangement of

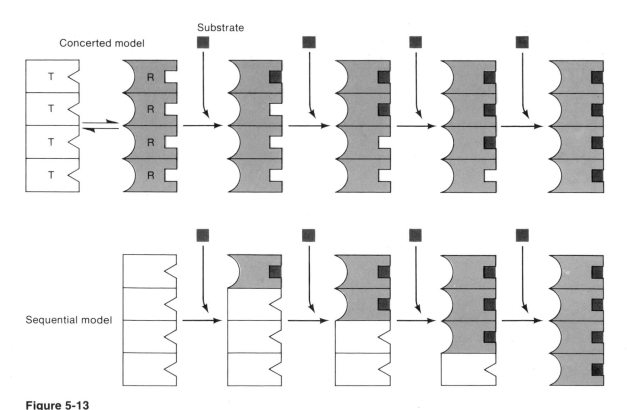

Figure 5-13
The major models for allostery, shown for the cases of activation of a tetramer. In the concerted model, the binding of one substrate molecule stabilizes the high affinity (R) form of all subunits. In the sequential model the binding of one substrate molecule induces that subunit to take the high-affinity form.

subunits is considered to be an essential component of the forces stabilizing the macromolecular aggregate. An R subunit can interact stably with another R form in the concerted model and two T subunits can aggregate, but an R and a T form cannot form a stable pair. As long as the subunits remain in contact, the stabilization of one R subunit causes conversion of *all* remaining T subunits to R subunits—that is, the symmetry of the protein aggregate is preserved. For instance, for a tetramer, the concerted model allows only TTTT and RRRR forms of the complex; TTTR, TTRR, and TRRR cannot exist. In the sequential model this notion of symmetry is discarded and it is assumed instead that the conversion of TTTT to RRRR occurs through steps such as TTTT→TTTR→TTRR→TRRR→RRRR, in which each step occurs as another molecule is bound.

The concerted model can be described by a simple mathematical formalism that relates the amount of A that is bound to the concentration of free A. The equations that describe the sequential model are considerably more complex. Both sets of equations can be found in several references given at the end of the chapter.

Experiments show that the behavior of some macromolecules is best explained by the concerted model while, for others, the sequential model is better. There are, however, many molecules for which neither model is satisfactory. These molecules seem to exist in more than two states and complex models are needed to account for their properties.

An Allosteric Enzyme: Aspartyl Transcarbamoylase

The best-understood example of an allosteric protein is the enzyme aspartyl transcarbamoylase (ATCase), which has sites for both an activator and an inhibitor. This enzyme catalyzes one step in the synthesis of the RNA precursor cytidine triphosphate (CTP) from glutamine. The overall reaction sequence is

$$\text{Glutamine} + CO_2 + \text{ATP} \longrightarrow \text{Carbamoyl phosphate}$$

$$\text{Aspartate} \searrow \quad \text{Aspartyl transcarbamoylase}$$

$$\text{CTP} \longleftarrow \text{UTP} \longleftarrow \text{UMP} \longleftarrow \text{Carbamoyl aspartate}$$

The enzyme consists of twelve subunits of two types. Six of these are catalytic subunits (C), which form two trimers. The remaining subunits are regulatory subunits (R); these form three dimers. (Note that the letter R refers to regulatory and not the R form discussed in the preceding section; this terminology, perhaps confusing, is standard.) Each R dimer is in contact with two C subunits, but each in a different trimer. This is

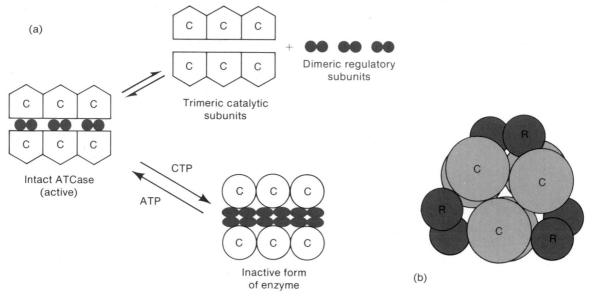

Figure 5-14
Aspartyl transcarbamoylase. (a) Its subunit structure (two trimeric catalytic subunits, labeled C, and three dimeric regulatory subunits, labeled R) and the conversion to an inactive form by CTP. (b) Three-dimensional arrangement of the subunits in the active molecule.

shown schematically in Figure 5-14(a). The actual arrangement of the subunits has been determined by x-ray crystallography and is shown in Figure 5-14(b). ATCase is much more complicated than the tetramer consisting of only one type of polypeptide, shown in Figure 5-13, so that its behavior should not be interpreted strictly according to either the concerted or sequential model. The allosteric principle is the same, though.

As an allosteric enzyme, ATCase behaves in the following way. The enzyme is stimulated by the RNA precursor ATP and inhibited by CTP, which is not the immediate product of the reaction but the ultimate product (Figure 5-15). The enzyme can be easily dissociated in the

Figure 5-15
Allosteric effects in aspartyl transcarbamoylase. ATP is an activator increasing the reaction rate; CTP is an inhibitor and reduces the reaction rate. The red curve results when neither ATP nor CTP is present. (After J. C. Gerhart. 1970. *Curr. Top. Cell Regul.*, 2: 275.)

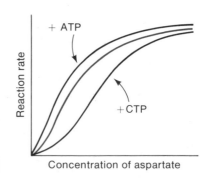

laboratory to yield the two C trimers and the three R dimers and these can be purified. When this is done, it is found that aspartate binds only to a trimer (one aspartate per C subunit); ATP and CTP each bind only to a dimer (one ATP binding site and one CTP per R subunit). Alone, a C trimer is able to synthesize carbamoyl aspartate but the activity is unaffected by the concentration of ATP or CTP. Thus activation and inhibition require that the R dimers be in contact with the C trimers. It has also been found that the R subunits undergo a change in shape when CTP is bound and that there is also a change in shape of the C subunits if the R dimers and the C trimers are associated, in exact accord with the principles of the allosteric effect.

A COMPLEX PROTEIN: IMMUNOGLOBULIN G

As has already been mentioned, many proteins consist of subunits, which may or may not be identical. In this section the structure of one such protein, immunoglobulin G, is described. It is not an allosteric protein. Immunoglobulin G has been selected as an example because its structure is well understood and its remarkable mode of synthesis is discussed in detail in Chapter 22.

The immunoglobulins (**antibodies**) are the proteins of the immune system. Their function is to interact with specific foreign molecules (**antigens**) and thereby render them inactive. This interaction is called the **antigen-antibody reaction.** The best-understood immunoglobulin is immunoglobulin G (**IgG**); other classes of immunoglobulins are IgA, IgM, IgD, and IgE. In this section, we shall examine only the IgG class, itself comprising several subclasses defined by slight structural differences.

Cleavage of the disulfide bonds of IgG yields two polypeptide chains whose molecular weights are about 25,000 and 50,000. The lighter polypeptide is called an **L chain;** the heavier one is an **H chain.** IgG is a tetramer containing two L chains and two H chains. A schematic structure of IgG is shown in Figure 5-16. Experimentally, IgG can be cleaved with the proteolytic enzyme papain, which causes each of the H chains to break, as shown in the figure, thus producing three separate subunits. The two units that consist of an L chain and a fragment of the H chain equal in mass to the L chain are called F_{ab} fragments (the subscript stands for "antigen-binding"). The third unit, consisting of two equal segments of the H chain, is called the F_c fragment.

Two sites on an IgG molecule can bind antigen. Each site is at the end of an F_{ab} segment, as shown in Figure 5-17. The F_c segment is not involved in antigen-antibody binding but is used in later processes needed to destroy the antigen.

The amino acid sequences of many subclasses of IgG molecules, each capable of combining with only a single kind of antigen molecule,

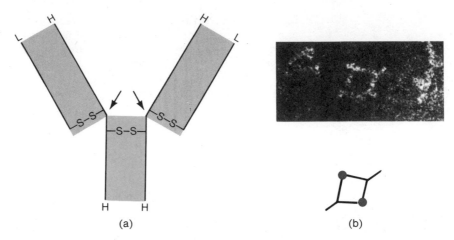

Figure 5-16
(a) The subunit structure of immunoglobulin G. There are two L subunits and two H subunits. Disulfide bonds join each L strand to an H strand and join the two H strands to one another. Treatment with papain cleaves the H strands at the arrows, yielding two F_{ab} units (shaded black) and one F_c unit (shaded red). (b) An electron micrograph showing two immunoglobulin G molecules joined at the antigen binding site. The joining is a result of binding two antigen molecules not visible in the micrograph but shown as red dots in the interpretive drawing (Courtesy of Ray Valentine.)

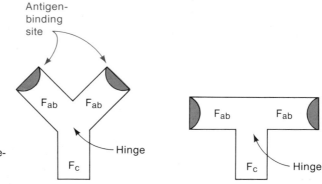

Figure 5-17
Immunoglobulin G is a Y-shaped molecule. It contains a hinge that gives it segmental flexibility.

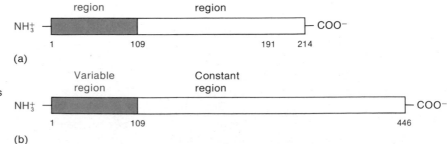

Figure 5-18
Structure of (a) the L chain and (b) the H chain of IgG showing the regions of homologous sequence. Numbers refer to amino-acid positions counted from the amino terminus.

have been determined. All immunoglobulins are structurally similar inasmuch as L and H chains both have what is called a **variable (V)** and a **constant (C)** region.

The V and C regions have separate functions in IgG. The V regions confer antigen-binding specificity, whereas the C regions are responsible for the overall structure of the molecule and for its recognition by other components of the immune system. One might expect, then, that the V regions of both the L and H chains would be closely associated and located in the antigen-binding regions of the IgG molecule. That this is indeed the case is shown in the schematic diagram shown in Figure 5-18.

The C regions of the L chains (C_L) of all types of IgG molecules have an identical amino acid sequence. Likewise, the C regions of all H chains (C_H) are identical for all IgG, though different from the C_L sequences. The V regions differ from one type of IgG to the next, however. The comparative sizes of the C_L, V_L, C_H, and V_H regions are shown in Figure 5-19.

Some of the similarities between amino acid sequences in different parts of an IgG molecule are quite striking. For example, the C region of the H chain can be divided into three segments C_H1, C_H2, and $C_{II}3$, whose amino acid sequences are quite similar though not identical to one another. Furthermore, the sequence of C_L resembles the C_H sequences, though generally they are different. The V_L and V_H sequences also are nearly the same. This means that the IgG molecule, which already has twofold symmetry, consists of four domains, each of

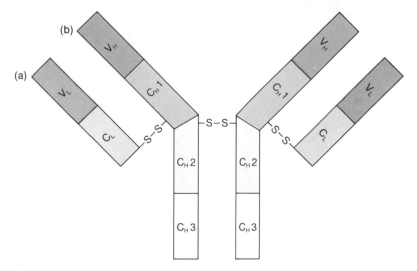

Figure 5-19
The arrangement and relative sizes of the variable and constant regions of the IgG (a) L chain and (b) H chain. The numbers refer to amino acids counted from the amino terminus. Note that the lengths of the variable regions are the same.

which has twofold symmetry, as shown in Figure 5-19, in which the intensity of the shading indicates two homologous regions. The symmetry of the molecules is made especially evident in Figure 5-20, which shows the three-dimensional structure of IgG. Note how in each domain the two components are somewhat wrapped around each other to form what is known as the **immunoglobulin fold.** The point of contact in the fold consists of two antiparallel β sheets between which are buried many hydrophobic amino acids.

Several other features of the IgG molecule should be noted, as they are often also found in other multisubunit proteins whose function is to bind other molecules. The most striking property is that many regions of symmetry exist. Although the kinds of symmetry differ from one molecule to the next, symmetry of some kind is a common element. For example, hemoglobin, which has four subunits, has several planes of symmetry that are utilized in arranging particular regions of the molecule for binding oxygen; the major muscle protein myosin has symmetric domains that interact with other muscle proteins; and the Cro protein made by *E. coli* phage λ, which binds to a symmetric base sequence in DNA, binds as a symmetric dimer (described in the following chapter). A second notable feature of IgG, common also to

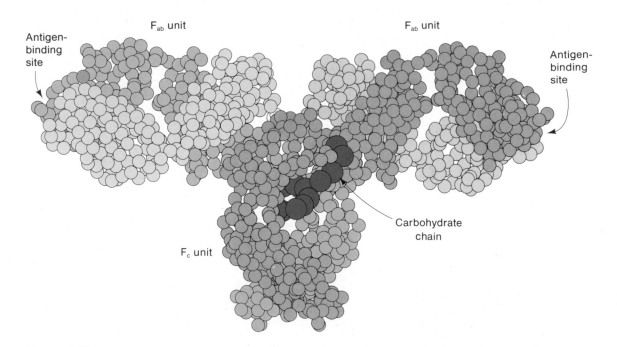

Figure 5-20
Schematic drawing of the three-dimensional structure of an IgG molecule. One of the H chains is shown in dark pink, the other in dark gray. One of the L chains is shown in light pink, the other in light gray. (After E. W. Silverton, M. A. Navia, and D. R. Davies. 1977. *Proc. Nat. Acad. Sci.,* 74: 5142.)

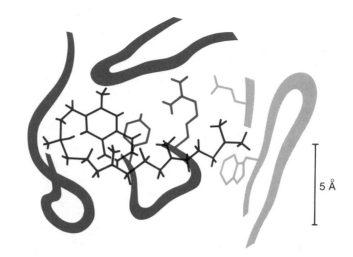

5 Å

Figure 5-21
Mode of binding of a derivative of vitamin K (shown in black) to an antibody-combining site. Amino acid residues from the H chain are shown in pink and those from the L chain in red. Note that the antigen (vitamin K) is bound by both the L and H chains. (After I. M. Amzel, R. Poljak, F. Saul, J. Varga, and F. Richards. 1974. *Proc. Nat. Acad. Sci.,* 71: 1427.)

other molecules designed to bind two identical molecules, is that the structure contains multiple subunits capable of providing binding sites. IgG normally binds two antigen molecules—one at each of the two V-region binding sites. This activity is important in the mechanism for ridding the organism of an antigen after it has been bound to IgG. A third element of the structure of IgG is the use of extended hydrophobic segments (as in the immunoglobulin fold) as a means of joining subunits. A fourth feature is the joining of regions in two different subunits to generate a binding site. This is shown clearly in Figure 5-21, in which an antigen that is a small molecule (in this case, vitamin K) is bound to one of the antigen binding sites.* The antigen fits neatly into a cleft formed where the H chain passes the end of the L chain on the IgG molecule, and interacts with amino acids in both chains.

ENZYMES

Enzymes are special proteins able to catalyze chemical reactions. Their catalytic power exceeds all manmade catalysts. A typical enzyme accelerates a reaction 10^8-to 10^{10}-fold, though there are enzymes that increase the reaction rate by a factor of 10^{15}. Enzymes are also highly specific in that each catalyzes only a single reaction or set of closely related reactions. Furthermore, only a small number of reactants, often only one, can participate in a single catalyzed reaction. Since nearly every biological reaction is catalyzed by an enzyme, these clearly require a very large number of distinct enzyme molecules.

*Vitamin K is not a substance against which an antibody would normally be made. However, by special techniques an animal's immune system can be tricked into making an antibody to such a small molecule. Experimentally, such antibodies are valuable because their antigen-binding sites are small and simple and more amenable to study than large complex sites.

Enzyme reactions are studied for a variety of reasons. For example, biochemists are interested in understanding detailed reaction mechanisms and therefore study the effect of pH, substrate modification, and inhibitors on reaction rates. The molecular biologist interested in regulation of enzyme synthesis or enzyme activity will choose a particular enzyme and measure the amount of enzyme activity as a function of various biological parameters. Enzymes are also useful laboratory tools inasmuch as they can be used to hydrolyze unwanted substances and prepare chemical compounds. Finally, the mechanisms by which complex processes occur, such as DNA and protein synthesis, often include a large number of enzymes and other factors acting together, and understanding the processes requires detailed knowledge of the enzymatic reactions.

The detailed mechanism of catalysis by particular enzymes is beyond the scope of this book. However, all enzymes have certain general features, the knowledge of which is important for understanding molecular biological phenomena. These features are described in this unit.

The Enzyme-Substrate Complex

In any reaction that is enzymatically catalyzed, one reactant always forms a tight complex with the enzyme. This reactant is called the **substrate** of the enzyme; in descriptive formulas, it is denoted S. The complex between the enzyme E and the substrate is called the enzyme-substrate or **ES complex.**

Initially the enzyme and the substrate are bound by weak bonds, as in Figure 5-11, though in a few cases a covalent bond forms. The site on the enzyme at which the substrate binds is the **active site.** The extraordinary selectivity in enzyme catalysis is almost entirely a result of specificity of enzyme-substrate binding. After the ES complex forms, the substrate is usually altered in some way that facilitates further reaction. When the substrate is in the altered state, the ES-complex is said to be active and is usually denoted (ES)*. The (ES)* complex then engages in one or a series of transformations, which result in conversion of the substrate to the product and dissociation of the product from the enzyme. The extent to which ES forms is determined by the strength of binding between E and S; this is called the **affinity** of E and S. A useful measure of affinity is the **Michaelis constant, K_M.** This is the substrate concentration at which the reaction rate of the enzyme (which increases with substrate concentration) is half-maximal; it is the same as the concentration at which half of the enzyme molecules in the solution have their active sites occupied by a substrate molecule.

For most enzymatic reactions, formation of ES is reversible in the sense that ES can dissociate, yielding free E and free S. Usually, dissociation of the ES-complex is more rapid than conversion of the complex to enzyme and product; when this is the case, the value of K_M is a measure of the strength of the ES binding. That is, a high value of K_M indicates weak binding, and a low value of K_M means strong binding. The strength of binding depends on several conditions, such as the presence of particular ions, the overall ion concentration, and sometimes the presence of inhibitors. Thus knowing the numerical value of K_M is often of use when one is trying to understand the mechanism of regulation of the activity of a particular enzyme. For most enzymes K_M ranges from 10^{-6} to 10^{-1} M, which shows that the affinity of E and S varies widely for different enzymes.

Theories of Formation of the Enzyme-Substrate Complex

Catalysis by an enzyme occurs in several steps—binding of the substrate, conversion to the product, and release of the product. The initial step, formation of the ES complex, is in principle easy to understand and is often considered in thinking about molecules. The subsequent rearrangments are chemical phenomena and will not be discussed in this book.

    There are two major theories of enzyme binding, the **lock-and-key** and **induced fit** models (Figure 5-22). In the lock-and-key model, the

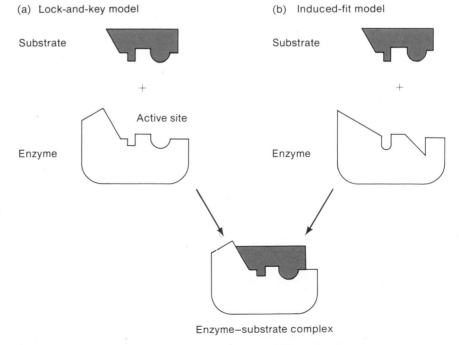

Figure 5-22
Two models for enzyme-substrate binding. (a) The lock-and-key model. The active site of the enzyme by itself is complementary in shape to that of the substrate. (b) The induced-fit model. The enzyme changes shape upon binding a substrate. The active site has a shape complementary to that of the substrate only after the substrate is bound.

(a) Lock-and-key model

Substrate

+

Active site

Enzyme

(b) Induced-fit model

Substrate

+

Enzyme

Enzyme–substrate complex

shape of the active site of the enzyme is complementary to the shape of the substrate. In the induced fit model, the enzyme changes shape upon binding the substrate and the active site has a shape that is complementary to that of the substrate only *after* the substrate is bound. For every enzyme-substrate interaction examined to date, one of these two models applies. It is often the case, though, that the substrate itself undergoes a small change in shape; in fact, the strain to which the substrate is subjected is often the principle mechanism of catalysis—that is, the substrate is held in an enormously reactive configuration.

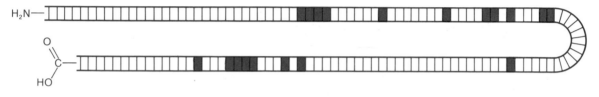

Figure 5-23
Schematic diagram of the amino acid sequence of lysozyme showing that the amino acids (red) in the active site are separated along the chain. Folding of the chain brings these amino acids together.

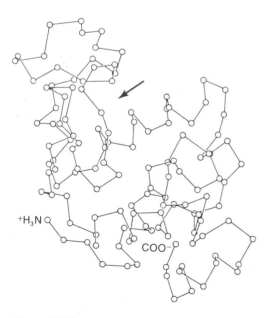

Figure 5-24
Three-dimensional structure of lysozyme. Only the α-carbon atoms are shown. The active site is in the cleft indicated by the arrow. (Courtesy of Dr. David Phillips.)

Figure 5-25
A space-filling molecular model of the enzyme lysozyme. The arrow points to the cleft that accepts the polysaccharide substrate. (C atoms are black; H, white; N, gray; O, gray with slots.) (Courtesy of John A. Rupley.)

Molecular Details of an Enzyme-Substrate Complex

The first detailed analysis of enzyme-substrate binding was carried out by using hen egg-white lysozyme. This enzyme cleaves certain bonds between sugar residues in some of the polysaccharide components of bacterial cell walls and is responsible for maintaining sterility within eggs. The amino acid sequence of lysozyme is shown in Figure 5-23. The 19 amino acids that are part of the active site are printed in red; it should be noticed that they form widely separated clusters along the chain. Only when the chain is folded do they come into proximity and form the active site. The folding of the chain is shown in Figure 5-24. The deep cleft indicated by the arrow is the active site. This is seen more clearly in the space-filling model shown in Figure 5-25. The substrate is a hexasaccharide segment that fits into the cleft and is distorted upon binding. The enzyme itself changes shape when the substrate is bound. A variety of interactions stabilize the binding. A few of the hydrogen bonds that form with a trisaccharide portion of the substrate are shown in Figure 5-26. Note that hydrogen bonds have formed both with side chain groups (for example, those of tryptophans 62 and 63 and aspartate 101) and with the peptide bonds (for example, bonds 59 and 107). There

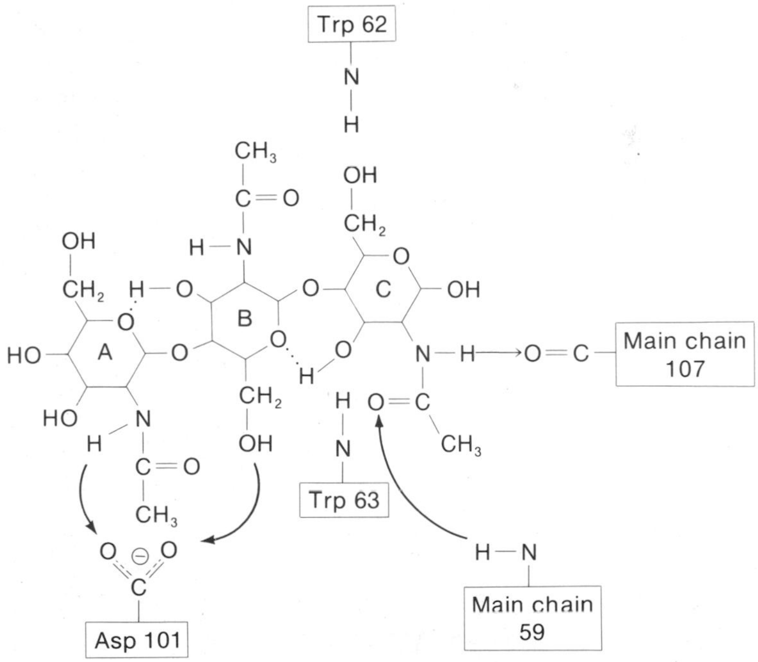

Figure 5-26
Hydrogen bonds between three amino-acid side chains and two main-chain peptide groups of lysozyme and three saccharide rings (A, B, C) of the enzyme substrate.

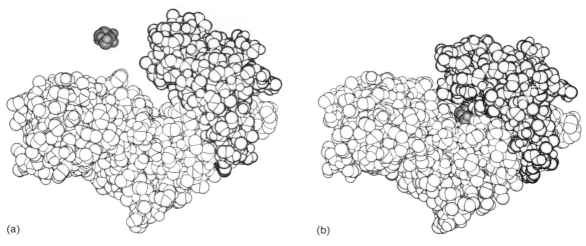

(a)

(b)

Figure 5-27
Structure of yeast hexokinase A. All atoms except hydrogen are shown separately. (a) Free hexokinase and its substrate, glucose. (b) Hexokinase complexed with glucose. Note the marked change in enzyme structure that accompanies glucose binding. (With permission, from Bennett, J. and T. A. Steitz. 1980. *J. Mol. Biol.*, 140: 211. Copyright: Academic Press Inc. [London] Ltd.)

are also numerous van der Waals contacts and a few polar interactions, neither of which is shown in the figure. Another enzyme, yeast hexokinase A, has been studied in order to examine what structural changes occur on substrate binding; these changes are now well documented. This is shown in the pair of space-filling models shown in Figure 5-27.

Dissociation of an Enzyme-Substrate Complex: Turnover

The rate at which a product forms from an ES complex is characterized by the average elapsed time between complex formation and dissociation of the product. This property is commonly expressed as the number of complexed substrate molecules converted to the product per enzyme molecule per second; this is called the **turnover number.** These considerations are summarized by the following statement in which E, S, and P refer to enzyme, substrate, and product respectively:

$$E + S \rightleftharpoons ES \rightleftharpoons (ES)^* \rightarrow E + P$$

DETECTION OF ENZYMATIC ACTIVITY

There are numerous ways to study enzyme reactions. Two of these procedures, optical assays and radioactivity assays, are very commonly used and are described next.

Optical Assays of Enzyme Activity

In an optical assay of an enzyme, some component in the reaction mixture is detected by the ability of that component to absorb light of a particular wavelength. Since it is always possible to find a range of concentrations of the substances in which the absorbance A and the concentration are proportional, the concentration of the substances can be measured. The optically absorbing substance can be a reactant, in which case A decreases with time, or a product, in which case A increases.

There are very few substances in cells that absorb visible light and not many more that absorb in the ultraviolet. Some examples of the latter are the nucleic acid bases, the amino acids tryptophan, tyrosine, histidine, and phenylalanine, and a few cofactors and vitamins. (When such substances are present, they may be assayed.) However, if the protein concentration of a mixture is very high, the ultraviolet absorption of the proteins may overwhelm the absorbance of the substance of interest. For this reason, biochemists have attempted to design substrates that generate a product that absorbs in a range of wavelengths for which there are no naturally occurring absorbing compounds. An example of such a substrate is o-nitrophenyl galactoside (ONPG); this is used to assay the enzyme β-galactosidase, which we will encounter frequently in this book.

The enzyme β-galactosidase catalyzes cleavage of lactose to form galactose and glucose:

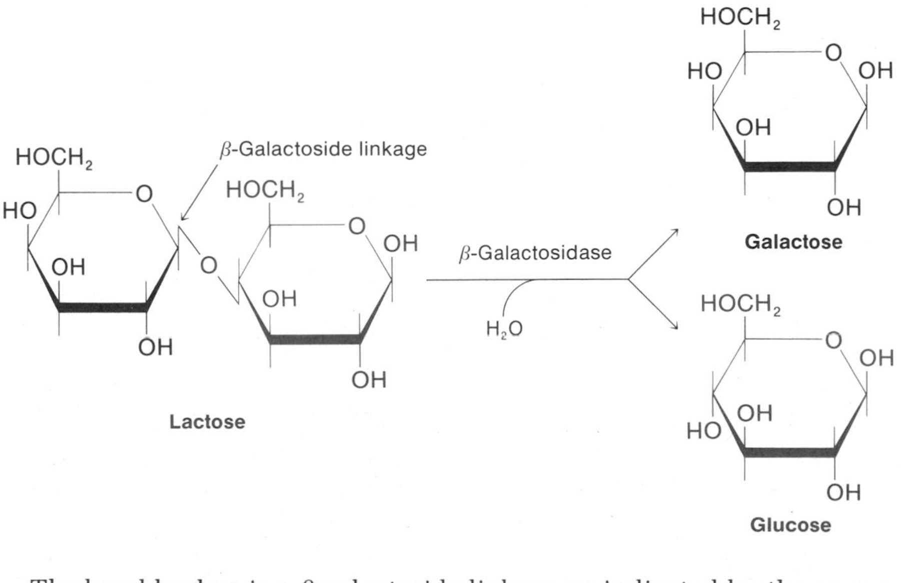

The bond broken is a β-galactoside linkage, as indicated by the arrow. This bond is also present in ONPG. The enzyme can hydrolyze ONPG, which is colorless, yielding galactose and o-nitrophenol, which is intensely yellow:

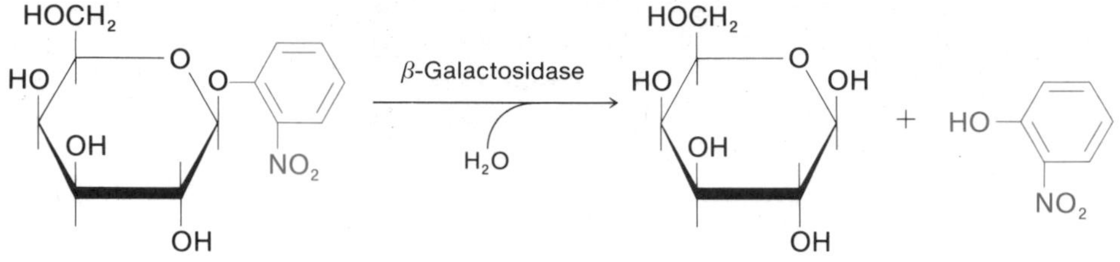

o-Nitrophenylgalactoside   Galactose   o-Nitrophenol
                                                  (yellow)

Thus, the activity of the enzyme can easily be followed by assaying the concentration of ONPG at a wavelength of 420 nm (blue light).

Radioactivity Assay for Enzymes

In a radioactivity assay a reactant that is radioactive is added to a reaction mixture and either the appearance of another radioactive substance or the loss of the radioactive reactant is measured. There is enormous variety in such assays. One of the most common assays is used to measure polymerization. This assay is based upon the following fact—all proteins and nucleic acids are insoluble in 1 M trichloracetic acid (TCA), whereas all amino acids and nucleotides are TCA-soluble. This property allows one to measure protein synthesis by adding a radioactive (e.g., ^{14}C) amino acid to a reaction mixture containing the other 19 nonradioactive amino acids and the appropriate enzymes and factors. After a period of time, TCA is added and the mixture is filtered and washed with TCA. The ^{14}C-labeled amino acid is soluble and passes through the filter but if ^{14}C-protein has been made, it will be precipitated and ^{14}C will be retained on the filter. Counting the filter-bound radioactivity is then a measure of the extent of reaction. Synthesis of DNA can also be measured in this way by using a mixture containing the four deoxynucleotide precursors of DNA, one of which is radioactive—for example, [^{14}C]thymidine triphosphate (TTP)—and other appropriate components of the mixture. After reaction, TCA is added and the mixture is filtered; [^{14}C]TTP passes through the filter but any DNA that has been synthesized is retained on the filter. This type of assay will be encountered again in this book.

Detection of Enzymatic Activity in a Crude Extract

It is often desirable, especially in preliminary studies, to be able to detect an enzymatic activity in a cell extract—that is, a solution obtained by breaking open a concentrated cell suspension. The interpretation of such experiments requires caution, as we shall see.

An enzyme is a catalyst and hence only alters the rate of reaction without affecting reaction equilibrium. This means that it accelerates both the forward and the back reaction by the same factor. This has an important consequence in the laboratory—namely, that an enzyme can be detected by assaying for the product of *either* of these reactions. Usually the forward reaction is the one that is detected because the formation of its product is favored by equilibrium. However, in choosing assay conditions in a cell extract, substances may be added to the assay mixture that can affect the equilibrium. Thus, if an enzymatic reaction in which B is converted to A is detected in a cell extract, one cannot be immediately sure that this is a biologically significant reaction because the assay conditions might be such that the biological reaction is being driven backwards. For example, the biological reaction might be $A + H^+ \rightarrow B + C$ but sufficient C might be present and the pH high enough in the extract to drive the reaction from right to left. As an example, some years ago an enzyme was discovered that catalyzed polymerization of polyribonucleotides; later, its biological role was shown to be that of a ribonuclease.

PREPARATION OF RADIOACTIVE PROTEINS

In many types of experiments it is desirable to label proteins with radioactivity. If the protein is of microbial origin, this is usually done by adding radioactive protein precursors to the growth medium. Any radioactive amino acid will suffice, but addition of some amino acids will result in the labeling of other molecules as well. For instance, glycine can be utilized as a general carbon source so that radioactivity will appear in a wide variety of compounds, and aspartate label can appear in nucleic acids because aspartate is in the pathway for purine biosynthesis. The amino acids leucine and phenylalanine cannot be converted to any macromolecule other than proteins; therefore, the ^{3}H- and ^{14}C-labeled forms are the precursors of choice for preparing radioactive proteins.

Sulfur-35 in the form of the sulfate ion is also a useful label since it is incorporated into the two sulfur-containing amino acids, methionine and cysteine. The single disadvantage of the use of ^{35}S is that sulfur is also contained in some RNA molecules (there is thiouridine in some transfer RNA molecules) and that sulfur-containing amino acids are not present in all proteins. The former problem can be eliminated by treating a radioactive sample with RNase to hydrolyze the RNA. The latter is usually only a problem if a specific protein that lacks sulfur is to be labeled.

Many proteins that are studied are derived from animal tissue and it is not possible to label these by feeding radioactive amino acids to an animal because the radioactivity will be diluted substantially by the normal food of the animal. (An unsuccessful attempt was made several years ago to prepare a radioactive cow.) The usual procedure is to purify the protein first and then make it radioactive by iodination. In an appropriate reaction mixture containing KI and $^{125}I_2$, iodination of tyrosine and histidine occurs and it is possible to make highly radioactive proteins. This procedure is generally useful but sometimes has the side effect that iodination eliminates enzymatic activity if tyrosine or histidine are in the active site of an enzyme.

REFERENCES

Anfinsen, C. G. 1973. "Principles that govern the folding of protein molecules." *Science,* 181, 223–230.

Baldwin, R. L. 1978. "The pathway of protein folding." *Trends Biochem. Sci.,* 3, 66–68.

Bernhard, S. 1968. *The Structure and Function of Enzymes.* Benjamin.

Cantor, C. R., and P. R. Schimmel. 1980. *Biophysical Chemistry. Part I. The Conformation of Biological Macromolecules.* W. H. Freeman and Co.

Capra, J. D., and A. B. Edmundson. 1977. "The antibody-combining site." *Scient. Amer.,* 236 (January), 50–59.

Dickerson, R. E., and I. Geis. 1980. *The Structure and Action of Proteins.* Harper & Row.

Edelman, G. M. 1973. "Antibody structure and molecular immunology." *Science,* 180, 830–840.

Fersht, A. 1977. *Enzyme Structure and Mechanisms.* W. H. Freeman and Co.

Freifelder, D. 1982. *Physical Biochemistry.* 2nd ed. W. H. Freeman and Co.

Haschemeyer, R. H., and A. E. V. Haschemeyer. 1973. *Proteins: A Guide to Study by Physical and Chemical Methods.* Wiley.

Kendrew, J. C. 1961. "The three-dimensional structure of a protein molecule." *Scient. Amer.,* 202 (December), 96–102.

Koshland, D. E. 1974. "Protein shape and biological control." *Scient. Amer.,* 229 (October), 52–64.

Lehninger, A. L. 1982. *Biochemistry.* Worth.

Neurath, H., and R. L. Hill. 1975. *The Proteins.* Academic Press.

Perutz, M. 1964. "The hemoglobin molecule." *Scient. Amer.,* 212 (November), 64–71.

Perutz, M. 1978. "Hemoglobin structure and respiratory control." *Scient. Amer.,* 239 (December), 92–125.

Phillips, D. C. 1966."The three-dimensional structure of an enzyme molecule." *Scient. Amer.* 214 (November), 78–84.

Schachman, H. K. 1974. "Anatomy and physiology of a regulatory enzyme— aspartyl transcarbamylase." *Harvey Lectures,* 68, 67–92. Academic.

Schultz, G. E., and R. H. Schirmer. 1979. *Principles of Protein Structure.* Springer-Verlag.

Silverton, E. W., M. A. Navia, and D. R. Davies. 1977. "Three-dimensional structure of an intact human immunoglobulin." *Proc. Nat. Acad. Sci.,* 74, 5140–5144.

Wold, F. 1971. *Macromolecules: Structure and Function.* Prentice-Hall.

Macromolecular Interactions and the Structure of Complex Aggregates

In the preceding chapter a great deal of information has been given about the structure of individual proteins and nucleic acids. We have also explained that many proteins contain two or more subunits, which may or may not be identical. This is true also for nucleic acids, though it is rarer and the number of subunits is usually quite small; for example, double-stranded DNA has two subunits and retrovirus RNA, which has more subunits than any other nucleic acid, contains four strands.

As will be seen throughout this book, the existence of interactions between different macromolecules is the rule rather than the exception and structural components both within individual cells and extra-cellular in organisms are invariably assemblies of macromolecules. For example, nucleic acids are often associated with proteins, as in chromosomes (which are DNA-protein complexes), extracellular viral nucleic acids are encased in protein shells, and bone and cartilage are complex assemblies of proteins and other macromolecules. Proteins can also interact with lipids to produce membranes such as those which separate the contents of a cell from the environment and which separate different intracellular components from one another. Finally, polysaccharides form extraordinarily complex structures such as the cell walls of bacteria and of plants.

The study of such complex structures and how they are formed is called structural biology. A few structures are almost completely understood and many more are actively being studied. In this chapter only a few structures will be described. These have been selected by two criteria—they are generally important or illustrate general principles, and their structures are reasonably well understood. The reader interested in exploring the subject further should consult the references given at the end of the chapter.

COLLAGEN—A MULTIPROTEIN ASSEMBLY

Collagen is the most abundant protein in mammals and probably the most common protein in the world. It is the major protein component of tendon, cartilage, bone, skin, and blood vessels. The principal function of collagen is to provide a fiber (Figure 6-1) with great tensile strength, and the molecule is neatly designed to accomplish this end. Collagen is of special interest for two reasons: (1) it is a multiprotein assembly in which subunits are joined to form a trimeric protein and (2) there is a higher organization in which the trimers are further joined to form a fiber. Collagen is synthesized by cells called **fibroblasts.** The collagen is

Figure 6-1
(a) An electron micrograph of a collagen fiber (large dark object at the right) which has been mechanically teased apart to release a single collagen fibril. The pattern of bands should be compared with Figure 6-6. These bands result from the binding of ions containing heavy metals to charged amino acids. A careful look shows that some bands are clustered to form darker units.
(b) Individual isolated fibrils on which metal has been deposited to enhance the major bands. (Courtesy of F. O. Schmitt.)
(c) An enlarged view of a stained collagen fibril that has been treated briefly with pepsin. Notice how the fibril is splayed at the bottom and shows the individual tropocollagen molecules. (Courtesy of Peter Davison.)

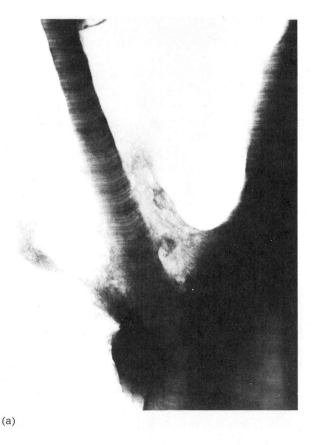

(a)

(b)

(c)

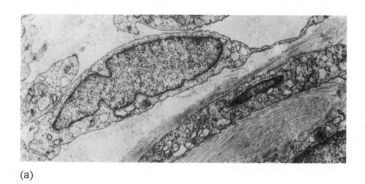

(a)

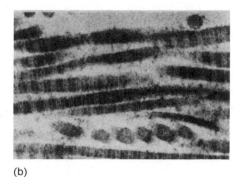

(b)

Figure 6-2
(a) A cross-section of connective tissue showing a fibroblast (large central object) and microfibrils in the intracellular space (fibers at lower right). (b) An enlarged view of the microfibrils. (Courtesy of H. Hecker.)

secreted to the extracellular space (Figure 6-2) as a precursor called **procollagen** and then assembled as follows:

Procollagen filaments (0–4 nm)
↓
Protofibrils (11–15 nm)
↓
Microfibrils (30–200 nm)
↓
Collagen fibrils (0.2–0.4 μm)
↓
Fibers (1–10 μm)
↓
Tendon, bone, cartilage, etc. (macroscopic).

Amino Acid Composition and Sequence of Collagen

The amino acid composition and sequence of the polypeptide chains from which collagen is assembled have several striking properties:

1. In each chain, which contains nearly 1000 amino acids, one-third of the amino acids are glycine. This value is much higher than that for a typical protein, in which about 5 percent of the amino acids is glycine; that is, the amino acid sequence can be represented as

   . . . Gly–A–B–Gly–C–D–Gly–E–F–Gly–G–H . . .

   in which A, B, C, . . . are amino acids other than glycine.

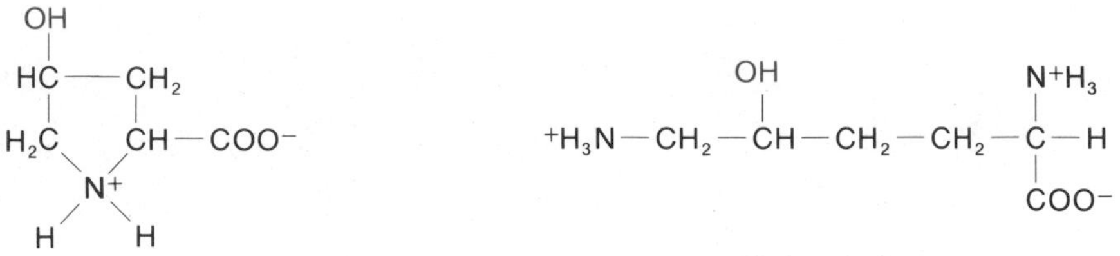

OH
|
HC———CH$_2$

H$_2$C   CH—COO$^-$
   N$^+$
  /   \
 H   H

**Hydroxyproline
(Hyp)**

OH   N$^+$H$_3$
| |
$^+$H$_3$N—CH$_2$—CH—CH$_2$—CH$_2$—C—H
                              |
                              COO$^-$

**Hydroxylysine
(Hyl)**

Figure 6-3
Formulas for hydroxyproline and hydroxylysine.

2. One-fourth of the amino acids are proline or a derivative, hydroxyproline (Figure 6-3). Proline is an amino acid somewhat infrequently encountered and hydroxyproline is not found in other proteins.
3. Many of the lysines are also hydroxylated (Figure 6-3).
4. The sequence glycine-proline-hydroxyproline occurs frequently in collagen, whereas in most proteins recurring amino acid sequences are unusual.

We will see that these amino acids and their arrangement are responsible for the more important properties of collagen.

Tropocollagen—A Triple-Stranded Helical Unit

The single-stranded monomer of collagen does not exist in nature except during one stage of collagen synthesis. Instead, three strands of this monomer are joined to form **tropocollagen.**

In part, the structure of tropocollagen is the result of the large amount of proline present in the individual polypeptide chains. In Chapter 3 it was pointed out that proline is an imino acid and thus fails to form a typical peptide bond. The bond it engages in is much less flexible and thereby causes each polypeptide strand to be rather extended. This extension exposes most of the potentially interactive groups, enabling the individual strands to interact by means of inter-strand hydrogen bonds; this generates a **triple-stranded helix** (Figure 6-4), which is the first stage in forming a fiber with great tensile strength. These hydrogen bonds and also van der Waals interactions pull the strands together to form a very tight complex in which the proline and most of the other side chains are external and the glycines are internal. Examination of the triple helix explains why glycine has evolved to be in every third position. The inner region of a triple helix is very crowded, as shown in Figure 6-5. The α-carbon atoms of glycine are shown in the cross-section of the triple helix; there is clearly no space

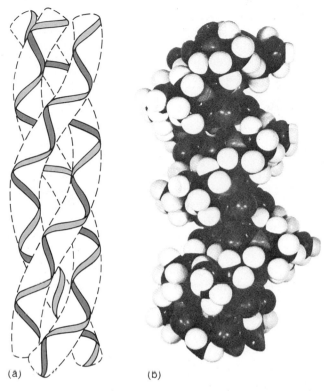

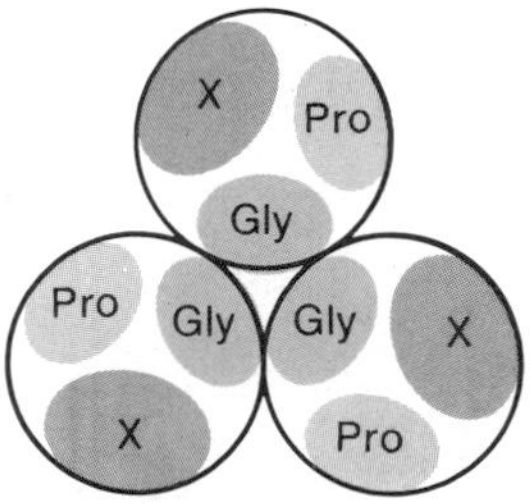

Figure 6-5
A schematic diagram of a cross section of triple-stranded collagen. Each heavy circle represents the outline of each α-helical strand. The crowding in the inner part of the triple helix is evident—glycine is the only amino acid whose side chain will fit. The shape labeled x can be any side chain.

(a)   (b)

Figure 6-4
(a) Model of the triple-stranded collagen helix. (b) A space-filling model of a segment of the collagen triple helix. (With permission, from Crick, F., and A. Rich. 1961. *J. Mol. Biol.,* 3: 483. Copyright: Academic Press Inc. [London] Ltd.)

for the bulky side chains of other amino acids. Thus the presence of glycine enables the structure to have maximum tightness and to increase its strength.

Interaction of Tropocollagen Units to Form a Fiber

In order to form a long fiber some means is needed to align tropocollagen molecules end to end. However, there is no direct way to form a strong end-to-end aggregate, as we will soon see. Tropocollagen contains numerous clusters of positively and negatively charged amino acids on its surface, which enables tropocollagen molecules to aggregate side to side. Having so many charges of both signs on a *single* chain would cause the molecule to fold back on itself were it not for the rigidity of the triple helix. These charges instead cause side-to-side aggregation. If the tropocollagen molecules were aligned precisely side to side, a broad unit would result whose length would equal the length of one tropocollagen molecule and whose width would be determined by the number of tropocollagen molecules in the unit. Such aggregates have been prepared in the laboratory and are shown in Figure 6-6. Patterns of dark bands are seen if the units are reacted with uranyl acetate, which binds to negative charges. If phosphotungstic acid, which binds to positive charges, is used as a stain, the band patterns are identical, which shows that the positively and negatively charged amino acids occupy approximately the same positions.

Another type of aggregate can be prepared in the laboratory. This structure is a long fiber that is easily disrupted. Its band pattern differs from that shown in the figure. An analysis of the band pattern shows that in the form in Figure 6-6 the tropocollagen molecules are in a parallel array; in long fibers, they are in an antiparallel array, as shown in Figure 6.7. Band patterns have also been observed with native collagen—that is, collagen that has not been formed in the laboratory from tropocollagen (Figure 6-1). These band patterns are different from those observed in synthetic material, which indicates that in natural collagen the tropocollagen molecules form neither parallel nor antiparallel arrays. An analysis of the band patterns of natural collagen shows that the tropocollagen molecules are aggregated side to side, but they are in a **quarter-staggered array** (Figure 6-8). This arrangement allows the generation of long fibers without the necessity for end-to-end aggrega-

Figure 6-6
Electron micrograph of disaggregated and reconstituted collagen in the so-called "segment-long-spacing" form, in which the monomers are parallel and joined side by side. The band pattern differs from that of native collagen shown in Figure 6-1(a). (Courtesy of Peter Davison.)

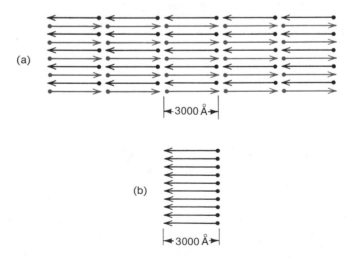

(a)

|←3000 Å→|

(b)

|←3000 Å→|

Figure 6-7
Diagrams showing two different aggregates of tropocollagen. (a) The fibrous-long-spacing form, in which the molecules are antiparallel and joined end to end. (b) The segment-long-spacing form, in which molecules are parallel.

tion. This staggering is actually a better way to generate strength. If collagen were formed by end-to-end joining of either of the arrays shown in Figure 6-6, one would expect the end-to-end joining of the aggregates to be cooperative. Thus, if a single end-to-end interaction were broken down (perhaps by accidental chemical attack), this would weaken adjacent pairs and possibly cause a cascading breakdown that could lead to cleavage of the fiber. However, in the quarter-staggered array, breakdown of the interaction between any pair of neighboring molecules does not weaken the fiber in any significant way.

In addition to the charge attraction just described, a second interaction strengthens the quarter-staggered array. This interaction utilizes the collagen-specific amino acids, hydroxyproline and hydroxylysine, which engage in covalent cross-links between adjacent polypeptide chains in tropocollagen and between adjacent tropocollagen chains (Figure 6-9). It is clear that many structural features of the collagen monomer are combined to form a macromolecular assembly that has great tensile strength.

Another structural feature of collagen plays a special functional role. Between the head and tail of each successive tropocollagen molecule in the quarter-staggered array is a gap (Figure 6-8). It is thought that this gap plays a role in bone formation. Bone is formed by the deposition of crystals of hydroxylapatite $(Ca_{10}(PO_4)_6(OH)_2)$ on a collagen fiber. The crystallization process can be carried out *in vitro*. It is found that initially microcrystals form on a collagen fiber at intervals of 670 Å, which is exactly the spacing of the gaps. Crystallization does not occur on the synthetic structures shown in Figure 6-7, neither of which possesses gaps.

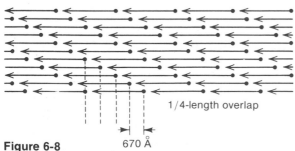

1/4-length overlap

670 Å

Figure 6-8
Schematic diagram showing the quarter-staggered array of tropocollagen in native collagen. The gap between each arrowhead and tail is about 400 Å. In reality, the structure is helical, as shown in Figure 6-4.

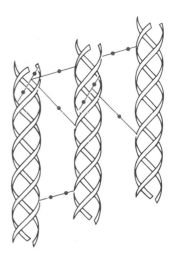

Figure 6-9
Intramolecular and intermolecular cross-links in collagen. (Courtesy of M. L. Tanzer.)

A COMPLEX DNA STRUCTURE: THE *E. COLI* CHROMOSOME

All genes of *E. coli*, and presumably of most bacteria, are contained on a single supercoiled circular DNA molecule. The total length of the circle is about 1300 μm. A cylindrical bacterium has a diameter and a length of about 1 and 3 μm respectively (Figure 6-10); clearly the bacterial DNA must be highly folded when it is in a cell.

Interaction of the DNA with a Scaffold Protein

When *E. coli* DNA is isolated by using a technique that avoids both breakage of the molecule and denaturation of proteins, a highly compact structure known as a bacterial chromosome or a **nucleoid** is found. This structure contains RNA, protein, and supercoiled DNA. An electron micrograph of the *E. coli* chromosome is shown in Figure 6-11. Two features of this structure should be noted: (1) the loops of DNA at the periphery are all supercoiled and (2) near the center of the chromosome is a dense region called the **scaffold.** The shape of the scaffold varies from one chromosome to the next but its length is fairly constant, ranging from 3 to 5 μm, or roughly one to two times the length of the bacterium. If the chromosome is treated with RNase (which degrades RNA to nucleotides), trypsin (which degrades proteins), or various detergents that break down protein-protein interactions, the chromosome expands markedly, though it remains much more compact than a free DNA molecule. As proteins are removed, the scaffold is disrupted and the chromosome goes through a series of transitions between forms having different degrees of compactness. The conclusion from these and other observations is that the chromosome is held in compact form by the binding of different regions of the DNA molecule to the scaffold and that the scaffold is a complex structure containing RNA and protein.

A Multiply-Looped DNA Structure

Treatment of a chromosome with very tiny amounts of a DNase producing single-strand breaks, followed by sedimentation of the treated DNA, gives some insight into the physical structure of the chromosome. In Chapter 4 it was explained that if a supercoiled DNA molecule receives one single-strand break, the strain of underwinding is immediately removed by free rotation about the opposing sugar-phosphate bond and all supercoiling is lost. Since a nicked circle is much less

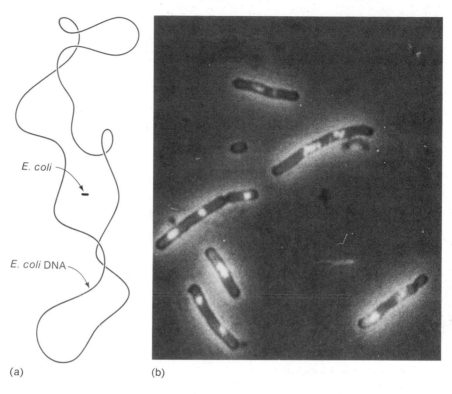

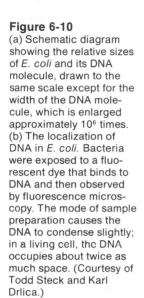

Figure 6-10
(a) Schematic diagram showing the relative sizes of *E. coli* and its DNA molecule, drawn to the same scale except for the width of the DNA molecule, which is enlarged approximately 10^6 times. (b) The localization of DNA in *E. coli.* Bacteria were exposed to a fluorescent dye that binds to DNA and then observed by fluorescence microscopy. The mode of sample preparation causes the DNA to condense slightly; in a living cell, the DNA occupies about twice as much space. (Courtesy of Todd Steck and Karl Drlica.)

(a)

(b)

Figure 6-11
Electron micrograph of the chromosome of *E. coli* attached to two fragments of a proteinaceous substance, which may be the cell membrane. This single molecule of double-helical DNA is intact and supercoiled. (From H. Delius and A. Worcel. 1974. *J. Mol. Biol.,* 82: 108. Copyright: Academic Press, Inc. [London] Ltd.)

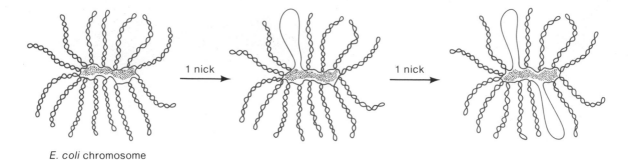

E. coli chromosome

Figure 6-12
A schematic drawing of the highly folded supercoiled
E. coli chromosome, showing only 15 of the 46 loops
attached to the scaffold and the opening of a loop
by a single-strand break (nick).

compact than a supercoil of the same molecular weight, the nicked circle sediments much more slowly. Thus, one single-strand break causes an abrupt decrease (by about 30 percent) in the s value. However, if one single-strand break is introduced by a nuclease into the E. coli chromosome, the s value decreases by only a few percent. Furthermore, a second break causes a second decrease in the s value, and subsequent breaks cause additional stepwise changes. After about forty-five breaks the form of the chromosome remains constant. This clearly indicates that free rotation of the entire DNA molecule does not occur when a single-strand break is introduced. The structure of the E. coli chromosome that has been deduced from these data is shown in Figure 6-12. The DNA is assumed to be fixed to the scaffold at 45 ± 10 positions, each of which prevents free rotation. Thus, there would be about 45 ± 10 supercoiled loops of DNA. One single-strand break would then cause one supercoiled loop to become open and each subsequent break would, on the average, open another loop. This notion has been confirmed in electron micrographs of chromosomes in which there are one or two nicks; structures such as that in Figure 6-11 but with one or two open loops are observed.

The enzyme DNA gyrase, which plays an important role in DNA replication (Chapter 8), is responsible for the supercoiling. If coumermycin, an inhibitor of E. coli DNA gyrase, is added to a culture of E. coli cells, the chromosome loses its supercoiling in about one generation time. Indirect measurements of the number of binding sites for gyrase on the chromosome indicate that there are roughly forty-five sites. The spatial distribution of these sites is not known but it is tantalizing to think that there may be one binding site in each supercoiled loop.

In the next section it will be seen that DNA is arranged in a much more complex way in eukaryotic cells.

CHROMOSOMES AND CHROMATIN

The DNA of all eukaryotes is organized into morphologically distinct units called **chromosomes** (Figure 6-13(a)). Each chromosome contains only a single enormous DNA molecule. For example, the DNA molecule in a single chromosome of the fruit fly *Drosophila* has a molecular weight greater than 10^{10} and a length of 1.2 cm. (Since the width of all DNA molecules is 20 Å or 2×10^{-7} cm, the ratio of length to width of *Drosophila* DNA has the extraordinary value of 6×10^{6}.) These molecules are much too long to be seen in their entirety by electron microscopy because, at the minimum magnification needed to see a DNA molecule, the field of view is only about 0.01 cm across. However, the DNA can be visualized by autoradiography (Figure 6-13(b)). The pattern of grains in autoradiographic film above a DNA molecule is 100 times wider than the DNA; thus, a much lower magnification can be used and the entire molecule will be included in the field of view.

A chromosome is much more compact than a DNA molecule and in fact a DNA molecule cannot spontaneously fold to form such a compact structure because the molecule would be strained enormously. Instead, DNA is made compact by a hierarchy of different types of folding, each of which is mediated by one or more protein molecules.

(a)

(b)

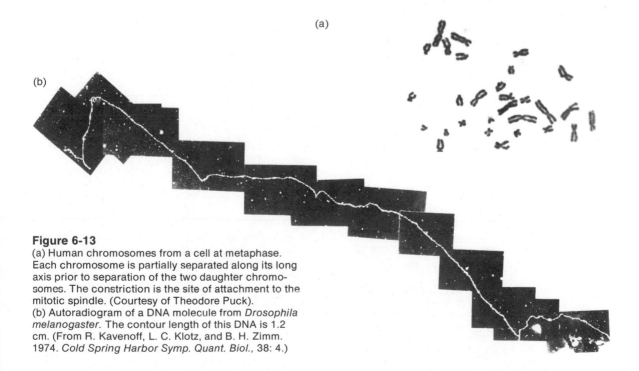

Figure 6-13
(a) Human chromosomes from a cell at metaphase. Each chromosome is partially separated along its long axis prior to separation of the two daughter chromosomes. The constriction is the site of attachment to the mitotic spindle. (Courtesy of Theodore Puck).
(b) Autoradiogram of a DNA molecule from *Drosophila melanogaster*. The contour length of this DNA is 1.2 cm. (From R. Kavenoff, L. C. Klotz, and B. H. Zimm. 1974. *Cold Spring Harbor Symp. Quant. Biol.*, 38: 4.)

Histones and Chromatin

The DNA molecule in a eukaryotic chromosome is bound to very basic proteins called **histones.** The complex comprising DNA and histones is called **chromatin.** There are five major classes of histones—H1, H2A, H2B, H3, and H4—whose properties are listed in Table 6-1. Histones have an unusual amino acid composition in that they are extremely rich in the positively charged amino acids lysine and arginine. The lysine-to-arginine ratio differs in each type of histone. (In the older literature H1 was called the lysine-rich histone and H3 and H4 were called arginine-rich.) The positive charge of the histones is one of the major features of the molecules, enabling them to bind to the negatively charged phosphates of the DNA. This electrostatic attraction is apparently the most important stabilizing force in chromatin since, if chromatin is placed in solutions of high salt concentration (e.g., 0.5 M NaCl), which breaks down electrostatic interactions, chromatin dissociates to yield free histones and free DNA. Chromatin can also be reconstituted by mixing purified histones and DNA in a concentrated salt solution and gradually lowering the salt concentration by dialysis. The result shows that no other components are needed to form chromatin.

Reconstitution experiments have been carried out in which histones from different organisms are mixed. Usually, almost any combination of histones works because, except for H1, the histones from dif-

Table 6-1
Types of histones

Type	Lys/Arg ratio	Number of residues	Mass (kdal)	Location
H1	20.0	215	21.0	Linker
H2A	1.25	129	14.5	Core
H2B	2.5	125	13.8	Core
H3	0.72	135	15.3	Core
H4	0.79	102	11.3	Core

Figure 6-14
The amino acid sequence of histone H4 from calf thymus. Histone H4 from pea seedlings has the same amino acid sequence except that residue 60 is isoleucine and residue 77 is arginine.

```
Ser-Gly-Arg-Gly-Lys-Gly-Gly-Lys-Gly-Leu-      10
Gly-Lys-Gly-Gly-Ala-Lys-Arg-His-Arg-Lys-      20
Val-Leu-Arg-Asp-Asn-Ile-Gln-Gly-Ile-Thr-      30
Lys-Pro-Ala-Ile-Arg-Arg-Leu-Ala-Arg-Arg-      40
Gly-Gly-Val-Lys-Arg-Ile-Ser-Gly-Leu-Ile-      50
Tyr-Glu-Glu-Thr-Arg-Gly-Val-Leu-Lys-Val-      60
Phe-Leu-Glu-Asn-Val-Ile-Arg-Asp-Ala-Val-      70
Thr-Tyr-Thr-Glu-His-Ala-Lys-Arg-Lys-Thr-      80
Val-Thr-Ala-Met-Asp-Val-Val-Tyr-Ala-Leu-      90
Lys-Arg-Gln-Gly-Arg-Thr-Leu-Tyr-Gly-Phe-     100
Gly-Gly                                        102
```

ferent organisms are very much alike. In fact, the amino acid sequences of both H3 and H4 are nearly identical (sometimes one or two of the amino acids differ) from one organism to the next. Histone H4 from the cow differs by only two amino acids from H4 from peas—arginine for lysine and isoleucine for valine (Figure 6-14)—which shows that the structure of histones has not changed in the 10^9 years since plants and animals diverged. Clearly histones are very special proteins indeed.

The Structural Hierarchy of Chromosomes

As a cell passes through its growth cycle, the structure of its chromatin changes. In a resting cell the chromatin is dispersed and fills the entire nucleus. Later, after DNA replication has occurred, the chromatin condenses about 100-fold and chromosomes form. Chromosomes have been isolated and gradually dissociated and chromosomes at various degree of dissociation have been observed by electron microscopy (Figure 6-15). The chromosome is first broken down into thick fibers of varying width. These are composed of fibers 25–30 nm wide, formed from smaller fibrils 10 nm wide, which appear like a string of beads. The beadlike structure is also seen when chromatin isolated from resting nuclei is examined (Figure 6-16). This string of beads is chromatin. The beads, which have a diameter of 100 Å, are connected by DNA strands 20 Å wide. The beadlike particles are an orderly aggregate of histones and DNA.

Nucleosomes

Prior to electron microscopic studies, the effect of various DNA endonucleases on chromatin also suggested that chromatin contains repeating units. Treatment of chromatin with micrococcal nuclease (which cannot attack DNA that is in contact with protein) yields a collection of small particles containing DNA and histones (Figure 6-17). If, after enzymatic digestion, the histones are removed, DNA fragments having roughly 200 base pairs or a multiple of 200 are found. After a long period of digestion the multiple-size units are not found and all of the DNA has the 200-base-pair-unit size. Fragments have also been isolated before removal of the histones and have been examined by electron microscopy; it has been found that a fragment containing $200n$ base pairs consists of n connected beads, indicating that there is a fundamental bead unit containing about 200 base pairs.

The beadlike particles are called **nucleosomes.** Each nucleosome is found to consist of one molecule of histone H1, two molecules each of

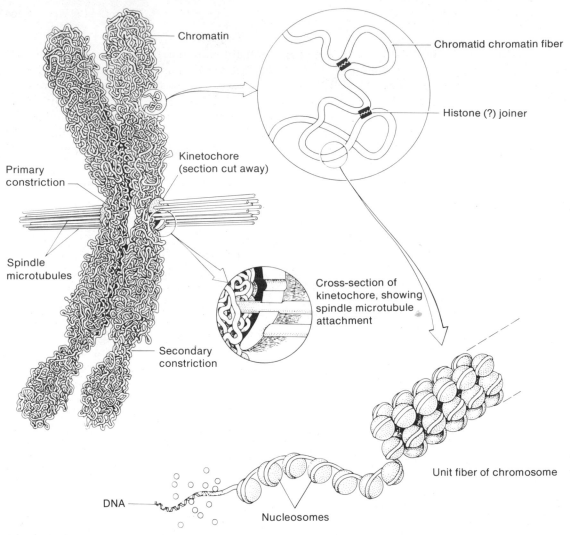

Figure 6-15
Drawing of a chromosome showing how it is broken down to smaller units.
(From S. Wolfe, 1981. *Biology of the Cell.* Wadsworth.)

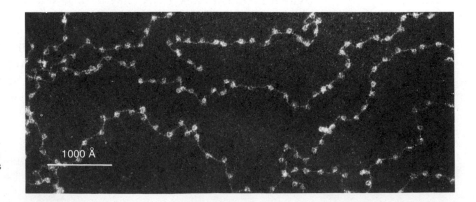

Figure 6-16
Electron micrograph of chromatin. The beadlike particles have diameters of nearly 100 Å. (Courtesy of Ada Olins.)

histones H2A, H2B, H3, and H4, and a DNA fragment. Treatment of the nucleosomes (which have been obtained by digestion of chromatin with pancreatic DNase) with the enzyme micrococcal nuclease gradually removes additional amounts of DNA. All histones remain associated with the DNA until the number of base pairs is less than 160, at which point H1 is lost. More bases can be removed but the number cannot be reduced to less than 140 base pairs. The structure that remains is called the **core particle.** It contains an octameric protein disc consisting of two

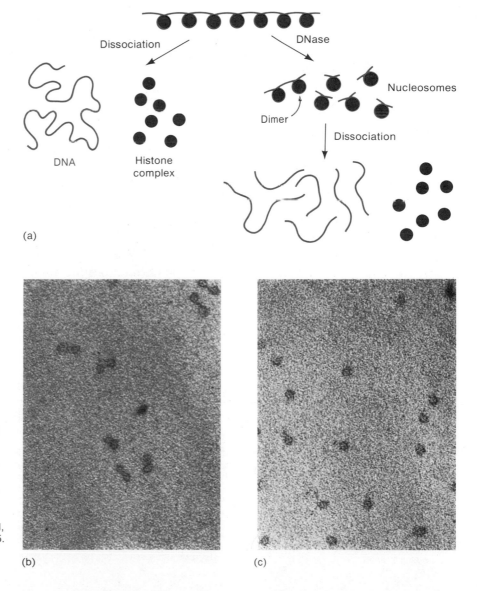

(a)

Figure 6-17
(a) The DNase-digestion method for production of individual nucleosomes and 200-base-pair fragments (b,c). (b) Electron micrographs of monomers and dimers. (c) Nucleosomes produced as described in part (a). (From J. T. Finch, M. Noll, and R. D. Kornberg. 1975. *Proc. Nat. Acad. Sci.,* 72: 3321.)

(b)   (c)

Figure 6-18
Schematic diagram of nucleo-some core particle. The DNA molecule is wound 1-3/4 turns around a histone octamer (2 mole-cules each of histones H2A, H2B, H3, and H4). Histone H1 (not shown) is bound to the linker DNA. Note that the two linker units point in the same direction.

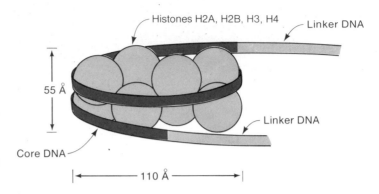

Figure 6-19
Schematic diagram of chromatin. Although not shown, the structure is likely to follow a zigzag path because the linkers leading to and from a single nucleosome are parallel.

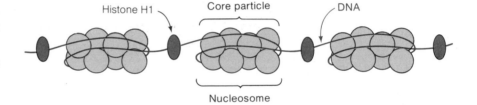

copies each of H2A, H2B, H3, and H4, around which the 140-base-pair segment is wrapped like a ribbon (Figure 6-18). Thus a nucleosome consists of a core particle and **linker DNA** to which H1 is attached. The overall structure of the chromatin fibril is that shown in Figure 6-19. Possibly, H1 also binds to adjacent core particles to make a more compact structure.

The binding of DNA and histone has been examined. About 80 percent of the amino acids in the histones are in α-helical regions. Preliminary data suggest that many of the extended α-helical regions lie in the larger groove of the DNA helix and that the complex is stabilized by an electrostatic attraction between the positively charged lysines and arginines of the histones and the negatively charged phosphates of the DNA.

The DNA content of nucleosomes varies from one organism to the next, ranging from 150 to 240 base pairs per unit. The core particles of all organisms have the same DNA content (140 base pairs), so that the observed variation results from different sizes of the linker DNA between the nucleosomes (namely, 10 to 140 base pairs). Suprisingly, in the same organism the lengths of the linkers may also vary from one tissue to another (for instance, brain versus liver). The significance of this variation is unknown.

Assembly of Nucleosomes: The 30-nm Fiber and the Solenoidal Model of Chromosome Structure

As we stated earlier, chromosomes have several orders of organization. The nucleosome is the first; in this structure a segment of DNA containing 150 to 240 base pairs and whose length is 510 to 726 Å when extended in solution is compacted into a cylinder about 110 Å across and 55 Å high. The next level of organization is the 30-nm fiber mentioned earlier. This fiber has been isolated and characterized by several techniques.

The currently accepted model for the chromosome is the solenoidal model shown in Figure 6-20. In this structure a linear beaded chain is arranged as a hollow helix that is 360 Å in diameter and that has a repeating unit of 110 Å. Other experiments suggest that these helices are arranged in a series of loops. The looped structure has the degree of compactness of DNA in a resting nucleus. Chromosomes are about 50–100 times as compact as the looped structure allows. Further compacting is probably provided by some type of scaffold (such as was seen in Figure 6-11 for the *E. coli* chromosome) consisting of nonhistone proteins and possibly RNA. Electron micrographs have been taken of metaphase chromosomes stripped of histones. The DNA is unfolded somewhat but, as shown in Figure 6-21, it remains tightly associated with the scaffold in a form having an enormous number of loops. Condensation and organization around the scaffold probably serve only to create a discrete unit that can be moved along the mitotic spindle prior to cell division.

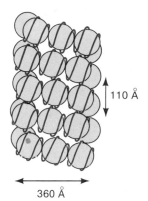

110 Å

360 Å

Figure 6-20
A proposed solenoidal model of chromatin. Six nucleosomes (shaded) form one turn of the helix. The DNA double helix (shown in red) is wound around each nucleosome. (After J. T. Finch and A. Klug. 1976. *Proc. Nat. Acad. Sci.*, 73: 1900.)

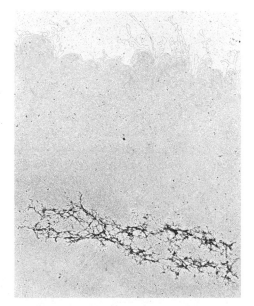

Figure 6-21
Electron micrograph of a segment of a human metaphase chromosome from which the histones have been chemically removed. The dense network near the bottom of the figure is the protein scaffold on which the chromosome is assembled. (Courtesy of Ulrich Laemmli.)

INTERACTION OF DNA AND A PROTEIN THAT RECOGNIZES A SPECIFIC BASE SEQUENCE

In the formation of chromatin the histones do not bind to specific base sequences but recognize only the general DNA structure. There are, however, numerous proteins that bind only to particular sequences of bases, and we will encounter many such proteins in this book. These proteins are of three types: (1) polymerases, which initiate synthesis of DNA and RNA from particular base sequences; (2) regulatory proteins, which turn on and off the activity of particular genes; and (3) certain nucleases, which cut phosphodiester bonds between a single pair of adjacent nucleotides contained in unique sequences of four to six nucleotides.

In the case of many DNA-binding proteins, the specific sequence of bases to which the protein binds is known. However, at present in only a few cases, one of which we now describe, do we understand the particular features of the DNA that are recognized and the mode of binding.

Binding of the Cro Protein to Phage λ DNA

The Cro protein made by *E. coli* phage λ is an important regulatory protein in the life cycle of the phage. Its biological role is not yet thoroughly understood, but that will not be of concern here. What is important is that it binds to a specific base sequence in a DNA molecule.

Cro is a small basic protein which contains 66 amino acids and binds to a sequence of DNA containing 17 base pairs. Both the amino acid sequence of the protein and the base sequence of the interaction region in DNA are known; this information has made it reasonable to attempt to work out the structural features of the interaction. At present, information about dual or matching sequences of this kind is known for only a few systems.

The three-dimensional structure of the Cro protein has been determined by standard techniques of x-ray diffraction and is shown in Figure 6-22. It is a simple structure (compared to larger proteins), consisting of three strands of antiparallel β sheet (residues 2–6, 39–45, and 48–55) and three α helices (residues 7–14, 15–23, and 27–36). Residues 63–66 are not shown in the figure as they possess no order (and have no bearing on the structure of the Cro protein or the Cro-DNA interaction).

Figure 6-23 shows the sequence of the seventeen bases in the Cro-binding site of λ DNA. Three procedures have been used to determine some of the points of contact between the protein and the DNA. In the first procedure, purified Cro protein and λ DNA are mixed using conditions that cause the Cro-DNA complex to form. The

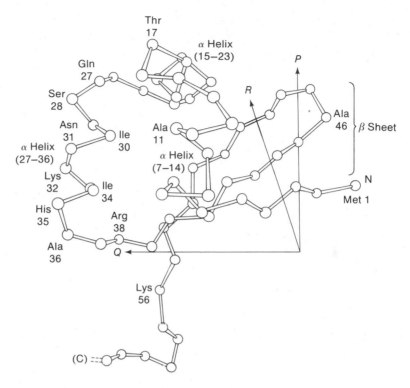

Figure 6-22
Three-dimensional structure of the λ Cro protein. Segments consisting of
α helices and β structure are indicated. The arrows P, Q, and R indicate the
three geometric axes referred to in other figures. The complete C-terminus
is not shown. (After W. F. Anderson et al., 1981. Nature, 290: 754.)

TATCACCGCAAGGGATA
ATAGTGGCGTTCCCTAT

Figure 6-23
The base sequence of
one of the λ Cro protein
binding sites. This se-
quence is called oR3. The
adenines and guanines
shown in Figure 6-24 are
printed in red.

Cro-DNA complex is treated with dimethyl sulfate, which causes the
addition of a methyl group to guanine in the N-7 position and to adenine
in the N-3 position of the purine ring. The complex is then dissociated
and the Cro protein is removed. Then, by direct analysis of the base
sequence of the DNA, the positions of the methylated guanines and
adenines are determined. Certain guanines and adenines are resistant to
methylation; presumably, in the complex, Cro is in contact with these
bases and prevents the dimethyl sulfate from approaching these bases.
In the second procedure, the Cro-DNA complex is treated with a reagent
that adds an ethyl group to phosphates, and the particular phosphates
that have become ethylated are then located. Again, certain phosphates
fail to be ethylated and these are presumably also in contact with the
protein. The first two procedures have assessed the extent to which Cro
inhibits the reaction with certain sites on the DNA. The third procedure
is designed to confirm the positions of the phosphates that are in
contact. In the third procedure, any phosphate can be altered—usually

one or two at a time—to determine which positions, when altered, inhibit binding of Cro. In this procedure ethylation is carried out, but the treatment is so brief that only one or two phosphates are ethylated at random in a given molecule. Cro is then added in excess so that any DNA molecule capable of forming a Cro-DNA complex is in such a complex. Some of the ethylated DNA molecules form a complex with Cro but others do not. The latter class—the free DNA molecules—is isolated in order to determine the positions of the ethylated phosphates in these molecules. Ethylation of these phosphates has interfered with the binding of Cro to the DNA; Cro would presumably bind to these phosphates if they were not ethylated. The positions of these phosphates must be in the contact region. Note that these positions are different from those ethylated by the second procedure. At the time of this writing, there is no information about contact of Cro to the cytosine and the thymine bases or to deoxyribose.

Figure 6-24 shows the Cro-binding segments of λ DNA drawn as a double helix. The guanines and adenines that cannot be methylated

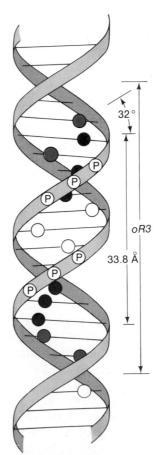

Figure 6-24
Points at which the *oR3* region of λ DNA makes contact with Cro. Phosphates are labeled P; guanines and adenines in contact are solid black and red circles; guanines and adenines not in contact are open black and red circles, respectively. The solid circles are the bases shown in red in Figure 6-23. Note that the bases in contact are only in the wider (major) groove of the DNA. The lengths and angles should be compared with Figure 6-25. (After W. F. Anderson *et al.*, 1981. *Nature,* 290: 754.)

when Cro is bound and the phosphates whose ethylation prevents binding are indicated. The potential methylation sites (the N-3 and N-7 atoms) of the bases face the reader. Since these bases and the phosphates shown are contact points, the figure illustrates that the contact points all lie on a single side of the helix, which, from a geometrical point of view, seems reasonable. Furthermore, the contact points are only in the major (wider) groove of the DNA. (This arrangement is also assumed by the λcI protein, which binds to λ DNA, and the *E. coli* CAP protein, which binds to certain regulatory sites in *E. coli* DNA and may be a common feature in all DNA-protein interactions.) Two additional points should be noted: (1) the adenines and guanines to which Cro is bound are on both strands (see also Figure 6-23); and (2) the contact points on the DNA are nearly symmetrically arranged.

An analysis of the arrangement of Cro protein molecules in a water-containing crystal indicates that the proteins are in an array having twofold symmetry. This array is shown in Figure 6-25. In

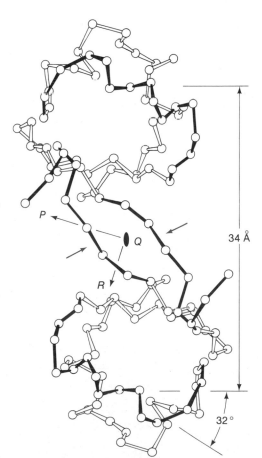

Figure 6-25
Arrangement of two Cro molecules in a protein crystal, viewed along the *x* axis of Figure 6-23. Regions of the protein backbone closest to the viewer and presumably in contact with DNA are drawn solid. These regions include two α helices 34 Å apart, inclined at 32°, and also a pair of extended strands (arrows) close to the *Q* axis. (After W. F. Anderson *et al.*, 1981. *Nature,* 290: 754.)

solution Cro sometimes exists as a dimer, and it is likely that together the two monomers have the same symmetry that the molecules have in a crystal. This observation is particularly significant because the Cro-binding DNA sequence is also somewhat symmetric, and such a symmetric array is a reasonable basis for binding to a nearly symmetric base sequence. Figures 6-26 and 6-27 show the presumed structure of the Cro-DNA complex. In Figure 6-26 the symmetry of the structure is readily apparent. As drawn, one monomer is above the other, and they interact near their carboxyl termini, where the molecules cross the Q

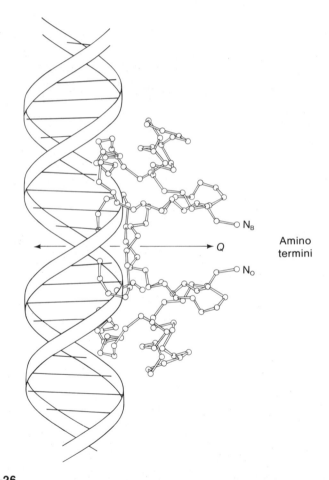

N_B

Q

N_O

Amino termini

Figure 6-26
Presumed structure of the Cro-*oR3* complex. The DNA is rotated 90° relative to that in Figure 6-24 so that the contact points are on the right side of the molecule. The amino termini of both Cro monomers are labeled N_O and N_B. The Q axis is that shown in Figures 6-22 and 6-25. A pair of α helices, related by two-fold symmetry, occupy successive major grooves of the DNA, and the two extended chains, indicated in Figure 6-25, run parallel to the axis of the DNA. (After W. F. Anderson *et al.*, 1981. *Nature,* 290: 754.)

axis. Figure 6-27 is a stereoscopic view of the complex in which the protein is oriented as in Figure 6-25. Note that the DNA molecule in Figure 6-27 is in the plane of the paper and that the protein molecule is behind the DNA. This view also shows that the binding takes place within one of the grooves (the major groove) of the helix, as was indicated in Figure 6-24.

A close look at the complex (for which there is not enough detail in the figures) would show that amino acids in each Cro monomer are situated in a way that they can form hydrogen bonds with the DNA bases, which confers binding specificity, and also interact electrostatically with the phosphate groups, which confers binding strength.

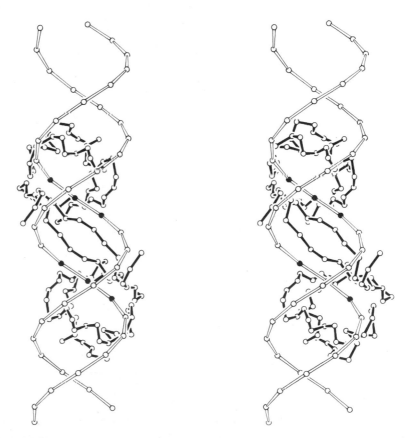

Figure 6-27
Stereo view of the Cro-DNA complex at an angle that the orientation of the Cro dimer is the same as that shown in Figure 6-25. The open circles on the DNA backbone indicate the phosphates; the solid circles indicate those phosphates whose ethylation hinders binding. By relaxing the eyes and focusing at infinity, the reader should be able to merge the drawings and see the stereo effect. (After W. F. Anderson *et al.*, 1981. *Nature*, 290: 754.)

Since the Cro-DNA complex is one of the first sequence-specific complexes whose structures have been worked out, it may be premature to make generalizations. However, several features may be common to all DNA-protein complexes in which binding occurs within a particular base sequence. In subsequent chapters it will be seen that most DNA base sequences recognized by specific proteins have some degree of symmetry. Furthermore, a great many of the sequence-specific DNA-binding proteins are multisubunit proteins. Thus it might be expected that the proteins will often form a symmetric array that facilitates recognition of a symmetric sequence. Preliminary evidence with several other DNA-binding proteins, which bind both specifically and nonspecifically, indicates that these proteins usually have two or three extended α-helical regions on the surface of the protein. Since one of the α helices of each Cro monomer lies within the major groove of the DNA and is responsible for much of the binding interaction, it has been suggested that insertion of an α helix into the major groove may be a general feature of DNA-protein interactions. Remember that such an interaction was also suggested for histone binding in chromatin.

BIOLOGICAL MEMBRANES

Biological membranes are organized assemblies consisting mainly of proteins and lipids. The structures of all biological membranes have many common features but small differences exist to accommodate the varied functions of different membranes.

Roles of Membranes in Biological Systems

There are many types of membranes, each having its own function. A fundamental function of membranes is to separate the contents of a cell from the environment. However, it is always necessary that cells take up nutrients from the surroundings; therefore, the enclosing membrane of a cell must be permeable. The permeability of the membrane must be selective though, so that the concentrations of compounds within a cell can be controlled; otherwise, intracellular concentrations would be inseparable from extracellular concentrations. Selective permeability is attained by means of restricting free passage of most intra- and extracellular substances and allowing transport of specific substances by molecular pumps, channels, and gates. External membranes also have a signal function. For instance, they often contain receptors for signal molecules such as hormones. The external membrane is also responsible for transmitting electrical impulses, as in nerve cells.

There are also numerous types of intracellular membranes. These membranes serve to compartmentalize various intracellular components and to provide surfaces on which certain molecules can be adsorbed prior to chemical reaction with other adsorbed molecules. Some internal membranes even contain enzyme systems. For example, photosynthesis takes place within the inner membrane of chloroplasts in plant cells.

Membrane Structure: The Basic Bilayer

Figure 6-28 shows an electron micrograph in which the membrane that encloses a red blood cell is seen in cross-section. The most notable feature of this membrane is that it consists of two layers—it is called a **membrane bilayer.** This structure is a consequence of the chemical

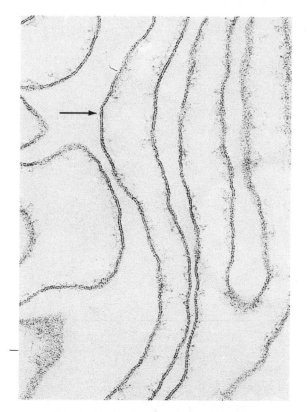

Figure 6-28
Electron micrograph of a preparation of plasma membranes from red blood cells. A single membrane is denoted by an arrow. (Courtesy of Vincent Marchesi.)

nature of the lipids that form the membrane. There are many membrane lipids of which the most prevalent are the phospholipids. The chemical structure of one phospholipid is shown in Figure 6-29. This molecule, like all phospholipids, has a central sugar moiety (glycerol in this case) linked to a highly polar group, phosphatidylcholine, and to two very nonpolar long hydrocarbon chains. Figure 6-30 shows a space-filling model of this molecule and a schematic drawing. The essential feature of the molecule is a polar "head" group to which hydrocarbon "tails" are attached. Such a molecule, which has distinct polar and nonpolar

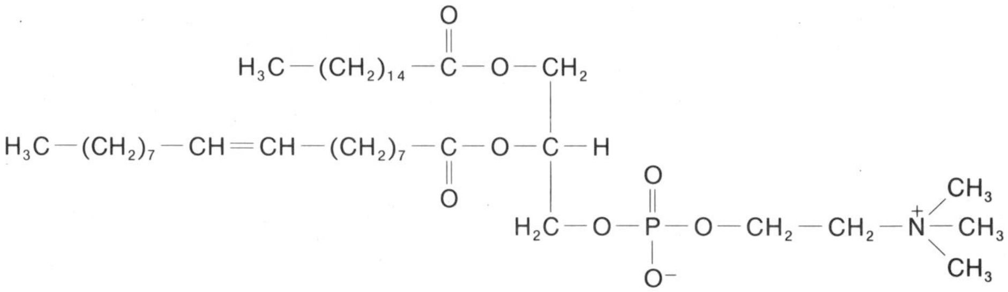

Figure 6-29
Structure of a phosphatidyl choline (1-palmitoyl-2-oleoyl-phosphatidyl choline).

$$H_3C-(CH_2)_{14}-\overset{\overset{O}{\|}}{C}-O-CH_2$$

$$H_3C-(CH_2)_7-CH=CH-(CH_2)_7-\underset{\underset{O}{\|}}{C}-O-\overset{}{\underset{}{C}}-H$$

$$H_2C-O-\underset{\underset{O^-}{|}}{\overset{\overset{O}{\|}}{P}}-O-CH_2-CH_2-\overset{+}{N}\underset{CH_3}{\overset{CH_3}{-}}CH_3$$

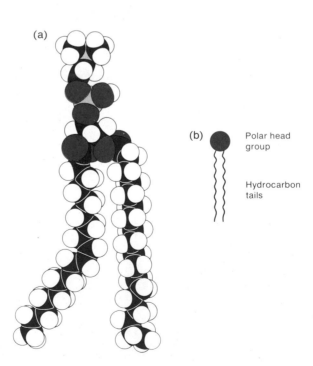

(a)

(b)   Polar head group

Hydrocarbon tails

Figure 6-30
(a) Space-filling model of a phosphatidyl choline molecule. (b) The essential features of a phospholipid or glycolipid molecule.

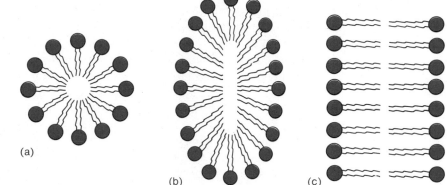

Figure 6-31
(a) Diagram of a section of a micelle formed from phospholipid molecules. (b) An ellipsoidal micelle. (c) Diagram of a section of a membrane bilayer formed from phospholipid molecules.

segments, is called **amphipathic.** When placed in water, amphipathic molecules tend to aggregate. This is because only the polar head is capable of interaction with the polar water molecules. The hydrocarbon tails are brought together by hydrophobic interactions—that is, they cluster because they are unable to interact with water. If the length of the hydrocarbon tail satisfies certain geometric constraints, a collection of amphipathic molecules forms a **micelle,** a spherical array of molecules in which the nonpolar tails form a hydrocarbon microdroplet enclosed in a shell composed of the polar heads (Figure 6-31(a)). The geometric constraints limit the size of a stable micelle and, as the size of the hydrocarbon tail increases, the array becomes ellipsoidal (Figure 6-31(b)) with the ratio of the lengths of the major axis to the minor axis of the ellipsoid increasing with length of the tail. Above a certain tail length, the only stable configuration is an ellipsoid whose major axis is "infinitely" long—that is, an extended lipid bilayer sheet, as shown in part (c) of the figure. The lipid bilayer can also be stabilized by van der Waals attractive forces between the hydrocarbon tails.

Vesicles

The ends of a lipid bilayer are unstable because the hydrocarbon chains are exposed to water. Thus lipid bilayers tend to close upon themselves to form hollow bilayer spheres known as **vesicles** (Figure 6-32). A lipid bilayer is a fluid structure that can be mechanically disrupted. Once disrupted, however, they spontaneously re-form so that contact between the exposed hydrocarbon tails and water is avoided. Often a vesicle is broken in more than one position, so that several vesicles form on the reclosing of fragments. An extended structure such as a complete cell membrane will usually form a large number of small vesicles, if it is extensively fragmented and allowed to re-form. This protocol allows one to study the permeability of a membrane in an elegant way (Figure 6-33). Either extended membranes or vesicles are suspended in a solution of a substance to be tested—for example, an amino acid. The suspension is then violently agitated, which breaks the membranes into small fragments or breaks the vesicles. When the agitation ceases, the vesicles re-form, but some amino acid molecules are trapped inside. The vesicles can then be collected and suspended in a solution lacking the amino acid. One can then determine whether in time the amino acid appears in the external fluid. Such passage will not occur if the membrane is impermeable to the amino acid. Otherwise, the rate of appearance of the amino acid is a measure of the permeability of the membrane to that molecule.

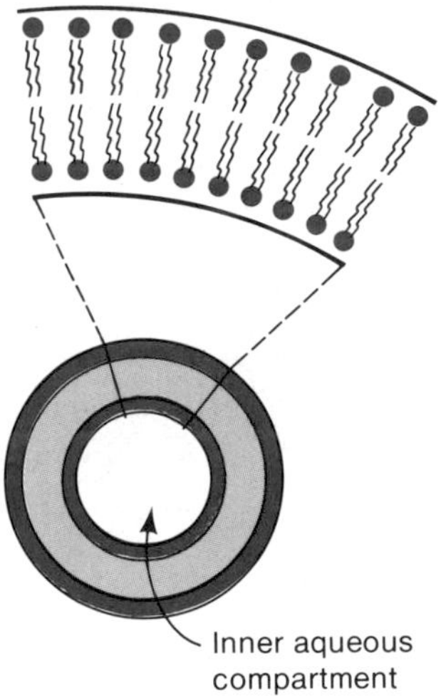

Figure 6-32
Schematic diagram of a lipid vesicle. The membrane is enlarged to show the double layer.

Inner aqueous compartment

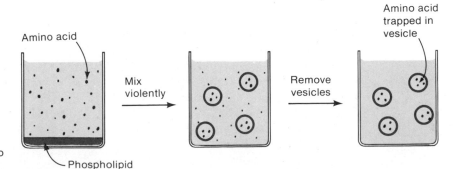

Figure 6-33
Method of preparation of a suspension of lipid vesicles containing amino acid molecules.

Modulation of the Basic Bilayer Structure: Permeability and Transport

Experiments of the sort just described, using synthetic vesicles made of a single phospholipid, indicate that the lipid bilayer is impermeable to all ions and highly polar molecules. How, then, do these molecules get in and out of cells? Naturally occurring biological membranes contain many protein molecules, and membranes having different functions contain different proteins. These proteins, which act in a variety of ways, are responsible for the transport of all molecules having polar regions and for most nonpolar molecules as well.

There are two classes of membrane proteins—the **integral membrane proteins,** which are contained wholly or in part within the membrane, and the **peripheral proteins,** which lie on the membrane surface and are bound to the integral proteins. A technique known as **freeze-fracture electron microscopy** shows that many protein molecules are present in membranes. This technique is shown schematically in Figure 6-34. A small volume of a suspension containing cells is frozen and then fractured. The line of fracture lies within some of the lipid bilayers in the suspension so that the surface exposed by the fracture is the inner surface of the bilayer. An electron micrograph of the plasma membrane of a red blood cell prepared by the freeze-fracture method is shown in Figure 6-35. The outer surface of the membrane is smooth but the inner surface is covered with many globular protein molecules.

Numerous physical measurements have shown that the integral membrane proteins can freely diffuse laterally throughout the bilayer. The degree of motion is determined by the fluidity of the lipid layer, which is in turn a function of the van der Waals interactions between the hydrocarbon tails of the particular phospholipids in the membrane.

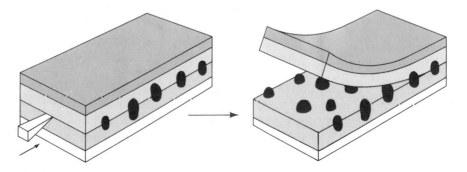

Figure 6-34
Technique of freeze-fracture electron microscopy applied to membranes. The membrane sample is frozen and struck with a sharp point in the direction of the cleavage plane. The membrane can then be split to show whether internal particles (black ellipses) are present.

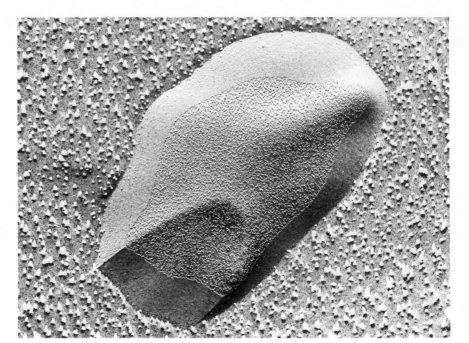

Figure 6-35
An electron micrograph of the plasma membrane of red blood cells, prepared by the freeze-etch technique. The interior of the membrane, which has been exposed by fracture of the membrane, contains numerous globular particles having a diameter of about 75 Å. These particles are termed integral membrane proteins. (Courtesy of Vincent Marchesi.)

This phenomenon was first noted by S. Jonathan Singer and Garth Nicholson and led to the **fluid mosaic model** of membrane structure (Figure 6-36). It was further shown that the proteins do not spontaneously rotate (flip-flop) from one side of the membrane to another. This is because the integral proteins also have polar and nonpolar regions as shown in the figure, and inasmuch as the external region is polar and the internal region is nonpolar, rotation would require passage of the polar region through the nonpolar center of the bilayer.

Since different proteins protrude from the two sides of the membrane, the membrane is an asymmetric structure having, if one were to take the point of view of a cell, an inside and an outside. It is this asymmetry that determines the direction of movement of molecules entering and leaving a cell.

There are many modes of transport of molecules across the bilayer and only a few will be mentioned. One mode makes use of channels through the membrane. These channels usually are passages through the integral membrane proteins; the passages will have an abundance of polar amino acids if their function is to allow transit of polar substances. Often the channels can be opened and closed by means of conformational changes of the membrane proteins. Other transport systems utilize chemical reactions that convert the substance to be transported to a molecule that can enter the membrane and then, after transit of the modified molecule, restore the original molecule at the

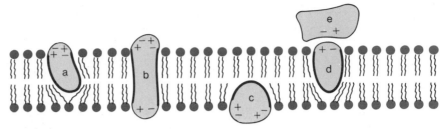

Figure 6-36
The structure of a membrane according to the fluid mosaic model. Four integral proteins (a)—(d) are embedded to various degrees into a lipid bilayer such that the hydrophobic surface of each protein (heavy lines) is in the membrane and the polar region (indicated by + and −) are external. Peripheral proteins (e.g., (e)) are on the surface and bound to a polar region of an integral protein. Integral proteins can drift laterally but cannot flip-flop.

Figure 6-37
A schematic model for active transport. (1) An extracellular molecule binds to a carrier protein. (2) As a result of binding, the carrier protein changes shape. (3) The carrier protein rotates. (4) The bound molecule is discharged to the intracellular space. (5) By means of an energy-requiring process the carrier rotates again. (6) The original shape of the carrier is restored.

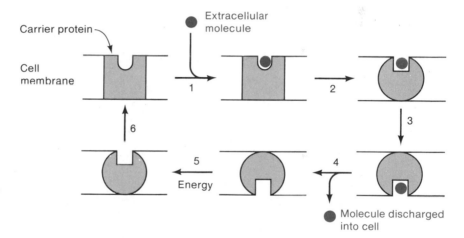

other side of the membrane; these chemical mechanisms are usually very complex, consume a great deal of energy, and are poorly understood. The interesting model shown in Figure 6-37 uses a carrier protein that can act like a revolving door. An integral carrier protein binds the molecule to be transported and undergoes a change in shape and charge distribution that allows the protein to rotate. During one of the rotations the molecule to be transported is released at the other side of the membrane. Upon release, the carrier protein undergoes a second conformational change that causes it either to lose the ability to rotate or to bind the transported molecule. This step is a necessary one because otherwise the carrier molecule would bring as many molecules into the cell as it would take out of the cell.

Transport of this sort is an energy-requiring process; a certain amount of evidence suggests that the energy is needed to restore the molecule to the "ready" state—that is, facing the right direction and having the right binding site.

THE BACTERIAL CELL WALL

Animal cells are enclosed by a single membrane of complex structure, as we have seen in the previous section. Bacteria, plants, and unicellular eukaryotes are encased in a multilayer structure called a **cell wall,** which includes the cell membrane as one of the layers. In molecular biological research with bacteria many simple operations are carried out that have profound effects on the bacterial cell wall and often the laboratory worker is unaware of these effects. It will be seen in this section that the cell wall contains several semisolid layers between which there are numerous freely diffusing molecules and that certain laboratory treatments destroy one or more of these layers, causing leakage of substances from one region to another or in and out of the cell.

Multiple Layers in the Cell Wall

Bacteria come in several shapes, each of which is determined by the arrangement of certain macromolecules in the bacterial cell walls. For this reason it is difficult to make a general statement about wall structures. Furthermore, there are two distinct classes of cell envelopes known as Gram-positive and Gram-negative. Operationally these forms are distinguished by their ability to take up a particular dense stain. In this section we will be concerned primarily with the cell wall of Gram-negative bacteria such as E. coli. Figure 6-38(a) shows an electron micrograph of a section of E. coli in which the multiple layers of the cell wall are clearly seen. Treatment of cells with various reagents, such as trypsin (which hydrolyzes protein), detergents (which remove lipids), and lysozyme, causes dissolution of particular layers. Electron microscopy of samples so treated allows one to identify a sensitive layer as being the outer, middle, or inner layer. Moreover, by dissolving one or two layers, the remaining material can be purified and analyzed chemically. The result of such studies is the model shown in Figure 6-39. The cell wall consists of two major segments, each called a membrane because of the presence of phospholipid. These two membranes, the inner and outer membrane, are separated by an aqueous region known as the **periplasmic space.** The outer membrane is a complex structure that is conveniently thought of as having two components—the protein-phospholipid membranous region and the **peptidoglycan layer.** Gram-positive bacteria differ from Gram-negative bacteria in that they lack the outer membrane (Figures 6-38(b) and 6-40).

The Peptidoglycan Layer

The peptidoglycan layer, which can be seen most clearly as the thick outer layer in Figure 6-38(b), is an unusual substance. It is a polymer

Figure 6-38
Two types of bacterial cell walls. (a) An electron micrograph of a thin section of *E. coli* (a Gram-negative bacterium) showing the multiple layers. (Courtesy of Jack Pangborn.) (b) An electron micrograph of a thin section of a Gram-positive bacterium, *Staphylococcus stapholyticus*. The thick outer layer is the peptidoglycan. (Courtesy of Harriet Smith.) (c) An electron micrograph of the cell wall of a Gram-positive bacterium prepared in a way that shows the cell wall structure more clearly than in (b). (Courtesy of J. Pate.) (d) An electron micrograph of the cell wall of a Gram-negative bacterium prepared in a way that shows the three layers more clearly than in (a). (Courtesy of T. D. Brock and S. F. Conti.) The segments in (c) and (d) are enlarged about three times more than in (a) and (b). See also Figures 6-39 and 6 40.

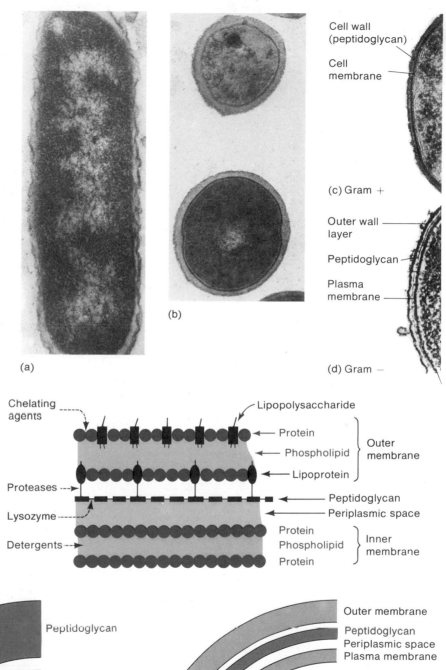

Figure 6-39
Schematic diagram of the cell wall of a Gram-negative bacterium. The sites of attack by chelating agents, proteases, lysozyme, and detergents are indicated by broken arrows.

Figure 6-40
The differences between the cell walls of (a) Gram-positive and (b) Gram-negative bacteria. See also Figure 6-38(c), (d).

consisting of both sugar and peptide units (Figure 6-41). Two sugars, N-acetylglucosamine (NAG) and N-acetylmuramic acid (NAM), alternate to form a linear polysaccharide chain. A pentaglycine peptide forms cross-links between the NAM residues to form the large sheetlike

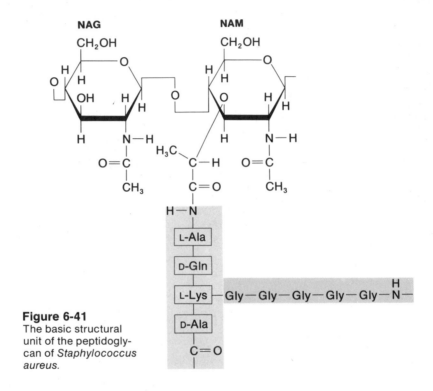

Figure 6-41
The basic structural unit of the peptidoglycan of *Staphylococcus aureus.*

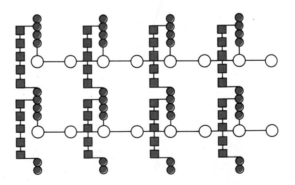

Figure 6-42
Schematic diagram of the peptidyglycan. The sugars are the large open circles, the tetrapeptides are small red circles, and the pentaglycine bridges are small solid squares. The cell wall is a single, enormous macromolecular sac because of contiguous cross-linking.

structure shown in Figure 6-42. The remarkable feature of peptidogly-can is that the sheet ultimately closes on itself to form one enormous saclike macromolecule which encloses the entire inner membrane and cytoplasm of the bacterium. The cross-linking is not always as regular as is shown in the figure; in fact, the particular variations of the cross-linking pattern determine the shape of each bacterial species.

Peptidoglycan is the site of attack of the enzyme lysozyme. Thus, since peptidoglycan is the only rigid component of the cell wall, a lysozyme-treated bacterium assumes a spherical shape (whether the bacterium is initially spherical or rodlike, or has any other shape); a cell treated in this way is called a **spheroplast.**

The antibiotic penicillin interferes with the synthesis of peptido-glycan. Thus, if a bacterium is allowed to grow in the presence of penicillin, it enlarges without growth of the strong peptidoglycan layer; the volume of the cell then exceeds that which can be held by the weak membranous layers and the cell bursts, which accounts for the lethal effect of penicillin.

The Inner Membrane and Permeability

The inner or cytoplasmic membrane has a structure much like the basic membrane structure described in the previous section. This membrane provides the major osmotic barrier and determines, for the most part, which molecules can enter and leave the cytoplasm. *Uncharged* molecules having a molecular weight less than 100 are usually able to pass freely through the membrane. However, larger uncharged mole-cules and all charged molecules require transport systems of the type we have discussed. Some charged molecules may never be able to enter a bacterium because of lack of the appropriate transport system. For example, no phosphorylated organic molecules (such as nucleotides) can enter a cell. Treatment of a cell with a good lipid solvent such as toluene or ether removes the lipid layer and thus also the permeability barrier. Cells treated in such a way are said to be **permeabilized.** They are unable to grow and divide but are able to carry out many biochemical reactions. Permeabilized cells have been extremely useful in the study of DNA synthesis since, in contrast with untreated cells, they are able to take up deoxynucleoside triphosphates, the immediate precursor of DNA. Permeabilizing is often also a convenient way to assay certain enzymes. For example, synthesis of the enzyme β-galac-tosidase has been widely studied as a means of understanding the regulation of gene expression (Chapter 14). It is assayed by permeabi-lizing cells with toluene prior to adding the charged substrate.

The cytoplasmic membrane also contains (or has tightly bound) enzymes. Most notable are the cytochromes and other enzymes that are important in the oxidative synthesis of ATP (oxidative phosphorylation), enzymes responsible for cell wall synthesis and lipid biosynthesis, and possibly DNA-synthesizing enzymes.

The Outer Membrane: Lipopolysaccharide, Lipoprotein, and Porin

The outer membrane is a very complex structure. It is basically a mass of closely associated phospholipids between two layers of proteins. Interspersed in the outer protein layer are many units of lipopolysaccharide, a very unusual molecule. The lipopolysaccharide (or LPS) consists of three covalently linked components—a complex phospholipid known as lipid A, a phosphorylated core oligosaccharide, and a polysaccharide known as the O side chain or O antigen (Figure 6-43). The orientation of the LPS in the outer membrane is shown in Figure 6-44. The lipopolysaccharide units have several protective functions. First, they provide a second permeability barrier that keeps out many

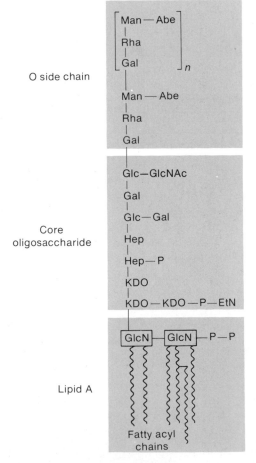

Figure 6-43
The three regions of the lipopolysaccharide molecule of *Salmonella typhimurium:* O side chain, core oligosaccharide, and lipid A. Abbreviations used: Abe, abequose; EtN, ethanolamine; Gal, galactose; Glc, glucose; GlcN, glucosamine; GlcNAc, *N*-acetylglucosamine; Hep, heptulose; KDO, 2-keto-3-deoxyoctonate; Man, mannose; Rha, rhamnose.

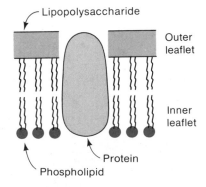

Figure 6-44
The arrangement of lipopolysaccharide molecules and phospholipids in the inner and outer leaflets, respectively, of the outer membrane of a bacterial cell wall.

large molecules. For example, *E. coli* is somewhat resistant to penicillin because the O side chains prevent penicillin from reaching the internal enzymes that synthesize peptidoglycan. Second, the LPS gives the cell a very hydrophilic surface, which decreases the ability of many protozoa, which feed on bacteria, to consume these cells. Many protozoa are surrounded by hydrophobic membranes that can fold around and envelop a hapless bacterium if the bacterium is not too heavily charged. Third, some LPS units are the sites of attachment of phages (which is, of course, detrimental to the bacterium). *E. coli,* like all other Gram-negative bacteria, contains LPS but, oddly, the strain of *E. coli* most commonly used in research, namely, *E. coli* K12, lacks O side chains. The LPS units are not covalently attached to any components of the outer membrane. Treatment of cells with the chelating agent ethylenediaminetetracetate (EDTA) causes release of about half the cellular LPS, suggesting that divalent cations (e.g., Mg^{2+} and Ca^{2+}), which are tightly chelated, might be involved in stabilizing the LPS in the outer membrane.

The inner side of the outer membrane contains the most abundant protein in the outer membrane—the **anchoring lipoprotein** (Figure 6-39). This small protein (consisting of 58 amino acids), which is almost completely α-helical and rodlike, contains three fatty acids covalently attached to the N-terminal amino acid (cysteine). These fatty acids cause the anchoring lipoprotein to be oriented so that the fatty acids are in the phospholipid layer. The C-terminal amino acid is lysine, which has a free amino group. Roughly one-third of the anchoring lipoprotein molecules are covalently linked via the ε-NH_2 group of the terminal lysine to the peptidoglycan. Thus, these lipoprotein molecules anchor the outer membrane to the peptidoglycan and thereby strengthen the cell envelope.

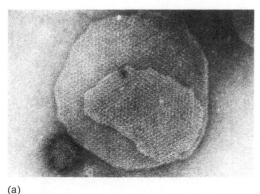

(a)

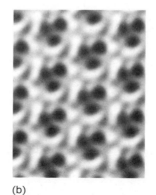

(b)

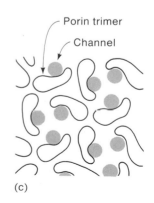

Porin trimer

Channel

(c)

Figure 6-45

The arrangement of the membrane protein porin in *E. coli.* (a) Negative-contrast electron micrograph of arrays of porin channels. (b) Computer-processed image of a sevenfold-greater-magnified portion of panel (a). (c) An interpretive drawing of the photograph in panel (b). (From A. C. Steven, B. Ten Heggeler, B. Müller, J. Kistler, and J. P. Rosenbusch. 1977. *J. Cell Biol.,* 72: 292.)

A final element in the outer membrane is a protein called **porin.** This protein forms trimers which pass through the phospholipid layer, providing pores through which small molecules can pass (Figure 6-45). Porin is responsible for the transport of small polar molecules such as glucose from the growth medium to the cell. (Disaccharides such as sucrose and lactose are too large for the porin channels and enter the cell via special transport systems.)

In view of the excluding properties of the outer membrane, one might wonder how lysozyme is able to degrade the peptidoglycan of Gram-negative bacteria. In fact, the Gram-negative bacteria are quite resistant to lysozyme compared to Gram-positive bacteria, which lack an outer membrane. Gram-negative bacteria must first be treated with EDTA and osmotic shock to remove the LPS and to weaken the outer membrane, in order to sensitize the cells to lysozyme.

The Periplasmic Space

Between the peptidoglycan layer and the inner membrane is an aqueous layer called the **periplasmic space.** This region becomes evident when the LPS is removed by EDTA or if the cell is subjected to a sudden decrease in osmotic pressure. (This often happens if cells are rapidly diluted from their growth medium into a dilute buffer.) These treatments cause a weakening of the outer membrane and the peptidoglycan layer, and subsequent release of certain proteins. Most of these proteins are nucleases and proteases and are thought to be used in digesting impermeable nutrients; for instance, in certain circumstances these proteins are excreted and enable a cell to use exogenous proteins and nucleic acids as food. Other molecules in the periplasmic space are proteins that bind specific ions, sugars, and amino acids, and are needed for transport of these substances across the inner membrane.

Osmotically Fragile Cells

Many mutants of E. coli that have defective membranes have been isolated in an attempt to learn more about particular components. One type of mutant, the osmotically sensitive mutant, is of particular value. Cells of these mutants have such weak walls that the cell literally explodes unless the cell is suspended in a hypertonic fluid, such as a buffer or a growth medium containing 20 percent sucrose. The laboratory worker can lyse these cells at will merely by diluting the sucrose-containing fluid, a procedure that is advantageous for some biochemical analyses. This mutation has also been introduced into potentially harmful bacteria that are of interest experimentally; the mutation constrains the bacterium to grow only in the laboratory in media containing 20 percent sucrose, and is thereby useful in preventing growth of a cell that might have escaped to the environment.

SELF-ASSEMBLY OF COMPLEX AGGREGATES

Several aggregates, consisting of two or more macromolecules and having well-defined structures, have been described. However, nothing has been said yet about how such aggregates form. Basically there are two possibilities:

1. The structures form spontaneously from the subunits without an additional source of energy; this is called **self-assembly.**
2. Aggregation is directed, possibly by enzymes or some template, and requires a source of energy.

Examples are known of aggregates formed in both ways. In this section one of the best understood examples of self-assembly is described— namely, the formation of a particle of tobacco mosaic virus.

Assembly of Tobacco Mosaic Virus

Tobacco mosaic virus (TMV) is a rodshaped virus 3000 Å long and 180 Å in diameter (Figure 6-46). There are 2130 identical subunits in the protein coat; these are closely packed in a helical array around a single-stranded RNA molecule that consists of 6390 nucleotides (Figure 6-46(b)). The RNA is deeply buried in the protein and each protein subunit interacts with three nucleotides.

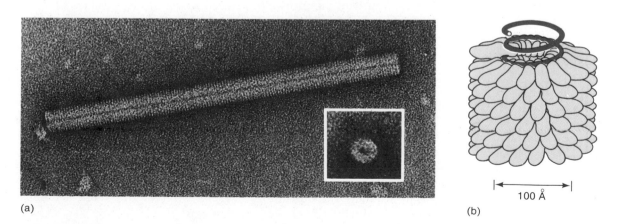

(a)

(b)

100 Å

Figure 6-46

(a) Electron micrograph of tobacco mosaic virus. The insert shows a cross-sectional view of a broken particle, in which the hollow central core can be seen. (Courtesy of Robley Williams.) (b) Model of a part of a tobacco mosaic virus particle, showing the helical array of protein subunits around a single-stranded RNA molecule. (After A. Klug and D. L. D. Caspar. 1960. *Advan. Virus Res.*, 7: 274.)

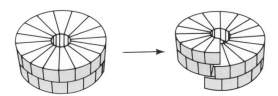

Figure 6-47
Schematic diagram showing the conversion of a
TMV protein disc into the helical "lock-washer"
form. (After A. Klug. 1972. *Fed. Proc.*, 31: 40.)

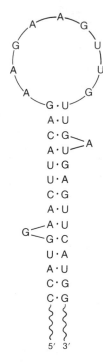

Figure 6-48
A portion of the initiation
region in TMV RNA for
the assembly of the virus
particle.

The simplest mechanism for the assembly of TMV would be step-wise addition of single protein subunits to the RNA. However, such a mechanism is unsatisfying logistically, inasmuch as the process would be a very slow one. About 17 coat subunits would have to be added to a flexible RNA molecule before the complex could close on itself, by forming a turn of the helix, and thereby acquire stability. This problem is solved if a stable complex of many subunits is first formed and then the complex is added to the RNA. The coat-protein subunits spontaneously aggregate to form a two-layered disc consisting of 34 subunits. Each layer of a disc is a ring of 17 subunits, which is nearly the same as the number of subunits ($16\frac{1}{3}$) in one turn of the TMV helix. The two-layered disc is a key intermediate in the assembly of TMV. An important property of the disc is that the subunits can slide over each other to form a two-turn helix; the disc looks like what it is called—a lock-washer (Figure 6-47).

The two-layer disc interacts rapidly with TMV RNA. A particular base sequence is specifically recognized by the disc and serves to initiate assembly of the virus. This initiation region was first isolated by adding a few discs to TMV RNA to coat the initiation region and then digesting the rest of the RNA with a nuclease. After such treatment the protected fragment always contained a common core of about 65 nucleotides bound very tightly and specifically to discs. The base sequence of this initiation region strongly suggests that it forms a hairpin structure with a base-paired stem and a loop (Figure 6-48). It is thought that the loop binds the first disc to start the assembly of the virus. The initiation loop is far from either end of the RNA—roughly 5300 nucleotides from the 5' end and 1000 nucleotides from the 3' end of the RNA. Two "tails" of RNA emerge from one end of a growing TMV particle (Figure 6-49). The length of the 3' tail is rather constant throughout most of the assembly process, whereas the 5' tail becomes shorter as the virus particle becomes longer.

The currently accepted model for the formation of TMV particles is shown in Figure 6-50. Assembly starts with the insertion of the initiation loop into the central hole of a two-layered protein disc. The loop binds

Figure 6-49
Electron micrograph of partially reconstituted tobacco mosaic virus particles. Note that two branches of the RNA emerge from each growing particle, indicating that assembly does not begin at a terminus of the RNA molecule. From the relative lengths of the branches the position on the RNA at which assembly begins can be determined. The length of the longest rod is about 2000 Å. (Courtesy of G. Lebeurier and A. Nicolaieff.)

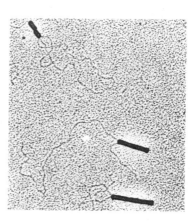

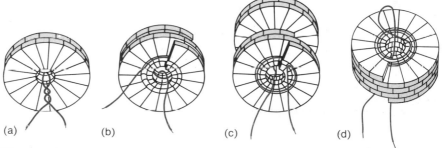

(a)   (b)   (c)   (d)

Figure 6-50
Model for the assembly of tobacco mosaic virus. (a) The initiation region of the RNA is looped into the central hole of the protein disc yielding (b) the helical "lock-washer" form. (c) Additional discs add to the looped end of the RNA. (d) One of the RNA tails is continually pulled through the central hole to interact with incoming discs.

to the first turn of the disc and the adjacent base-paired stem opens. This interaction transforms the disc into the helical lock-washer form and traps the viral RNA. The viral helix is now started. Another disc then adds to the newly formed loop of RNA that protrudes from the central hole. A new loop is formed after the addition of each disc by the drawing up of the 5' tail through the central hole of the growing virus particle. The virus is completed when the 3' tail becomes coated, which occurs by an unknown mechanism.

REFERENCES

Anderson, W. F., D. H. Ohlendorf, Y. Takeda, and B. A. Matthews. 1981. "Structure of the Cro repressor from bacteriophage λ and its interaction with DNA." *Nature*, 290, 754–758.

Bretscher, M. S. 1973. "Membrane structure: some general principles." *Science*, 181, 622–624.

Butler, P. J. G., and A. J. Klug. 1978. "The assembly of a virus." *Scient. Amer.*, 239 (November), 62–69.

Capaldi, R. A. 1974. "A dynamic model of cell membranes." *Scient. Amer.*, 230 (March), 26–33.

Dupraw, E. J. 1970. *DNA and Chromosomes*. Holt, Rinehart, and Winston.

Eyre, D. R. 1980. "Collagen: molecular diversity in the body's protein scaffold." *Science*, 207, 1315–1322.

Fox, C. F. 1972. "The structure of cell membranes." *Scient. Amer.*, 226 (February), 30–47.

Fox, C. F., and A. Keith. 1972. *Membrane Molecular Biology*. Sinauer.

Ghuysen, J. M. 1977. "Biosynthesis and assembly of bacterial cell walls." In G. Poste and G. L. Nicholson (eds.), *Cell Surface Reviews. Vol. IV. Membrane Assembly and Turnover*. Elsevier.

Gomperts, B. D. 1977. *The Plasma Membrane*. Academic Press.

Gross, J. 1961. "Collagen." *Scient. Amer.*, 202 (May), 120–129.

Kaback, H. R., and J. S. Hong. 1973. "Membranes and transport." *CRC Crit. Reviews Microbiol.*, 2, 333–357.

Kornberg, R. D. 1977. "Structure of chromatin." *Ann. Rev. Biochem.*, 46, 931–951.

Leive, L. (ed.) 1973. *Bacteria, Membranes and Walls*. Marcel Dekker.

Lodish, H. F., and J. E. Rothman. 1979. "The assembly of cell membranes." *Scient. Amer.*, 240 (January), 48–63.

Nikaido, H., and T. Nakae. 1979. "The outer membrane of Gram-negative bacteria." *Adv. Microbiol. Physiol.*, 14, 163–250.

Ramachandran, G. N., and A. H. Reddi. (eds.) 1976. *Biochemistry of Collagen*. Plenum.

Rich, A., and F. H. C. Crick. 1961. "The molecular structure of collagen." *J. Mol. Biol.*, 3, 483–506.

Rothfield, L. I. (ed.) 1971. *Structure and Function of Biological Membranes*. Academic Press.

Sharon, N. 1969. "The bacterial cell wall." *Scient. Amer.*, 220 (May), 92–98.

Singer, S. J., and G. L. Nicholson. 1972. "The fluid mosaic model of the structure of cell membranes." *Science*, 175, 720–723.

Stryer, L. 1981. *Biochemistry*. W. H. Freeman and Co.

Tanford, C. 1980. *The Hydrophobic Effect. Formation of Micelles and Biological Membranes*. 2nd Ed. Wiley-Interscience.

Traub, J. W., and K. A. Piez. 1971. "The chemistry and structure of collagen." *Adv. Prot. Chem.*, 25, 243–352.

Weissman, G., and R. Claiborne. (eds.) 1975. *Cell Membranes*. H. P. Publishing.

7 The Genetic Material

The genetic material of any organism is the substance that carries the information determining the properties of that organism. Furthermore, it is responsible for transferring the genetic information from parent to progeny. In almost all organisms the genetic material is DNA; the exceptions are a few bacteriophages and numerous plant and animal viruses in which the genetic material is RNA.

Before the birth of modern molecular biology it was widely believed that the genetic material consisted wholly of protein molecules. This idea was based on the concept that of all of the known intracellular molecules, only proteins were sufficiently complex and variable in structure and in chemical properties to contain the great variety of information required of the genetic substance. DNA was not considered a possible genetic material, though DNA was known to be present only in the nuclei of cells, because, for a variety of reasons now only of historical interest, DNA was thought to be too simple to be a gene and was assumed instead to be some structural component of chromosomes. Two experiments, which are described later in the chapter, reversed this view and, in so doing, helped to create the science of molecular genetics.

Once the significance of DNA and its remarkable structure (as described in Chapter 4) was appreciated, a variety of experiments were performed that led to what is known as the Central Dogma—that is, a statement of the way in which the information contained in DNA is translated into protein structure. This concept is described in the following section.

THE CENTRAL DOGMA

What is commonly called the Central Dogma is the statement that

DNA is copied to make RNA and RNA is the template for protein synthesis.

This statement is based on the recognition that (1) amino acids do not bind to DNA molecules, so a cell cannot produce a protein molecule with a particular amino acid sequence simply by aligning amino acids along a DNA molecule; and (2) in the long run, there is much less risk of damage to a cell's single DNA molecule if it is not used directly as a template for the synthesis of each protein molecule. Considering these two points, the Central Dogma states that an intermediate molecule, now known to be a **messenger RNA molecule (mRNA),** is copied from the DNA (Figure 7-1) and then used repeatedly for protein synthesis. This statement is usually written as

DNA $\longrightarrow$ mRNA $\longrightarrow$ protein.

THE GENETIC CODE AND THE ADAPTOR HYPOTHESIS

There are only four bases in DNA and twenty amino acids in proteins, so some combination of the bases is needed to specify a particular amino acid. The set of such combinations is called the **genetic code**; this is discussed in detail in Chapter 12. A base sequence corresponding to a particular amino acid is a **codon.**

The surface distinctions among the twenty amino acids are neither specific enough nor sufficiently large for patterns of bases to be recognized by means of van der Waals forces or hydrogen-bonding;

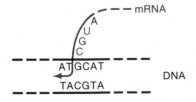

Figure 7-1
Schematic diagram showing how mRNA is copied from DNA. The upper strand of the DNA molecule is being copied and the mRNA molecule is elongating by addition of nucleotides at the end of the molecule denoted by the arrow. As the head of the arrow progresses, the mRNA strand is released at the other end, so that at any instant only a few RNA bases are hydrogen-bonded to DNA.

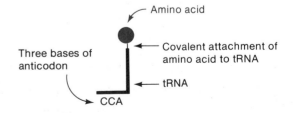

Figure 7-2
A schematic drawing of a tRNA molecule to which an amino acid (red) is covalently linked and which has an anticodon CCA corresponding to a codon UGG.

thus, recognition requires another intermediate—an adaptor RNA molecule. This adaptor, which we know as **transfer RNA (tRNA),** contains a site to which a particular amino acid is attached, and a base sequence called the **anticodon,** which hydrogen-bonds via complementary base pairing to a given codon (Figure 7-2). Thus the process of protein synthesis may be stated briefly as follows:

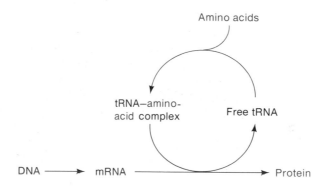

In outline, the mechanism of protein synthesis can be depicted as in Figure 7-3. An mRNA molecule is copied from a DNA molecule and then is laid onto the surface of a protein-synthesizing unit. Elsewhere, appropriate tRNA–amino-acid complexes have formed and these are then aligned along the mRNA molecule. Peptide bonds are made between successive amino acids; finally, the tRNA–amino-acid bond is broken and the completed protein is removed. Details of these processes, which are oversimplified in the figure, are presented in Chapters 11 through 13.

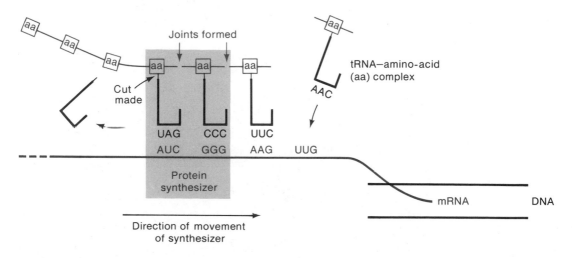

Figure 7-3
A diagram showing how a protein molecule is synthesized according to the Central Dogma. See text for details.

IDENTIFICATION OF DNA AS THE GENETIC MATERIAL

The Transformation Experiments

The development of the current idea that DNA is the genetic material began with an observation in 1928 by Fred Griffith, who was studying the bacterium responsible for human pneumonia—i.e., *Streptococcus pneumoniae* or *Pneumococcus*. The virulence of this bacterium was known to be dependent on a surrounding polysaccharide capsule that protects it from the defense systems of the body. This capsule also causes the bacterium to produce smooth-edged (S) colonies on an agar surface. It was known that mice were normally killed by S bacteria (Figure 7-4(a)). Griffith then isolated a rough-edged (R) colony mutant, which proved to be both nonencapsulated and nonlethal (Figure 7-4(b)). He subsequently made a significant observation—namely, whereas both R and heat-killed S were nonlethal, a mixture of live R and heat-killed S

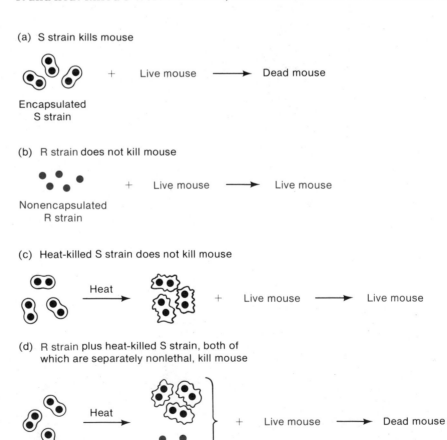

(a) S strain kills mouse

Encapsulated
S strain   +   Live mouse ⟶ Dead mouse

(b) R strain does not kill mouse

Nonencapsulated
R strain   +   Live mouse ⟶ Live mouse

(c) Heat-killed S strain does not kill mouse

Heat   +   Live mouse ⟶ Live mouse

(d) R strain plus heat-killed S strain, both of
which are separately nonlethal, kill mouse

Heat   +   Live mouse ⟶ Dead mouse

Figure 7-4
The Griffith experiment showing conversion of a nonlethal bacterial strain to a lethal form by a cell extract.

was lethal (Figure 7-4(c,d)). Furthermore, the bacteria isolated from a mouse that had died from such a mixed infection were only S—i.e., the live R had somehow been replaced by or **transformed** to S bacteria. Several years later it was shown that the mouse itself was not needed to mediate this transformation because when a mixture of R and heat-killed S was grown in a culture fluid, living S cells were produced. A possible explanation for this surprising phenomenon was that the R cells restored the viability of the dead S cells; but this idea was eliminated by the observation that living S cells grew even when the heat-killed S culture in the mixture was replaced by a cell *extract* prepared from broken S cells, which had been freed from both intact cells and the capsular polysaccharide by centrifugation (Figure 7-5). Hence, it was concluded that the cell extract contained a **transforming principle,** the nature of which was unknown.

The next development occurred some 15 years later when Oswald Avery, Colin MacLeod, and Maclyn McCarty partially purified the transforming principle from the cell extract and demonstrated that it was DNA. These workers modified known schemes for isolating DNA and prepared samples of DNA from S bacteria. They added this DNA to a live R bacterial culture; after a period of time they placed a sample of the S-containing R bacterial culture on an agar surface and allowed it to grow to form colonies. Some of the colonies (about 1 in 10^4) that grew were S type (Figure 7-5). To show that this was a permanent genetic change, they dispersed many of the newly formed S colonies and placed them on a second agar surface. The resulting colonies were again S type.

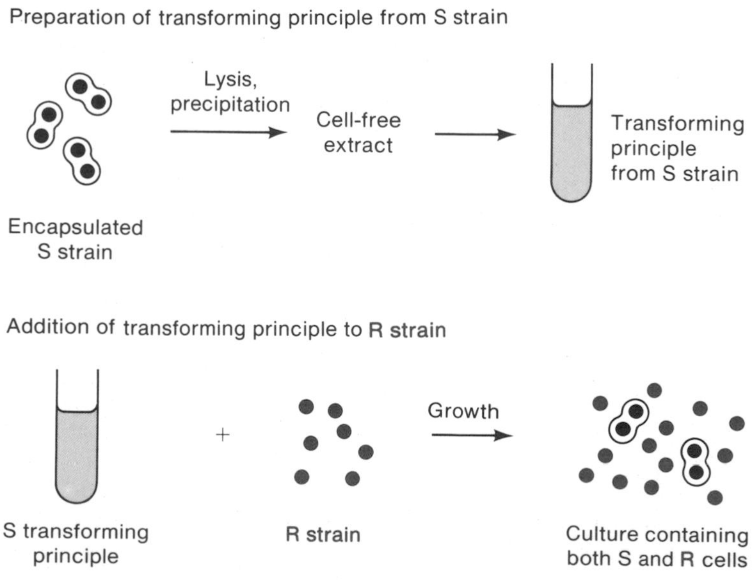

Preparation of transforming principle from S strain

Lysis, precipitation

Cell-free extract

Transforming principle from S strain

Encapsulated S strain

Addition of transforming principle to R strain

S transforming principle   +   R strain   Growth   Culture containing both S and R cells

Figure 7-5
The transformation experiment.

If an R colony arising from the original mixture was dispersed, only R bacteria grew in subsequent generations. Hence the R colonies retained the R character, whereas the transformed S colonies bred true as S. Because S and R colonies differed by a polysaccharide coat around each S bacterium, the ability of purified polysaccharide to transform was also tested, but no transformation was observed. Since the procedures for isolating DNA then in use produced DNA containing many impurities, it was necessary to provide evidence that the transformation was actually caused by the DNA alone.

This evidence was provided by the following four procedures.

1. Chemical analysis showed that the major component was a deoxyribose-containing nucleic acid.
2. Physical measurements showed that the sample contained a highly viscous substance having the properties of DNA.
3. Experiments demonstrated that transforming activity is not lost by reaction with either (a) purified proteolytic (protein-hydrolyzing) enzymes—trypsin, chymotrypsin, or a mixture of both—or (b) with ribonuclease (an enzyme that depolymerizes RNA).
4. It was demonstrated that treatment with materials known to contain DNA-depolymerizing activity (DNase) inactivated the transforming principle.

The fourth point, which is the most critical, is worth examining further since Avery, MacLeod, and McCarty—unable to purchase purified DNase as we now can— had to prove it was the DNase activity in the various fluids tested that destroyed the transforming activity.

Some of the data they used to make this argument are shown in Table 7-1 and Figure 7-6. In Table 7-1 the results of incubating transforming principle with five crude enzyme preparations are shown. For each preparation the activity of three enzymes—phosphatase, tributyrin esterase, and DNA depolymerase—was measured and correlated with the ability to inactivate transforming principle. For example, since rabbit bone phosphatase contained phosphatase and esterase activity but failed to prevent transformation, these enzymes could not be responsible for the inactivation. Instead, loss of transformation correlated with the presence of DNA depolymerase, but since the preparations were impure and probably contained many other enzyme activities, these results could not be taken as proof. Figure 7-6 shows the loss of viscosity of the transforming principle produced by dog and rabbit sera that were pretreated by heating them to different temperatures. Since the viscosity was presumably due to DNA, the viscosity decrease was taken as a measure of the DNase activity and the loss (upon heating) of the ability of the sera to reduce viscosity was presumably a result of loss of the DNase activity as heat inactivated the enzyme. These heated sera were also tested for the ability to inactivate

Table 7-1

The inactivation of transforming principle by crude enzyme preparations

	Enzymatic activity			
Crude enzyme preparations	Phos-phatase	Tributyrin esterase	Depolymer-ase for deoxyribo-nucleate	Inacti-vation of trans-forming principle
Dog intestinal mucosa	+	+	+	+
Rabbit bone phosphatase	+	+	−	−
Swine kidney phosphatase	+	−	−	−
Pneumococcus autolysates	−	+	+	+
Normal dog and rabbit serum	+	+	+	+

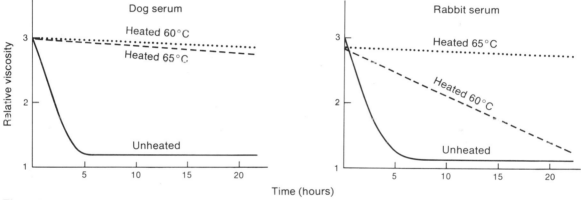

Figure 7-6

Relative viscosity of transforming principle as a function of time of incubation in three samples of dog and rat sera: (1) unheated, (2) heated for 30 minutes at 60°C, (3) heated for 30 minutes at 65°C. The decrease in viscosity is owing to depolymerization of DNA. When the viscosity is constant, the DNase in the particular serum must have been inactivated. (From O. Avery, C. MacLeod, and M. McCarty, 1944. *J. Expt. Med.*, 79: 137–158.)

transforming principle, and it was found that the temperature dependence of depolymerizing activity followed that of inactivation of transformation. The final point was that the addition of sodium fluoride to any preparation that had the ability both to inactivate transformation and to depolymerize DNA always simultaneously removed both activities. Thus, as Avery, MacLeod, and McCarty stated in the publication of their findings in 1944, "the fact that transforming activity is destroyed only by preparations containing depolymerase for desoxyribonucleic acid and the further fact that in both instances the enzymes concerned are inactivated at the same temperature and inhibited by fluoride provide additional evidence for the belief that the active principle is a nucleic acid of the desoxyribose type." The caution with which they stated their inference should be noted.

Avery, MacLeod, and McCarty avoided stating explicitly that DNA was *the* genetic material and concluded at the end of their work only that nucleic acids have "biological specificity, the chemical basis of which is as yet undetermined." The problem they faced in persuading the scientific community to accept their conclusion was that, whatever the genetic material was, it had to be a substance capable of enormous variation in order to contain the information carried by the huge number of genes. At the time, DNA was considered to be a tetranucleotide and this notion certainly seemed incompatible with any strong contention that DNA should be considered the sole component of the genetic material. Furthermore, the general consensus was that genes were made of chromosomal protein—an idea that, in the course of a 40-year period, had logically evolved from the recognition that protein composition and structure varied greatly between organisms. For this reason the Avery-MacLeod-McCarty transformation work had little initial impact and those people who supported the genes-as-protein theory posed the following two alternative explanations for the team's results: (1) the transforming principle might not be DNA but rather one of the proteins invariably contaminating the DNA sample; (2) DNA somehow affected capsule formation directly by acting in the metabolic pathway for biosynthesis of the polysaccharide and permanently altering this pathway.

Point 1 had already been discounted in the original work because the experiments showed insensitivity to proteolytic enzymes and sensitivity to DNase. However, since the DNase was not a pure enzyme and its activity against the transforming principle was only inferred by correlating it to other activities against DNA in the same preparation, the possiblity could not be eliminated conclusively. Rollin Hotchkiss repeated the transformation experiment and, five years later, using a DNA sample whose protein content was only 0.02 percent, found that this extensive purification did not reduce the transforming activity. This result supported the view of Avery, MacLeod, and McCarty but still did not prove it. The second alternative, however, was cleanly eliminated, also by Hotchkiss, with an experiment in which he transformed a penicillin-sensitive bacterial strain so that it acquired penicillin resistance. Since penicillin resistance is totally distinct from the rough-smooth character of the bacterial capsule studied by the Avery team, this showed that the transforming ability of DNA was not limited to capsule synthesis. Interestingly enough, most biologists still remained unconvinced that DNA was the genetic material and it was not until Erwin Chargaff showed in 1950 that a wide variety of chemical structures in DNA were possible—thus allowing biological specificity—that this idea was thoroughly accepted.

The Chemical Experiments

The DNA-as-a-tetranucleotide hypothesis had arisen from the belief that DNA contained equimolar quantities of adenine, thymine, guanine, and cytosine. This incorrect conclusion was formulated for two reasons. First, in the chemical analysis of DNA, the technique used to separate the bases before identification did not resolve them very well, so that the quantitative analysis was poor. Second, the DNA analyzed was usually isolated from eukaryotes (and in eukaryotes the four bases are nearly equimolar) or from bacteria whose DNA happened to have nearly equal amounts. Using the DNA of a wide variety of organisms, Chargaff applied the newly developed technique of paper chromatography to separate the bases and ultraviolet absorbance measurements of the separated bases to determine the concentrations of each. He thereby showed that the molar concentrations of the bases could vary widely. Thus it was demonstrated that DNA could have variable composition, a primary requirement for a genetic substance. Upon publication of Chargaff's results, the tetranucleotide hypothesis quietly died and the DNA-as-gene idea began to catch on. Shortly afterward, Alfred Mirsky and Hans Ris (in one laboratory) and R. Vendrely and A. Boivin (an another) separately found that, for a wide variety of organisms, somatic cells have twice the DNA content of germ cells, a characteristic to be expected of the genetic material, given the tenets of classical chromosome genetics. Although it could apply just as well to any component of chromosomes, once this result was revealed, the hereditary nature of DNA rapidly became the fashionable idea. Although no additional data had been released by Avery, MacLeod, and McCarty, objections to their work were no longer heard.

The Blendor Experiment

An elegant confirmation of the genetic nature of DNA came from an experiment with *E. coli* phage T2. This experiment, known as the **blendor experiment** because a kitchen blendor was used as a major piece of apparatus, was performed by Alfred Hershey and Martha Chase, who demonstrated that the DNA injected by a phage particle into a bacterium contains all of the information required to synthesize progeny phage particles.

A single particle of phage T2 consists of DNA (now known to be a single molecule) encased in a protein shell (Figure 7-7(a,b)). The DNA is the only phosphorus-containing substance in the phage particle; the proteins of the shell, which contain the amino acids methionine and cysteine, have the only sulfur atoms. Thus, by growing phage in a

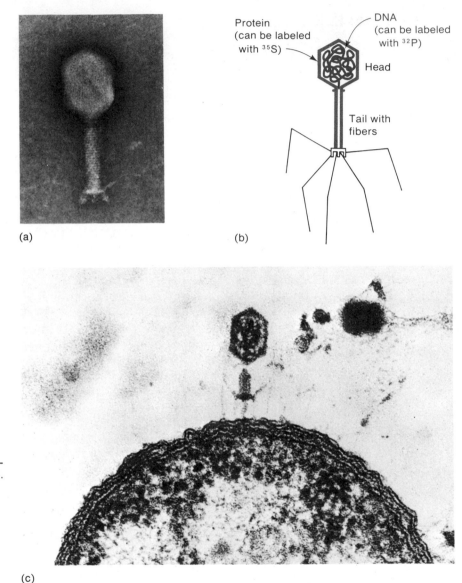

Figure 7-7
An electron micrograph of T4 phage. Note the detailed structure of the tail. (Courtesy of M. Wurtz.) (b) An interpretive drawing. (c) Phage T2 adsorbed to the surface of *E. coli.* (Courtesy of Lee Simon and Thomas Anderson.)

nutrient medium in which radioactive phosphate ($^{32}PO_4^{3-}$) is the sole source of phosphorus, phage containing radioactive DNA can be prepared. If instead the growth medium contains radioactive sulfur as $^{35}SO_4^{3-}$, phage containing radioactive proteins are obtained. If these two kinds of labeled phage are used in an infection of a bacterial host, the phage DNA and the protein molecules can always be located by their radioactivity. Hershey and Chase used these phage to show that ^{32}P but not ^{35}S is injected into the bacterium.

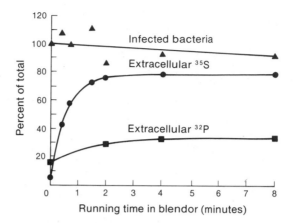

Figure 7-8
(a) Data from the blendor experiment. The ordinate presents the percent of ^{35}S and ^{32}P removed from bacteria infected with radioactively labeled T2 phage and the percent of infected bacteria surviving as infective centers, after the times of agitation in a blendor shown on the abscissa. (After A. D. Hershey and M. Chase. 1952. *J. Gen. Physiol.*, 36: 39.)

Each phage T2 particle has a long tail by which it attaches to sensitive bacteria (Figure 7-7(c)). Hershey and Chase showed that an attached phage can be torn from a bacterial cell wall by violent agitation of the infected cells in a kitchen blendor. Thus, it was possible to separate an adsorbed phage from a bacterium and determine the component(s) of the phage that could not be shaken free by agitation — presumably, those components had been injected into the bacterium.

In the first experiment ^{35}S-labeled phage particles were adsorbed to bacteria for a few minutes. The bacteria were separated from unadsorbed phage and phage fragments by centrifuging the mixture and collecting the sediment (the pellet), which consisted of the phage-bacterium complexes. These complexes were resuspended in liquid and blended. The suspension was again centrifuged, and the pellet, which now consisted almost entirely of bacteria, and the supernatant were collected. It was found that 80 percent of the ^{35}S was in the supernatant and 20 percent was in the pellet (Figure 7-8). The 20 percent of the ^{35}S that remains associated with the bacteria was shown some years later to consist mostly of phage tail fragments that adhered too tightly to the bacterial surface to be removed by the blending. A very different result was observed when the phage population was labeled with [32]P. In this case 70 percent of the ^{32}P remained associated with the bacteria in the pellet after blending and only 30 percent was in the supernatant. Of the radioactivity in the supernatant roughly one-third could be accounted

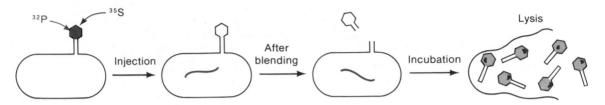

Figure 7-9
The blendor experiment. The parental DNA molecule is drawn in red. During incubation the DNA replicates and progeny phage are produced. Some, but not all, of the parental DNA appears in these progeny. Although not known to Hershey and Chase, progeny DNA molecules engage in genetic recombination resulting in dispersal of parental DNA among the progeny phage, as shown.

for by breakage of the bacteria during the blending. (The remainder was shown some years later to be a result of defective phage particles that could not inject their DNA.) When the pellet material was resuspended in growth medium and reincubated, it was found to be capable of phage production. Thus, the ability of a bacterium to synthesize progeny phage is associated with transfer of ^{32}P, and hence of DNA, from parental phage to the bacteria (Figure 7-9).

Another series of experiments, known as **transfer experiments,** supported the interpretation that genetic material contains ^{32}P but not ^{35}S. In these experiments progeny phage were isolated, after blending, from cells that had been infected with either ^{35}S- or ^{32}P-containing phage and the progeny were then assayed for radioactivity, the idea being that some parental genetic material should be found in the progeny. It was found that no ^{35}S but about half of the injected ^{32}P was transferred to the progeny. This result indicated that though ^{35}S might be residually associated with the phage-infected bacteria, it was not part of the phage genetic material. The interpretation (now known to be correct) of the transfer of only half of the ^{32}P was that progeny DNA is selected at random for packaging into protein coats and that all progeny DNA is not successfully packaged.

The Question of RNA as a Genetic Substance

The transformation and blendor experiments settled once and for all the question of the chemical identity of the genetic material. The absolute generality of the conclusion remained in question, though, because several plant and animal viruses were known to contain single-stranded RNA and no DNA, and common baker's yeast was thought to be free of DNA. The true nature of yeast was determined when improved techniques for isolating DNA were developed; yeast contains so much RNA compared to DNA that in the original assays the

DNA was overlooked. The observation of DNA-free RNA-containing phage and virus particles was understandable shortly afterwards as the role of RNA in the flow of information from gene to protein became clear—that is, that DNA stores genetic information for protein synthesis and the pathway from DNA to protein always requires the synthesis of an RNA intermediate which is copied from a DNA template. Thus, an organism that lacks DNA utilizes the base sequence of RNA both for storage of information and as a template from which the amino acid sequence of proteins can be obtained. RNA does not serve these two functions as efficiently as the DNA→RNA system, but we will see later in this chapter how it manages.

In the following section we discuss the properties that should be possessed by the genetic material and indicate how DNA satisfies the criteria.

PROPERTIES OF THE GENETIC MATERIAL

The genetic material must have the following properties:

1. Ability to store genetic information and to transmit it to the cell as needed.
2. Ability to transfer its information to daughter cells with minimal error.
3. Physical and chemical stability, so that information is not lost.
4. Capability for genetic change, though without major loss of parental information.

We will see in the following sections that DNA is particularly suited to be the genetic material.

Storage and Transmission of Genetic Information by DNA

The information possessed and conveyed by the DNA of a cell is of several types:

1. The sequence of amino acids in every protein synthesized by the cell.
2. A start and a stop signal for the synthesis of each protein.
3. A set of signals that determine whether a particular protein is to be made and how many molecules are to be made per unit time.

This information is contained in the sequence of DNA bases. The amino acid sequence and the start and stop signals are not obtained directly from the DNA base sequence but via RNA intermediates. That is, DNA serves as a template for synthesis of specific RNA molecules called messenger RNA. The base sequence of the mRNA is complementary to that of one of the DNA strands and it is from the mRNA sequence that the amino acid sequence is translated. In particular, the base sequence is "read" in order in groups of three bases called **triplets** or **codons** and each group corresponds to a particular amino acid or to a start or stop codon. This two-stage process has the advantage that the DNA molecule neither has to be used very often nor has to enter the protein-synthetic mechanism. Since protein synthesis occurs continually, one can appreciate why DNA evolved with bases having groups capable of forming hydrogen bonds. First, by specific patterns of hydrogen-bonding—that is, base-pairing—the base sequence of DNA is easily transcribed into an RNA molecule by means of an enzymatic system that adds a base to the polymerizing end of an RNA molecule only if that base can hydrogen-bond with the base being copied. This system of transcription is discussed in detail in Chapter 11. The use of hydrogen-bonding enables the cell to use less genetic material to hold information than if van der Waals forces had been selected in the course of evolution; this is because van der Waals forces are so weak that the error frequency in transmitting information would be much greater than with hydrogen bonds unless more bases (or other molecules) were used for complementary binding.

The process of base recognition is facilitated by the breathing (or localized denaturation) of DNA. That is, while the base pairs are temporarily broken, a polymerizing enzyme (RNA polymerase) can penetrate the helix and read the base sequence, as must be necessary for RNA synthesis.

Specific base sequences are also recognized by particular protein molecules, which are regulatory elements, that determine whether a protein is to be made. These proteins bind directly to sequences of six or more base pairs but do not require that the base pairs separate, as they bind in external grooves of the DNA molecule (Chapter 6). These regulatory proteins recognize specific configurations of electrons and bind by means of electrostatic and van der Waals forces, hydrophobic interactions, and hydrogen bonds. One example of such binding between a base sequence and a regulatory protein—namely, phage λ DNA and the protein produced by the *cro* gene—has been described in Chapter 6. Other elements—namely, repressors and the cyclic AMP receptor protein—are discussed in detail in Chapter 14.

Transmission of Information from Parent to Progeny

When a cell divides, each daughter cell must contain identical genetic information; that is, each DNA molecule must become two identical molecules each carrying the information that was contained in the parent molecule. This duplication process is called **replication.** Once again, the value of bases capable of hydrogen-bonding is apparent. That is, to make a replica, the replication system need only require that the base being added to the growing end of a new chain be capable of hydrogen-bonding to the base being copied.

Chemical Stability of DNA and of Its Information Content

In long-lived organisms a single DNA molecule may have to last one hundred years or more. Furthermore, the information contained in the molecule is passed on to successive generations for millions of years with only small changes. Thus, DNA molecules must have great stability.

The sugar-phosphate backbone of DNA is extremely stable. The C—C bonds in the sugar are resistant to chemical attack under all conditions other than strong acid at very high temperatures. The phosphodiester bond is a little less stable and can be hydrolyzed at room temperature at pH 2, but this is not a physiological condition. In considering the stability of the phosphodiester bond, one can see immediately why 2'-deoxyribose rather than ribose is the constituent sugar in DNA. At alkaline pH RNA hydrolyzes to free nucleotides exceedingly rapidly (see Chapter 4). Even at pH 8 RNA molecules are gradually fragmented by phosphodiester-bond breakage. The reaction mechanism, which is called a β-elimination reaction, requires that a hydroxyl group be present on the sugar. Except for the 3'-OH terminal nucleotide, there is only a single OH group in the RNA backbone, namely, the 2'-OH group. Thus, in DNA, which uses 2'-deoxyribose, there is no free OH group and hydrolysis by means of this pathway is not possible. This has the effect that at room temperature and pH-range 7–9 there is no detectable hydrolysis of DNA (less than one bond broken per million bonds per month).

An alteration in the chemical structure of a base definitely means loss of genetic information. In a cell there are certainly a large number of chemical compounds that can attack a free base. In considering this problem one can see immediately the value of the double helix. One aspect of this is that the molecule is redundant in the sense that identical information is contained in both strands; that is, the base sequence of one strand is complementary to that of the other strand. In fact, there exist in cells elegant repair systems that can remove an altered base and then by reading the sequence on the complementary strand, can replace the correct base. (Repair will be discussed in detail in Chapter 9.) The more important aspect of this double-helical structure is that the nature of the bases and the duplex structure of DNA provides extreme protection against chemical attack. The bases are hydrophobic rings having charged groups that contain the genetic information. It is therefore the charged groups that need protection. Hydrogen-bonding between each of these groups provides the first line of defense. The hydrophobicity of the bases causes stacking and this vastly reduces the area of the base that is vulnerable to attack. Hydrogen-bonding of the two strands places all of the stacked bases into one gigantic cylindrical cluster in the central region of which are the bases. DNA stacks so extensively that water is almost completely excluded from a stacked array of bases. Since any potentially harmful compound must usually be water-soluble, these molecules will find it difficult to come into close contact with the "dry" stack of bases. Certainly the likelihood of reaching the buried charged groups is small.

The N-glycosylic bond, which joins a DNA base to the 1'-carbon of a sugar, is stable except at extremes of pH. The bond would be less stable if the bases were not rings. Methylation of the purine ring on the 7-nitrogen decreases chemical stability enormously; this is the mode of action of the highly mutagenic and destructive nitrogen and sulfur mustards used in chemical warfare in World War I—namely, any of these various mustards provides an alkyl group that attaches to the 7-nitrogen of guanine. The guanine then falls off the sugar, leaving a hydroxyl group on the 1'-carbon, allowing phosphodiester hydrolysis to occur.

The bases themselves, with the exception of cytosine, are very stable. At a very low rate, however, cytosine is deaminated to form uracil:

Cytosine   Uracil

Deamination is a disastrous change because the deamination product, uracil, pairs with adenine rather than with guanine. This has two effects: (1) an incorrect base will appear in mRNA and (2) an adenine instead of a guanine will occur in newly replicated DNA strands (Figure 7-10). There exists an intracellular system, which will be described in Chapter 8, for removing uracil from DNA and replacing it with thymine. The necessity for eliminating a uracil formed by deamination explains why DNA utilizes thymine and not uracil as the base complementary to adenine. If uracil were the normal DNA base, there would be no way to distinguish a correct uracil from an incorrect uracil produced by deamination of cytosine. By using thymine, the cell follows the rule:

Always remove a uracil from DNA because it is unwanted.

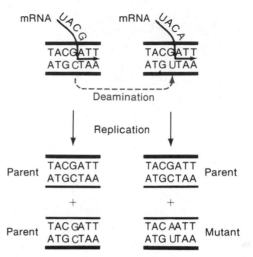

Figure 7-10
The effect of deamination of cytosine to form uracil in the base sequences of mRNA and of two daughter DNA molecules. The C→U transition is shown in red in the uppermost panel. "Parent" and "mutant" refer to base sequences before and after deamination. Newly replicated DNA is indicated by a thin line.

It is not obvious why RNA uses uracil and not thymine nor why DNA evolved with cytosine rather than with some base that would not deaminate. This may have been an evolutionary accident. It is possible, though, that the original DNA and RNA molecules both contained cytosine and uracil because these were the only pyrimidines available in the primordial sea. Cells then gained the ability to methylate uracil to form thymine because this provides a cell with a criterion for eliminating the result of a cytosine→uracil conversion. It is significant that the final step in the synthesis of thymidylic acid is the methylation of deoxyuridylic acid. The thymidylic acid is then converted to the triphosphate needed for incorporation into DNA (see Chapter 8).

The Ability of DNA to Change: Mutation

All of the information contained in a cell resides in its DNA. Thus, if a cell is to be able to improve through time—that is, to evolve—then the base sequence of its DNA must be capable of change. Furthermore, the new sequence must persist so that progeny cells will have the new property. The process by which a base sequence changes is called **mutation.** There are two main mechanisms of mutation:

1. A **chemical alteration** of the base that gives it new hydrogen-bonding properties and causes a new base to be present in a newly replicated daughter molecule.
2. A **replication error** by which an incorrect base or an extra base is accidentally inserted in the daughter molecules.

On the average, mutational changes are deleterious and lead to cell death. Therefore, it is important that not too many mutations occur in a single DNA molecule because otherwise the rare advantageous alteration would always occur in a cell destined to die by virtue of lethal mutations. Thus, the mutation rate must be low or controlled. This is accomplished in two ways. First, the hydrophobic, water-free core of the DNA molecule reduces the accessibility of the DNA to attacking molecules, as we discussed earlier. Second, the cell has evolved several repair mechanisms (discussed in Chapter 9) for correcting alterations and replication errors. These repair systems are not entirely efficient though and allow mutations to occur at a rate that is very low but useful in the long run. These mutations are, as we have said, usually deleterious, so that it is important that the parental information is not lost. Such loss is prevented in two ways: (1) The other members of the species retain the parental base sequence, and (2) a double-stranded molecule is redundant. Normally only one strand is altered, and DNA replicates in such a way that after cell division the DNA molecule in each daughter cell receives only one of the parental single strands. Thus, it is possible for one of the daughter DNA molecules to be normal and the other to be mutant; the mutant may not be able to survive but the cell with the parental base sequence will. If the mutant is better

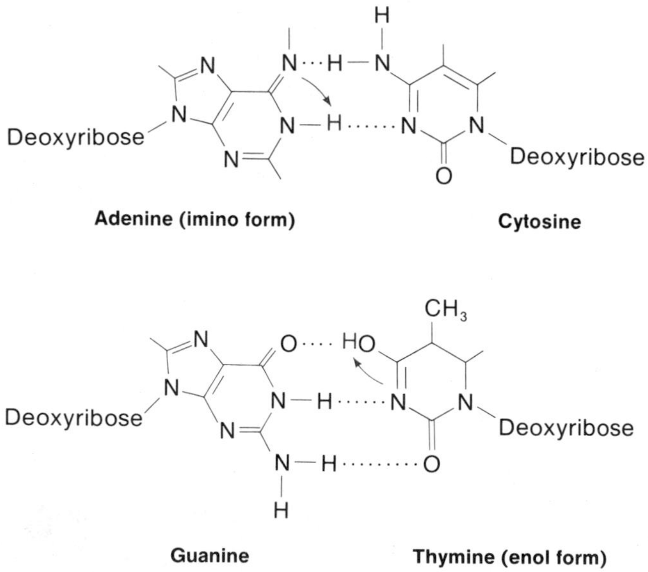

Figure 7-11
Base-pairing between the rare imino form of adenine and cytosine and the enol form of thymine and guanine. The red H is the one that has moved from the more common position. Compare with Figure 4-2, which shows the standard base pairs.

equipped to survive and multiply than the parent organisms or any other member of the species, then after a great many generations, Darwin's principle of survival of the fittest will lead to ultimate replacement of the parental genotype by the mutant phenotype in nature.

The second mechanism of mutation—replication errors—follows from the tautomeric properties of the bases adenine and thymine. The rare imino form of adenine can form a stable hydrogen bond with cytosine, and the enol form of thymine can pair with guanine (Figure 7-11). It may be the case that in the primordial seas nucleic acid molecules having other bases also arose spontaneously but only those having tautomeric bases survived because their mutability allowed adaptation to changing circumstances and evolution.

RNA AS THE GENETIC MATERIAL

In the preceding sections the properties of DNA that suit it to be the genetic material have been described. RNA is clearly not as suitable but nonetheless has survived in some organisms even though it lacks beneficial features of the base protection and the redundancy associated with a double helix (though one virus and one phage are known to have double-stranded RNA). It is noteworthy that RNA is not present as the genetic material in cellular organisms, for which chemical stability is required, but only in phages and viruses. In these particles the RNA molecule is protected from the environment by a nearly insoluble protein shell, as will be seen in Chapters 15 and 21. Furthermore, RNA molecules spend most of their time as inert particles, replicating only infrequently in host organisms. When they do replicate, they do so very rapidly, so that an enormous number of progeny particles are produced in a short period of time and sheer numbers compensate for the lesser chemical stability of RNA. Thus the RNA phages and viruses have evolved special compensatory features that enable them to survive despite their deficiencies relative to DNA-containing organisms.

REFERENCES

Avery, O. T., C. M. MacLeod, and M. McCarty. 1944. "Studies on the chemical nature of the substance inducing transformation of pneumococcal types." *J. Exp. Med.,* 79, 137–158.

Cairns, J., G. S. Stent, and J. D. Watson (eds). 1966. *Phage and the Origins of Molecular Biology.* Cold Spring Harbor Laboratory.

Crick, F. H. C. 1958. "On protein synthesis." *Symp. Soc. Expt. Biol.,* 12, 138–163.

Freifelder, D. (ed.) 1978. *The DNA Molecule.* W. H. Freeman and Co.

Hershey, A. D., and M. Chase. 1952. "Independent function of viral protein and nucleic acid in growth of bacteriophage." *J. Gen. Physiol.* 36, 39–56.

Judson, H. 1979. *The Eighth Day of Creation.* Simon and Schuster.

Mirsky, A. E. 1968. "The discovery of DNA." *Scient. Amer.,* 218 (June), 78–92.

Olby, R. 1974. *The Path to the Double Helix.* University of Washington Press.

Portugal, F. H., and J. S. Cohen. 1977. *A Century of DNA: A History of the Discovery of the Structure and Function of the Genetic Substance.* MIT Press.

Stent, G. S., and R. Calendar. 1978. *Molecular Genetics.* W. H. Freeman and Co.

Watson, J. D. 1968. *The Double Helix.* Athenaeum.

8 DNA Replication

Genetic information is transferred from parent to progeny organisms by a faithful replication of the parental DNA molecules. Usually the information resides in one or more double-stranded DNA molecules. Some bacteriophage species contain single-stranded instead of double-stranded DNA. In these systems replication consists of several stages in which single-stranded DNA is first converted to a double-stranded molecule, which then serves as a template for synthesis of single strands identical to the parent molecule. Viruses containing single- and double-stranded RNA molecules are also known; these organisms use several different modes of replication, some of which include double-stranded DNA as an intermediate. The modes of replication of each of these types of molecules differ in detail, though certain fundamental features are common to each mode.

Replication of double-stranded DNA is a complicated process that is not completely understood. This complexity is, at least in part, an acknowledgement of the importance of the following facts: (1) replication requires a supply of energy to unwind the helix; (2) single-stranded DNA tends to form intrastrand base pairs; (3) a single enzyme can catalyze only a limited number of physical and chemical reactions; (4) several safeguards have evolved that are designed both to prevent replication errors and to eliminate the rare errors that do occur; and (5) both circularity and the enormous size of DNA molecules impose geometric constraints on the replicative system, and how these fit into the system has to be understood. To add to the difficulty for the researcher, there is not a unique mode of replication common to all organisms having double-stranded DNA.

This chapter begins by examining a few general features of the replication process but then, out of necessity, we will plunge into complicated topics. An attempt has been made to explain the sequence of events in a logical order; however, the reader will probably find that a full understanding of some stages of replication can only be had after understanding certain phenomena associated with other stages described elsewhere in the chapter.

THE BASIC RULE FOR
REPLICATION OF ALL NUCLEIC ACIDS

All genetically relevant information contained in any nucleic acid molecule resides in its base sequence, so the prime role of any mode of replication is to duplicate the base sequence of the parent molecule. The specificity of base pairing—adenine with thymine and guanine with cytosine—provides the mechanism used by all replication systems. Furthermore,

1. Nucleotide monomers are added one by one to the end of a growing strand by an enzyme called a **DNA polymerase.**
2. The sequence of bases in each new or **daughter strand** is complementary to the base sequence in the old or **parent strand** being copied—that is, if there is an adenine in the parent strand, a thymine will be added to the end of the growing daughter strand when the adenine is being copied.

In the following section we consider how the two strands of a daughter molecule are physically related to the two strands of the parent molecule.

THE GEOMETRY OF DNA REPLICATION

The production of daughter DNA molecules from a single parental molecule gives rise to several topological problems, which result from the helical structure and great size of typical DNA molecules and the circularity of most DNA molecules. These problems and their solutions are described in this section.

Semiconservative Replication of Double-Stranded DNA

The purpose of DNA replication is to create daughter DNA molecules that are identical to the parental molecule. Two modes of replication can be distinguished. These differ in whether or not the two single strands of the parent molecule become rearranged after one round of replication.* These are the **semiconservative** and **conservative** modes (Figure 8-1).

In the semiconservative mode, first proposed by Watson and Crick, each parental DNA strand serves as a template for one new or daughter strand and as each new strand is formed, it is hydrogen-bonded to its parental template (Figure 8-2). Thus, as replication proceeds, the parental double helix unwinds and then rewinds again into two new double helices, each of which contains one originally parental strand and one newly formed daughter strand. At the time this was proposed,

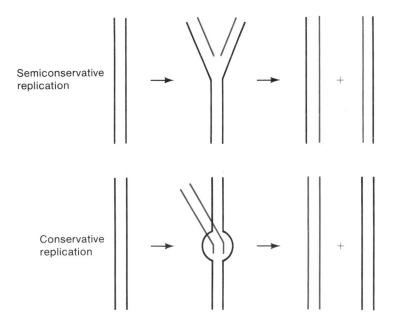

Figure 8-1
Two ways in which DNA can replicate. In semiconservative replication, the parental strands (black) separate, each daughter molecule receiving one parental and one newly synthesized daughter strand (red). In conservative replication the parental strands remain entwined, so that after one round of replication, one of the daughter molecules contains only the parental strands (double black lines) and the other only daughter strands (double red lines).

*The term "round of replication" refers to a single transit of the replication system along a DNA molecule.

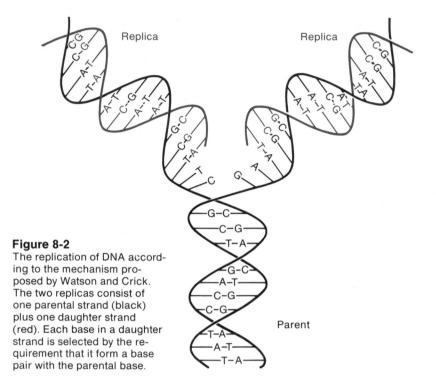

Figure 8-2
The replication of DNA according to the mechanism proposed by Watson and Crick. The two replicas consist of one parental strand (black) plus one daughter strand (red). Each base in a daughter strand is selected by the requirement that it form a base pair with the parental base.

DNA denaturation was not understood and strand separation was thought to be impossible. The idea of inseparability was based on an incorrect calculation of the time required for a helix to unwind (an enormous value was calculated) and a misinterpreted experiment that suggested that the molecular weight of DNA was not halved upon denaturation. In addition, since the length of a single bacterial DNA molecule is about 1.3 mm and the length of a typical bacillus is about 0.003 mm, the idea of strand separation with total disentanglement of the daughter strands seemed preposterous.

These considerations provided the impetus for proposing the conservative mode. In the conservative mode the two strands of the parental molecule remain entwined before and after replication and unwind at the replication site only as much as is sufficient to allow the base sequence there to be read by the polymerizing enzyme. From this reading a single daughter molecule is produced; the parental molecule remains intact (Figure 8-1). Thus, in conservative replication, complete unwinding of the parental helix is not necessary.

The correctness of the semiconservative model was demonstrated by the following experiment.

The Meselson-Stahl Experiment

The conservative and semiconservative modes differ in the composition of daughter molecules—in the conservative mode, of any two "daughter" DNA molecules one is in fact also the original "parent" molecule, and in the semiconservative mode, the parental strands are distributed equally between the two daughter molecules, one parental strand forming one of the two strands in each daughter. Matthew Meselson and Franklin Stahl developed a simple method by which the parental and daughter strands could be distinguished. A culture of bacteria was grown for many generations in a growth medium containing ^{15}N-labeled NH_4Cl as the sole source of nitrogen (called a "heavy medium"). In this way the parental DNA molecules were labeled with the heavy isotope ^{15}N, thereby increasing the density of the DNA. The cells were then transferred to a medium containing the common isotope of nitrogen, namely ^{14}N (light medium). At various times after the transfer, samples of cells were collected and the DNA was isolated. The DNA molecules were fragmented during isolation; Figure 8-3 represents the composition of the fragments at one and two generations after the density shift. In the semiconservative mode, after one generation all daughter molecules would have one ^{15}N strand and one ^{14}N strand (this is called a **hybrid** molecule); hence all daughter molecules would have the same density (hybrid density)—namely, midway between that of

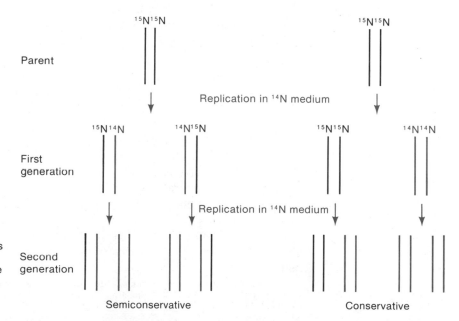

Figure 8-3
Distribution of parental ^{15}N-labeled strands (black) during two rounds of replication in ^{14}N (red) medium, according to the semiconservative and conservative modes of replication.

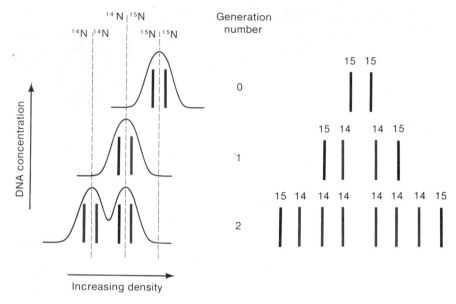

Figure 8-4
The type of data obtained in the Meselson-Stahl experiment. *E. coli* were grown for many generations in ¹⁵N medium (black) and then at zero time transferred to ¹⁴N medium (red). After one generation, all DNA had hybrid density. DNA of less than hybrid density is not seen before the first generation, because the chromosome is fragmented into about 200 pieces during isolation. At the right, the state of the isolated DNA at various times is shown.

[¹⁵N¹⁵N]- and [¹⁴N¹⁴N]DNA molecules.* In the conservative mode both [¹⁵N¹⁵N]DNA and [¹⁴N¹⁴N]DNA molecules would result; hence, the two first-generation daughter molecules would differ in density. After two generations the predicted distributions in density also differ, one from the other, as shown in the figure. Meselson and Stahl used density gradient centrifugation in CsCl (see Chapter 3) to measure the densities of the molecules present as a function of time after the change from heavy to light medium. Their results, shown in Figure 8-4, that all DNA had a hybrid density after one round of replication, indicated that the semiconservative mode is correct.

A second experiment confirmed the structure of the [¹⁴N¹⁵N]DNA found after one generation. In this experiment the hybrid DNA was denatured by heating to 100°C and then centrifuged in CsCl. The heated DNA yielded two bands having the densities of denatured (single-stranded) [¹⁴N]- and [¹⁵N]DNA (Figure 8-5), proving that the DNA of hybrid density did in fact consist of one ¹⁴N strand and one ¹⁵N strand.

*Double-stranded DNA molecules containing either ¹⁴N or ¹⁵N in *both* strands are denoted [¹⁴N¹⁴N] DNA and [¹⁵N¹⁵N]DNA, respectively. A molecule containing ¹⁴N in one strand and ¹⁵N in the other is denoted [¹⁴N¹⁵N]DNA.

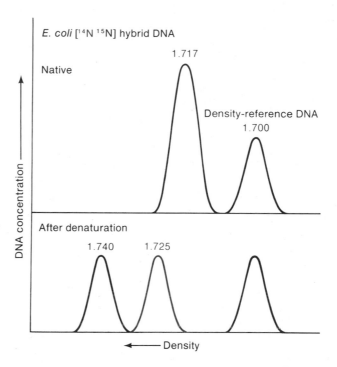

Figure 8-5
The density of native, double-stranded, first-generation $^{14}N^{15}N$ hybrid DNA molecules and of single polynucleotide strands derived from them by denaturation. Reference DNA of density 1.700 g/cm³ was added to both denatured (lower panel) and undenatured (upper panel) E. coli hybrid DNA, and the two DNA samples were centrifuged to equilibrium in a cesium chloride density gradient. The ordinate indicates the concentration of DNA in the density gradient.

Now that semiconservative replication is established, we are faced with the problem of how to explain unwinding of the helix. In a later section the energetics of unwinding will be considered. In the following section the geometric problem of unwinding is discussed.

The Geometry of Semiconservative Replication

Unwinding a double helix in semiconservative replication presents a mechanical problem. Either the two daughter branches at the Y-fork shown in Figure 8-2 must revolve around one another or the unreplicated portion must rotate. If the molecule were fully extended in solution, there would be no problem and rotation of the unreplicated portion would be the simpler motion, as there would be less friction with the solvent. However, since the molecule is 600 times longer than the cell that contains it (so that it must be repeatedly folded, as was shown in Chapter 6), such rotation is unlikely. A simple solution would be to make a single-strand break in one parental strand ahead of the growing fork; this would enable a small segment of the unreplicated region to rotate, thereby eliminating the geometric problem (Figure 8-6).

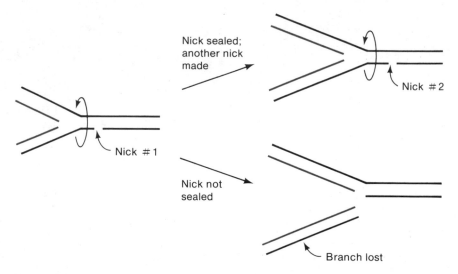

Figure 8-6
Mechanism by which a nick ahead of a replication fork allows rotation. If the nick is not sealed, a newly formed branch is lost.

The only requirement would then be to re-form the broken bond. However, this repair would have to occur before the replication fork had passed the nick; otherwise a daughter strand would be lost, as shown in the figure.

Most DNA molecules replicate as circular structures. This introduces a geometric problem that is more severe than that just described. This problem is solved by an enzyme, DNA gyrase, which will be discussed in the next section.

Geometry of the Replication of Circular DNA Molecules

Most bacteriophage species contain linear DNA molecules which, by a variety of mechanisms, circularize after infection. (An exception is *E. coli* phage T7, which will be discussed later in this chapter.) The DNA molecules of the few bacteria examined to date are also circular. The first demonstration that *E. coli* DNA replicates as a circle came from two elegant experiments performed by John Cairns. In both experiments, cells were grown in a medium containing [³H]thymine so that all DNA synthesized after transfer to the radioactive medium would be radioactive. The DNA was isolated without fragmentation and placed on photographic film. Each ³H-decay exposed one grain in the film and after several months there were enough grains to visualize the DNA with a microscope; the pattern of black grains on the film located the

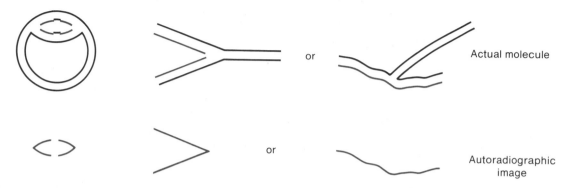

Figure 8-7

Patterns of grains in an autoradiogram of a partially replicated circle or linear molecule. The red indicates radioactive DNA. The upper row shows the actual molecule. The lower row shows the pattern of radioactivity that would be observed for each molecule in the upper row. Even though the labeled portions of the molecule are not physically continuous, the grain pattern is continuous because the separation of the discontinuous segments is exceedingly small.

molecule. When Cairns grew the bacteria for a fraction (about one-tenth) of a generation in radioactive medium, he observed closed patterns of grains; this is characteristic of replication of a circle because when autoradiograms of linear molecules are made, linear (probably a bit curved) arrays of grains are observed (Figure 8-7). When the bacteria were grown for slightly less than two generations in radioactive medium, branched circular patterns were seen. An autoradiogram and its interpretation are shown in Figure 8-8. Five years later, electron micrographs of circularly replicating molecules from other organisms were obtained (Figure 8-9). A replicating circle is schematically like the Greek letter θ (theta), so this mode of replication is usually called θ replication; likewise, a θ molecule is sometimes also called a Cairns molecule.

The unwinding problem in θ replication is formidable because lack of a free end makes rotation of the unreplicated portion impossible. An advancing nick, as in Figure 8-6, or a swivel at the replication origin (the point of initiation of replication, as proposed by Cairns) would solve the problem. These suggestions, though not quite correct, anticipate the correct mechanism.

As replication of the two daughter strands proceeds along the helix, in the absence of some kind of swiveling the nongrowing ends of the daughter strands would cause the entire unreplicated portion of the molecule to become overwound (Figure 8-10). This in turn would cause *positive* supercoiling (see Chapter 4) of the unreplicated portion. This supercoiling obviously cannot increase indefinitely because, if it were to do so, the unreplicated portion would become coiled so tightly that no further advance of the replication fork would be possible. This

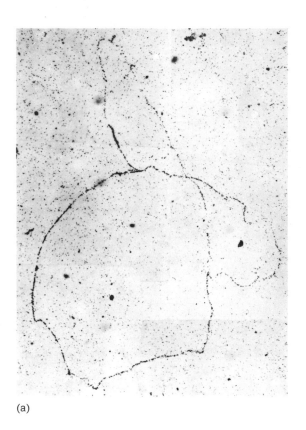

(a)

Figure 8-8

(a) Autoradiogram of the intact replicating chromosome of an *E. coli* bacterium that has been allowed to incorporate [³H]thymine into its DNA for slightly less than two generation periods. The continuous lines of dark grains were produced by electrons emitted during a two-month storage period by decaying ³H atoms in the DNA molecule. (From J. Cairns, *Cold Spring Harbor Symp. Quant. Biol.* 1963. 28: 44). (b) Interpretation of the autoradiogram of part (a). (i) Original molecule. (ii) One of two identical daughter molecules produced after one generation. (iii) Two daughter molecules present after slightly less than two generations. (iv) How (iii(b)) would appear in an autoradiogram. The heavy line has twice the grain density of the dashed line. (v) How (iv) would appear when arranged like the molecule in part (a).

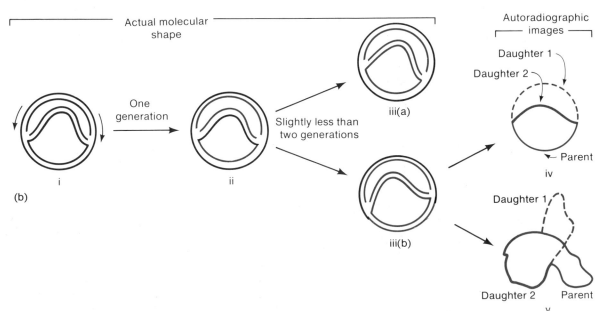

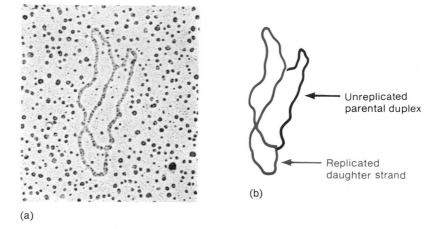

Figure 8-9
Electron micrograph of a ColE1 DNA molecule (molecular weight = 4.2 × 10⁶) replicating by the θ mode. The parental and daughter segments are shown in the drawing. (Courtesy of Donald Helsinki.)

(a)

(b)

Unreplicated parental duplex

Replicated daughter strand

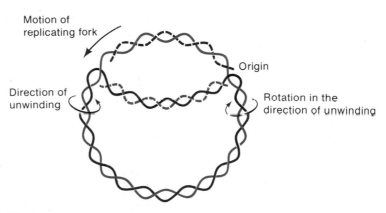

Motion of replicating fork

Origin

Direction of unwinding

Rotation in the direction of unwinding

Figure 8-10
Drawing showing that the unwinding motion (curved arrows) of the daughter branches of a replicating circle lacking positions at which free rotation can occur causes overwinding of the unreplicated portion.

topological constraint ought to be avoidable by the simple nicking-sealing cycle just described but, in general, the twists are removed in another way. As discussed in Chapter 4, most naturally occurring circular DNA molecules are *negatively* supercoiled. Thus, initially the overwinding motion is no problem because it can be taken up by the underwinding already present in the negative supercoil. However, after about 5 percent of the circle is replicated, the negative superhelicity is used up and the topological problem arises.

Most organisms contain one or more enzymes called **topoisomerases.** These enzymes can produce a variety of topological changes in

DNA; the most common are production of negative superhelicity and the removal of superhelicity. In *E. coli* there is an enzyme called **DNA gyrase** (Eco topoisomerase II) which is able to produce negative superhelicity, and it is DNA gyrase that is responsible for removing the positive superhelicity generated during replication. The evidence for this point comes from experiments using drugs that inhibit DNA gyrase (e.g., coumermycin, nalidixic acid, oxolinic acid, and novobiocin) and from the study of gyrase mutants. Addition of any of these drugs to a growing bacterial culture inhibits DNA synthesis; in a phage-infected cell they prevent supercoiling of injected phage DNA molecules (that is, the molecules circularize but do not supercoil). Proof that these effects are due to inhibition of DNA gyrase comes from isolating a strain of *E. coli* that can grow normally in the presence of these antibiotics. With this strain the antibiotics have no effect on either DNA replication or supercoiling. If DNA gyrase is isolated from both wild-type *E. coli* and the mutant *E. coli* strain and each form is tested for binding of the antibiotics, it is found that only the wild-type enzyme binds the antibiotic. (This approach of probing for an activity with an inhibitor and comparing *in vitro* and *in vivo* results is used in many problems in molecular biology.)

DNA gyrase does have a nicking-sealing activity but this activity cannot by itself introduce negative superhelical turns. How twisting is done is explained in the following section.

Mechanism of Action of DNA Gyrase

If a circular DNA molecule is exposed to DNA gyrase and ATP, the DNA molecule undergoes a series of structural transitions, each of which introduces *two* superhelical twists into the molecule, until the molecule is so tightly coiled that no further twisting is possible. *The number of superhelical twists in a particular molecule is always an even number.* This fact gave the first clue to understanding how gyrase works.

The basic model for gyrase activity is called the **sign inversion model** and is shown in Figure 8-11. Let us first examine molecules I and II of the figure. If a single gyrase molecule were to bring together the two sites in molecule I indicated by the arrows and, in so doing, twist the molecule in the direction shown by the circular arrow, a cross-over region or a **node** would be produced. By definition this node would be a positive node. Because the molecule is circular and the base pairs are most stable when maintaining their 12° rotation with respect to one another in the double helix, the lower part of molecule II apparently twists in the opposite sense, generating a negative node. Molecule II would be stable as long as gyrase remained bound to the positive node.

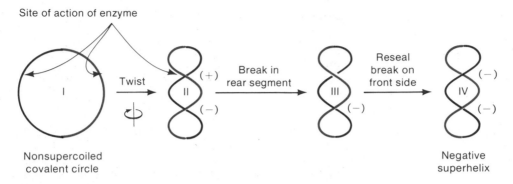

Figure 8-11

The sign inversion model for superhelix formation. The left and right halves of the circle are in different colors only to facilitate seeing whether one segment passes in front of or behind another segment. + and − refer to positive and negative nodes respectively. Note that the enzyme acts only on the upper node.

At this point the net twisting of the circular DNA molecule would still be zero. In molecule III the portion of the upper node that is farthest from the viewer would then be cleaved (a double-strand break) and moved *in front of* the other portion, which would convert the positive node to a negative node. Sealing of this double-strand break would yield molecule IV, which would have two negative twists.

This overall model, which in principle is the one that occurs, has the following requirements: (1) Gyrase must be capable of binding to two DNA segments, which may be quite distant, and it must have at least two binding sites. (2) The two free ends in molecule III must be held together; otherwise the circle would completely open up and become a linear molecule. (3) A mechanism is needed to pass the two free ends from one side of the unbroken strand to the other, and this motion requires energy.

The first requirement is met by the mode of binding of DNA to gyrase. The DNA molecule is wrapped around the gyrase in a manner resembling (but not identical to) the coiling of DNA around a histone octamer in a nucleosome (Chapter 6). Wrapping brings separated regions into apposition.

The second requirement is fulfilled by the subunit structure of the enzyme. Gyrase consists of four subunits, two α and two β subunits. In the cleavage shown in molecule III the 5′-P terminus of each end is covalently joined to one α subunit. This linkage has two important effects. First, the two ends are attached to a single gyrase molecule and therefore cannot fly apart. Second, the energy gained in breaking the phosphodiester bonds is stored in the DNA−5′-P−gyrase linkage, so that no activation (such as triphosphate formation) is needed for rejoining the ends after sign inversion.

The third requirement is satisfied by the fact that the β monomers bind adenosine triphosphate (ATP) and the binding induces a conformational change that results in movement of the two α subunits with respect to one another. Note that cleavage of ATP is not needed for that movement—the binding energy is the source of energy. After rejoining of the broken ends, ATP is hydrolyzed to yield adenosine diphosphate (ADP) and inorganic phosphate (P_i). The ADP, which does not bind to gyrase, then falls off of the β subunit, causing the enzyme to return to the original configuration, ready to start the cycle again and introduce two more twists. In Chapter 2 we saw that in metabolic reactions the free energy of hydrolysis of ATP is often used to drive reactions to completion. In the gyrase reaction the role of ATP is quite different in that hydrolysis merely serves to convert bound ATP to a molecule that cannot bind to the enzyme. The role of ATP as a mediator of cyclic conformational changes is found in many biochemical processes in which molecules or parts of molecules move with respect to one another. We will see a striking example of this in Chapter 13 when we discuss the mechanism by which a growing protein chain moves along a ribosome; in that case, guanosine triphosphate (GTP) is used.

The details of the supercoiling reaction are not yet resolved, though an understanding should be forthcoming shortly. However, a model has recently been proposed that includes the known properties of DNA gyrase. This model, which is shown in Figure 8-12, may not be correct in detail but is probably near enough to the correct mechanism to be worthy of careful study. The structure at I shows a complex between a portion of a DNA molecule and the $\alpha_2\beta_2$ tetramer of gyrase. The DNA is wrapped in a right-handed superhelix in such a way that the segment to be cleaved (red) and the segment to be passed from one side of the node to the other (pink) are on opposite sides of the gyrase. In structure II the shaded segment has been cleaved. The cuts are known to be staggered by a distance of four base pairs so that each free end has a 5'-P group at the end of a four-base single-stranded segment. Each 5'-P group is linked to an α subunit. The subunits of gyrase are thought to be arranged so that there is a hole in the center of the tetramer. This hole is essential for the supercoiling reaction. After cleavage to form structure II, gyrase undergoes a shape change (caused by binding of ATP), also shown in structure II, and this change allows the unbroken pink segment to pass between the α subunits, as in the structure at III. The enzyme then closes, which brings the ends of the broken segment together to be sealed (as in structure IV) and releases the unbroken segment through a gap formed at the back of the gyrase (structure V). ADP is then released and the gyrase is restored to the form shown in structure I, ready to start the cycle again.

The breaking and passing of DNA through the gyrase molecule explains two other reactions that are catalyzed by gyrase—namely, catenation and decatenation (Figure 8-13). **Catenation** is the linking of

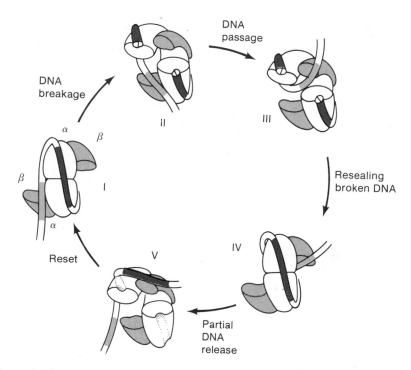

Figure 8-12

Model for negative supercoiling of DNA by gyrase. Step I represents a gyrase molecule consisting of two α and two β (shaded) monomers with a section of a circular duplex DNA wrapped in a right-handed superhelix. The solid red DNA molecule contains the site of double-strand breakage at its middle. The DNA, beginning about 85 base pairs from the cleavage site, is to be passed through the transient gap. The enzyme contains a hole (concealed in step I) that is exposed by opening the molecule about a hinge between the α subunits (as in step II). To effect supercoiling, a right-handed DNA loop in step I is converted into a left-handed one (step IV) by the following steps: II, DNA breakage and opening of the enzyme; III, passing of DNA through the gap by exchange of the stippled DNA between β protomers; and IV, closing the enzyme and resealing the broken DNA. This sequence of steps introduces two negative supercoils. Step V represents a way of resetting the DNA to complete the cycle; the resealed DNA section is released, followed by reopening of gyrase to allow escape of the passed DNA section. (From A. Morrison and N. R. Cozzarelli, 1981. *Proc. Nat. Acad. Sci.*, 78: 1416.)

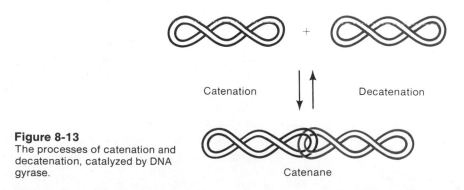

Figure 8-13

The processes of catenation and decatenation, catalyzed by DNA gyrase.

two circular DNA molecules to form a chain and decatenation is the reverse process. One catenane will produce two DNA circles. These two reactions can be carried out if the two units both bind to a single gyrase molecule in such a way that a segment of each circle can pass through a segment of the other. Decatenation is probably important in DNA synthesis because, when a circle replicates, the products under certain conditions are two catenated circles; this will be discussed later.

In DNA replication DNA gyrase can relieve positive supercoiling by introducing two negative twists each time two positive twists arise. The procedure would probably be most effective if the gyrase bound just ahead of the growing fork, though whether there is such a binding site is not known with certainty.

So far, we have discussed the geometric problems of DNA replication. Now we shall enter the complex realm of the enzymatic mechanism of replication.

ENZYMOLOGY OF DNA REPLICATION

The enzymatic synthesis of DNA is a complex process, primarily because of the need for high fidelity in copying the base sequence and for physical separation of the parental strands. The number of steps that must be completed is far too great to be accomplished by a single enzyme and, in fact, about twenty proteins are known at present to be necessary. Thus, in an effort to provide some understanding with a minimum of confusion, each step in the process will be treated separately. We will consider the basic chemistry of polymerization, the source of the precursors, the problems raised by the chemistry of polymerization, the means of initiating and terminating synthesis, and the mechanisms for eliminating replication errors.

The Polymerization Reaction and the Polymerases

In 1957, Arthur Kornberg showed that in extracts of *E. coli* there exists a DNA polymerase (now called **polymerase I** or **pol I**). This enzyme was able to synthesize DNA from four precursor molecules—namely, the four deoxynucleoside 5′-triphosphates (dNTP), dATP, dGTP, dCTP, and dTTP—as long as a DNA molecule to be copied (a **template** DNA) was provided. Neither 5′-monophosphates nor 5′-diphosphates, nor 3′-(mono-, di-, or tri-) phosphates can be polymerized—only the 5′-triphosphates are substrates for the polymerization reaction; soon we will see why this is the case. Some years later, it was found that pol I, though playing an essential role in the replication process, is not the

O⁻ structure in margin (deoxynucleoside triphosphate)

**Deoxynucleoside
triphosphate**

major polymerase in *E. coli*; instead, the enzyme responsible for advance of the replication fork is polymerase III or pol III.* Pol III also exclusively uses 5′-triphosphates as precursors and requires a DNA template before polymerization can occur. Pol I and pol III have many features in common and, in fact, a few types of DNA molecules replicate by using only pol I. The overall chemical reaction catalyzed by both DNA polymerases is:

$$\text{Poly(nucleotide)}_n\text{-3}'\text{-OH} + \text{dNTP} \rightarrow \text{Poly(nucleotide)}_{n+1}\text{-3}'\text{-OH} + \text{PP}$$

in which PP represents pyrophosphate cleaved from the dNTP.

We have already mentioned (Chapter 2) that nucleoside monophosphates cannot be added to the end of a growing DNA strand because the free energy of hydrolysis of a phosphodiester bond is large and negative; hydrolysis and not polymerization is the preferred direction of reaction. The rate of spontaneous hydrolysis of these bonds is very low but it does occur; as a consequence, most cells contain an auxiliary enzyme called a ligase, which seals nicks that arise spontaneously. (We will see shortly that this is not the main function of ligase.) In a nucleoside triphosphate, the strong negative charge on adjacent phosphate groups (see margin) mutually repel one another and hence weaken the P—O bonds; this repulsion is reflected in the rather large negative value of the free energy of hydrolysis for these substances, which tends to compensate for the large negative free energy of formation of the phosphodiester bond in DNA. However, the value of the total free energy for the polymerization reaction written above is still slightly greater than zero ($+0.5$ kcal/mole = about 2400 joule/mole), so equilibrium still lies to the left as written. Indeed, if excess pyrophosphate (one of the products of cleavage of the triphosphate) and a polymerase are added to a solution containing a partially replicating DNA molecule, the polymerase catalyzes depolymerization. In order to drive the reaction to the right, pyrophosphate must be removed, and this is done by a potent **pyrophosphatase,** a widely distributed enzyme that breaks down pyrophosphate to inorganic phosphate. Hydrolysis of pyrophosphate has a large negative free energy, so that essentially all of the pyrophosphate is rapidly removed.

The energetics of the polymerization reaction explains why polymerization has not evolved to make use of diphosphate precursors, which also have a large, negative free energy of hydrolysis. If the diphosphates were used as precursors, the free energy change would

---

*The numbers I and III refer only to the order in which the enzymes were first isolated, not to their relative importance. Another polymerase, pol II, has also been isolated from *E. coli*. It plays no role in DNA replication. Its biological function is unknown; mutant cells lacking pol II grow normally in the laboratory. For convenience, the abbreviation "pol" will be used in this chapter to refer to specifically named polymerases.

still be positive for the polymerization reaction and the product would be inorganic phosphate, which cannot be broken down further, but this is not easily removed by known biochemical reactions. Hence, if the polymerization depended on diphosphate precursors, no straightforward way would exist to drive the reaction in the direction of polymerization.

In the following secti ..., ...e synthesis of nucleoside triphosphates is described. This section may be omitted, but, if used, the reader should pay special attention to the synthesis of thymidine triphosphate (TTP), because it introduces a complication that will be described later.

Source of Precursors

There are two distinct pathways for the synthesis of the DNA precursors—the **de novo pathway** and the **salvage pathway.** In the latter, free bases, nucleosides, and nucleotides obtained either by degradation of nucleic acids or from the growth medium are built up to nucleoside monophosphates by a variety of reactions. For further information, see the references at the end of the chapter.

In the *de novo* pathway, ribonucleoside monophosphates are synthesized from phosphoribosylpyrophosphate, amino acids, CO_2, and NH_3 (rather than from free bases). The nucleoside monophosphates (NMP) are then converted to nucleoside diphosphates (NDP) by enzymes called **kinases.** The kinases are specific for each base but nonspecific with respect to ribose or deoxyribose. The reaction involves cleavage of ATP and is:

$$NMP + ATP \longrightarrow NDP + ADP$$

The NDP are then converted to their deoxy form by an enzyme, **ribonucleotide reductase.** This enzyme is not base-specific: it acts on all ribo-NDP:

$$\text{Ribo-NDP} \xrightarrow{\substack{\text{Ribonucleotide} \\ \text{reductase}}} \text{Deoxy-NDP}$$

The enzyme **nucleoside diphosphate kinase,** which is neither base- nor sugar-specific, then forms the triphosphate. Because of the lack of base specificity of both this enzyme and ribonucleotide reductase, the nucleotide dUTP, which is not an immediate precursor for either DNA or RNA, is synthesized. As will be seen later, bacteria devote considerable effort toward preventing incorporation of uracil (U) into DNA and removing the few molecules of uracil that do become part of the DNA.

The synthesis of TTP requires special consideration since thymidine is unique to DNA. Thymidine monophosphate (TMP) is formed by methylation of dUMP; the reaction is catalyzed by the enzyme **thymid-ylate synthetase** and the methyl group by which uracil and thymine differs is obtained from tetrahydrofolate:

$$
\text{dUTP} \longrightarrow \text{dUMP} + \text{tetrahydrofolate} \xrightarrow[\text{synthetase}]{\text{Thymidylate}} \text{TMP} + \text{dihydrofolate}
$$
$$
\searrow
$$
$$
\text{PPi}
$$

Recognizing this mode of synthesis has been important for at least two reasons:

1. Mutants lacking thymidylate synthetase (thy^- mutants) have been isolated. Exogenous thymine or thymidine, which can be built into TMP by the salvage pathway, must be provided to these mutants in order for them to grow. Since there are no salvage enzymes that convert thymine to any compound other than thymidine and TMP, addition of radioactive thymine or thymidine to growth medium allows radio-labeling of DNA without labeling any other compound. This technique has been important in the study of replication *in vivo*. In eukaryotes it is often not possible to obtain thy^- mutants. However, addition of fluorouracil deoxyriboside (FUDR), a thymidine analogue in which fluorine replaces the methyl group, inhibits thymidylate synthetase. FUDR is phosphorylated and then forms a ternary complex with the enzyme and tetrahydrofolate, thereby inactivating the enzyme.

2. Inhibitors of tetrahydrofolate synthesis also introduce a require-ment for exogenous thymine or thymidine, without which the cells die. A drug that inhibits the conversion of dihydrofolate to tetrahydrofo-late—methotrexate—is commonly used in cancer chemotherapy.

Properties of Polymerases I and III

Pol I and pol III have many features in common. Both enzymes can only polymerize deoxynucleoside 5'-triphosphates and can do so only while copying a template DNA. Furthermore, polymerization can only occur by addition to a **primer**—that is, an oligonucleotide hydrogen-bonded to the template strand and whose terminal 3'-OH group is available for reaction (that is, a "free" 3'-OH group). The meaning of a primer is made clear in Figure 8-14, which depicts six potential template molecules; of these, only three can be said to be active—(c), (e), and (f)—each of which has a free 3'-OH group. The lack of activity with (d) and the direction of synthesis with (e) and (f) indicate that nucleotides do not add to a free 5'-P group. The lack of any synthesis with (a) or (b) indicates that addition to a 3'-OH group cannot occur if there is nothing to copy. Thus, we draw two conclusions:

1. Both a primer with a free 3'-OH group and a template are needed.
2. Polymerization consists of a reaction between a 3'-OH group at the end of the growing strand and an incoming nucleoside 5'-triphosphate. When the nucleotide is added, it supplies another free 3'-OH group. Since each DNA strand has a 5'-P terminus and a 3'-OH terminus, strand growth is said to proceed in the 5'→3' (5'-to-3') direction.

These two points are summarized in Figure 8-15.

Pol I has several activities other than its ability to polymerize; some of these are nevertheless important during the polymerization process. Notably, polymerase functions as a 3'→5' exonuclease, a 5'→3' exonuclease, and an endonuclease, and can perform nick translation and strand displacement. These functions are described in the following sections.

The 3'→5' Exonuclease Activity

Occasionally polymerases add a nucleotide to the 3'-OH terminus that cannot hydrogen-bond to the corresponding base in the template strand. This may be purely a mistake or may result from the tautomerization of adenine and thymine discussed in Chapter 10. Inasmuch as such a nucleotide alters the information content of the daughter DNA molecule, it is reasonable that cells would evolve mechanisms for correcting such errors. All cells do, in fact, possess numerous repair systems that can excise a damaged (chemically altered) or an unpaired base (see Chapter 9). In the case of a damaged base, the distinction between the correct base (i.e., the undamaged member of the pair) and the altered base is clear. However, if a standard base (i.e., A, T, G, or C) incapable of pairing is inserted, these repair systems have no simple way of distinguishing the correct from the incorrect. Thus it is essential that the unpaired base be removable while its incorrectness is recognizable—namely, as an unpaired base at the 3'-OH terminus of a growing strand.

Pol I responds to an unpaired terminal base by terminating polymerizing activity, because the enzyme requires a primer that is correctly hydrogen-bonded. When such an impasse is encountered, a 3'→5' exonuclease activity, which may be thought of simply as pol I running backwards or in the 3'→5' direction, is stimulated, and the unpaired base is removed (Figure 8-16). After removal of this base, the exonuclease activity stops, polymerizing activity is restored, and chain growth begins again. This exonuclease activity is called the **proofreading** or **editing function** of pol I.

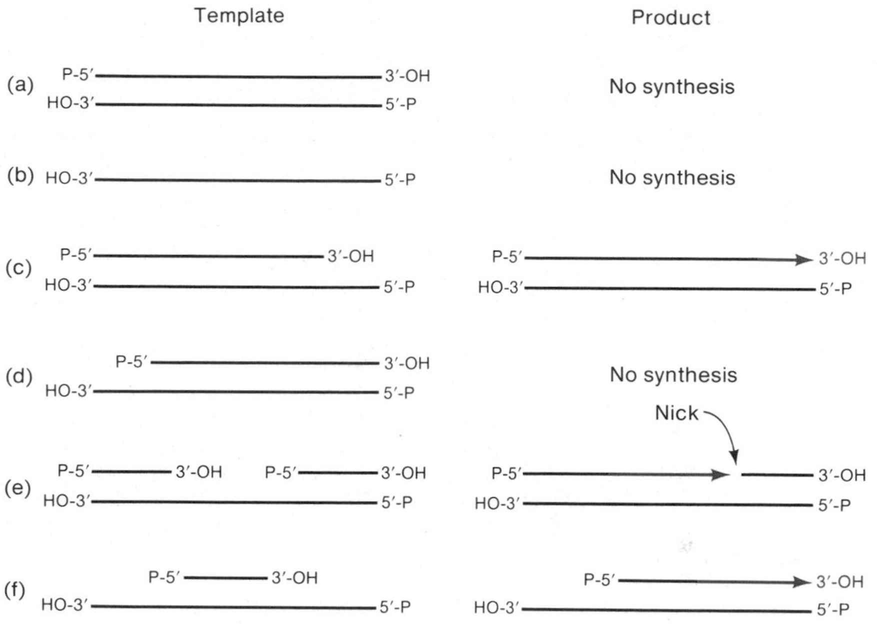

Template
Product
(a)
P-5′
3′-OH
HO-3′
5′-P
No synthesis
(b)
HO-3′
5′-P
No synthesis
(c)
P-5′
3′-OH
HO-3′
5′-P
P-5′
3′-OH
HO-3′
5′-P
(d)
P-5′
3′-OH
HO-3′
5′-P
No synthesis
Nick
(e)
P-5′
3′-OH
P-5′
3′-OH
HO-3′
5′-P
P-5′
3′-OH
HO-3′
5′-P
(f)
P-5′
3′-OH
HO-3′
5′-P
P-5′
3′-OH
HO-3′
5′-P

Figure 8-14
The effect of various templates used in DNA polymerization reactions. A free 3′-OH on a hydrogen-bonded nucleotide at the strand terminus and a non-hydrogen-bonded nucleotide at the adjacent position on the template strand is needed for strand growth. Newly synthesized DNA is red.

5′-P
A
G
T
A • T
T • A
C • G
G • C
A • T
P
3′-OH
5′
P—P—P
3′-OH
Template
Primer
3′-OH
5′-P
3′
5′

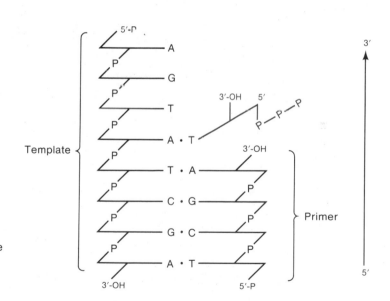

Figure 8-15
Schematic diagram of a replicating DNA molecule showing the distinction between template and primer and the meaning of 5′→3′ synthesis.

5′
3′
G T A A G T C G
T C A G C
C
3′
5′
Site of hydrolysis by
3′ → 5′ exonuclease

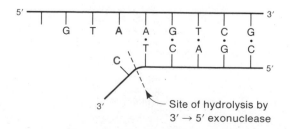

Figure 8-16
The 3′→5′ exonuclease activity of DNA polymerase I showing the site of hydrolysis. The C that is removed (red) does not base-pair with the A being copied (red).

The 5′→3′ Exonuclease Activity

Figure 8-17 shows several substrates and the products of the 5′→3′ exonuclease activity of pol I. From these results five features of the 5′→3′ exonuclease can be deduced.

1. Nucleotides are removed from the 5′-P terminus only; although not shown in the figure, they are removed one by one.
2. More than one nucleotide can be removed by successive cutting.
3. The nucleotide removed must have been base-paired.
4. The nucleotide removed can be either of the deoxy- or the ribo-type.
5. Activity can be at a nick as long as there is a 5′-P group.

We will see shortly that the main function of the 5′→3′ exonuclease activity is to remove ribonucleotide primers.

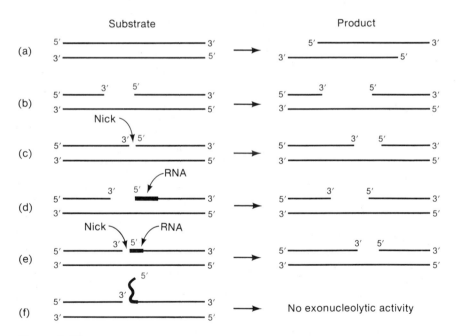

Figure 8-17
Several substrates and products after activity of the polymerase I 5′→3′ exonuclease. The 5′ terminus attacked by the enzyme is shown in red. (a) 5′-Terminal nucleotides are removed from each end of the molecule. (b) A gap is enlarged by cleavage at a 5′ terminus. (c) A gap is formed by removal of a 5′-terminal nucleotide at a nick. The gap is then enlarged as in (b). (d) RNA (heavy line) at the boundary of a gap is removed by cleavage at the 5′ terminus of the RNA. (e) A gap is formed from a nick at a DNA-RNA boundary by removal of 5′-P-terminal ribonucleotides. (f) A non-hydrogen-bonded 5′ terminus is resistant to the exonuclease; there is an endonucleolytic activity, as described in Figure 8-20.

Nick Translation

The $5' \rightarrow 3'$ exonuclease activity at a single-stranded break (a nick) can occur simultaneously with polymerization. That is, as a 5'-P nucleotide is removed, a replacement can be made by the polymerizing activity (Figure 8-18). Since pol I cannot form a bond beween a 3'-OH group and a 5'-monophosphate, the nick moves along the DNA molecule in the direction of synthesis. This movement is called **nick translation.**

Strand Displacement

Experimental conditions can be chosen so that polymerization will occur at a nick (see nick translation, above) without concomitant $5' \rightarrow 3'$ exonuclease activity. The growing strand then displaces the parental strand (Figure 8-18). This is thought to be an important step in the mechanism of genetic recombination (Chapter 18). Of all *E. coli* polymerases known to date, pol I is the only one capable of carrying out an unaided displacement reaction. In other strand displacement reactions, auxiliary proteins are required and ATP is cleaved to fuel the unwinding of the helix; this will be discussed when the events at a replication fork are described.

A bizarre reaction often accompanies strand displacement *in vitro*. This is called **template-switching** (Figure 8-19) and probably has no biological significance.

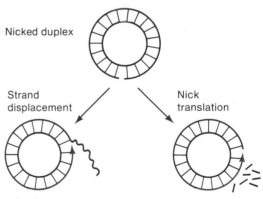

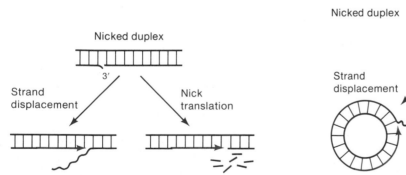

Figure 8-18
Strand displacement and nick translation on linear and circular molecules. In nick translation a nucleotide is exonucleolytically removed for each nucleotide added. The growing strand is shown in red.

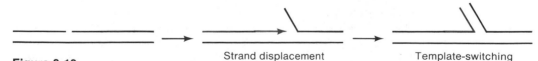

Strand displacement   Template-switching

Figure 8-19
Strand displacement followed by template-switching.

An Endonuclease Activity

The $5' \rightarrow 3'$ exonuclease activity is also capable of an endonuclease activity, as shown in Figure 8-20. An endonucleolytic cut is made between two base pairs that follow a 5'-P-terminated segment of unpaired bases. This is unimportant in normal replication but is probably a stage in a major excision-repair system (Chapter 9).

Polymerase III

E. coli pol III is a very complex enzyme. In its most active form it is associated with eight other proteins to form the **pol III holoenzyme,** occasionally termed pol III. The term holoenzyme refers to an enzyme that contains several different subunits and retains some activity even when one or more subunits is missing. The smallest aggregate having enzymatic activity is called the **core enzyme.** The activities of the core enzyme and the holoenzyme are usually very different. Genes encoding five of the subunits have been identified; these are called *dnaE, dnaN, dnaQ, dnaX,* and *dnaZ.* The *dnaE* protein possesses the major polymerizing activity but each of the subunits, except for the *dnaQ* protein is essential for replication. Pol III shares with pol I a requirement for a template and a primer but its substrate specificity is much more limited. For instance, pol III cannot act at a nick nor is it active with single-stranded DNA primed by either a DNA or RNA nucleotide fragment (Figure 8-21). (Why it is deficient in these ways is not known.) The principal activity *in vitro* is on gapped DNA in which the gap is less than 100 nucleotides long. Such a gap is akin to the state of the DNA at a replication fork—that is, the parental strands are separated and bear short single-stranded regions ahead of the growing daughter chain (Figure 8-2). Pol III cannot carry out strand displacement either, and another system is needed to unwind the helix in order that a replication fork will be able to proceed (this will be discussed in detail in a later section). The enzyme, like pol I, possesses a $3' \rightarrow 5'$ exonuclease activity which performs the major editing function in DNA replication. This function is carried out by the *dnaE* subunit, which is also the major polymerizing subunit, as we have just mentioned. The *dnaQ* subunit plays an important role in editing, also, but the biochemical basis of the role is not yet known; it probably interacts with the *dnaE* subunit. The principal evidence supporting the view that it is involved in providing fidelity to the replication process is that bacteria containing a mutation in the *dnaQ* gene have a somewhat higher mutation frequency. Pol III also possesses a $5' \rightarrow 3'$ exonuclease activity; however, the enzyme acts only on single-stranded DNA so that it cannot carry out nick translation. The biological role of the $5' \rightarrow 3'$ exonuclease activity of pol III is unknown at present.

Although pol III holoenzyme is the major replicating enzyme in *E.*

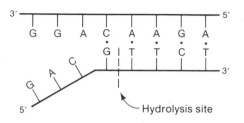

Figure 8-20
The endonuclease activity of polymerase I.
Base-pairing does not occur between the
bases shown in red.

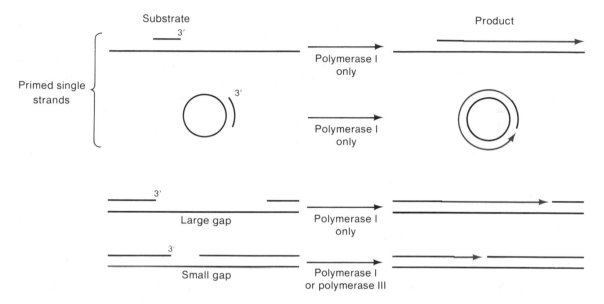

Figure 8-21
The activities of polymerase I and polymerase III holoenzyme on four substrates, showing that polymerase III
acts only on small gaps. Newly synthesized DNA is
shown in red. Note that, in the product, the growing
daughter strand is never covalently joined to the single-
stranded DNA ahead of the red arrowhead.

coli, much less is known about it than about pol I, because it is a more
complex enzyme and somewhat unstable when isolated, and it has been
discovered more recently. Study of pol III is currently an active field of
research.

Later in this chapter, when the events at the growing fork catalyzed
by the *E. coli* replication system are examined, it will be seen that pol I
and pol III holoenzyme are both essential for *E. coli* replication.*
However, a requirement for two polymerases is not common to all
organisms; for instance, *E. coli* phage T4 synthesizes its own polymerase
and this enzyme is capable of carrying out all necessary polymerization
functions.

*When mutants of *E. coli* lacking pol I were first isolated and found to be viable, it was concluded that
pol I is a nonessential enzyme. However, the explanation for the viability of such mutants is that all
known pol I mutants have a residual activity and make some functional enzyme.

The Direction of Chain Growth

All known polymerases (both DNA and RNA polymerases) are capable of chain growth in only the 5′→3′ direction; that is, the growing end of the polymer must have a free 3′-OH group. It is possible for the following reasons that the enzymes evolved in this way to facilitate editing. If 3′→5′ growth were to occur, the growing strand would also be terminated with a 5′-triphosphate and the 3′-OH group of the incoming nucleotide would react with it. Chemically this is certainly acceptable but since the bonds formed contain only a single phosphate, an editing function would leave a free 5′-monophosphate. In order for chain growth to proceed, an enzymatic system would be needed to enter the replication fork and convert the monophosphate to a triphosphate. There is already a great deal going on in the replication fork, so that it would seem more economical for the cell to require 5′→3′ growth exclusively. However, the observation that chain growth proceeds in only one direction introduces what is probably the greatest complication in the entire replication process; this will be described shortly.

DNA Ligase

Figure 8-21 indicates that neither replication from a primed circular single strand nor gap filling results in a continuous daughter strand. Discontinuity results because no known polymerase can join a 3′-OH and a 5′-monophosphate group. The joining of these groups is accomplished by the enzyme **DNA ligase,** which functions in replication and other important processes.

E. coli DNA ligase can join a 3′-OH group to a 5′-P group as long as both are termini of adjacent base-paired deoxynucleotides—the enzyme cannot bridge a gap (Figure 8-22).

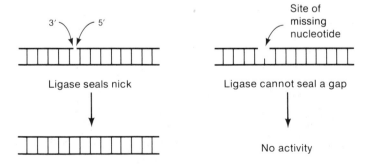

Figure 8-22
The action of DNA ligase. A nick having a 3′-OH and a 5′-P terminus is sealed (left panel). If one or more nucleotides are absent, the gap cannot be sealed.

In the usual polymerization reaction, the activation energy for phosphodiester bond formation comes from cleaving the triphosphate. Since DNA ligase has only a monophosphate to work with, it needs another source of energy. It obtains this energy by hydrolyzing either ATP or nicotine adenine dinucleotide (NAD); the energy source depends upon the organism from which the DNA ligase is obtained. The *E. coli* DNA ligase uses NAD.

DISCONTINUOUS REPLICATION

In the model of replication shown in Figure 8-2 both daughter strands are drawn as if replicating continuously. However, no known DNA molecule replicates in this way—instead,

One of the daughter strands is made in short fragments, which are then joined together.

The reason for this mechanism and the properties of these fragments are described in the following sections.

Fragments in the Replication Fork

As we have just seen, pol I and pol III can add nucleotides only to a 3'-OH group. Examination of the growing fork indicates that if both daughter strands grew in the same overall direction, only one of these strands would have a free 3'-OH group; the other strand would have a free 5'-P group because the two strands of DNA are antiparallel (Figure 8-23). Thus one of the following must be true:

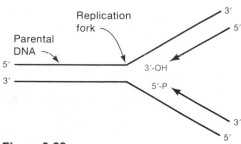

Figure 8-23
The termini (red) that would be present in a replication fork if both strands were to grow in the same overall direction.

1. There is another polymerase that can add a nucleotide to the 5′ end; that is, it would catalyze strand growth in the 3′→5′ direction. If the 5′ end were a monophosphate, this polymerase would probably use monophosphates as DNA precursors and would need an energy-supplying cofactor such as ATP. Alternatively, the 5′ end could be a triphosphate, as discussed in the preceding section, and the hypothetical polymerase would add a triphosphate.

2. The two strands both grow in the 5′→3′ direction but from opposite ends of the parental molecule, as shown in Figure 8-24. If this were correct, a significant fraction of the unreplicated molecule would have to be single-stranded.

3. The two strands both grow in the 5′→3′ direction at a single growing fork and hence do not grow in the same direction along the parental molecule. Two simple ways to accomplish this are shown in Figure 8-25. Note that both models predict that some newly made DNA consists of fragments. The model in panel (b) also states that some of the newly made DNA should spontaneously renature after denaturation. That is, if the DNA in the fork has not been cut at the fork, and if it were freed from the parental strand by denaturation at high temperature, upon cooling it would spontaneously form a double-stranded hairpin because the daughter strand contains complementary base sequences. In lab jargon, it is said that such a molecule is nondenaturable.

Many experiments showed that (i) there are no triphosphates in DNA, (ii) only a very small fraction (about 0.05 percent) of bacterial DNA is single-stranded at any time, and (iii) that newly synthesized DNA is completely denaturable. Furthermore, no polymerase that catalyzes 3′→5′ growth has ever been detected. Thus, if the proposals made are mutually exclusive, a model such as that in Figure 8-25(a) must be correct (and indeed it is). We call the fragments defined by the model **precursor fragments.** They are also often called nascent DNA.

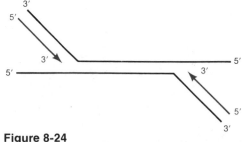

Figure 8-24
One way to replicate an antiparallel DNA molecule by means of 5′→3′ chain growth. Daughter strands are shown in red.

Detection of Fragments

In 1968, Reiji Okazaki demonstrated that in *E. coli* newly synthesized DNA consists of fragments that later attach to one another to generate continuous strands. The presence of fragments supported the discontinuous replication model of Figure 8-25(a). Okazaki did two experiments to demonstrate this. In the first experiment, [³H]dT was added to a growing bacterial culture in order to label the new DNA strands with radioactivity, and 30 seconds later the cells were collected and all of their DNA was isolated. This is called a **pulse-labeling experiment.** The DNA was then sedimented in alkali, which causes strand separation. The type of data obtained (Figure 8-26(a)) showed that the most recently made ("pulse-labeled") DNA sediments very slowly in comparison with the single strands obtained from the parental DNA (even though these strands are usually broken in the course of isolation); from the s value it was estimated that the fragments of pulse-labeled DNA range in size from 1000 to 2000 nucleotides, whereas the isolated parental DNA is usually 20 to 50 times as large. In the second experiment, the bacteria were pulse-labeled for 30 seconds; then, the [³H]dT was replaced with nonradioactive dT, and the bacteria were allowed to grow for several minutes. This is called a **pulse-chase experiment** and it allows one to examine the current state of molecules synthesized at an earlier time. Okazaki observed that the s value of the radioactive material increased with time of growth in the nonradioactive medium. These experiments are presented and interpreted in Figure 8-26 in terms of the discontinuous replication model shown in Figure 8-25(a). Apparently the joining of the pulse-labeled fragments to the growing daughter strands caused the label to sediment with the bulk of the DNA.

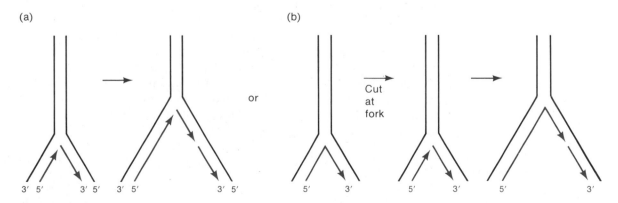

Figure 8-25
Two ways to accommodate 5'→3' polymerization with a single overall direction of movement of the fork. (a) The discontinuous mode and (b) the "knife and fork" mode. Daughter strands are shown in red.

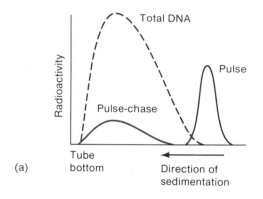

Figure 8-26
(a) The type of data obtained by alkaline sedimentation of pulse-labeled DNA (black) and pulse-chased DNA (red). Total DNA is the sum of the nonradioactive and radioactive DNA, as might be indicated by the optical absorbance. The *s* value of the sedimenting material increases from right to left.
(b) The location of radioactive DNA (red) at the time of pulse-labeling and after a chase. The radioactive molecules present in alkali are shown. The fragmentation resulting from removal of uracil (see later in text) accounts for the fact that all pulse-labeled DNA has a low *s* before the chase.

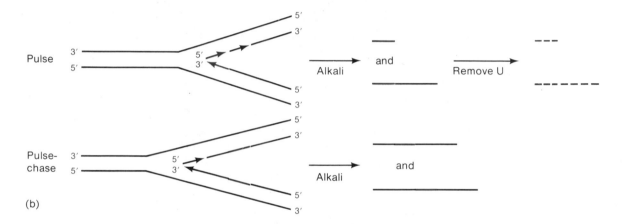

These fragments, which are widely known as **Okazaki fragments,** have the properties predicted by the discontinuous model; that is, they are initially small and then become large as they are attached to previously made DNA. However, the model predicts that only half of the radioactivity should be found in small fragments whereas the data shown in Figure 8-26(a) indicate that all of the newly synthesized DNA consists of fragments. This result should be surprising because there is no reason why the DNA of the 3'-OH-terminated strand should be synthesized discontinuously. The answer is that this strand is made continuously and is fragmented after synthesis. Why this occurs is explained in the following section.

Uracil Fragments

Earlier it was explained that both TTP and dUTP are present in cells and that pol III (and pol I, as well) cannot distinguish dUTP from TTP

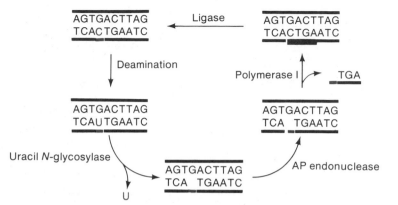

Figure 8-27
Scheme for repair of cytosine deamination. The same mechanism could remove a uracil that is accidentally incorporated.

because both are equally able to form hydrogen bonds with adenine. Thus, some dUTP is incorporated into DNA to make A · U pairs. In order to minimize this incorporation, E. coli possesses an enzyme, **dUTPase,** that converts dUTP to dUMP, which can no longer be incorporated into DNA. However, the action of dUTPase is not perfectly efficient; some dUTP survives and is incorporated.

In Chapter 7 the deamination of cytosine to uracil was described. After one round of replication such a deamination would lead to replacement of a G · C pair by an A · U pair, which would become an A · T pair after another round of replication. Since this would be mutagenic, cells have evolved a mechanism for replacing the unwanted U by a C. The first step of this repair cycle (Figure 8-27) is removal of the uracil by the enzyme **uracil N-glycosylase.** This enzyme cleaves the N-glycosylic bond and leaves the deoxyribose in the backbone. A second enzyme, **AP endonuclease,*** makes a single cut, freeing one end of the deoxyribose. This is followed by removal of the deoxyribose (probably by a second AP endonuclease that acts on the other side of the apurinic site), after which pol I fills the gap with the correct nucleotides. (This sequence, endonuclease—enlargement—polymerase, is an example of a general repair mechanism called excision-repair, which will be described in the next chapter.)

Uracil N-glycosylase cannot distinguish between a uracil arising by deamination and a uracil resulting from incorporation of dUTP into DNA. Thus, whenever a dUTP molecule is incorporated, uracil N-glycosylase goes to work. The steps of repair following the activity of this enzyme are fairly slow; hence, after removal of uracil, the phosphodiester bond can be hydrolyzed in alkali (see Chapter 4). Thus, all newly synthesized DNA will appear to be fragmented. The question then is

---

*AP stands for apurinic acid, a polynucleotide from which purines have been removed by hydrolysis of the N-glycosylic bonds.

whether any of the fragments observed in Figure 8-26(a) are true precursor fragments arising as suggested in Figure 8-25(a). The answer comes from studies with two bacterial mutants, one lacking dUTPase (dut^-) and the other lacking uracil-N-glycosylase (ung^-). The relevant experiments show the following:

1. In a dut^- mutant, the Okazaki fragments are smaller than in a dut^+ mutant because dUTP is not hydrolyzed; there is more uracil in the DNA and fragments result when uracils are excised.

2. In an ung^- mutant, roughly half of the newly made DNA consists of fragments, because this mutant lacks the ability to excise uracil, so no fragments are generated. The size of these ung-independent fragments is about the same as that of the ung-dependent ones so that there is about 1 incorporated dUTP per 1000 nucleotides or about 1 uracil per 250 thymines in newly made DNA.

3. An $ung^- dut^-$ double mutant behaves like an ung^- mutant.

These results argue that about half of the Okazaki fragments are true precursor fragments and that the discontinuous-synthesis model in Figure 8-25(a) is probably correct. Further support for this model comes from high-resolution electron micrographs of replicating DNA molecules, showing a short single-stranded region on one side of the replication fork (Figure 8-28). This region results from the fact that synthesis of the discontinuous strand is initiated only periodically; perhaps a particular base sequence or some other signal is required for initiation. In fact, the 3'-OH terminus of the continuously replicating strand is always ahead of the discontinuous strand. This had led to the use of the convenient terms, **leading strand** and **lagging strand,** for the continuously and discontinuously replicating strands, respectively (Figure 8-29).

The experiments just described have been carried out with *E. coli.* (The electron micrograph is of phage λ DNA.) Discontinuous replication is apparently a universal phenomenon, as it has been observed with numerous bacterial species, bacteriophages, animal viruses, and eukaryotes. In eukaryotes, the fragments are only about 100–200 bases long (1000–2000 for bacteria); however, whether these smaller fragments represent the actual size of the precursor fragment or simply a higher level of incorporation of dUTP is not known since ung^- mutants of eukaryotes have not yet been isolated.

In the next sections we examine how synthesis of a precursor fragment is initiated. In order to do so, we must first consider the general question of initiation and priming of replication. This will also put us in a position to understand (1) how fragments are attached to one another, and (2) the role of pol I in replication of *E. coli* DNA.

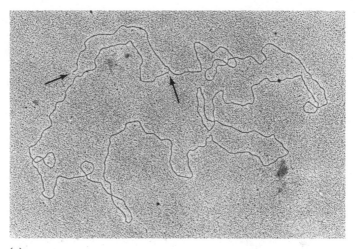

(a)

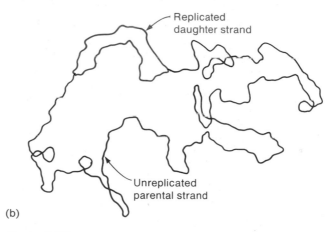

Replicated
daughter strand

Unreplicated
parental strand

(b)

Figure 8-28
(a) A replicating θ molecule of phage λ DNA. The arrows show the
two replicating forks. The segment between each pair of thick lines
at the arrows is single-stranded DNA; note that it appears thinner and
lighter. (b) An interpretive drawing. (Courtesy of Manuel Valenzuela.)

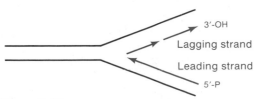

3'-OH

Lagging strand

Leading strand

5'-P

Figure 8-29
A growing fork showing the direction of growth
of the leading and lagging strands (both in red).

The RNA Terminus of Precursor Fragments

Pol III cannot lay down the first nucleotide to initiate chain growth but requires a primer. There are two ways that might be envisioned to produce a primer: (1) A pool of oligonucleotide fragments might exist that, by binding to parental strands, can provide a free 3'-OH end which pol III can extend. (2) A polymerizing enzyme, distinct from pol III, synthesizes a primer oligonucleotide, which is then extended by pol III; this is the correct explanation.

In *E. coli* initiation of synthesis of the leading strand and of the precursor fragments of the lagging strand occurs by distinct mechanisms, possibly because initiation of leading strand synthesis requires priming of double-stranded DNA whereas, in initiation of synthesis of a precursor fragment, single-stranded DNA is primed (that is, the strand to be copied is already unwound). Analysis of the mechanism of priming will be examined shortly. At this point, the most important feature of priming is that in every case examined so far, the primer for both leading and lagging strand synthesis is a short RNA oligonucleotide that consists of 1 to 60 bases; the exact number depends on the particular organism. This RNA primer is synthesized by copying a particular base sequence from one DNA strand and differs from a typical RNA molecule in that after its synthesis *the primer remains hydrogen-bonded to the DNA template.* In bacteria two different enzymes are known that synthesize primer RNA molecules—**RNA polymerase,** which is the same enzyme that is used for synthesis of most RNA molecules such as messenger RNA, and **primase** (the product of the *dnaG* gene). Experimentally, these enzymes can be distinguished *in vivo* by their differential sensitivities to the antibiotic rifampicin—RNA polymerase, but not primase, is inhibited by the antibiotic. In *E. coli,* initiation of leading strand synthesis is rifampicin-sensitive, presumably because RNA polymerase is used; however, initiation of precursor fragment synthesis is resistant to the drug, as it uses primase. In all cases, the growing end of the RNA primer is a 3'-OH group to which pol III can easily add the first deoxynucleotide; the 5'-P end of the RNA chain, which remains free, has a 5'-triphosphate group. Thus, a precursor fragment has the following structure, while it is being synthesized:

PPP-5' ━━━━━━━━━━━━━━━━━━━━━━━━━ 3'-OH
RNA   DNA

The Joining of Precursor Fragments

The precursor fragments are ultimately joined to yield a continuous strand. This strand contains no ribonucleotides so that assembly of the lagging strand must require removal of the primer ribonucleotides,

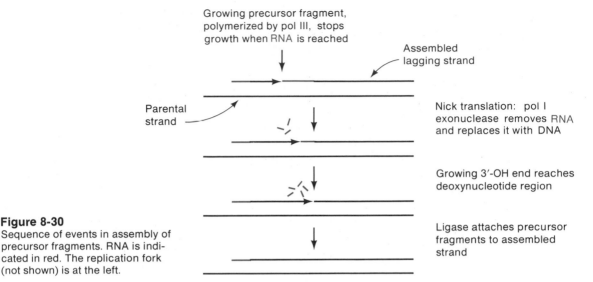

Growing precursor fragment, polymerized by pol III, stops growth when RNA is reached

Assembled lagging strand

Parental strand

Nick translation: pol I exonuclease removes RNA and replaces it with DNA

Growing 3′-OH end reaches deoxynucleotide region

Ligase attaches precursor fragments to assembled strand

Figure 8-30
Sequence of events in assembly of precursor fragments. RNA is indicated in red. The replication fork (not shown) is at the left.

replacement with deoxynucleotides, and then joining. In *E. coli* the first two processes are accomplished by DNA pol I and joining is catalyzed by DNA ligase. How this is done is shown in Figure 8-30. Pol III extends the growing strand until the RNA nucleotide of the primer of the previously synthesized precursor fragment is reached. Pol III can go no further since its 5′→3′ exonuclease is inactive on base-paired DNA; it cannot join a 5′-triphosphate at the terminus of a polymer (i.e., on the primer) to a 3′-OH group on the growing strand and it cannot carry out strand displacement. Thus pol III dissociates from the DNA, leaving a nick. *E. coli* DNA ligase cannot seal the nick because a triphosphate is present; even if an additional enzyme could cleave the triphosphate to a monophosphate, DNA ligase would be inactive when one of the nucleotides is in the ribo form. However, pol I works efficiently at a nick as long as there is a 3′-OH terminus. In this case the enzyme carries out nick translation, probably proceeding into the deoxy section, but there DNA ligase can compete with pol I and seal the nick. Thus, the precursor fragment is assimilated into the lagging strand. By this time, the next precursor fragment has reached the RNA primer of the fragment just joined and the sequence begins anew.

It seems as if everything might be simpler if the primer were made by a special DNA polymerase that could initiate strand growth. Hence it is interesting to ask why the primer is a ribonucleotide and not simply DNA. (In fact, *in vitro* it is possible to adjust experimental conditions so that primase polymerizes deoxynucleotides.) The answer to this question is not known with certainty; however, when considering a "why" question, it is important to think about the problems faced by a system and the consequences of an alternative mechanism. We have seen that great effort goes into maintaining high fidelity of replication. It seems likely that if an error is made in the first few bases of a primer, an editing function that detects unpaired bases could not correct it; this is because stacking provides little stabilization of the double-stranded structure formed by a very short oligonucleotide, so hydrogen-bonding is very weak. Thus, the error frequency probably cannot be reduced in the initial portion of the primer by editing. It may also be the case that, for a variety of reasons, priming is an error-prone process. If there is no good way to correct these errors, an effective solution is to discard the primer and recopy the DNA segment by a minimal-error process. This is, of course, exactly what is done by the $5' \rightarrow 3'$ exonuclease of pol I. Thus, the advantage of using RNA as a primer is that it is easily recognized by pol III (it is chemically different from DNA) as a signal to stop synthesis so that pol I can nick-translate; nick-translation is, of course, a minimal-error process because the $3' \rightarrow 5'$ editing function remains active. The reader should bear in mind that this argument is hypothetical; it does, however, seem reasonable.

Synthesis of precursor fragments follows synthesis of the leading strand. In the next section, how the leading strand advances into the parental double helix is described. Once this is understood, we can return to the question of how RNA primer synthesis gets started—a surprisingly complicated process.

EVENTS IN THE REPLICATION FORK

DNA replication requires not only an enzymatic mechanism for adding nucleotides to the growing chains but also a means of unwinding the parental double helix. In this section we will see that these are distinct processes and that the unwinding mechanism is closely related to the initiation of synthesis of precursor fragments.

Advance of the Replication Fork and the Unwinding of the Helix

The pol III holoenzyme cannot carry out strand displacement, as has already been discussed, because it is unable to unwind the helix. In order for a helix to be unwound, hydrogen bonds and hydrophobic interactions must be eliminated, and this requires energy. Pol I is able to utilize both the free energy of hydrolysis of the triphosphate group and the binding energy of forming a new hydrogen bond to unwind the parental molecule as it synthesizes the leading strand. No other polymerase can do that; usually a helix must first be unwound in order for a DNA polymerase to advance. Helix-unwinding is accomplished by enzymes called **helicases.** The helicase active in *E. coli* DNA replication is believed to be a protein or enzyme encoded by the *rep* gene and called the Rep protein, though this has not yet been proved. The Rep protein hydrolyzes ATP and utilizes the free energy of hydrolysis in an unknown way to unwind the helix, hydrolyzing two ATP molecules per base pair broken.

In *E. coli* the pol III holoenzyme synthesizing the leading strand is not immediately behind the advancing Rep protein (Figure 8-31). Thus, the Rep protein leaves in its wake two single-stranded regions, a large one on the strand to be copied by precursor fragments and a smaller one just ahead of the leading strand. In order to prevent the single-stranded regions from annealing or from forming intrastrand hydrogen bonds, the single-stranded DNA is coated with *E. coli* single-strand binding protein (**ssb protein**). This is one of a class of binding proteins that bind tightly to both single-stranded DNA and to one another. As pol III holoenzyme advances, it must displace the ssb protein so that base pairing of the nucleotide being added can occur. It is not known, at present, whether pol III holoenzyme carries out this displacement by itself or whether it is aided by the protein called the *dnaB* product. This protein also moves along a DNA molecule fueled by ATP hydrolysis. The DnaB protein will be described in more detail when initiation of synthesis of precursor fragments is discussed.

Some replication systems possess a single protein that is both a helicase and a ssb protein. This is true of *E. coli* phage T4, which synthesizes a protein called the gene-32 protein. The 32-protein, which is also described in Chapter 5, binds very tightly to single-stranded DNA and exceedingly tightly to itself. Its binding energy is great enough to unwind the helix. Phage T4 makes its own DNA polymerase and this enzyme is able to displace the 32-protein. Thus the events at the growing fork of T4 DNA are probably those shown in Figure 8-32.

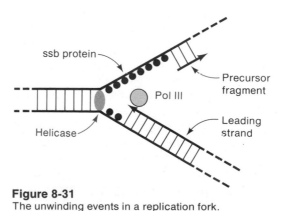

Figure 8-31
The unwinding events in a replication fork.

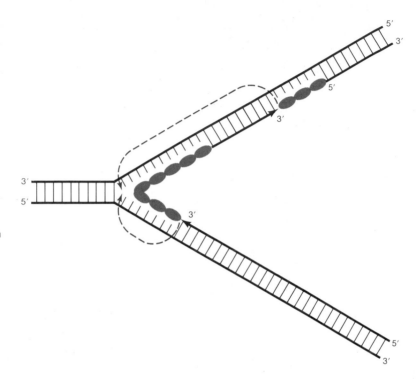

Figure 8-32
The action of the T4 gene-*32* protein (red ellipses) in unwinding T4 DNA and stabilizing the unwound regions. The protein is displaced by T4 DNA polymerase and probably binds again at the unwinding end of the protein aggregate. The red dashed arrows show that the protein is displaced by growth of the daughter strands and the positions at which proteins bind again in the replication fork.

The mechanism of initiation of replication of precursor fragments is closely connected to both the unwinding of the helicase and the action of the ssb protein. This is described in the following section.

Initiation of Replication from a Single-Stranded Template: Initiation of Precursor Fragments

A direct study of the mechanism of initiation of the synthesis of precursor fragments has not yet been carried out because experimentally the problem is very complex. Instead, attention has been directed to several *E. coli* phages having single-stranded DNA, in which the first step in replication is the conversion of single-stranded circular DNA to a double-stranded covalent circle. Since some of the phages studied use *E. coli* enzymes exclusively, it is thought that they can serve as a model for understanding initiation of precursor fragment synthesis.

The simplest mechanism of initiation seen so far occurs with the single-stranded circular DNA obtained from phage G4. This DNA molecule has a small palindromic segment which forms a double-stranded hairpin. This structure and part of the replication sequence are shown in Figure 8-33. The initial step, which is common to the

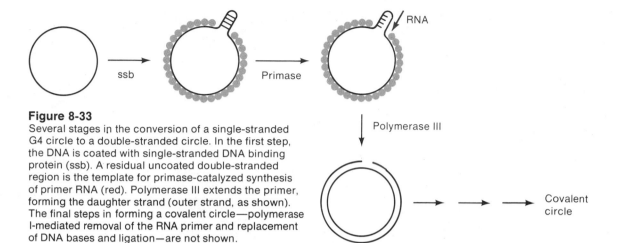

Figure 8-33
Several stages in the conversion of a single-stranded G4 circle to a double-stranded circle. In the first step, the DNA is coated with single-stranded DNA binding protein (ssb). A residual uncoated double-stranded region is the template for primase-catalyzed synthesis of primer RNA (red). Polymerase III extends the primer, forming the daughter strand (outer strand, as shown). The final steps in forming a covalent circle—polymerase I-mediated removal of the RNA primer and replacement of DNA bases and ligation—are not shown.

replication of all single-stranded molecules, is the coating of all single-stranded regions with ssb protein. The double-stranded region remains uncoated and is the site of binding of primase, which synthesizes a small segment of RNA. Pol III holoenzyme then extends this RNA primer and, displacing the ssb protein, proceeds to replicate around the circle. When the 5′-P terminus is reached, pol I replaces the pol III holoenzyme and removes the RNA, replacing it with DNA. Ligase seals the nick and replication is complete. Note that with G4 DNA, initiation begins at a unique sequence and that only ssb protein and dnaG primase are needed to synthesize the primer; thus, primer synthesis is rifampicin-resistant.

The single-stranded circular DNA of phage M13 is somewhat more complicated. It also has a hairpin loop that is not coated by ssb protein, and this loop is the priming region. However, initiation of primer synthesis requires several unidentified factors, and the RNA primer is synthesized by *E. coli* RNA polymerase rather than primase. Thus, primer synthesis is sensitive to rifampicin.

The most complicated system studied so far and the one that is most closely related to *E. coli* DNA synthesis is that of the circular single-stranded DNA of phage φX174. This DNA molecule does not have a hairpin loop. The primer is synthesized by primase (thus, synthesis is rifampicin-resistant). The first step in primer synthesis is the formation of a complex called a **preprimosome,** which contains six proteins: the n, n′, n″, i, DnaB, and DnaC proteins. The protein n′ binds to single-stranded DNA and acquires a bound ATP molecule, and then primase joins the preprimosome, forming a unit called the **primosome.** The protein n′ uses the energy of ATP hydrolysis to move the primosome along the DNA until a priming site, which is chosen at random, is found

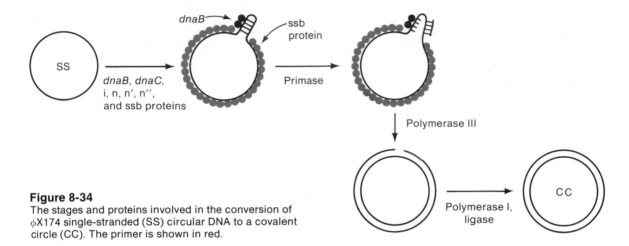

Figure 8-34
The stages and proteins involved in the conversion of
φX174 single-stranded (SS) circular DNA to a covalent
circle (CC). The primer is shown in red.

(Figure 8-34). Although the details are unclear, the DnaB protein alters the structure of the DNA at the site that has been selected, and this alteration enables primase to initiate synthesis of an RNA primer. Replication then proceeds as with G4 DNA—namely, by pol III holoenzyme, pol I, and ligase, in sequence. The important characteristics of the prepriming reaction are: the lack of a specific sequence at which initiation occurs; the use of a complex of six proteins; the requirement for ATP; and synthesis of RNA by primase.

A large number of single-stranded DNA molecules have now been studied and all are replicated by one of the three modes just described. It is thought that these may be the only modes, at least, in *E. coli* systems. The initiation of synthesis of precursor fragments in *E. coli* is known from *in vivo* studies to require primase, DnaB and DnaC proteins, and ssb protein. No information is available yet about the other proteins, though it is hypothesized that the initiation of synthesis of precursor fragments uses the same mechanism as described for φX174. Since each precursor fragment must be initiated, it is assumed that the primosome moves along the DNA following the motion of the replication fork, allowing primase to synthesize primers repeatedly at various intervals. In the following section this hypothesis and other information are combined to generate a reasonable picture of the events at a replication fork.

A Summary of Events at the Replication Fork

Figure 8-35 summarizes the proposed events that occur in or near a replication fork in *E. coli*. A helicase, driven by ATP hydrolysis and probably aided by binding of ssb protein, unwinds the helix. Tempera-

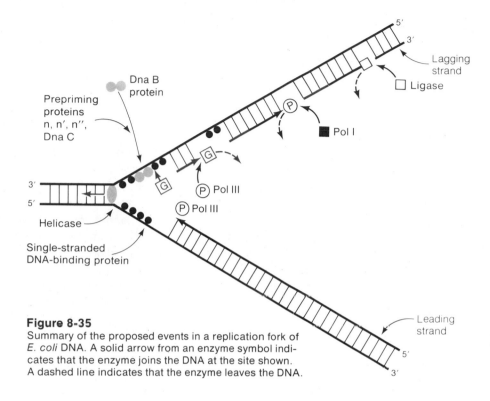

Figure 8-35
Summary of the proposed events in a replication fork of
E. coli DNA. A solid arrow from an enzyme symbol indi-
cates that the enzyme joins the DNA at the site shown.
A dashed line indicates that the enzyme leaves the DNA.

ture-sensitive mutants of the DnaB protein fail to synthesize any DNA at
all at restrictive temperatures, so that the DnaB protein may also
participate in some way in the unwinding process. The unpaired bases
are coated with ssb protein. The leading strand advances along one
parental strand by nucleotide addition catalyzed by the pol III holoenzyme. The DnaB protein complex moves along the other parental
strand, prepriming it so that primase will synthesize a primer RNA. Pol
III holoenzyme adds nucleotides to the primer, thereby synthesizing a
precursor fragment. This synthesis continues up to the primer of the
preceding precursor fragment; at this point pol I replaces pol III
holoenzyme; by nick-translation the RNA is removed and replaced by
DNA. Once the RNA is gone, DNA ligase seals the nick, thereby joining
the precursor fragment to the lagging strand.

The advance of pol III holoenzyme is continually delayed by $3' \rightarrow 5'$
exonuclease editing. Furthermore, uracil residues which are accidentally incorporated into DNA are removed by uracil N-glycosylase. The
deoxyribose is removed by AP endonuclease, the gap is (often) enlarged
and then filled by pol I, and then ligase makes the final seal.

Table 8-1
Replication genes and proteins either isolated from *E. coli* or characterized
by the existence of mutants

Gene	Function or gene product
dnaA	Initiation of a new round of replication in *E. coli.*
dnaB	Chain elongation. Component of primosome.
dnaC	Initiation of a new round of replication and of precursor fragments; binds to *dnaB* protein. Component of primosome.
dnaD	Same as *dnaC* gene; designation no longer in use.
dnaE	Polymerizing and editing activity of poly III.
dnaF	Ribonucleotide reductase.
dnaG	Primase for precursor fragments.
dnaH	Similar to *dnaA*; poorly characterized.
dnaI	Chain elongation; details unknown.
dnaJ	Initiation; details unknown.
dnaK	Initiation; details unknown.
dnaN	Chain elongation; subunit of pol III holoenzyme.
dnaP	Initiation; details unknown; subunit of pol III holoenzyme.
dnaQ	Fidelity; subunit of pol III holoenzyme.
dnaX	Chain elongation; subunit of pol III holoenzyme.
dnaZ	Chain elongation; subunit of pol III holoenzyme.
polA	Polymerase I. Chain elongation. Its $5' \rightarrow 3'$ exonuclease removes primer RNA. Required to join a precursor fragment to the lagging strand.
gyrA	α subunit of DNA gyrase, which introduces negative superhelical twists into nonsupercoiled DNA and removes positive twists in the unreplicated portion of replicating DNA. Inactivated by nalidixic acid.
gyrB	β subunit of DNA gyrase (see above). Inactivated by coumermycin.
lig	DNA ligase. Seals nicks; required to join a precursor fragment to the lagging strand.
dut	dUTPase.
rep	Helicase. Presumably unwinds DNA at the replicating fork.
ung	Uracil *N*-glycosylase; removes uracil from DNA.
rpoA,B,C,D	Subunits of RNA polymerase; synthesizes primer in some systems.
ssb	Single-stranded DNA-binding protein. Required for advance of the replication fork, priming, and all chain growth. Prevents formation of inter- and intrastrand hydrogen bonds.
Not defined	Proteins i, n, n′, n″. Prepriming of ϕX174 DNA synthesis and presumably of precursor fragments. Components of the primosome.

The advance of the replication fork continues until replication is completed. How termination is accomplished is described in a later section.

The reader must surely be impressed with the complexity of this process. However, in some systems replication is somewhat simpler. For instance, with E. coli phage T7, which is discussed in detail in Chapter 15, priming is accomplished by a phage RNA polymerase and all replication is performed by a single DNA polymerase encoded in the phage DNA; none of the known E. coli enzymes are needed. Phages T4 and λ replicate in a somewhat more complicated manner than T7 but still the mechanism is far simpler than E. coli replication. Why should the process in E. coli be so complex? This may be a consequence of the requirement for fidelity. The larger the DNA molecule, the greater is the probability that a replication error will be made during a round of replication. Furthermore, a bacterial DNA is replicated only once per generation; a phage replicates many times in an infected cell, so that one defective replica would not significantly affect the successful outcome of the infection. Thus, in order to reduce the error frequency to a tolerable level, organisms with a larger DNA molecule and a single replication event need more replication proteins, since each protein can be designed to minimize or correct a particular type of error.

So far, a large number of E. coli replication genes have been mentioned. These genes and their properties, when known, are listed in Table 8-1.

We now leave our analysis of the replication fork and examine how the fork itself is created at the outset; what is the process by which synthesis of double-stranded DNA is initiated? This is one of the least understood subjects of replication especially because it may occur in several ways; it is described in the following section.

INITIATION OF SYNTHESIS OF THE LEADING STRAND

All known double-stranded DNA molecules initiate a round of replication at a unique base sequence, called the **replication origin** or **ori.** The sequence is specific to each organism, though there is one example in which two organisms have the same sequence.

Initiation can occur in two ways—**de novo initiation,** in which the leading strand is started afresh, and **covalent extension,** in which the leading strand is covalently attached to a parental strand. In this section we discuss de novo initiation; covalent extension is discussed in a later section in which rolling circle replication is described. A particular topological configuration associated with de novo initiation is also described in this section.

De Novo Initiation in E. coli

De novo initiation remains poorly understood. The single feature common to all bacterial and phage systems examined to date is the requirement for an RNA primer synthesized by an RNA polymerizing enzyme. Additional proteins are also needed. For instance, initiation of *E. coli* DNA synthesis requires the products of the genes *dnaA, dnaH, dnaJ, dnaK,* and *dnaP* (Table 8-1). The biochemical activity of these gene products is, however, not known at present.

A feature that has been observed in several phage systems is the requirement for a phage-encoded protein that makes a nick in one strand at the origin. Two examples are the φX174 gene-*A* product and the gene-*O* and gene-*P* proteins of phage λ. How the nicking is connected to the initiation event is not known. In one system studied so far, a ribonuclease, RNase H, active on RNA that is hydrogen-bonded to DNA, is required. It is possible that this nuclease activity terminates primer synthesis and allows a switch from RNA synthesis to chain elongation by pol III holoenzyme.

All known DNA molecules, with only few exceptions, replicate as circles and hence initiate within the helix. Even those molecules that replicate as linear molecules initiate within the helix rather than at one end. Thus, because of the requirement for 5′→3′ synthesis, initiation of all DNA molecules creates a replication "bubble" consisting of one single-stranded and one double-stranded branch. This is called a D loop and its properties will be examined in the next section.

D Loops

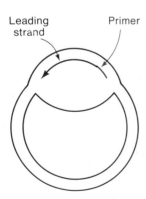

Leading strand | Primer

Figure 8-36
A circular DNA molecule with a D loop.

In *de novo* initiation, synthesis of the leading strand precedes that of the lagging strand. Thus, before synthesis of the first precursor fragment begins, a replication bubble exists that consists of one double-stranded branch, made up of one parental strand paired with the leading strand, and one single-stranded branch, which is the unreplicated, second parental strand (Figure 8-36). Since the leading strand displaces the unreplicated parental strand, the bubble is called a **displacement loop** or **D loop.** Such a configuration is ordinarily a transient one, existing only until synthesis of precursor fragments begins—which may require that the leading strand release a specific sequence (in single-stranded form) that can be used for prepriming. However, in certain circumstances, namely in a replication system that does not employ DNA gyrase to relieve topological constraints, a D loop may be long lived.

In the initial stages of replication of a naturally occurring circular DNA molecule, advance of the replication fork does not require the presence of DNA gyrase, because such a circular DNA molecule initially is negatively supercoiled and the negative twists compensate for the positive turns introduced by movement of the fork. However, once the fork has advanced sufficiently that the negative twists are used up, a topological constraint-relieving system such as gyrase is needed. The superhelix density of all naturally occurring DNA molecules is 0.05 twists per turn of the helix. This means that, if positive superhelicity is forbidden, the leading strand could move roughly 5 percent of the distance along a circular, negatively superhelical molecule before gyrase would be needed. A replication complex cannot in fact move along a double helix if the movement would produce more than one or two turns of overwinding. Thus, if neither gyrase nor any other unwinding system were present, replication would cease when the unreplicated portion had lost its supercoiling. The molecular configuration of the bubble would depend upon whether the first precursor fragment had been initiated; if it had not, the structure in Figure 8-36 would result.

Such a structure is seen in animal cell mitochondrial DNA (Figure 8-37). The molecular weights of these molecules in most species is about 10^7 and the molecular weight of a precursor fragment is typically about 10^6, so the lack of double-stranded DNA in both branches is consistent with the fact that only 4 percent of the DNA (or 2×10^5 molecular weight units of the leading strand) is replicated. The fact that 70 percent of all replicating mitochondrial DNA molecules are in the D-loop configuration indicates that indeed replication has ceased at this point.

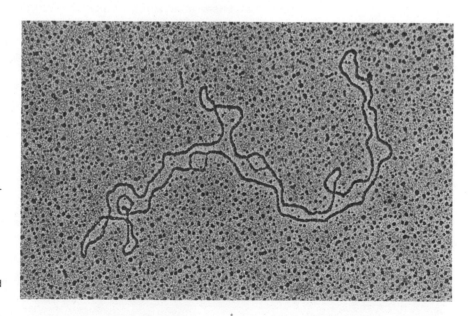

Figure 8-37
An electron micrograph of a dimer of mouse mitochondrial DNA showing diametrically opposing D loops. The single strand in each D loop appears thinner than the double-stranded DNA. The total length of the molecule is 10 μm. (Courtesy of David Clayton.)

Exactly what happens to allow replication to proceed is not known precisely. Mitochondrial DNA replication takes about one hour, which is about 100 times slower than bacterial DNA. After synthesis is complete, another 40 minutes elapse before the daughter molecules are supercoiled. These results are interpreted as follows. It is assumed that either a gyraselike enzyme, which is very inefficient, is present, or a simple nicking-sealing system is present. The system does not work continuously, so there are repeated pauses in replication to remove topological constraints; the system may only be active in response to slightly overwound DNA.

Loops with the appearance of D loops have also been seen in the DNA of E. coli phage λ. These loops may not be related to replication for three reasons: (1) they are rarely, if ever, at the replication origin; (2) often many D loops are found on one molecule—three to five is a common number and 35 loops were once seen on a single molecule; (3) the double-stranded branch, in some cases, appears to contain RNA. The significance of these structures is unknown.

Initiation by Covalent Extension: Rolling Circle Replication

There are numerous instances in which, in the course of replication, a circular phage DNA molecule gives rise to linear daughter molecules in which the base sequence of the DNA present in the phage is repeated numerous times, forming a concatemer (Figure 8-38). These concatemers are usually an essential intermediate in phage production. Likewise, in bacterial mating, a linear DNA molecule is transferred by a replicative process from a donor cell to a recipient cell. Both phenomena are consequences of initiation by covalent extension, an event that gives rise to a replication mode known as **rolling circle replication.**

Consider a duplex circle in which, by some initiation event, a nick is made having 3'-OH and 5'-P termini (Figure 8-39). Under the influence of a helicase and ssb protein a replication fork can be generated. Synthesis of a primer is unnecessary because of the 3'-OH group, so leading-strand synthesis can proceed by elongation from this terminus. At the same time, the parental template for lagging-strand synthesis is displaced. The polymerase used for this synthesis is apparently pol III

XYZABC   XYZABC   XYZABC   XYZABC

Figure 8-38
A concatemer consisting of the repeating unit ABC . . . XYZ. Note that the definition of concatemer does not make any requirements about the terminal sequences.

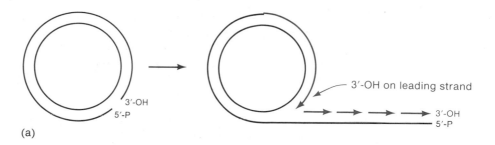

(a)

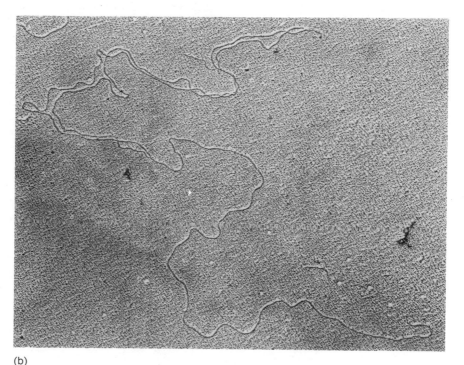

Figure 8-39
(a) Rolling circle or σ replication. Newly synthesized DNA is shown in red. (b) An electron micrograph of a rolling circle of phage λ DNA isolated from phage-infected *E. coli*. The length of the branch is 15.2 μm. (Courtesy of Marc Better.)

(b)

holoenzyme. (In this circumstance we can ignore the fact that the enzyme cannot ordinarily carry out a displacement reaction, because in this case, displacement is a result of a coupling of helicase, ssb protein, and pol III holoenzyme.) The displaced parental strand is replicated in the usual way by means of precursor fragments. The result of this mode of replication is a circle with a linear branch; it resembles the Greek letter *sigma* and is called σ replication or rolling circle replication.

There are four significant features of rolling circle replication:

1. The leading strand is covalently linked to the parental template for the lagging strand.
2. Before precursor fragment synthesis begins, the linear branch has a free 5'-P terminus.

3. Rolling circle replication continues unabated, generating a concatemeric branch.
4. The circular template for leading strand synthesis never leaves the circular part of the molecule.

A variant of the rolling circle mode, called **looped rolling-circle replication,** is used to generate a progeny single-stranded circle from a double-stranded circular template. For phage ϕX174 this is accomplished in the following way (Figure 8-40). A phage protein called the gene-*A* protein makes a nick at the origin and in the process becomes covalently linked to the newly formed 5′-P terminus. Using the Rep and ssb proteins and pol III holoenzyme, chain growth occurs from the 3′-OH group, displacing the broken parental strand (which we call the (+) strand). This strand becomes coated with ssb protein and is not replicated. Leading strand synthesis continues until the origin is reached. At this point, the *A*-gene product binds to the 3′-OH group of the (+) strand and, using the energy obtained from the original nicking event, it joins the 3′-OH and 5′-P groups of the (+) strand, dissociates, and attaches to the newly synthesized (+) strand. This process can continue indefinitely, generating numerous circular (+) strands. Note

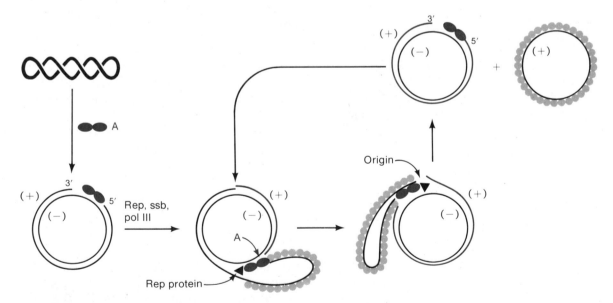

Figure 8-40

A diagram of looped rolling-circle replication of phage ϕX174. The gene-*A* protein nicks a supercoil and binds to the 5′ terminus of a strand (known as the + strand) whose base sequence is the same as that of the DNA in the phage particle. Rolling circle replication ensues to generate a daughter strand (red) and a displaced + single strand that is coated with ssb protein and still covalently linked to the A protein. When the entire + strand is displaced, it is cleaved from the daughter + strand and circularized by the joining activity of the A protein. The cycle is ready to begin anew. Note that the − strand is never cleaved.

that in looped rolling-circle replication, the displaced strand never exceeds the length of the circle, in contrast with ordinary σ replication.

It is interesting that the first hint of a rolling circle mechanism came from a genetic experiment. It was observed that in a very long mating between an *E. coli* Hfr donor cell and a recipient cell the sequence of genes transferred as a function of time was *ABC . . . XYZAB . . . .* That is, *A* and *B,* the first genes transferred, entered the recipient a second time, immediately after *Z*, the final gene, had been transferred. We will see many examples in this book of molecular phenomena uncovered by simple genetic experiments.

BIDIRECTIONAL REPLICATION

In this section we examine the events following a D-looplike initiation step in which chain growth occurs continuously. It will be seen that unless special rules are imposed, replication will be bidirectional—that is, there will be two replication forks.

Somewhat after initiation of synthesis of the leading strand at the origin, the first precursor fragment is synthesized. This is shown in Figure 8-41-I, in which the overall direction in which the replicating fork moves is counterclockwise. In our earlier analysis of lagging-strand replication, it was noted that synthesis of each precursor fragment is terminated when the growing end reaches the primer of the previously synthesized fragment. However, in the case of the first precursor fragment, there can be no previously made fragment. We may consider two possible events: (1) there is a termination signal for synthesis of the first fragment, or (2) there is no such signal and the precursor fragment continues to grow. In the second case, the precursor fragment is

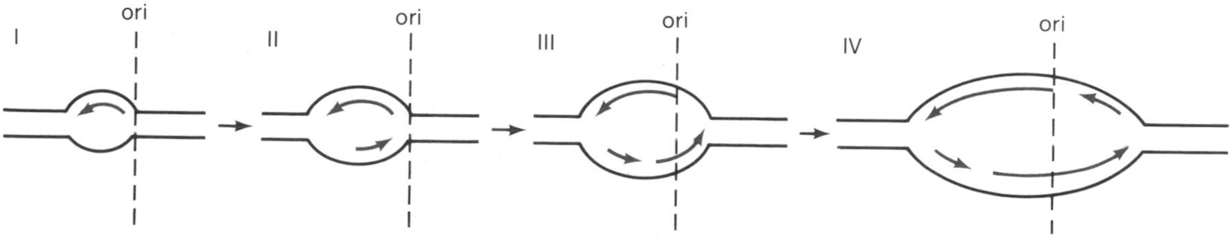

Figure 8-41
The formation of a bidirectionally enlarging replication bubble. I. The leftward-leading strand starts at *ori*. II. The leading strand has progressed far enough that the first rightward precursor fragment begins. III. The leftward-leading strand has progressed far enough that the second rightward precursor fragment has begun. The first rightward precursor fragment has passed *ori* and has become the rightward-leading strand. IV. The rightward-leading strand has moved far enough that the first leftward precursor fragment has begun. There are now two complete replication forks.

equivalent to a leading strand for a second replication fork, moving clockwise, as shown in the figure. Clockwise replication requires the synthesis of precursor fragments in the second replication fork but this can be achieved by the standard mechanism. The result of these events is that the DNA molecule will have two replication forks moving in opposite directions around the circle. This is called **bidirectional replication.** If alternative (1) occurs, there is a single fork and replication is unidirectional. There is no particular reason to have such a stop signal and, in fact, bidirectional replication has the advantage of halving the time required to complete the process. Bidirectional replication was first detected with E. coli phage λ and it has since been observed with many other phages, and in bacteria and plasmids as well as in animal cells. Unidirectional replication has also been observed but less frequently.

The method used to study the direction of replication was developed by Ross Inman and Maria Schnös and is called **denaturation mapping.** It is done as follows. If a DNA molecule is heated to a temperature at which melting is just detected, single-stranded bubbles form in the regions having very high A + T content. If formaldehyde is added and the DNA is cooled, the bubbles persist. DNA treated in this way can be observed by electron microscopy and the position of the bubbles can be noted (Figure 8-42). These positions are constant from one molecule to the next (Figure 8-43). When the technique is applied to a replicating molecule, the bubbles serve as fixed reference points against which the positions of a branch-point can be plotted. Figure 8-44 shows the kinds of molecules that would be expected for unidirectional and bidirectional replication of a hypothetical molecule. Note that (1) the relation between the positions of the bubbles and the branch points differ for the two modes and (2) a replication fork is identified by its changing distance from a bubble.

In unidirectional replication, one branch point remains at a fixed position with respect to the bubbles; this position defines the replication origin. In bidirectional replication both branch points move with respect to the bubbles so that each branch point is a replication fork. If both replication forks move at the same rate, the origin is always at the midpoint of each branch of the replication loop. With phage λ it is found that the midpoints of all replication loops (that is, loops of all sizes) are at the same position with respect to the denaturation bubbles; this indicates that the two growing forks move at the same rate and also locates the origin. It has also been observed that there is one single-stranded region in each replication fork at opposite ends of each double-stranded branch of the loop; this is expected of a replication fork since each fork has a lagging strand.

A small fraction of λ DNA molecules have been observed to replicate unidirectionally, some clockwise and some counterclockwise. The significance of this observation is unknown.

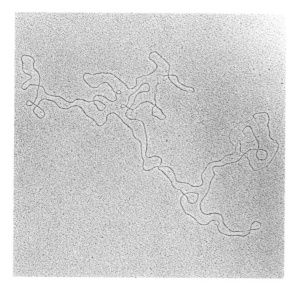

Figure 8-42
An electron micrograph of a partially denatured circular phage λ DNA molecule. With this method of sample preparation the single strands are very thin and faint compared to the double-stranded DNA. (Courtesy of Manuel Valenzuela.)

(a)   Map

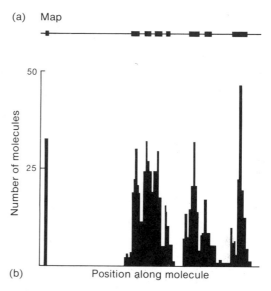

Position along molecule

Figure 8-43
A denaturation map of *E. coli* phage λ DNA showing (a) the size and positions of most denaturation bubbles and (b) the number of molecules whose strands are separated at the positions shown, observed in a single experiment.

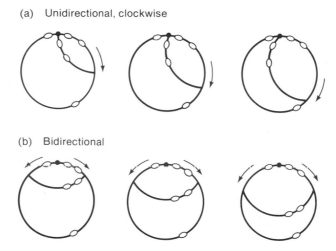

(a)   Unidirectional, clockwise

(b)   Bidirectional

Figure 8-44
A diagram showing the relative positions of branch points and a replication bubble for a DNA molecule replicating (a) unidirectionally and clockwise or (b) bidirectionally. The arrows indicate the direction of movement of the replication fork and the dots indicate the replication origin.

Bidirectional replication has been widely observed with phage, bacterial, and plasmid DNA. A small number of phages and plasmids use the unidirectional mode exclusively, so that a stop signal must exist that prevents chain growth of the first precursor fragment past the origin. One plasmid uses a surprising variant of the bidirectional mode in that initiation of movement of one fork occurs at a much later time than the primary initiation event.

Bidirectional replication is the major mode of chain growth with eukaryotes. However, since eukaryotic DNA is almost always fragmented in the course of isolation, a different technique was required for observing the replication forks of eukaryotic DNA. The method used, which employed autoradiography, is shown in Figure 8-45(a). A culture of living cells was pulse-labeled with [³H]thymidine, the DNA was isolated, and then autoradiograms were prepared. Except in the rare case of labeling at the time of initiation, a replication bubble will become radioactive only adjacent to the fork. Remembering that only radioactive DNA is visualized in an autoradiogram, we can expect that a unidirectionally replicating bubble will appear as a labeled V, whereas in bidirectional replication both forks are labeled and this will appear as two V's whose arms point at one another. Figure 8-45(b) shows an actual autoradiogram that demonstrated bidirectional replication in animal cells.

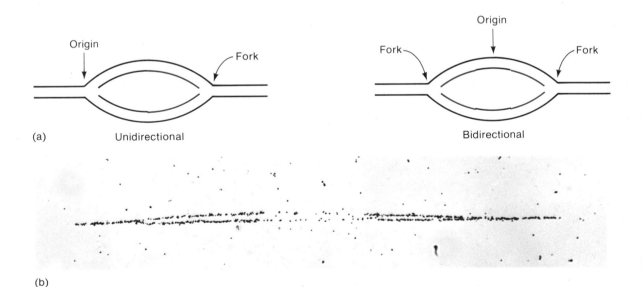

(a) Unidirectional Bidirectional

(b)

Figure 8-45
(a) Expected patterns of labeling for unidirectional and bidirectional replication. Radioactive DNA is shown in red. (b) Actual pattern of labeling, as seen in an autoradiogram. (From D. M. Prescott and P. Kuempel. 1973. *Meth. Cell Biol.,* 7: 147.) Note the very faint pattern of grains between the labeled V sections. This occurs because the cells had been grown for one generation in a medium having a very small amount of [³H]thymidine prior to pulse-labeling; this was done to allow the strands connecting the V's to be visible.

TERMINATION OF REPLICATION

We have discussed initiation of a round of replication by formation of a replication fork and the events that occur while the fork moves along a parental DNA molecule. A round of replication is completed when a molecule is totally replicated. The termination event(s) is not well understood and must differ in circular and linear molecules; these two cases are considered in the following sections.

Termination of Replication of a Circle

In a unidirectionally replicating molecule, replication terminates at the origin. In a bidirectionally replicating molecule this clearly is not the case. There are two possible modes of termination: (1) there is a defined termination sequence, or (2) two growing points collide and termination occurs wherever the collision point happens to be. In both cases, termination might occur exactly halfway around the circle (at the antipode of the origin). Both termination modes have been observed. In several plasmids it has been observed by denaturation mapping that one growing fork stops at a fixed position before the antipode is reached; the other fork advances until this termination site is reached. Why this first fork stops moving is not known. Possibly, a termination protein sits at this point or some base sequence signals dissociation of the replication machinery.

With *E. coli* phage λ, termination is by simple collision. To prove this, a λ variant having a large genetic deletion was studied. If there is a genetically determined termination site, which happens to be at the antipode of the wild-type phage DNA, then in the deletion mutant, this site would be moved nearer to the origin and one replication fork would stop before reaching the new antipode. However, both forks have been observed to continue on, indicating that there is no fixed termination site in λ DNA.

Termination has a topological problem. When double-stranded circular DNA replicates semiconservatively, the result is a pair of circles that are linked as in a chain. Such a structure is called a **catenane.** Catenated molecules have been observed in some systems, but there is no evidence to indicate that they result from replication. At any rate, ultimately two separated circles do result. The mechanism for this separation is not known, though it is possible, on paper at least, to use a topoisomerase such as gyrase to perform this separation. Little is known about this process.

Termination of the replication of a linear DNA molecule lacks this topological problem. However, a more serious problem is occasioned by termination, as will be seen in the next section.

Termination of Phage T7 DNA, a Linear DNA Molecule

E. coli T7 DNA replicates as a linear molecule. The origin is located 17 percent of the total distance from the left end of the molecule and replication is bidirectional (Figure 8-46). Initially there is a single replication bubble (molecule I), and when the leftward fork reaches the terminus, the molecules assume a Y form (molecule III). A problem in termination can be seen by examining the synthesis of the final precursor fragment in each fork (Figure 8-47). Assuming that priming could occur at each terminus, we can see that two linear molecules would result, each having a single terminal RNA segment (only one is shown in the figure). Ribonucleases exist that can remove this RNA but once it has been removed, either a 5'-OH or a 5'-P group (depending on the particular RNase) would remain at the ends of the molecule, and no known polymerase can act at these termini. The molecules are nevertheless completed and no RNA remains. How this is accomplished is not known, but the following hypothesis, which accounts for all of the facts, has been made.

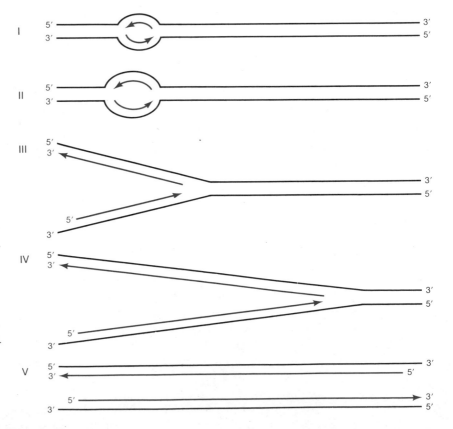

Figure 8-46
Origin and direction of replication of T7 DNA. Bidirectional replication starts at a point about 17 percent from the left end. Because the two forks move at equal rate, the left end is finished first. Note that the region at the 3' end of each template strand remains unreplicated. The daughter strands are shown in red. The arrows show the direction of chain growth but the precursor fragments are drawn as if already joined to the lagging strand.

T7 DNA has a stretch of several hundred nucleotides that is repeated at each end of the DNA molecule (Figure 8-48). This is called **terminal redundancy** and it means that unreplicated bases at the right and left ends of the daughter molecule are complementary to free single-stranded sequences at the opposite end of the molecule. It is hypothesized that these sequences on two separate molecules pair, and the excess single-stranded material is removed by a DNase, as shown in

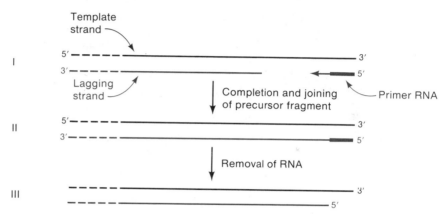

Figure 8-47
Events at the terminus of a replicating linear molecule. This is an expanded view of the upper molecule of row V in Figure 8-46. I. Synthesis of the final precursor fragment begins by extension from an RNA primer. II. The precursor fragment in I is joined to the lagging strand. III. RNA is removed, leaving a 5'-P-terminated gap that cannot be filled by any known polymerase.

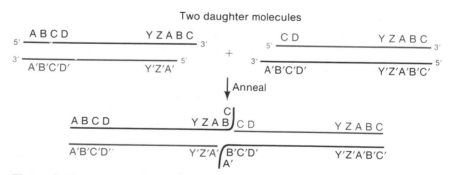

Figure 8-48
A hypothesis for the generation of concatemers from two incomplete daughter molecules shown in the lowermost section of Figure 8-47. Note that the color coding of the single strands is the same as in Figure 8-47. Annealing of the daughter molecules is made possible by the terminally redundant sequence ABC. The excess unmatched single strands bearing A' and C are presumably removed by a DNase. When the dimer replicates, two daughter molecules result that are terminally incomplete, as in Figure 8-47. These can anneal to form tetramers. This model was first proposed by J. D. Watson in 1972 (*Nature New Biol.*, 239: 197).

the figure. The result is a dimer called a **concatemer.** Each unit of the dimer contains a replication origin so that the dimer can also replicate. The result of this round of replication is a pair of daughter dimers, each of which has complementary single-stranded termini. Once again, reannealing and trimming can occur to yield a tetramer. This process can continue indefinitely to form a gigantic linear concatemer. At any time linear T7 DNA molecules can be cut out of this structure. Huge concatemers of T7 DNA have been isolated from phage-infected cells, which lends support to this hypothesis.

It may be the case that one reason DNA molecules evolved to be circular is to avoid termination problems.

METHYLATION OF DNA AND MISMATCH REPAIR

In Chapter 4 the existence of methylated bases, particularly cytosine and adenine, was mentioned. The methylation of adenine, which in *E. coli* is accomplished by the product of the *dam* gene, has been studied carefully. Two important observations are that the methylated adenine is normally contained in the sequence G-A-T-C, indicating that there is sequence specificity, and that there is a gradient of methylation along a newly synthesized daughter strand, the least methylation occurring at the replication fork. The parental strand is always uniformly methylated. Different methylation of parental and daughter strands during replication is important for the fidelity of replication.

As has already been described, both pol I and pol III occasionally catalyze incorporation of an "incorrect" base, which cannot form a hydrogen bond with the template base in the parental strand; such errors are usually corrected by the editing function of these enzymes. However, the integrity of the base sequence of DNA is so important that a second system exists for correcting the occasional error missed by the editing function. This correction system is called **mismatch repair.** In mismatch repair a pair of non-hydrogen-bonded bases is recognized as incorrect and a polynucleotide segment is excised from one strand, thereby removing one member of the unmatched pair. The resulting gap is filled in by pol I, which presumably uses this "second chance to get it right" to form only correct base pairs; then the final seal is made by DNA ligase.

If it is to correct but not create errors, the mismatch repair system must be able to distinguish the correct base in the parental strand from the incorrect base in the daughter strand. Studies with *dam*⁻ (methylation-defective) mutants gave a clue as to how this distinction is made. For any genetic locus the mutation frequency in a *dam*⁻ mutant is much higher than in a *dam*⁺ bacterium. This indicates that incorrectly incorporated bases are less frequently corrected in a *dam*⁻ mutant than in the wild type. The reason for this is that the mismatch repair system recognizes the degree of methylation of a strand and preferentially excises nucleotides from the undermethylated strand. The daughter strand is always the undermethylated strand, as its methylation lags somewhat behind the moving replication fork, whereas the parental strand is fully methylated, having been methylated in the previous round of replication.

The mismatch repair system is very important for another reason—namely, because certain unusual molecules called base analogues (see Chapter 10) can be incorporated into DNA without being recognized by the editing function. An example of this is a base having both an enol and a keto form in equilibrium and in which only one of these forms can hydrogen-bond with a standard DNA base. We can imagine a situation in which the enol-keto equilibrium lies far toward the keto form and only the enol form can base-pair with (let us say) a cytosine. Substitution of this base for a guanine would be very rare but nonetheless this base might be incorporated into the DNA during a brief period in which the base is in the enol form. Since in that form the odd base would nevertheless be able to pair, the editing function would not recognize it as incorrect. Afterwards, however, the base would tend to be predominately in the keto form so there would be a mismatched base pair most of the time. Since methylation lags somewhat behind the advancing fork, this incorrect base would be in an undermethylated strand and this would be excised and corrected by the mismatch repair system.

REGULATION OF BACTERIAL DNA SYNTHESIS AND REINITIATION OF A PARTIALLY REPLICATED DNA MOLECULE

In *E. coli* the time required for a single cell to double in size and divide depends upon the rate of production of useful energy and precursors. At 37° (if glucose is provided as the sole carbon source and all other nutrients are simple inorganic compounds—that is, if the cells are in a glucose-minimal medium) it takes about 45 minutes for a cell to replicate. If succinate is the sole carbon source, ATP is synthesized more slowly and the doubling time is 70 minutes. With even poorer carbon sources, the doubling time can be increased to ten hours. In a glucose medium, the time required to replicate the bacterial DNA is 40 minutes; that is, initiation is delayed by a few minutes following completion of a round of replication. Surprisingly, in succinate medium, the replication time is still 40 minutes, so the time between successive rounds of replication is 30 minutes. In a medium in which the doubling time is ten hours, the replication time is increased by only a few minutes. These facts indicate that the initiation of DNA replication must be regulated.

The regulation of DNA synthesis in bacteria is poorly understood but certainly utilizes numerous specific proteins and at least one type of RNA molecule. It has been proposed that the ratio of the total amount of protein and DNA may be important because inhibition of protein synthesis by amino acid starvation inhibits initiation of a new round of replication but allows synthesis of a partially replicated DNA molecule to go to completion. It is also possible that particular initiation proteins must accumulate to critical concentrations.

If *E. coli* is grown in a complex nutrient broth, the doubling time of the bacteria is reduced below 45 minutes; for example, in very rich media a 21-minute doubling time is observed. However, it has been shown that replication still takes 40 minutes—an apparent paradox.

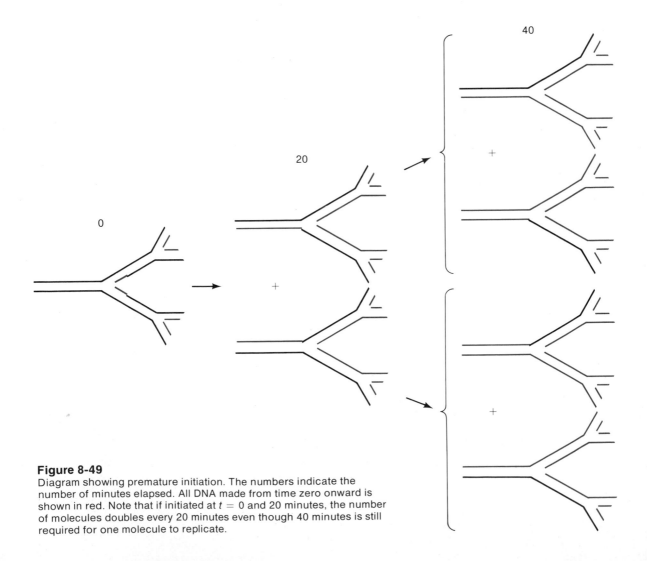

Figure 8-49
Diagram showing premature initiation. The numbers indicate the number of minutes elapsed. All DNA made from time zero onward is shown in red. Note that if initiated at *t* = 0 and 20 minutes, the number of molecules doubles every 20 minutes even though 40 minutes is still required for one molecule to replicate.

This observation is explained by the phenomenon of **premature initiation.** That is, in very rich media, following initiation and subsequent polymerization, a second initiation event occurs before replication is complete. Figure 8-49 shows how a second initiation event at the time replication is half complete allows segregation of two daughter molecules to occur at twice the normal rate. The mechanism underlying premature initiation is not known.

Premature initiation is not confined to bacteria. For instance, several phage species, which make replicas very rapidly, do so by premature initiation. This is usually detected by electron microscopy. All phages do not do this though; for instance, it is very rare with *E. coli* phage λ.

One of the most striking examples of regulation is seen when studying hybrid plasmid DNA molecules having two origins. These molecules are made in the laboratory with recombinant DNA techniques (Chapter 20) by joining two different circular molecules each having its own origin. The hybrid molecule is put into *E. coli* (by DNA transformation) and allowed to replicate. When replicating molecules are isolated, it is invariably found that replication has been initiated at only one, and often a particular, origin. The reasons for this selectivity are explained in Chapter 17.

REPLICATION OF EUKARYOTIC CHROMOSOMES

The replication of eukaryotic chromosomes presents many problems not found in the prokaryotes because of the enormous size of eukaryotic chromosomes and the geometric complexity imposed by the organization of the DNA into nucleosomes. How these problems are handled is described in this section.

Multiple Forks in Eukaryotic DNA

The rate of movement of a replication fork in *E. coli* is about 10^5 base pairs per minute. In eukaryotes the polymerases are much less active and the rate ranges from 500 to 5000 base pairs per minute. Since a typical animal cell contains about fifty times as much DNA as a bacterium, the replication time of an animal cell should be about 1000 times as great as that of *E. coli* or about 30 days. However, the duration of the replication cycle is usually several hours and this is accomplished by having multiple initiation sites. For instance, the DNA of the fruit fly *Drosophila* has about 5000 initiation sites, each separated by about 30,000 bases, and each site replicates bidirectionally. The number of sites is regulated in a way that is not understood. For example, in the round of replication following fertilization of *Drosophila* eggs, the

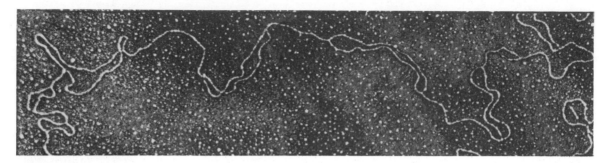

Figure 8-50
Replicating DNA of *Drosophila melanogaster* showing many replicating eyes. The molecular weight of the seg-ment shown is roughly 20×10^6. (Courtesy of David Hogness.)

number of initiation points reaches 50,000, and it takes only 3 minutes to replicate all of the DNA. An example of a fragment of this rapidly replicating *Drosophila* DNA is shown in Figure 8-50.

The enormous number of growing forks in eukaryotic cells is reflected in the number of polymerase molecules. In *E. coli* there are between 10 and 20 molecules of pol III holoenzyme. However, a typical animal cell has 20,000 to 60,000 molecules of polymerase α, which is believed to be the major polymerase.

Replication of Chromatin

The replication of double-stranded DNA proceeds through both a polymerization step (nucleotide addition) and a dissociation step (strand separation). The replication of chromatin, which is the form that DNA has in eukaryotes, proceeds through an additional dissociation step—namely, dissociation of DNA and histone octamers—and a histone-DNA reassociation step (see Chapter 6 for a discussion of the structure of the octamers contained in nucleosomes). In chromatin, DNA is wrapped around a histone octamer to form a nucleosome and if the DNA were never unwrapped from the histone spool, severe geometric problems would arise at the growing fork. Moreover, after DNA dissociates from the histones, newly formed DNA must rejoin with the nucleosomal octamers so that each daughter molecule will be organized into nucleosomes, just as the parent was. Clearly some mechanism is needed for regulating the histone-DNA dissociation and reassociation reactions.

Examination of replication forks in DNA that has not been deproteinized during isolation indicates that nucleosomes form very rapidly after replication. For example, Figure 8-51 shows that all portions of a replication eye have the beadlike appearance characteristic of nucleosomes.

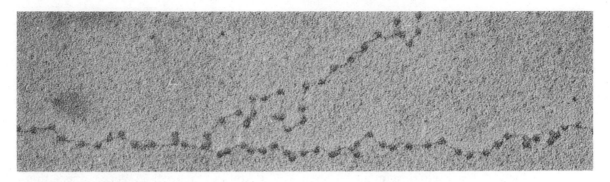

Figure 8-51
A replicating fork showing nucleosomes on both branches. The diameter of each particle is about 110 Å. (Courtesy of Harold Weintraub.)

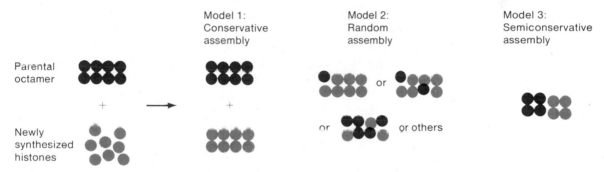

Figure 8-52
Three models for the assembly of newly made histone octamers.

The synthesis of histones occurs simultaneously with DNA replication—that is, histones are made in the cell as they are needed, so that the cell does not contain an appreciable amount of unassociated histone molecules. In light of this, we would like to know whether newly synthesized histones mix with parental histones in the octamers associated with daughter DNA molecules. Three distinguishable models have been considered (Figure 8-52):

Model 1. Parental histone octamers are conserved. That is, the octamers do not dissociate.

Model 2. The parental histone octamers totally dissociate to monomers. The parental monomers mix at random with new histone monomers, re-form octamers, and then rejoin the DNA molecule.

Model 3. The parental histone octamers dissociate not to monomers but to tetramers (or possibly dimers), which then mix with new histones prior to reassembly of nucleosomes.

These alternatives have been distinguished by a density labeling experiment. Cells were grown in a light medium ($^{12}C^{14}N$) for a long time and then grown for one hour in a heavy medium ($^{13}C^{15}N$) also containing a tritiated (^3H) amino acid so as to make the histone radioactive. Octamers were then isolated and their densities determined by equilibrium centrifugation in a cesium formate density gradient. (Cesium formate is much like CsCl but for technical reasons it is more useful for the analysis of the density of proteins.) The radioactive octamers, which contain newly synthesized histones, were found in only a single position in the density gradient, namely, the positions at which all histone octamers are heavy. If model 1 of Figure 8-52 were correct, there would be a single such radioactive peak at the high density. In contrast, model 2 would yield a broad band having a range of densities from light to heavy. If model 3 were correct, a single band of hybrid density would result. Only model 1 proved to be consistent with the experimental results. Thus, whereas DNA replication is semiconservative, production of histone octamers is conservative.

The result just stated does not give any information about how the conserved histone octamers are arranged on the daughter strands. Two possibilities are shown in Figure 8-53. In the first panel the parental octamers are distributed between both daughter strands, which would imply that the octamers dissociate completely from the DNA during replication. In the second arrangement all parental octamers are located on one daughter strand. These alternatives were distinguished by repeating the density shift just described, producing chemical cross-links between adjacent octamers, and then separating the cross-linked octamers from the DNA. The experiment was based on the fact that if the number of cross-links is small, only a few adjacent nucleosomes would join, forming 16-unit polymers (two linked octamers). The 16-unit polymers could be isolated and their densities were determined. The two possibilities are shown in the figure. The results indicated that

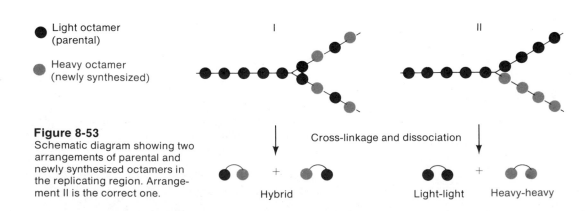

Figure 8-53
Schematic diagram showing two arrangements of parental and newly synthesized octamers in the replicating region. Arrangement II is the correct one.

all 16-unit polymers were either heavy-heavy or light-light. Thus, all parental octamers have adjacent locations—that is, they occupy the same daughter molecule. A second important experiment also led to this conclusion. As we have said, histones are made as they are needed. If cycloheximide, an inhibitor of protein synthesis, is added to growing cells, DNA replication continues in those cells in which replication had begun prior to addition of the inhibitor; however, no histones, which are proteins, are made. Replicating chromatin obtained from cyclohex-imide-treated cells has been observed by electron microscopy. Molecules of the type shown in Figure 8-54 are seen. Each fork in an eye has naked DNA in one branch; the other branch has the parental conserved octamers. This suggests that the parental octamers do not dissociate from the parental DNA strand during DNA replication.

  The final question is whether the parental octamers are associated with a particular strand (leading or lagging strand). So far, it has been possible to answer this question for only one system—an animal virus, SV40, whose DNA is also organized in nucleosomes (which may not be true of all viral DNA). In this system the parental octamers are associated exclusively with the leading strand.

Figure 8-54
Replication of chromatin in the presence of cycloheximide, an inhibitor of protein synthesis. (a) A schematic diagram of an eye. (b) Replicating chromatin from cyclohexi-mide-treated cells. Left panels: two replication forks. Right panel: interpretive drawings. Particles have a cross-section of 110 Å. (Courtesy of Harold Weintraub.)

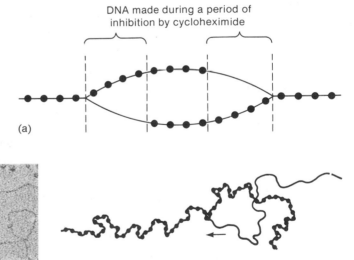

(a)

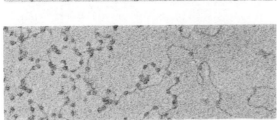

(b)

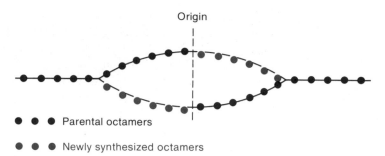

Figure 8-55
The arrangement of parental and newly synthesized octamers in an eye,
when parental octamers are associated with the leading strand and
newly synthesized octamers are associated with the lagging strand.

Since DNA replication in each eye is bidirectional and the parental
octamers are always associated with leading strand synthesis, then each
branch of an eye is covered with a long tract of parental octamers and a
long tract of new octamers. The dividing point between these tracts is
the replication origin of each eye, as shown in Figure 8-55.

REFERENCES

Abdel-Monem, M., and H. Hoffman-Berling. 1980. "DNA unwinding enzymes."
Trends Biochem. Sci., 5, 128–130.

Alberts, B., and R. Sternglantz. 1977. "Recent excitement in the DNA replication
problem." *Nature,* 269, 655–661.

Cold Spring Harbor Laboratory. 1979. Replication and recombination. *Cold
Spring Harb. Symp. Quant. Biology.* Vol. 43.

DePamphilis, M., and P. Wassarman. 1980. "Replication of eukaryotic chromosomes: a close-up of the replication fork." *Ann. Rev. Biochem.,* 49, 627–666.

Geider, K., and H. Hoffman-Berling. 1981. "Proteins controlling the helical
structure of DNA." *Ann. Rev. Biochem.,* 50, 233–260.

Gellert, M. 1981. "DNA topoisomerases." *Ann. Rev. Biochem.,* 50, 879–910.

Gilbert, W., and D. Dressler. 1968. "DNA replication: the rolling circle model."
Cold Spring Harb. Symp. Quant. Biol., 33, 473–478.

Kornberg, A. 1979. "Aspects of DNA replication." *Cold Spring Harbor Symp.
Quant. Biol.,* 43, 1–9.

Kornberg, A. 1980. *DNA Replication.* Also, *1982 Supplement.* W. H. Freeman
and Co.

Lehman, I. R. 1974. "DNA ligase: structure, mechanism, and function." *Science,*
186, 790–797.

McHenry, C., and A. Kornberg. 1977. "DNA polymerase III holoenzyme of *E.
coli*: purification and resolution into subunits." *J. Biol. Chem.,* 252, 6478–
6484.

McKnight, S. L., and O. L. Miller. 1977. "Electron microscopic analysis of
chromatin replication in the cell blastoderm of a *Drosophila melanogaster*
embryo." *Cell,* 12, 795–804.

Meselson, M., and F. W. Stahl. 1957. "The replication of DNA in *E. coli*." *Proc. Natl. Acad. Sci.,* 44, 671–687.

Morrison, A., and N. R. Cozzarelli. 1981. "Contact between DNA gyrase and its binding site on DNA: features of symmetry and symmetry revealed by protection from nuclease." *Proc. Nat. Acad. Sci.,* 78, 1416–1420.

Ogawa, T., and T. Okazaki. 1980. "Discontinuous DNA replication." *Ann. Rev. Biochem.,* 49, 421–428.

Prescott, D. M., and P. L. Kuempel. 1972. "Bidirectional replication of the chromosome in *E. coli*." *Proc. Nat. Acad. Sci.,* 69, 2842–2845.

Riley, D., and H. Weintraub. 1979. "Conservative segregation of parental histones during replication in the presence of cycloheximide." *Proc. Nat. Acad. Sci.,* 76, 328–332.

Rowen, L., and A. Kornberg. 1978. "Primase, the *dnaG* protein of *E. coli*: an enzyme which starts DNA chains." *J. Biol. Chem.,* 253, 758–764.

Schnös, M., and R. Inman. 1970. "Position of branch points in replication of lambda DNA." *J. Mol. Biol.,* 51, 61–69.

Seidman, M. M., A. J. Levine, and H. Weintraub. 1980. "The asymmetric segregation of parental nucleosomes during chromosome replication." *Cell,* 18, 439–449.

Tomizawa, J. 1978. "Replication of colicin E1 plasmid DNA *in vitro*." In *DNA Synthesis: Present and Future,* edited by I. Molineux and K. Kohiyama. Plenum, pp. 797–826.

Tomizawa, J., and G. Selzer. 1979. "Initiation of DNA synthesis in *E. coli*." *Ann. Rev. Biochem.,* 48, 999–1034.

Tye, B. K., P. Nyman, I. R. Lehman, S. Hochhauser, and B. Weiss. 1977. "Transient accumulation of Okazaki fragments as a result of uracil incorporation into nascent DNA." *Proc. Nat. Acad. Sci.,* 74, 154–157.

Valenzuela, M., D. Freifelder, and R. Inman. 1976. "Lack of a unique termination site in lambda DNA replication." *J. Mol. Biol.,* 102, 569–578.

Watson, J. D., and F. H. C. Crick. 1953. "Genetic implications of the structure of desoxyribonucleic acid." *Nature,* 171, 964–967.

Weissbach, A. 1977. "Eukaryotic DNA polymerases." *Ann. Rev. Biochem.,* 46, 25–47.

Wickner, S. H. 1978. "DNA replication proteins of *E. coli*." *Ann. Rev. Biochem.,* 49, 421–428.

9 Repair

There is no single molecule whose integrity is as vital to the cell as DNA. Thus, in the course of hundreds of millions of years there have evolved efficient systems for correcting occasional replication errors. Two of these have already been discussed in the previous chapter—namely, the editing function of both polymerase I and polymerase III and the uracil N-glycosylase system. DNA is also subject to damage in many ways, by environmental agents such as radiation as well as by intracellular chemicals; thus, it is understandable that repair systems exist for eliminating a variety of altered structural features of DNA. In this chapter we examine what is recognized by the repair systems and how the correct structure is restored. We begin by describing the principal kinds of damage that can occur in a DNA molecule.

ALTERATIONS OF DNA MOLECULES

There are three distinct mechanisms for altering the structure of DNA: (1) base substitutions during replication, (2) base changes resulting from the inherent chemical instability of the bases or of the N-glycosylic bond, and (3) alterations resulting from the action of other chemicals and environmental agents. These mechanisms are responsible for the occurrence of the following defects:

1. An incorrect base in one strand that cannot form hydrogen bonds with the corresponding base in the other strand. This defect can result from a replication error that by chance is not corrected by the editing function; presumably this failure is rare. A more common event is spontaneous deamination of cytosine to uracil, followed by conversion of the uracil to thymine in subsequent rounds of replication. Less frequent than the cytosine→uracil→thymine transition is occasional deamination of adenine to form hypoxanthine, which forms base pairs with cytosine instead of thymine. Hypoxanthine N-glycosylase corrects this alteration. Deamination will not be discussed further in this chapter as its mechanisms of repair have already been explained in Chapter 8 (see Figure 8-27).

2. Missing bases. The N-glycosylic bond of a purine nucleotide is spontaneously broken at physiological temperatures, though at a very low rate. This process is called **depurination** because the purine is lost from the DNA. The rate of spontaneous depurination is about one purine removed per 300 purines per day at pH 7 and 37°C, which amounts to about 10^4 purines per day in a mammalian cell and 0.25 purines per day per generation time for a bacterium. The rate is increased as the pH is lowered or as the temperature is elevated. Noxious chemicals called **alkylating agents** (Figure 9-1), once used in chemical warfare and now used in cancer treatment, react primarily with guanine, adding an aklyl group to N-7 of the purine ring. This alkylation weakens the N-glycosylic bond and in time this bond is hydrolyzed, resulting in depurination.

Since breakage of the N-glycosylic bond is the first step in the N-glycosylase repair systems, restoration of the correct structure uses the apurinic acid (AP) nuclease pathway already shown in Figure 8-27.

3. Altered bases. Bases can be changed into strikingly different compounds by a variety of chemical and physical agents. For instance, ionizing radiation (such as the β particles emitted by naturally occurring radioisotopes or laboratory x rays) can break purine and pyrimidine rings and can cause several types of chemical substitutions—the most frequent substitutions are made in thymine; a well-studied product, which is important in repair processes, is 5,6-dihydroxydihydrothymine (denoted DHDT). Free radicals produced in many meta-

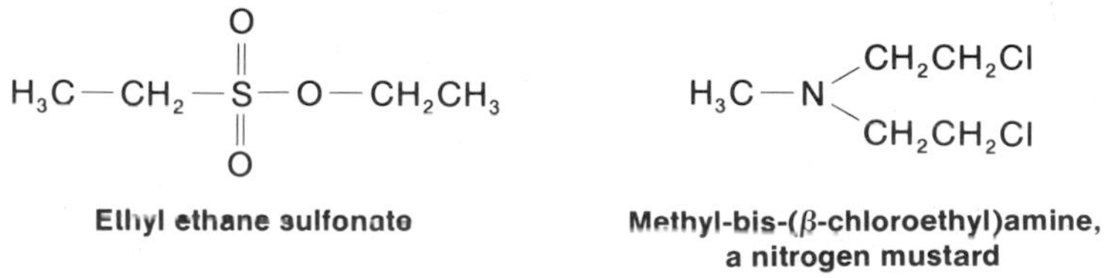

$$H_3C - CH_2 - \overset{\overset{\displaystyle O}{\|}}{\underset{\underset{\displaystyle O}{\|}}{S}} - O - CH_2CH_3$$

Ethyl ethane sulfonate

$$H_3C - N \overset{\displaystyle CH_2CH_2Cl}{\underset{\displaystyle CH_2CH_2Cl}{\diagdown}}$$

Methyl-bis-(β-chloroethyl)amine, a nitrogen mustard

Figure 9-1
Alkylating agents.

bolic reactions can also cause a variety of significant changes. The best-studied altered base is the dimer formed by two pyrimidines as a result of ultraviolet radiation. The most prominent of these dimers is the thymine dimer shown in Figure 9-2. The significant effects of the presence of thymine dimers are the following: (1) the DNA helix becomes distorted as the thymines, which are in the same strand, are pulled toward one another (Figure 9-3); and (2) as a result of the distortion, hydrogen-bonding to adenines in the opposing strand, though possible (because the hydrogen-bonding groups are still present), is significantly weakened; this weakening causes inhibition of advance of the replication fork, as will be discussed in a later section.

4. *Single-strand breaks.* A variety of agents can break phosphodiester bonds. Among the more common chemicals are peroxides, sulfhydryl-containing compounds (for example, cysteine) and metal ions such as Fe^{2+} and Cu^{2+}. Ionizing radiation produces strand breaks

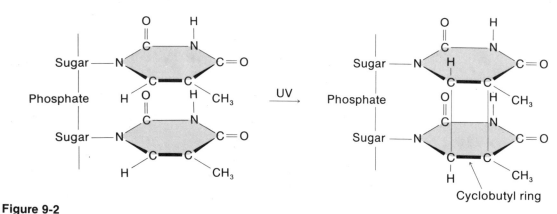

Figure 9-2
Structure of a cyclobutylthymine dimer. Following ultraviolet (UV) irradiation, adjacent thymine residues in a DNA strand are joined by formation of the bond shown in red. Although not drawn to scale, these bonds are considerably shorter than the spacing between the planes of adjacent thymines, so that the double-stranded structure becomes distorted. The shape of the thymine ring also changes as the C=C double bond of each thymine is converted to a C—C single bond in each cyclobutyl ring.

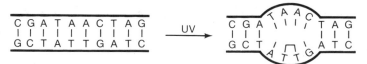

Figure 9-3
Distortion of the DNA helix caused by two thymines moving closer together when joined in a dimer. The dimer is shown as two joined lines.

both by the action of secondary electrons produced by a β particle or an x-ray photon passing nearby and by production of free radicals in water (e.g., $\cdot$OH) that can attack the bond. In the case of energetic radiation the bases adjacent to the broken bond are usually damaged also. DNases that are present in cells probably also make frequent and possibly accidental phosphodiester scissions.

5. *Double-strand breaks.* If a DNA molecule receives a sufficiently large number of randomly located single-strand breaks, two breaks may be situated opposite one another, resulting in breakage of the double helix. Actually, the breaks need not be directly opposite one another to constitute a double-strand break but can be separated by a few base pairs (Figure 9-4). Double-strand breaks also form in a single event. This can result from exposure to highly ionizing radiation (for example, x rays, naturally occurring cosmic rays, electron beams, γ rays, and radiation beams of various kinds used in radiotherapy). When highly ionizing radiation passes through water, large clusters of electron pairs and free radicals form in volumes of about 5 Å in diameter. These clusters cause local, multiple alterations. Double-strand breaks arising in this way are invariably accompanied by adjacent base damage and rarely are repaired.

6. *Cross-linking.* Some antibiotics (for example, mitomycin C) and some reagents (the nitrite ion) can form covalent linkages between a base in one strand and an opposite base in the complementary DNA strand. This prevents strand separation during DNA replication and also causes a local distortion of the helix.

Each of the defects just described can be corrected by one of several repair systems.

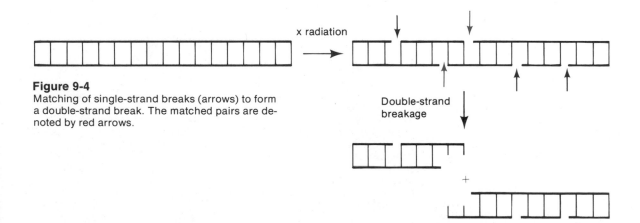

Figure 9-4
Matching of single-strand breaks (arrows) to form a double-strand break. The matched pairs are denoted by red arrows.

Repair was first observed and is best understood in bacteria. It is a widespread and probably universal phenomenon, however, and is now well documented in many microorganisms and in mammalian cells. In the following section we show how repair is recognized both biologically and chemically.

BIOLOGICAL INDICATION OF REPAIR

The existence of repair was first suggested by several observations of the effect of ultraviolet light on the ability of bacteria to form colonies. Once it was realized that DNA is the genetic material, it was expected that ultraviolet light in a wavelength range absorbable by DNA would be able to cause chemical damage and hence either death or mutagenesis in an irradiated cell. The means of investigating this possibility is to obtain a **survival curve.** This is done in the following way.

Samples are drawn at intervals from a population of bacteria irradiated by ultraviolet light. The samples are plated on nutrient agar and the colonies that grow are counted. The proportion of cells able to form colonies is plotted as a function of ultraviolet dose; it is found that colony-forming ability decreases as the dose increases. An example of a survival curve is shown in Figure 9-5.

There are many types of repair. The first type that was recognized is called **photoreactivation**—that is, the survival of various ultraviolet-irradiated bacteria is higher if the bacteria are exposed to intense visible light (for example, sunlight) after ultraviolet irradiation but before plating (Figure 9-6(a,b)). Thus, the visible light eliminates some of the damage introduced by the ultraviolet light.

Figure 9-5
A typical ultraviolet-light survival curve for a bacterium. Initially, the curve is fairly flat because initial damage does not cause killing; this could either mean that several lethal hits are needed to kill a cell or that, at low doses, a great deal of the damage is repaired. Note that the y axis is logarithmic.

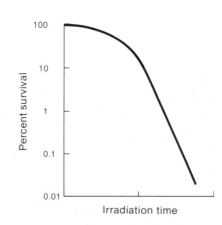

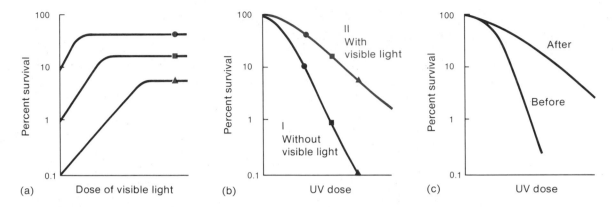

Figure 9-6

Types of repair. (a) Increase in survival of three different samples of ultraviolet-light-irradiated bacteria as a function of the dose of visible light. This is called photoreactivation. (b) A pair of survival curves showing the effect of postirradiation with visible light. Curve I consists of points taken from the y axis of part (a) and curve II is a plot of the red points taken from the plateaus in part (a). (c) Survival curves before and after incubation in buffer following ultraviolet-light irradiation. This is called liquid-holding recovery.

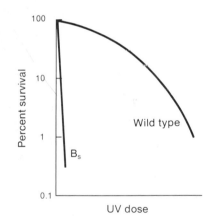

Figure 9-7

Survival curves showing the great sensitivity to ultraviolet light of the mutant *E. coli* B_s.

Another type of repair is **liquid-holding recovery.** If ultraviolet-irradiated cells are held in a nonnutrient but warm buffer for several hours before plating, the surviving fraction for a particular dose is increased (Figure 9-6(c)). Thus, it seemed that delaying cell growth or perhaps merely DNA replication created additional time for repair to occur. Furthermore, the liquid-holding recovery seemed to take place without light from any source. Proof that recovery resulted from two distinct repair processes—a light-dependent one (photoreactivation) and a light-independent one—came from the isolation of an extraordinarily ultraviolet-sensitive *E. coli* mutant called B_s (Fig. 9-7). The mutant will not recover its growth potential after ultraviolet irradiation when held in a buffer in the dark but it does photoreactivate normally. The isolation of another mutant called pho^-, which cannot photoreactivate but can undergo liquid-holding recovery, confirms the conclusion that there are two repair systems. Liquid-holding recovery is now known to be a special case of a general phenomenon called **dark repair.** The chemical mechanisms of both photoreactivation and dark repair are described in a later section.

Other mutants have been found that indicate the existence of other repair systems. For example, the alk^- mutant of *E. coli* is more sensitive than wild-type bacteria to alkylating agents but has normal sensitivity to ultraviolet and x rays. Thus, there must be a repair system that recognizes alkylated bases. *E. coli rorA*$^-$ also is sensitive to x rays but not to ultraviolet light, indicating that there is a system that specifically repairs x-ray damage. Ras^- mutants are abnormally sensitive to both x rays and ultraviolet light, which suggests one of the following: (1) Another repair system exists that can act on both types of damage; (2) there are several kinds of x-ray and ultraviolet-light damage, one product is common to both agents, and a single system in ras^+ mutants repairs this product; and (3) an enzymatic step is common to both repair mechanisms and the *ras* gene specifies that step.

It should be noted that the existence of these mutants exemplifies how the isolation of a mutant indicates that a function exists.

It might seem surprising that cells should have evolved systems for repairing damage caused by ultraviolet light, which is not an agent commonly encountered in nature by most cells. To understand this, one must remember that the properties of cells reflect ancient history. During the period when life was evolving in the primordial seas, it is likely that the earth was not enveloped by a stratospheric ozone layer. Without this layer a very high flux of ultraviolet light would have reached the surface of the earth and would have been very damaging to primitive organisms. Thus, acquiring the ability to repair ultraviolet damage had great survival value and this ability has persisted for eons because, as will be seen shortly, the dark repair system is also able to repair other types of damage. How photoreactivation has persisted is less clear since it seems to be active only against products formed by ultraviolet irradiation.

IDENTIFICATION OF THE THYMINE DIMER AS A PRIMARY LESION THAT IS REPAIRABLE

If purified DNA whose thymines are ^{3}H-labeled is ultraviolet-irradiated and then acid-hydrolyzed to obtain free bases, [^{3}H]thymine dimers can be isolated and the mole fraction of thymine recoverable as dimers will increase with ultraviolet dose. Thymine dimers can also be isolated from purified [^{3}H]DNA obtained from ultraviolet-irradiated cells. The number of dimers per cell in a population of ultraviolet-irradiated bacteria, of which 50 percent of the bacteria survive, is many thousands—far more than are necessary to kill the cell. This number is very large because dimers are repairable, as shown by the following experimental results:

1. A population of bacteria is given a single dose of ultraviolet light and then irradiated with high-intensity visible light to induce photoreactivation. After various doses of visible light, the DNA is isolated and hydrolyzed, and the number of thymine dimers per cell is measured. It is found that the number of thymine dimers per cell decreases continually with increasing doses of visible light. Thus, photoreactivation, which increases survival, converts thymine dimers back to two thymine monomers.

2. A population of bacteria is ultraviolet-irradiated and then incubated for various periods of time in a nonnutrient buffer (that is, it is subjected to liquid-holding recovery). During this period the number of thymine dimers present in the DNA continually decreases and at the same time, thymine dimers appear both in the buffer and in the intracellular fluid. Thus, during the dark repair process, thymine dimers are excised from the DNA.

In the following sections the biochemical mechanisms for photoreactivation, dark repair, and two other processes will be described.

BIOCHEMICAL MECHANISMS FOR REPAIR OF THYMINE DIMERS

There are four major pathways for repair, which can be subdivided into two classes—light-induced repair (photoreactivation) and light-independent repair (dark repair). The latter can be accomplished by three distinct mechanisms: (1) excision of the damaged bases (**excision repair**), (2) reconstruction of a functional DNA molecule from undamaged fragments (**recombinational repair**), and (3) disregard of the damage (**SOS repair**). The chemical mechanisms for each of these repair processes are described in this section.

Photoreactivation

Photoreactivation is an enzymatic cleavage of thymine dimers activated by visible light (300–600 nm). An enzyme called the photoreactivating or **PR enzyme** has been isolated from almost all cells, from bacteria to

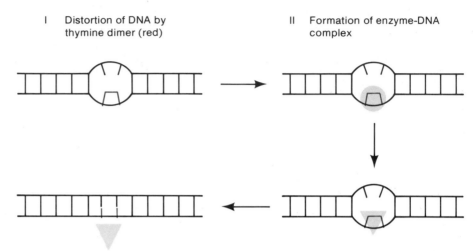

I   Distortion of DNA by
    thymine dimer (red)

II   Formation of enzyme-DNA
     complex

IV   Release of enzyme

III   Absorption of visible light
      and activation of enzyme

Figure 9-8
Scheme for enzymatic
photoreactivation of a
thymine dimer.

animals. The PR enzyme itself does not absorb light nor does it bind any light-absorbing compound. In a way that is not known the enzyme-DNA complex nevertheless absorbs light and uses the light energy to cleave the C—C bonds of the cyclobutyl rings shown in Figure 9-2. The activity of the PR enzyme is shown in Figure 9-8. First, the enzyme binds specifically to a thymine dimer (cytosine dimers and cytosine-thymine dimers also are formed by ultraviolet-irradiation but much less frequently than thymine dimers; the PR enzyme is also active against these dimers). If light is absorbed, the thymine dimer is monomerized (the chemical mechanism is unknown); finally, the PR enzyme, which only binds to thymine dimers in DNA, dissociates.

Excision Repair

Excision repair is a multistep enzymatic process. The four steps, which are often summarized as **cut—patch—cut—seal,** are shown in Figure 9-9. In the first step, an **incision** step, a repair endonuclease recognizes the distortion produced by a thymine dimer and makes a single cut in the sugar-phosphate backbone ahead of the dimer. At the incision site there is a 5′-P group on the side of the cut containing the dimer and a 3′-OH group on the other side. The 3′-OH group is recognized by polymerase I, which then synthesizes a new strand while displacing a DNA segment consisting of about twenty nucleotides and carrying the thymine dimer. This segment is excised in part by the 5′→3′ exonuclease activity of polymerase I and partly by the associated endonuclease activity. Other

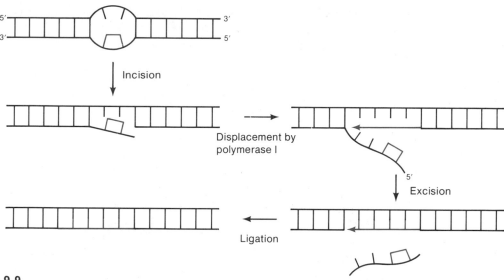

Figure 9-9
Scheme for excision repair of a thymine dimer by the cut–patch–cut–seal mechanism. The thymine dimer and the displacing segment synthesized by polymerase I are both shown in red.

exonucleases can also carry out the excision step; it has been found that excision still occurs in a mutant lacking the 5′→3′ exonuclease of polymerase I and its associated endonuclease activity. The excised fragment is ultimately degraded to single nucleotides plus a thymine dimer deoxynucleotide by the combined activity of numerous scavenging exo- and endonucleases. The final step of the repair process is joining of the newly synthesized segment to the original strand by DNA ligase.

The incision activity of *E. coli* is determined by three genes called *uvrA*, *uvrB*, and *uvrC*. The products of the *uvrA* and *uvrB* genes are two subunits of a protein complex (called ultraviolet-endo I) that has endonuclease activity. The UvrC product is necessary for maximum endonuclease activity *in vivo*, but its precise role is not yet known. Proof that the Uvr system is responsible for excision repair *in vivo* comes from studies with mutants. Mutation in any of these genes renders *E. coli* exceedingly sensitive to killing by ultraviolet light (Figure 9-10) and when *uvrA*⁻ and *uvrB*⁻ mutants are irradiated and then incubated, thymine dimers are not found in the intracellular fluid. That these phenomena are due to a loss of endonucleoytic activity was shown by the following results. An analysis of the formation of single-strand breaks in ultraviolet-irradiated DNA (performed by measurement of the size of the DNA obtained when the isolated DNA is sedimented in alkali, which denatures the DNA) shows that breaks occur in the DNA of *uvr*⁺ bacteria but not in that of *uvrA*⁻ and *uvrB*⁻

mutants. Breaks are also found in the DNA of *uvrC*⁻ mutants but the number appearing per unit time is smaller.

The Uvr system is able to repair lesions other than thymine dimers. These lesions have in common either the displacement of bases, as in thymine dimer formation, or the addition of bulky substituents on the bases. It is guessed that the incision enzyme recognizes a helix distortion.

The bacterial excision-repair system is also active on phage DNA. If a phage suspension is irradiated with ultraviolet light and then plated on a bacterial lawn, the ability of the phage to produce plaques decreases with dose; if the phage DNA is labeled with [³H]thymidine, [³H]thymine dimers are formed. Repair is evident from the fact that if a *uvr*⁺ bacterium is infected with such an irradiated phage, thymine dimer excision occurs; there is no excision if the host is a *uvr*⁻ mutant and the fraction of the population that survives is lower than the fraction observed with a *uvr*⁺ host. Thus, the thymine dimer is a major lesion in ultraviolet-light-irradiated phages and its excision by the bacterial system enables an irradiated phage to survive. Some phages even possess their own repair systems and can thus excise thymine dimers even when a *uvr*⁻ host is infected.

Incision enzymes have been isolated from many sources. Two particularly well-studied ultraviolet-repair enzymes are those of *E. coli* phage T4 and of the bacterium *Micrococcus luteus*. The incision enzymes of yeast and of many mammalian cells have also been characterized in detail. The mammalian systems seem to be more complicated and require a larger collection of enzymes.

Many human diseases may result from inability to carry out excision repair. The best studied, xeroderma pigmentosum, is a result of mutations in genes that encode the ultraviolet-light incision system. Cells cultured from patients with the disease are killed by much smaller ultraviolet doses than are cells from a normal person. Furthermore, their ability to remove thymine dimers from their DNA is very much reduced. Patients with this disease develop skin lesions when exposed to sunlight and commonly develop one of several kinds of skin cancer.

Recombinational Repair

Figure 9-10 shows that an *E. coli uvrA*⁻ mutant is much more sensitive to ultraviolet light than is the wild type. However, it can be shown (see References) that an ultraviolet dose yielding an average of one lethal event per *uvrA*⁻ cell produces about 300 thymine dimers per cell, none of which is excised. If we assume that one thymine dimer is sufficient to be lethal, then there must be some way to eliminate the consequences of the other thymine dimers. Of course, an alternative (but incorrect)

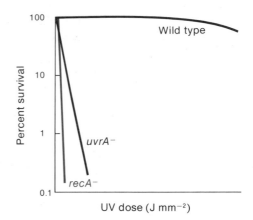

Figure 9-10
Survival curves of *E. coli* showing
the sensitization to ultraviolet light
resulting from the *uvrA⁻* and *recA⁻*
mutations.

explanation might be that it takes many thymine dimers to kill a
bacterium. Evidence that there is another repair system comes from
examination of cells mutant in the gene *recA*, a gene that is essential for
genetic recombination in E. *coli* (hence the symbol *rec*); as mentioned in
Chapter 1, in an Hfr × F⁻ bacterial mating, no recombinants are found
if the F⁻ cell is *recA⁻*. An ultraviolet irradiation survival curve for a
recA⁻ mutant shows that it is very ultraviolet-sensitive (Figure 9-10),
which suggests that the *recA* gene must be a component of a repair
system. Excision of thymine dimers occurs normally in a recA⁻ mutant
so that *recA*-mediated repair clearly differs from excision repair.
Furthermore, this system is apparently more effective than excision
repair, inasmuch as a *recA⁻* mutant is even more ultraviolet-sensitive
than a *uvrA⁻* mutant (Figure 9-10). The existence of two repair systems
is confirmed by a quantitative analysis of the survival curve of a
uvrA⁻recA⁻ double mutant, which is more ultraviolet-sensitive than
either of the single mutants and for which the number of thymine
dimers per lethal event is roughly one. It has been established that the
Uvr system is responsible for removal of many of the thymine dimers
whereas the *rec* system eliminates the dire effects of most of those that
are not removed.

In order to discuss the mechanism of recombinational repair, it is
necessary to know the effect of a thymine dimers on DNA replication.
When polymerase III reaches a thymine dimer, the replication fork fails
to advance. A thymine dimer is still capable of forming hydrogen bonds
with two adenines because the chemical change in dimerization does
not alter the groups that engage in hydrogen bonding. However, the
dimer introduces a distortion into the helix and when an adenine is
added to the growing chain, polymerase III reacts to the distorted region
as if a mispaired base had been added; the editing function then
removes the adenine. The cycle begins again—an adenine is added and

Figure 9-11

Blockage of replication by thymine dimers (represented by joined lines) followed by re-starts several bases beyond the dimer. The black region is a segment of ultraviolet-light-irradiated parental DNA. The red region represents synthesis of a daughter molecule from right to left. The daughter strand contains gaps.

then it is removed; the net result is that the polymerase is stalled at the site of the dimer. (The same effect would occur if, instead of a dimer, radiation or chemical damage resulted in formation of a base to which no nucleoside triphosphate could base pair.) Evidence that such a phenomenon occurs after ultraviolet irradiation is the existence of an ultraviolet-light-induced idling process—that is, rapid cleavage of deoxynucleoside triphosphates to monophosphates without any net DNA synthesis (i.e., without advance of the replication fork). A cell in which DNA synthesis is permanently stalled cannot complete a round of replication and does not divide. Protein synthesis continues but, without cell division, a long filamentous cell called a snake results; this growth leads to cell death. However, in a rec^+ cell, which may be either uvr^+ or uvr^-, the inhibition of DNA synthesis is usually not permanent. (If the ultraviolet dose is very large, the Rec system may itself be damaged, in which case DNA synthesis is permanently inhibited.) There are two different ways in which DNA synthesis manages to get going again—**postdimer initiation** and **transdimer synthesis.** These are responsible for **recombination repair** and **SOS repair,** respectively. In this section recombination repair is described; SOS repair is discussed in the succeeding section.

One way to deal with a thymine dimer block is to pass it by and initiate chain growth beyond the block (Figure 9-11). Such postdimer initiation does occur after a pause of about five seconds per thymine dimer but the mechanism, which appears to involve unprimed reinitiation, is unknown. The result of this process is that the daughter strands have large gaps, one for each unexcised thymine dimer. There is no way to produce viable daughter cells by continued replication alone, because the strands having the thymine dimer will continue to turn out gapped daughter strands, and the first set of gapped daughter strands would be fragmented when the growing fork enters a gap. However, by a recombination mechanism called **sister-strand exchange** proper double-stranded molecules can be made.

The essential idea in sister-strand exchange is that a single-stranded segment free of any defects is excised from a "good" strand on the homologous DNA segment at the replication fork and somehow inserted into the gap created by excision of a thymine dimer (Figure

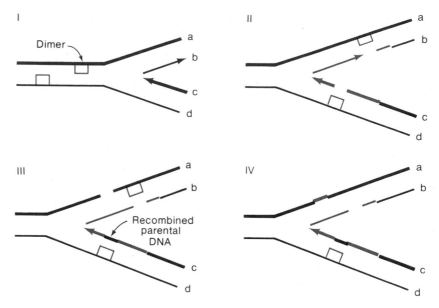

Figure 9-12
Recombinational repair. I. A molecule containing two thymine dimers (red boxes) in
strands a and d is being replicated. II. By postdimer initiation, a molecule is formed
whose daughter strands b and c have gaps. If repair does not occur, in the next round
of replication, strands a and d would yield gapped daughter strands, and strands b and
c would again be fragmented. III. By sister-strand exchange a continuous segment of
a is excised and inserted into strand c. (In a second round of replication, strand c
would be a template for synthesis of functional DNA.) IV. The gap in a is next filled.
Such a DNA molecule would probably engage in a second sister-strand exchange in
which a segment of c would fill the gap in b. The gap in c would then be filled. In this
way strand c also becomes a functional template. DNA synthesized after irradiation is
shown in red. Heavy and thin lines are used for purposes of identification only.

9-12). The combined action of polymerase I and DNA ligase joins this
inserted piece to adjacent regions, thus filling in the gap. The gap
formed in the donor molecule by excision is also filled in completely by
polymerase I and ligase. If this exchange and gap filling are done for
each thymine dimer, two complete daughter single strands can be
formed and each can serve in the *next* round of replication as a template
for synthesis of normal DNA molecules. Note that the system fails if two
dimers in opposite strands are very near one another because then no
undamaged sister strand segments are available to be excised. The
molecular details of recombinational repair are not known with
precision, so that the model shown in Figure 9-12 must be considered to
be a working hypothesis that is at present consistent with the facts.

Recombinational repair is an important mechanism because it
eliminates the necessity for delaying replication for the many hours that

would be needed for excision repair to remove all thymine dimers. It may also be the case that some kinds of damage cannot be eliminated by excision repair—for example, alterations that do not cause helix distortion but do stop DNA synthesis. Recombination repair has been demonstrated in several bacteria, but it is not known whether it occurs in animal cells.

Recombinational repair also occurs with ultraviolet-irradiated phages of some types; this is indicated by the fact that, if a population of a phage that lacks its own repair system is ultraviolet-light irradiated, fewer phage will plate on a $recA^-$ host than on a $recA^+$ host.

Since recombinational repair occurs after DNA replication, in contrast with excision repair, it is often called **postreplicational repair.**

SOS Repair

SOS repair is a **bypass system** that allows DNA chain growth across damaged segments at the cost of fidelity of replication. It is an error-prone process; even though intact DNA strands are formed, the strands are often defective. (The principle involved is that survival with some loss of information is better than no survival at all.) SOS repair is not understood particularly well at present but is thought to invoke a relaxation of the editing system in order to allow polymerization to proceed across a dimer (transdimer synthesis) despite the distortion of the helix. In the previous section the idling caused by the response of the editing system to a dimer was described. When the SOS system is activated, apparently the editing system does not react to the distortion, the growing fork advances, and two adenines are placed in the growing chain at the sites specified by the thymines in the dimer. An early hypothesis about the mechanism of SOS repair suggested that the polymerase placed any bases in the growing chain at the damaged site and that this random replacement was the cause of mutagenesis. However, recently an important observation has been made about the base sequence in the daughter strand—namely, that incorrect nucleotides are indeed occasionally added to the growing chain but the mistakes are not necessarily at the sites of the adenines that would normally be added if the DNA were not damaged. The current hypothesis is that damaged DNA induces an error-prone replication system that has less proofreading activity (if any) than the normal replication system. This lower-fidelity system causes *all* newly made DNA strands to have a higher-than-normal number of mispaired bases. It seems likely that polymerase III is modified in some way (possibly by loss or alteration of a subunit needed for high-fidelity replication) so it can continue chain growth without being stalled at damaged sites

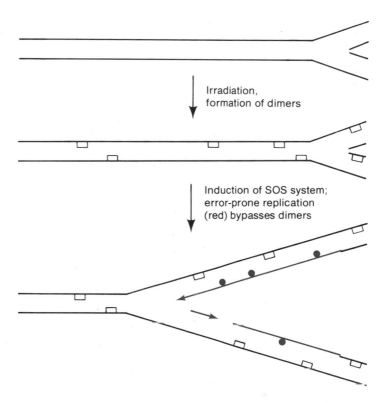

Figure 9-13
SOS repair. A DNA molecule in an early stage of replication is irradiated with ultraviolet light and thymine dimers are formed. The SOS system is induced and all subsequent replication (shown in red) has a higher-than-usual number of misincorporated bases (red dots). Some of these bases can be removed by the mismatch repair system.

Irradiation, formation of dimers

Induction of SOS system; error-prone replication (red) bypasses dimers

(Figure 9-13). Because of the presence of mispaired bases, most progeny will be mutants (although the base-pair mismatch repair system presumably can correct many of these errors.) How long this error-prone replication continues is not known. The SOS repair system is thought to be a major cause of ultraviolet-induced mutagenesis.

SOS repair also requires an active *recA* gene, though the reason for this is not precisely known. Some comments about the role of this gene will be made later in this section.

SOS repair (and probably recombinational repair as well) differs from excision repair in that SOS repair is *induced* as a result of the damage to the DNA. That is, the responsible enzymes are not present until after a cell has been damaged. The best evidence for this point comes from an analysis of the repair of ultraviolet-light-inactivated phage by irradiated bacterial host cells.

Most phages do not need a fully functional (or even viable) host to produce progeny phage particles. For instance, a study of the result of infecting ultraviolet-irradiated E. *coli* with phage λ showed that, if the host is irradiated with a dose of ultraviolet light yielding 10 percent survival of the ability to form colonies, there is no loss of the ability to support phage infection. However, if the λ particles are also irradiated,

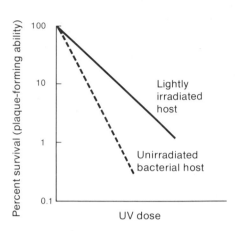

Figure 9-14
W reactivation of ultraviolet-light-irradiated phage λ. The dashed line shows the survival curve (for plaque-forming ability) obtained when λ phage irradiated with various doses of ultra-violet light are plated on unirradiated bacteria. The solid line represents survival of plaque-forming ability, when ultraviolet-light-irradiated λ are plated on lightly irradiated bacteria.

the irradiated *E. coli* cells are better able to support growth of the irradiated λ than are unirradiated cells. That is, the survival of a ultraviolet-irradiated λ is higher on an irradiated host than on an unirradiated host (Figure 9-14). This phenomenon, which is called **UV-reactivation** or **W reactivation** (for Jean Weigle, who discovered it), clearly involves a repair process and occurs only if the host is Rec+; the phenomenon does not require the *uvr* genes. Although more phage survive, the surviving population contains a higher proportion of phage mutants when the irradiated host is used; that the mutation frequency increases suggests that the repair mechanism makes use of an error-prone replication system. It is now thought that this phenomenon is the SOS system in action. If so, the SOS system has been activated or turned on by ultraviolet-irradiation of the host.

A hint at what is going on in W reactivation comes from a similar effect that has been observed with a temperature-sensitive *dnaB⁻* mutant. Heating the mutant for a short while (which inhibits DNA synthesis, as is explained in Chapter 8) prior to infection with ultravio-let-irradiated λ also increases the survival of the phage. This phenome-non is also mimicked if a thymine-requiring mutant bacterial host is briefly starved for thymine, which stops DNA synthesis. These obser-vations have given rise to the idea, as yet unproved, that it is inhibition of host DNA synthesis by thymine dimers that turns on the SOS repair system. Thus, it would be the first dimer encountered by polymerase III in the replication fork that would induce the SOS system, and all further (downstream) dimers could be bypassed.

Several bacterial genes are involved in the SOS pathway. Of these the best understood are the *recA* and *lexA* genes, the *lexA* gene encoding a regulator of the synthesis of the *recA* product. The *recA*-gene product has three main biochemical activities: one of these is its recombinational activity, which functions in an unknown way; another is a protease activity; and a third is a single-stranded-DNA binding activity. At least two of these activities are important for SOS repair, as we will soon see.

Although the regulation of gene activity is discussed primarily in Chapter 14, it is a useful introduction at this point to consider why the SOS system should normally be shut down, and also the mechanism by which it is switched on. Clearly, if cells have evolved an elegant editing function for maintaining high fidelity in replication, it is reasonable that they should have means of making all error-prone replicative processes inactive. This inactivation is accomplished by means of the *lexA* gene product. When this protein is present (and it normally is), the genes (most of which are unknown) that encode the SOS repair enzymes are made inactive in the sense that very little mRNA for those genes can be synthesized. Furthermore, in an unirradiated cell, or in any cell in which DNA synthesis is occurring, the *recA* protein lacks protease activity. However, when DNA replication is prevented, the small amount of *recA* product that is present is converted to an active protease. (The activation requires single-stranded DNA and possibly an unknown small molecule; the former may be either at a replication fork or in gaps produced by postdimer reinitiation.) This proteolytic activity of the *recA* protein functions only on selected proteins; one of these is the *lexA* protein, which is cleaved into two fragments. The fragmented *lexA* protein is no longer able to prevent mRNA production; therefore the proteins of the SOS system are synthesized. When repair is complete, the *recA* product reverts to an inactive form; *lexA* protein, which has continually been synthesized and cleaved, is no longer cleaved and hence is able to turn off the synthesis of the SOS system. It has recently been suggested that the DNA-binding activity of the *recA* protein may also be required—possibly as a substitute for the ssb (single-strand-binding) protein at the site of a thymine dimer.

Summary of the Repair of Thymine Dimers

Four repair systems have been described—excision repair, photoreactivation, recombination, and SOS repair. These systems differ both in the source of energy—light in the case of photoreactivation and ATP hydrolysis for the other three—and in the mechanisms used to allow functional daughter DNA molecules to be formed. Excision repair and photoreactivation both repair the template, recombination repair forms a new template, and SOS repair ignores the damage and forms uninterrupted daughter strands, despite the presence of damaged segments of the template. Figure 9-15 summarizes these pathways. It should be noted that photoreactivation is the only mechanism in which there is no interruption of the sugar-phosphate backbone and that SOS repair is the only process that leads to mutation.

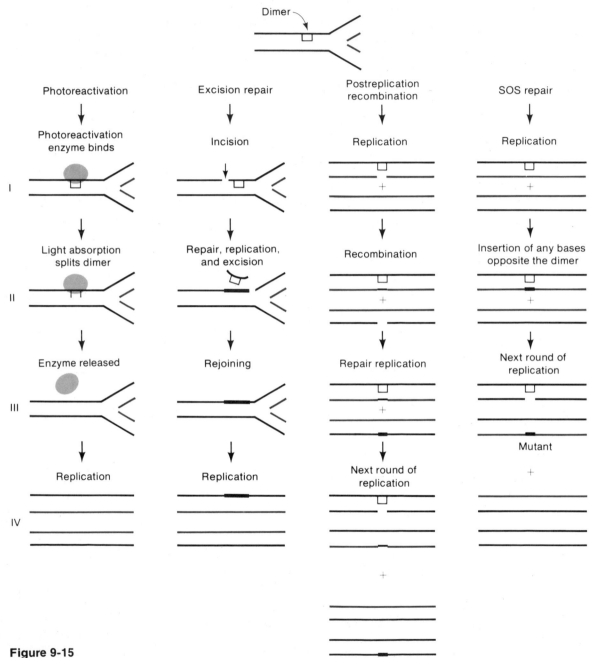

Figure 9-15
The four major mechanisms for repair of thymine dimers.
The red lines show the single strands (and their completed forms) in the advancing fork of the uppermost
panel. DNA made by repair replication is depicted by a
heavy line.

REPAIR OF CROSS-LINKS

Cross-links have been experimentally introduced into bacterial DNA in two ways—by addition of antibiotics such as mitomycin C, and by photochemical mechanisms by addition of the substance 8-methoxy-psoralen to the suspending buffer and then irradiation with 360 nm light (Figure 9-16). The survival of these bacteria has been plotted against time of exposure to a particular concentration of mitomycin C or to a dose of visible light following psoralen treatment. With both treatments *E. coli uvr⁻* and *recA⁻* mutants are much more sensitive than wild-type *E. coli*, so apparently wild-type *E. coli* has evolved a repair system for cross-links, which utilizes *uvr* and *recA* genes. How repair of cross-links does occur is not at all known at present. Simple excision repair cannot suffice since each strand lacks a copiable base (Figure 9-17). A proposal, some of whose features are probably correct, is shown in Figure 9-18.

Repair of cross-links must also occur in mammalian cells; cultured cells obtained from patients with Fanconi's anemia (a bone marrow disease) are much more sensitive to psoralen-induced killing than are cells obtained from a normal human.

Figure 9-16
Molecular structure of psoralen and the mechanism by which it binds to DNA, forming bonds to single strands and interstrand cross-links. Psoralen molecules are shown in red. Psoralen intercalates into the DNA and, when irradiated, forms bonds to individual bases. Occasionally, two bonds will form to bases in the *same* strand (intrastrand cross-links), but more commonly, for stereochemical reasons, a psoralen molecule forms an interstrand cross-link between bases in adjacent pairs in different strands, as shown in the figure. The base pairs are drawn as interrupted horizontal lines to emphasize the interstrand nature of the cross-link.

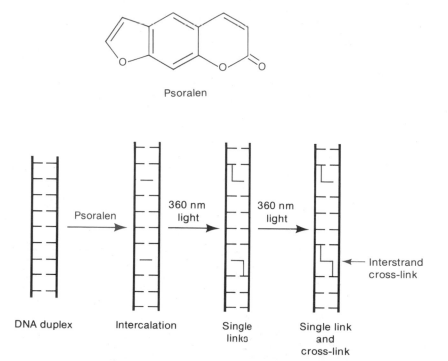

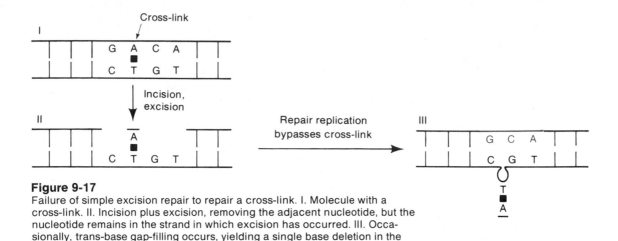

Figure 9-17
Failure of simple excision repair to repair a cross-link. I. Molecule with a cross-link. II. Incision plus excision, removing the adjacent nucleotide, but the nucleotide remains in the strand in which excision has occurred. III. Occasionally, trans-base gap-filling occurs, yielding a single base deletion in the "repaired" strand.

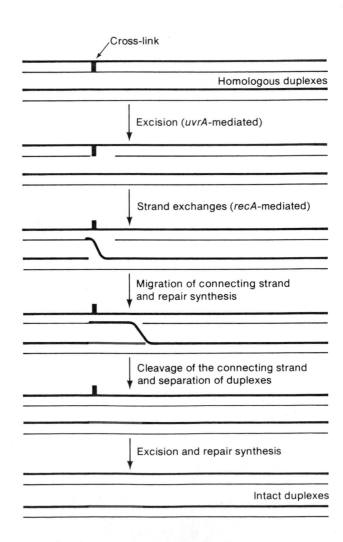

Figure 9-18
One possible mechanism for repair of cross-links in which genetic recombination plays a role. A cross-linked nucleotide is excised from one strand and nucleases enlarge the gap. By a recombinational event a homologous strand from another DNA molecule is inserted into the gap. This inserted segment is the template in a second repair step, which consists of excision of the cross-linked nucleotide from the other strand followed by ordinary repair replication.

REPAIR OF DAMAGE BY IONIZING RADIATION

In this section we consider repair of single-strand and double-strand breaks and the excision of one type of an altered base, namely, 5,6-dihydroxydihydrothymine (DHDT).

The repair of single-strand breaks introduced by ionizing radiation has been carefully studied. A few of the single-strand breaks have free 3'-OH and 5'-P groups without any other alterations and are repaired by DNA ligase. However, the efficiency of repair of the break is very much reduced in mutants lacking polymerase I (*polA⁻* mutants) which suggests that polymerase I-mediated repair replication is needed. Cells mutant only in the 5'→3'-exonuclease activity of polymerase I are also defective in this repair. The reason is that base damage invariably has occurred near each single-strand break.

The repair of base damage, including that which is not adjacent to breaks, has been examined by studying the removal of DHDT. This process is not as well understood as thymine-dimer excision but in both *E. coli* and in mammalian cells an endonuclease activity has been found that makes cuts near DHDT (and other altered bases). This enzyme is not the thymine dimer-incision nuclease inasmuch as *uvrA⁻* cells are not sensitive to x rays and the endonuclease activity is not reduced in a *uvrA⁻* cell. The incision event is followed by excision by the 5'→3' exonuclease of polymerase I, after which repair replication mediated by polymerase I and DNA ligase occurs. This process is probably required, at least in part, for repair of single-strand breaks accompanied by base damage. A DHDT-*N*-glycosylase was isolated from *E. coli* in 1981. Cells that are *recA⁻* are also extraordinarily sensitive to x rays; thus, recombination repair apparently plays an important role in the repair of x-ray damage, though at the present time its role is unknown.

Once again, evidence from cultured cells obtained from diseased humans shows that repair of x-ray damage occurs in mammalian cells. In this case the cells are obtained from an inherited retinoblastoma, a cancer of the retina; these cells are hypersensitive to x rays and unable to repair single-strand breaks efficiently.

There is little solid evidence at present for the direct repair of double-strand breaks. Since there is no obvious way to bring together two freely moving ends, the few supporting observations have been viewed with suspicion. Since it is believed that all DNA molecules, bacterial included, are associated with structural proteins, as in the DNA of mammalian chromosomes, it is questionable whether there really are two freely moving ends at the site of a potential break. Thus, the view is that when repair is observed, the two ends have been held adjacent to one another by these proteins. Further experiments are needed to resolve this question. In yeast there is good evidence for a complex mechanism for repair of double-strand breaks that requires the action of the recombination pathway.

WHY DO DAMAGED CELLS DIE?

We have discussed a variety of repair systems that seem to be able to fix everything, yet cells still are killed by various forms of radiation and toxic chemicals. Why? There are many answers to this question. One trivial reason is that there are probably some alterations that are not repairable—for instance, a base might be altered to yield a replication block that cannot be bypassed by postreplicational or SOS repair. Also, damaged regions may be opposite one another or so clustered that there is nothing to use as a template for repair. A second possibility is that the repair systems themselves can be damaged; this is not unlikely since it takes a very large amount of damage before lethality ensues in a wild-type organism. The repair systems might also be saturated by such large amounts of damage that some lethal imbalance in the protein/DNA ratio occurs. Finally, some of the repair systems may be sufficiently error-prone that mutations sometimes occur in essential genes.

REFERENCES

Grossman, L. 1974. "Enzymes involved in the repair of DNA." *Adv. Radiat. Biol.,* 4, 77–130.

Grossman, L., A. Braun, R. Feldberg, and I. Mahler. 1975. "Enzymatic repair of DNA." *Ann. Rev. Biochem.,* 44, 19–44.

Hanawalt, P. C., P. K. Cooper, A. K. Ganesan, and C. A. Smith. 1979. "DNA repair in bacteria and mammalian cells." *Ann. Rev. Biochem.,* 48, 783–836.

Hanawalt, P. C., E. C. Friedberg, and C. F. Fox. 1978. *DNA Repair Mechanisms.* Academic Press.

Hanawalt, P. C., and R. B. Setlow. 1975. *Molecular Mechanisms for Repair of DNA.* Plenum.

Haynes, R. H., and B. A. Kunz. 1982. "DNA repair and mutagenesis in yeast." *In* J. Strathern, E. Jones, and J. Broach (eds.), *Molecular Biology of the Yeast Saccharomyces: Life Cycle and Inheritance,* pp. 371–414. Cold Spring Harbor Laboratory.

Little, J. W., S. H. Edmiston, L. Z. Pacelli, and D. W. Mount. 1980. "Cleavage of the E. *coli lexA* protein by the *recA* protease." *Proc. Nat. Acad. Sci.,* 77, 3225–3229.

Nichols, W. W., and D. G. Murphy. 1977. *DNA Repair Processes.* Symposia Specialists, Miami.

Photochemical and Photobiology Reviews. 1976–present. Plenum. Short review articles on many aspects of repair appear every year.

Roberts, J. J. 1978. "The repair of DNA modified by cytotoxic, mutagenic, and carcinogenic chemicals." *Adv. Radiat. Biology.,* 6, 212–436.

Setlow, R. B. 1966. "Cyclobutane-type pyrimidine dimers in polynucleotides." *Science,* 153, 379–380.

Ward, J. 1975. "Molecular mechanisms of radiation-induced damage to nucleic acids." *Adv. Radiat. Biol.,* 5, 182–240.

10 Mutagenesis, Mutations, and Mutants

In previous chapters the use of mutants has been encountered repeatedly. In this chapter we explain what a mutant actually is, how a mutant is created, and how it is detected. We begin with the terminology.

TERMINOLOGY

The terminology used in discussing mutants is somewhat confusing because of the similarity of four words—**mutant, mutation, mutagen,** and **mutagenesis.** These terms have the following meanings.

    Mutant refers to the genetic state of the organism. That is, if one of the ensemble of characteristics (the **phenotype**) that comprises a so-called normal organism is different from the "wild-type" character, then that organism is in that respect said to be a mutant. Thus, for *E. coli,* which in nature is able to utilize galactose as a carbon source, its wild-type state is Gal$^+$, and a Gal$^-$ cell, unable to utilize galactose, is a mutant. The terminology "wild type" is a misnomer, however, and often does not refer to the state of the organism in nature because many organisms are defective (mutant) in some particular respect, even in their most common natural state. For example, *E. coli* isolated from nature is usually unable to utilize lactose (it is Lac$^-$), yet the Lac$^+$ phenotype is invariably called wild type and Lac$^-$ is called mutant.

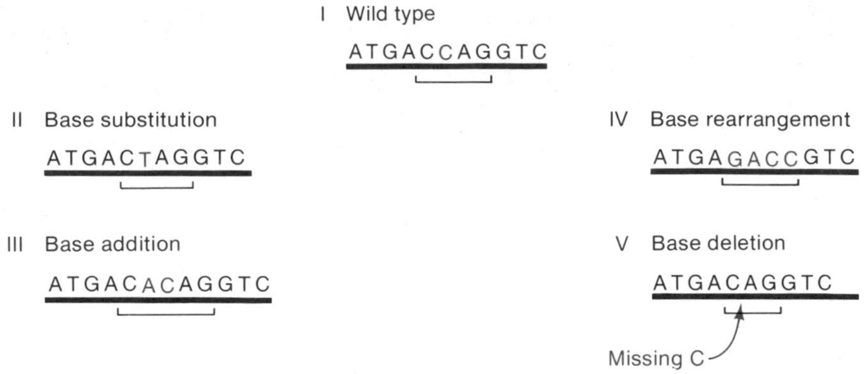

I   Wild type

ATGACCAGGTC

Figure 10-1
Four types of mutations.
Only the base sequence
in one DNA strand is
shown. Changes are
shown in red. The hori-
zontal brackets indicate
the affected segment.

II   Base substitution

ATGAC_TAGGTC

IV   Base rearrangement

ATGAGACCGTC

III   Base addition

ATGACACAGGTC

V   Base deletion

ATGACAGGTC

Missing C

Generally speaking (but not always), when referring to biochemical
properties, the + form is called wild type and a − form is a mutant. The
same points apply to nutritional requirements, such as a requirement
for an exogenous supply of arginine—that is, the ability to carry out
arginine biosynthesis (Arg⁺) is considered to be wild type and the
inability to do so (arginine-requiring or Arg⁻) is mutant; this convention
is independent of the genotype found in nature.

Mutation refers to any structural alteration of DNA that is present
in a mutant and that gives rise to the mutant phenotype. That is, an
organism is a Leu⁻ mutant because it has a mutation in the *leu* gene. A
mutation is always a change in the base sequence of DNA. The most
common change is a substitution, addition, rearrangement, or a deletion
of one of more bases (Figure 10-1).

A **mutagen** is a physical agent or a chemical reagent that causes
mutations to occur. For example, nitrous acid reacts with some DNA
bases, changing their identity and hydrogen-bonding properties; thus, it
is a mutagen.

Mutagenesis is the process of producing a mutation. If it occurs in
nature without the addition of a mutagen, it is called **spontaneous
mutagenesis** and the resulting mutants are **spontaneous mutants.** If a
mutagen is used, the process is **induced mutagenesis.** Unfortunately, the
term mutation is sometimes used when mutagenesis is meant.

TYPES OF MUTATIONS AND THEIR NOTATION

Mutations can be categorized in several ways. One distinction is based
on the nature of the change—specifically, on the number of bases
changed. Thus, we may distinguish a **point mutation,** in which there is
only a single changed base pair, from a multiple mutation, in which two
or more base pairs differ from the wild-type sequence. A point mutation
may be a **base substitution,** a **base insertion,** or a **base deletion,** but the
term most frequently refers to a base substitution.

A second distinction is based on the consequence of the change in terms of the amino acid sequence that is affected. For example, if there is an amino acid substitution, the mutant is a **missense** mutation. If the substitution produces a protein that is active at one temperature (typically 30°C) and inactive at a higher temperature (usually 40–42°C), the mutation is called a **temperature-sensitive** or Ts mutation. According to the most recent convention, a gal^- temperature-sensitive mutation is written gal^-(Ts). In Chapter 12 it will be seen that sometimes there is no amino acid that corresponds to the new base sequence; in that case, termination of synthesis of the protein occurs at that point and the mutation is called a **chain termination mutation** or a **nonsense mutation.** There are three kinds of nonsense mutations, each representing one of three particular base sequences that do not correspond to any amino acid. They are called **amber** (Am), **ochre** (Oc), and **opal** (no symbol agreed on). A gal^- amber mutant and a gal^- ochre mutant would be designated gal^-(Am) and gal^-(Oc), respectively.

When mutants are isolated in the laboratory, they are usually collected in large numbers. Normally each mutation is given a number that states the chronological order in which it was "isolated." Thus, the fourth gal^- mutation is written $gal4$. If $gal82$ and $gal223$ were Am and Ts mutations, respectively, the designations would be $gal82$(Am) and

Table 10-1

Summary of notation used to designate bacterial mutants and mutations

Phenotype or genotype	Designation*
Phenotype	
Lacking in ability to make a substance; the substance must be supplied	Sub$^-$
Possessing ability to make a substance	Sub$^+$
Resistance to an antibiotic	Ant-r
Sensitivity to an antibiotic	Ant-s
Genotype	
Wild-type gene for making a substance	*sub$^+$*
Mutant gene for making a substance	*sub$^-$*†
Mutant *subA* gene	*subA$^-$*
Mutation in *subA* gene having a temperature-sensitive phenotype	*subA$^-$*(Ts)
Mutation in *subA* gene having an amber phenotype	*subA$^-$*(Am)
Mutation number 63 in *subA* gene	*subA63*
Gene for resistance or sensitivity to a particular antibiotic	*ant*
Genotype conferring resistance to an antibiotic	*ant-r*
Genotype conferring sensitivity to an antibiotic	*ant-s*

* The arbitrary abbreviations "sub" and "ant" mean substance and antibiotic in this table.

†The notation *sub* for *sub$^-$* is widespread. Because of possible ambiguity, the superscript minus is used throughout this book.

gal223(Ts). Even though there is a "standard" notation, some variation is still seen. For example, gal3(Ts) and gal5(Am) are frequently seen written as gal ts3 and gal am5, respectively.

Often a phenotype (which is always capitalized), such as the ability to synthesize tryptophan (Trp$^+$), is determined by several genes encoding all of the enzymes in the biosynthetic pathway; such genes are usually distinguished by capital letter suffixes, as trpA, trpE, and so forth. Sometimes the suffix abbreviates the enzymatic activity as in galK (the gene for galactokinase). Mutation 103 in the trpA gene is written trpA103. The notation of bacterial mutants is summarized in Table 10-1.

ISOLATION OF MUTANTS

Mutants are usually isolated from large populations of organisms and some means is needed to distinguish a mutant and a wild-type organism. Furthermore, the mutants may constitute an extremely small fraction of the total population, so that prior to searching for a mutant it is necessary to increase the mutant fraction. These two procedures—detection, or screening, and enrichment—are described in this section.

Detection of Mutants

Mutants can be detected in several ways. Let us first examine two simple procedures used to detect bacterial mutants that are either resistant to an antibiotic or unable to utilize a particular sugar. These procedures are the following:

1. A streptomycin-resistant (Str-r) mutant is found simply by plating a large number of bacteria on agar containing streptomycin. Only a Str-r cell can form a colony. This method applies to the isolation of mutants resistant to any chemical compound that can be incorporated into agar.

2. Sugar-utilization mutants are isolated by means of color indicator plates. Consider an agar prepared from a complex broth that contains a very high concentration of all of the amino acids, the sugar lactose, and the two pH-sensitive dyes, eosin and methylene blue; this is called **EMB agar.** A portion of a culture containing both Lac$^+$ and Lac$^-$ cells is put on the agar. The Lac$^+$ cells will preferentially use the lactose as a source of energy and, in metabolizing lactose oxidatively, will excrete H$^+$ ions. Thus, the pH of the medium surrounding the cells decreases, causing the dyes that permeate the colony to become a deep purple. The Lac$^-$ cells cannot utilize the lactose but are able to use amino acids such as glycine as a source of energy. However, in the

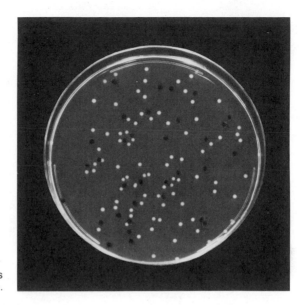

Figure 10-2
E. coli Lac⁺ (black) and Lac⁻ (white) colonies growing on EMB agar. In color the agar is a deep purple, Lac⁺ colonies are purple, and Lac⁻ colonies are white.

degradation of an amino acid, NH_3 is released, so that there is a local increase in pH around these cells. This decolorizes the dyes and produces a white colony (Figure 10-2). Thus, on EMB agar Lac⁺ colonies are purple and Lac⁻ colonies are white. These color differences occur for all metabolizable sugars and for many other carbon sources as well. Other indicator plates are also used—two of the more common types employ MacConkey agar (Lac⁺ is red, Lac⁻ is yellow) and tetrazolium agar (Lac⁺ is white, Lac⁻ is red).

The sugar-utilization mutants just described can grow in a broth lacking the sugar, if another carbon source is provided. If we seek a leucine-requiring (Leu⁻) mutant, we cannot use an agar lacking leucine, however, because the desired mutant will not grow. Instead, the elegant technique of **replica-plating** will solve the problem. When a mixture of 20 Leu⁺ cells and 5 Leu⁻ cells is plated on a minimal agar supplemented with leucine (we call this the master plate), 25 colonies will form. Ordinarily, a second petri dish is prepared that contains minimal agar lacking leucine (a test plate) and, by using 25 sterile toothpicks, a tiny portion of each Leu⁺ colony can be transferred from the master plate to the test plate; only 20 of the transfers will form a colony and, if one keeps track of the initial colonies transferred, the five Leu⁻ colonies on the master plate will be identified. By contrast, in the replica-plating procedure all 25 colonies can be transferred from the master plate to test plates in a single step, in the following way. A circular piece of sterile velvet is pressed onto the leucine-containing master plate and then picked up. A portion of each colony is retained in the hairs of the cloth.

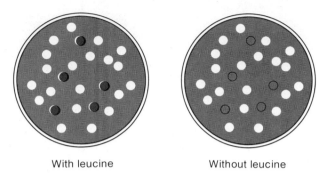

With leucine   Without leucine

Figure 10-3
The effect of replica plating 20 Leu⁺ (white) colonies and
5 Leu⁻ (red) colonies. The positions of the Leu⁻ colonies
are unoccupied in the plate at the right, which identifies the
Leu⁻ colonies in the plate at the left. In reality both Leu⁺
and Leu⁻ colonies would have the same color (white) on the
plate at the left.

The velvet is then pressed onto the surface of two fresh plates, one with leucine and one without leucine, and then colonies are allowed to grow. Leu⁺ colonies form on both plates, but the Leu⁻ cells fail to grow on the leucine-deficient agar, and the Leu⁻ colonies are identified as those which grow only on the agar containing leucine (Figure 10-3). The particular advantage of the replica-plating technique is that a very large number of colonies can be screened rapidly. For example, typically one would be seeking not five mutants in twenty-five but one in 10^4 and this can be done by replica-plating from fifty plates having two hundred colonies on each plate.

Replica-plating can also be used to isolate temperature-sensitive mutants. This is done by forming colonies at 30°C and then transferring the colonies to two plates which are incubated at either 30°C or 42°C. A colony that forms on the plate incubated at 30°C but is absent from the 42°C plate contains a temperature-sensitive mutation. Replica-plating is widely used to select mutants of all colony-forming microorganisms such as yeast and algae.

Phage mutants are detected in two ways. "Plaque-morphology" mutants are found because they look different from wild-type plaques. For example, some mutants make plaques that are turbid, or clear, or have a halo, or are a combination of these. Conditional mutants—that is, mutants that are either temperature-sensitive or host-dependent (forming plaques on one host but not on another)—are commonly used and can also be found by replica-plating. With a temperature-sensitive mutant, plaques are formed at 30°C and then transferred to two plates already seeded with bacteria, and incubated at either 30°C or 42°C. A temperature-sensitive mutant forms a plaque on the 30°C plate but not on the 42°C plate. Similarly, a host-dependent mutant can be detected by plating on a permissive host and replica-plating onto either a permissive or a nonpermissive host.

Enrichment for Mutants

If a population of bacteria that has not been mutagenized is screened for the presence of Leu$^-$ mutants, the number found will be less than one per 10^7 cells. With 200 colonies per plate, 5×10^4 plates will have to be replicated in order to find one Leu$^-$ mutant. With induced mutagenesis the frequency can be increased, somewhat facilitating the task of isolating mutants. However, it is often the case that induced mutagenesis does not increase the mutant fraction to more than one mutant per 10^5 cells, so that 500 replica plates will need to be screened to find a single mutant. In that case, mutant enrichment procedures are used that favor the growth of the mutant over the wild type. One such procedure is the **penicillin selection technique.** This method, which is applicable only to bacteria, is based upon the fact that penicillin kills bacteria by interfering with the synthesis of the bacterial cell wall. When penicillin is added to a culture of growing bacteria, cell wall synthesis stops but growth (enlargement) of the cell continues until the cell explodes. A Leu$^-$ cell in a growth medium lacking leucine cannot synthesize protein and thus cannot enlarge, so that it is not killed by penicillin. Consider a culture containing one Leu$^-$ cell per 10^6 Leu$^+$ cells growing in medium containing leucine. The culture is filtered and the cells collected on the filter are resuspended, this time in a growth medium containing penicillin but *lacking leucine.* The Leu$^+$ cells continue to grow and after one hour about 99 percent of them begin to explode; in contrast, the Leu$^-$ cells, which cannot grow, are unharmed. Incubation of the culture is stopped at this point, because the exploded cells release their contents to the medium and the pool of leucine released from these cells would allow the Leu$^-$ cells to grow, and therefore also to be killed. The treated culture at this time consists of one Leu$^-$ cell per 10^4 cells and the mutant could be found with fifty replica plates. The cells could also be passed through another enrichment cycle; there would then be one Leu$^-$ cell per 100 Leu$^+$ cells, and the mutant cell would be found with fewer plates.

Detection of Mutants in Diploid Cells

In earlier chapters it was pointed out that an essential step in the advance of molecular biology was the use of mutants to simplify the study of a system. For example, RNA synthesis can be studied in a bacterial cell that is not simultaneously synthesizing proteins, by using a Leu$^-$ mutant and a growth medium lacking leucine. Also, the existence of mutations that result in overproduction of a particular enzyme indicates that in the wild type the synthesis of the enzyme is normally regulated; the general properties of the unregulated mutant may give insight into the mechanism of regulation. The use of mutants has not generally been practical with eukaryotic cells, except when a particular eukaryote has a haploid phase, as will be seen.

Most mutations are recessive; in a diploid cell the mutation cannot be expressed to give a mutant phenotype. In order to have a mutant phenotype a mutation in both copies of the gene is needed and it is usually quite difficult to achieve that state. Following mutagenesis the probability of a mutation in a particular gene—say, A—might be one mutation per 10^3 to 10^4 cells (10^{-3} to 10^{-4} per cell). Thus, since mutagenesis is a random process, the probability of occurrence of two A^- mutations in one cell is the square of the probability of the single mutation—namely, 10^{-6} to 10^{-8} per cell. Thus, isolation of the double mutant would require screening of at least 10^6 cells. With animal cells the problem of obtaining mutants is compounded by the fact that even if the mutant fraction were not so small, there is no general method to isolate most mutants, because colony formation is inefficient and there is no equivalent to replica-plating. Some mutants are available, though. For example, most animal cells already are auxotrophic for several amino acids, so that protein synthesis, at least, can be inhibited by withholding a required amino acid from the growth medium. In addition, it has been possible to isolate mutants resistant to certain drugs, e.g., azaguanine, by adding the drug to a growing culture exposed to a mutagen and selecting survivors. These mutants have been useful in some types of experiments. However, mutants have not yet been of great value in animal cell biology. In Chapter 22 we shall see that an important breakthrough in the study of eukaryotes has been the development of recombinant DNA techniques.

The situation is quite different with eukaryotic microorganisms that have a stable, replicating haploid phase. This is because, with haploid cells, the genotype directly affects the phenotype, so that mutants are detected easily. The unicellular eukaryotic microorganisms, such as yeast, are as easy to work with as bacteria; for example, they form colonies on agar and can be replica-plated. In yeast, induced mutagenesis can raise the fraction that is mutant to 0.1 percent of the population, so that even though there is no mutant enrichment technique comparable to penicillin selection, the mutants can be easily isolated simply by direct testing of 1000 colonies. An additional advantage of the unicellular eukaryotic organisms is that haploid cells can be mated to form diploid cells and in these cells genetic recombination can occur. The diploid cells can then be stimulated to produce haploid cells, which can then be reisolated; in many of the haploid cells several mutations may be present. With yeast, thousands of single and multiple mutants of many types are available for a variety of investigations. The mutant approach has also been widely used with the unicellular protozoan *Tetrahymena* and the alga *Chlamydomonas*.

BIOCHEMICAL BASIS OF MUTANTS

A mutant may be defined as an organism in which either the base sequence of DNA or the phenotype has been changed. These definitions are the same (except for the case of a silent mutation, which will be discussed shortly) since the base sequence of DNA determines the amino acid sequence of a protein. The chemical and physical properties of each protein are determined by its amino acid sequence, so that a single amino acid change is capable of inactivating a protein. This was first demonstrated by Vernon Ingram, who found that the mutant hemoglobin molecule obtained from patients with sickle-cell anemia differed from normal hemoglobin only in that a single glutamic acid in the normal protein is replaced by valine in the mutant.

From the discussion of protein structure in Chapter 5 it is easy to understand how an amino acid substitution can change the structure, and hence, the biological activity, of a protein. For instance, consider a hypothetical protein whose three-dimensional structure is determined entirely by an interaction between one positively charged amino acid (for example, lysine) and one negatively charged amino acid (aspartic acid). A substitution of methionine, which is uncharged, for the lysine would clearly destroy the three-dimensional structure, as would substitution of histidine, which is positively charged, for aspartic acid. Similarly, a protein might be stabilized by a hydrophobic cluster, in which case substitution of glutamine (polar) for leucine (nonpolar) would also be disruptive.

An amino acid substitution does not always create a mutant. For instance, a hydrophobic cluster might be virtually unaffected by a replacement of one leucine by another nonpolar amino acid such as isoleucine. Similarly, a negatively charged amino acid might successfully substitute for another negatively charged amino acid. When an amino acid substitution has no detectable effect on the phenotype of a cell, it is called a **silent mutation.** There is also another type of silent mutation, namely, a base change without an amino acid alteration; this is a result of the redundancy of the genetic code, which will be explained in Chapter 12.

The shapes of proteins are determined by such a variety of interactions that sometimes an amino acid substitution is only partially disruptive. For instance, an isoleucine might substitute successfully for leucine and be silent, but replacement with a more bulky amino acid such as phenylalanine might cause subtle stereochemical changes, though a hydrophobic cluster is preserved. This could be manifested as a reduction, rather than a loss, of activity of an enzyme. For example, a bacterium carrying such a mutation in the enzyme that synthesizes adenine might grow very slowly (but it would grow) unless adenine is provided in the growth medium. Such a mutation is called a **leaky mutation**; these mutations are not particularly useful for most genetic studies.

Generally speaking, the following types of amino acid substitutions are expressed as nonleaky mutations: polar to nonpolar, nonpolar to polar, change of sign of a charge, small side chain to bulky side chain, sulfhydryl to any other side chain, hydrogen-bonding to non-hydrogen-bonding, any change to or from proline (which changes the shape of the polypeptide backbone), and any change in a substrate-binding site.

So far, we have discovered mutants that arise from amino acid substitutions. Three other types of alterations occur that invariably eliminate activity of the protein totally. One of these is the deletion of bases, which causes one or more amino acids to be absent in the completed protein. A second is the frameshift, which will be discussed shortly. In a frameshift all amino acids starting from a particular point (the site of the mutation) are different, so that the mutant contains a protein that can be extensively different from the wild-type protein. The third type is a chain termination mutant, in which a base change results in the production of a signal for stopping protein synthesis. In such a mutant the protein chain prematurely terminates at the site of mutation so that the mutant protein is a fragment of the wild-type protein.

In the following section we will see how an altered base sequence can arise.

MUTAGENESIS

The production of a mutant requires that a change occur in the base sequence. This can occur spontaneously by replication errors or can be stimulated to occur in five main ways: (1) removal of an incorrectly inserted base is prevented; (2) a base is inserted that tautomerizes and allows a substitution to occur in subsequent replication; (3) a previously inserted base is chemically altered to a base having different base-pairing specificity; (4) one or more bases are skipped during replication; or (5) one or more extra bases are inserted during replication. In the following sections we describe mutagens that act by one or more of these mechanisms and address the question of spontaneous mutagenesis. The information that follows is also presented in Table 10-2.

Table 10-2
Types of mutagens

Mutagen	Mode of action	Example	Consequence
Base analogue	Substitutes for a standard base during replication and causes a new base pair to appear in daughter cells in a later generation	5-Bromouracil	A · T → G · C, and G · C → A · T
		2-Aminopurine	A · T → G · C
Chemical mutagen	Chemically alters a base so that a new base pair appears in daughter cells in a later generation	Nitrous acid	G · C → A · T, and A · T → G · C
		Hydroxylamine	G · C → A · T
		Ethyl methane sulfonate (EMS)	G · C → A · T, G · C → C · G, and G · C → T · A
		NNG	Same as EMS
		Ultraviolet light	All single base-pair changes are possible.
Intercalating agents	Addition or deletion of one or more base pairs	Acridines	Frameshifts
Mutator genes	Excessive insertion of incorrect bases or lack of repair of incorrectly inserted bases	———	All single base-pair changes are possible.
None	Spontaneous deamination of 5-methylcytosine (MeC)	———	G · MeC → A · T

Note: Italicized changes in base pairs are transversions; those that are not italicized are transitions.

Base-Analogue Mutagens

By a base analogue one means a substance other than a standard nucleic acid base which can be built into a DNA molecule by the normal process of polymerization. Such a substance must be able to pair with the base on the complementary strand being copied or the 3′→5′ editing function will remove it. However, if it can tautomerize or if it has two modes of hydrogen-bonding, it will be mutagenic.

The substituted base 5-bromouracil (BU) is an analogue of thymine inasmuch as the bromine has about the same van der Waals radius as the methyl group of thymine (Figure 10-4). In subsequent rounds of replication BU functions like thymine and primarily pairs with adenine. Thymine can sometimes (but rarely) assume an enol form that is

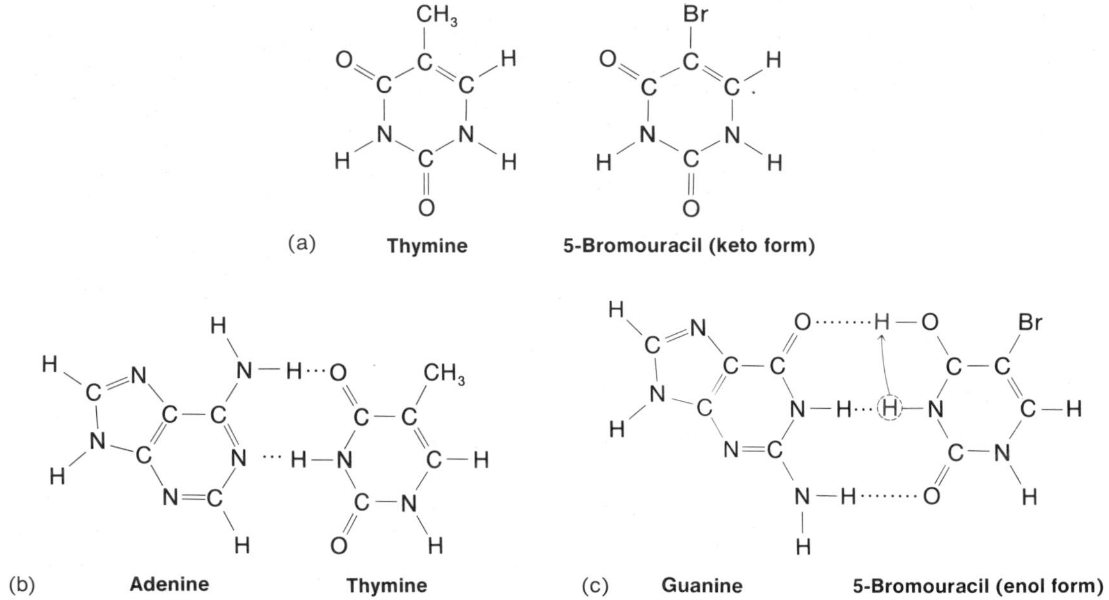

(a)   **Thymine**   **5-Bromouracil (keto form)**

(b)   **Adenine**   **Thymine**   (c)   **Guanine**   **5-Bromouracil (enol form)**

Figure 10-4
Mutagenesis by 5-bromouracil. (a) Structural formulas of thymine and 5-bromouracil. (b) A standard adenine-thymine base pair. (c) A base pair between guanine and the enol form of 5-bromouracil. The red H in the dashed circle shows the position of the H in the keto form.

capable of pairing with guanine, and this conversion occasionally gives rise to mutants in the course of replication. The mutagenic activity of 5-bromouracil stems from a shift in the keto-enol equilibrium caused by the bromine atom; that is, the enol form exists for a greater fraction of time for BU than for thymine. Thus, if BU replaces a thymine, in subsequent rounds of replication, it occasionally generates a guanine, which in turn specifies cytosine, resulting in formation of a G · C pair (Figure 10-5). BU can also induce a change from G · C to A · T. The enol form is actually sufficiently prevalent that BU is sometimes (but infrequently) incorporated into DNA in that form. When that occurs, BU is acting as an analogue of cytosine rather than thymine. However, even though it may become part of the DNA by temporarily having the base-pairing properties of cytosine, the keto form is the predominant form, so that in subsequent rounds of replication BU will usually pair like thymine. Thus, a G · C pair, which, as a result of an incorporation error, is converted to a G · BU pair, ultimately becomes an A · T pair, as shown in Figure 10-5(b). Thus, BU can stimulate a transition from A · T to G · C as well as from G · C to A · T.

Note that it takes two rounds of replication in order to generate a new base pair. Similarly, a mutant progeny cell would not appear until

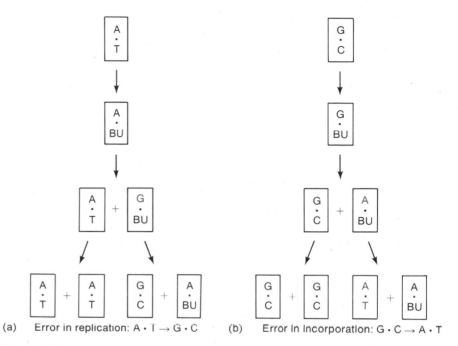

(a)   Error in replication: A · T → G · C   (b)   Error In Incorporation: G · C → A · T

Figure 10-5

Two mechanisms of 5-bromouracil- (or BU)-induced mutagenesis. (a) During replication, BU, in its usual keto form, substitutes for T and the replica of an initial A · T pair
becomes an A · BU pair. In the first mutagenic round of replication the BU, in its rare
enol form, pairs with G. In the next round of replication, the G pairs with a C, completing the transition from an A · T pair to a G · C pair. (b) During replication of a G · C
pair a BU, in its rare enol form, pairs with a G. In the next round of replication the
BU is again in the common keto form and it pairs with A, so that the initial G · C pair
becomes an A · T pair. The replica of the A · BU pair produced in the next round of
replication is another A · T pair.

two generations had elapsed; one says that it takes two generations for
the mutation to be "expressed."

Both base-pair changes induced by BU maintain the original purine
(Pu)-pyrimidine (Py) orientation. That is, the original and the altered
base pairs both have the orientation Pu · Py—namely, A · T and G · C. If
the original pair was T · A, the altered pair would be C · G—that is,
Py · Pu for both the original and the altered pairs. A base change that
does not change the Py · Pu orientation is called a **transition**. Base
analogue mutations are always transitions. Later we will see changes
from Pu · Py to Py · Pu and Py · Pu to Pu · Py; when such a change of
orientation occurs, the mutation is called a **transversion.**

Another useful base analogue is 2-aminopurine (AP), which substitutes primarily for adenine (Figure 10-6). AP does not readily tautomerize but is able to form base pairs with both thymine and cytosine.

The pairing with cytosine is weak because it forms only a single hydrogen bond but it is strong enough to allow occasional incorporation of a cytosine during subsequent replication. Thus, after two rounds of replication there is a transition from an A • T pair to a G • C pair (Figure 10-7).

2-Aminopurine   2-Aminopurine

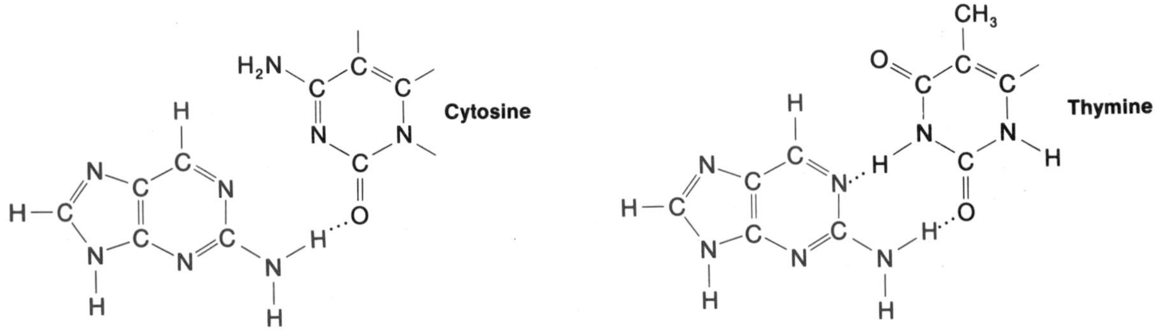

Figure 10-6
The mode of pairing of 2-aminopurine, an analogue of adenine, with both cytosine and thymine.

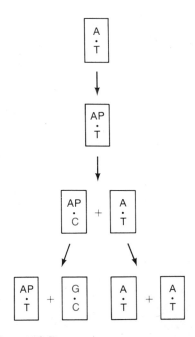

Figure 10-7
Transition induced by substitution of 2-aminopurine (AP) for A. In a later round of replication AP pairs with C, which after one more replication cycle yields a G·C pair.

Chemical Mutagens

By a chemical mutagen is meant a substance that can alter a base that is already incorporated in DNA and thereby change its hydrogen-bonding specificity. Four very powerful chemical mutagens are nitrous acid (HNO_2), hydroxylamine (HA), ethylmethane sulfonate (EMS), and N-methyl-N'-nitro-N-nitrosoguanidine (NNG). The chemical structures of these are shown in Figure 10-8.

Nitrous acid primarily converts amino groups to keto groups by oxidative deamination. Thus cytosine, adenine, and guanine are converted to uracil (U), hypoxanthine (H), and xanthine (X), respectively. These bases can form the base pairs U · A, H · C, and X · C. Therefore, the changes are G · C→A · T and A · T→G · C as cytosine and adenine respectively are deaminated. Since G and X both pair with C, the conversion G→X does not cause mutations to occur in this way (though X is involved in rare mutant formation by another mechanism). These changes are summarized in Figure 10-9. Note that these changes are all transitions also.

Hydroxylamine reacts specifically with cytosine and converts it to a modified base that pairs only with adenine, so that a G · C pair ultimately becomes an A · T pair. The chemistry of the alteration is complex.

EMS and a related substance ethylethane sulfonate (EES) are alkylating agents that react primarily with guanine and to some extent with adenine. The addition of an alkyl group to N-7 of the purine ring has two effects, both of which involve the formation of a quaternary nitrogen at that position. In one case, the quaternary group stimulates ionization of the ring in the position shown in Figure 10-10. In the ionized form the alkylated guanine can pair with thymine instead of cytosine, which results in a transition from a G · C pair to an A · T pair. The alkylated purine also has a labile N-glycosylic bond, which hydrolyzes easily to yield a depurinated site. Such a site is potentially repairable by the apurinic acid endonuclease (Chapter 9) but apparently replication sometimes precedes this repair. When replication occurs, any base can be inserted. Thus, after a second round of

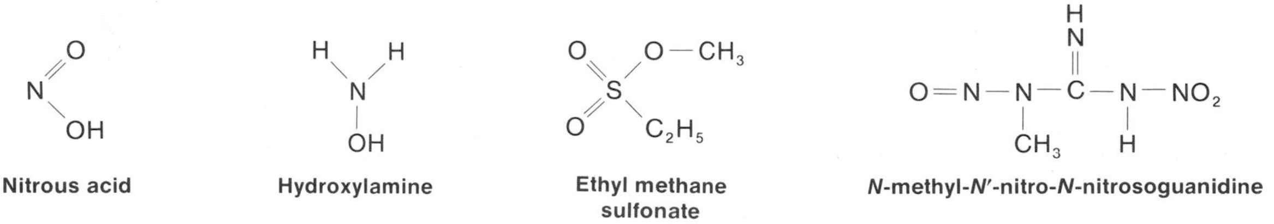

Nitrous acid   **Hydroxylamine**   **Ethyl methane sulfonate**   **N-methyl-N'-nitro-N-nitrosoguanidine**

Figure 10-8
Structures of four chemical mutagens.

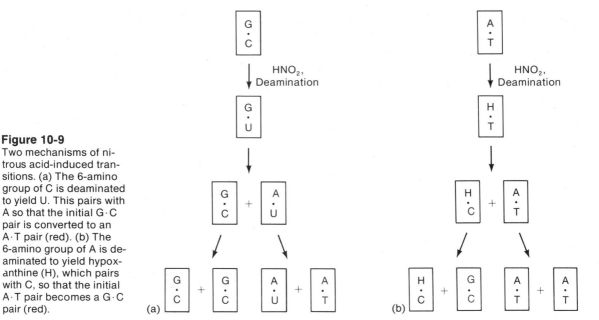

Figure 10-9
Two mechanisms of nitrous acid-induced transitions. (a) The 6-amino group of C is deaminated to yield U. This pairs with A so that the initial G·C pair is converted to an A·T pair (red). (b) The 6-amino group of A is deaminated to yield hypoxanthine (H), which pairs with C, so that the initial A·T pair becomes a G·C pair (red).

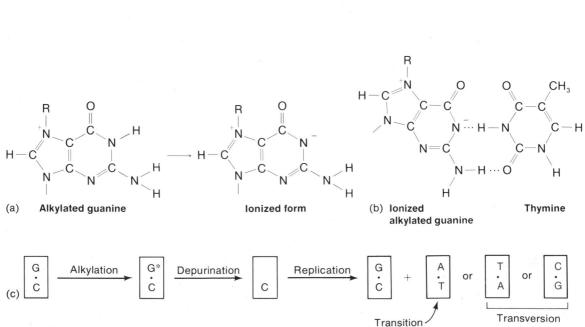

Figure 10-10
Mechanism of mutagenesis by an alkylating agent. (a) Ionization of alkylated guanine. (b) Hydrogen bonding of ionized alkylated guanine with thymine. (c) Depurination followed by random insertion of a base pair during replication.

replication the original G · C pair can be converted to a C · G, A · T, or T · A pair (Figure 10-10). This is the first example we have seen of a transversion. Treatment of phages with low pH buffers or at high temperature also produces depurinations (see Chapter 4). On replication of phages treated in this way, numerous transversions occur, in agreement with the replication error mechanism just suggested.

NNG is an unusual and extraordinarily powerful mutagen that acts by alkylation. It is possible to produce a population that is 1 percent mutant for any particular gene. Bacteria carrying an NNG-induced mutation are usually found to be multiply mutant in the sense that there are often additional mutations in the same gene and in other genes. For example, an NNG-induced *gal⁻* mutant is frequently *pro⁻* also. The distribution of the mutations is extraordinary—the mutations are clustered in genes that are adjacent in the genetic map. The explanation for this phenomenon is that NNG exerts its most powerful effect in a replication fork. Thus, in each bacterium, mutations are clustered in the region of the DNA that was traversed by the replication fork during exposure to the mutagen. Since the mutagenesis is usually performed in a nonnutrient buffer in which replication proceeds exceedingly slowly, the mutations that are formed are very near one another. NNG has the advantage that mutants are easily found, but has the disadvantage that a desired mutation must usually be separated from unwanted nearby mutations by subsequent genetic manipulations in the laboratory.

Ultraviolet Irradiation

Ultraviolet light is a fairly potent mutagen. In *E. coli* the number of mutants induced by ultraviolet light can be reduced by exposure to visible light (photoreactivation), which implicates the thymine dimer and other pyrimidine dimers. When a RecA⁻ bacterium is ultraviolet-irradiated, no mutants result; from this it is apparent that ultraviolet-induced mutagenesis requires a DNA repair mechanism—either recombination repair or the SOS system (see Chapter 9). It is generally thought that the system responsible for the mutagenesis is the error-prone SOS repair system.

Mutagenesis by Intercalating Substances

Acridine orange, proflavine, and acriflavine (Figure 10-11), which are substituted acridines, are planar, three-ringed molecules whose dimensions are roughly the same as those of a purine-pyrimidine pair. In aqueous solution, these substances form stacked arrays and are also able to stack with a base pair; this is done by insertion between two base

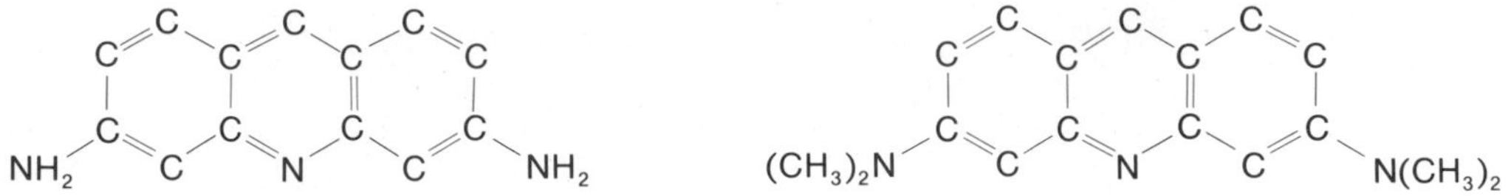

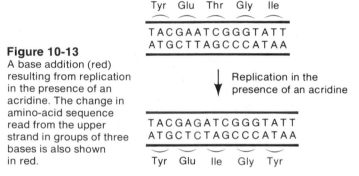

Figure 10-11
Structures of two mutagenic acridine derivatives.

Figure 10-12
Separation of two base pairs (shown in red) by an intercalating agent.

Figure 10-13
A base addition (red) resulting from replication in the presence of an acridine. The change in amino-acid sequence read from the upper strand in groups of three bases is also shown in red.

pairs, a process called **intercalation.** Since the thickness of the acridine molecule is approximately that of a base pair and because the two bases of a pair are normally in contact, the intercalation of one acridine molecule causes adjacent base pairs to move apart by a distance equal to that of the thickness of one base pair (Figure 10-12). This has bizarre effects on the outcome of DNA replication, though the mechanism of action of the mutagen is not known. When DNA containing intercalated acridines is replicated, additional bases appear in the sequence (Figure 10-13). The usual addition is a single base, though occasionally two bases are added. Deletion of a single base also occurs but this is far less common than base addition. Mutations of this sort are called **frameshift mutations.** This is because the base sequence is read in groups of three bases when it is being translated into an amino acid sequence and the addition of a base changes the reading frame (Figure 10-13). This will be discussed in greater detail in the section on reversion.

Mutagenesis by Insertion of Long Segments of DNA (Transposable Elements)

E. coli, and many other organisms as well, contain long DNA segments (hundreds to thousands of base pairs long) that are mobile and that are called **transposable elements.** In a complex way to be described in Chapter 19, a transposable element replicates; one replica remains at the original insertion site and the other replica is inserted in another region of the chromosome. This process of insertion of a replica at a second site is called **transposition.** When transposition occurs, the sequence frequently inserts itself into a bacterial gene, thereby mutating that gene. An *E. coli* phage called Mu (for mutator) also inserts its DNA, at one stage of its life cycle, into the bacterial chromosome at random sites, and thereby generates mutations. The insertion mechanism of Mu is similar to that of transposition. Insertion sequences and Mu will be discussed in detail in Chapter 19.

Mutator Genes

In *E. coli* there are genes that in a mutant state cause mutations to appear very frequently in other genes throughout the genetic map. These genes are called **mutator genes.** This is a misnomer, because the function of each gene is probably to keep the mutation frequency low; that is, it is only when the product of a mutator gene is itself defective that there is widespread production of mutations. With two exceptions the mechanisms by which these genes act are unknown.

Polymerases make occasional errors and correct these errors by means of $3' \rightarrow 5'$-exonuclease editing functions. An alteration in the region of the polymerase molecule that reduces or eliminates the editing function constitutes a mutator. Of course, mutations are not being created—it is only that random errors are not being corrected.

A second mutator gene is the *dam* gene described in Chapter 9. This gene is responsible for the methylation of DNA. If a misincorporated base is missed by the $3' \rightarrow 5'$ editing function, a mismatch repair system comes into play and removes one member of the pair. The greater methylation of the parental strand compared to the methylation of the newly synthesized daughter strand instructs the mismatch repair system to remove the incorrect base from the daughter strand. In a *dam⁻* mutant there is no methylation of the parent (or any other) strand and the mismatch repair system frequently excises the parental (correct) base and inserts a base that can pair with the daughter (incorrect) base. Note that this gives rise to a mutation that does not require additional replication for expression.

MUTATIONAL HOT SPOTS

It has already been pointed out that in a particular protein some amino acids can be replaced by others without changing the phenotype. Furthermore, in Chapter 12 it will be seen that all base substitutions do not lead to amino acid changes. Thus, it should be expected that within a particular gene, mutations will not be distributed randomly but should be confined to certain sites. An average gene contains 1000–2000 base pairs and changes in about half of these will produce amino acid substitutions. Of these substitutions perhaps half again will produce mutant proteins and one might expect about 250–500 distinct mutational sites in a typical gene. Furthermore, if several thousand mutations are independently isolated, each of the distinct sites might be represented by more or less the same number of isolates; however, this is not the case.

The most detailed study of the distribution of mutations in a single gene was made by Seymour Benzer, using the *rII* gene of *E. coli* phage T4. Suspensions of T4 phage were treated with a variety of mutagens and thousands of *rII* mutants were isolated and mapped. The spatial distribution of the observed mutations is far from random—certain sites are altered with very high frequency, as shown in Figure 10-14. A similar phenomenon has been observed with the *lacZ* gene and the *trpE* gene of *E. coli* and in a few other systems. A site at which the number of mutations isolated vastly exceeds the number at other sites is called a **hot spot.**

There are probably many explanations for hot spots. For example, with chemical mutagenesis certain bases might be in regions of the DNA that breathe more often and hence are more susceptible to attack. Certain sequences might also be more prone to accept a base analogue or to produce an error during SOS repair. However, there are a few cases in which the explanation seems clear.

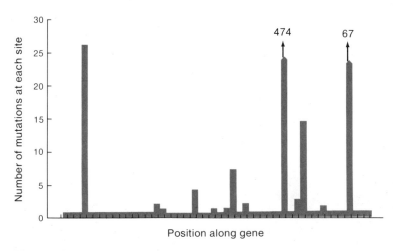

Figure 10-14
A portion of the T4 *rII* gene showing the number of mutations isolated at each site. (From S. Benzer. 1961. *Proc. Nat. Acad. Sci.,* 47: 410.)

About 5 percent of the cytosines in a typical DNA molecule are in the methylated form—5-methylcytosine (MeC). The role of MeC (and other methylated bases as well) is not clearly understood in most cases, though one role is described in Chapter 20. At any rate, they are not harmful and do not change the hydrogen-bonding properties of the base—that is, MeC pairs with guanine just as cytosine does. MeC is also subject to alteration by spontaneous deamination, as discussed earlier for cytosine. When cytosine is deaminated, uracil is formed and this is removed by uracil N-glycosylase. However, when MeC is deaminated, the result is 5-methyluracil, which is another name for thymine; therefore the G · MeC pair becomes a G · T pair, which in subsequent replication yields an A · T pair. Note that the mismatch repair system (Chapter 9) is capable of converting the G · T pair back to a correct G · C pair. However, spontaneous deamination can occur in a nonreplicating DNA molecule (e.g., in a resting cell or in a phage), and both strands will be equally methylated by the *dam*-gene product (Chapter 9). Thus the mismatch repair system receives no signal indicating that the G · C pair is the correct one and could just as well convert the G · T pair to an A · T pair. Thus the mutation frequency can be very high at a MeC site. A given G · MeC→A · T transition will of course only produce a mutant if the change causes an amino acid substitution that affects the activity of the gene product. Such MeC sites do not occur very often, so hot spots should not be particularly frequent. Direct determination of the base sequence of several genes and of hot spot mutants has shown that, indeed, MeC accounts for most of the hot spots for spontaneous mutagenesis.

This phenomenon probably also explains some of the hot spots seen in nitrous acid mutagenesis. Nitrous acid is also a deaminating agent. The C→U and A→H deaminations will often be repaired by uracil- and hypoxanthine-N-glycosylase, respectively. The G→X change does not change the base-pairing specificity, but the 5-MeC→T transition should be highly mutagenic.

DELETIONS, DUPLICATIONS, AND REARRANGEMENTS

At very low frequency gross genetic changes occur that must be classified as mutations. These phenomena—deletions, duplications, and sequence rearrangement—often involve thousands or even tens of thousands of bases.

Deletions of a few hundred to a few thousand base pairs arise spontaneously, but such deletions can also be stimulated by cross-linking agents. Presumably, large segments of cross-linked DNA are discarded in some way in order that separation of daughter cells can occur. It seems likely that some recombinational process is needed.

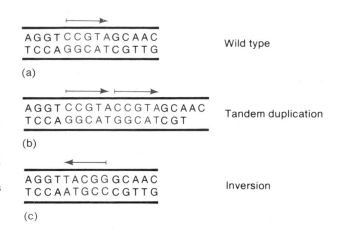

Figure 10-15

A section of a DNA molecule demonstrating a tandem duplication and an inversion of a sequence of five bases (red). The arrows indicate an arbitrary direction for reading the base sequence and are given only to facilitate viewing the sequence. In a real organism, duplicated and inverted sequences may contain many thousands of bases.

In a duplication (or triplication, which sometimes occurs), a base sequence repeats and the identical sequences appear in tandem (Figure 10-15(b)). The duplicated sequence is typically so long that many genes are included. Special techniques are required to isolate and characterize a genetic duplication.

A typical gene rearrangement is an inverted sequence; that is, a segment of DNA has in effect been cut out and reinserted with a reversed orientation (Figure 10-15(c)). A rearrangement will usually have a mutant phenotype with respect to the two genes at which the interchange has occurred; if the genes are nonessential for growth, the mutations will not be recognized.

The mechanisms of formation of duplications and rearrangements are not known—presumably, they entail errors in replication, recombination, or both.

Gene duplications are rather unusual in microorganisms. However, they are very common with animal cells, as shown by C_0t analysis (Chapters 4 and 23). In most cases, the significance of the duplications is not known, though often in amphibian cells, gene multiplication seems to arise when multiple copies are needed to make a particularly large amount of a gene product.

One type of gene duplication accounts for drug resistance in animal cells. For instance, most animal cells are killed by the drug methotrexate, an antitumor agent that inhibits the reduction of tetrahydrofolate. Resistant cell lines have been found and these frequently have multiple copies of the gene responsible for tetrahydrofolate synthesis. In this way the concentration of the enzyme inhibited by methotrexate is rendered so high that active enzyme is always present. Thus, even in the presence of the inhibitor, the concentration of the tetrahydrofolate-synthesizing enzyme is so high that active enzyme is always present.

REVERSION

So far we have discussed changes from the wild-type to the mutant state. The reverse process, in which the wild-type phenotype is regained, also occurs; this process is called **back mutation, reverse mutation,** or, most commonly, **reversion.** One way that the wild-type phenotype may be restored is to regain the wild-type genotype (that is, the wild-type base sequence). However, this is not always what happens because, as will be seen shortly, reversion can occur in several ways. We first examine how reversion is detected.

Detection of Revertants

If 10^4 Leu$^-$ bacteria are plated on agar lacking leucine, no colonies will form, but if 10^7 bacteria are plated, a few colonies will appear. If one of these colonies is collected and suspended in a buffer, and a few hundred cells of the suspension are plated, each cell will form a new colony because the colony consists of bacteria that make their own leucine; these bacteria are called **revertants.** Notice that nothing is said about whether the genotype is leu^+ but only that the cells exhibit the Leu$^+$ *phenotype*—that is, they can grow in the absence of leucine. Only if it can be shown that the *genotype* is also leu^+, however, can the cells be said to be *leu* revertants.

Reversion is a result of spontaneous mutagenesis and thus it is a nearly random process. It is important to realize that revertants are not formed as a result of the absence of leucine but rather that they preexisted in the population and were merely selected by growth without leucine. It is possible, on the other hand, in the laboratory to stimulate reversion by mutagenesis, as we will see shortly.

Some mutations do not revert at a detectable frequency. When this is the case, there are usually two possible reasons: (1) The mutant is a double mutant—that is, two base sequences, which are usually not adjacent, are changed; since the probability of correcting two defects is the product of the probabilities of correcting each defect separately, the frequency of reversion of a double mutant is ordinarily too low to be detected. (2) The mutation is a deletion of many consecutive bases; the probability of replacement of several bases is nil. The lack of the ability to induce reversion by any mutagen is a useful criterion for the presence of a multiple mutant or a large deletion.

Second-Site Revertants

The most useful type of revertant in molecular biological studies has been a kind that does not faithfully recreate the wild-type base sequence. In these revertants the reversion does not occur at the site of the mutation but instead entails a mutation at a second site. Such mutations are often called **second-site** or **suppressor mutations.** In some cases the reverse mutation does not even occur within the mutated gene, and we may distinguish between intragenic and extragenic reversion. One example of extragenic reversion will be discussed shortly, but the major treatment of this phenomenon is in Chapter 12.

The somewhat large value of the frequency of reversion makes one suspect that spontaneous reversion does not always result in a restoration of the wild-type base sequence. In a population of about 10^9 leu^+ cells, only about 100 leu^- mutants are typically present, and the mutations creating these mutants will in general be distributed over roughy 100 different sites in the DNA. In a second population there may also be 100 mutants; some may be mutant at sites present in the first population but most will be located at other sites. In many populations there will be, roughly speaking, mutants representing about 500 different sites. Thus, if we were to insist that a reversion were to occur at one particular site (in order to revert a particular mutant to the wild-type base sequence), then on the average only $100/(500 \times 10^9)$ or one cell in 5×10^9 would be mutated at that site. If base changes occur at random and a change from any one base pair to one of the other three is equally probable, then the original base sequence would be restored in about one cell in 1.5×10^{10} cells. However, it is usually found that in a population derived from a single Leu^- mutant, about one in 10^8 cells is Leu^+. For any particular mutant, the number just stated can vary widely, but it is frequently observed that the fraction of the mutant population with the revertant phenotype is much too high to be explained by a return to the wild-type base sequence. The explanation is simply that reversion events occurring at many different sites can produce the same phenotype. This has been confirmed by biochemical data in which it has been shown that the amino acid sequence in a revertant is often not the wild-type sequence and that the original mutational amino acid substitution is still present. The following discussion shows that this result is not unexpected.

Consider a hypothetical protein containing 97 amino acids whose structure is determined entirely by an ionic interaction between a positively charged $(+)$ amino acid at position 18 and a negative one $(-)$ at position 64 (Figure 10-16). If the $(+)$ amino acid is replaced by a $(-)$ amino acid, the protein is clearly inactive. Three kinds of reversion events would restore activity (Figure 10-16(a)): (1) The original $(+)$ amino acid could be put back. (2) A different $(+)$ amino acid could be put at position 18. (3) The $(-)$ amino acid at position 64 could be replaced by a $(+)$ amino acid; this second-site mutation would restore the activity of the protein. A possibility which would not generally work but which might work in a specific case is to insert a $(+)$ amino acid at position 17 or 19.

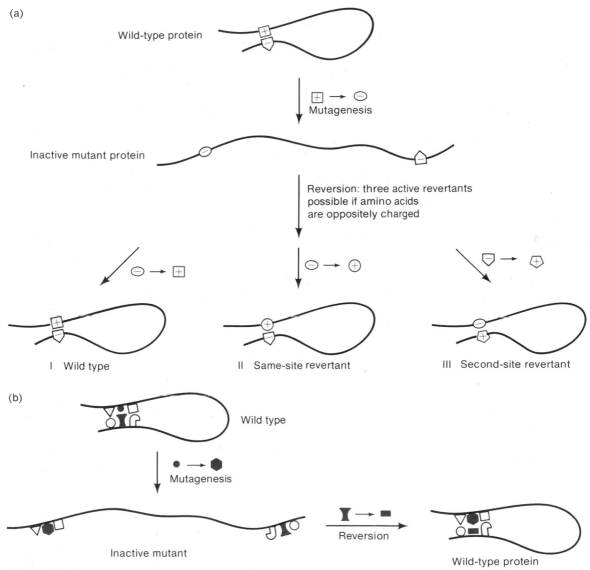

Figure 10-16
Several mechanisms of reversion. In panel (a) the charge of one amino acid is changed and the protein loses activity. The activity is returned by (I) restoring the original amino acid, or (II) by replacing the (−) amino acid by another (+) amino acid, or (III) by reversing the charge of the original (−) amino acid. In each case the attraction of opposite charges is restored. In panel (b) the structure of the protein is determined by interactions between six hydrophobic amino acids. Activity is lost when the small circular amino acid is replaced by the bulky hexagonal one and is restored when space is made by replacing the convex amino acid by the small rectangle.

Figure 10-16(b) shows another, but more complicated, example of intragenic reversion. In this case the structure of a protein is maintained by a hydrophobic interaction. The replacement of an amino acid with a small side chain by a bulky phenylalanine changes the shape of that region of the protein. A second amino acid substitution providing space for the phenylalanine could restore the protein structure.

The analysis of second-site amino acid substitution has been an important aid in determining the three-dimensional structure of proteins because the following rule is often obeyed:

> If a substitution of amino acid A by amino acid X, which creates a mutant, is compensated for by a substitution of amino acid B by amino acid Y, then A and B are either three-dimensional neighbors or are both contained in two interacting regions.

Revertants are sometimes found to be temperature sensitive; though a second-site mutation can restore activity, the new interaction that replaces the original interaction destroyed by the mutation may not be as strong as the wild-type interaction and hence more easily disrupted by thermal motion.

Second-Site Revertants of Frameshift Mutations

Revertants of frameshift mutations almost always occur at a second site. It is of course possible that a particular added base could be removed or a particular deleted base could be replaced by a spontaneous event, but this would not occur very often. Second-site reversion of a frameshift mutation has two requirements illustrated in Figure 10-17: (1) the

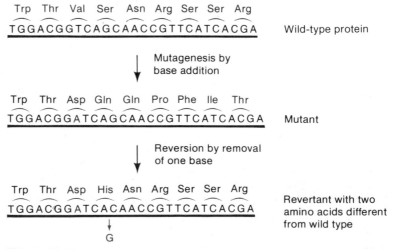

Figure 10-17
Reversion by base deletion from an acridine-induced base-addition mutant.

reverting event must be very near the original site of mutation, so that very few amino acids are altered between the two sites; and (2) the segment of the polypeptide chain in which both changes occur must be able to withstand substantial alterations. Frameshift mutations will be considered in greater detail in Chapter 12, when the genetic code is examined.

Reversion at the Mutational Site Stimulated by Mutagens

There is one instance in which reversion may frequently occur at the mutational site—namely, when reversion of a base substitution mutation is stimulated by certain mutagens that themselves cause base substitution. We have seen that substitution of 5-bromouracil (BU) for thymine or 2-aminopurine for adenine often causes a change from an $A \cdot T$ pair to a $G \cdot C$ pair. If a cell containing a 5-bromouracil-induced mutation is mutagenized by growth in the presence of 2-aminopurine, reversion is stimulated (that is, the reversion frequency in 2-aminopurine is greater than the spontaneous reversion frequency) presumably because the frequency of second-site base substitutions is greater. The reversion frequency is even greater if the mutagen is hydroxylamine, which causes $G \cdot C \rightarrow A \cdot T$ transitions. It is thought that the increased stimulation comes about because the hydroxylamine not only causes second-site reverting mutations but also sometimes restores the original $A \cdot T$ pair by altering the mutant $G \cdot C$ pair. Similarly, hydroxylamine-induced mutants are stimulated to revert by hydroxylamine but more so by growth in medium containing 5-bromouracil or 2-aminopurine. In summary, base substitution mutations can be reverted by any base-substituting mutagen but reversion frequency is greater if there is the potential for restoring the original base sequence.

It should be noted that growth in acridines, which cause frameshift mutations, would not generally stimulate reversion of base substitution mutations. On the other hand, frameshift mutations are stimulated to revert by acridines but not by base-substituting mutagens. Whether a mutation can be reverted by a base substitution mutagen or by an acridine is a useful criterion for identifying a base substitution or a frameshift mutation, respectively.

Extragenic Reversion

Extragenic reversion refers to a mutational change in a second gene that eliminates or suppresses the mutant phenotype. One type of suppression, which occurs when two proteins interact, has already been examined (Chapter 5) and occurs in the following way. A mutation in the binding site of protein A prevents the protein from interacting with protein B. Another mutation in the binding site of protein B alters this binding site so that the mutant B protein can bind to the mutant A protein; thus, the interaction between the two proteins is restored. The occurrence of extragenic reversion of this kind is an important indicator of the interaction between two proteins and is a splendid example of a genetic result with profound implications for molecular structure.

A second type of extragenic reversion has the remarkable property that the second-site mutation not only eliminates the effect of the original mutation but also suppresses mutations in many other genes as well. This type will be examined in detail in Chapter 12.

Reversion as a Means of Detecting Mutagens and Carcinogens

The detection of a mutation conferring antibiotic resistance entails a positive selection in that only the mutants can grow in agar containing the antibiotic. Detecting a nutritional mutant such as Leu$^-$ is a negative process in the sense that all cells can grow on agar lacking the nutrient except the mutant. We have seen that the negative selection is not straightforward, requiring enrichment procedures and replica plating. On the other hand, the isolation of a Leu$^+$ revertant is a positive procedure. Furthermore, the frequency of reversion of a nutritional mutant is usually higher than that of production of antibiotic resistance, so that positive selection for reversion of a nutritional mutant is a more convenient way of measuring mutation frequencies and the effectiveness of various mutagens. This principle has led to the development of a technique for detecting potential carcinogens—that is, substances that induce cancer in experimental animals.

Many carcinogens are able to bind to DNA or to alter DNA bases and thereby act as mutagens with bacteria. Together these facts led to the hypothesis that most carcinogens are mutagens and that mutagenesis is one way to induce cancer. A very simple test for mutagenesis was established—namely, a carcinogen was put in agar, known numbers of a mutant bacterium were plated, and the number of revertants was measured. When the number of revertants exceeded that observed in the absence of the carcinogen, the substance was considered mutagenic. This test fails for many known carcinogens, however, because some substances are not by themselves carcinogenic but are converted to an active carcinogen by some metabolic step that occurs in animals but not in bacteria. Such a conversion system has been found in an extract of the rat liver and is used in the following test for carcinogens developed by Bruce Ames.

A set of histidine-requiring (His⁻) mutants of the bacterium *Salmonella typhimurium,* which contain either a base substitution or a frameshift mutation, are used for tests of reversion to His⁺. The frequency of spontaneous His⁺ revertants is low in this mutant but revertants are readily produced in one or more of these mutants by most known mutagens. Agar is prepared containing (1) a very small amount of histidine, sufficient to initiate growth of individual cells but not enough for colony formation (needed because two rounds of replication are required for a mutant to be expressed, as was discussed earlier), and (2) a known carcinogen or substance to be tested. A small amount of an extract of rat liver and about 10^8 His⁻ cells are spread on the agar (plate A). The same number of cells is applied to another plate (B) which lacks the carcinogen; the number of colonies appearing on plate B is usually about 5 to 10 and these are the spontaneous revertants. When a known carcinogen is present, more colonies are found on plate A. The number of colonies on plate A depends on the concentration of the substance being tested and, for a known carcinogen, correlates roughly with its known effectiveness as a carcinogen.

The Ames test has now been used with thousands of substances and mixtures (such as industrial chemicals, food additives, pesticides, hair dyes, etc.) and numerous unsuspected substances have been found to stimulate reversion in this test. This does not mean that the substance is definitely a carcinogen but only that it has a high probability of being so. The final proof of carcinogenicity always comes from testing for tumor formation in laboratory animals. However, the Ames test delimits the number of substances that have to be tested in animals since to date no substance that fails the Ames test causes tumors. The summary of these findings is that

Most carcinogens are mutagens but not all mutagens are carcinogens.

Until proof to the contrary is obtained, however, all substances that do stimulate reversion by this test should be treated with caution.

Another test, the Devoret test, which is based on a slightly different principle, will be described in Chapter 16.

REFERENCES

Ames, B. W. 1979. "Identifying environmental chemicals causing mutations and cancer." *Science,* 204, 587–593.

Ames, B. W., W. E. Durston, E. Yamasaki, and F. D. Lee. 1973. "Carcinogens are mutagens: a simple test system combining liver homogenates for activation and bacteria for detection." *Proc. Nat. Acad. Sci.,* 70, 2381–2385.

Brenner, S., L. Barnett, F. H. C. Crick, and L. Orgel. 1961. "The theory of mutagenesis." *J. Mol. Biol.,* 3, 121–124.

Budowsky, E. I. 1976. "The mechanism of the mutagenic action of hydroxylamine." *Prog. Nucl. Acid Res. Molec. Biol.,* 16, 125–187.

Drake, J. W. 1970. *The Molecular Basis of Mutation.* Holden-Day.

Drake, J. W., and R. H. Baltz. 1976. "The biochemistry of mutagenesis." *Ann. Rev. Biochem.,* 45, 11–38.

Freese, E. 1971. "Molecular mechanisms of mutation." *Chem. Mutagens,* 1, 1.

Hayes, W. 1968. *The Genetics of Bacteria and Their Viruses.* Ch. 13. Halsted.

Heidelberger, C. 1975. "Chemical carcinogens." *Ann. Rev. Biochem.,* 44, 79–121.

Orgel, L. E. 1965. "The chemical basis of mutation." *Adv. Enzym.,* 27, 289–346.

Stent, G. S., and R. Calendar. 1978. *Molecular Genetics.* Ch. 12. W. H. Freeman and Co.

Streisinger, G., Y. Okada, J. Emrich, J. Newton, A. Tsugita, E. Terzhagi, and M. Inouye. 1966. "Frameshift mutations and the genetic code." *Cold Spring Harb. Symp. Quant. Biol.,* 31, 77–84.

11 Transcription

Gene expression is accomplished by the transfer of genetic information from DNA to RNA molecules and then from RNA to protein molecules. RNA molecules are synthesized by using the base sequence of one strand of DNA as a template in a polymerization reaction that is catalyzed by enzymes called DNA-dependent RNA polymerases or simply **RNA polymerases.** The process by which RNA molecules are initiated, elongated, and terminated is called **transcription.**

Two aspects of transcription must be considered—(1) enzymology, and (2) the signals that determine at what points on a DNA molecule transcription begins and stops. The enzymatic mechanism of transcription has been fairly well understood for many years. The recognition sites were extremely difficult to study until the development a few years ago of recombinant DNA techniques (Chapter 20), which allow the isolation of particular segments of a larger DNA molecule, and of a procedure for determining the base sequence of DNA. In every case cited in this chapter, information about recognition sites has been obtained by a combination of these techniques.

Throughout this chapter and elsewhere in this book we will be stating base sequences of both DNA and RNA molecules. For DNA there is no reason to give the sequences in both strands (which would be redundant), so a convention has been adopted by which only the sequence of the **noncoding** DNA strand is stated. This strand has been chosen since, for a transcribed region, the base sequence on this strand is that of the mRNA synthesized from that region. This convention has the advantage that the base sequences corresponding to amino acids (the codons) and to start and stop signals are in their standard forms and thus easily recognizable.

We begin the treatment of transcription by describing the characteristics of the enzymatic mechanism of polymerization of RNA.

ENZYMATIC SYNTHESIS OF RNA

In this section we describe the basic features of the polymerization of RNA, the identity of the precursors, the nature of the template, the properties of the polymerizing enzyme, and the mechanisms of initiation, elongation, and termination of synthesis of an RNA chain.

Basic Features of RNA Synthesis

The essential chemical characteristics of the synthesis of RNA are the following:

1. The precursors in the synthesis of RNA are the four ribonucleoside 5'-triphosphates (rNTP) ATP, GTP, CTP, and UTP. On the ribose portion of each NTP there are two OH groups—one each on the 2'- and 3'-carbon atoms (Figure 11-1).
2. In the polymerization reaction a 3'-OH group of one nucleotide reacts with the 5'-triphosphate of a second nucleotide; a pyrophosphate is removed and a phosphodiester bond results (Figure 11-2). This is the same reaction that occurs in the synthesis of DNA.
3. The sequence of bases in an RNA molecule is determined by the base sequence of the DNA. Each base added to the growing end of the RNA chain is chosen by its ability to base-pair with the DNA strand used as a template; thus, the bases C, T, G, and A in a

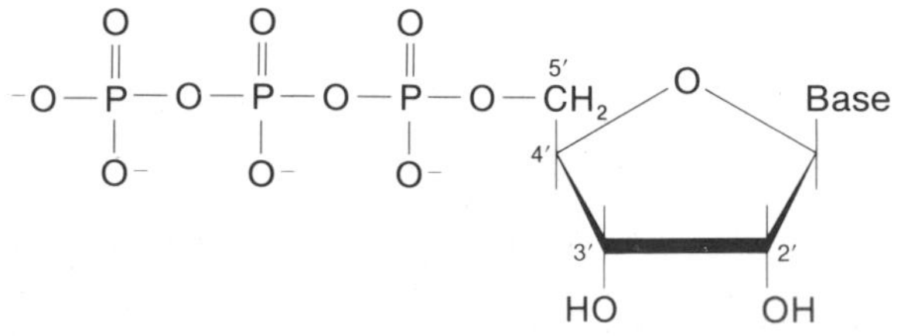

Figure 11-1
A ribonucleoside 5'-triphosphate. The 2'-OH group shown in red is replaced by an H in a *deoxy*nucleotide.

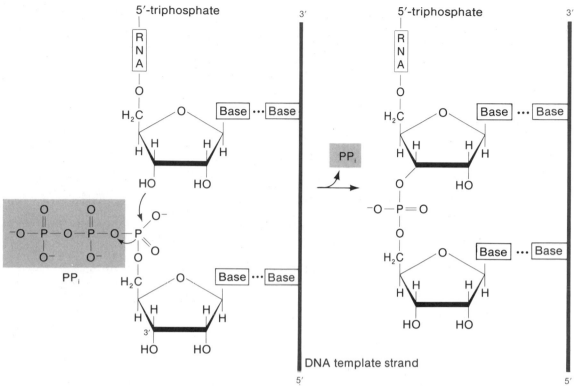

Figure 11-2

Mechanism of the chain-elongation reaction catalyzed by RNA polymerase. The red arrow joins the reacting groups. The pyrophosphate group (shaded in red) and the red hydrogen atom do not appear in the RNA strand. The DNA template and the RNA strands are antiparallel, as in double-stranded DNA.

DNA strand cause G, A, C, and U, respectively, to appear in the newly synthesized RNA molecule.

4. The DNA molecule being transcribed is double-stranded, yet in any particular region only one strand serves as a template. The meaning of this statement is shown in Figure 11-3.

5. The RNA chain grows in the 5'→3' direction: that is, nucleotides are added only to the 3'-OH end of the growing chain—this is the same as the direction of chain growth in DNA synthesis. Furthermore, the RNA strand and the DNA template strand are antiparallel to one another.

6. RNA polymerases, in contrast with DNA polymerases, are able to initiate chain growth; that is, no primer is needed.

7. Only ribonucleoside 5'-triphosphates participate in RNA synthesis and the first base to be laid down in the initiation event is a triphosphate. Its 3'-OH group is the point of attachment of the

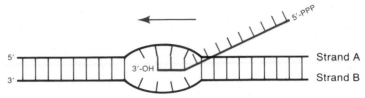

Figure 11-3
An RNA strand (shown in red) is copied only from strand A of a segment of a DNA molecule. No RNA is copied from strand B in that region of the DNA molecule. However, elsewhere, for example, in a different gene, strand B might be copied; in that case, strand A would not be copied in that region of the DNA. The RNA molecule is antiparallel to the DNA strand being copied and is terminated by a 5'-*tri*phosphate at the nongrowing end. The red arrow shows the direction of RNA chain growth.

subsequent nucleotide. Thus, the 5' end of a growing RNA molecule terminates with a triphosphate (Figure 11-3).

The overall polymerization reaction may be written as

$$n\text{NTP} + \text{XTP} \xrightarrow[\text{Mg}^{2+}]{\text{DNA, RNA-P}} (\text{NMP})_n^{-}\text{XTP} + n\text{PP}_i$$

in which XTP represents the first nucleotide at the 5' terminus of the RNA chain, NMP is a mononucleotide in the RNA chain, RNA-P is RNA polymerase, and PP_i is the pyrophosphate released each time a nucleotide is added to the growing chain. The Mg^{2+} ion is required for all nucleic acid polymerization reactions.

Most of the features just described and other properties of RNA synthesis have been determined by experiments using the following simple assay for polymerization. RNA is insoluble in trichloracetic acid (TCA) and forms a precipitate, but the nucleotides are soluble in TCA. Thus, a reaction mixture is prepared containing DNA, Mg^{2+}, RNA polymerase (which may be in a crude extract), three nonradioactive NTP, and one radioactive NTP labeled either in the base with ^3H or ^{14}C or in the phosphate attached to the ribose (the α phosphate) with ^{32}P. Following an incubation period, TCA is added and the mixture is filtered. If RNA has been synthesized, radioactive material will be retained on the filter. The amount of radioactivity on the filter is proportional to the amount of RNA synthesized.

The synthesis of RNA consists of four discrete stages: (1) binding of RNA polymerase to a template at a specific site, (2) initiation, (3) chain elongation, and (4) chain termination and release. A discussion of these stages is facilitated if the structure of RNA polymerase is understood. In the following section we will describe the properties of the RNA polymerase of *E. coli,* which is the best understood RNA polymerase; the properties of other bacterial RNA polymerases are quite similar. The RNA polymerases of mammalian cells will be described in a later section.

E. coli RNA Polymerase

E. coli RNA polymerase consists of five subunits—two identical α subunits and one each of types β, β′, and σ. The molecular weights of the proteins are 36,500, 150,000, 160,000, and 82,000 for α, β, β′, and σ respectively, so that the total molecular weight is 465,000. This is one of the largest enzymes known (Figure 11-4). *E. coli* mutants containing altered subunit structures have been isolated and these mutants have defined the genes *rpoA* (α), *rpoB* (β), *rpoC* (β′), and *rpoD* (σ). Other proteins have been found associated with RNA polymerase on occasion but their functions are unknown.

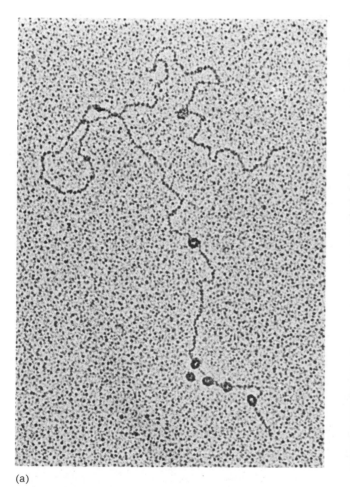

(a)

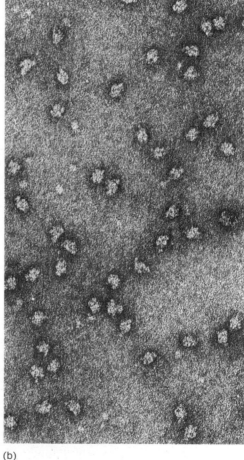

(b)

Figure 11-4

(a) *E. coli* RNA polymerase molecules bound to phage DNA. (b) *E. coli* RNA polymerase holoenzyme viewed by negative-contrast electron microscopy (× 270,000). (Courtesy of Robley Williams.)

The σ subunit dissociates from the enzyme easily and in fact does so during one stage of polymerization. The term **core enzyme** is used to describe the σ-free unit—namely, $\alpha_2\beta\beta'$. The complete enzyme, $\alpha_2\beta\beta'\sigma$ is called the **holoenzyme.** We will use the name RNA polymerase when the holoenzyme is meant.

RNA polymerase is sufficiently large that it can come into contact with many DNA bases simultaneously. An estimate of the size of the region of the DNA where contact is made is obtained by selectively degrading adjacent DNA bases with DNase, a procedure known as the **DNase protection method.** RNA polymerase is bound to DNA; then a DNA endonuclease is added to the mixture. This nuclease degrades most of the DNA to mono- and dinucleotides but leaves untouched DNA segments covered by RNA polymerase; these segments usually contain about 40 base pairs and are presumably protected from degradation by the RNA polymerase. However, since the mechanism of the steric interference is not known, the *exact* size of the region of contact cannot be deduced from this experiment; only an estimate can be made. For example, if the nucleotides at the edge of the contact region are susceptible to the enzyme, the size will be underestimated, but if a few nucleotides adjacent to the contact region are insensitive to the nuclease, the size will be overestimated.

Specific points of contact within the contact region can be identified with another protective method—this one, the **dimethyl sulfate protection method** (used to study the DNA–Cro-protein interaction examined in Chapter 6). Dimethyl sulfate methylates adenine and guanine (not cytosine and thymine) but cannot react with these bases if they are in close contact with a protein. When a protein molecule binds to a long sequence of DNA bases, many bases are not in close contact with the protein because the surfaces of the protein and DNA molecules are not completely complementary in shape. Thus, the adenines and guanines in the actual contact region will be unavailable for methylation; the sites of contact can be identified by determining the positions of nonmethylated adenine and guanine in the endonuclease-protected region. One might wonder why the entire contact region is resistant to DNase yet only specific contact points are resistant to methylation. This is explained by the different relative sizes of the DNase and dimethyl sulfate molecules. The separation of a DNA base that is in the contact region (but not in direct contact) from RNA polymerase is usually only a few angstrom units; this space is sufficient for a small molecule such as dimethyl sulfate to approach the base but not great enough for a bulky enzyme molecule to reach the adjacent phosphodiester bond. The principles involved in the DNase protection and dimethyl sulfate protection techniques are shown in Figure 11-5.

At any moment many RNA molecules have to be synthesized within a cell. Thus, one *E. coli* cell contains between 3000 and 6000 RNA

Figure 11-5
A schematic diagram of the analysis of the interaction of RNA polymerase (RNA-P) with DNA. Identification of (a) the protected region (left path of arrows) and (b) of some of the bases in contact with the enzyme (right path of arrows). For clarity, only one strand of DNA is shown, though the experiment is performed with double-stranded DNA. To locate the binding sequence the base sequence of the protected fragment is compared to that of a larger segment of DNA. Contact points are identified by comparing the methylation patterns obtained when RNA-P is either present or absent.

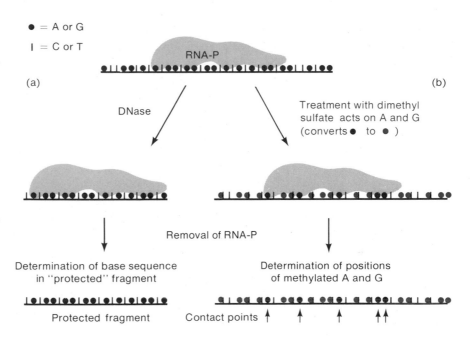

polymerase molecules; the number is greater when cells are growing rapidly.

Site Selection: I. The Promoter

The first step in transcription is binding RNA polymerase to a DNA molecule. Binding occurs at particular sites called **promoters,** which are specific sequences of 20–200 bases at which several interactions occur. (A promoter is also frequently defined as a region protected by RNA polymerase from digestion by endonucleases.) The existence of promoters was first demonstrated by the isolation of a particular class of Lac⁻ mutations in *E. coli.* These mutations not only eliminate gene activity but also are noncomplementable (because they are *cis*-acting) and prevent synthesis of the RNA transcript of the *lac* gene. These mutations are called **promoter mutations.**

Several events must occur at a promoter. RNA polymerase must recognize a specific DNA sequence, attach in a proper configuration, open the DNA to gain access to the bases to be copied, and then initiate synthesis. These events are guided by the base sequence of the DNA, the polymerase σ subunit (without which the promoter is not recognized) and, for some promoters, by auxiliary proteins. The details of these events are not yet known, but the process can be broken down into three

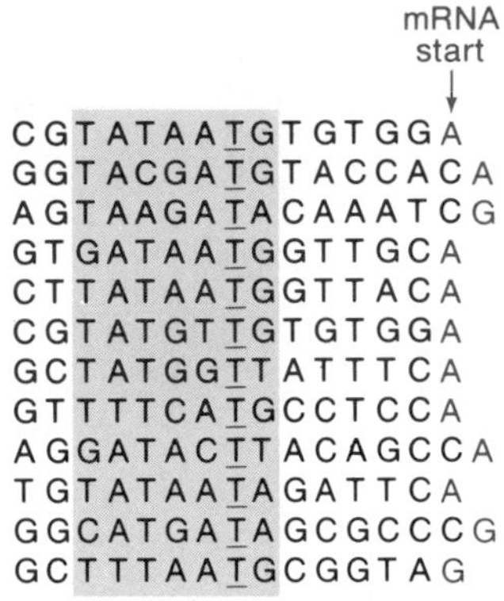

mRNA
start
↓

C G T A T A A T G T G T G G A
G G T A C G A T G T A C C A C A
A G T A A G A T A C A A A T C G
G T G A T A A T G G T T G C A
C T T A T A A T G G T T A C A
C G T A T G T T G T G T G G A
G C T A T G G T T A T T T C A
G T T T T C A T G C C T C C A
A G G A T A C T T A C A G C C A
T G T A T A A T A G A T T C A
G G C A T G A T A G C G C C C G
G C T T T A A T G C G G T A G

Figure 11-6
Segments of the noncoding strand of protected regions from various genes showing the common sequence of seven bases (red) known as the Pribnow box. The start point for mRNA synthesis is shown. The "conserved" T is underlined.

mRNA
start
↓

−40   −30   −20   −10   0
•   •   •   •   •
G G C A C C C C A G G C T T T A C A C T T T A T G C T T C C G G C T C G T A T G T T G T G T G G A A T T G
              ↓ ↓ ↓   ↓   ↓ ↓
              Δ A A   A   A A

Figure 11-7
A region of the noncoding strand of the promoter for the *lac* gene showing six mutations (red arrows) that affect promoter activity; Δ means a base deletion. The Pribnow box is shaded in red. Many base changes are known; all are either in or near the Pribnow box or are clustered around base −35 and thus define an important site (see page 377).

parts—(a) template binding at a polymerase recognition site, (b) movement to an initiation site, and (c) establishment of what is termed an open-promoter complex (shown schematically later in Figure 11-9). The approach to elucidating these steps for many genes has been to isolate the DNA segment (the promoter) that is protected by RNA polymerase from DNase digestion, determine the base sequence in the segment, and look for common features in the sequences (Figure 11-6). The specific sites of contact are also determined by the dimethyl sulfate protection method. This is important because one might expect that the specific contact sites would be in the regions common to all promoters.

The RNA molecules synthesized *in vitro* from each of these promoter regions must also be sequenced if one wishes to identify the initiation sequence, which is the sequence of the first few bases that are transcribed; this sequence is just the complement of the bases at the 5′ terminus of the RNA molecule. Additional information is obtained by determining the sequence of bases in promoters having mutations that either eliminate initiation *in vivo* or change the requirements for initiation (Figure 11-7). The rationale is that if a base change affects promoter activity, that base must be contained in the promoter. This technique has allowed researchers to identify the bases in the protected segment that are actually part of the promoter. So far, 46 promoters have been sequenced.

Site Selection: II. The Pribnow Box

Figure 11-6 shows portions of several promoter sequences in *E. coli* and *E. coli* phages (each promoter sequence is recognized by *E. coli* RNA polymerase) and their important features. In a region from five to ten bases to the left of the first base copied into mRNA is the right end of a sequence called the **Pribnow box.** All sequences found in Pribnow boxes are considered to be variants of a basic sequence TATAATG. The underscored T, at base 6 in the Pribnow box, from six to nine bases to the left of the first base transcribed (the distance depending on the distance from the Pribnow box to the transcription start point), is present in all promoters sequenced to date. It is called the "conserved T" and different sequences are usually compared by aligning conserved T's vertically, as shown in the figure. In 35 of 46 known Pribnow boxes in *E. coli*, the first two bases are TA; the variants, TG, CA, GA, and TC, retain one of the two TA bases. The Pribnow box is thought to be the sequence that orients RNA polymerase, so that synthesis proceeds from left to right (as the sequence is drawn), and the region at which the double helix opens to form the open-promoter complex (see below).

Before enough sequences were known that the conserved T was recognizable, the first base transcribed was chosen as a reference point and numbered zero. The direction of transcription was called "downstream"; all "upstream" bases, which are not transcribed, were given negative numbers starting from the zero reference. The Pribnow box is enclosed between -13 and -4, depending on the particular promoter. This numbering convention has become standard.

There are several mutations in the Pribnow box, two of which are shown in Figure 11-7, that prevent initiation of transcription. These mutations clearly indicate the importance of this sequence. Other bases outside of the Pribnow box are important too, as indicated by the other mutations shown in the figure.

Site Selection: III. The -35 Sequence

Examination of the complete sequence of the region protected by RNA polymerase indicates that for many (but not all) promoters, there is a second important region, to the left of the Pribnow box, whose sequences in different promoters have common features (Figure 11-8).

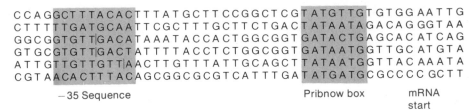

C C A G G C T T T A C A C T T T A T G C T T C C G G C T C G T A T G T T G T G T G G A A T T G
C T T T T T G A T G C A A T T C G C T T T G C T T C T G A C T A T A A T A G A C A G G G T A A
G G C G G T G T T G A C A T A A A T A C C A C T G G C G G T G A T A C T G A G C A C A T C A G
G T G C G T G T T G A C T A T T T T A C C T C T G G C G G T G A T A A T G G T T G C A T G T A
A T T G T T G T T G T T A A C T T G T T T A T T G C A G C T T A T A A T G G T T A C A A A T A
C G T A A C A C T T T A C A G C G G C G C G T C A T T T G A T A T G A T G C G C C C C G C T T

−35 Sequence   Pribnow box   mRNA
                                                                   start

Figure 11-8

Base sequences in the noncoding strand of six different RNA polymerase-protected regions showing the similarity between the −35 sequences. In each case, mutations that eliminate promoter activity have been found in the −35 sequence. The vertical lines indicate the HindII cuts mentioned in the text. The Pribnow boxes rather than the mRNA start points are aligned.

This sequence, which is called the **−35 sequence** and typically contains nine bases, is thought to be the initial site of binding of the enzyme. Evidence for this notion comes from the following experiment. RNA polymerase is removed from the protected fragment and the fragment is purified. If fresh RNA polymerase is then added, binding will occur, indicating that the binding site is on the fragment. However, if the fragment is first treated with a restriction nuclease (Chapter 20) called HindII , which makes a double-strand break at the sites indicated in the figure by the lines, RNA polymerase can no longer bind; presumably, the binding site is destroyed by the nuclease. Thus, RNA polymerase is thought to bind first at the leftmost side of the protected region and then to the Pribnow box. How it moves from one site to the next is not known. A theory that the enzyme "slides" along the DNA was popular at one time; it has not been ruled out but is considered to be unlikely. Another possibility, which has some experimental support, is that the σ subunit binds first to a recognition site at the left in a highly specific interaction and then, owing to the great size of the enzyme, the appropriate region of the polymerase can come in contact with the Pribnow box region (Figure 11-9). Once bound to the Pribnow box, the polymerase then dissociates from the leftmost recognition site.

The **open-promoter complex** is a highly stable complex and is the active intermediate in chain initiation. In this complex a local unwinding ("melting") of the DNA helix occurs starting about ten base pairs from the left end of the Pribnow box and extending to the position of the first transcribed base. This melting is necessary for pairing of the incoming ribonucleotides. The base composition of the Pribnow box sequence (A+T-rich) renders the DNA strand susceptible to denaturation. Presumably RNA polymerase itself induces this conformational change.

The promoters discussed in this section are classified as **high-level** or **strong promoters.** There are also weak promoters in which recognition by RNA polymerase is poor. The number of RNA molecules

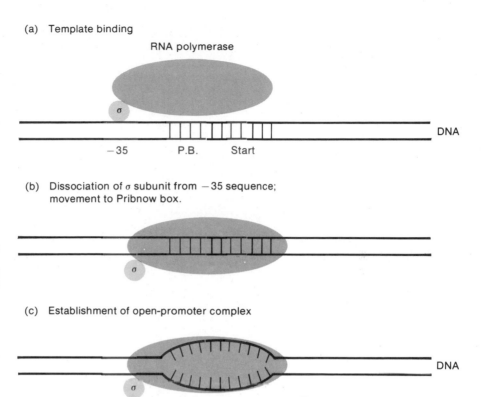

(a) Template binding

RNA polymerase

σ

DNA

−35   P.B.   Start

(b) Dissociation of σ subunit from −35 sequence; movement to Pribnow box.

σ

(c) Establishment of open-promoter complex

σ

Open complex

DNA

Figure 11-9
A proposed scheme for the binding of RNA polymerase to a promoter to form an open-promoter complex. Regions of the DNA molecule important for binding are shown in red. The shape of RNA polymerase is idealized for schematic purposes. The enzyme covers the region from bases −45 to +15, and the unpaired region in (c) extends from (roughly) −12 to +2. The enzyme is shown in contact with both strands because the strands are actually wrapped around one another in a helical array; however, true binding occurs only to bases in the coding strand. "P.B." indicates the Pribnow box.

synthesized per unit time from genes with weak promoters is much less than from a strong promoter with the result that fewer protein molecules are made per unit time by genes with weak promoters. Promoter strength is one factor which determines the number of copies of each protein molecule present in the cell. In most cases examined so far the difference between weak and strong promoters lies in the structure of the −35 region.

Site Selection: IV. The CAP Site

Some promoters totally lack the common −35 sequence—for example, the λpre, galP, and araBAD promoters. These are active only in the presence of positive effector molecules (see Chapters 14 and 16); for example, the λpre promoter is active only when the λ cII protein is present. The mechanisms of action of these effectors are not well understood, though a study of the lac promoter suggests that they bind to a site in the −50 to −30 region and, by a mechanism that differs from that

```
AATGTGAGTTAGCTCACTCA
TTACACTCAATCGAGTGAGT
```

Figure 11-10
Base sequence of both strands of the
principal CAP protein binding site
(site 1) of the *E. coli lac* promoter
showing the axis of the rotational or
inverted repeat symmetry (vertical
line) and the symmetric elements
(shaded in red).

Figure 11-11
A hypothetical arrange-
ment of two CAP protein
molecules and RNA poly-
merase on the *lac* pro-
moter. I and II are the
CAP sites mentioned in
the text.

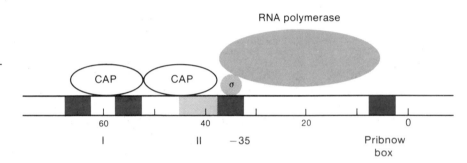

used when a common −35 sequence is present, they stimulate binding
of RNA polymerase to the Pribnow box. The *lac* promoter has a −35
sequence but does not bind RNA polymerase well unless a factor
known as the cyclic AMP (cAMP) receptor protein (**CAP**) is present.
There are two CAP-cAMP binding sites in the *lac* promoter, one in the
−70 to −50 segment (site I) and another in the −50 to −40 segment (site
II). Site I contains an inverted repeat sequence (that is, the base
sequence on one DNA strand is of the form ABCD . . . D′C′B′A, in
which a prime denotes a complementary base), which seems to be
characteristic of most strong binding sites (Figure 11-10). Site II is a very
weak binding site but when a CAP-cAMP complex is bound to site I, the
ability of CAP-cAMP to bind to site II is very much enhanced. Once site
II is occupied, RNA polymerase binds tightly to the −35 region and
thereby to the Pribnow box. This stimulation of binding of RNA
polymerase by CAP has an important regulatory role in many systems
and will be discussed in Chapter 14. It is also of significance with
respect to the function of the Pribnow box. In the absence of CAP, RNA
polymerase does bind to the Pribnow box, but a closed-promoter
complex results; that is, dissociation of the DNA strands in the initiation
region does not occur. The sequence of the box is TATGTTG—the
underlined G may increase the thermal stability so that RNA polymerase cannot induce melting. The binding of the CAP-cAMP complex is

believed to destabilize the G + C-rich region next to site I, lowering the melting temperature of the entry site for RNA polymerase in the Pribnow box region and thereby facilitating the establishment of an open-promoter complex (Figure 11-11). Support for this view comes from the sequence of a mutation in the *lac* promoter that enables transcription to occur very efficiently in the absence of CAP. The Pribnow box for this mutant (the mutant is called *uv5*), has two base changes G→A and T→A, and has the sequence TATAATG. With this promoter an open-promoter complex forms without CAP. Presumably the loss of the internal G · C pair reduces the thermal stability of the sequence and enables RNA polymerase to dissociate the helix in the absence of CAP-cAMP.

RNA Chain Initiation

Once the open-promoter complex has formed, RNA polymerase is ready to initiate synthesis. RNA polymerase contains two nucleotide binding sites called the **initiation site** and the **elongation site.** The initiation site binds only purine triphosphates, namely ATP and GTP, and one of these (usually ATP) is the first nucleotide in the chain. (UTP has been observed but it is rare.) Thus, the first DNA base that is transcribed is usually thymine. The initiating nucleoside triphosphate binds to the enzyme in the open-promoter complex and forms a hydrogen bond with the complementary DNA base (Figure 11-12). The elongation site is then filled with a nucleoside triphosphate that is selected strictly by its ability to form a hydrogen bond with the next base in the DNA strand. The two nucleotides are then joined together, the first base is released from the initiation site, and initiation is

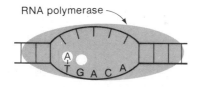

I. RNA polymerase binds to promoter, slides into place, and forms an open complex. ATP in initiation site binds to T on coding strand.

II. A NTP is added to the elongation site and is covalently linked to the A.

III. RNA polymerase moves over to the next DNA base. The initiating dinucleotide is released. A NTP enters the elongation site and is covalently linked to the dinucleotide. Then movement of RNA polymerase continues.

Figure 11-12
A scheme for initiation of RNA synthesis. The enzyme is drawn smaller than in Figure 11-9 and without the σ factor. Only the bases in the lower strand are recognized.

Promoter	Sequence
E. coli galP1	5′–G TA T G– 3′
E. coli araBAD	5′–G T A TG –3′
Phage φ X174 gene *A*	5′–G TT T A –3′
Virus SV40	5′–G T TT A –3′

Figure 11-13
Segments of the coding strand of DNA surrounding the base that is the template for the
first base of RNA in four different promoters.
Note that the choice of the base for initiation
is not related simply to the surrounding base
sequence. The coding strand (not the noncoding strand) sequence is given here in
order to show the bases recognized by RNA
polymerase.

complete. The dinucleotide remains hydrogen-bonded to the DNA. The
elongation phase then begins when the polymerase releases the base
and then moves along the DNA chain.

The β subunit of RNA polymerase is involved in forming the first
internucleotide linkage. The antibiotic rifamycin SV binds tightly to the
β subunit of RNA polymerase and prevents the initiation site from being
filled. Once the dinucleotide has formed, the binding of the β subunit to
the initiation site is unnecessary. The related drug **rifampicin** allows
formation of the dinucleotide but blocks the movement of the enzyme.
However, if a third nucleotide has been added to the chain, rifampicin is
without further effect. Thus, rifamycin SV and rifampicin are inhibitors
of chain initiation but not chain growth. These drugs have been used
extensively in investigations of transcription as they can be added to a
growing culture for *in vivo* studies or to a reaction mixture for *in vitro*
studies and will inhibit the initiation of new chains very rapidly. Since
it takes only a few seconds to complete a growing RNA chain, RNA
synthesis is rapidly terminated.

How the first base copied is selected is not known. It is always
within six to nine bases of the conserved T of the Pribnow box, but
neither the precise position nor the identity of adjacent bases shows any
obvious pattern. This is shown in Figure 11-13.

Chain Elongation

After several nucleotides (eight is the best estimate) are added to the
growing chain, RNA polymerase changes its structure and loses the σ
subunit. Thus, most elongation is carried out by the core enzyme (Figure

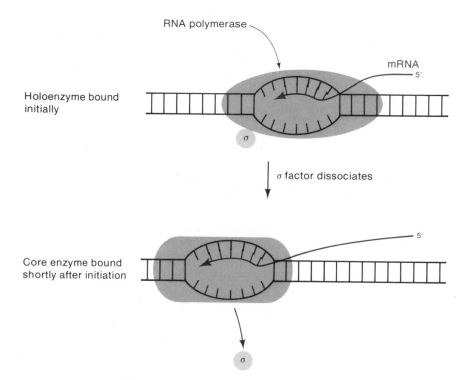

Figure 11-14
Diagram of a transcription bubble shortly after transcription begins and, at a slightly later stage, when the core enzyme has changed shape and the σ subunit has dissociated from the core enzyme.

11-14). The core enzyme moves along the DNA, binding a nucleoside triphosphate that can pair with the next DNA base and opening the DNA helix as it moves. The open region extends only over a few base pairs; that is, the DNA helix recloses just behind the enzyme.

The newly synthesized RNA is released from its hydrogen bonds with the DNA as the helix re-forms. Treatment of a transcribing system with pancreatic RNase, an enzyme that cleaves single-stranded RNA but not a DNA-RNA hybrid, shows that about ten RNA bases are protected from nuclease action. Whether these bases are hydrogen-bonded to DNA or this is simply steric interference by the massive core enzyme is not known. It has been suggested that only one or two RNA bases may actually be paired with DNA. The small amount of DNA-RNA hybrid that exists during transcription contrasts with the long (up to 50 bases) segment that is present during the priming of DNA synthesis (Chapter 8).

A peculiarity of the chain elongation reaction is that it does not occur at a constant rate—that is, synthesis stops or slows down when particular regions of a DNA molecule are being transcribed. This reduction in rate is called a **pause** and neither its significance nor its cause is understood. In the few instances in which the base sequence of the RNA is known it has been observed that a pause occurs about eight to

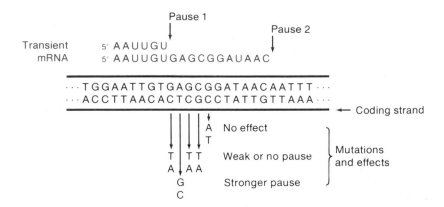

Figure 11-15
A segment of the DNA of the *E. coli lac* system. The positions of two pauses and the transient mRNA molecule (red) are shown. Mutations that affect pause 2 but not pause 1 are shown. Those that weaken the pause are G·C→A·T changes that would tend to destabilize DNA. One G·C→A·T change has no effect; it is in a region so rich in G + C that the change has little effect on stability. An A·T→G·C, which should stabilize the segment, increases the pause.

ten bases after a particularly G +C-rich region (Figure 11-15). Furthermore, in mutant forms of sequences in which there is a G·C→A·T change, the pause is often eliminated if the transition occurs in a position that would reduce the stability of the hydrogen-bonded region, as shown in the figure. The tentative interpretation of this observation is that chain elongation requires that the RNA just synthesized dissociate from the DNA template and that the G +C-rich region form a more stable DNA-RNA hybrid, thereby halting the advance of the enzyme until removal of the RNA is completed. A possible role for the pause will be given in the following section.

Termination and Release of Newly Synthesized RNA

Termination of RNA synthesis occurs at specific base sequences within the DNA molecule. Twenty termination sequences have so far been determined and each has the characteristics shown in Figure 11-16. There are three important regions:

1. First, there is an inverted-repeat base sequence containing a central nonrepeating segment; that is, the sequence in one DNA strand would read like ABCDEF-XYZ-F′E′D′C′B′A′ in which A and A′, B and B′, and so on, are complementary bases. Thus, this sequence is capable of intrastrand base pairing, forming a "stem-and-loop" configuration in the RNA transcript and possi-

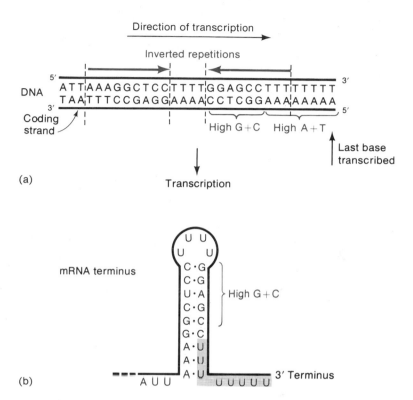

(a)

Figure 11-16
Base sequence of (a) the DNA of the *E. coli trp* operon at which transcription termination occurs and of (b) the 3′ terminus of the mRNA molecule. The inverted repeat sequence is indicated by reversed red arrows. The mRNA molecule is folded to form a stem-and-loop structure thought to exist. The relevant regions are labeled in red; the terminal sequence of U's in the mRNA is shaded in red.

(b)

bly in the DNA strands. It is possible though not yet proved that the stem-and-loop serves a purpose independent of its role in termination—namely, to render the newly synthesized RNA resistant to degradation by RNase II, an intracellular enzyme that is inactive against double-stranded RNA.

2. The second region is near the loop end of the putative stem (sometimes totally within the stem) and is a sequence having a high G + C content.

3. A third region (sometimes absent) is a sequence of A · T pairs (which may begin in the putative stem) that yields in the RNA a sequence of six to eight uracils often followed by an adenine.

The transcript does not terminate at a unique site: in a collection of RNA molecules initiated from a single promoter, some end with five U's, some with six U's, some with six U's and an A, and so on. The final base is always 16–24 bases from the midpoint of the terminating sequence.

How these features lead to termination is not known, but the currently accepted hypothesis is that termination occurs when a sequence is reached at which there is a pause in the advance of the RNA

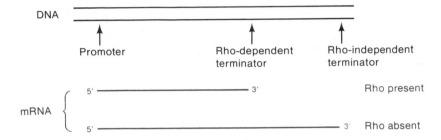

Figure 11-17
The effect of Rho on termination of RNA synthesis. This effect is readily demonstrated *in vitro*. If Rho is added to a reaction mixture *after* RNA polymerase has passed the Rho-dependent terminator, RNA synthesis continues until the Rho-independent terminator is reached. The red lines represent the completed mRNA molecules.

chain. In particular, it is thought that RNA polymerase somehow responds to this slowdown by initiating an unknown sequence of events that leads to termination. The evidence that supports this hypothesis is that a pause usually precedes termination and that mutations in certain termination sequences will eliminate both termination and pauses. It seems that there are several pauses and that the first pause is in the inverted repeat region of the sequence, because certain mutations in this region reduce the efficiency of termination. Furthermore, the length of the first pause in wild-type sequences increases with $G + C$ content of this region, which suggests that the DNA itself contains a stem and loop in each strand (or perhaps that this configuration is in equilibrium with the standard double-helical configuration). A second pause is presumably in the high-$G + C$ region. This pause may occur because RNA polymerase has difficulty traversing the region or, more likely, because the newly synthesized RNA cannot easily dissociate from the DNA template. The mechanism by which the polymerase is released from the DNA is unknown but may be a function of the poly(U) sequence, which binds very weakly to the adenine sequence in the DNA.

The process of termination includes (1) cessation of RNA chain elongation, (2) release of newly formed RNA, and (3) release of the RNA polymerase from the DNA. There are two kinds of termination events—those that are self-terminating (dependent on the DNA base sequence only) and those that require the presence of a termination protein called **Rho** (Figure 11-17).* Both types of events occur at specific but distinct base sequences. Rho, an oligomeric protein, does not bind to the core polymerase or to the holoenzyme and only binds very weakly to DNA. It binds tightly to RNA and, when bound, acquires a powerful ATPase

*Rho was originally called the ρ factor; ρ (Greek rho) was a symbol for release.

activity. The action of Rho is very poorly understood and at present several mechanisms are possible. For example, some sequences require Rho in order to cause the RNA polymerase to pause at the putative stem-and-loop configuration. With other sequences, synthesis stops without Rho but Rho is required to prevent chain growth from continuing after a delay. At one terminator, synthesis stops completely but Rho is needed to release the RNA. Clearly, Rho has a multifaceted activity.

Some terminators do not have an absolute requirement for Rho but are more efficient in its presence. However, even with this group there are different modes of action. For instance, the 6S mRNA of *E. coli* phage λ sometimes fails to terminate if Rho is absent but always terminates if Rho is present; in contrast, the λ *oop* RNA is always terminated but releases more rapidly when Rho is present.

Rho-dependent termination is very important in regulatory processes, as will be seen in later chapters, because the activity of termination can be switched off by regulatory proteins called **antiterminators.** Termination is also associated with some aspects of protein synthesis, in that often termination is a consequence of inhibition of protein synthesis. This will be described in Chapter 14 when attenuators are discussed.

Mutations sometimes occur that generate a new termination site well before the natural termination site; a shorter mRNA is then produced. Genetic mapping techniques indicate that these mutations have the following characteristics. They are localized at a single site, but if the site is in one gene of a polygenic mRNA molecule, they reduce the activity of all genes downstream from the mutation. These mutations represent one of a class called **polar mutations.** (Other types of polar mutations are described in Chapters 13 and 15.) The explanation for this type of polarity is that the downstream genes are not transcribable because the mutation has activated a premature termination site. The first polar mutation of this type to be understood was one that generated a Rho-dependent termination site (Figure 11-18). When mapped, this mutation was found to occur in the first gene of a cluster of three whose products are components of the pathway for utilizing galactose; the resulting mutant was Gal⁻ and the products of genes 2 and 3 were not made. The connection between this mutation and Rho was made evident when a temperature-sensitive mutation in the *rho* gene was introduced into the bacterial strain carrying the *gal⁻* polar mutation. When the temperature was raised from 34°C to 42°C, Rho was inactivated and the activities of genes 2 and 3 were restored. The bacterial strain of course remained Gal⁻, however, because the mutation had caused a deleterious amino acid substitution to be made in gene 1.

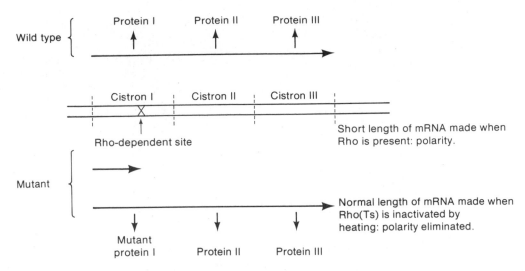

Figure 11-18
The effect of a polar mutation on mRNA synthesis in a polycistronic system. When premature termination is prevented by eliminating Rho activity, the polar effect is gone, though a mutant protein is still made. Usually, the mutation is slightly upstream from the Rho-dependent site.

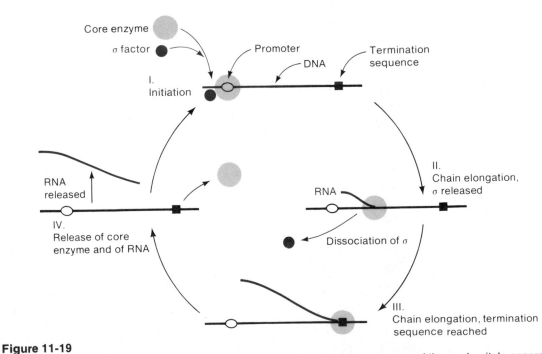

Figure 11-19
The transcription cycle of *E. coli* RNA polymerase showing dissociation of the σ subunit shortly after chain elongation begins, dissociation of the core enzyme during termination, and re-formation of the holoenzyme from the core enzyme and the σ subunit. In general, initiation of a second round of transcription does not utilize the same core enzyme and σ subunit used in the first round.

The final step in the termination process is dissociation of the core enzyme from the DNA. Following this event, the core enzyme interacts with a free σ subunit to re-form the holoenzyme. Thus there is a σ cycle—σ falls off the holoenzyme after elongation begins and rejoins the core enzyme after the core enzyme falls off the DNA (Figure 11-19).

CLASSES OF RNA MOLECULES

There are three major classes of RNA molecules—messenger RNA (mRNA), ribosomal RNA (rRNA), and transfer RNA (tRNA). All are synthesized from DNA base sequences, which are termed RNA genes, but they have very different functions in protein synthesis, as will be seen in this and the next chapter. There are also significant differences between the structures and modes of synthesis of the RNA molecules of prokaryotes and eukaryotes, though the basic mechanisms of their functions are nearly the same. The greatest amount of information has been obtained from studies with bacteria and bacterial cell extracts, so that it is here that we begin. Transcription in eukaryotes and the structure and synthesis of eukaryotic RNA molecules are discussed in a later section.

Structure of Messenger RNA

The base sequence of a DNA molecule determines the amino acid sequence of every polypeptide chain in a cell, though amino acids have no affinity for DNA. Thus, instead of a direct pairing between amino acids and DNA, a multistep process is used in which the information contained in the DNA is converted to a form in which amino acids can be arranged in an order determined by the DNA base sequence. This process begins with the transcription of the base sequence of one of the DNA strands (the **coding strand** or **sense strand**) into the base sequence of an RNA molecule and it is from this molecule—messenger RNA—that the amino acid sequence is obtained by the protein-synthesizing machinery of the cell. (The DNA strand that is not transcribed is called the **antisense strand**.) As we will see in Chapter 12, the base sequence of the mRNA is then read in groups of three bases (a group of three is called a **codon**) from a start codon to a stop point, with each codon corresponding either to one amino acid or a stop signal.

A DNA segment corresponding to one polypeptide chain plus the start and stop signals is called a **cistron** and a mRNA encoding a single polypeptide is called monocistronic mRNA. It is very common for a mRNA molecule to encode several different polypeptide chains; in this case it is called a **polycistronic mRNA** molecule. The cistrons contained in polycistronic mRNA often correspond to the proteins of a metabolic pathway. For example, in *E. coli* the three proteins required to metabolize galactose are synthesized from a single mRNA molecule and the ten enzymes needed to synthesize histidine are encoded in another mRNA molecule. The use of polycistronic mRNA is an economical way for a cell to regulate synthesis of related proteins in a coordinated way. For example, the usual way to regulate synthesis of a particular protein is to control the synthesis of the mRNA molecule that encodes it. With a polycistronic mRNA molecule the synthesis of several related proteins—in similar quantities and at the same time—can be regulated by a single signal.

The sizes of mRNA molecules vary within a broad range. The smallest proteins contain about 50 amino acids; correspondingly, 150 nucleotides (3 per amino acid) would be needed in a monocistronic mRNA molecule. A protein of more typical size would have 300 to 600 amino acids, corresponding to 900 to 1800 mRNA bases. In prokaryotes polycistronic mRNA is actually more common than monocistronic mRNA, and a typical mRNA size is therefore from about 3000 to 8000 bases; a few enormous molecules such as the histidine mRNA, which encodes ten proteins, contain 12,000 nucleotides.

In addition to cistrons and start and stop sequences for translation, other regions in mRNA are significant. For example, translation of a mRNA molecule (that is, protein synthesis) seldom starts exactly at one end of the RNA and proceeds to the other end; instead, initiation of synthesis of the first polypeptide chain of a polycistronic mRNA may begin hundreds of nucleotides from the 5'-P terminus of the RNA. The section of nontranslated RNA before the coding regions is called a **leader;** in some cases, the leader contains a regulatory region (called an **attenuator**) that determines the rate of protein synthesis. Untranslated sequences are found at both the 5'-P and the 3'-OH termini and a polycistronic mRNA molecule may contain intercistronic sequences (**spacers**) hundreds of bases long.

We have previously stated that mRNA is transcribed from only one DNA strand at a given place along a gene. It is important to realize, though, that not all mRNA molecules are synthesized from the same DNA strand. That is, in an extended segment of a DNA molecule, mRNA would be seen to be growing in both directions (Figure 11-20).

When discussing RNA synthesis, the following terminology is often used. The two strands of DNA are called *l* and *r* strands; *l* and *r,* which stand for left and right, refer to directions of synthesis of the mRNA when the DNA strands are depicted in a standard but arbitrary way (horizontally). That is, rightward-moving mRNA is transcribed from the *r* strand. According to this convention, since the growth of mRNA proceeds from the 5' end to the 3' end, and since a DNA-RNA hybrid is

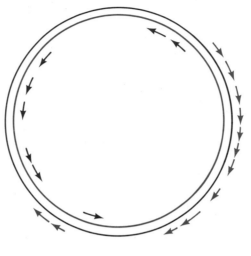

Figure 11-20
A physical map of *E. coli* phage T4 showing many identified clockwise and counterclockwise transcripts (arrows). The black mRNA molecules are transcribed from the red DNA strand and the red mRNA molecules are transcribed from the black DNA strand.

Figure 11-21
A drawing showing the definitions of the *l* strand and the *r* strand of a DNA molecule. The arrows represent transcripts.

an antiparallel structure, the *r* strand is drawn with the 3'-OH terminus at the left, as shown in Figure 11-21.

Although mRNA is synthesized as a single-stranded molecule, it has numerous double-stranded regions. (These regions have not been seen by electron microscopy but the evidence for their existence is that the A_{260} of an RNA solution increases on heating and that the molecule can be cleaved by RNases active only on double-stranded RNA.) Some of the double-stranded regions may reflect the termination signals on the DNA, as we have just seen, or serve as signals for certain stages of protein synthesis, as will be seen in the next chapter. Another function of the double-stranded segments will become evident when we discuss posttranscriptional modification (processing) of RNA.

The Lifetime of mRNA

An important characteristic of prokaryotic mRNA is that its lifetime is short compared to other types of prokaryotic RNA molecules. For bacteria the half-life of a typical mRNA molecule is a few minutes

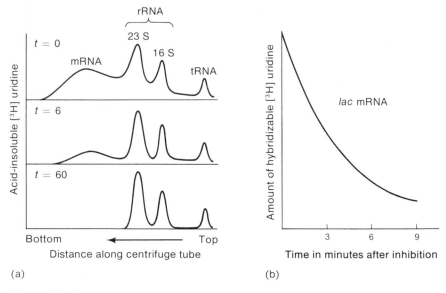

Figure 11-22
Two ways to demonstrate the short lifetime of mRNA. (a) Analysis of total mRNA. Bacteria are incubated in the presence of [³H]uridine to make radioactive RNA. At $t = 0$ the bacteria are transferred to a medium containing an inhibitor of transcription. Samples are taken at various times—for example, 0, 6, and 60 minutes. RNA is then isolated and fractionated by sucrose gradient centrifugation; sedimentation proceeds from right to left. Four classes of RNA are seen. The stable rRNA and tRNA molecules persist for a long time. The mRNA is degraded and gradually disappears. (b) Analysis of a specific mRNA molecule—the one coded by the *E. coli lac* genes. Bacterial RNA is labeled as in part (a). Purified RNA is hybridized to denatured phage λ DNA containing the *E. coli* lactose-metabolism *lac* genes. Only mRNA encoded by the *lac* genes will hybridize to λ*lac* DNA. The half-life of the *lac* mRNA can be measured directly from this analysis and is approximately three minutes.

(Figure 11-22). This feature, which may seem terribly wasteful, has an important regulatory function. If a protein is no longer required, a cell need only turn off synthesis of the mRNA that encodes the protein; soon afterwards, none of that particular mRNA will remain, and synthesis of the protein no longer occurs. Of course, this also means that in order to maintain synthesis of a particular protein, the mRNA molecules encoding these proteins must be synthesized continuously. Continuous synthesis is small payment by the cell for the ability to regulate the synthesis of specific proteins. When eukaryotes are discussed, it will be seen that eukaryotes have special mechanisms for increasing the lifetime of the particular mRNA molecules that encode continually needed proteins and decreasing the lifetime of those no longer needed.

The short lifetime of bacterial mRNA is one criterion used to identify mRNA in prokaryotes. The usual experiment to determine whether a particular molecule or class of molecules is mRNA is the **pulse-chase experiment** (Chapter 8). RNA is labeled briefly by growing bacteria in the presence of [^{3}H]uridine; bacteria are then grown in a medium containing no [^{3}H]uridine and a high concentration of nonradioactive uridine. A stable RNA molecule will be present through many generations (Figure 11-22). However, in a molecule of mRNA, the amount of acid-insoluble radioactivity (that is, RNA) will decrease with a half-life of two to three minutes.

Degradation of mRNA proceeds enzymatically primarily from the 5′-P terminus. The enzymology is not completely understood but the enzymes appear to be RNA exonucleases, though in a few cases endonucleases are also involved. During interaction with other components of the protein-synthesizing machinery, mRNA receives some protection from degradation. Stable mRNA, such as that found in specialized eukaryotes (for instance, the hemoglobin mRNA of reticulocytes or the silk fibroin mRNA of the silkworm) is protected from degradation in additional ways—for example, by modification of the 5′ terminus to render it resistant to exonuclease action. This will be discussed in greater detail later in the chapter.

Stable RNA: Ribosomal RNA and Transfer RNA

During the synthesis of proteins genetic information is supplied by messenger RNA. RNA also plays other roles in protein synthesis. For example, proteins are synthesized on the surface of an RNA-containing particle called a **ribosome;** these particles consist of three classes of ribosomal RNA (**rRNA**), which are stable molecules (Figure 11-22(a)), and a large number of proteins which have various functions. Amino acids do not line up against the mRNA template independently during protein synthesis but are aligned by means of a set of about fifty adaptor RNA molecules called transfer RNA (**tRNA**), also a stable species. Each tRNA molecule is capable of "reading" three adjacent mRNA bases (a codon) and placing the corresponding amino acid at a site on the ribosome at which a peptide bond is formed with an adjacent amino acid. Neither rRNA nor tRNA is used as a template. The roles of ribosomes and of tRNA molecules will be explained in detail in the following two chapters. In this chapter we are concerned only with their synthesis, which, as we will see, involves posttranscriptional modification of a remarkable sort.

The synthesis of both rRNA and tRNA molecules is initiated at a promoter and completed at a terminator sequence and, in this respect, their synthesis is no different from that of mRNA. However, the following three properties of these molecules indicate that neither rRNA nor tRNA molecules are the **primary transcripts** (immediate products of transcription):

1. The molecules are terminated by a 5′-monophosphate rather than the expected triphosphate found at the ends of all primary transcripts.
2. Both rRNA and tRNA molecules are much smaller than the primary transcripts (the transcription units).
3. All tRNA molecules contain bases other than A, G, C, and U and these "unusual" bases (as they are called) are not present in the original transcript.

All of these molecular changes are made after transcription by a process called **posttranscriptional modification** or, more commonly, **processing.** In the following sections we will consider processing of prokaryotic rRNA and tRNA. Eukaryotic processing is discussed later.

Processing of tRNA Molecules

The best-understood tRNA molecule from the point of view of its synthesis is the *E. coli* tRNA$_I^{Tyr}$ molecule, a molecule that reads the codon GAU and brings the amino acid tyrosine to the site on the ribosome at which polymerization occurs. The molecule contains 85 nucleotides and its base sequence is known (see Chapter 12). In *E. coli* there are two copies of the tRNA$_I^{Tyr}$ gene (that is, two copies of the DNA from which tRNA$_I^{Tyr}$ is transcribed) and the copies are adjacent. Each of these genes, whose base sequences are identical, consists of about 350 (not 85) nucleotides separated by a "spacer" of 200 nucleotides (Figure 11-23). The two genes are transcribed as a single RNA molecule that is cut up after transcription is complete. In order to simplify the study of the synthesis of tRNA$_I^{Tyr}$, genetic techniques have been used to create a transcription unit containing only a single gene. This unit contains 350 nucleotides. The transcription start site is 41 nucleotides before (upstream from) the 5′ end of the tRNA base sequence and a Rho-dependent stop site is 224 nucleotides downstream from the 3′ terminus of the tRNA. This transcript is processed by a series of steps shown in Figure 11-24, which may be grouped into the following three stages:

1. *Formation of the 3′-OH terminus.* This process involves the action of an endonuclease that recognizes a hairpin loop (I in the figure) and an exonuclease that recognizes the three-base sequence CCA. After endonuclease digestion at site 1, the seven bases upstream are removed

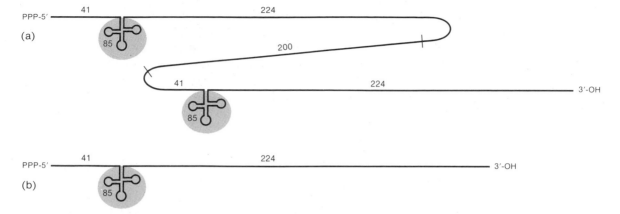

Figure 11-23

The *E. coli* tRNA$_I^{Tyr}$ gene showing (a) the complete transcript with two adjacent identical tRNA segments and the spacer region and (b) the single genetic unit used to study processing. The numbers indicate the number of bases in each segment of the transcript. The tRNA sequences are shaded in red.

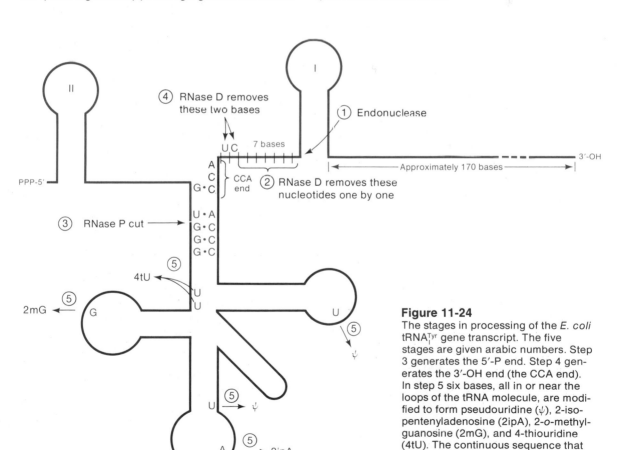

Figure 11-24

The stages in processing of the *E. coli* tRNA$_I^{Tyr}$ gene transcript. The five stages are given arabic numbers. Step 3 generates the 5'-P end. Step 4 generates the 3'-OH end (the CCA end). In step 5 six bases, all in or near the loops of the tRNA molecule, are modified to form pseudouridine (ψ), 2-isopentenyladenosine (2ipA), 2-o-methylguanosine (2mG), and 4-thiouridine (4tU). The continuous sequence that forms the final tRNA molecule is given in black.

by an exonuclease called RNase D (step 2). (This enzyme may be identical to another processing enzyme called RNase Q.) This enzyme initially stops two bases short of the CCA terminus, though it later removes these two bases after the 5′ end is processed. This leaves a molecule called pre-tRNA that is easily isolated from *E. coli* and that has the structure

$$P\text{-}5'-(41 \text{ bases})-\text{tRNA}-(2 \text{ bases})-3'\text{-OH}.$$

2. *Formation of the 5′-P terminus.* The 5′-P terminus is formed by an enzyme called RNase P, which is believed to be responsible for generation of the 5′-P terminus of all *E. coli* tRNA molecules. The evidence for this comes from studies with an *E. coli* mutant in which RNase P is inactive at 42°C. When this mutant is grown at 42°C, large RNA molecules accumulate that contain tRNA sequences plus a hairpin loop (II in the figure) at the 5′ terminus. These are tRNA precursors and contain the sequences for more than thirty different tRNA molecules. RNase P removes the excess RNA from the 5′ end of a precursor molecule by an endonucleolytic cleavage (step 3) that generates the correct 5′ end and a single fragment. (The fragment is later degraded by exonucleases and the nucleotides are re-used.) It seems that RNase P does not recognize a specific base sequence at the cleavage site or anywhere else, but instead recognizes the overall three-dimensional configuration of the tRNA molecule with its several hairpin loops and then makes a cut at just the right place. The evidence for this statement is that sequence alterations (mutations) in the segment to be removed do not affect the cleavage unless the alteration causes extensive disruption of the stem-and-loop arrangement. Once the 5′-P terminus has been formed, RNase D removes the two 3′-terminal nucleotides (step 4), leaving a tRNA molecule having the correct length.

3. *Production of the modified bases.* The final modification is to produce the altered bases in the tRNA (step 5). Highly specialized enzymes that act only on bases at specific sites in tRNA produce the necessary changes. In the case of tRNA$_I^{Tyr}$, two uridines are converted to pseudouridine (ψ), two uridines to two 4-thiouridines (4tU), one guanosine to 2′-o-methylguanosine (2mG) and one adenosine to isopentenyladenosine (2ipA), as shown in the figure. These and other modified bases are formed in most tRNA molecules.

Although the modifications described are for the artificial system containing only one copy of the tRNA$_I^{Tyr}$ gene, it is assumed that the same events also occur in the natural system, containing two copies of the gene.

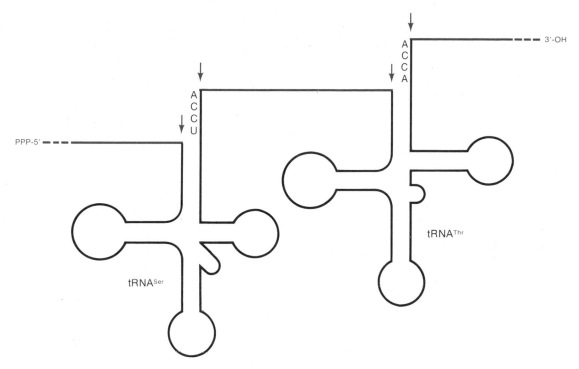

Figure 11-25
Generation of tRNA^Ser and tRNA^Thr from a single transcript of *E. coli* genes. A portion of the transcript that contains both tRNA^Ser and tRNA^Thr is shown here. The segments in red are removed by cutting at the sites indicated by the arrows.

Multiple copies of a particular tRNA molecule are commonly found in a single transcription unit.* For example, there are four copies of one of the tRNA^Leu molecules in its precursor molecule. The occurrence of different tRNA molecules in a single transcript is also frequent. For example, one tRNA^Ser and one tRNA^Thr are present in a single unit in *E. coli;* how the two tRNA molecules are formed is shown in Figure 11-25. *E. coli* phage T4 contains the information for the synthesis of eight different tRNA molecules, each of which exists in at least two copies, and two other stable RNA molecules of unknown function. These are all cleaved from one gigantic transcript containing more than 1000 bases.

In the following section in which processing of ribosomal RNA is described, we will see that some tRNA molecules are even obtained from the transcript that contains rRNA.

*This phenomenon occurs for several reasons. For example, multiple copies of a gene allow a greater rate of synthesis of a particular tRNA molecule than if only a single copy is made of a gene; they also permit cells to generate suppressors (Chapter 12).

Processing of Ribosomal RNA in *E. coli*

Bacterial ribosomes contain three kinds of rRNA (see Chapter 13 for details). Each kind is designated by its sedimentation coefficient, expressed in svedberg units (S), which is an empirical and convenient way to describe its size. The three types of molecules are called 5S rRNA, 16S rRNA, and 23S rRNA; they contain 120, 1541, and 2904 nucleotides, respectively. These molecules, plus several tRNA molecules, are contained in a continuous transcript having more than 5000 nucleotides whose base sequence is almost completely known. At present seven distinct primary transcripts containing rRNA have been isolated; they differ by the identity of the tRNA molecules and the location of the tRNA sequences with respect to the rRNA sequences. A diagram of one of these transcripts is shown in Figure 11-26. This transcript contains four different tRNA molecules and the segments are, from the 5′ end to the 3′ end,

16S rRNA–tRNA^{Ile}–tRNA^{Ala}–23S rRNA–5S RNA–tRNA^{Asp}–tRNA^{Trp}.

No particular pattern exists for the relative positions of the rRNA and tRNA molecules in the various primary transcripts.

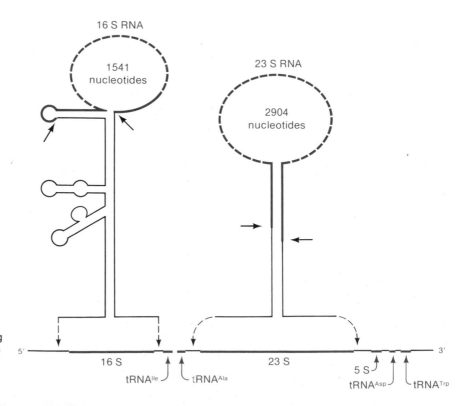

Figure 11-26
A schematic diagram of one of the *E. coli* rRNA transcripts from which 5S, 16S, and 23S rRNA molecules are excised. The regions containing the 16S rRNA and 23S rRNA molecules are shown in expanded form above the line designating the transcript. The arrows indicate the termini of the 16s rRNA and 23s rRNA molecules.

A transcript such as the one just shown is cut, usually while it is being synthesized, by several enzymes acting in sequence; the complete set of enzymes has not yet been identified for any transcript, though. The first enzyme is apparently RNase III, a nuclease that cleaves double-stranded RNA. The evidence for this is that some copies of the complete uncut transcript can be recovered from a mutant that lacks RNase III. Following the RNase III cuts, other enzymes complete the processing of the rRNA; RNase P is required for some of the cuts around the tRNA segments.

The sequence of processing events is not the same in all rRNA transcripts or in all organisms, but the basic pattern of excision of all rRNA components from a single precursor seems to be a general phenomenon. The reason for this system, which appears somewhat wasteful, is that it enables a constant ratio of the 16S, 23S, and 5S RNA molecules to be maintained. Since one molecule of each of these three is present in a ribosome and since the molecules are used nowhere else in the cell, efficiency would demand that if each were transcribed separately, some means would be needed to maintain a 1:1:1 ratio.

Processing of bacterial mRNA has not been observed, but two cases are known in which phage mRNA is processed. One case is the mRNA of *E. coli* phage T7, which is cleaved by RNase III into five distinct mRNA molecules. The other is *E. coli* phage ϕ80 for which one mRNA molecule is cleaved by RNase P. (Phage T7 will be discussed in detail in Chapter 15.) Eukaryotic mRNA is extensively processed; this is described later in this chapter.

Summary of Processing in Prokaryotes

Although a great deal remains to be learned about the processing of prokaryotic RNA, the following facts seem to be well established.

1. All stable RNA molecules are processed; mRNA molecules, with the exception of those of phages T7 and ϕ80, are unstable and are not processed.
2. About ten nucleases account for all of the cuts. The endonucleases always generate a 5'-P and a 3'-OH group.
3. The 3'-OH ends are generally formed by exonucleases, though 5S rRNA may be an exception.
4. Many processing enzymes are not sequence-specific but recognize large structural features. Although many known RNases make sequence-specific cuts, none is a processing enzyme.
5. RNase III can be replaced by tRNA processing enzymes when RNase III is absent.
6. Bases other than A, U, G, and C are formed by enzymatic modification of bases already present in otherwise completed molecules.

TRANSCRIPTION IN EUKARYOTES

The basic features of the transcription and the structure of mRNA in eukaryotes are similar to those in bacteria. However, there are five notable differences, which are the following:

1. Eukaryotic cells contain three classes of nuclear RNA polymerase and these are responsible for the synthesis of different classes of RNA.
2. Many mRNA molecules are very long lived.
3. Both the 5′ and 3′ termini are modified; a complex structure called the **cap** is found at the 5′ end and a long (up to 200 nucleotides) sequence of polyadenylic acid—poly(A)—is found at the 3′ end.
4. The mRNA molecule that is used as a template for protein synthesis is usually about one-tenth the size of the primary transcript. During mRNA processing intervening sequences called **introns** are excised and the fragments are rejoined.
5. All eukaryotic mRNA molecules are monocistronic.

These points are illustrated in Figure 11-27, which shows a schematic diagram of a typical eukaryotic mRNA molecule and how it is produced. Clearly the production of mRNA in eukaryotes is not a simple matter of transcribing the DNA.

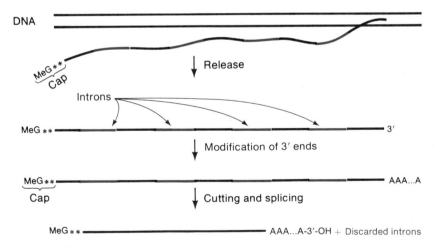

Figure 11-27
Schematic drawing showing production of eukaryotic mRNA. The primary transcript is capped before it is released. Then, its 3′-OH end is modified, and finally the intervening regions are excised. MeG denotes 7-methylguanosine and the two asterisks indicate the two nucleotides whose riboses are methylated.

Initiation and termination in eukaryotes are not well understood and no meaningful comparison can be made with prokaryotes. The problem is that, to date, *in vitro* synthesis of the mRNA of specific genes has not been achieved except for synthesis of 5S RNA from the DNA of oocytes. This transcript has a unique length, so that, presumably, defined initiation and termination sequences are recognizable. The principal information about initiation of mRNA synthesis comes from sequencing of several DNA sequences shown to be complementary to mRNA molecules made *in vivo*. Data are limited, but it appears that a base sequence (similar to the Pribnow box in *E. coli*) that is 25–30 bases upstream from the transcriptional start site is required for chain initiation. This sequence—TATAAATA—which is known as the **TATA box** (pronounced "tah-tah") or the Hogness box, is flanked by two G+C-rich regions. It is likely that there is another sequence important for initiation that is 30–50 bases farther upstream. There is no information available about termination nor is there any evidence for termination factors like Rho.

In the following sections some of the features listed above are examined more closely.

RNA Polymerases of Eukaryotes

The first RNA polymerase isolated was obtained from mammalian cells. However, difficulties in purifying what proved years later to be several enzymes, and the successes in the molecular biology of *E. coli* hampered further study of eukaryotic polymerases. Even at this time knowledge of the eukaryotic polymerases seems primitive compared to that of *E. coli* RNA polymerase.

The three major classes of eukaryotic RNA polymerases are denoted I, II, and III; they can be distinguished by the ions required for their activity, the optimal ionic strength, and their sensitivity to inhibition by various antibiotics. All are found in the eukaryotic nucleus. Minor RNA polymerases, which have not yet been studied in detail, are found in mitochondria and chloroplasts. The locations and products of the nuclear RNA polymerases are listed below.

	Class I	Class II	Class III
Location:	Nucleolus	Nucleoplasm	Nucleoplasm
Product:	rRNA	mRNA	tRNA, 5S RNA

Note that RNA polymerase II is the enzyme responsible for all mRNA synthesis.

The enzymatic activities in crude extracts all result from proteins of large molecular weight, generally about 500,000. Each enzyme consists of two large subunits ($M = 117$ to 214×10^3) and four to eight small subunits. The functions of the individual subunits are not known. The class II and class III enzymes are inhibited by α-amanitin, the toxic product of toadstool mushrooms, and are identified by their sensitivity to this substance. Each of the enzymes can be isolated in a variety of chromatographically and electrophoretically distinct forms, the significance of which is unknown. The biochemical reaction catalyzed by the eukaryotic RNA polymerases is the same as that catalyzed by E. coli RNA polymerase.

Structure of 5′ and 3′ Termini of Eukaryotic mRNA Molecules: "Caps" and "Tails"

The 5′ terminus of a eukaryotic mRNA molecule is blocked by the formation of the 5′-5′-pyrophosphate linkage of a methylated guanosine derivative, 7-methylguanosine (7-MeG), to either 2′-o-methylguanosine (2′-oMeG) or 2′-o-methyladenosine (2′-oMeA). The structural unit

$$(7\text{-MeG})-5'\text{-PPP-}5'-(2'\text{-oMe(G/A)})-3'\text{-P-}5'-\text{nucleoside-}3'\text{-P}$$

in which P and PPP refer to mono- and triphosphate groups respectively, has no free 5′ terminus and is said to be "capped." The structure of a generalized 5′ terminus of a mRNA molecule is shown in Figure 11-28. In yeast and in the slime molds, the most abundant cap structure is that designated cap 0, in which 7-MeG is added to the 5′ terminus and no further methylation occurs. This type of structure is not found in animal cells. In viral mRNA and most animal cell mRNA, methylation of ribose also occurs on the second nucleotide, giving the structure known as cap 1. In some cases, methylation also occurs in the third nucleotide, giving the cap 2 structure. It seems likely at present that each mRNA molecule has a specific cap structure.

Capping occurs shortly after initiation of synthesis of the mRNA, possibly before RNA polymerase II leaves the initiation site, and precedes all excision and splicing events. The biological significance of capping has not yet been unambiguously established but it is believed that it is required for efficient protein synthesis. Capping may function to protect the mRNA from degradation by nucleases and to provide a feature for recognition by the protein-synthesizing machinery (the ribosome).

Most, but not all, animal mRNA molecules are terminated at the 3′ end with a poly(A) tract, which is added to the primary mRNA by a nuclear enzyme, poly(A) polymerase. The adenylate residues are not added to the 3′ terminus of the primary transcript. Transcription normally passes the site of addition of poly(A), so that an endonucleolytic cleavage must occur before the poly(A) is added. A base sequence— AAUAAA—10 to 25 bases upstream from the poly(A) site is a component of the system for recognizing the site of poly(A) addition. Interestingly, some primary transcripts contain two or more sites at which poly(A) can be added. The differentially terminated mRNA molecules usually play different roles in the life cycle of the particular organism. The length of the poly(A) segment can be from 20 to 200 nucleotides. The significance of the poly(A) terminus is unknown at present, but it is

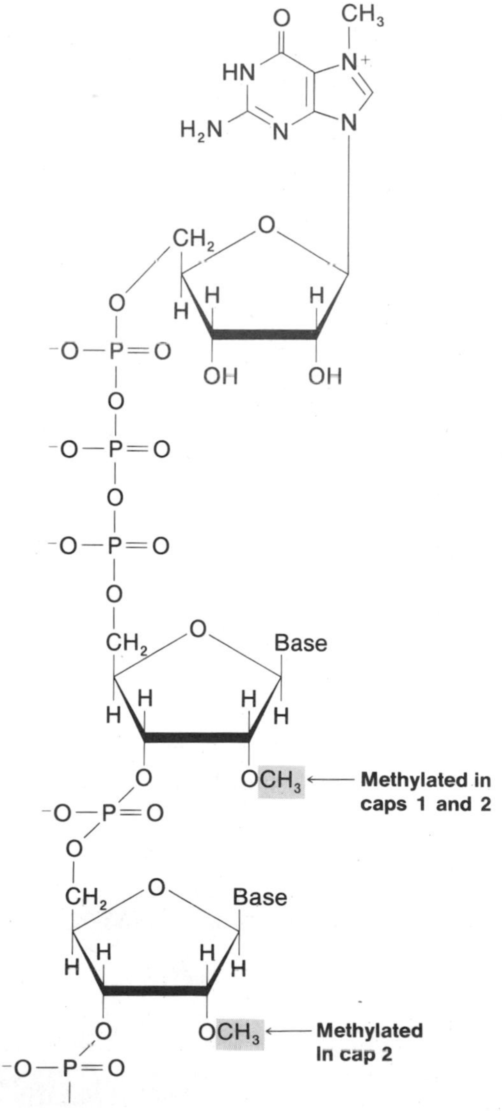

Figure 11-28
Structure of caps at the 5′ ends of eukaryotic mRNA molecules. There are three types of caps (0, 1, and 2). All caps contain 7-methylguanosine (shown in red) attached by a pyrophosphate linkage to the 5′ end of the mRNA molecule. In cap 0, none of the riboses are methylated. In cap 1, one is methylated and in cap 2, two are methylated.

believed that it increases the stability of mRNA; a possible role in attachment of mRNA to intracellular membranes has also been proposed. Some cellular mRNA molecules lack poly(A), so its presence cannot be obligatory for successful translation. The splicing events described in the next section always follow addition of poly(A); however, if polyadenylation is inhibited, splicing occurs normally. Thus, poly(A) has no recognizable role in the further processing of mRNA. After addition of poly(A) and transport of the mRNA to the cytoplasm the poly(A) tail is gradually shortened; in time, no poly(A) remains and the mRNA is degraded. The role of poly(A) will be discussed further in the chapter on viruses (Chapter 21).

RNA Splicing

A characteristic of almost all eukaryotic mRNA molecules is the presence of untranslated intervening sequences (**introns**) that interrupt the coding sequence and are excised from the primary RNA transcript (Figures 11-27 and 11-29). Discovery of the existence of introns has been one of the major benefits of the recombinant DNA technology (without which the intron might not otherwise be known) and is probably the most baffling new phenomenon of the past decade. For many of the surprises in molecular biological research in the past, usually a few hours' thought reveals a reason why a certain phenomenon "has to be." However, after two years, the reason for the existence of introns is as mysterious as ever.

In the processing of mRNA the amount of discarded RNA ranges from 50 percent to nearly 90 percent of the primary transcript. The remaining segments (**exons**) are joined together to form the finished mRNA molecules. The excision of the introns and the formation of the final mRNA molecule by joining of the exons is called **RNA splicing.** The 5′ segment (the cap) of a primary transcript is never discarded and hence is always present in the completely processed molecule; the 3′ segment is also usually retained but cases are known in which the poly(A) tail and an additional small terminal segment are removed. Thus the number of exons is usually one more than (but sometimes equal to) the number of introns.

Figure 11-29
A diagram of the conalbumin primary transcript and the processed mRNA. The seventeen introns, which are excised from the primary transcript, are shown in black.

Primary transcript

Excise introns

mRNA

One must be sure to appreciate the difference between introns and the regions excised from the prokaryotic primary transcripts containing tRNA and rRNA sequences. In the latter, the excised sequences always *surround,* but do not interrupt, each tRNA and rRNA molecule. However, introns are segments that are removed from *within* the coding sequence and that *must* be removed in order to obtain the correct amino acid sequence of the encoded protein.

The existence and the positions of introns in a particular mRNA molecule are readily demonstrated by electron microscopic examination of a DNA-RNA hybrid formed from a strand of the DNA that is transcribed and the fully processed mRNA molecule. The example of adenovirus mRNA and the corresponding DNA are shown in Figure 11-30, along with an interpretation of the data.

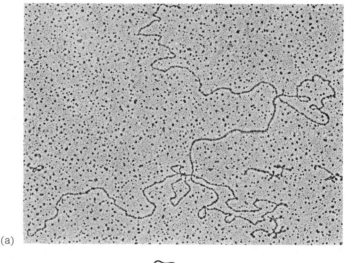

(a)

Figure 11-30
RNA splicing with adenovirus. (a) An electron micrograph of a DNA-RNA hybrid obtained by annealing a single-stranded segment of adenovirus DNA with fully processed capped and polyadenylated mRNA encoding the major virion protein. The loops are single-stranded DNA. (Courtesy of Tom Broker and Louise Chow.) (b) An interpretive drawing. RNA and DNA strands are shown in red and black respectively. Four regions do not anneal—three single-stranded DNA segments corresponding to the introns and the poly(A) tail of the mRNA molecule.

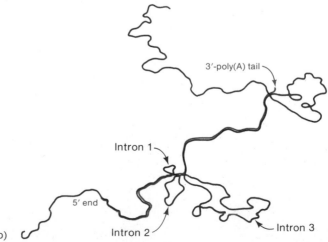

(b)

The mechanism of cutting and splicing is not understood, but a battery of endonucleases and RNA ligases are known to be required. At the present time, putative processing enzymes have been detected in yeast. The remarkable feature of the splicing reaction is that cuts are made at unique positions in transcripts that contain thousands of bases. The demand in fidelity of the excision and splicing reaction is extraordinary, for if an error of one base were made, the correct reading frame would be destroyed. This fidelity must presumably be achieved by recognition of particular base sequences at or near the intron-exon junction.

Base-sequence studies of the regions surrounding 90 different introns indicate that usually two nearly identical repeated sequences of two to six bases can be found at each end of an intron (Figure 11-31). The most striking point about these sequences is that, though the positions of the cuts that are made have not been identified for all introns, it is possible to find two sites (at which cutting might occur) in the sequence such that the excised segment (the intron) would have the sequence GU at the 5' end and AG at the 3' end, as shown in the figure. Possible sequences have been observed in such diverse organisms as chicken, mouse, rabbit, and an animal virus (infecting monkeys), and their existence is taken to indicate a general requirement (sometimes called Chambon's rule) that:

The base sequence of an intron begins with GU and ends with AG.

Exactly how these hypothetical sites of cutting are recognized is unknown. It is very unlikely that these short sequences carry all of the information for splicing specificity; discovery of additional signals is anticipated.

Figure 11-31
Base sequences surrounding the termini of several introns. The intron is shown in black. The terminal GU and AG are underlined. The related terminal sequences are boxed.

Protein		Intron
β-Globin, intron 1	...G C A G G U U G	U C A G G C U G...
β-Globin, intron 2	...C A G G U G A	A C A G U C U C...
Igλ, intron 1	...U C A G G U C A	G C A G G G G C...
SV40 early T antigen	...U A A G G U A A	U U A G A U U C...
Ovalbumin, intron 2	...U A A G G U G A	A C A G G U U G...
Ovalbumin, intron 3	...U C A G G U A C	U C A G U C U G...
Ovalbumin, intron 4	...G C C A G U A A	A C A G G A G A...
Ovalbumin, intron 5	...A A U G G U A A	A A A G G A A A...
Ovalbumin, intron 6	...U G A G G U A U	C C A G C A A G...
Ovalbumin, intron 7	...G C A G G U A U	G C A G C U U G...

A few complete introns have been sequenced and no special features (e.g., symmetry, repeated sequences, etc.) have been noted. One current hypothesis is that a very small RNA molecule anneals with distant base sequences on the primary transcript to bring exons together during the cutting and splicing event (Figure 11-32). Molecules of this type have recently been identified, but whether these molecules function as hypothesized has not yet been determined.

The number of introns per mRNA molecule varies considerably (Table 11-1). Furthermore, within a particular mRNA molecule the introns are widely distributed and have many different sizes (Figure 11-29), but introns are invariably longer than exons.

Splicing occurs in the nucleus after capping and addition of poly(A). The existence of splicing explains the fact the nucleus contains an enormous number of different RNA molecules, whose size distribution is very great. All animal mRNA molecules are monocistronic, so that one might expect the mRNA to have a small range of sizes. However, the amount of RNA that is discarded as a result of excision and splicing ranges widely, so the sizes of the primary transcripts cover a range of molecular weights from about 6×10^4 to 3×10^6. Further-

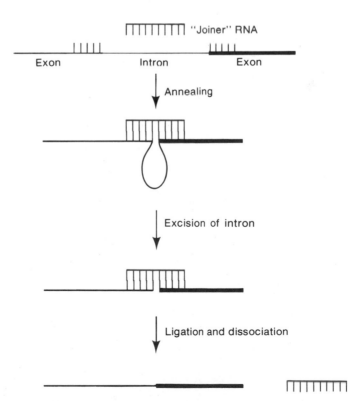

Figure 11-32
One hypothesis for the mechanism of excision of introns and the joining of exons.

Table 11-1
Translated eukaryotic genes in which introns have been demonstrated

Gene	Number of introns
α-Globin	2
Immunoglobulin L chain	2
Immunoglobulin H chain	4
Yeast mitochondria cytochrome *b*	6
Ovomucoid	6
Ovalbumin	7
Ovotransferrin	16
Conalbumin	17
α-Collagen	52

Note: At present the genes for histones and for interferon are the only known translated genes in the higher organisms that do not contain introns.

more, introns are excised one by one, and apparently ligation occurs before the next intron is excised, so the number of different nuclear RNA molecules present at any instant is huge. Translation does not occur until processing is complete, so that before processing was recognized, the nucleus appeared to contain a significant amount of possibly useless RNA; these molecules were given the general name **heterogeneous nuclear RNA** (HnRNA), a term that is still used. After processing is completed, the mature mRNA is transported to the cytoplasm to be translated; it is not known why the primary transcript and partially processed molecules are not transported.

The biological significance of splicing is unclear, for it would seem easier to synthesize mRNA molecules directly without splicing. Theories of why splicing evolved abound but only a few organisms are known in which splicing serves an obvious purpose. For instance, the virus SV40, whose DNA is so small that it can only encode a few genes, uses one of its mRNA molecules twice, once before splicing and then again after a rather small number of bases has been removed. In this way two different proteins are encoded in a particular DNA segment. This type of economy has only been observed in the small animal viruses (see Chapter 21 for further discussion), which, without splicing, would have an inadequate amount of DNA for their necessary functions. (Another example of splicing will be discussed in Chapter 21 when mRNA production in adenovirus is described.) However, the use of splicing to increase coding capacity does not provide an adequate explanation for its ubiquity in animal cells for most eukaryotic cells have more than enough DNA. A few types of mRNA molecules are not spliced—for example, the transcript of the gene encoding α-interferon. At the other extreme is the transcription unit for collagen, which contains 52 introns.

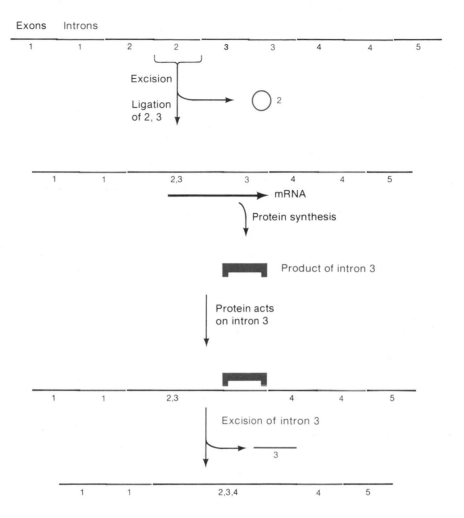

Figure 11-33
An outline of the program for excising two particular introns from the mRNA of the yeast cytochrome *b* gene.

So far, there is only one known example of a eukaryotic cell in which information for protein synthesis is encoded in an intron (Figure 11-33). In the yeast cytochrome *b* gene, splicing begins by excision of an intron as a circle. (That is, not only are the ends of the exon joined but also the ends of the intron are joined.) Once this intron is excised, the two adjacent exons and the following intron (intron 3) comprise a coding unit that is translated. The product that is translated from the coding unit is a protein that is needed for excision of intron 3. This type of mechanism suggests that splicing may sometimes serve a fundamental role in the regulation of gene expression in animal cells, an idea that will be pursued in Chapter 22.

One manifestation of the role of splicing in gene expression is exemplified by the commitment of an antibody-producing cell to production of a particular antibody. This will also be discussed in Chapter 22. Another interesting example of the role of splicing in cell differentiation may be seen in two lines of mouse cells—one is a stem cell line (that is, a cell capable of differentiation) and the other a line of already differentiated cells obtained from the same tissue. The virus SV40 is capable of infecting both cell lines and of being transcribed. When the differentiated cell line is infected, progeny SV40 viruses are synthesized, but in the stem cell line, no viruses are produced; the defect is that the SV40 transcripts are not processed. Thus, it seems that differentiation is associated with the ability to process certain classes of transcripts. No other examples of this are known at present.

An example of a possible requirement for the presence of introns for gene expression is also known—namely, in the β-globin gene. By genetic engineering techniques two forms of an animal virus (again SV40, but this is of no signficance) have been prepared that contain either the intact globin gene or a modified globin gene that lacks the introns. (The latter was prepared by making a DNA copy of the fully processed globin mRNA and inserting this copy into the viral DNA.) Cells were separately infected with these viruses and globin production was examined. It was observed that globin is synthesized only in the cells having the gene that contains the introns. Whether splicing itself is needed or one or more of the introns contains some information necessary for gene expression is not known.

The Transcription Unit Concept

In prokaryotes a mRNA molecule can encode one or more proteins and all proteins can be translated. This is not the case for eukaryotes for, as will be discussed in detail in Chapter 13, only one protein molecule can be translated from a particular mRNA molecule. Nonetheless, in eukaryotes many of the primary transcripts contain information for synthesizing more than one protein. It has been useful in eukaryotic molecular biology to define a segment of DNA that can be transcribed as a **transcription unit** and to distinguish a **simple transcription unit**—one that carries information for only one protein—from a **complex transcription unit**—one that carries information for more than one protein molecule. If only one protein molecule can be translated from a single mRNA molecule, how can one say that a complex transcription unit encodes two or more proteins? As we will see in Chapter 22, a primary transcript can be altered so that it terminates at one of several poly(A)-addition sites, generating distinct mRNA molecules, and also that a particular transcript with a poly(A) tail can be processed in different ways (that is, different introns can be removed), producing mRNA molecules having totally different coding sequences. This ability has important consequences for regulation in eukaryotes.

Eukaryotic rRNA Genes

Eukaryotic ribosomes contain four RNA molecules. They are characterized by their sedimentation coefficients and are referred to as 5S, 5.8S, 18S, and 28S rRNA molecules (see Chapter 13 for details). The organization of the genes for these RNA species differs from that observed in prokaryotes.

In the nucleolus of a typical animal cell several hundred copies of a DNA sequence (ribosomal DNA or rDNA) encode the 18S, 5.8S, and 28S rRNA molecules (in that order). The 5S rRNA genes are located outside the nucleolus. In several species of *Xenopus*, the African clawed toad, which are the best understood systems, the three ribosomal genes comprise a 40S transcription unit (Figure 11-34) from which three RNA segments are excised by processing enzymes giving rise to 18S, 28S, and 5.8S rRNA molecules. These transcription units are separated by large untranscribed spacer segments, as shown in the figure. The repeating unit has a molecular weight of about 8×10^6.

The gene for the 5S rRNA, which contains 120 base pairs, is contained in a separate transcription unit consisting of the 120 base-pair transcribed region and an untranscribed segment of about 600 base pairs (Figure 11-35). The 5S rRNA genes are also arranged in tandem—

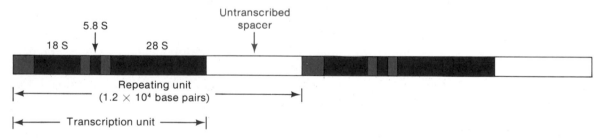

Figure 11-34
Organization of the genes in the 40S precursor of *Xenopus* rRNA. The solid colors represent the transcribed region. The red regions are removed from the transcript by the processing enzymes. The black regions are the three rRNA genes.

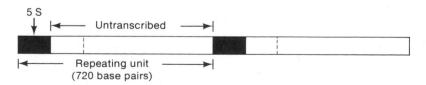

Figure 11-35
Organization of the genes for the 5S rRNA of *Xenopus*. Only the black regions, which are 120 units long, are transcribed. Between each black region and the dashed line is a segment whose base sequence is very near that of the 5S rRNA; however, oddly it is not transcribed.

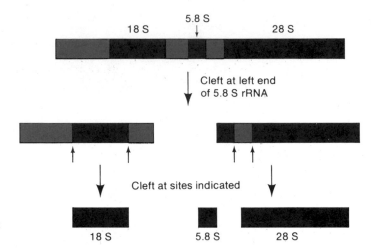

Figure 11-36
Representative stages in the formation of mammalian rRNA from a primary transcript.

there are several enormous units containing several thousand copies of the gene. Altogether, there are 24,000 copies of the 5S RNA gene! During oogenesis (development of an egg) there is an increase in the number of copies of the 40S unit until in the mature egg there are 2×10^6 copies of the unit, representing 75 percent of the total DNA of the cell.

All eukaryotic rRNA transcripts require processing, which occurs by events similar to those observed in prokaryotes—namely, trimming of the 5′ and 3′ termini and excision of the unwanted regions of the transcript. Processing of eukaryotic rRNA has been most widely studied in cultured mammalian cells. The sequence of events is shown in Figure 11-36. Several endonucleases are involved, one of which closely resembles *E. coli* RNase III. Figure 11-36 shows that internal segments are excised from primary rRNA transcripts, but these regions are not referred to as introns because, by definition, an intron is a segment within a gene. In fact, most eukaryotic rRNA genes do not contain introns, though an exception occurs in the fruit fly *Drosophila melanogaster*. This organism contains 285 copies of the rRNA genes of which roughly one-third of the copies contains introns; however, the intron-containing genes are not transcribed.

Formation of Eukaryotic tRNA Molecules

Eukaryotic tRNA molecules are also excised from large transcripts (called pre-tRNA transcripts), which may contain one or more tRNA sequences. The enzymology of processing is not yet fully understood, though an enzyme similar to *E. coli* RNase P (which is utilized for processing of tRNA) has been purified from mouse cells. The best information at present comes from studies with yeast.

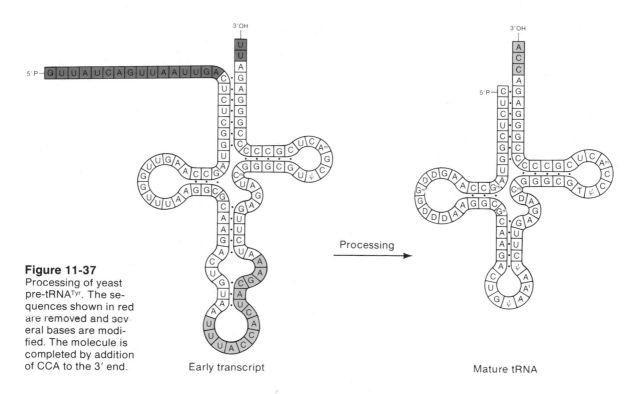

Figure 11-37
Processing of yeast pre-tRNA^Tyr. The sequences shown in red are removed and several bases are modified. The molecule is completed by addition of CCA to the 3′ end.

Early transcript

Processing

Mature tRNA

Many, but not all, yeast tRNA genes contain introns of 10–20 bases; thus, not only external processing (as in prokaryotes) but splicing is necessary to produce mature yeast tRNA (Figure 11-37), in contrast with yeast rRNA. The processing mechanism in yeast also differs from that in *E. coli* in that the 3′-terminal CCA sequence is not present in the primary transcript but is added as part of the processing operation.

In discussing RNA splicing of mRNA molecules it was pointed out that probably all introns are terminated by a GU at one end and an AG at the other. This is not true of the introns of tRNA; in this case, there is no base sequence common to all excised regions.

The tRNA-processing enzymes may be the same in organisms that are evolutionarily quite distant. This is suggested by the following experiment. DNA fragments containing a yeast tRNA gene were microinjected into a *Xenopus* egg to determine whether a toad cell could produce normal yeast tRNA. The striking result was that the DNA was not only transcribed but that mature yeast tRNA was produced. Since yeast and toads presumably evolved hundreds of millions of years apart, the fact that the specificity of the processing system has been conserved for a very long time indicates that processing must serve some very important function in eukaryotic gene expression.

MEANS OF STUDYING INTRACELLULAR RNA

Several procedures are used to study RNA metabolism *in vivo*. In most of the techniques, radioactive RNA is prepared by adding ^{3}H-labeled or ^{14}C-labeled uridine to growth medium. However, this also produces radioactive DNA, because uridine can be metabolized to cytosine and thymine; this is significant when studying RNA metabolism because, when RNA is isolated from cells, it is usually contaminated with DNA. The contaminating DNA can be removed, though, by treatment of the sample with pancreatic DNase, for this enzyme degrades the DNA to mononucleotides and small oligonucleotides, which can be easily separated from the RNA.

In the following three sections two types of methods for studying intracellular RNA molecules are presented. In the first section, procedures are described that are used to fractionate various species of RNA, which can then be studied separately. This type of analysis is especially useful in studying rRNA and tRNA because the number of species is small. In the second section, methods used principally in the study of mRNA are described; in these, different RNA molecules are not separated but detected in a mixture by specific DNA-RNA hybridization. A variant of the hybridization procedure—the **Southern transfer method**—in which DNA molecules are fragmented and fractionated and then hybridized to RNA, is so important and in such widespread use that it is described separately in the third section.

Separation of RNA Molecules

In early studies of RNA metabolism the various species of RNA were distinguished by zonal sedimentation in a sucrose gradient. The 16S and 23S rRNA molecules were easily recognized by this procedure because they sediment in discrete narrow bands (shown earlier in Figure 11-22). The tRNA (4S) forms an obvious band near the top of the centrifuge tube, but the 5S RNA is difficult to resolve because it sediments quite near the tRNA, which is in great excess. Thus, 5S RNA was first identified in samples of purified ribosomes, which contain only 5S, 16S, and 23S RNA, by dissociating the ribosomes to free the RNA and then sedimenting this RNA mixture. Messenger RNA also presented some problems because the range of size of mRNA is large so that there is no distinct zone of mRNA. Instead it forms a diffuse band superimposed on the 16S and 23S rRNA bands (Figure 11-22), and this makes it difficult to identify the mRNA unambiguously. The problem of identifying mRNA was solved by the pulse-chase procedure, which clearly distinguishes mRNA from the stable RNA molecules. For example, in a 30-second pulse with [^{14}C]uridine, all RNA species are labeled but after a chase with nonradioactive [^{12}C]uridine for a few minutes, radioactivity

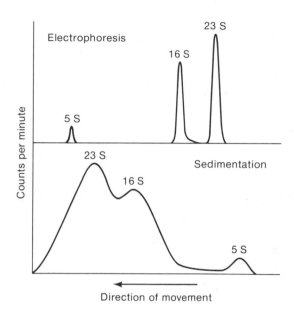

Figure 11-38
Separation of 5S, 16S, and 23S rRNA by gel electrophoresis in polyacrylamide and by sedimentation through a sucrose density gradient. Note how much narrower the electrophoretic bands are than the sedimentation zones. Also, the relative positions of the three classes of molecules are reversed in the two separation techniques.

remains only in the stable RNA molecules such as tRNA and rRNA. Thus, mRNA is identified as the fraction whose lifetime is only a few minutes. The lifetime of eukaryotic mRNA is often many hours, so this procedure is usually not feasible; studies of eukaryotic mRNA (with the exception of yeast, whose mRNA is short-lived) had to await the development of the recombinant DNA technology (Chapter 20).

The resolution of the various RNA components is considerably improved when they are separated by electrophoresis in polyacrylamide, as shown in Figure 11-38.

DNA-RNA Hybridization

A great deal of information about the stable classes of RNA has been obtained from experiments in which the amount of each species was measured by sedimentation or electrophoresis. However, these methods are less useful in the study of mRNA because they are not capable of resolving particular mRNA species. The most informative experiments with mRNA have utilized hybridization procedures, of which the most useful application is filter hybridization, described in Chapter 4 (Figure 4-15). Briefly, single-stranded DNA containing a particular gene is bound to a filter, which is subsequently incubated with an extract containing radioactive RNA. If a DNA-RNA hybrid forms, radioactivity is retained on the filter even after extensive washing.

In performing hybridization one needs a pure sample of a DNA molecule that contains a sequence complementary to that of the RNA molecule of interest. Furthermore, in a typical hybridization experiment 0.1–2 μg of DNA is needed. When studying phage or viral mRNA, the necessary DNA is readily available because it can be obtained from purified phage and viruses; and since it is straightforward to prepare several milligrams of phage or viral DNA, the required amounts present no problem. In studying bacterial mRNA it is usually not feasible to use the total bacterial DNA because any particular gene would consist of only about 0.03 percent of the total DNA. This small percentage gives rise to two problems—first, regions of the DNA other than the gene of interest may have some degree of homology with the mRNA, and second, in order to have the necessary quantity of the DNA of the gene, 0.3–6 *milligram* of DNA would be needed, which vastly exceeds the capacity of a reasonably sized nitrocellulose filter to hold DNA. Furthermore, with total bacterial DNA a particular mRNA would not be detected if it were in a mixture containing many bacterial mRNA molecules. For bacteria the solution to this problem is to "clone" the gene—in other words, to create a new DNA molecule, which can be easily isolated, which consists mostly of sequences nonhomologous to the mRNA of interest, and which contains the particular gene as a major component. A standard procedure is to use either specialized genetic techniques or the recombinant DNA technology (Chapter 20) to insert the gene into the DNA of *E. coli* phage λ. Thus, simply by preparing a phage DNA sample one has purified DNA, roughly 3 percent of which is the gene sequence, the remainder being phage DNA. If necessary, by quite simple techniques this sequence can be excised from the phage DNA and purified. A DNA molecule such as a phage, viral, or plasmid DNA carrying a foreign DNA sequence to be used in a hybridization experiment is termed a **probe.**

In studying eukaryotic mRNA the concentration problem just described is even more formidable than with bacterial DNA since, owing to the large amount of DNA in a typical eukaryote, a gene sequence may be only 0.0001 percent of the total DNA. Once again, the recombinant DNA technology comes to the rescue because bacterial plasmids, phages, and animal viruses containing eukaryotic DNA can all be prepared. The methods needed to form and identify these recombinant molecules are more complex than with bacterial DNA and are described in detail in Chapter 20. Nonetheless, probes containing many different cloned genes have been prepared and the use of these probes in eukaryotic DNA biology makes the detection of a particular eukaryotic mRNA as straightforward as that of any phage mRNA molecule.*

In the following sections several examples of the use of DNA-RNA hybridization are described.

---

*In fact, the development of probes by recombinant DNA techniques has been a major breakthrough in the study of eukaryotes in recent years.

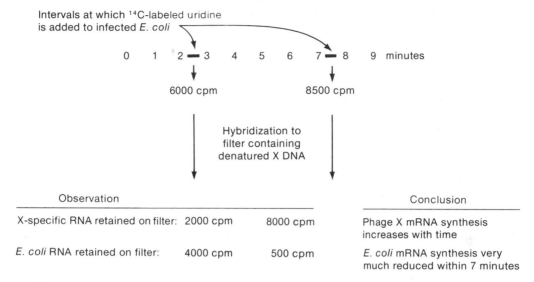

Figure 11-39
Use of hybridization to monitor phage mRNA synthesis and shutdown of *E. coli* RNA synthesis during infection with phage X. See text for a description of the experiment.

Detecting Phage X mRNA

Figure 11-39 shows the use of filter hybridization to compare the amounts of phage X mRNA synthesized in *E. coli* during the third and eighth minute after infection and to study the effect of phage infection on *E. coli* RNA synthesis. Phage X mRNA and *E. coli* RNA of all classes become labeled after incubation of the infected cells in the presence of [^{14}C]uridine; however, only phage X mRNA will be retained on a filter to which phage X DNA has been bound. During the third minute after infection [^{14}C]uridine is incorporated into RNA at the rate of 6000 cpm/minute;* hybridization indicates that the synthesis of phage X mRNA occurs at a rate of 2000 cpm/minute. During the eighth minute after infection the rate of incorporation increases to 8500 cpm/minute; now 8000 cpm/minute is hybridizable and is phage X mRNA. Thus the rate of phage X mRNA synthesis increases fourfold (from 2000 to 8000 cpm/minute). This simple experiment also illustrates the shutdown of *E. coli* RNA synthesis after phage infection, from $6000 - 2000 = 4000$ cpm/minute to $8500 - 8000 = 500$ cpm/minute in the two time intervals.

---

*The notation "cpm" symbolizes counts of radioactivity per minute in which "minute" refers to the *counting* time. In the term cpm/minute, the word minute refers to the *labeling* time.

Distinguishing Two Classes of mRNA by Hybridization Competition

This example describes a technique that enables one to determine both the time at which particular classes of mRNA are synthesized and the lengths of the lifetimes of these molecules. It will be followed by two examples in which modifications of the technique are presented that allow the transcribed gene to be physically located.

Hybridization competition is a variation of the basic filter hybridization technique that allows one to determine whether two different cell extracts contain the same mRNA. Consider one extract containing [^{3}H]lac mRNA that has been hybridized to a filter bearing lac DNA. We further assume that the amount of lac mRNA added is sufficient to hybridize with every lac DNA sequence on the filter and that this amounts to 1 μg of [^{3}H]lac mRNA with an activity of 1000 cpm being hybridized. If 1 μg of nonradioactive lac mRNA from a second extract were mixed with the [^{3}H]lac mRNA prior to hybridization, the nonradioactive mRNA would compete with the [^{3}H]lac mRNA for the limited number of complementary sequences in the DNA. Since binding of [^{1}H]mRNA and [^{3}H] mRNA is equally probable and since 1 μg of mRNA is the maximum amount of mRNA that can be retained, the filter will have only 500 cpm. Similarly, if 3 μg of [^{1}H]mRNA were added, only $1/(3 + 1) = 1/4 \times 1000 = 250$ cpm would be found on the filter, as shown in Figure 11-40(a). (Note that if an extract obtained from cells unexposed to lactose were allowed to compete with the [^{3}H]lac mRNA, 1000 cpm would be measured on the filter because the extract would contain no lac mRNA, as stated in the foregoing example. These data are plotted in the hybridization competition curve shown in Figure 11-40(b).

This example can now be extended to a system in which two different mRNA molecules are examined at the same time. Consider a culture of bacteria infected with a phage Q and in which two different mRNA molecules are made; mRNA$_1$ is made at a constant rate from 1–10 minutes after infection and mRNA$_2$ is made from 5–20 minutes after infection. RNA is pulse-labeled for one minute in three different time intervals—namely, 3–4, 6–7, and 15–16 minutes after infection—and the labeled RNA is isolated immediately after the labeling period. In addition, nonradioactive RNA is isolated at 4, 7, and 16 minutes after infection of different cultures. Nine sets of hybridization reactions are prepared, each with denatured Q DNA bound to the filter. The data are shown in Figure 11-41 and can be analyzed by the following steps:

1. Binding of the 3–4-minute RNA is totally inhibited if sufficient 3–4-minute and 6–7-minute unlabeled DNA is added but is not inhibited at all by 16-minute RNA. This shows that the 3–4-minute RNA sequences are present at 7 minutes but gone by 16 minutes (conclusion 1).

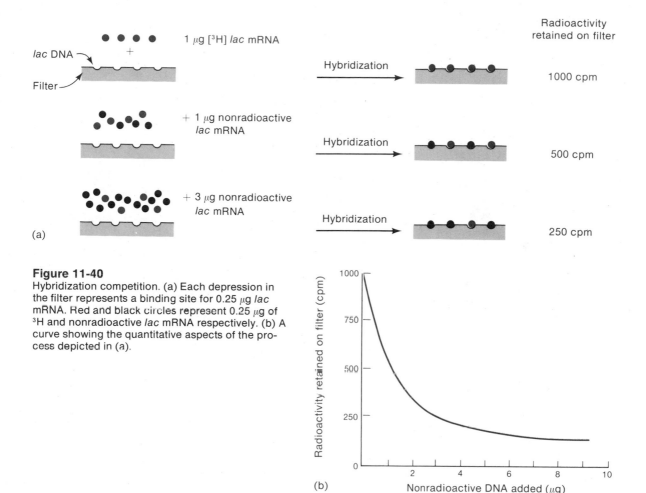

Figure 11-40
Hybridization competition. (a) Each depression in the filter represents a binding site for 0.25 μg *lac* mRNA. Red and black circles represent 0.25 μg of ³H and nonradioactive *lac* mRNA respectively. (b) A curve showing the quantitative aspects of the process depicted in (a).

2. The binding of 6–7-minute RNA can be totally inhibited by 7-minute RNA but only partially inhibited by 4-minute and 16-minute RNA. Thus, there must be at least two kinds of mRNA labeled from 6–7 minutes (conclusion 2). One of these is the mRNA that is also labeled from 3–4 minutes and another is still being synthesized at 16 minutes.
3. A quantitative analysis suggests that there are only two species of mRNA labeled in the 6–7-minute interval: note that 40 percent of the 6–7-minute RNA is inhibited by 4-minute RNA; thus 40 percent of the 6–7-minute RNA must in fact be 3–4-minute RNA (conclusion 3). Also, the remainder, or 60 percent, of the

6–7-minute RNA is inhibited by 16-minute RNA, so that all molecules labeled in the period from 6–7 minutes contain sequences that are present either at 4 minutes or 16 minutes (conclusion 4).

4. If we assume that there is only one type of molecule labeled in the period from 3–4 minutes and another type present at 16

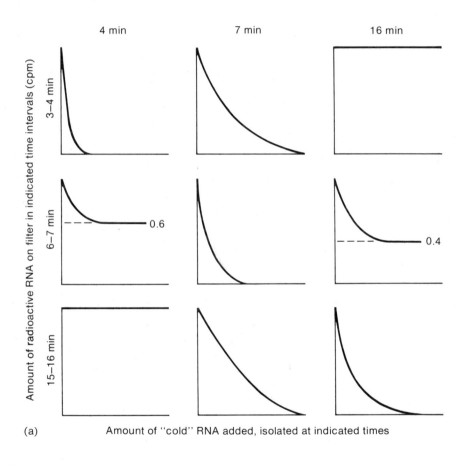

(a)

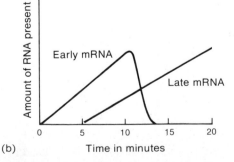

(b)

Figure 11-41
The hybridization competition data discussed in the text. (a) The data. (b) An analysis of the data.

minutes, we can conclude that only two types of mRNA are being examined. The early mRNA is present at least from 3–7 minutes but is no longer around at 16 minutes (conclusion 5); the later mRNA is not present at 4 minutes, so that its transcription begins some time between 4 and 6 minutes and it is still present at 16 minutes (conclusion 6). Additional labeling in other time intervals and competition with extracts obtained at other times could define the time of synthesis more closely and would give evidence for other classes of mRNA molecules, if they were to exist.

This method has been very useful in determining the timing of synthesis of various classes of RNA in phage and viral infections. A more elegant procedure is the following.

Hybridization with Defined Fragments of DNA

By using recombinant DNA techniques (Chapter 20) researchers can break DNA molecules into defined fragments containing particular genes or clusters of genes, and these fragments can easily be purified and bound to filters. Thus, labeled mRNA can be hybridized to DNA bound to a collection of filters and in this way shown to be transcribed from particular genes. This technique is especially useful also in analyzing sequential transcription such as that which occurs in the phage infections just described. Consider a phage whose DNA is broken up into five segments, A, B, C, D, and E, as shown in Figure 11-42(a). These

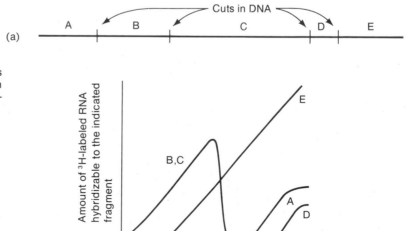

Figure 11-42
Transcription of specific regions of DNA assayed by hybridization with fragments produced by cutting the DNA at specific places. RNA is pulse-labeled in various time intervals and hybridized separately to each of the fragments. The data provide the temporal program of transcription. For example, transcription of the B–C region begins at time zero and terminates about ten minutes later, whereas transcription of the E region begins at five minutes and continues indefinitely at a constant rate. Compare to Figure 11-40(b).

different DNA segments can be bound to filters separately and used to analyze the composition of phage mRNA. The mRNA is again labeled in one-minute intervals at different times after infection and hybridized to each filter. Figure 11-42(b) shows that mRNA is synthesized from genes in segments B and C from 0 to 10 minutes, from segment E starting at 5 minutes, and in segments A and D (mRNA molecules not recognized in the competition experiment) from 15 minutes on. (Of course, the data would normally not be so unambiguous, because frequently the cuts would be within a transcriptional unit.) Additional information can be obtained from the following technique.

Hybridization with Specific Strands

The *l* and *r* strands of most DNA molecules can be physically separated. The separation technique is based on the facts that pyrimidines (C and T) are often present in one strand of DNA in tracts about ten bases long and that the number of pyrimidine tracts is not usually the same in both strands. If denatured DNA is mixed with a polyribopurine (a mixed polymer of inosine and guanine, poly(I,G), is a favorite), the ribopolymer binds to each strand. Since the density of RNA in CsCl is much greater than that of DNA, the DNA-poly(I,G) hybrids have a higher density than the parent DNA strands. In addition because the number of pyrimidine tracts in the *l* and *r* strands differ, the amount of polymer bound to each strand differs so that the *l* and *r* strands have distinct densities and can be separated into two bands by sedimentation to equilibrium in CsCl (Figure 11-43). The material in each band can be

Figure 11-43

Separation of *r* (black) and *l* (red) strands of DNA. The poly(I,G) molecule has a density much greater than that of single-stranded DNA and hence increases the density of the single strands by binding to sequences of cytosine and thymine (pyrimidine tracts). Since the individual single strands usually have different numbers of pyrimidine tracts, the complexes with poly(I,G) will have different densities and therefore come to rest at separate positions when centrifuged to equilibrium in a CsCl density gradient. The complexes are removed from the CsCl and the poly(I,G), which is a polyribonucleotide, is digested with an RNase. The purified single strands of DNA are generally placed on a nitrocellulose filter and used in hybridization experiments. The denser and less dense strands are also frequently called the H (heavy) and L (light) strands.

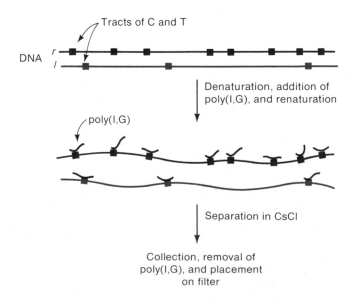

Figure 11-44
(a) Hybridization to *r*- and *l*-strand DNA. The mRNA is that used in the experiments described in Figure 11-41 and 11-42. (b) The four transcripts (shown in red) are shown on the map of Figure 11-42(a). The circled numbers indicate the order in which the mRNA molecules are made. The length of each arrow defines the extent of the region in which transcription has been mapped—not the actual length of mRNA molecules. For example, non-transcribed segments of any size might also be present in region 1.

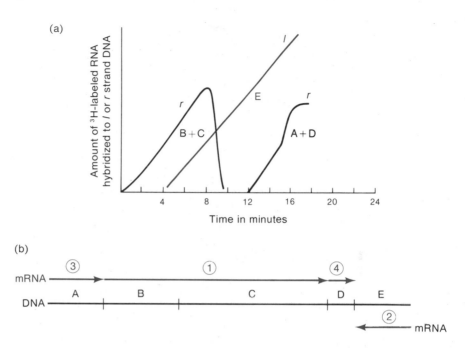

collected, the ribopolymer removed, and the single strands of DNA bound to filters. The mRNA molecules labeled as in the foregoing example can then be hybridized to *l*-strand and r-strand filters. Typical hybridization results might be those shown in Figure 11-44.

The Southern Transfer Procedure

The Southern transfer procedure (developed by Edwin Southern and sometimes called **blotting**), a method for performing hybridization to particular DNA segments, avoids the necessity of purifying the DNA fragments obtained by treatment of DNA by restriction endonucleases.

At present the best way to separate DNA fragments from one another is by electrophoresis through agarose gels. A specific fragment can be isolated by cutting out of a gel a portion that contains the fragment of interest. A variety of procedures, most of which are cumbersome and tedious, are available for recovering the DNA molecule from the gel. If hybridization is to be performed, the fragment must be bound to a nitrocellulose filter. In the Southern transfer technique a collection of fragments is handled in such a way that all fragments are transferred from a gel to a sheet of nitrocellulose in a single step, significantly simplifying the entire process.

The Southern transfer technique is carried out as follows (Figure 11-45). DNA is enzymatically fragmented and then electrophoresed

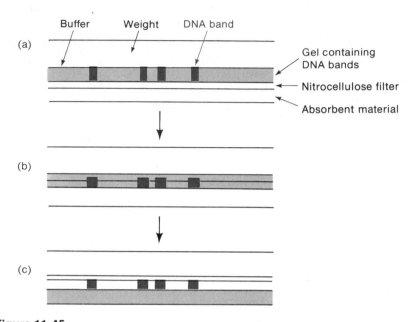

Figure 11-45
The Southern transfer technique. (a) A stack—consisting of a weight, a gel, a filter, and absorbent material—at the time the weight is applied. (b) A later time—the weight has forced the buffer (shaded area), which carries the DNA, into the nitrocellulose. (c) The lowest layer has absorbed the buffer but not the DNA, which remains bound to the nitrocellulose.

through an agarose gel. Following electrophoresis the gel is soaked in a denaturing solution (usually NaOH), so that all DNA in the gel is converted to single-stranded DNA, which is needed for hybridization. A large sheet of nitrocellulose paper is placed on top of several sheets of ordinary filter paper; the gel, which is typically in the form of a broad flat slab, is then placed on the nitrocellulose filter and covered with a glass plate to prevent drying. A weight is then placed on the top of the stack and the liquid is squeezed out of the gel. The liquid passes downward through the nitrocellulose filter. Denatured DNA binds tightly to nitrocellulose (Chapter 3); the remaining liquid passes through the nitrocellulose and is absorbed by the filter paper. DNA molecules do not diffuse very much, so that if the gel and the nitrocellulose are in firm contact, the positions of the DNA molecules on the filter are identical to their positions in the gel. The nitrocellulose filter is then dried in vacuum, which insures that the DNA remains on the filter during the hybridization step. The dried filter is then moistened with a very small volume of a solution of ^{32}P-labeled RNA, placed in a tight-fitting plastic bag to prevent drying, and held at a temperature suitable for renaturation (usually for 16–24 hours). The filter is then removed, washed to remove unbound radioactive molecules, dried, and autoradiographed with x-ray film. The blackened positions of the film indicate the locations of the DNA molecules whose DNA base sequences are complementary to the sequences of the added radioactive molecules. Since usually the genes contained in each fragment are known, specific mRNA molecules can be identified. The degree of blackening of the film is easily measured quantitatively and is proportional to the amount of RNA that has hybridized. Thus, the amount of mRNA transcribed from each region of a DNA molecule can be measured. Note that by this technique many different mRNA molecules can be studied simultaneously.

The Southern transfer method has many other uses, particularly for DNA-DNA hybridization analysis.

REFERENCES

Abelson, J. 1979. "RNA processing and the intervening sequence problem." *Ann. Rev. Biochem.,* 48, 1035–1069.

Adhya, S., and M. Gottesman. 1978. "Control of transcription termination." *Ann. Rev. Biochem.,* 47, 967–996.

Altman, S. 1981. "Transfer RNA processing enzymes." *Cell,* 23, 3–4.

Bujard, H. 1980. "The interaction of *E. coli* RNA polymerase with promoters." *Trends Biochem. Sci.,* 5, 274–278.

Chambon, P. 1981. "Split genes." *Scient. Amer.,* 244 (May), 60–71.

Crick, F. 1979. "Split genes and RNA splicing." *Science,* 204, 264–271.

Danchin, A., and A. Ullmann. 1980. "The coordinate expression of polycistronic operons in bacteria." *Trends Biochem. Sci.,* 5, 51–53.

Erkmann, V. A. 1976. "The structure and function of 5S and 5.8S RNA." *Prog. Nucleic Acid Res. Molec. Biol.,* 18, 45–90.

Freifelder, D. 1982. *Physical Biochemistry.* W. H. Freeman and Co. Chapters 7 and 9 have detailed descriptions of hybridization techniques and the Southern transfer method.

Lathe, R. 1978. "RNA polymerase of *E. coli.*" *Current Topics Microbiol. Immunol.,* 83, 37–92.

Losick, R., and Chamberlin, M. 1976. *RNA Polymerase.* Cold Spring Harbor Laboratory.

Miller, O. L. 1973. "The visualization of genes in action." *Scient. Amer.,* 228 (March), 34–42.

Perry, R. P. 1976. "Processing of RNA." *Ann. Rev. Biochem.,* 45, 605–630.

Shatkin, A. J. 1976. "Capping of eukaryotic mRNA." *Cell,* 9, 645–653.

Spiegelman, S. 1964. "Hybrid nucleic acids." *Scient. Amer.,* 210 (May), 48–56.

12 Translation

I. THE INFORMATION PROBLEM

The synthesis of every protein molecule in a cell is directed by intracellular DNA. There are two aspects to understanding how this is accomplished—the **information** or **coding problem** and the **chemical problem.** By the information problem is meant the mechanism by which a base sequence in a DNA molecule is translated into an amino acid sequence of a polypeptide chain. The chemical problem refers to the actual process of synthesis of the protein: the means of initiating synthesis; linking together the amino acids in the correct order; terminating the chain; releasing the finished chain from the synthetic apparatus; folding the chain; and, often, postsynthetic modification of the newly synthesized chain. The overall process is called **translation.**

Translation is a complex process requiring the participation of about two hundred distinct macromolecules. Our approach in Chapters 12 and 13 is to outline the general features of the process and then to discuss the various stages in some detail. The structures of two major components—namely, the ribosome and transfer RNA (tRNA)—are actively being studied at present; these topics will be discussed in detail in sections that may be omitted by the beginning student.

OUTLINE OF TRANSLATION

Protein synthesis occurs on intracellular particles called **ribosomes.** These particles, which in prokaryotes consist of three RNA molecules and about 55 different protein molecules, contain the enzymes needed to form a peptide bond between amino acids, a site for binding the mRNA, and sites for bringing in and aligning the amino acids in preparation for assembly into the finished polypeptide chain. Amino acids themselves are unable to interact with the ribosome and cannot recognize bases in the mRNA molecule. Thus there exists the collection of carrier molecules described in the previous chapter, **transfer RNA (tRNA).** These molecules contain a site for amino acid attachment and a region called the **anticodon** that recognizes the appropriate base sequence (the **codon**) in the mRNA. Thus, proper selection of the amino acids for assembly is determined by the positioning of the tRNA molecules, which in turn is determined by hydrogen-bonding between the anticodon of each tRNA molecule and the corresponding codon of the mRNA. There are two sites on a ribosome for binding tRNA molecules, the **P site** and the **A site.**

In outline, protein synthesis consists of several stages (Figure 12-1):

1. An initiation complex forms in which the P site is filled with an initiator tRNA molecule (bearing methionine) and in which the mRNA is correctly positioned so that a start codon matches up with the initiator tRNA.
2. The A site is occupied by an amino acid–tRNA complex (called **aminoacyl-tRNA**); the particular aminoacyl-tRNA that is chosen is determined by the codon on the mRNA that is adjacent to the start codon.
3. The first amino acid (the one on the initiator tRNA) is transferred to and covalently joined to a second amino acid (the one attached to the tRNA in the A site) while both are still attached to tRNA.

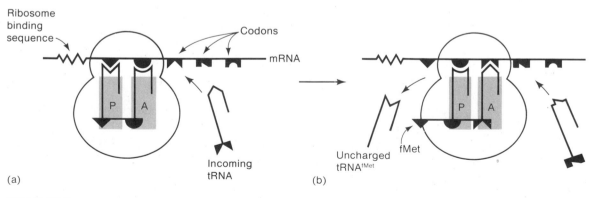

Figure 12-1
Schematic diagram of the beginning of polypeptide synthesis. In (a) the P site is occupied by fMet-tRNA^fMet and the A site is occupied by the tRNA that matches the second codon. In (b) the mRNA has advanced by one codon. The fMet is separated from its tRNA and leads the new polypeptide chain. The ribosome binding sequence is a base sequence in the mRNA that binds to the ribosome.

4. The first tRNA molecule is then ejected from the P site.
5. The aminoacyl-tRNA molecule in the A site advances to the P site.
6. The aminoacyl-tRNA corresponding to the third codon occupies the A site.
7. The process continues until a stop codon in the mRNA occupies the A site. Then, the completed polypeptide chain is released from the ribosome.

Clearly there must be many different tRNA molecules, because each amino acid must be brought in to the ribosome in a way that ensures that it corresponds to the base sequence of the mRNA. Thus, there are specific tRNA molecules that correspond to each amino acid.

The formation of aminoacyl-tRNA is catalyzed by a set of enzymes called **aminoacyl-tRNA synthetases.** Each enzyme recognizes specific sites (a recognition site and an amino acid attachment site) on each tRNA molecule as well as one of the 20 amino acids, so that the appropriate amino acid will be attached to the correct tRNA molecule.

In this chapter we shall be examining the various stages of the information problem in detail.

THE GENETIC CODE

By the **genetic code** one means the collection of base sequences (codons) that correspond to each amino acid and to stop signals for translation.

Since there are 20 amino acids, there must be more than 20 codons to include signals for starting and stopping the synthesis of particular protein molecules. If one assumes that all codons have the same number of bases, then each codon must contain at least three bases. The argument for this conclusion is the following. A single base cannot be a codon because there are 20 amino acids and only four bases. Pairs of bases also cannot serve as codons because there are only $4^2 = 16$ possible pairs of four bases. Triplets of bases are possible because there are $4^3 = 64$ triplets, which is more than adequate. We will see shortly that the code is indeed a triplet code and that all 64 codons carry information of some sort. In many cases, several codons designate the same amino acid—that is, the code is **redundant.**

Two Types of Codes

There are several ways in which a codon could be read from a mRNA molecule. The two most important alternatives originally considered are the **overlapping and nonoverlapping codes** (Figure 12-2). In an

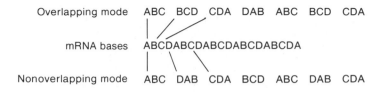

Figure 12-2
Two ways of reading the bases of a mRNA molecule. In the overlapping mode the seven codon triplets are read from nine bases; in the nonoverlapping mode 21 bases are needed.

overlapping code each base serves as the first base of some codon; in a nonoverlapping code, each base is used in only one codon. The overlapping code is certainly more economical in that a particular region of the DNA can carry the information for the synthesis of three times as many proteins as a nonoverlapping code can. The overlapping code has a particular disadvantage, though, in that a single mutagenic base change could alter as many as three amino acids, whereas only one would be altered with a nonoverlapping code. The overlapping code could have the biological effect of increasing the fraction of mutations that are lethal and thereby reducing the probability of evolutionary improvement by constraining the amino acid sequence. The two kinds of codes were distinguished experimentally by examining the amino acid sequence of wild-type and mutant proteins made from particular viral genes. In each case observed it was found that the mutant protein was altered by one and not three amino acids. Thus, the code was shown to be of the nonoverlapping type.*

We have argued that the genetic code must be at least a three-letter code but have not ruled out codes having more than three letters. A great deal of chemical evidence, which will be described shortly, clearly identified the sequences of three bases specifying particular amino acids but, prior to this work, a brilliant genetic experiment was performed that clearly indicated that the code is a **triplet code.** This experiment, characteristic of the classic period of molecular biology, is described because it shows the great power of genetic arguments in molecular biology.

A Genetic Argument for a Triplet Code

The molecule proflavin (an acridine derivative) binds tightly to a DNA molecule. In 1961, before the triplet code had been determined, it was

*We will see later that some phages and viruses have overlapping genes. The codons do not overlap but the genes themselves do. Since the overlapping genes are not in the same phase, a mutation usually causes an amino acid change in two different proteins.

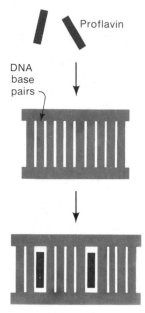

Figure 12-3
A segment of a DNA molecule showing the intercalation of proflavin. In normal DNA there is virtually no space between the base pairs. Proflavin, whose thickness equals that of a base pair, pushes the base pairs apart and thereby lengthens the DNA molecule. Proflavin binds tightly to DNA because it stacks between the bases. The proflavin-induced distortion of the parent DNA strand interferes with replication and can lead to two types of mutants in the daughter strands. In one instance an extra base is added, leading to an insertion mutation; in the second a base is omitted, leading to a deletion mutation.

believed that proflavin binds by intercalating between the DNA bases (Figure 12-3), though there was no proof of this idea. Proflavin is also highly mutagenic if it is applied during a replication cycle. That is, treatment of a phage suspension with proflavin is not mutagenic, but if proflavin is added to a phage-infected culture of bacteria, mutant phage progeny are produced.

Acridine-induced mutants (see Chapter 10) have one property that distinguishes them from other kinds of mutants—namely, they do not revert to wild type by treatment with base-substitution mutagens; they might, of course, have been deletions, but this possibility was ruled out by the finding that reversion of an acridine-induced mutant could be achieved if such a mutant was further cultured in a growth medium containing mutagenic acridines. It was far too improbable that the base sequence of the wild-type parent would be restored in mutant progeny by further mutagenesis, so it was concluded that acridine mutants were of a new type, and the key to understanding these mutants was that they arose only if the phage replicated in the presence of a substance thought to intercalate. As an explanation of how acridine mutations are caused, therefore, it was proposed that intercalation of an acridine molecule moves two adjacent bases apart by the thickness of a single base, and that during replication an additional base is occasionally inserted when DNA polymerase meets an already intercalated molecule (Figure 12-4). Thus, acridine-induced mutants were in fact base-addition mutants. (We will see shortly that base deletion also occurs.)

Clearly, a base addition must be mutagenic, because it upsets the phase of reading of the code for a specific protein: every amino acid downstream from the added base will be different, as shown in the figure. It was then thought that reversion of an acridine-induced mutant should be explainable as the removal of an extra base; however, the efficiency with which acridine-induced mutants reverted was sufficiently high that it seemed unlikely that the particular additional base giving rise to a mutation would itself be removed during the reversion process. Ultimately, an understanding of this phenomenon, as well as proof of the three-letter nature of the code, came from a study of a collection of acridine-induced mutants in the gene called *rIIB* of *E. coli* phage T4.

Figure 12-4
Change in the amino acid sequences of a protein caused by addition of an extra base.

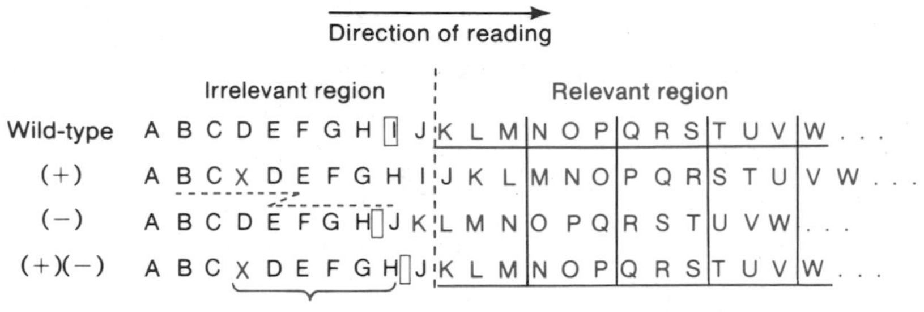

Direction of reading →

	Irrelevant region	Relevant region

Wild-type   A B C D E F G H I J K L M N O P Q R S T U V W . . .

(+)   A B C X D E F G H I J K L M N O P Q R S T U V W . . .

(−)   A B C D E F G H I J K L M N O P Q R S T U V W . . .

(+)(−)   A B C X D E F G H I J K L M N O P Q R S T U V W . . .

Figure 12-5
A schematic diagram of the T4*rIIB* gene showing the irrelevant (black) and relevant (red) regions. A, B, C, and so forth represent consecutive bases. In the (+) mutant, base X is inserted between C and D. In the (−) mutant base I is deleted. The dashed red line shows the recombinational event that generated the (+)(−) recombinant. In (+)(−) the horizontal red brace denotes the sequence in the irrelevant region that differs from the wild-type sequence. The relevant regions are the same in wild type and (+)(−) and in each case are in the same reading frame—namely, beginning at the eleventh letter. The reading frame in the relevant region is indicated by the vertical black line.

The T4 *rIIB* gene can be subdivided, genetically, at least, into two regions that we term *irrelevant* and *relevant* (Figure 12-5). The irrelevant region is defined by the fact that it is very difficult to obtain base substitution mutants in this region and, in fact, the region can be deleted without full loss of function of the gene. (This means that the irrelevant region of the *rIIB* protein can tolerate many amino acid changes without loss of function.) However, by proflavin mutagenesis, many mutations in the irrelevant region can be found, and these mutations are reverted with high frequency by growth in the presence of an acridine. Normally, if two phage strains, each carrying a mutation in the same gene, are crossed by mixed infection of a bacterium, some wild-type phage progeny arise by genetic recombination between the two genetic loci. However, in the experiments to investigate acridine-induced mutagenesis, it was found that when these mutants were crossed, wild-type phage progeny did not always arise. In fact, acridine-induced mutants could be placed in two distinct classes, which were arbitrarily termed (+) and (−). These classes were distinguished by the following properties:

1. A cross between two (+) mutants did not yield wild-type phage.
2. A cross between two (−) mutants did not yield wild-type phage.
3. A cross between one (+) and one (−) mutant did yield wild-type phage.

These results were interpreted in the following way. The (+) mutants were thought to have one added base and the (−) mutants to lack one base. A double mutant of the type (+)(+) or (−)(−) was thought to have two additional bases or to lack two bases, respectively. In both cases

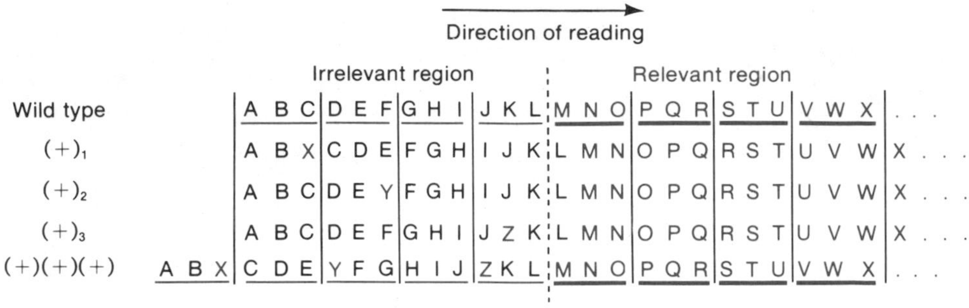

Direction of reading →

		Irrelevant region		Relevant region	

Wild type	A B C	D E F	G H I	J K L	M N O	P Q R	S T U	V W X	. . .
(+)₁	A B X	C D E	F G H	I J K	L M N	O P Q	R S T	U V W	X . . .
(+)₂	A B C	D E Y	F G H	I J K	L M N	O P Q	R S T	U V W	X . . .
(+)₃	A B C	D E F	G H I	J Z K	L M N	O P Q	R S T	U V W	X . . .
(+)(+)(+)	A B X	C D E	Y F G	H I J	Z K L	M N O	P Q R	S T U	V W X	. . .

Figure 12-6
Diagram showing that combining three base additions (X, Y, Z, shown in red) restores the reading frame in the relevant region (heavy red underline), if the code is a triplet code. An extra amino acid is present in the irrelevant region of (+)(+)(+). Vertical black lines separate the codons.

reading of the code would be shifted two bases out of phase and a functional protein would not be made. However, in a (+)(−) double mutant, the advanced reading frame *following* the (+) locus would be incorrect, but the correct reading frame would be restored at the (−) locus. Between the two mutant sites, the amino acid sequence would not be that of the wild-type phage, but nonetheless the (+)(−) and (−)(+) phage strains would be functional, because both mutant loci were in the irrelevant region. These interpretations proved correct. The point is that *as long as the reading frame is restored before the relevant region is reached, a functional protein can be produced.* By this reasoning a reversion event of a (+) mutant must often be the production of a (−) mutation at a second site.

The fact that a double (+)(+) mutant never had the wild-type phenotype meant that the code could not be a two-letter code; if it were a two-letter code, the reading frame would be restored in this combination. The critical experiment to test for a triplet code was the construction of a triple mutant. As expected, the triple mutants (+)(+)(+) and (−)(−)(−) had the wild-type phenotype whereas the mixed triples, (+)(+)(−) and (+)(−)(−), were still mutant. How the combination of three mutants of the same type can yield a wild-type phage is shown in Figure 12-6.

This elegant experiment provided strong evidence for the triplet code. In later experiments using the lysozyme gene of phage T4, the amino acid sequence of the wild-type protein and acridine-induced single mutants were compared and it was found that indeed there was a site after which all amino acids are changed in the mutant. The amino-acid sequence of double mutants of the type (+)(−) also showed that a short sequence of amino acids in the double mutant differed from that of wild-type, as expected, if (+) and (−) mutants were not in adjacent codons. Once the genetic code was worked out, it was shown also that many (−) mutants were actually (+)(+) mutants—that is, two bases had been added rather than a single base deleted. Interestingly, when this occurs, the two added bases are adjacent to one another.

Elucidation of the Base Composition of the Codons

The first step in the elucidation of the genetic code was taken in 1961 when Marshall Nirenberg and Heinrich Matthei incubated a mixture of ribosomes, tRNA molecules, radioactive amino acids, a cell fraction (of unknown composition) known to be required for protein synthesis, and the synthetic polynucleotide polyuridylic acid (poly(U)), and obtained synthesis of polyphenylalanine. From this simple result and previous knowledge that the code is a triplet code, it was immediately proved that the codon UUU corresponds to the amino acid phenylalanine. Variations on this basic experiment made it possible to identify all U-containing codons, as will be seen in this section.

Poly(U) and other synthetic polynucleotides are prepared by incubation of ribonucleoside diphosphates with the enzyme polynucleotide phosphorylase in a reaction that does not need a template. If uridine 5′-diphosphate (UDP) is the only added nucleotide, the result is poly(U). However, if UDP and guanosine 5′-diphosphate (GDP) are added, polymers are made containing both uracil and guanine. The base sequences of such polymers are random, so in a collection of mixed polymers it is possible to calculate the frequency of occurrence of each codon. The result of such a calculation for polymers prepared from a mixture that is 76 percent UDP and 24 percent GDP is shown in Table 12-1. Note that there are four frequency classes for the eight possible triplets. If this polymer mixture is used in 20 different assays, each containing all 20 amino acids of which only 1 in each assay is radioactive,

Table 12-1
Expected frequency of triplets in a random copolymer of U (0.76) and G (0.24)

Triplet	Probability	Relative frequency
UUU	$.76 \times .76 \times .76 = .439$	100
UUG	$.76 \times .76 \times .24 = .139$	31.6
UGU	$.76 \times .24 \times .76 = .139$	31.6
GUU	$.24 \times .76 \times .76 = .139$	31.6
UGG	$.76 \times .24 \times .24 = .0438$	10.0
GUG	$.24 \times .76 \times .24 = .0438$	10.0
GGU	$.24 \times .24 \times .76 = .0438$	10.0
GGG	$.24 \times .24 \times .24 = .0138$	3.1

Table 12-2
Amino acid incorporation using a random copolymer
of U (0.76) and G (0.24) as mRNA

Amino acid	Relative amount incorporated	Inferred codon composition
Phenylalanine	100	3U
Valine	37	2U, 1G
Leucine	36	2U, 1G
Cysteine	35	2U, 1G
Tryptophan	14	1U, 2G
Glycine	12	1U, 2G

the frequency with which each amino acid is included in newly synthesized protein molecules can be measured. The results of such an experiment are shown in Table 12-2. Note that the incorporation frequencies also fall into several classes. The smallest class, corresponding to GGG, is apparently absent, but it will be seen that this is because both GGG and GGU are glycine codons and the latter contribution obscures the former. Comparison of the data in the two tables suggests that the codons corresponding to valine, leucine, and cysteine contain two U's and one G, whereas those corresponding to tryptophan and glycine contain one U and two G's.

Another variation of the basic experiment gives more information. For instance, if a single guanine is added to the terminus of a poly(U) chain, it is found that the polyphenylalanine is often terminated by leucine. Thus, UUG must be the leucine codon. If many guanines are added at the terminus, the polyphenylalanine is sometimes terminated by glycine, indicating that GGG corresponds to glycine (Figure 12-7). At low frequency leucine and tryptophan are also found at the carboxyl terminus of the polyphenylalanine. These must come from the codons UUG and UGG at the transition point between U and G. Since it had already been shown that UUG is the leucine codon, this observation also identified UGG as the codon for tryptophan.

Similar experiments were carried out with mixed polymers of uracil, adenine, and cytosine—poly(U,A) and poly(U,C)—to identify the

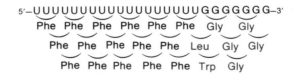

Figure 12-7
Polypeptide synthesis from UUUU . . . UUGGGGGGG
in three different reading frames showing the origin
of the incorporation of glycine, leucine, and
tryptophan.

remaining U-containing codons. Other homopolymers were also studied. Poly(A) results in the formation of polylysine, which indicates that AAA is a lysine codon and allows experiments with mixed polymers of the type just described to be performed to identify the A-containing codons. Similarly, poly(C) was found to produce polyproline. Poly(G) fails to act as a synthetic mRNA because it forms a triple-stranded helix and cannot be translated.

Elucidation of the Base Sequence of the Codons

Other elegant experiments were used to identify the codons that could not be identified by the procedure just described and to confirm the assignments already known. For instance, H. Ghobind Khorana synthesized a series of ribopolymers of a sequence of repeating pairs. One example of such a polymer is poly(AC)—i.e., . . . ACACAC . . .* This polymer will be read in groups of threes as . . . ACA CAC ACA CAC ACA CAC . . . so that the protein made will consist of two alternating amino acids. This polymer yields

. . . Thr-His-Thr-His-Thr-His . . .

From other experiments it was known that threonine has a (2A,1C) codon and histidine has a (2C,1A) codon, so that the codon assignments must be Thr = ACA and His = CAC.

Similar experiments using poly(UG)—i.e., . . . UGUGUGUGUG . . .—yield the polyamino acid

. . . Cys-Val-Cys-Val-Cys-Val-Cys . . .

from which the assignments Cys = UGU and Val = GUG were made.

Polymers were also made by Khorana in which a triplet is repeated: for example, the polymer UACUACUACUAC . . . , which can be read in one of three ways:

U A C U A C U A C U A C U A C U A C U A C U A C U

Tyr   Tyr   Tyr   Tyr   Tyr . . . .

    Thr   Thr   Thr   Thr   Thr   Thr . . . .

       Leu   Leu   Leu   Leu   Leu   Leu   Leu . . .

---

*Remember the convention for abbreviating copolymers. Poly(X,Y) is a polynucleotide in which bases X and Y are randomly arranged; poly(XY) is the alternating copolymer XYXYXY. . . . Also, in any polynucleotide ABC . . . XYZ, the direction of the chain is P-5'-ABC . . . XYZ-3'-OH, or, using the p notation, pApBpC . . . pZ.

As expected, three polyamino acids—polytyrosine, polythreonine, and polyleucine—were made.

Nirenberg and Philip Leder also discovered that synthetic trinucleotides stimulate the binding of tRNA molecules to ribosomes. For example, if a mixture is prepared consisting of ribosomes, all of the tRNA molecules (to which amino acids, one of which is radioactive, are already bound), and the trinucleotide UUU, and then the ribosomes are isolated, it is found that only phenylalanyl-tRNA is associated with the ribosomes in any appreciable amount. This type of experiment was repeated with all 64 trinucleotides and in most cases unambiguous results were obtained; 50 codons were confirmed in this way.

Identification of the Stop Codons

The stop codons were also identified by using polynucleotides having a defined sequence. For example, with the polymer poly(UAUC), translation can be started at any of four sites:

UAU CUA UCU AUC UAU CUA UCU AUC . . .

U AUC UAU CUA UCU AUC UAU CUA UCU . . .

UA UCU AUC UAU CUA UCU AUC UAU CUA . . .

CUA UCU AUC UAU CUA UCU AUC UAU . . .

However, only a single type of protein is made—that containing the repeating sequence Tyr-Leu-Ser-Ile . . .—with any of the four amino acids starting the chain. This is consistent with the codon assignments and shows that in this synthetic system, translation can begin at any site on the synthetic messenger.

The polymer poly(GUAA) gives a very different result. This polymer can be read as:

GUA AGU AAG UAA GUA AGU AAG . . .

G UAA GUA AGU AAG UAA GUA AGU . . .

GU AAG UAA GUA AGU AAG UAA GUA . . .

AGU AAG UAA GUA AGU AAG UAA . . .

However, it is found that large polypeptides are not synthesized. In fact, only two peptides are found—the tripeptide Val-Ser-Lys and the dipeptide Ser-Lys. From the known codon assignments—namely, Val = GUA, Ser = AGU, and Lys = AAG—these peptides can be lined up in four reading frames:

1   GUA AGU AAG UAA GUA AGU AAG . . .

     Val   Ser   Lys   ?

2   G UAA GUA AGU AAG UAA GUA AGU . . .

     ?   Val   Ser   Lys   ?

3   GU AAG UAA GUA AGU AAG UAA GUA . . .

     Lys   ?

4   AGU AAG UAA GUA AGU AAG UAA . . .

     Ser   Lys   ?

Note that synthesis stops in each case just before a UAA codon, which indicates that UAA must be a stop codon. Later it was found that there is no tRNA that can bind to UAA, which is the reason that this is a stop codon.

Similar experiments showed that poly(AUAG) yielded only the tripeptide Ile-Asp-Arg and the dipeptide Asp-Arg, which provided proof that the codon UAG is also a stop codon. Furthermore, with poly(UGA) only polymethionine and polyaspartic acid are synthesized, showing that UGA is also a stop codon.

As a laboratory joke, the three stop codons were given the following names—UAG, **amber** codon; UAA, **ochre** codon; UGA, **opal** codon—which are now in common use. We shall use these terms also.

Each of the codon assignments, including those for the stop codons, has been amply confirmed by direct sequencing of DNA molecules that encode proteins whose amino acid sequences are known. For example, there are instances in which a wild-type protein of known sequence, containing lysine—let us say, ABC-Lys- . . . KLM—is converted to the mutant fragment ABC . . . ; this clearly indicates a mutation from the lysine codon AAG to the stop codon UAG.

As will be seen in Chapter 13, in protein synthesis each codon is recognized by a tRNA molecule having a base sequence that can form hydrogen bonds with the codon. The convention used to describe the anticodon base sequence requires special mention. In Chapter 3 it was pointed out that base sequences are by convention always stated in the $5' \to 3'$ direction. That is, the codon AUG is accurately written 5'-AUG-3'. Thus the anticodon corresponding to AUG is CAU (5'-CAU-3') and not UAC, because of the antiparallel nature of all double-stranded DNA and RNA segments. Actually, many base pairs other than the standard A · U and G · C pairs are used in codon-anticodon pairing; this will be discussed in a later section.

Identification of the Start Codons

In the Nirenberg system, protein synthesis can start at any base in the polymer used to direct the synthesis. However, *in vivo* protein synthesis cannot begin at any base in an RNA molecule. Instead, a start codon is needed. The codon AUG is the most commonly used start codon; in a few instances, the codon GUG is used. In all DNA molecules whose base sequences have been compared to amino acid sequences, the AUG codon appears in the same reading frame as the base sequence corresponding to a particular protein. Special signals, which we will describe shortly, designate a particular AUG codon as a start codon and thereby define the reading frame; if the signal is not received, the codon is simply read as an internal codon corresponding to methionine. When used to initiate polypeptide chain growth in prokaryotes, the AUG codon causes initiation of a polypeptide chain with a modified amino acid, N-formylmethionine (fMet); in eukaryotes, unmodified methionine is used.

Universality of the Code

The experiments with the Nirenberg and Khorana systems that have been described used components isolated from the bacterium *E. coli*. These experiments have been repeated with ribosomes and tRNA molecules obtained from many species of bacteria, yeast, plants, and animals, including mammals. With the exception of mitochondria (discussed later in this chapter) the same codon assignments can be made for all organisms that have been examined. Thus, the genetic code is considered to be universal.

A summary of the genetic code is shown in Table 12-3.

Redundancy of the Code

We have now accounted for four codons as signals—the three stop codons and one start codon. The remaining 60 codons, as well as internal and start AUG, all correspond to amino acids. Since there are only 20 amino acids, there are many cases in which several codons direct the insertion of the same amino acid into a protein chain. Thus, the genetic code is highly **redundant,** as shown in Table 12-3.

Examination of the codon assignments shows that the redundancy is not random, in that multiple codons corresponding to a single amino acid are not distributed haphazardly throughout Table 12-3. In fact, with the exception of serine, leucine, and arginine, all **synonyms**

Table 12-3
The "universal" genetic code

First position (5' end)	Second position				Third position (3' end)
	U	C	A	G	
U	Phe	Ser	Tyr	Cys	U
	Phe	Ser	Tyr	Cys	C
	Leu	Ser	Stop	Stop	A
	Leu	Ser	Stop	Trp	G
C	Leu	Pro	His	Arg	U
	Leu	Pro	His	Arg	C
	Leu	Pro	Gln	Arg	A
	Leu	Pro	Gln	Arg	G
A	Ile	Thr	Asn	Ser	U
	Ile	Thr	Asn	Ser	C
	Ile	Thr	Lys	Arg	A
	Met	Thr	Lys	Arg	G
G	Val	Ala	Asp	Gly	U
	Val	Ala	Asp	Gly	C
	Val	Ala	Glu	Gly	A
	Val	Ala	Glu	Gly	G

Note: The boxed codons are used for initiation. GUG is very rare.

(codons corresponding to the same amino acid) are in the same box—that is, the codons differ by only the third base. For example GGU, GGC, GGA, and GGG all code for glycine. Furthermore, redundancy is the rule—that is, there are multiple codons except for tryptophan and methionine.

Certain features of the redundancy are quite regular—for example:

1. Pairs of codons of the general form XYC and XYU always code for the same amino acid.
2. The pairs XYG and XYA usually code for the same amino acid.

The structural basis for this phenomenon will become evident later when we examine codon-anticodon binding.

What is the biological significance of the redundancy? The simplest explanation is that it minimizes the deleterious effects of mutations. If redundancy were not the case, then 44 of the 64 codons would not code for an amino acid—that is, they would be stop codons. Thus, two-thirds of the base changes could not possibly lead to improvement of an organism and the evolutionary process would be seriously limited.

It has been proposed that originally the code was a two-letter code and proteins contained only 15 different amino acids. This would allow one codon to serve as a stop signal. The third base might have been used either as a spacer, part of the alignment system, or to stabilize binding to the ribosome.

Genetic Confirmation of Codon Assignments

The codon assignments shown in Table 12-3 are completely consistent with all chemical observations and have been confirmed independently by a comparison of the amino acid sequences of wild-type proteins against base-substitution mutant proteins. In every case the amino acid substitution can be accounted for by observing the codons corresponding to the two differentiating amino acids. For example, in tobacco mosaic virus coat protein a Pro→Ser substitution has been observed. This can be accounted for by the changes CCC→UCC, CCU→UCU, CCA→UCA, and CCG→UCG. Similarly, in mutant hemoglobins the substitutions of a particular glutamic acid by both valine and lysine have been observed. This is consistent with GAA as the glutamic acid codon and changes to GUA and AAA for valine and lysine, respectively.

Protein fragments have also been isolated from cells carrying chain termination mutations. These are also easily explained as a mutation from a sense codon to a stop codon—for a change from the tyrosine codon UAC to the stop codon UAG.

THE DECODING SYSTEM

We have now determined the relation between the base sequence of the coding region of a mRNA molecule and the amino acid sequence of the protein translated from that molecule. However, as we have already mentioned, the amino acids do not by themselves line up along the mRNA molecule. This decoding operation—that is, the conversion of the base sequence within a mRNA molecule to an amino acid sequence of a protein—is accomplished by a system consisting of two different types of molecules. One type is a set of small RNA molecules called **transfer RNA (tRNA)** and the other is a set of enzymes, the **aminoacyl synthetases.** The structures and the modes of action of these molecules are described in the sections that follow.

Transfer RNA and the Aminoacyl Synthetases

There are many different tRNA molecules in a particular cell. Each tRNA molecule has several important regions, three of which are of concern to us at this point. One of these regions is a contiguous sequence of three bases that can hydrogen-bond by base-pairing to each codon; this sequence is the **anticodon.** A second site is the **amino acid attachment site;** the amino acid that binds to this site corresponds to the particular codon in mRNA that forms base pairs with the anticodon of the tRNA. In order that the aminoacyl synthetase match the amino acid and the anticodon correctly, the enzyme must be able to distinguish one tRNA molecule from another. Thus each tRNA molecule must have a **recognition region.** Each of these regions will be discussed in detail shortly.

The name of the amino acid that is linked to a particular tRNA molecule by a specific aminoacyl synthetase is also appended when naming both types of molecules. Thus, one says that leucyl-tRNA synthetase attaches leucine to tRNA^{Leu}. Because of the redundancy of the code, there is sometimes (but not always) more than one tRNA molecule for a particular amino acid; these may be denoted tRNA$_I^{Leu}$ and tRNA$_{II}^{Leu}$ or sometimes tRNA$_{UUA}^{Leu}$ and tRNA$_{CUA}^{Leu}$. A tRNA molecule and its corresponding aminoacyl synthetase are called **cognates;** the terms cognate tRNA or cognate synthetase are also used when referring to one member of a pair.

The Cloverleaf Structure of tRNA

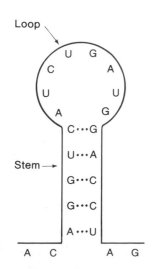

All tRNA molecules are small, single-stranded nucleic acids ranging in size from 73 to 93 nucleotides. The base sequence of tRNA^{Ala} from the yeast *Saccharomyces cerevisiae* was determined by Robert Holley and coworkers in 1965 and was a *tour de force* at the time. It was noticed that several base sequences could form extended double-stranded regions and that it is possible to fold the molecule to form a coverleaf structure in which open loops are connected to one another by double-stranded stems (Figure 12-8). In the nearly two decades that have passed since Holley and his team made their determination, almost 200 different tRNA molecules from bacteria, yeast, plants, and animals have been sequenced. For each sequence base pairing can be arranged so the polynucleotide chain can be folded into the cloverleaf configuration; study of many of these molecules by physical methods has shown in each case that this is the correct configuration. By careful comparison of these sequences certain features have been found to be common to almost all of the molecules. This has led to the idea of a "standard" tRNA molecule consisting of 76 nucleotides arranged in a cloverleaf form (Figure 12-9). By convention, the nucleotides are numbered 1 through 76 starting from the 5'-P terminus. This 76-base sequence is sufficiently fundamental that in tRNA molecules having more than 76 bases the additional ones can usually be recognized as additions to the standard sequence. The most common additions follow standard bases

Figure 12-8
An example of a stem and a loop.

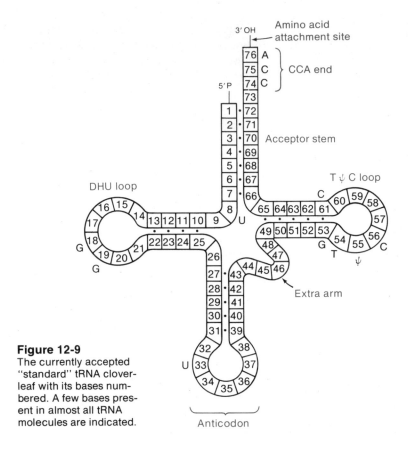

Figure 12-9
The currently accepted "standard" tRNA cloverleaf with its bases numbered. A few bases present in almost all tRNA molecules are indicated.

17, 20, and 47; the "extra" bases are numbered 17:1, 17:2, 20:1, 20:2, 47:1, 47:2, and so forth.

The standard tRNA molecule has the following features:

1. Bases in positions 8, 11, 14, 15, 18, 19, 21, 24, 32, 33, 37, 48, 53, 54, 55, 57, 58, 60, 61, 74, 75, and 76 are invariant—that is, they are the same in nearly all tRNA molecules whose sequences are known at present.

2. The 5'-P terminus is always base-paired. It is thought that this contributes to the stability of tRNA.

3. The 3'-OH terminus always is a four-base single-stranded region having the base sequence XCCA-3'-OH in which X can be any base. This is called the **CCA** or **acceptor stem.** The adenine in the CCA sequence is the site of attachment of the amino acid by the cognate synthetase.

4. There are many so-called "modified" bases in tRNA. A few of these, dihydrouridine (DHU), ribosylthymine (rT), pseudouridine (ψ), and inosine (I) occur frequently and in particular regions. Others are found in a variety of positions. In some cases the substituents of the purine and pyrimidine rings are so numerous and complex that the base is called "hypermodified." In only a few cases is the significance of these unusual bases known.

5. There are three large single-stranded loops. The lowermost or **anticodon loop** contains seven bases. The anticodon occupies positions 34 through 36. It is almost always preceded by two pyrimidines (usually base 33 is uridine) and followed by a modified purine. Thus the anticodon loop has the general sequence

$$5'\text{-Py}-\text{U}-\text{XYZ}-\text{Pu (modified)}-\text{Variable base}$$
Anticodon

The loop containing bases 14 through 21 is called the **DHU loop.** It is not constant in size in different tRNA molecules but may contain up to three extra bases between bases 17 and 18 and one to three extra bases following base 19. The loop containing bases 54 through 60 almost always contains the sequence TψC and is called the **TψC loop.**

6. There are four double-stranded regions called stems (or, sometimes, arms). The stems have no invariant bases and often contain base pairs other than G · C and A · T—for example, G · U pairs are found in many molecules. The three stems attached to each loop are given the name of the corresponding loop, as, for example, the anticodon stem.

7. An additional loop, containing bases 44 through 48, is also present. In the smallest tRNA molecules it contains four bases, lacking base 47, whereas in the largest tRNA molecule it contains 21 bases with 16 bases present between bases 47 and 48. This highly variable loop is called the **extra arm;** its function is not understood.

Three-Dimensional Structure of tRNA

Although the tRNA molecule is conveniently depicted as a planar cloverleaf, its three-dimensional structure is actually quite different. This has been shown by x-ray crystallographic analysis of yeast tRNA[Phe]. Figure 12-10(a) shows the skeletal model of this tRNA mole-

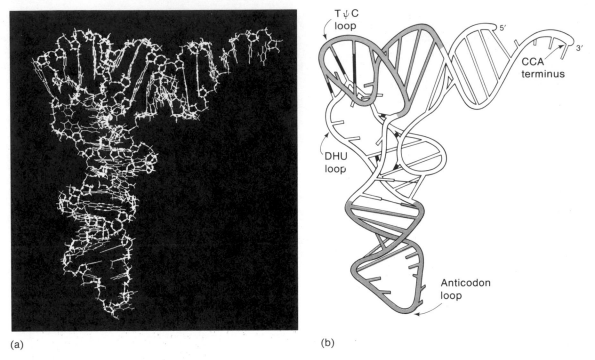

(a)   (b)

Figure 12-10
(a) Photograph of a skeletal model of yeast tRNA^{Phe}. (b) Schematic diagram of the three-dimensional structure of yeast tRNA^{Phe}. (Courtesy of Dr. Sung-Hou Kim.)

cule; an interpretive drawing is in panel (b) of the figure. The major features of the three-dimensional structure are the following:

1. The base pairs proposed in the cloverleaf model do exist.

2. The molecule is folded in such a way that there are two helical double-stranded branches, each about 60 Å long and mutually perpendicular. One branch consists of the acceptor stem and the TψC stem; the other consists of the stems of the DHU and anticodon loops. Thus, the molecule is L-shaped.

3. The CCA acceptor terminus is at one end of the L and the anticodon loop is at the other end. The amino acid and the anticodon are separated by about 80 Å.

4. The TψC loop, which is thought to interact with ribosomal RNA, is at the corner of the L, maximally separated from the CCA end and the anticodon loop.

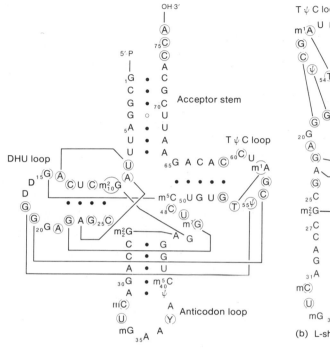

(a) Cloverleaf arrangement

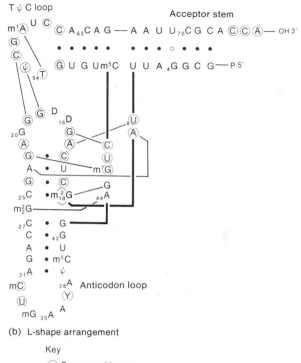

(b) L-shape arrangement

Key

○ Conserved bases
() Semiconserved bases
₃₈ Subscript as prefix: position of base in tRNA
m As prefix: methylated base

Figure 12-11

Nucleotide sequence of yeast tRNA^Phe in the (a) cloverleaf and (b) L arrangements. The bases that are strongly conserved are circled; less conserved bases are in parentheses. The bases that form tertiary hydrogen bonds are connected by thin lines. Notice that most of the conserved bases are localized in the middle of the L and are involved in forming the tertiary hydrogen bonds. The solid circles indicate hydrogen-bonded base pairs. The open circle is a nonstandard base pair.

5. All bases except five (16, 17, 20, 47, and 76) are stacked, even though half of the bases are not in double-stranded helical regions. Even the bases of the single-stranded 3′-terminal segment (except for the terminal adenine) are stacked.

6. There are nine base pairs that are not shown in the cloverleaf model. These are called **tertiary base pairs** because they are responsible for the three-dimensional folding. Only one of these, a G · C pair, is a standard Watson-Crick pair. Furthermore, most of these pairs are formed from bases that are conserved in all tRNA molecules, as shown in Figure 12-11. This fact suggests that all tRNA molecules may have the same general structure. The idea of a universal structure is supported by the observation that different tRNA molecules can co-crystallize,

forming a uniform crystal; a low-resolution x-ray crystallographic analysis of these mixed crystals does not indicate that two types of molecules are present and yields molecular dimensions that are the same as those of a single kind of molecule. This suggests that the two tRNA molecules in a mixed crystal have common structural features. Since mixed crystals containing many different tRNA molecules have been tested with similar results, it is likely that all tRNA molecules are structurally very similar.

7. In addition to the tertiary base pairs described in item 6, the tertiary structure is stabilized by hydrogen bonds between bases and different ribose units and between ribose units in separate nucleotides. Details of these interactions are not yet known, though they often include hydrogen bonding with the 2'-OH group of ribose. The need for a 2'-OH group, which is absent in 2'-deoxyribose, may be the reason that cells evolved with transfer RNA rather than transfer DNA.

Attachment of an Amino Acid to a tRNA Molecule

When an amino acid is attached to a tRNA molecule, the tRNA is said to be **aminoacylated** or **charged.** This is designated in several ways. For example, if the amino acid is serine, the charged tRNA would be written seryl-tRNA or seryl-tRNA^{Ser}. There are cases in which a tRNA molecule is mischarged. If tRNA^{Leu} were mischarged with serine, this would be written seryl-tRNA^{Leu}. The term "uncharged tRNA" refers to a tRNA molecule lacking an amino acid.

There are two reasons that cells have evolved with acylated tRNA as an intermediate in protein synthesis. The first and obvious reason is to align the amino acids according to the order of the codons. The second reason is a thermodynamic one—namely, that peptide-bond cleavage, rather than peptide-bond formation, is thermodynamically favored. This situation is reversed by the simple expedient of first forming an amino acid ester with a 2'-OH or 3'-OH group of the 3'-terminal nucleotide of tRNA (Figure 12-12). Thus, the amino acid ester is the activated intermediate of protein synthesis.

Acylation is accomplished in two steps both of which are catalyzed by an aminoacyl synthetase. In the first, or activation, step an aminoacyl AMP is synthesized in a reaction between an amino acid and ATP:

$$^+H_3N-\overset{\overset{\displaystyle H}{|}}{\underset{\underset{\displaystyle R}{|}}{C}}-\overset{\overset{\displaystyle O}{\|}}{C}-O^- + ATP \rightleftharpoons {}^+H_3N-\overset{\overset{\displaystyle H}{|}}{\underset{\underset{\displaystyle R}{|}}{C}}-\overset{\overset{\displaystyle O}{\|}}{C}-O-\overset{\overset{\displaystyle O}{\|}}{\underset{\underset{\displaystyle O^-}{|}}{P}}-Ribose-Adenine + PP_i$$

Pyrophosphate

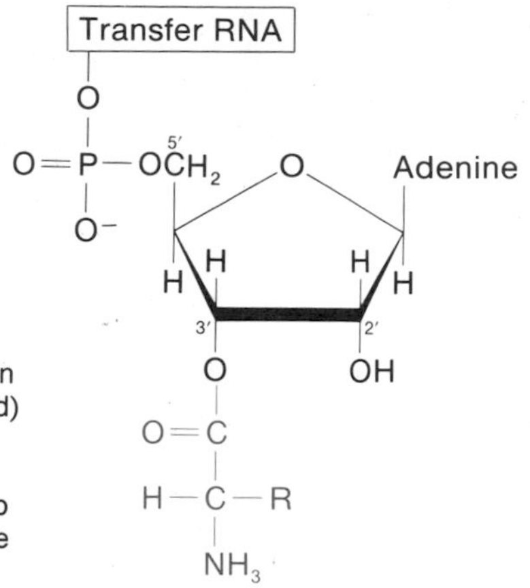

Figure 12-12
Mode of attachment of an amino acid (shown in red) to a tRNA molecule. The amino acid is covalently linked to the 3'-OH group of the terminal adenosine of tRNA.

In the second, or transfer, step the aminoacyl-AMP reacts with tRNA to form acylated tRNA and AMP:

$$\text{Aminoacyl-AMP} + \text{tRNA} \longrightarrow \text{Aminoacyl-tRNA} + \text{AMP}$$

The transfer reaction is an equilibrium reaction in which neither direction predominates. As in DNA synthesis, the reaction is driven to the right by hydrolysis of pyrophosphate, so the overall reaction is

$$\text{Amino acid} + \text{ATP} + \text{tRNA} + \text{H}_2\text{O} \longrightarrow \text{Aminoacyl-tRNA} + \text{AMP} + 2\text{P}_i$$

The Aminoacyl Synthetases

There is at least one and usually only one aminoacyl synthetase for each amino acid. For a few amino acids specified by more than one codon, more than one synthetase exists. By 1982, 19 synthetases corresponding to 17 amino acids had been purified. Since there must be a synthetase for the remaining 3 amino acids, the total number of synthetases is at least 22. The enzymes are clearly very specific and must be able to recognize both a particular tRNA molecule and a particular amino acid. Because of the great structural similarities between all tRNA molecules, one would expect the synthetases to be similar, but this is not the case. There are three classes of synthetases: (1) those consisting of a single polypeptide chain, (2) those having two identical chains, and (3) those containing four chains of two types, with two copies of each type. Furthermore, the molecular weights of the subunit range from 33,000 to

113,000 and the molecular weights of the holoenzyme range from 54,000 to 270,000. The synthetases are definitely a heterogeneous group of proteins. Understanding the significance of this heterogeneity must await purification of more synthetases and detailed studies of the acylation reaction mechanisms of each enzyme, which may differ from one molecule to the next.

Recognition of a tRNA Molecule by Its Cognate Synthetase

Each synthetase must be able to recognize its cognate tRNA molecule.

The recognition region on the tRNA molecule spans the entire molecule rather than being a localized site like the codon binding and amino acid attachment. Two experimental approaches—**photoactivated cross-linking** and **RNase-protection**—have been used to determine the regions of a tRNA molecule that are in contact with a synthetase. In both methods the synthetase and the tRNA molecule are first allowed to bind to one another. In the cross-linking technique (Figure 12-13), the synthetase-tRNA complex is irradiated with ultravi-

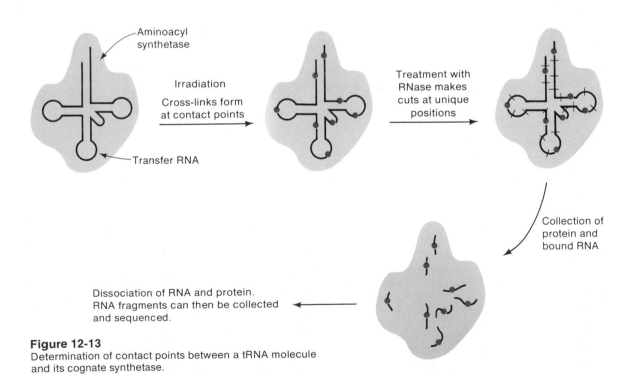

Figure 12-13
Determination of contact points between a tRNA molecule and its cognate synthetase.

olet light, which causes chemical bonds (cross-links) to form between bases and amino acids in the protein. The cross-links can occur only if the bases are within 2 or 3 Å of an amino acid. After irradiation, the tRNA molecule is fragmented and oligonucleotides bound to proteins are identified by electrophoresis. Each oligonucleotide fragment comes from a particular (and known) segment of the tRNA molecule. In the RNase-protection method the complex is treated with a collection of RNases that digest all of the RNA that can be reached by the enzymes. Points of contact between the tRNA molecule and the synthetase are protected from the nucleases. The results of such studies with several synthetase-tRNA complexes indicate that there are many points of contact, including the acceptor stem, the anticodon loop, and the extra arm. All contact points are on the same side of the three-dimensional structure. The odd nucleosides are rarely engaged in contact; instead, they are probably involved in attachment to the ribosome. There are several examples in which a synthetase can bind to a tRNA molecule that is not its cognate, and again, there are many points of contact. Aminoacylation does not occur in these pairs because the CCA terminus is not positioned at the active site of the enzyme (Figure 12-14). Thus, it seems that the specificity of aminoacylation is not derived exclusively from binding ability but also from the three-dimensional structure of each tRNA molecule, which serves to place the 3′-terminal adenine at the aminoacylation site of the synthetase.

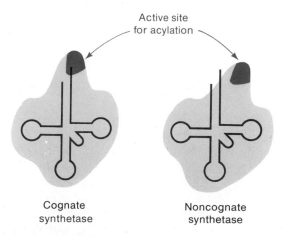

Cognate
synthetase

Noncognate
synthetase

Figure 12-14
Diagram showing how an aminoacyl synthetase can bind to a noncognate tRNA yet be unable to charge the tRNA molecule. The tRNA and the aminoacyl synthetase molecules are not drawn to scale.

Error Correction in Aminoacylation

The location of each amino acid in a growing polypeptide is determined only by the codon-anticodon interaction. This is known from the following experiment. A sample of tRNA can be acylated in a reaction mixture containing all of the synthetases, 19 nonradioactive amino acids, and [^{14}C]cysteine. The only charged tRNA that will be radioactive is [^{14}C]cysteyl-tRNA^{Cys}. The charged tRNA molecules are then hydrogenated with Raney nickel, which results in the conversion of [^{14}C]cysteine to [^{14}C]alanine; therefore, the only radioactive tRNA molecules are [^{14}C]alanyl-tRNA^{Cys}. These molecules are then used in an *in vitro* protein-synthesizing system containing hemoglobin mRNA as the only mRNA molecule. The synthesized hemoglobin is next isolated and its amino acid sequence determined. When this is done, it is found that [^{14}C]alanine is present at sites normally occupied by cysteine; thus the protein-synthesizing apparatus does not examine the identity of each amino acid but looks only at the tRNA molecule. This means that the burden of producing the correct amino acid sequence is on the synthetase.

Several amino acids are structurally similar and it is to be expected that synthetases might make occasional mistakes. If the error frequency is high, it seems reasonable to expect that an editing mechanism, as we saw with DNA synthesis, has evolved. Valine and isoleucine constitute such a possibly ambiguous pair of amino acids and, in fact, isoleucyl-tRNA synthetase forms valyl-AMP, which remains bound to the synthetase, at a frequency of about one per 225 activation events. This would mean that 1/225 of all isoleucine positions in proteins could contain a valine. For a typical protein containing 500 amino acids of which 25 were isoleucines, about one copy in nine could be altered. Since there are at least ten known examples of misacylation, there could be an error in almost every molecule. This does not occur, though.

The editing mechanism that corrects the valine-isoleucine error is a hydrolytic step in which valyl-AMP is cleaved and removed from the enzyme. The hydrolysis is carried out by the isoleucyl-tRNA synthetase itself. Interestingly, the signal that activates the hydrolytic function is the attempted binding of valine to tRNA^{Ile}. The number of times this editing system fails and valyl-tRNA^{Ile} is formed is about 1 in 800. Thus, the overall error frequency—that is, the fraction of isoleucine sites occupied by valine—is $(1/225)(1/800) = 1/180,000$. If all possible amino acid misacylations occur at this frequency, only about 0.17 percent of the proteins would be defective. A similar case, in which methionyl-tRNA synthetase forms threonyl-AMP and homocysteyl-AMP, has also been analyzed.

In examining the instances in which an incorrect complex is formed, it has been noticed that the side chain of the incorrect amino acid is never larger than that of the correct amino acid. These observations have been put together in a scheme that predicts what misacylations will occur and when hydrolytic editing will occur. This scheme is called the **double-sieve mechanism.** According to this mechanism, amino acids larger than the correct amino acid are never activated because they are too large to fit into the active site of the synthetase. Thus, the size of the active site forms a sort of sieve. Those amino acids that are smaller than the correct one can be activated, though at a lower rate since they fit less well into the active site. It is further proposed that the hydrolytic site is too small for the correct amino acid; only any of eleven amino acids that are smaller than the correct one can be removed by hydrolysis. Thus, the hydrolytic site forms the second sieve. There are a few cases in which two amino acids have the same size and nearly the same shape—for example, valine and threonine; indeed, threonine is activated by valine-tRNA synthetase. However, threonine is removed by tRNA-stimulated hydrolysis. This is explained in terms of obvious chemical differences between the two side chains—that the threonine hydroxyl group forms a hydrogen bond that draws the amino acid into the hydrophilic hydrolytic site.

At present, the available data is consistent with the double-sieve model. Such predicted editing functions as would be needed to correct activation of glycine by alanine-tRNA synthetase and of serine and valine by threonine-tRNA synthetase, are actively being sought.

Is the acylated tRNA also proofread? The previously described experiment with Raney nickel suggests that it is not. The tRNA molecule is sufficiently complex and capable of such structural variation that pairing between a tRNA molecule and a noncognate synthetase should be very weak. (Remember that misacylation occurs because the amino acids are not so greatly different from one another.) Indeed, from the binding constants obtained from studies of correct and incorrect complexes, the frequency of mispairing can be calculated as roughly one mispair per 10^7 pairs, which does not require a special editing function. To date, there is no evidence for tRNA proofreading, and it is likely that it does not occur. In a later section (on missense suppression) we will see examples of high frequency mispairing occurring with mutant tRNA molecules.

THE CODON-ANTICODON INTERACTION

The fidelity of protein synthesis is determined by two features of the system—correct charging of a particular tRNA molecule by the cognate aminoacyl synthetase and the codon-anticodon pairing. The latter is the subject of this section.

Presence of Inosine in the Anticodon Loop

Once it was established that 61 of the 64 possible codons represent amino acids, it was expected that there would be 61 distinct tRNA molecules. This view was originally supported by the finding that when leucyl-tRNA molecules are purified and tested in the *in vitro* protein-synthesizing system it was found that $tRNA_I^{Leu}$ induces incorporation of leucine with poly(U,C) but not with poly(U,G), whereas $tRNA_{II}^{Leu}$ responds to poly(U,G) but not to poly(U,C). Clearly, $tRNA_I^{Leu}$ pairs with a codon carrying C but not G (actually CUU) and $tRNA_{II}^{Leu}$ pairs with a codon carrying G but not C (UUG). However, the one-codon-one-tRNA idea was demolished by the properties of the $tRNA^{Ala}$ of yeast, sequenced by Holley; this molecule responds to three of the four GCX alanine codons, namely, GCU, GCC, and GCA. Study of the sequence of the non-hydrogen-bonded regions in the tRNA molecule indicated that there is only one place in the molecule at which the base sequence is 5'-GC-3'. Thus, assuming that the anticodon consists of three adjacent bases, these two bases must be part of the anticodon.

However, this assignment requires an anticodon sequence of 5'-IGC-3' (Figure 12-15) in which I denotes the unusual nucleoside

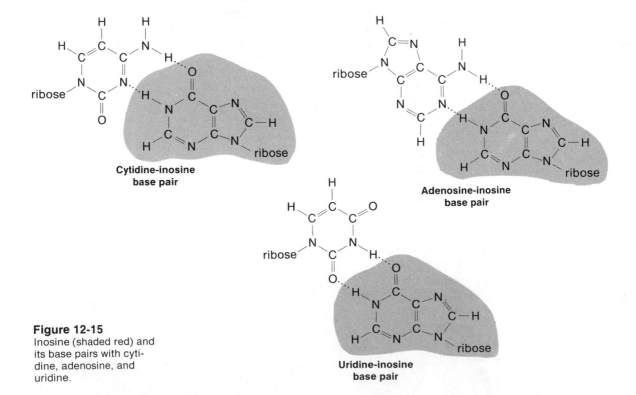

Figure 12-15
Inosine (shaded red) and its base pairs with cytidine, adenosine, and uridine.

Cytidine-inosine base pair

Adenosine-inosine base pair

Uridine-inosine base pair

inosine.* Hence, since the tRNA molecule responds to the three codons GCU, GCC, and GCA, then inosine either does not form hydrogen bonds or it can pair with each of the bases U, C, and A. If inosine fails to form hydrogen bonds, then tRNA[Ala] ought to respond to GCG, which is not the case. This suggests that inosine does form hydrogen bonds but not of the strict Watson-Crick type, as shown in the figure.

The chemical origin of inosine in the anticodon is rather interesting. Sequence analysis of a large number of tRNA molecules has shown that no anticodon starts with A. Whenever A appears in the first position in the unmodified tRNA transcript, it is deaminated by an enzyme called anticodon deaminase to hypoxanthine (the base of inosine). All inosine in tRNA molecules is formed in this way.

Redundancy as a Response of One Anticodon to Several Codons: The Wobble Hypothesis

The pattern of the redundancy of the code suggests that something is missing in the explanation of codon-anticodon binding; the most striking aspect of the redundancy is that with only a few exceptions, the identity of the third codon base appears to be unimportant. That is, XYA, XYB, XYC, and XYD are usually synonyms.

In 1965, Francis Crick made a proposal, known as the **wobble hypothesis,** that explains the fact that some tRNA molecules respond to several codons and also provides insight into the pattern of redundancy of the code. Up to that time it was generally assumed that no base pair other than $G \cdot C$, $A \cdot T$, or $A \cdot U$ would be found in a nucleic acid. This is true of DNA because the regular helical structure of double-stranded DNA imposes two steric constraints: (1) two purines cannot pair with one another because there is not enough space for a planar purine-purine pair, and (2) two pyrimidines cannot pair because they cannot reach one another. Crick proposed that since the anticodon is located within a single-stranded RNA loop, the codon-anticodon interaction might not require formation of a structure with the usual dimensions of a double helix. By model-building he showed that the steric requirements were less stringent at the third position of the codon; by allowing a little play in the structure (this play is called wobble), Crick demonstrated that other base pairs can exist between codon and anticodon. He required, first, that the first two base pairs be of the standard type in order to maximize stability and, second, that the third base pair not produce as much distortion as a purine-purine pair might cause. He included inosine in his model because it was known to be in the anticodons of several tRNA molecules, and he proposed that the base pairs

*Note that inosine is not a base but a nucleoside, specifically, ribosylhypoxanthine.

Table 12-4
Allowed pairings according to the wobble hypothesis

Third position codon base	First position anticodon base*
A	U,I
G	C,U
U	G,I
C	G,I

*A is not allowed; see 1(c) below.

listed in Table 12-4 are possible in the third position of the codon.

The possibility of forming the four base pairs shown in the table $-A \cdot I, U \cdot I, C \cdot I$, and $G \cdot U$—explains how one tRNA molecule can respond to several codons. This is shown in the following paragraph.

There are two major species of yeast tRNA^{Ala}. The one sequenced by Holley responds to the codons GCU, GCC, and GCA. Its anticodon is IGC, which is consistent with the entries in Table 12-4 and shows that only inosine can pair with U, C, and A. (Remember the convention for naming the codon and the anticodon—*always with the 5′ end at the left.* Thus the codon 5′-GCU-3′ is matched by the anticodon 5′-IGC-3′). Similarly, yeast tRNA$_{\text{II}}^{\text{Ala}}$ responds only to GCG; there are two possible anticodons, CGC and UGC, because both C and U can bond to G. If the anticodon were UGC, tRNA$_{\text{II}}^{\text{Ala}}$ would respond to both GCG and GCA, which is not the case; thus, the anticodon cannot be UGC. If it were CGC, the only codon recognized would be GCG, which is the case. Thus, the anticodon must be CGC, as indeed it is.

The most striking achievement of the wobble hypothesis is that it explains the arrangement of synonyms in the code. The following points can be made (compare Figure 12-16 and Table 12-4):

1. The codons XY<u>C</u> and XY<u>U</u> are always synonyms.
   a. If the anticodon to XY<u>C</u> is <u>G</u>Y′X′, this anticodon can also pair with XY<u>U</u>, because G can pair with U in the third position of a codon.
   b. If the anticodon is <u>I</u>Y′X′, then pairing also occurs with XY<u>U</u>, XY<u>A</u>, and XY<u>C</u>. Thus, no anticodon can pair only with XY<u>C</u> and not with XY<u>U</u> or only with XY<u>U</u> and not with XY<u>C</u>.
   c. If the anticodon to XY<u>U</u> were <u>A</u>Y′X′, then XY<u>U</u> would pair with two tRNA molecules (having anticodons <u>G</u>Y′X′ and <u>I</u>Y′X′). However, A is not found in the first position of an anticodon (except in the mitochondrial anticodon for glycine), because the enzyme anticodon deaminase acts at this position to convert A to I.

2. The codon XY<u>A</u> has the anticodons <u>I</u>Y′X′ or <u>U</u>Y′X′.

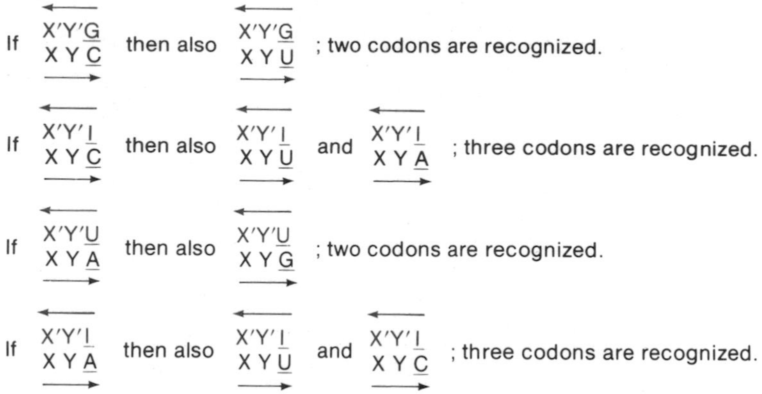

Figure 12-16
The explanation of synonyms given by the wobble hypothesis. In each case the codon is black and the anticodon is red. Codons in the same horizontal row are synonyms. The arrows show the 5'→3' direction for both codon and anticodon. The bases of the "wobble pair"—that is, the third base of the codon and the first base of the anticodon—are underlined.

If $\overset{\longleftarrow}{\underset{\underset{\longrightarrow}{X\ Y\ \underline{C}}}{X'Y'\underline{G}}}$ then also $\overset{\longleftarrow}{\underset{\underset{\longrightarrow}{X\ Y\ \underline{U}}}{X'Y'\underline{G}}}$; two codons are recognized.

If $\overset{\longleftarrow}{\underset{\underset{\longrightarrow}{X\ Y\ \underline{C}}}{X'Y'\underline{I}}}$ then also $\overset{\longleftarrow}{\underset{\underset{\longrightarrow}{X\ Y\ \underline{U}}}{X'Y'\underline{I}}}$ and $\overset{\longleftarrow}{\underset{\underset{\longrightarrow}{X\ Y\ \underline{A}}}{X'Y'\underline{I}}}$; three codons are recognized.

If $\overset{\longleftarrow}{\underset{\underset{\longrightarrow}{X\ Y\ \underline{A}}}{X'Y'\underline{U}}}$ then also $\overset{\longleftarrow}{\underset{\underset{\longrightarrow}{X\ Y\ \underline{G}}}{X'Y'\underline{U}}}$; two codons are recognized.

If $\overset{\longleftarrow}{\underset{\underset{\longrightarrow}{X\ Y\ \underline{A}}}{X'Y'\underline{I}}}$ then also $\overset{\longleftarrow}{\underset{\underset{\longrightarrow}{X\ Y\ \underline{U}}}{X'Y'\underline{I}}}$ and $\overset{\longleftarrow}{\underset{\underset{\longrightarrow}{X\ Y\ \underline{C}}}{X'Y'\underline{I}}}$; three codons are recognized.

    a. If the anticodon is I̲Y'X', then XY A̲ is synonymous with both XY U̲ and XY C̲.

    b. If the anticodon is U̲Y'X', then XY A̲ is synonymous with XY G̲.

Thus, *every codon ending with A is redundant.* This point also explains the number of different tRNA molecules that correspond to a particular amino acid. For example, in boxes in Table 12-3 in which there are four codons for one amino acid, there is one tRNA molecule that recognizes codons ending only in G and a second tRNA that recognizes XYA, XYC, and XYU, as seen with the pair of yeast tRNA^Ala molecules described on the opposite page.

    3. The codons AUG (Met) and UGG (Trp) are nonredundant (they are both single entries in the universal code table). Figure 12-16 and Table 12-4 indicate that a codon XY G̲ can have an anticodon C̲Y'X' that responds to no other codon. However, when this occurs, the codons XY A̲, XY C̲, and XY U̲ must be matched by an I̲Y'X' anticodon, because if the codon XY A̲ were matched by U̲Y'X', this anticodon would also pair with XY G̲, which would then be redundant. The alternative to anticodon U̲Y'X' is I̲Y'X', in which case XY A̲, XY C̲, and XY U̲ are redundant. This pattern is true of AUG (Met) and the triply redundant set of Ile codons (AUU, AUC, and AUA).

    4. There is no tRNA having an anticodon complementary to UGA. The anticodon for the Cys codons UGU and UGC could be GCA or ICA. If the anticodon were ICA, the codon UGA would also be a Cys codon. By having GCA as the Cys anticodon, UGA is reserved as a nonsense codon; that is, it cannot be recognized by any existing tRNA molecule.

    5. Finally, UAU and UAC for Tyr, and UAA and UAG for Stop, are explained by having GUA as the Tyr anticodon and supposing that no tRNA has an anticodon UUA.

In the examples considered so far, most of the redundancy has been explained by the presence of the modified base inosine in the anticodon. This arrangement is true for yeast but not for all other organisms. In *E. coli* another modified base has recently been discovered in the wobble position of tRNA^{Val} (it probably exists in other organisms, too). This base, uridine 5-oxyacetic acid, can pair with U, A, and G. Another modified base in this position in some organisms is 2-thiouridine, which pairs only with A. Five other modified bases have also been observed in the first position of the anticodon; their pairing characteristics vary. Apparently, the first base in the anticodon can be modified to produce either less or greater base-pairing specificity.

The reader must not be led to the conclusion at this point that, because of the varying kinds of redundancy, the codon assignments vary from one organism to the next. This is certainly not the case (except, as will be described later, for mitochondria). The differences in the first anticodon positions merely determine the number of different tRNA molecules corresponding to a particular amino acid.

The reader is urged to refer constantly to Tables 12-3 and 12-4 while reviewing these paragraphs.

These findings can be generalized as follows:

Every tRNA molecule reads either one, two, or three codons, the number depending on whether the first base of the anticodon, which is sometimes called the wobble base, is C (1 codon), U or G (2 codons), or a modified base (3 codons). The redundancy of the code is caused in part by wobble in the pairing of the third codon base.

The modified bases have probably evolved in the wobble position because they maximize the number of codons that can be read by a single tRNA molecule.

Stability of the Codon-Anticodon Interaction

The stability of a double-stranded DNA or RNA molecule consisting of two complementary trinucleotides is very low. In a solution of ordinary ionic composition such a molecule would be quite rare at room temperature because there would be virtually no stabilization by base stacking—that is, the two terminal base pairs in the molecule would tend to be unstacked, leaving the central pair very weakly bound. How, then, can one account for the high fidelity (i.e., the low-error frequency) of codon-anticodon binding? Several factors contribute to the stability of the interaction: (1) the folding of the tRNA molecule and the compactness of the anticodon loop is such that the anticodon bases are stacked; (2) beyond the 5' side of the anticodon of all tRNA molecules is a U, which suggests that the U is involved in anticodon function—presumably in the codon-anticodon interaction; (3) the codon and the anticodon normally pair when both tRNA and mRNA are bound to a ribosome. Presumably, in the bound state the codon and the anticodon are positioned in such a way that the binding is maximized. Support for this view comes from the very weak binding *in vitro* of a tRNA molecule and a free trinucleotide that is complementary to the anticodon.

Minor tRNA Molecules

In *E. coli* there are several instances in which chemically distinct tRNA molecules having the same anticodon have been isolated. Usually one of the forms exists in a much greater amount. The less-frequent forms are called **minor tRNA molecules.** In some cases the minor form differs from the usual form only in the absence of the terminal 3'-adenosine; it is possible that these minor forms are not truly distinct species but result from exonucleolytic loss of the adenosine when the molecules are isolated. However, most of the minor tRNA molecules have distinct base sequences and clearly differ from the major form. The role of the minor tRNA molecules is unknown; we will see later that mutant minor tRNA molecules sometimes serve as intergenic suppressors.

THE SPECIAL PROPERTIES OF THE PROKARYOTIC INITIATOR tRNA^fMet

The initiator tRNA molecule in prokaryotic organisms—tRNA^fMet—has several biological properties that distinguish it from all other types of tRNA molecules:

1. The tRNA is acylated at the outset with an amino acid—methionine—that is subsequently further modified prior to binding to the ribosome. That is, tRNA^fMet is charged with methionine by methionyl-tRNA synthetase. (This same enzyme also charges tRNA^Met with methionine.) The methionine of charged tRNA^fMet is immediately recognized by another enzyme, tRNA methionyl transformylase, which transfers a formyl group from N-10-formyltetrahydrofolate (fTHF) to the NH_3^+ group of the methionine to form N-formylmethionine (fMet):

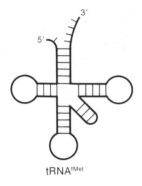

tRNA^{fMet}

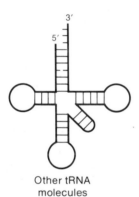

Other tRNA
molecules

Figure 12-17
A notable difference be-
tween *E. coli* tRNA^{fMet} and
other *E. coli* tRNA molecules—that is, the 5′-
terminal base is not
hydrogen-bonded.

Transformylase does not catalyze formylation of methionyl-tRNA^{Met}, so there must be structural differences between tRNA^{Met} and tRNA^{fMet}.

2. Ribosomes contain two tRNA binding sites, the A site and the P site. After several preliminary steps, a formylated, charged tRNA^{fMet} molecule enters the P site without prior occupation of the A site, whereas all other aminoacyl-tRNA molecules bind initially to the A site and move to the P site *only after a peptide bond has been formed*. Thus, some structural feature (or more than one) must be present in tRNA^{fMet} that is absent in all other tRNA molecules. The most striking difference is evident from the base sequence—that is, in the tRNA^{fMet} molecules of all prokaryotes so far examined the 5′-terminal base is not hydrogen-bonded to the opposite base in the acceptor stem (Figure 12-17). An x-ray crystallographic analysis of the structure of *E. coli* tRNA^{fMet} has revealed the following. The overall structure of tRNA^{fMet} is very similar to other tRNA molecules. Nevertheless, a hydrogen bond, which in most tRNA molecules occurs between the uracil adjacent to the anticodon and a ribose in the anticodon loop, is absent in tRNA^{fMet}. This gives the anticodon loop a more open structure. There are also small differences in the acceptor stem and the DHU loop. However, too little is known at present to be sure that one or more of these features is entirely responsible for the biological differences.

The codon-anticodon interaction of tRNA^{fMet} differs from that of other tRNA molecules in an interesting way. We have already pointed out the ambiguity in pairing of the third codon base (the wobble codon). With *E. coli* tRNA^{fMet}, there is some wobble in the *first* codon base. That is, though the major codon recognized by tRNA^{fMet} is AUG, tRNA^{fMet} also interacts significantly with the valine codon GUG and weakly with the leucine codons UUG and CUG. Possibly this wobble results from the lack of the hydrogen bond in the anticodon loop, as just mentioned. In fact, all four codons are used as start codons and are recognized by tRNA^{fMet}.

Eukaryotic initiator-tRNA molecules differ from the prokaryotic initiator molecule in several ways. The most striking difference is that whereas eukaryotic organisms produce both a normal tRNA^{Met} and an initiator tRNA, which is also charged with methionine, the methionine does not undergo formylation. That is, in eukaryotes, the first amino acid in a growing polypeptide chain is Met and not fMet. The codon for both kinds of tRNA molecules in eukaryotes is AUG just as is true for prokaryotes. Another important difference between eukaryotic and prokaryotic initiator-tRNA molecules is the presence of the sequence AUC or AψC instead of TψC in the TψC loop. The eukaryotic initiator-tRNA molecules are like the prokaryotic tRNA^{fMet} in lacking hydrogen-bonding of the 5′-terminal base.

At this point, the reader might be perplexed by two questions. (1) How is an AUG start codon distinguished from an internal AUG codon for Met? This question will be answered when initiation of protein synthesis is discussed in Chapter 13. (2) How did Nirenberg and Matthei successfully synthesize proteins on templates such as poly(U) and poly(U,C), which clearly could not have had an AUG codon? The answer in this case concerns the Mg^{2+} concentration in the reaction mixture that they used. As will be seen in Chapter 13, in the usual sequence of events in protein synthesis a $tRNA^{fMet}$ molecule interacts with a mRNA molecule and a free 30S ribosome, and then a 50S ribosome joins the complex. When the Mg^{2+} concentration is 0.005 to 0.01 M, the joining of the 30S and 50S subunits occurs only if a start codon (AUG or GUG) is present. However, Nirenberg used 0.02 M $MgCl_2$; at this concentration of Mg^{2+} free 30S and 50S ribosomal subunits join together spontaneously, and the resulting particle is able to bind mRNA and any tRNA molecule whose anticodon matches a codon in the mRNA. Thus, in 0.02 M $MgCl_2$ the specificity is lost and initiation can occur at any codon.

TRANSFER RNA GENES

In Chapter 15 we shall see that there is a tendency in bacteria and in phages for related genes to be clustered. This is definitely not true of the tRNA genes, which are scattered throughout the chromosome; even genes that make isoacceptor-tRNA (different tRNA molecules that bind the same amino acid) are rarely adjacent. For instance, $tRNA_I^{Tyr}$ and $tRNA_{II}^{Tyr}$ are at minute 27 and minute 88 of the 100-minute E. coli map (Figure 12-18). Clustering of some tRNA genes does occur; when this is the case, the tRNA molecules are usually components of a single transcript that is processed. For example, $tRNA_{II}^{Tyr}$, $tRNA_{II}^{Glu}$, $tRNA_{II}^{Thr}$, $tRNA_{IV}^{Thr}$, and $tRNA_{II}^{Gly}$ are excised from a single transcript.

According to the wobble theory, only 32 tRNA molecules are needed to respond to 61 different codons; however, the actual number of tRNA genes is much greater, as shown in Table 12-5. There are two reasons for this great multiplicity of tRNA genes: the existence of (1) minor tRNA molecules and (2) duplicate copies of a particular gene. For instance, in E. coli there are two identical copies of the $tRNA_I^{Tyr}$ gene and both are in a single transcription unit. Even more striking is $tRNA_I^{Leu}$ of E. coli, for which there are four (or possibly five) copies in one transcription unit and at least one more copy in another transcription unit. At present, the genes for at least nine different tRNA molecules are duplicated to various extents in E. coli and this may be true of other tRNA molecules as well. Duplication is not always the case, though, because $tRNA^{Trp}$ is transcribed from only one gene.

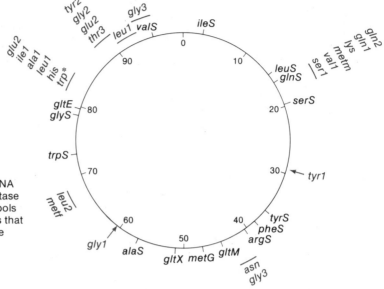

Figure 12-18
Genetic map of *E. coli* showing tRNA genes (red) and aminoacyl synthetase genes (black). A stack of red symbols outside a red curved line indicates that those tRNA genes are at nearly the same location.

Table 12-5
Number of tRNA genes in various organisms or particles

Organism	Number of tRNA genes*
Mitochondria	22
E. coli	62
Yeast	300
Drosophila (fruit fly)	600
Xenopus (toad)	8000

*These are numbers per haploid set of chromosomes. The *E. coli* number is a minimum value. The others are approximate.

The arrangement of the duplicated genes shows no particular pattern. For example, some transcripts are found to contain multiple copies of tRNA genes A and B in which the gene order is AABB, whereas for other tRNA genes the order is CDCD. The reason for the existence of tandem duplications is not clear. However, they must serve an important immediate (as opposed to evolutionary) function because they are stable; when tandem duplication of genes other than tRNA genes is artificially accomplished in *E. coli* in the laboratory, the duplication is rapidly lost by intergenic genetic recombination. This is not the case for the duplicated tRNA genes, which have probably been stable for millions of generations. In the next section we will see that suppression is facilitated by tRNA gene duplication. It is unlikely, however, that the ability to form a suppressor is a sufficient driving force for maintaining a duplication, for this ability confers an advantage on subsequent generations rather than on an existing cell.

SUPPRESSORS

In Chapter 10 conditional mutations were described. These are mutations that yield the wild-type phenotype in certain circumstances (genetic background, temperature, and so forth) and a mutant phenotype in other conditions. Of particular interest in this section are the **suppressor-sensitive mutations,** which behave like the wild type when a suppressor molecule is present. An important example of this phenomenon is a phage mutant that can grow in one strain of bacteria (denoted Sup$^+$) but fails to grow in other strains. Suppressor-sensitive mutations are of two main types: **nonsense** or **chain termination** mutations and **missense** or **base substitution** mutations (see Chapter 10). The nonsense type will be considered first.

Genetic Detection of a Nonsense Suppressor

Chain termination mutations are very common, for they can arise in many ways. For example, a single base change in any of the codons AAG, CAG, GAG, UCG, UUG, UGG, UAC, and UAU can give rise to the amber chain termination codon UAG. If such a mutation occurs within a gene, a mutant protein with little or no function will result because there is no tRNA molecule whose anticodon is complementary to UAG. Thus, only a fragment of the wild-type protein is produced and this usually fails to function unless the mutation is very near the carboxyl terminus of the protein.

In certain bacterial mutants a chain termination mutation does not cause termination; this can be seen in the following simple system. Consider a bacterium B that has mutated with the result that chain growth terminates at a Tyr site in the *lac* gene, so that the bacterium is genotypically and phenotypically Lac$^-$ (Figure 12-19). We now seek a Lac$^+$ revertant of B by mutagenizing a population of B, allowing several generations of growth, and plating the culture on agar selective for Lac$^+$ colonies. Three classes of revertants will be found, as shown in the figure. In class I the chain termination mutation is reversed by a base substitution mutation that converts UAG back to UAC. In this class the complete protein chain will be present, the tyrosine will be restored at the correct position, and the protein will have the wild-type base sequence. In other words, there will be perfect reversion to wild type. In class II a new amino acid, serine, is present in the now complete protein chain, and the base sequence is likewise altered from the original wild-type sequence. In this case the chain termination mutation will be converted to a silent mutation that has a wild-type or near wild-type phenotype. In the figure this has happened as a result of a base substitution mutation that converts UAG to UCG, the serine codon, and the substitution of serine for tyrosine does not markedly alter function;

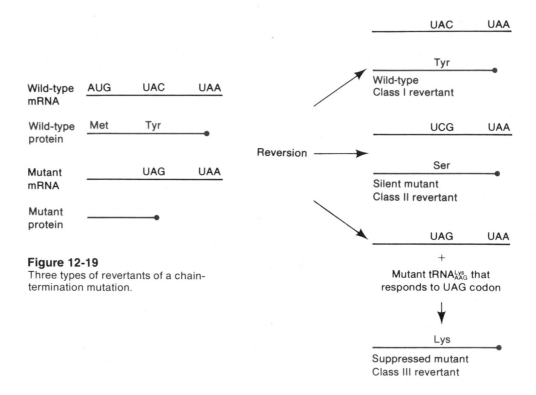

Figure 12-19
Three types of revertants of a chain-termination mutation.

that is, serine is an acceptable amino acid at this position in the protein. The class III revertants are those of interest for this discussion. Those revertants produce a complete polypeptide chain usually by supplying some other amino acid at the mutant site without in any way altering the original mutant base sequence—that is, the codon remains UAG. Class III revertants also typically have the property that a mutant phage that has a chain termination mutation in some essential protein is able to form a plaque on a class III revertant. The revertant cell has gained the ability to ignore (or **suppress,** to use the genetic term) UAG-type chain termination mutations. Thus, somehow the mutant UAG codon is being translated into an amino acid.

Nonsense Suppressor tRNA

The molecular explanation for the suppression phenomenon described in the foregoing paragraphs is that the class III revertant, which is called a suppressor-containing bacterium or a **suppressor mutant,** contains an altered tRNA molecule—one that has the anticodon CUA, which can pair with UAG. This mutant tRNA is called a **suppressor tRNA.**

How can such a mutant tRNA molecule arise? Since a tRNA

molecule is simply the product of a tRNA gene, these mutants occur in exactly the same way as any other mutant—by an error in replication or repair. It is important to realize that a suppressor tRNA does not arise in response to a chain termination mutant; rather, it is formed spontaneously or as a result of mutagenesis, and researchers have selected it from a population of bacteria containing both it and a chain termination mutation by using a growth medium in which only cultures of cells having a wild-type phenotype can form a colony.

What has been mutated in the production of a suppressor tRNA? Clearly it must be a normal tRNA gene. Thus, in the example given previously, a tRNA^{Lys} molecule whose anticodon is CUU, has been altered to have the anticodon CUA, which can hydrogen-bond to the codon UAG.

Inasmuch as a single base change is sufficient to alter the complementarity of an anticodon and a codon, there are (at most) eight tRNA molecules having a complementary anticodon that, with a single changed base, will also suppress a UAG codon. Thus, the following amino acids (whose codons are also indicated) can be put at the site of a chain termination codon: Lys (AAG), Gln (CAG), Glu (GAG), Ser (UCG), Trp (UGG), Leu (UUG), and Tyr (UAC and UAU). Note that these are the same amino acid codons that can be altered by mutation to form a UAG site.

Suppressors also exist for chain termination mutants of the ochre (UAA) and opal (UGA) type. These too are mutant tRNA molecules whose anticodons are altered by a single base change.

In conventional notation, suppressors are given the genetic symbol *sup* followed by a number (or occasionally a letter) that distinguishes one suppressor from another. A cell lacking a suppressor is designated *sup0* and *sup⁻*. Some of the *E. coli* suppressors are listed in Table 12-6.

Table 12-6
Properties of several *E. coli* suppressors

Genetic designation		Codons suppressed
Current	*Obsolete**	
sup1 †	*supD*	UAG
sup2 †	*supE*	UAG
sup3 †	*supF*	UAG
sup4	*supC*	UAA, UAG
sup9	--------	UGA, UGG
sup36	--------	AGA, AGG
supβ	*supL*	UAA, UAG

*These symbols were in use before a standard notation was adopted but are still seen.

†The symbols *supI*, *supII*, and *supIII* are still occasionally used for *sup1*, *sup2*, and *sup3*.

Biochemical Features of Nonsense Suppression

Several features of nonsense suppression should be recognized:

1. Not every UAG suppressor can restore a functional protein by suppressing each UAG chain termination mutation. Thus, a UAG codon produced by mutating the leucine UUG codon might be suppressed by a suppressor tRNA that inserts tyrosine, serine, or tryptophan, but might not be able to tolerate a substitution by the electrically charged amino acids lysine, glutamine, or glutamic acid.

2. Suppression may be incomplete, for two reasons. First, replacement of leucine by tyrosine or serine may restore activity sufficient for survival of colony formation but the activity may be subnormal because the amino acids are not hydrophobic. Second, the nonsense codon is usually translated as sense by a suppressor tRNA only part of the time (that is, chain termination is still fairly frequent), so that the concentration of the protein produced is lower than that of the wild-type protein.

3. An ochre suppressor also suppresses the amber codon, but an amber suppressor does not always suppress an ochre codon. This is a result of wobble. An ochre (UAA) suppressor must have the anticodon UUA. Because G·U is a wobble pair, interaction of UUA with UAG (amber) is possible with G·U in the wobble (third) position. An amber suppressor may have the anticodons CUA or UUA. The UUA anticodon can pair with the ochre UAA but the CUA cannot (see Table 12-4). Thus, ochre mutants are not suppressed by amber suppressors.

4. A cell can survive the presence of a suppressor only if the cell also contains two or more copies of the tRNA gene. Clearly if a tRNA^{Ser} molecule that reads the UCG codon is mutated, then the UCG can no longer be read as a sense codon. This will lead to chain termination wherever UCG occurs and a cell harboring such a mutant tRNA molecule will fail to terminate virtually every protein made by the cell. However, as was mentioned earlier, there are multiple copies of most tRNA molecules; moreover, there are also minor tRNA molecules having the same anticodons as the major molecules. Thus, in any living cell containing a suppressor tRNA, there must always be an additional copy of a wild-type tRNA that can function in normal translation. The exception to this generalization is the tRNA^{Trp} molecule of *E. coli*. There is only a single copy of this tRNA gene, so that an amber suppressor that inserts tryptophan is not possible in an ordinary bacterium. Interestingly, in partial diploids in which a second copy of the tRNA^{Trp} gene is contained in an F plasmid, a tryptophan-inserting amber suppressor has been isolated. This suppressor has the peculiar property that it allows some glutamine as well as tryptophan to be inserted. The

reason is that the anticodon change in the mutant slightly alters the recognition by glutaminyl-tRNA synthetase, which occasionally acylates the tRNA^{Trp} molecule with glutamine.

5. Some tRNA molecules can mutate to yield either an amber or an ochre suppressor. For example, tRNA^{Tyr}, which has the anticodon QUA (Q is a modified base called queosine), can mutate to CUA to be an amber (UAG) suppressor and to UUA to be an ochre (UAA) suppressor.

6. Although suppression of chain termination is typically a result of an anticodon mutation, it can also result from an alteration elsewhere in the molecule. One example is the opal *sup9* suppressor, in which a base change *in the DHU stem* somehow alters the specificity of codon-anticodon binding. This suppressor is an especially interesting one since it is an alteration of the tRNA^{Trp} molecule of which there is only a single copy. Clearly, some mechanism is needed to allow translation of the tryptophan codon if a cell containing *sup9* is to survive. In fact, the *sup9* suppressor retains the ability to respond to the tryptophan codon in addition to the UGA codon. Furthermore, it does not recognize the UGA codon with high efficiency (it is a so-called **weak suppressor**) but at a sufficient rate to allow completion of the prematurely terminated protein at a level adequate for survival. Why the *sup9* suppressor is able to respond to both UGG and UGA with the anticodon CCA is not known. Presumably the alteration in the DHU stem affects the orientation of the tRNA molecule on the ribosome and produces ambiguous pairing between the codon and the anticodon.

Normal Termination in the Presence of a Suppressor tRNA

If a cell contains an amber suppressor, then proteins terminated by a single UAG codon will not be completed and the existence of a suppressor tRNA should be lethal. There are two ways that this problem can be avoided:

1. Protein factors active in termination (see Chapter 13) respond to chain termination codons even though a tRNA molecule that recognizes the codon is present, i.e., suppression is weak. For example, if the probability of recognition is only x percent, then only x percent of the prematurely terminated mutant proteins would be completed and only x percent of normally terminated proteins would be ruined by lack of termination.

2. Normal chain termination may utilize pairs of distinct termination codons such as the sequence UAG–UAA. Thus, the existence of a UAG suppressor would not prevent termination of a doubly terminated protein.

It is likely that both of these mechanisms are responsible for the existence of viable suppressor-containing organisms. Suppression of potentially lethal UAG and UGA mutations is very efficient; termination is prevented more than 50 percent of the time. (Note that the *activity* of the suppressed mutant may not always be as high as 50 percent; this is because the particular amino acid insertion is not always well tolerated by the protein rather than because of inefficient suppression of chain termination.) A partial explanation for the viability of organisms containing UAG and UGA suppressors came initially from studies of the base sequence of the RNA phages R17 and MS2; it is clear from the amino acid sequence of the phage proteins and the base sequence of the RNA molecule that one of the proteins is terminated by paired chain termination codons. The codons are not always adjacent, though, and may be separated by several bases; the number of separating bases is always a multiple of three. It may be the case that duplication of chain termination codons is a frequent occurrence in many organisms. This is not a satisfactory explanation because, in these same phages, there are genes terminated by a single UAG codon yet the phages are able to grow normally in a host cell containing a UAG suppressor.

Only a limited number of base sequences of complete genes are known. However, it seems clear that chain termination is accomplished most commonly by a single UAA codon. This is consistent with the observation that UAA suppressors are typically very inefficient, preventing termination of mutationally induced UAA codons between 1 and 5 percent of the time. Furthermore, cells containing UAA suppressors are generally unhealthy, growing rather slowly compared to cells lacking any suppressor or containing either a UAG or UGA suppressor. Presumably, several percent of the normal proteins, perhaps those needed critically, are damaged by not being terminated at the right time. The reader can see that an understanding of how cells survive having suppressors is incomplete. Presumably this reflects a lack of understanding of how normal chain termination occurs (see Chapter 13). It is likely that unknown factors are active in normal chain termination; some of these factors may be the secondary structure of the mRNA or special ribosomal features that reduce the recognition of natural termination signals by suppressors. Some support for the existence of such factors is the following: there are numerous instances of leaky chain termination mutations—that is, a UAG codon within a gene and in the correct reading frame, at which termination does not always occur. In lab jargon, it is said that **read-through** occurs past the mutant site. Furthermore, mutations in ribosomal proteins sometimes increase the efficiency of termination of leaky mutations. In order to explain the growth of suppressor-containing bacteria, it has even been proposed that the chain termination triplets play little or no role in natural termination. This hypothesis has, however, been ruled out by the observation that base changes in a natural chain termination codon abolish termination. For example, in hemoglobin a U→C change in the UAA stop codon of the α chain results in insertion of glutamine at that site and addition of thirty amino acids to the carboxyl end of the protein before another stop codon is encountered. Thus, if termination requires information other than the presence of the UAA codon, that information is not by itself sufficient to allow termination to occur.

Clearly, understanding of termination must await the accumulation of more information, especially base sequence data.

Suppression of Missense Mutations

Suppression can also occur for missense mutations. For example, a protein in which valine (nonpolar) has been mutated to aspartic acid (polar), resulting in loss of activity, is restored to the wild-type phenotype by a missense suppressor that substitutes alanine (nonpolar) for aspartic acid. Such a substitution can occur in three ways: (1) a mutant tRNA molecule may recognize two codons, possibly by a change in the anticodon loop; (2) a mutant tRNA molecule can be recognized by a noncognate aminoacyl synthetase and be misacylated; and (3) a mutant synthetase can charge a noncognate tRNA molecule. Examples of each of these classes of suppressors are known. Before considering these examples, it must be pointed out that suppression of missense mutations is necessarily inefficient. If a suppressor that substitutes alanine for aspartic acid worked with, say, 20 percent efficiency, then in virtually every protein molecule synthesized by the cell at least one aspartic acid would be replaced, which is a situation that a cell could not possibly survive. The usual frequency of missense suppression is about 1 percent. The missense suppressor that inserts glycine at an AGA codon is 25 percent efficient. However, the AGA codon does not occur very frequently, so damage to normal proteins is minimal. In this way a small amount of a functional, essential protein is made and thereby a mutant cell is able to survive. Missense suppression still, however, introduces a significant number of defective proteins of all other types and, as a result, a cell carrying a missense suppressor usually grows slowly and is generally unhealthy. The principle that applies to these cells is that it is better to be sick than dead.

Missense Suppression Caused by an Altered Anticodon

An altered anticodon can produce a missense suppressor without affecting the specificity with which the cognate synthetase charges the tRNA molecule. A well-studied example is that of tRNA^{Gly}. There are three forms of this molecule, each carrying a different anticodon. The tRNA$_I^{Gly}$ has the anticodon CCC and is responsible for translating the glycine codon GGG; tRNA$_{II}^{Gly}$ has the anticodon UCC and hence pairs with GGA and, by wobble, with GGG. Since both tRNA$_I^{Gly}$ and tRNA$_{II}^{Gly}$ both respond to GGG, tRNA$_I^{Gly}$ can be altered in the anticodon without the cell losing the ability to read the GGG. A mutant tRNA$_I^{Gly}$ has been isolated in which the anticodon is changed from CCC (Gly) to CCU (Arg). This mutant is a missense suppressor—it is acylated with glycine which is inserted at AGG, an arginine codon. The tRNA molecule is a minor one; in most cases, AGG is still read as arginine. However, occasionally it is read as glycine and this may restore the function of a protein having an arginine at a mutant site. The suppression would be especially effective if the arginine site arose by alteration of a site that initially coded for glycine. There are many examples of missense suppressors arising in this way.

Missense Suppression by Mischarging of a Mutant tRNA

Single base changes in the acceptor stem and a few other locations increase the frequency of incorrect mischarging by synthetases. For example, base changes at six different positions (Figure 12-20) of the acceptor stem of E. coli tRNA$_I^{Tyr}$ stimulate occasional mischarging with glutamine by glutaminyl-tRNA synthetase.

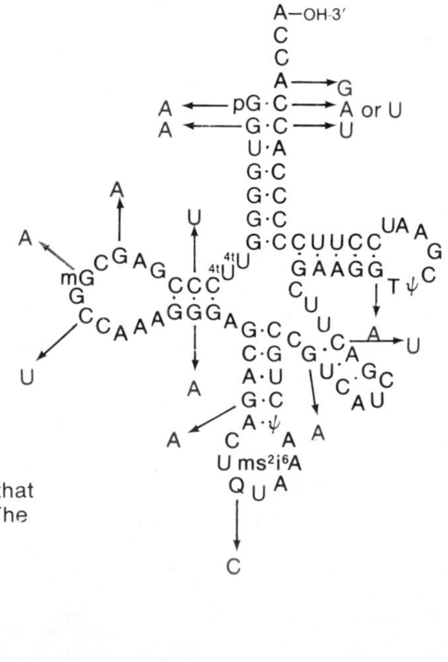

Figure 12-20
Mutations in tRNA$_I^{Tyr}$ that cause mischarging. The superscripts indicate modified bases.

Changes in the anticodon also induce misacylations that lead to rather complicated patterns of suppression. For example, the *sup3* tRNA$_I^{Tyr}$ molecule is primarily charged with tyrosine but the anticodon change that allows recognition of the UAG codon stimulates occasional misacylation with glutamine.

Mutant Synthetases as Suppressors

No mutant tRNA synthetase has yet been found that causes misacylation of a wild-type tRNA. However, recently it has been shown that the occasional charging of *sup9*-tRNA$_I^{Tyr}$ (described in the previous section) by glutamine is caused by a mutant glutaminyl-tRNA synthetase. This suggests that other misacylating mutant synthetases will be found in the future.

Frameshift Suppressors

When a frameshift mutation occurs—for example, a base addition—all downstream codons are read incorrectly and the entire amino acid sequence following the mutation is incorrect. There is no way that either a missense or a nonsense suppressor could allow synthesis of an active protein. Restoration of activity is possible only by recreating the original reading frame.

In bacteria three types of spontaneous revertants of frameshift mutations occur (Figure 12-21). The most straightforward type is that in which the added base is simply removed. A second type is an intragenic second-site suppressor in which either a nearby base is removed or two extra bases are added. In both cases, the reading frame is restored and as long as the additional amino acids between the mutant and revertant sites can be tolerated, protein activity returns. Sometimes this type of reversion requires the presence of a chain termination suppressor. (Figure 12-21(a), revertant II). The third type of reversion is produced by a true extragenic suppressor, some of which are mutant tRNA molecules having a four-letter anticodon. An example of such a suppressor is a mutant tRNA^{Gly} in which the anticodon loop possesses an extra C—the anticodon contains the sequence CCCC. It is possible that this allows pairing with a sequence of four G's, though other explanations have also been given. For instance, it has been proposed that the additional base is not engaged in pairing but is merely a spacer that forces the next codon read to be displaced by one base.

Single-base deletions are also intragenically suppressible—possibly because some tRNA molecules may pair with only two bases of the codon. This hypothesis derives from the fact that almost all single-base

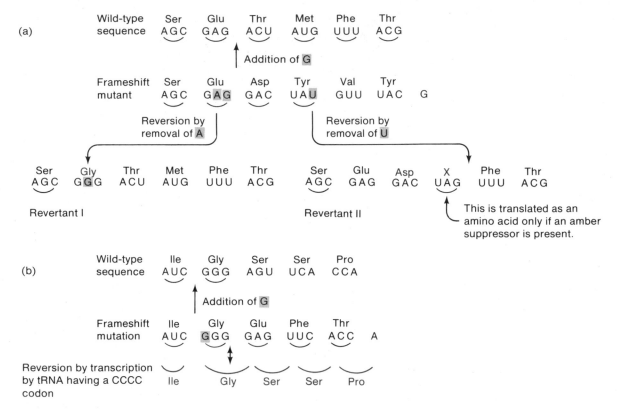

Figure 12-21
Reversion of frameshift mutants. (a) Reversion by eliminating a base, to restore the reading frame. In revertant I
there is a Gly (red) where there is a Glu in the wild-type
protein. Revertant II is functional only if the UAG codon
is read by an amber suppressor; amino acids Asp and X
replace Thr and Met, respectively. (b) Reversion by
means of a suppressor that can read a four-base unit.

deletion mutations are somewhat leaky, possibly because various
normal tRNA molecules occasionally (or perhaps always?) engage in
two-base pairing. Other explanations have been given for frameshift
suppression; few of the ideas are supported by good data. Basically, the
mechanism of frameshift suppression is poorly understood.

Temperature-Sensitive (Ts) Suppressors

A collection of Ts suppressors of amber mutations has been isolated.
These are able to suppress amber mutations at 30°C but not at 42°C.*
Most are tRNA molecules in which the acceptor stem has been altered,

*Note that these are not suppressors of Ts mutations.

which allows the tRNA to undergo significant changes in shape at 42°C. The Ts suppressors have not been studied in detail but are instead used as an experimental tool. For instance, synthesis of a protein containing an amber mutation proceeds normally at 30°C in a bacterium having a Ts suppressor. However, synthesis can be immediately stopped merely by raising the temperature sufficiently. Of even greater value is the fact that the number of functional molecules depends on temperature, so that the effect of the internal concentration of a molecule can be studied merely by growth of the cells at several temperatures; in this way the internal concentration of a protein can be varied systematically.

Suppressor-enhancing Mutations

As we mentioned, the efficiency of nonsense suppression is never 100 percent and is in fact quite low. This is especially true of Ts suppressors at slightly elevated temperatures. If a cell containing both an amber mutant of an essential gene and a Ts amber suppressor is grown at a somewhat elevated temperature, most cells grow poorly, forming small colonies. In a population of such cells, a few cells form large colonies. Analysis of the cells in these colonies shows that they contain secondary mutations called **suppressor-enhancing mutations** (sue). The sue mutations are also active with temperature-insensitive suppressors, increasing the efficiency of suppression between three- and tenfold. There are several known sue mutations—called sueA, sueB, and sueC. They are distinct mutations: cells containing more than one of these mutations are more effective than those containing a single mutation. For example, sueA, sueB, and sueC mutations enhance suppression 3.3-, 10-, and 3.3-fold respectively, yet the double mutations sueA sueB and sueB sueC enhance suppression 13.3- and 27-fold, respectively.

The molecular mechanism of sue mutations is not understood at all.

Suppressors in Yeast and Their Use in Animal Cell Research

At present, interest in yeast suppressors centers on their practical use in mammalian cell genetics. For a variety of reasons, which we shall not go into at this point, it would be valuable to have amber mutants of animal proteins. Mutants of animal cells have been isolated but few are known to be chain termination mutants because suppressors have not yet been isolated in animal cells. However, purified amber suppressor tRNA of yeast has been injected by a microsyringe into individual mutant mammalian cells. For some mutations a period of suppression lasts several hours. Ultimately the yeast tRNA is destroyed and since there is no synthesis of this suppressor tRNA, the mutant phenotype returns.

This technique allows the identification of amber mutations. Yeast tRNA is used rather than *E. coli* tRNA because yeast is a eukaryote and its tRNA molecules function more efficiently on the ribosomes of the mammalian cell than does *E. coli* tRNA.

Role of Genetics in the Study of tRNA

At the present time, most of the information about tRNA structure has been derived from base sequence analysis and powerful physical techniques such as x-ray crystallography and infrared and nuclear magnetic resonance spectroscopy. Once the base sequence of yeast tRNA^{Phe} was elucidated, the cloverleaf model, which we have already seen, seemed to be obvious. Evidence supporting a cloverleaf structure was provided by simple genetics experiments many years before the physical techniques were applied to the problem. Genetic methods have also provided most of the information about the organization and multiplicity of tRNA genes. It is instructive to look closely at the genetic procedure.

The genetic approach began with the isolation of a derivative of a phage φ80 that was selected to carry the gene for the *sup3* tRNA. With standard genetic methods, a φ80 mutant was prepared that had two amber mutations. This mutant was able to form a plaque on a lawn of bacteria containing the *sup3* gene but not on a *sup0* bacterium and formed in the following way. Phage φ80 is a lysogenic phage (Chapter 16) and inserts its DNA into the *E. coli* chromosome next to the *sup3* gene. Sometimes when the phage DNA is excised from the chromosome, it takes with it some of the adjacent chromosomal DNA. A phage carrying chromosomal DNA in this way is called a **transducing particle.** How transducing particles form is shown in Figure 16-23 (Chapter 16). A transducing particle is described by stating the bacterial genes that are carried; thus a φ80 phage with the *sup3* gene is written φ80*sup3*. A φ80*sup3* particle is very easy to isolate if the particle also contains an amber mutation, because the *sup3* gene enables the particle to plate on a *sup0* bacterium (Table 12-7). The fact that the particle can grow in a Sup⁻ bacterium indicates that active *sup3* tRNA is transcribed from the phage DNA.

When φ80*sup3* infects a bacterium, an enormous amount of *sup3* tRNA is made, because the phage DNA replicates and provides many gene copies, which can be transcribed. This is a useful procedure for isolating primary transcripts because these transcripts are enriched about 100-fold compared to an uninfected cell. Transducing particles carrying almost every known *E. coli* suppressor tRNA gene have been isolated and have provided researchers with a means of obtaining highly purified specific tRNA molecules.

Table 12-7
Plaque-forming ability of a ϕ80 mutant and its *sup3*-transducing variant on two bacterial strains

Bacteria	Plaque-forming ability		
	Phage ϕ80	Phage ϕ80(Am)	Phage ϕ80(Am)sup3
sup⁻	Yes	No	Yes
sup3	Yes	Yes	Yes

One of the first observations made with this particle was that both the *sup3* tRNA and the wild-type tRNA^{Tyr} molecule, whose mutant form is *sup3* tRNA, are made in large and equal quantities. This indicated that *E. coli* normally contains at least two copies of the tRNA^{Tyr} gene and that only one of these is a mutant in a *sup3*-containing strain. The implications of this finding have already been discussed. When the primary tRNA transcript was examined, it was found that both tRNA^{Tyr} and *sup3* tRNA are transcribed as a single large RNA molecule and that each tRNA molecule is excised from the primary transcript. This was the first evidence for RNA processing.

A second type of experiment provided supporting evidence for the cloverleaf model. A ϕ80*sup3* variant (no longer containing any amber mutations) was mutagenized in order to obtain a particle with a mutant, nonfunctional *sup3* tRNA. (Note that whereas *sup3* tRNA is a tRNA^{Tyr} with a mutation in the anticodon, the tRNA is still functional; the mutation that is being described now was selected to eliminate *all* function of the *sup3* tRNA molecule.) Such a mutant was isolated in the following way. The mutagenized phage population was plated on a bacterial host carrying a *lac*⁻(Am) mutation on agar containing lactose and the indicator 5-bromo-4-chloro-3-indole-β-galactoside. Lac⁺ cells form blue colonies and Lac⁻ cells make white colonies on this agar. With a Lac⁻ indicator bacterium a normal (nontransducing) ϕ80 phage yields a colorless plaque on this agar. However, ϕ80*sup3* suppresses the *lac*⁻ mutation, so that ϕ80*sup3* forms blue plaques on this agar. A ϕ80*sup3* carrying a mutation (*sup*⁻) in the *sup3* tRNA will be unable to suppress the bacterial *lac*⁻ mutation and will form white plaques. Many white plaque formers were isolated and used in subsequent infections in order to obtain large quantities of the *sup*⁻ mutant tRNA molecules. The sequence of many of these mutant tRNA molecules was determined. It was found that often a base in one of the presumed stem regions of the cloverleaf was altered in such a way that a base pair in the cloverleaf stem could no longer be formed, suggesting that the presumptive base pair is essential for tRNA function. Several particles

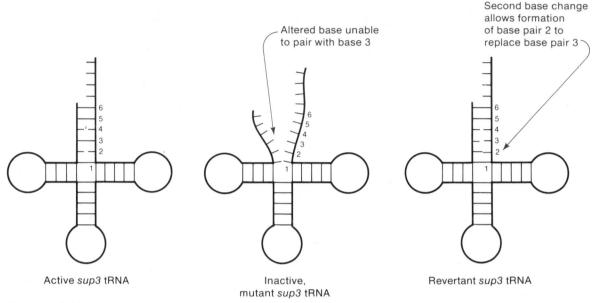

Figure 12-22
Study of reversion of mutant *sup3* tRNA, indicating that there is base-pairing in the acceptor stem.

carrying such mutations were mutagenized again and revertants in which the *sup3* tRNA molecule were once again active were isolated by the blue plaque test. The base sequences of these revertants were determined and it was found that in each revertant a second base had changed that restored pairing (Figure 12-22). The original base pair was not required to restore activity—any pair would do. From these data one could infer that certain bases that were rather distant in the polynucleotide chain had to be paired, if the tRNA molecule is to be functional. Since these bases were in each case in the presumptive stem region, strong support was given for the cloverleaf model. Several years later, physical studies confirmed these base-pair assignments.

SELECTION OF THE CORRECT AUG CODON FOR INITIATION

The sequence AUG serves both as an initiation codon and the codon for methionine. Since methionine occurs within protein chains, there must be some signal in the base sequence of the mRNA that identifies particular AUG triplets as start codons. The precise nature of the signal is not known, though we have some idea. Most of the available information comes from determining the base sequence of several bacterial mRNA molecules.

The Initiation Signal of a Prokaryotic mRNA

We will see in Chapter 13 that in the presence of charged tRNA^{fMet} and several proteins required for initiation, a ribosome and a mRNA molecule form a stable complex. If no other tRNA molecules are present, protein synthesis is stalled at this initiation stage. If the complex is treated with pancreatic RNase, an enzyme that hydrolyzes RNA only if it is free and single-stranded, all of the RNA is hydrolyzed except for that which is bound to the ribosome. This protected fragment can be isolated and its base sequence determined. This type of experiment has been performed with phage mRNA molecules, which encode proteins whose amino acid sequences are known (Figure 12-23.) The protected RNA fragment is usually 30 to 40 nucleotides long and always contains an AUG (or, infrequently, a GUG) codon. This codon is known in each case to be a start codon because the adjacent bases (in the 3′ direction) constitute the codons for the initial amino acids of the proteins. The AUG is usually 20 to 30 bases from the 5′ end, so that at least some of the preceding sequence (the **leader**) is recovered. The most notable feature of the leader region is that the sequence

AGGAGGU

(or a variant) often precedes the initiator triplet and is a fixed distance ahead of it.* This sequence is thought to define the start codon by providing a binding site for the ribosome, as we will see below. In a few cases the spacer region can form a hairpin. This feature was at one time thought to be the recognition site; however, there are enough protected sequences that lack this self-complementarity that this view is no longer held.

In 1975, an attractive hypothesis was put forward by J. Shine and L. Dalgarno to explain recognition of the initiation codon. The hypothesis

5′ AAUCUUGGAGGCUUUUUUAUGGUUCGUUCU 3′   Phage φX174 gene-*A* protein

5′ UAACUAAGGAUGAAAUGCAUGUCUAAGACA 3′   Phage Qβ replicase

5′ UCCUAGGAGGUUUGACCUAUGCGAGCUUUU 3′   Phage R17 gene-*A* protein

5′ AUGUACUAAGGAGGUUGUAUGGAACAACGC 3′   Phage λ gene-*cro* protein

| Pairs with 16S RNA | Pairs with tRNA^{fMet} |

Figure 12-23
Sequences of initiation regions for protein synthesis in four phage mRNA molecules.

*At the time of this writing—1982—this sequence has been found in 74 different mRNA molecules.

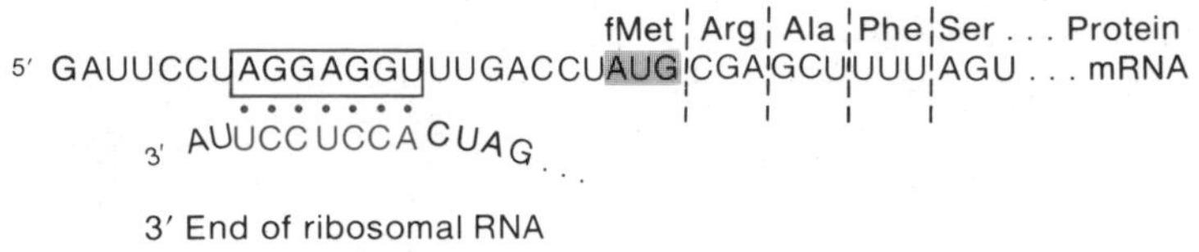

```
                                        fMet ¦ Arg ¦ Ala ¦Phe¦Ser . . . Protein
       5′ GAUUCCU AGG AGGU UUGACCU AUG CGA GCU UUU AGU . . . mRNA
                    • • • • • •                    ¦    ¦    ¦    ¦
          3′ AUUCC UCCA C UA G . . .
```

3′ End of ribosomal RNA

Figure 12-24
Base-pairing between the purine-rich region (in box) in the initiator region
of mRNA and the complementary region (red) near the 3′ end of rRNA.
The AUG start codon is shaded.

has not yet been proved rigorously but its general notions are widely
believed. The 3′-terminal sequence of the 16S rRNA of several bacteria
is

5′-PyACCUCCUUA-3′

in which Py can be any pyrimidine. The sequence is very rich in
pyrimidines, and it has been proposed that the purine-rich leader
sequence AGGAGGU forms moderately stable base pairs with the
pyrimidine-rich sequence in 16S RNA (Figure 12-24). This pairing is
thought to bring the AUG codon to a start site on the ribosome. The
evidence, other than the apparent sequence homology just described,
which supports this hypothesis is the following:

1. An intact 3′-OH terminus of 16S RNA is essential for initiation of
protein synthesis.

2. Certain ribosomal proteins (soon to be described) involved in
initiation are located in the ribosome near the 3′ terminus of the 16S
RNA molecule.

3. Antibiotics that inhibit initiation bind to protein components
right next to 3′-terminal sequences of 16S RNA.

4. Mutations that reduce the efficiency of translation (that is, the
number of protein molecules made per unit time) have been isolated
and many of these are base changes in the Shine-Dalgarno sequence.

5. Random copolymers rich in adenine and guanine inhibit initiation with natural mRNA molecules. Presumably, within these polymers
are sequences that bind to the 16S RNA molecule and prevent binding
of the mRNA molecule to the ribosome.

6. Direct binding of the initiator region of an mRNA molecule and
the 3′-terminal region of the 16S RNA molecule has been detected. This
was done in the following experiment. A 30-nucleotide fragment of the

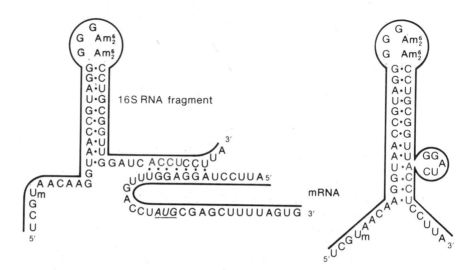

(a) Complex between mRNA and the colicin fragment   (b) Free colicin fragment

Figure 12-25
Evidence for binding of mRNA to a complementary sequence in 16S rRNA. The AUG
start codon is underlined. The four red hydrogen bonds in panel (b) are broken in the
formation of the complex in panel (a). The four bases shown in red undergo base-
pairing changes in forming the complex.

mRNA molecule encoding the initiator region for the synthesis of the A
protein of *E. coli* phage R17 was isolated and allowed to form an
initiation complex with *E. coli* ribosomes. A substance called **colicin E3,**
which makes a single endonucleolytic cut 59 nucleotides from the 3'
terminus of 16S RNA, was then added to the complex. The mixture was
treated with a detergent that dissociates ribosomes and releases RNA.
The RNA that was released is a mRNA-rRNA hybrid molecule (Figure
12-25(a)). In this hybrid the mRNA and the rRNA fragments were held
together by hydrogen bonds. The paired sequences are those shown in
Figure 12-24. Another result indicates that the mRNA must actually
invade the rRNA structure. Consider the molecule labelled "colicin
fragment" in Figure 12-25(b). This molecule is obtained if the ribosomes
are treated with colicin E3 before addition of the mRNA fragment. Note
that the hydrogen bonds shown in red in panel (b) are broken when the
complex in panel (a) forms and that the four bases shown in red in panel
(b) are paired with bases in mRNA, as shown in panel (a). Thus, the
binding of mRNA causes a small structural change in the rRNA.

Further support for the Shine-Dalgarno hypothesis comes from a
study of the efficiency by which the R17 A protein is synthesized with
ribosomes obtained from other bacterial species. The 3'-terminal se-

quences of the 16S RNA molecules of six bacterial species were determined and it was observed that the length and strength (number of G · C versus A · U pairs) of the base-paired regions are not the same. The amount of A protein synthesized with those different ribosomes correlates with the stability of the base-paired region. This suggests that a major factor in the frequency of intiation may be the strength of mRNA-16S RNA base-pairing.

In Chapter 4 it was pointed out that short homologous sequences of bases have low thermal stability because of reduced stacking stabilization. Thus, one might wonder whether the sequence of seven base pairs shown in Figure 12-25(a) is stable. The mRNA-16S RNA complex is in fact less stable than double-stranded DNA in that it is completely melted at 55°C; however, most bacteria rarely encounter temperatures above 45°C. There may also be other stabilizing tertiary interactions such as those in tRNA molecules. Furthermore, in the initiation complex with the ribosome, there are numerous protein factors (which will be discussed in Chapter 13) that probably provide additional stabilization.

Although the Shine-Dalgarno sequence is essential for recognition of an AUG start codon, it is not sufficient: several examples have been observed of Shine-Dalgarno sequences adjacent to AUG triplets that are not used as codons. The additional features of the mRNA base sequence needed for initiation are not yet known. Possibly, the protein factors mentioned in the preceding paragraph bind to specific sites on the mRNA.

Repeated Initiation in Polycistronic mRNA

Many prokaryotic mRNA molecules are polycistronic—that is, they contain sequences specifying the synthesis of several proteins. Thus, a polycistronic mRNA molecule must possess a series of start and stop codons. If a mRNA molecule encodes three proteins the minimal requirement would be the sequence

Start, protein 1, stop—start, protein 2, stop—start, protein 3, stop.

Actually, such an mRNA molecule is probably never so simple in that the leader sequence preceding the first start signal may be several hundred bases long and there is usually a sequence called a **spacer** of from 5 to 20 bases between one stop codon and the next start codon. Thus the structure of a tricistronic mRNA is more typically that shown in Figure 12-26.

We have already mentioned that stop codons sometimes occur in pairs. However, in addition to this pairing that occurs in the same

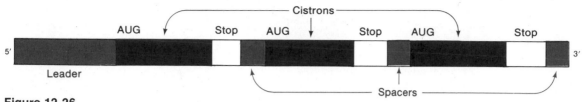

Figure 12-26
Arrangement of cistrons and untranslated regions (red) in a typical polycistronic mRNA molecules.

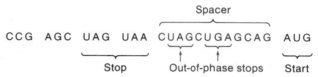

Figure 12-27
A portion of a typical spacer sequence. Usually the sequence is longer than that shown.

reading frame, there may be one or more out-of-phase stop codons in the spacer. For instance, the type of sequence shown in Figure 12-27 might occur. It is thought that the out-of-phase stops provide insurance against two possibilities—(1) on occasion the reading frame may "slip," or (2) a frameshift mutation might alter the reading frame. Having stop codons in all reading frames in the spacer region allows the translational system to have a fresh start at the following AUG.

The mechanism for initiating synthesis of the first protein molecule in a polycistronic mRNA is no different from that in a monocistronic mRNA. However, in a polycistronic mRNA, if the second protein is to be made, protein synthesis must start again after the stop codon of the first protein is reached. The mechanism by which this occurs will be taken up in the next chapter. At this point it is sufficient to state that either the start codon of the second protein is so near the preceding stop codon that synthesis reinitiates before the ribosome and the mRNA dissociate or a second initiation signal locks the protein-synthesizing system onto the second cistron.

Varying Rates of Initiation of Different Cistrons in a Polycistronic mRNA

In general, the number of copies of each protein synthesized from a polycistronic mRNA is not the same. There are several explanations for this phenomenon, one of which is certainly the variability of the strength of binding of the ribosome to the initiation sites for each cistron. An interesting case is one in which some initiation sites are

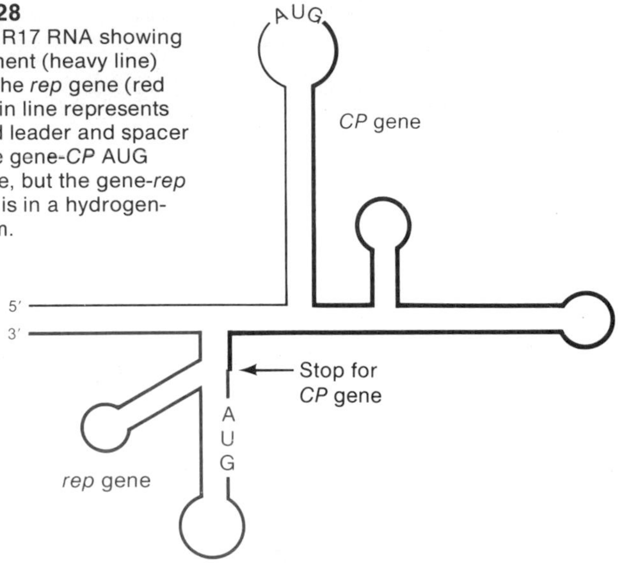

Figure 12-28
A portion of R17 RNA showing the *CP* segment (heavy line) and part of the *rep* gene (red line). The thin line represents untranslated leader and spacer regions. The gene-*CP* AUG codon is free, but the gene-*rep* AUG codon is in a hydrogen-bonded stem.

available only at certain times for pairing to the 16S rRNA. This can be seen with *E. coli* phage R17, which is an RNA phage whose nucleic acid is a polycistronic mRNA encoding three proteins that are made in very different amounts. This RNA molecule has a great deal of internal hydrogen-bonding. These hydrogen bonds have the effect that whereas the initiation site of the *CP* gene—the phage coat-protein gene—is available to the protein-synthesizing apparatus at all times, that of the *rep* gene is not (Figure 12-28). Thus, at the time of phage infection, ribosome binding occurs only at the initiation site of the *CP* gene. How, then, is the Rep protein ever made? As translation of the *CP* region of the mRNA proceeds, the secondary structure of this region is disrupted. The termination codon for the *CP* gene is so near the initiation site for the *rep* gene that this site is made available just about when the CP protein is being terminated; thus, synthesis of the Rep protein is initiated. Note that if a chain termination mutation were present in the *CP* gene well before the normal termination codon, initiation of the Rep protein could not occur because the base-paired region would not be opened. Mutations of this sort have been observed.

At first the CP and Rep proteins are synthesized at equal rates, suggesting that the initiation regions are recognized with equal efficiency. However, later, much more CP is made than Rep, because the CP protein binds to the *rep* site and thereby blocks initiation from that site. Like any binding interaction, CP molecules are continually binding to and leaving the *rep* site. As more CP protein is synthesized, the fraction of the time that the *rep* site is occupied increases and the number of initiation events per unit time decreases. In this way, synthesis of the Rep protein continually decreases, whereas synthesis of CP continues unabated. This is useful since many thousands of CP molecules are needed for synthesis of the hundreds of phage produced by an infected cell, whereas Rep is the replication enzyme, which is needed in only catalytic amounts. This is a beautiful example of translational regulation. We will see many examples of regulation of gene activity in Chapter 14.

Initiator tRNA and Ribosome Binding Sites in Eukaryotes

The basic pattern of protein synthesis in eukaryotes is quite similar to that for prokaryotes. In Chapter 13 it will be seen that the differences are mostly in detail. However, the mechanism by which the ribosome selects the correct reading frame in eukaryotic mRNA is quite different from that which has been described for prokaryotes. The statements that follow come from studies of synthesis of iso-1-cytochrome c of yeast; it is thought that most of the information applies equally well to other eukaryotic cells.

1. In eukaryotes the initiating tRNA molecule carries Met and not fMet. There are both a true $tRNA^{Met}$ molecule and an initiator $tRNA^{Met}$ molecule, as in prokaryotes, but in eukaryotes the latter is denoted $tRNA_f^{Met}$. The initiator tRNA responds only to the AUG codon and never to GUG, because a modified sulfur-containing base in the anticodon loop of $tRNA_f^{Met}$ prevents binding to GUG. Thus, in eukaryotes the start codon is always AUG.

2. In eukaryotes all mRNA molecules are monocistronic. Furthermore, eukaryotic ribosomes cannot translate all of the cistrons of a prokaryotic polycistronic mRNA. In fact, if bacterial polycistronic mRNA is added to an *in vitro* protein-synthesizing system that uses eukaryotic ribosomes, only the cistron nearest to the 5′ terminus is translated (Figure 12-29). The lack of polycistronic mRNA in eukaryotes agrees with genetic mapping results, which indicate that related gene products are usually in different parts of the same chromosome or on different chromosomes, quite different from prokaryotes, in which related genes are often adjacent.

3. In prokaryotes, of the many AUG triplets scattered throughout a mRNA molecule, one is selected to establish the reading frame and to initiate protein synthesis. The one that is chosen is always very near a particular ribosome binding site on the mRNA molecule—namely, the Shine-Dalgarno sequence. However, even though the 3′ termini of

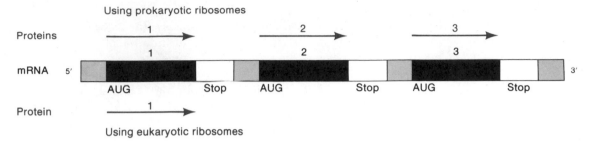

Figure 12-29

Difference in the products translated from a tricistronic mRNA molecule by the ribosomes of prokaryotes and eukaryotes. The prokaryotic ribosome translates all of the cistrons but the eukaryotic ribosome translates only one cistron—the one nearest the 5′ terminus of the mRNA. Translated sequences are in black, stop codons are white, and the leader and spacers are shaded.

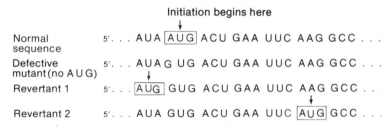

Figure 12-30

A segment of the yeast iso-1-cytochrome *c* mRNA. Initiation normally begins at the arrow. The defective mutant has lost the normal AUG start codon and cannot initiate. Revertants 1 and 2 of the defective mutant have each gained a new AUG codon and can initiate from either of these codons, which indicates that initiation need not occur next to a unique sequence. Base changes are shown in red; start codons are boxed.

prokaryotic 16S rRNA and eukaryotic 18S rRNA are nearly homologous, there is no evidence for a unique ribosome binding sequence in eukaryotic mRNA molecules. Instead, the initiator AUG codon is, with only two exceptions (see below), the AUG triplet that is nearest to the 5′ terminus of the mRNA molecules. The main evidence for this point is that there are many mutations in the yeast iso-1-cytochrome *c* gene, and these create a new AUG triplet nearer to the 5′ terminus than the wild-type initiator is, and initiation begins at the new AUG codon (Figure 12-30). Furthermore, if the normal AUG codon is eliminated by mutation, the next AUG sequence downstream (in the 3′ direction from the natural AUG start codon) serves as an initiator codon, which shows that a particular adjacent sequence is not necessary for initiation. This conclusion does not eliminate the possibility that the adjacent base sequence modulates the efficiency of initiation. In fact, a few mutations

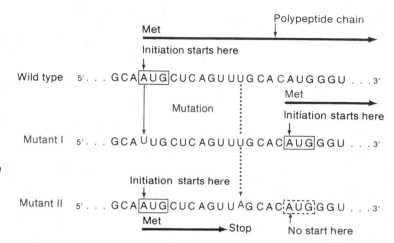

Figure 12-31
Demonstration that initiation does not occur in eukaryotes after a stop codon. In mutant I the AUG start codon is gone, so initiation occurs at the next AUG codon. In mutant II, a UAG stop codon is present after the natural start codon. The following AUG codon cannot be used. The heavy arrows represent polypeptide chains. Mutations are in red. Usable stop codons are in solid boxes; the nonusable stop codon is in a dashed box.

that are a few bases in the 5′ direction from the normal initiation codon of the yeast iso-1-cytochrome *c* gene reduce the efficiency of initiation two- or three-fold. (There are also two instances known in which an AUG codon upstream from the normal initiation codon is not used, which shows that the situation is not yet fully understood.)

4. Initiation cannot occur at an AUG triplet that follows a stop codon. Evidence for this statement is given in Figure 12-31. This deficiency explains why only the 5′-proximal cistron of a polycistronic mRNA can be translated.

An additional observation made with rat rRNA—namely, that the 3′ terminus lacks the sequence found in prokaryotic rRNA that is complementary to the Shine-Dalgarno sequence—indicates that eukaryotes have developed an alternative mechanism for recognizing mRNA.

These and many related results have led to a theory of initiation in eukaryotes known as the **scanning hypothesis** (Figure 12-32). The idea is that a ribosomal subunit attaches to the mRNA at or near the 5′ end of a mRNA molecule (possibly at the 5′ cap sequence) and then drifts along the molecule in the 5′→3′ direction until it encounters an AUG sequence. At this point the initiation complex to be described in Chapter 13 forms and the reading frame is established. When a stop codon is reached, the ribosome leaves the mRNA, which explains why an AUG following a stop codon cannot serve for initiation.

The scanning hypothesis is supported by the fact that certain circular eukaryotic mRNA molecules (circularized in the laboratory), which can be translated *in vitro* with *E. coli* ribosomes, are not

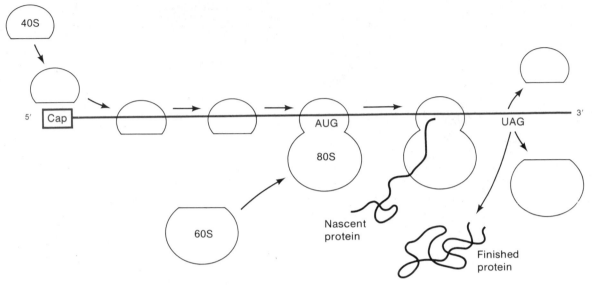

Figure 12-32
The scanning hypothesis. A 40S ribosome subunit attaches at or near the cap and drifts in the 3′ direction. It stops at an AUG codon and is joined by a 60S ribo-somal subunit to form an active 80S ribosome. The 80S ribosome dissociates when reaching a stop codon.

translated if eukaryotic ribosomes are used; if, however, the RNA is nicked (to linearize the molecule), *in vitro* translation with eukaryotic ribosomes occurs normally.

OVERLAPPING GENES

In all that has been said so far about coding and signal recognition, an implicit assumption has been that the mRNA molecule is scanned for start signals to establish the reading frame and that reading then proceeds in a single direction within the reading frame. The idea that several reading frames might exist in a single segment was not considered until recently. Actually, this possibility was supposedly eliminated in the early 1940's when it was noted by George Beadle and Edward Tatum in *Neurospora crassa* that mutations did not affect more than one gene. This view, called the one-gene–one-enzyme hypothesis, has through the years been supported by an enormous number of observations and justified by noting that, if there were overlapping reading frames, several constraints would be placed on the amino acid sequences of two proteins translated from the same portion of mRNA.

However, because the code is highly redundant, the constraints are actually not so rigid. For instance, each of the base sequences shown below,

I. UCU CCU GCA AUU CGU A

II. UCC CCA GCG AUC CGC A

III. UCA CCG GCC AUA CGA A

specifies the amino acid sequence Ser-Pro-Ala-Ile-Arg, yet in the reading frame in which the shaded C is the first codon position, the amino acid sequences are

I. Leu-Leu-Gln-Phe-Val
II. Pro-Gln-Arg-Ser-Ala
III. His-Arg-Pro-Tyr-Glu

Further inspection of the codon assignments shows that there are several hundred sequences of five amino acids that can be obtained by a one-base displacement of a sequence that otherwise translates as Ser-Pro-Ala-Ile-Arg. Thus, the constraint resulting from multiple reading frames is not as great as was initially thought. Clearly, if there are three reading frames, a single DNA segment would be utilized with maximal efficiency. The disadvantage of such a system is that evolution may be slowed, because single-base-change mutations would be deleterious more often than if there were a unique reading frame. Nonetheless, organisms have evolved that use overlapping reading frames. To date, the only examples are small viruses and the smallest phages.

The *E. coli* phage ϕX174 contains a single strand of DNA consisting of 5386 nucleotides whose base sequence is known. If a single reading frame were used, at most 1795 amino acids could be encoded in the sequence and, if we take 110 as the molecular weight of an "average" amino acid, at most 197,000 molecular weight units of protein could be made. However, the phage makes eleven proteins and the total molecular weight of these proteins is 262,000. This paradox was resolved when it was shown that translation occurs in several reading frames from three mRNA molecules (Figure 12-33). For example, the sequence for protein B is contained totally in the sequence for protein A but translated in a different reading frame. Similarly, the protein E sequence is totally within the sequence for protein D. Protein K is initiated near the end of gene *A*, includes the base sequence of B, and terminates in gene *C*; synthesis is not in phase with either gene *A* or gene *C*. Of note is protein A' (also called A*), which is formed by reinitiation within

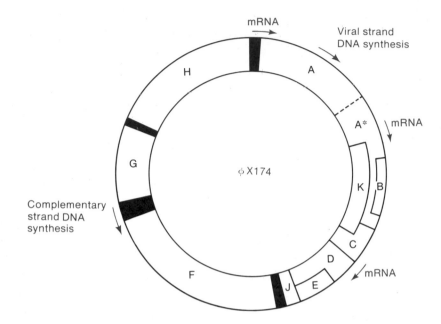

Figure 12-33
Genetic map of phage
φX174 showing the start
points for mRNA synthe-
sis and the boundaries of
the individual protein
molecules. The solid re-
gions are spacers.

gene A and in the same reading frame, so that it terminates at the stop codon of gene A. Thus, the amino acid sequence of A′ is identical to a segment of protein A. In total, five different proteins obtain some or all of their primary structure from shared base sequences in φX174. This phenomenon, known as **overlapping genes,** has been observed in the related phage G4 and in the small animal virus SV40.

It should be realized that the single structural feature responsible for gene overlap is the location of each AUG initiation sequence.

THE GENETIC CODE OF MITOCHONDRIA

Mitochondria are small particles contained in the cells of all multi-cellular and many unicellular eukaryotes. Their principal function is to generate ATP and other energy-storage compounds. Their structure and composition are somewhat unusual; a single mitochondrion is a highly convoluted membrane system containing a variety of enzymes used in generating energy, a set of tRNA molecules that are not found elsewhere in the cell, and a circular DNA molecule having a molecular weight of about 10^7. This DNA molecule encodes most of the mitochondrial enzymes and is the template for synthesis of all of the mitochondrial tRNA molecules.

In 1979, after 16 years of believing in the universality of the genetic code, the world of molecular biology was jolted by the discovery that the genetic code of mitochondria is not the same as the "universal" code. In 1980 it was found that several mitochondrial codes exist for different organisms. The high point of this study came in April of 1981 when the complete base sequence of human mitochondrial DNA was reported—a sequence consisting of 16,569 base pairs! In this sequence were found the genes for 12S and 16S ribosomal RNA, 22 different tRNA molecules, three subunits of the enzyme cytochrome oxidase (whose amino acid sequence is known), cytochrome *b,* and several other enzymes. A few mitochondrial mRNA molecules were also isolated and sequenced, so that their sites of transcription could be located on the DNA. Comparing the amino acid sequences of several proteins, the base sequences of the mRNA molecules, and the DNA confirmed the fact that the mitochondrial genetic code is not the same as the universal code, a point that has profound implications for the evolution of mitochondria.

The human mitochondrial code is shown in Table 12-8; entries given in red are those that differ from the universal code (compare to

Table 12-8
The genetic code of human mitochondria

First position (5′ end)	Second position				Third position (3′ end)
	U	C	A	G	
U	Phe	Ser	Tyr	Cys	U
	Phe	Ser	Tyr	Cys	C
	Leu	Ser	Stop	Trp	A
	Leu	Ser	Stop	Trp	G
C	Leu	Pro	His	Arg	U
	Leu	Pro	His	Arg	C
	Leu	Pro	Gln	Arg	A
	Leu	Pro	Gln	Arg	G
A	Ile	Thr	Asn	Ser	U
	Ile	Thr	Asn	Ser	C
	Met	Thr	Lys	Stop	A
	Met	Thr	Lys	Stop	G
G	Val	Ala	Asp	Gly	U
	Val	Ala	Asp	Gly	C
	Val	Ala	Glu	Gly	A
	Val	Ala	Glu	Gly	G

Note: the red entries are found in mitochondria but not elsewhere. Boxed codons are used as start codons. The mitochondrial codes of other organisms exhibit further differences.

Table 12-3). The differences are truly striking in that most are in the initiation and termination codons. That is, in mitochondria

1. UGA codes for tryptophan and not for termination.
2. AGA and AGG are termination codons rather than codons for arginine.
3. AUA and AUU are initiation codons, as well as AUG. Both AUA and AUG also code for methionine. AUU also codes for isoleucine, as in the universal code.
4. AUA codes for methionine (and initiation, as shown in point 3) instead of isoleucine.

The number of mitochondrial tRNA molecules is 22 which is less than the minimum number (32) needed to translate the universal code. This is possible because in each of the fourfold redundant sets—for example, the four alanine codons, GCU, GCC, GCA, and GCG—only one tRNA molecule (rather than two, as explained earlier) is used. In each set of four tRNA molecules the base in the wobble position of the anticodon is U or a modified U (not I). It is not yet known whether this U is base-paired in the codon-anticodon interaction or the U manages to pair weakly with each of the four possible bases. For those codon sets that are doubly redundant—for example, the two histidine codons, CAU and CAC—the wobble base always forms a $G \cdot U$ pair, as in the universal code.

The structure of the human mitochondrial tRNA molecule is also different from that of the standard tRNA molecule (except for mitochrondrial $tRNA_{UU\_}^{Leu}$). The most notable differences are the following:

1. The universal sequence $CT\psi C\_A$ is lacking in mitochondrial tRNA.
2. The "constant" seven-base-pair sequence of the $T\psi C$ loop varies from three to nine bases.
3. The invariant bases U8, A14, G15, G18, G19, and U48 of the standard tRNA molecule are not invariant in mitochondrial tRNA.

In standard tRNA molecules, each of these bases is involved in forming bonds that produce the folded L-shaped molecule. Thus, the mitochondrial tRNA molecule seems to be stabilized by fewer interactions. The three-dimensional configurations of these molecules are not known with certainty; possibly they differ from the standard L-shape and mitochondrial tRNA engages in a different type of interaction with the ribosome than standard tRNA molecules do.

Examination of the base sequence of thirteen different reading frames in the DNA molecule indicates that in seven of these there are no chain termination codons. These reading frames are shown in panel (a) of Figure 12-34. Note that each of these genes but one is immediately adjacent to a tRNA gene. In each case, just before the tRNA gene or the

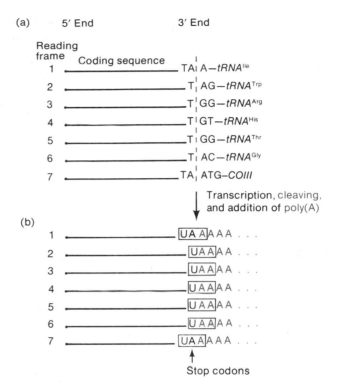

Figure 12-34
A model for termination of translation of mitochondrial mRNA.

COIII gene there is a thymine in the DNA in the first position of a codon in the reading frame. If the tRNA or *COIII* RNA segment were removed by processing enzymes and poly(A) were added, translation would run into the poly(A) tail, with the result that these proteins would end with a long tract of polylysine and termination would occur at the 3′ end of the poly(A) tail. The polylysine segments do not exist, so that a simple hypothesis shown in panel (b) of the figure has been proposed to explain termination. If the processing enzyme cuts the mRNA molecule at the site indicated by the vertical red dashed line and poly(A) is then added to the 3′ end, a UAA chain termination codon would be created, as shown. In two cases, a UA would appear to the left of the cleavage site and in the other case a U would appear there. This hypothesis is now being tested by sequencing of the processed mRNA molecule.

Most DNA molecules contain long noncoding segments (spacers) between genes. In mitochondrial DNA there are either a few bases or none. In each case there is a start codon (AUG, AUA, or AUU) at the 5′ end of the mRNA molecule or within a few bases of it. This raises the question of the nature of the ribosome recognition site, since mitochondrial mRNA contains no Shine-Dalgarno sequence or its equivalent. As has already been discussed for initiation in eukaryotes, the start codon that is recognized is always the one nearest the 5′ end of the mRNA molecule. This is also true of mitochondrial mRNA. However, other mRNA molecules usually commence with a short leader sequence that is thought to be responsible for binding to the ribosome. Such a leader is not present in the mitochondrial mRNA molecules. The most plausible explanation is that the 5′ terminus of a mitochondrial mRNA molecule is able to bind to the ribosome.

There is one final surprising point about translation of mitochondrial mRNA. We have just seen that initiation resembles the eukaryotic mode, and of course mitochondria are found only in eukaryotes. However, proteins are initiated with fMet, the prokaryotic initiator amino acid.

REFERENCES

Altman, S. (ed.) 1978. *Transfer RNA.* MIT Press.

Anderson, S., A. T. Bankier, B. G. Barrell, M. H. L. de Bruin, A. R. Coulson, J. Drouin, I. C. Operon, D. P. Nierlich, B. A. Roe, F. Sanger, P. H. Schreier, A. J. H. Smith, R. Staden, and I. G. Young. 1981. "Sequence and organization of the human mitochondrial genes." *Nature,* 290, 457–465.

Celis, J. E., and J. D. Smith. 1979. *Nonsense Mutations and tRNA Suppression.* Academic Press.

Clark, B. F. C., 1977. "Correlation of biological activities with structural features of transfer RNA." *Prog. Nucleic Acid Res. Mol. Biol.,* 20, 1–19.

Cold Spring Harbor Laboratory. 1966. *The Genetic Code.* Vol. 31, Symposium on Quantitative Biology.

Crick, F. H. C., L. Barnett, S. Brenner, and R. J. Watts-Tobin, 1961. "General nature of the genetic code for proteins." *Nature,* 192, 1227–1232.

Fersht, A. 1980. "Enzymatic editing mechanisms in protein synthesis and DNA replication." *Trends Biochem. Sci.,* 5, 262–265.

Goodman, H. M., J. Abelson, A. Landy, S. Brenner, and J. D. Smith. 1968. "Amber suppression: a nucleotide change in the anticodon of tyrosine transfer RNA." *Nature,* 217, 1019–1024.

Jukes, T. 1978. "The amino acid code." *Adv. Enzym.,* 47, 375–432.

Khorana, H. G. 1968. "Nucleic acid synthesis in the study of the genetic code." In *Nobel Lectures: Physiology or Medicine:* Vol. 4 (1973). American Elsevier.

Nirenberg, M. 1963. "The Genetic Code." *Scient. Amer.,* March. pp. 80–94.

Rich, A., and S. Kim. 1978. "The three-dimensional structure of transfer RNA." *Scient. Amer.,* January. pp. 52–62.

Schimmel, P., and A. G. Redfield. 1980. "Transfer RNA in solution: selected topics." *Ann. Rev. Biophys. Bioeng.,* 9, 181–221.

Schimmel, P. R., and D. Soll. 1979. "Aminoacyl-tRNA synthetases; general features and recognition of transfer RNA's." *Ann. Rev. Biochem.,* 48, 601–648.

Sherman, F. 1982. "Suppression in the yeast *Saccharomyces cerevisiae.*" In *The Molecular Biology of the Yeast Saccharomyces.* J. Strathern, E. Jones, and J. Broach, eds. Cold Spring Harbor Laboratory.

Soll, D., J. Abelson, and P. Schimmel (eds.) 1980. *Transfer RNA.* Volumes 1 and 2. Cold Spring Harbor Laboratory.

13 Translation

II. THE MACHINERY AND THE CHEMICAL NATURE OF PROTEIN SYNTHESIS

In the preceding chapter we described how the information in a mRNA molecule is converted to an amino acid sequence having specific start and stop points. We called this the informational problem. In this chapter we examine the chemical problem of attachment of the amino acids to one another. We begin with the structure of the ribosome—a complex nucleoprotein particle on which assembly occurs—and the physical role played by the ribosome in protein synthesis.

RIBOSOMES

A ribosome is a multicomponent structure that contains several of the enzymatic activities needed for protein synthesis and serves to bring together a single mRNA molecule and charged tRNA molecules in the proper position and orientation so that the base sequence of the mRNA molecule is translated into an amino acid sequence. The chemical composition of the ribosome of prokaryotes is well known, though there are still some questions about the composition of the ribosomes of eukaryotes. Both, however, share the same general structure and organization. Studies of the physical structure of the particles and the sites pertinent to protein synthesis are still incomplete, but are progressing. The properties of bacterial ribosomes are constant over a wide range of species. The E. coli ribosome is best understood and, though it structure differs slightly from that of eukaryotic ribosomes, it serves a useful model for discussion of all ribosomes; thus, we begir describing its chemical composition.

Chemical Composition of Prokaryotic Ribosomes

A prokaryotic ribosome is a nucleoprotein particle consisting of two subunits. The intact ribosome and the subunits are, for historical reasons, named by stating their sedimentation coefficients. Thus the intact particle is called a **70S ribosome** (because its s value is 70 svedbergs) and the subunits, which are unequal in size and composition, are termed **30S** and **50S particles.**

A 70S ribosome consists of one 30S particle and one 50S particle. In the laboratory the stability of the 70S ribosome is determined by the Mg^{2+} concentration of its environment. In 0.0005 M $MgCl_2$ the ribosome is almost completely dissociated into its subunits, whereas in 0.005 M $MgCl_2$ (roughly the concentration within bacteria) there is very little dissociation. Thus, the following equation describes the equilibrium:

$$\text{30S particle} + \text{50S particle} \xrightleftharpoons[\text{Low } [Mg^{2+}]]{\text{High } [Mg^{2+}]} \text{70S ribosome}$$

At physiological concentrations of the Mg^{2+} ion the 70S ribosome is the preferred form and

The 70S ribosome is the form active in protein synthesis.

The molecular weights of the E. coli particles are

70S ribosome:	2.7×10^6
50S particle:	1.8×10^6
30S particle:	0.9×10^6

These values are characteristic of the values of all prokaryotic ribosomes.

Both the 30S and the 50S particles can be further dissociated into RNA and protein molecules under appropriate conditions (Figure 13-1). The composition of each particle is the following:

30S particle:	one 16S rRNA molecule + 21 different proteins
50S particle:	one 5S rRNA molecule + one 23S rRNA molecule + 32 different proteins

The proteins of the 30S particle are termed S1, S2, . . . , S21 (S for "small particle"); there is one copy of each of these protein molecules in the 30S particle. The proteins of the 50S particle are denoted by a capital L (for large particle).

Each 50S particle contains one copy of each protein molecule except for proteins L7 and L12, of which there are four copies in total in a 50S particle, and L26, which is not a true component of the 50S subunit. The number of copies of L7 relative to those of L12 depends on the

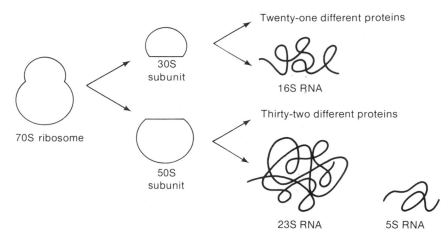

Figure 13-1
Dissociation of a prokaryotic ribosome. The con-
figuration of two over-
lapping circles will be
used throughout this
chapter, for the sake of
simplicity. The correct
configuration is shown in
Figure 13-4.

bacterial growth rate. These protein units are distinct molecules with
two exceptions. Protein L7 is a terminal-N-acetylated form of L12 and
the protein designated L8 is a complex consisting of one molecule of L10
to which L7 and L12 are bound. This complex results from a functional
association that probably exists in the ribosome; sometimes the complex does not dissociate during isolation. Furthermore, a protein
designated L26 (of which only one molecule per five 50S particles is
observed) is actually protein S20 of the 30S particle; occasionally, when
a 70S particle is dissociated, S20 leaves the 30S particle and becomes
associated with a 50S particle. The total number of chemically distinct
protein molecules in a 70S ribosome is 21 + 32 + 2 (the extra L7 and
L12) = 55. Most of these are very basic proteins containing up to 34
percent basic amino acids (arginine + lysine). This basicity probably
accounts in part for their strong association with the RNA, which is
acidic. The amino acid sequences of most of the ribosomal proteins are
known.

The sizes of the RNA components are the following:

5S rRNA:	120 nucleotides
16S rRNA:	1541 nucleotides
23S rRNA:	2904 nucleotides

The base sequence of each of these molecules is known. Roughly 70
percent of the bases in each RNA molecule are internally paired to form
a highly complex structure with numerous stems and loops. The
three-dimensional structures of these molecules have not yet been
elucidated.

A summary of the structure of the *E. coli* 70S ribosome is shown in
Figure 13-1.

Chemical Composition of Eukaryotic Ribosomes

Although the basic features of eukaryotic ribosomes are similar to those of bacterial ribosomes, all eukaryotic ribosomes are somewhat larger than those of prokaryotes. They contain a greater number of proteins (about 80) and an additional RNA molecule. The biological significance of the differences between prokaryotic and eukaryotic ribosomes is unknown.

A typical eukaryotic ribosome has an s value of about 80 and consists of two subunits, a **40S** and a **60S particle.** These sizes may vary by as much as ± 10 percent from one organism to the next, in contrast with bacterial ribosomes, whose sizes are nearly the same for all bacterial species examined. The best-studied eukaryotic ribosome is that of the rat liver. The components of the subunits of eukaryotic ribosomes are:

40S particle:	one 18S rRNA molecule + about 30 proteins
60S subunit:	one 5S, one 5.8S, and one 28S rRNA molecule + about 50 proteins

The 5.8S, 18S, and 28S rRNA molecules of eukaryotes correspond functionally to the 5S, 16S, and 23S molecules of bacterial ribosomes. The bacterial counterpart to the eukaryotic 5S rRNA is very likely present as part of the 23S rRNA.

The molecular weights of the 80S ribosome of the rat liver and its 40S and 60S subunits are:

80S ribosome:	4.3×10^6
60S particle:	2.8×10^6
40S particle:	1.4×10^6

The precise number of proteins found in the eukaryotic ribosome is uncertain at present because several of the proteins are not always found in purified subunits; thus, it is unclear whether they are ribosomal proteins that have been accidentally removed during purification or nonribosomal proteins that sometimes contaminate the samples.*

Eukaryotic ribosomal proteins are highly basic, like the E. coli ribosomal proteins. Study of eukaryotic ribosomal proteins has been hampered by their insolubility in ordinary buffers and by the difficulty in obtaining an adequate supply of material. The sequencing of the individual proteins and their characterization by physicochemical methods require about two grams of total ribosomal protein. This can be obtained from a few hundred liters of bacteria, an amount which can

---

*The problem of ascertaining that components associated after isolation are also associated *in vivo* is ubiquitous in all biology. Usually a great many additional experiments are needed before the question can be answered.

easily be grown and purified in two days. However, only about 3 mg of ribosomal protein can be obtained from an adult rat liver, so nearly 1000 rats are needed for only a few experiments. As we will see throughout this book, lack of availability of material is a common problem in studying the molecular biology of eukaryotes. However, recombinant DNA techniques (Chapter 20) such as cloning should eventually overcome problems of yield.

Comparison of *E. coli* and Eukaryotic Ribosomes

A comparison of the properties of *E. coli* and eukaryotic ribosomes is given in Table 13-1. We have already stated that there are more components in eukaryotic ribosomes (both RNA and proteins) than in

Table 13-1
Characteristics of the ribosomes of prokaryotes and eukaryotes (e.g., rat liver)

Property	Prokaryote*	Eukaryote (rat liver)†
Small subunit		
s value (total)	30S	36.9S (termed the 40S subunit)
Molecular weight	0.9×10^6	1.44×10^6
s value (RNA)	16S	18S
Molecular weight (RNA)	0.51×10^6	0.7×10^6
Proteins	21	≈ 30
Range of molecular weight	8300–25,800	11,200–41,500
Total molecular weight	0.39×10^6	0.74×10^6
Large subunit		
s value	50S	56.3S (termed the 60S subunit)
Molecular weight (total)	1.8×10^6	2.8×10^6
s value and molecular weights of corresponding RNA molecules	5S: 40,000 23S: 0.98×10^6	5S: 39,000 5.8S: 51,000 28S: 1.7×10^6
Total molecular weight of RNA	1.02×10^6	1.79×10^6
Proteins	32	≈ 50
Range of molecular weight	5300–24,600	11,500–41,800
Total molecular weight	0.78×10^6	1.05×10^6

*The values given are for *E. coli;* they vary very little from one prokaryote to the next.
†The values given may vary from one eukaryote to the next and may be as much as 15 percent greater.

prokaryotic ribosomes; it should be noted also that in general the eukaryotic components have a higher molecular weight.

The difference in size of eukaryotic and prokaryotic ribosomes has given rise to a great deal of speculation. Two points of view have been taken—namely, to explain why eukaryotic ribosomes are larger or, on the other hand, why prokaryotic ribosomes are smaller.

In the first view, one attempts to account for the need of additional components in terms of additional functions. For example, since eukaryotic mRNA lacks a Shine-Dalgarno sequence (see Chapter 12), initiation is presumed to be more complex in eukaryotes; we shall see shortly that indeed it is, and more proteins are needed, though the known extra proteins are not ribosomal proteins, but soluble factors. Also, it will be seen in Chapters 14 and 22 that regulation of synthesis of particular proteins in eukaryotes is frequently accomplished by regulation of translation, and this requires additional proteins. Later in this chapter we will also see that eukaryotic ribosomes must be capable of binding to cytoplasmic membranes, and this may require specific ribosomal components.

In the second point of view, one assumes that originally all ribosomes were large and that there was a greater selective pressure for bacteria to evolve a more efficient ribosome because (1) during rapid growth 30 percent of all cellular protein is ribosomal, and (2) it would be advantageous to combine the functions of several different proteins in a single molecule. A decision about the validity of either viewpoint must await greater understanding of the function of each component.

Reconstitution of *E. coli* Ribosomes

A major breakthrough in understanding bacterial ribosomal structure came about when Masayasu Nomura and his colleagues succeeded in reconstituting the *E. coli* ribosome from a mixture of purified ribosomal RNA and protein molecules. By using this technique it is possible to determine (1) whether a component is necessary in order to have a functional ribosome and (2) to identify the function of each component. The most important initial finding was that assembly of a 30S particle requires nothing other than the 16S RNA molecule and the 21 type-S proteins, and that a reconstituted particle has all the biological functions of the natural particles. Thus, no nonribosomal components (for example, enzymes or special factors) are needed to form the intact particle *in vitro*. The same observation was made for the 50S particle. Thus, the formation of a ribosome *in vivo* is a self-assembly process.

The RNA-protein interaction is a highly specific one. For instance, mRNA plus ribosomal proteins will not interact to form a compact particle (though a few proteins do bind weakly).

The relatedness of the ribosomes of prokaryotes is shown by the fact that the 16S rRNA of E. coli can be substituted by 16S rRNA of other bacteria in a reconstitution experiment. However, reconstitution will not occur if the 18S rRNA of a eukaryote (e.g., yeast) is mixed with the proteins of a bacterial 30S particle, or if bacterial 16S rRNA and eukaryotic proteins are mixed. This is one of many known examples of a molecular difference between prokaryotes and eukaryotes.

Reconstitution experiments have been carried out with mixtures that lack one protein at a time in order to determine whether all proteins are essential and to elucidate the sequence of events during assembly. In these experiments it has been found that sometimes an apparently intact particle forms, but the particle is unable either to combine with the complementary particle to form a 70S ribosome or to perform certain steps in protein synthesis. When mixtures of various combinations of proteins are studied, it is observed that there are stable intermediates. For example, there is a stable structure containing 16S rRNA and five proteins (S4, S8, S15, S18, and S20). Additional proteins added singly do not bind to this structure. However, a mixture of S5, S7, S12,

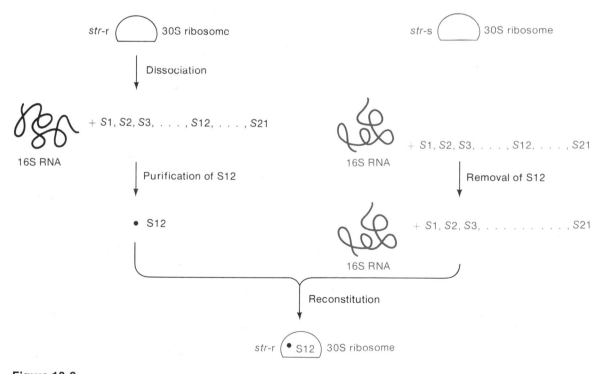

Figure 13-2

A reconstitution experiment showing that protein S12 is the determinant for resistance to streptomycin. Only the 30S particle is shown; to form a functional ribosome, the 50S particle is added after reconstitution.

S16, and S18 cause a transition to a second stable form in which all eleven proteins are bound.

Reconstitution experiments can also give information about the functions of individual proteins. However, the analysis of such experiments is complicated by the fact that often there is not a one-to-one correspondence between a particular protein and a particular function—that is, several proteins may be required for a particular function, or a protein may participate in two different functions. A particularly useful procedure is the reconstitution of mixtures obtained from wild-type and mutant proteins. For instance, the antibiotic streptomycin is an inhibitor of protein synthesis. Bacterial mutants that are streptomycin-resistant (Str-r) have been isolated and with these mutants protein synthesis is not inhibited by streptomycin *in vitro* or *in vivo*. If a mixture is prepared consisting of wild-type 16S rRNA and all of the wild-type S-proteins with the substitution of S12 from a streptomycin-sensitive (Str-s) bacterium for the S12 from Str-r bacterium, the reconstituted ribosomes are not inhibited by streptomycin (Figure 13-2). Thus, protein S12 determines the response to the drug. One might be tempted to conclude that S12 is the binding site for streptomycin. However, there are other possible interpretations of the data. For example, S12 might be responsible for arranging two other proteins to form a streptomycin binding site and a mutant S12 would hence prevent formation of the site. Another possibility is that S12 is adjacent to the binding site and that a mutant S12 sterically hinders binding.

The technique of reconstitution with mutant proteins has been used to examine a variety of drug-resistance mutations. It has been quite useful because if the particular protein molecule can be shown to bind the antibiotic, and if the step in protein synthesis inhibited by the antibiotic can be determined, this information can be used to identify one of the functions of that protein molecule in protein synthesis.

Physical Structure of Prokaryotic Ribosomes

Figure 13-3 shows an electron micrograph of a field of *E. coli* 70S ribosomes, and its two kinds of subunits. In any electron micrograph, individual particles are viewed from a variety of angles depending on how the particle has been deposited on the surface of the sample holder. By studying the features of many 30S, 50S, and 70S particles in many orientations, it is possible to reconstruct a three-dimensional model of the ribosome. Several models have been proposed, but the asymmetric subunit model shown in Figure 13-4 is generally agreed to fit the data on protein distributions best, especially the results from neutron diffraction described in the following section.

Several methods have been used to locate the various protein molecules with respect to one another and with respect to the RNA molecules. Some of these procedures are the following:

1. *Identification of proteins that bind to particular rRNA molecules.* Many of the proteins that bind to rRNA molecules have been identified by affinity chromatography. In this technique a purified sample of a particular rRNA molecule (for example, 16S rRNA) is chemically coupled via the 3′ terminus to small solid beads of Sepharose, a commercially available polysaccharide polymer commonly used in chromatography. A hollow tube (a chromatographic column) is

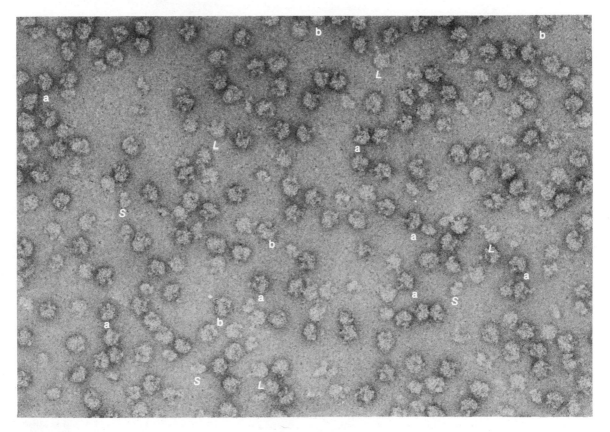

Figure 13-3
A field of ribosomes from *E. coli*. Some of the ribosomes are oriented as in the models shown in Figure 13-4. The letters a and b refer to the orientations in panels (a) and (b) respectively of Figure 13-4. A few ribosomal subunits that also lie in the field are identified by the letters *S* and *L*, which stand for small and large. The field has been negatively stained for electron microscopy; each ribosome is defined by an outline of the salt of a heavy metal, which in this case is uranium. The electron micrograph is adapted with the permission of the author from James A. Lake, "The Ribosome," in *Scientific American,* August 1981, pp. 84–97.

(a)

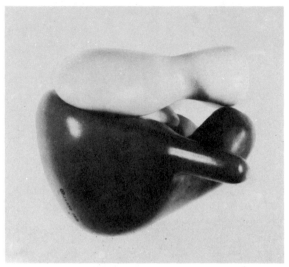

(b)

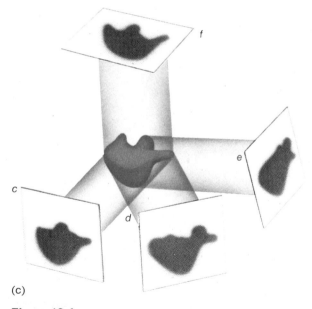

(c)

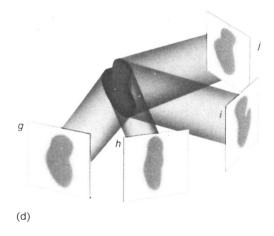

(d)

Figure 13-4

Three-dimensional model of the *E. coli* ribosomes showing the relative orientations of the subunits (the small subunit is light, the large subunit is dark) in two views. Ribosomes from plants, animals, and other bacteria all have a similar design, though they differ in some details. The model was deduced for the ribosome by finding distinctly different images of the large subunit (c) and the small subunit (d) in fields of subunits that had been negatively stained for electron microscopy. The letters *c, d, e,* and *f* identify images of the large subunit; *g, h, i,* and *j* identify images of the small subunit. Each image is a two-dimensional projection of a subunit that results from directing a beam of electrons through a three-dimensional structure and the uranium salt that surrounds it. Electron micrographs do not show the surface of a ribosome; thus, the three-dimensional shape of the ribosome is inferred from the multiple projections. (Reprinted with permission of the author from James A. Lake, "The Ribosome," in *Scientific American,* August 1981, pp. 84–97.

filled with these RNA-containing beads and a mixture of ribosomal proteins is passed through the column. Proteins that bind to the rRNA molecule stick to the beads, whereas those that fail to bind pass through the column. The bound proteins can be removed ("eluted") from the column by washing the column with a concentrated salt solution. The proteins can then be identified by the standard gel electrophoretic methods used to distinguish various ribosomal proteins. The number of bound proteins detected in this way is a minimum value because some proteins that are bound to RNA in a ribosome may fail to bind to the RNA when this is coupled to the Sepharose. Such failure occurs because, during the coupling process, the RNA molecules are often denatured and, though some base pairs certainly re-form, the overall structure of the coupled molecule may not be the same as the native molecule; hence, some binding sites may not be present in the coupled RNA. Moreover, some binding sites may be composed of portions of two different rRNA molecules and would necessarily be absent in a column prepared with a single type of RNA.

In the following procedure the proteins are tested one by one for binding.

2. *Protection of particular regions of RNA by specific proteins.* A particular type of rRNA molecule is mixed with a single purified ribosomal protein and then the mixture is treated with RNase. If the protein fails to bind, the rRNA is totally digested by the enzyme. If there is binding, all of the RNA except for the region bound to the protein is digested. This method not only indicates that binding occurs, but also, since the rRNA molecules have been totally sequenced, it is possible to identify the base sequence to which binding occurs. (This is the same procedure that was encountered in determining the sequences of promoters in DNA and ribosome binding sites in mRNA.) By using this procedure the following results have been obtained:

RNA	Bound proteins
5S	L5, L18, L25
16S	S4, S7, S8, S15, S17, S20
23S	L1, L2, L3, L4, L6, L7, L10, L11, L13, L14, L16, L20, L23, L24

Since the amino acid sequences of most of these proteins are known, it is also possible to identify the region of the protein that is bound to the RNA. This is done by carrying out chemical reactions that attack specific amino acids. The amino acids that are resistant to attack when the protein is bound to the RNA are considered to be in the contact site.

3. *Linkage of proteins by cross-linking reagents.* Certain chemicals cause a covalently linked bridge **(cross-link)** to form between two ε-amino groups of lysine. If a ribosomal subunit (30S or 50S) or an intact ribosome is treated with these reagents, and the particle is then dissociated, pairs of covalently linked neighboring proteins form, which can be separated from one another by gel electrophoresis. Each pair is then exposed to a reagent that cleaves the cross-link and the component proteins are identified by gel electrophoresis—that is, by comparison of the electophoretic diagram with that obtained with purified proteins (Figure 13-5). It is important to realize that the production of a cross-link between two proteins does not mean that the protein molecules are in physical contact, because the cross-link itself has length and need only span the distance between the two molecules. The cross-linking agent most commonly used (2-iminothiolane) produces a linker that is 15 Å long. Additional information about the distance between the proteins can be obtained by use of reagents producing shorter cross-links. For example, if two protein molecules are joined by a 15 Å linker, but not by a 10 Å linker, the molecules must be separated by a distance between 10 and 15 Å.

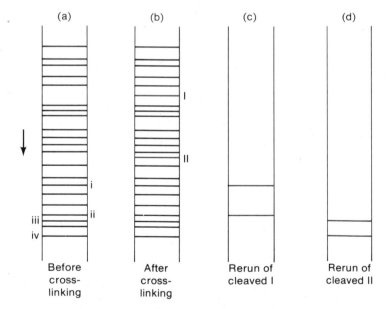

Figure 13-5
The principle of the cross-linking technique to identify proteins of the 30S ribosome that are in contact. Panel (a) shows the result of gel electrophoresis of the 21 proteins of a 30S ribosome. Each band represents a particular protein. Panel (b) shows a hypothetical distribution obtained with proteins isolated from a 30S ribosome that had been exposed to a cross-linking agent. Two new bands, I and II, are seen. If the material in band I is isolated and the cross-link is cleaved, gel electrophoresis shows two bands (panel (c)) whose positions match those of band i and ii of panel (a); thus, I = i + ii. A similar experiment with band II is shown in panel (d); here, II = iii + iv.

4. *Localization of proteins on the ribosome surface by immuno-electron microscopy.* Many of the proteins on the surface of each particle have been localized by immuno-electron microscopy. In these experiments an antibody is first prepared by injecting a purified ribosomal protein into a rabbit. The antibody, which is a large protein, reacts only with the specific protein that elicited its production, and an antibody can be prepared for each protein. In these experiments an antibody directed against a single ribosomal protein is mixed with either a 30S or 50S particle and the sample is then viewed by electron microscopy. Each antibody molecule has two identical binding sites (it is bivalent), as was shown in Figure 5-16 (Chapter 5). Thus, the antibody binds to its target protein on the surface of two identical ribosomal subunits, thereby forming a dimer (Figure 13-6). The point of contact of the two monomers is the same for both monomers and is the location of the particular protein. Since the orientation of the 30S ribosome can be determined (as in Figure 13-4), the position of the particular protein can be localized by looking at particles in various orientations. These procedures have been repeated with antibodies directed against many ribosomal proteins. Figure 13-7 shows the location of ribosomal proteins on the 30S and 50S particles determined by this procedure.

5. *Localization by neutron scattering.* Neutron scattering is a complex technique whose methodology and interpretation is beyond the scope of this book. Like x-ray diffraction, it can be used to localize individual atoms, though at present it cannot in practice yield an absolute structure. Its application to the study of ribosomes is based on

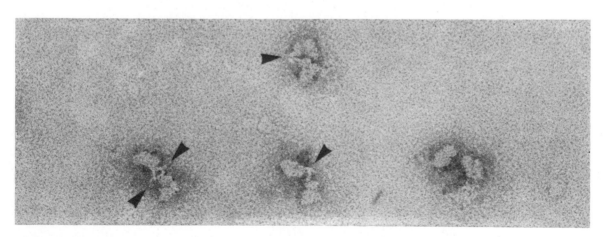

Figure 13-6
A field of 30S ribosomal subunits reacted with antibodies against protein S14. The IgG antibodies are indicated by arrows. Protein S14 is exposed, so it is possible for two IgG molecules to bind simultaneously to the ribo-somal subunit. (From Lake, Prendergast, Kahan, and Nomura. 1974. *Proc. Natl. Acad. Sci. USA,* 71: 4688–4692.)

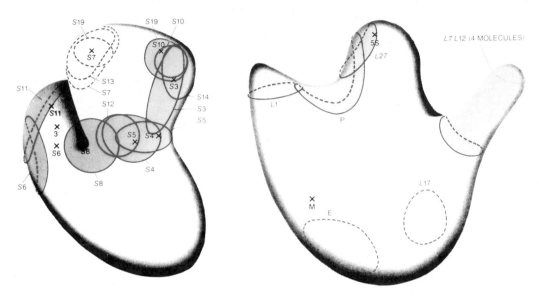

Figure 13-7
Map of proteins in the ribosome shows their location in the small subunit (left) and the large subunit (right). Ribosomal proteins are indicated by a prefix, *S* or *L*, indicating a small or a large subunit protein and a number referring to each protein. For example, *S*3 is a small subunit protein and *L*27 is a large subunit protein. The patches were mapped by immunoelectron microscopy; they are the sites on the surface of the ribosome where antibodies bind to the protein against which they are directed. Each protein makes one such appearance at the surface of a subunit, except for protein *S*19, which evidently makes two appearances, and protein *S*4, which is under *S*5 and *S*12. The crosses indicate locations mapped by neutron diffraction. Five additional sites are shown. They are M, the site at which the ribosome can be anchored to the intracellular membrane; 5S, the location of an RNA that forms part of the large subunit; 3′, an end of the ribosomal RNA denoted 16S rRNA; P, the site at which the ribosome joins successive amino acids to make a protein chain; and E, the site where the newly synthesized protein chain emerges from the ribosome. (Reprinted with permission of the author from James A. Lake, "The Ribosome," in *Scientific American,* August 1981, pp. 84–97.)

the fact that ^{1}H and ^{2}H nuclei scatter neutrons in very different ways. In practice, deuterated ribosomal proteins are prepared either by growth of bacteria in deuterated growth medium or by prolonged incubation of purified ribosomes in $[^2\text{H}]_2\text{O}$ (in which ^{1}H-^{2}H exchange occurs). Individual deuterated proteins are then isolated and purified. A ribosomal subunit is then reconstituted by using two different deuterated proteins—for example, $S2$ and $S6$—all the other being nondeuterated S proteins. The reconstituted particle is then subjected to neutron scattering. Analysis of the pattern of scattering yields the distance between the deuterated protein molecules. By using many different deuterated pairs, the three-dimensional protein map shown superimposed on Figure 13-7 has been derived.

The interface between the 30S and 50S particles is of special interest because, as we will see shortly, the particles associate at one step of protein synthesis and then dissociate at a later step. This interface has been studied by several procedures. If the cross-linking technique is applied to the 70S ribosome, some (S protein)-(L protein) pairs form, which identifies the proteins in the two particles that contact one another. Interparticle contact between proteins and RNA molecules has also been examined. For example, proteins such as S11 and S12 (from the 30S particle) protect 23S RNA (from the 50S particle) from digestion by RNase. Observing the interference of aggregation by antiprotein antibodies is also a useful method. Thus, if a particular S protein is in the interface region, an antibody directed against it will prevent formation of the 70S complex. This method has also been used with 50S particles and antibodies directed against L proteins.

Physical Structure of Eukaryotic Ribosomes

The techniques just described have also been used to study the structure of eukaryotic ribosomes. There are no major differences between the structures of prokaryotic and eukaryotic ribosomes apart from the larger number of proteins in eukaryotic ribosomes, of which some contain phosphorylated serine residues. In fact the proteins in eukaryotic ribosomes resemble those of prokaryotes quite closely; in several cases there is such close correspondence in the amino acid sequences that an antibody directed against a particular prokaryotic protein can interact with the analogous eukaryotic protein. A noteworthy example is the E. coli protein pair L12 and its N-acetylated derivative L7, which closely resemble the rat liver 60S-subunit proteins L40 and L41. These proteins have weak, but nonetheless constant sequence homologies. The yeast L40-L41 pair can also substitute for L7 and L12 in reconstituted E. coli ribosomes.*

The interpretation of these and other similarities is that the properties of the ribosomal proteins that are most important in ribosome function have been conserved throughout evolution. The similarities also suggest that prokaryotic and eukaryotic ribosomes did not arise independently, but that either one evolved from the other or both evolved from a common ancestor.

Chloroplasts and **mitochondria** (eukaryotic intracellular organelles) also possess ribosomes. These particles are similar to, but distinct from, the major eukaryotic type of ribosome and more closely resemble the bacterial type. For instance, in reconstitution experiments, the RNA

---

*Earlier we noted that a eukaryotic ribosomal protein cannot interact with a bacterial rRNA molecule. This exchange of yeast and E. coli proteins does not contradict this statement because neither the L7 nor the L12 proteins bind to RNA. Although the rat liver proteins L40 and L41 can substitute for E. coli L7 and L12 in E. coli ribosomes, the resultant activity is not as great as when the yeast pair is used. Thus the L7-L12 and L40-L41 pairs share not only common structural features, but also common functions.

Table 13-2
Sedimentation coefficients of the ribosomal subunits and the ribosomes of the mitochondria of various eukaryotic organisms

Organism	Sedimentation coefficient		
	Small subunit	Large subunit	Ribosome
Ascomycetes (e.g., fungi)	30–40	50	70–75
Ciliate protozoa (e.g., *Paramecium, Tetrahymena*)	35	55	80
Higher plants	40	60	77–80
Animals	30–35	40–45	55–66

molecules of these organelles cannot be interchanged either with each other or with the RNA of the cellular ribosomes.

Both the size and chemical composition of chloroplast ribosomes are very much like those of bacteria, a fact which is thought to have some evolutionary significance. The structure of mitochondrial ribosomes varies enormously from one organism to the next, as can be seen in the s values (and hence the size) of the subunits and the intact particles, as shown in Table 13-2. Possibly, these differences are related to the differences in the mitochondrial genetic code, as described in Chapter 12.

PROTEIN SYNTHESIS

Protein synthesis can be divided into three stages—(1) **polypeptide chain initiation,** (2) **chain elongation,** and (3) **chain termination.** The main features of the initiation step are binding of mRNA to the ribosome, selection of the initiation codon, and binding of the charged tRNA bearing the first amino acid (namely, fMet). In the elongation stage there are two processes—joining together two amino acids by peptide bond formation and moving the mRNA and the ribosome with respect to one another so that each codon can be translated successively. This movement is called **translocation.** In the termination stage the completed protein is dissociated from the synthetic machinery and the ribosomes are released to begin another cycle of synthesis. These three stages will be described separately, first in an overview and then in some detail. However, before entering this complex subject two simpler features of protein synthesis will be described—namely, the direction of reading of mRNA and the direction of polypeptide chain growth.

Direction of Polypeptide Chain Growth

Every protein molecule has an amino terminus and a carboxyl terminus. Synthesis begins at the amino terminus; in a protein having the sequence NH_2-fMet-Ser-Ala-Glu-Ala-COOH, the fMet is the starting amino acid and Ala is the last amino acid added to the chain. The growth of the polypeptide chain from the amino terminus to the carboxyl terminus was first demonstrated in rabbit reticulocytes involved in the synthesis of hemoglobin and has since been confirmed for all organisms studied—bacteria, yeast, protozoans, plants, and animals.

Direction of Translation of mRNA

Translation of mRNA molecules occurs in the $5'\rightarrow3'$ direction. The meaning of this statement is made clear by examining the original experiment that proved the point. In this experiment (Figure 13-8) the synthetic polynucleotide

$$\text{P-5'-AAAAAA- . . . -AAC-3'-OH}$$

was used as a mRNA in an *in vitro* protein-synthesizing system not requiring an AUG start codon. The codons AAA and AAC correspond to lysine and asparagine respectively. The protein that is translated from this mRNA is

$$\text{H}_2\text{N-Lys-Lys- . . . -Asn-COOH}$$

Since the protein is synthesized from the amino terminus to the carboxyl terminus, the direction of mRNA translation is $5'\rightarrow3'$. If the direction had been $3'\rightarrow5'$, the protein would have been

$$\text{H}_2\text{N-Gln-Lys- . . . -Lys-COOH}$$

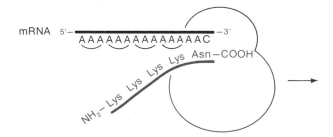

Figure 13-8
A ribosome translating a synthetic mRNA molecule, showing the direction of movement of the ribosome and the direction of polypeptide chain growth.

An Overview of Protein Synthesis

Protein synthesis in bacteria begins by the association of one 30S particle (not the 70S ribosome), a mRNA, a charged tRNA^{fMet}, three proteins known as **initiation factors,** and guanosine 5′-triphosphate (GTP). These molecules comprise the **30S preinitiation complex** (Figure 13-9). Following formation of the 30S preinitiation complex, a 50S subunit joins to the 30S particle to form a 70S initiation complex (Figure 13-9). This joining process requires hydrolysis of the GTP contained in the 30S preinitiation complex. There are two tRNA binding sites on the 50S subunit. (There is no definite proof of this statement, but all experimental evidence is compatible with the idea.) These sites are called the **aminoacyl** or **A site** and the **peptidyl** or **P site;** each site consists of a collection of segments of *L* proteins and 23S rRNA. The 50S subunit is positioned in the 70S initiation complex so that the tRNA^{fMet}, which was previously bound to the 30S preinitiation complex, occupies the P site of the 50S subunit. Positioning tRNA^{fMet} in the P site fixes the position of the anticodon of tRNA^{fMet} so as to allow it to pair with the initiator codon in the mRNA. Thus, *the reading frame is unambiguously defined upon completion of the 70S initiation complex.**

The A site of the 70S initiation complex is available to any tRNA molecule whose anticodon can pair with the codon adjacent to the initiation codon. However, entry to the A site by the tRNA requires a helper protein called an **elongation factor, EF** (specifically **EF-Tu**) (Figure 13-10). After occupation of the A site a peptide bond between fMet and the adjacent amino acid can be formed. Once it was thought that the blockage of the NH_2 group of fMet by the formyl group was responsible for peptide bond formation between the COOH group of fMet and the NH_2 group of the adjacent amino acid. However, in eukaryotic systems the starting amino acid is Met and not fMet and protein synthesis proceeds in the correct direction. Presumably, the relative orientation of the two amino acids in the A and P sites determines the linkage that is made.

The peptide bond is formed by an unusual enzyme complex called **peptidyl transferase.** The active site of peptidyl transferase consists of portions of several proteins of the 50S subunit. As the peptide bond is formed, the fMet is cleaved from the tRNA^{fMet} in the P site (Figure 13-11).

After the peptide bond forms, an uncharged tRNA occupies the P site and a dipeptidyl-tRNA is in the A site. At this point three

---

*If the 30S preinitiation complex and the 70S initiation complex formed only with tRNA^{fMet}, the Nirenberg technique for the determination of codon assignment would not have worked. However, at Mg^{2+} concentrations above the physiological intracellular value the system becomes less fastidious. Under these conditions the ribosome will bind to any base sequence on the mRNA and, for initiation, the P site can be occupied by any tRNA whose anticodon can pair with the mRNA base sequence that is present at the P site.

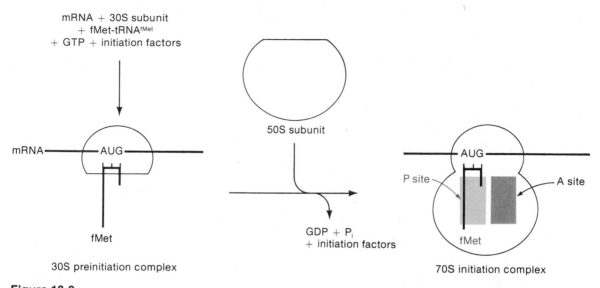

Figure 13-9
Early steps in protein synthesis: in prokaryotes: formation of the 30S preinitiation complex and of the 70S initiation complex. Details are shown in Figure 13-14.

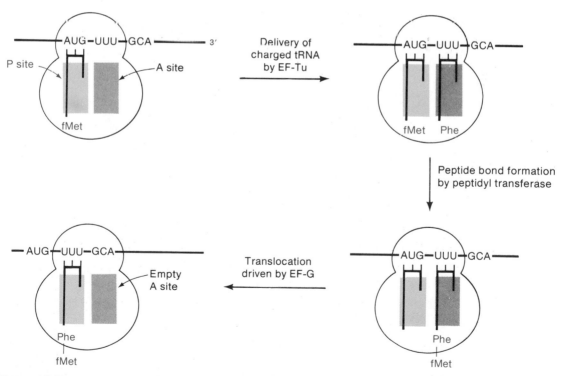

Figure 13-10
Elongation phase of protein synthesis: binding of charged tRNA, peptide bond formation, and translocation.

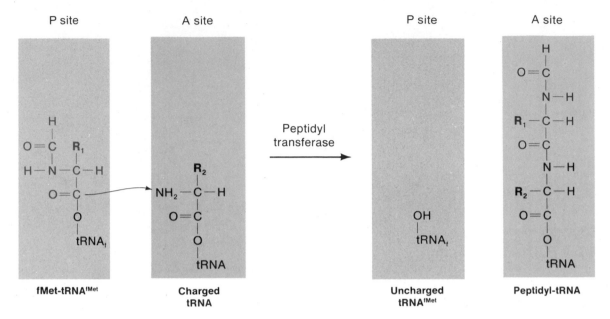

Figure 13-11
Peptide bond formation.

movements, which together comprise the translocation step, occur: (1) the deacylated tRNA^{fMet} leaves the P site, (2) the peptidyl-tRNA moves from the A site to the P site, and (3) the mRNA moves a distance of three bases in order to position the next codon at the A site (Figure 13-10). The translocation step requires the presence of another elongation protein **EF-G** and hydrolysis of GTP. The movement of the mRNA by three bases is probably dependent on the movement of the tRNA from the A site to the P site and, in fact, it is likely that mRNA translocation is a consequence of the tRNA motion. Supporting this view is the fact that the distance that mRNA moves is not fixed, whereas the translocation distance is normally three bases; if a glycyl frameshift suppressor having an extra base in the anticodon loop is present, the displacement in the translocation step is four bases instead of three.

After translocation has occurred, the A site is again available to accept a charged tRNA molecule having a correct anticodon. If a tRNA^{fMet} molecule, whose anticodon is the same as that of tRNA^{Met} molecule, were to enter the A site (because an internal AUG site were present), protein synthesis would stop because a peptide bond cannot form with the blocked NH$_2$ group of fMet. However, inasmuch as the factor EF-Tu is needed to facilitate tRNA entry into the A site, this misadventure is prevented, since EF-Tu cannot bind to tRNA^{fMet}.

When a chain termination codon is reached, there is no aminoacyl tRNA that can fill the A site and chain elongation stops. However, the polypeptide chain is still attached to the tRNA occupying the P site.

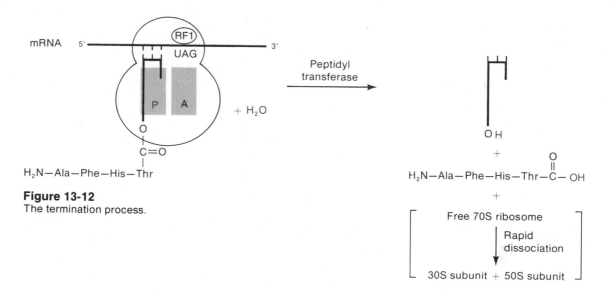

Figure 13-12
The termination process.

Release of the protein is accomplished by release factors (RF), proteins that in part respond to chain termination codons. There are two such release factors in *E. coli*—RF1, which recognizes the UAA and UAG codons, and RF2, which recognizes UAA and UGA.* Why the number of release factors is not one (i.e., useful for all codons) or three (one for each stop codon) is not known. Each release factor forms an activated complex with GTP; this complex binds to a termination codon and alters the specificity of peptidyl transferase. In the presence of release factors peptidyl transferase catalyzes the reaction of the bound peptidyl moiety with water rather than with the free aminoacyl-tRNA (Figure 13-12). Thus, the polypeptide chain, which has been held in the ribosome solely by the interaction with the tRNA in the P site is released from the ribosome. The 70S ribosome dissociates into 30S and 50S subunits and the system is ready to start synthesis of a second chain.

In the following three sections details of initiation, translocation and termination are described. The reader for whom the overview is sufficient should proceed to the section entitled "The Role of GTP."

Initiation in Prokaryotes

Initiation of protein synthesis consists of several stages. The first stage, formation of a preinitiation complex, requires the cooperative interaction of three nonribosomal proteins called **initiation factors.** In bacteria these factors are designated IF1, IF2, and IF3. They are not required

---

*The symbol RF1 for release factor should not be confused with RFI (replication form I), a term often used to designate a supercoiled DNA molecule present early in the replication cycle of some phages.

when using synthetic polynucleotides as mRNA at a high Mg^{2+} concentration, but are essential with natural mRNA molecules and at physiological Mg^{2+} concentrations.

At the intracellular concentration of Mg^{2+}, the 70S ribosome is rarely dissociated, yet formation of the preinitiation complex requires a free 30S particle. This particle is supplied by IF1 and IF3, whose primary function is to shift the equilibrium toward dissociation. IF1 does not itself cause dissociation, but increases the rate of dissociation and association (Figure 13-13), which increases the period of exposure of an IF3 binding site on the 30S ribosome and in the region that contacts the 50S particle. Thus, whenever a 70S ribosome becomes dissociated, an IF3 may bind to this site tightly, thereby preventing reassociation. IF2 can then join this complex to form a unit now consisting of one 30S particle, IF1, IF2, and IF3, and called the preinitiation complex (Figure 13-13). Cross-linking studies indicate that IF3 binds very close to the 3'-OH terminus of 16S rRNA, and there is also evidence that IF3 may have the ability to denature base-paired regions—in particular, the 3'-OH terminus of 16S rRNA. The denatured 3'-OH terminus of the 16S rRNA also includes the Shine-Dalgarno sequence needed to bind mRNA (Chapter 12). At this point the exact sequence of events is unclear. However, IF2 mediates binding of one molecule of GTP and charged $tRNA^{fMet}$. This addition only occurs if a formyl group is present, which is shown by two experimental results: (1) binding does not occur if $tRNA^{fMet}$ is acylated with Met, but does occur once the Met is formylated, and (2) misacylated $tRNA^{fMet}$ binds if the amino acid is formylated (for example, formylvalyl-$tRNA^{fMet}$ will bind). IF3 then leaves the complex when the $tRNA^{fMet}$ binds. It is not known whether mRNA joins the complex before or after binding of the $tRNA^{fMet}$ and it is thought that binding can occur in either order. Binding between the mRNA and the Shine-Dalgarno sequence of the 16S rRNA is aided in some way by the ribosomal proteins S1 and S12.

The loss of IF3 during binding of charged $tRNA^{fMet}$ frees the site on the 30S particle that binds to the 50S particle. Thus, the 50S particle binds to the complex and, in so doing, causes IF1 to fall off of the 30S particle (Figure 13-14). The P and A sites then form by a conjunction of S and L proteins. The P site is occupied by $tRNA^{fMet}$, but IF2 occupies the A site. IF2 must vacate the A site if a second charged tRNA molecule is to enter that site. In a way that is not understood, a GTPase is activated in the A site and the GTP bound to IF2 is hydrolyzed to yield GDP and P_i; IF2 cannot remain bound without the GTP, so the IF2 leaves the A site; this departure is aided by IF1 in an unknown way. The positioning of the fMet in the P site is determined by interactions of fMet and $tRNA^{fMet}$ mediated by specific ribosomal proteins. For instance, certain cross-linking agents produce linkages between fMet and the proteins L2 and L27.

The ribosome is now ready for the elongation stage.

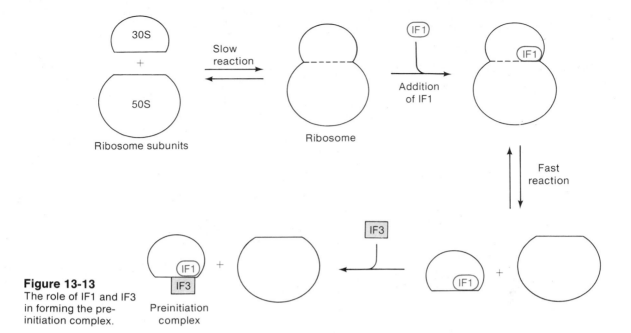

Figure 13-13
The role of IF1 and IF3 in forming the pre-initiation complex.

30S

+

50S

Ribosome subunits

Slow reaction

Ribosome

IF1

Addition of IF1

IF1

Fast reaction

IF3

+

IF1

+

IF1
IF3

Preinitiation complex

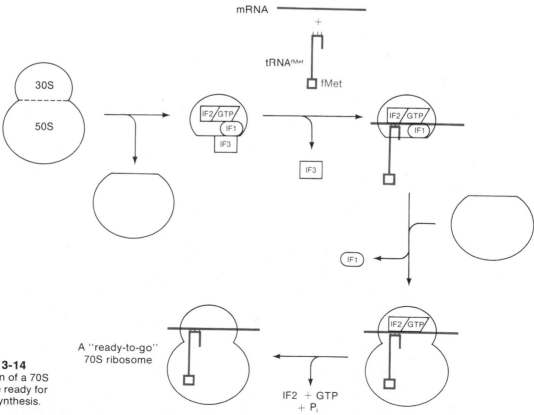

Figure 13-14
Formation of a 70S ribosome ready for protein synthesis.

mRNA

+

tRNA^fMet

fMet

30S

50S

IF2/GTP
IF1
IF3

IF3

IF2/GTP
IF1

IF1

IF2/GTP

A "ready-to-go" 70S ribosome

IF2 + GTP + P_i

Chain Elongation in Prokaryotes

Three nonribosomal proteins—the elongation factors, EF-Tu, EF-Ts, and EF-G—are required for chain elongation. These important proteins together comprise about 5–10 percent of the soluble protein of the cell.

EF-Tu is a carrier protein needed for the binding of charged tRNA to the A site of an active 70S ribosome. It acts in a multistep process that begins with the interaction of GTP with EF-Tu to form a binary complex (Figure 13-15). The binding of GTP causes a conformational change in EF-Tu, which allows a charged tRNA molecule to bind to the binary complex to form a ternary complex. In the ternary complex, EF-Tu is situated near the CCA terminus of the acceptor stem of the tRNA molecule. The anticodon loop is, however, completely free for pairing with the codon.

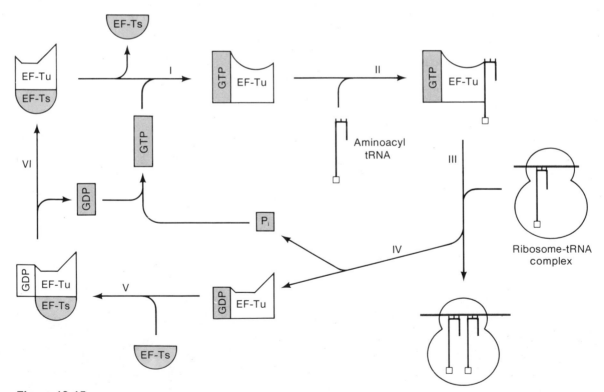

Figure 13-15

The cycle of the elongation factor EF-Tu. The steps shown are the following: I. Formation of active EF-Tu. II. Formation of ternary complex. III. Transfer of tRNA to a ribosome. IV. Release of GDP-EF-Tu. V. Binding of EF-Ts. VI. Formation of EF-Tu–EF-Ts.

The ternary complex is able to bind to the A site. There is some evidence that, as a result of codon-anticodon binding, the aminoacyl-tRNA undergoes a small conformational change that exposes the TψCG sequence in the TψC loop. This is thought to allow pairing of the tRNA to a sequence AAGC in the 5S RNA and that this pairing stabilizes binding to the A site. Once the ternary complex is in the A site, a ribosomal protein having GTPase activity causes hydrolysis of the GTP bound to EF-Tu to yield bound GDP and free P_i.

The GDP-(EF-Tu) complex lacks the ability of the GTP-(EF-Tu) complex to bind to acylated tRNA, and its dissociation from the tRNA leaves the acylated tRNA in the A site. Removal of EF-Tu likewise is necessary for the formation of a peptide bond; the evidence for this requirement is that the addition of substances that either prevent GTPase activity or prevent removal of EF-Tu also inhibit peptide bond formation. Presumably, until EF-Tu is removed, the charged tRNA is not positioned correctly for bond formation to occur.

GDP binds more tightly to EF-Tu than does GTP. Hence, some means is required either to regenerate GTP in the binary complex or to dissociate the complex. The latter is done by the factor EF-Ts, which displaces the GDP, forming an (EF-Ts)-(EF-Tu) intermediate. That is,

$$\text{GDP-(EF-Tu)} + \text{EF-Ts} \rightleftharpoons \text{(EF-Tu)-(EF-Ts)} + \text{GDP}$$

The released GDP joins the internal pool of metabolic intermediates and is reconverted to GTP. The EF-Ts is displaced from the (EF-Ts)-(EF-Tu) complex when GTP-(EF-Tu) re-forms. That is,

$$\text{(EF-Ts)-(EF-Tu)} + \text{GTP} \rightleftharpoons \text{GTP-(EF-Tu)} + \text{EF-Ts}$$

EF-Ts is unable to bind to this complex, presumably because it does not recognize EF-Tu after the GTP-induced conformational change. In summary, the sole function of EF-Ts is to regenerate GTP-(EF-Tu) from GDP-(EF-Tu) by intermediate formation of (EF-Ts)-(EF-Tu); that is, the net reaction is

$$\text{GDP-(EF-Tu)} + \text{GTP} \xrightleftharpoons{\text{EF-Ts}} \text{GTP-(EF-Tu)} + \text{GDP}$$

The system has now reached the stage at which both the P and A sites are occupied and a peptide bond can form. This simple reaction is unusually complex and poorly understood. The problem in understanding this reaction is that, though there is a peptidyl transferase enzymatic activity, no such enzyme has been isolated. The reaction occurs in a region of the 50S ribosomal subunit known as the **peptidyl transferase center.** Presumably, portions of the polypeptide chain of at least ten different 50S ribosomal proteins contribute to formation of the active site for this enzymatic activity. The overall chemical reaction in forming a peptide bond is shown in Figure 13-11.

Formation of the peptide bond is accompanied by cleavage of the bond connecting fMet and tRNA^{fMet}. Deacylated tRNA binds very poorly to the P site, so this tRNA leaves the ribosome immediately after peptide bond formation. In addition, the binding of the peptidyl-tRNA molecule that is now situated in the A site is weakened. The binding of peptidyl-tRNA to the P site is always strong, so there should be a tendency for the peptidyl-tRNA to move from the A site to the P site. This movement, which is called **translocation,** does occur, but is not dependent on random dissociation and reassociation (a random dissociation-reassociation mechanism could lead to premature termination). A third elongation factor, EF-G, controls this process. As in several previous binding events, the process in which EF-G participates begins by the formation of a complex with GTP. This is then followed by hydrolysis of GTP and release of GDP. There are numerous theories that attempt to explain translocation and the role of EF-G. However, at present there is no solid evidence to support any of these theories and the mechanism of translocation remains obscure.

As already mentioned, the elongation factors account for 5–10 percent of all bacterial protein. This large percentage is needed to maintain a high rate of protein synthesis, since the factors are used every time a peptide bond is formed. The actual amount of these factors depends on the growth medium and increases with the growth rate. Whether the growth rate determines their number or vice versa is unknown. Although the apparent need for the factors in protein synthesis does not exceed one per 70S ribosome, the number of molecules of each type per ribosome are 1, 1, and about 10 for EF-Ts, EF-G, and EF-Tu, respectively; the reason for the excess EF-Tu is not known. It is possible that EF-Tu has other functions in the cell; for instance, it is a major protein in the cell membrane. Like ribosomal proteins, the elongation factors differ little from one bacterial species to the next and are fully interchangeable in all prokaryotic *in vitro* protein-synthesizing systems.

Chain Termination

The presence of chain termination codons in the A site stimulates, in an unknown way, the appearance of release factors RF1 and RF2, each of which responds to a particular codon. A third factor, RF3, is also involved in termination; however, all that is known about RF3 is that it binds GTP and somehow enhances the binding of RF1 and RF2 to the ribosome. Chain termination is not a well-understood process.

The Role of GTP

GTP, like ATP, is an energy-rich molecule. Generally, when such molecules are hydrolyzed, the free energy of hydrolysis is used to drive reactions that otherwise are energetically unfavorable. This does not seem to be the case in protein synthesis. A review of the reaction sequence indicates that the role of GTP is to facilitate binding of protein factors either to tRNA or to the ribosome. Furthermore, hydrolysis of GTP to GDP and P_i always precedes dissociation of the bound factor. Comparison of the structure of the free factor and the factor-GTP complex indicates that the factor undergoes a slight change in shape when GTP is bound. Thus, GTP is an allosteric effector—namely, a small molecule that can induce a shape change in a macromolecule by binding to it. Since it is easily hydrolyzed by various GTPases, the use of GTP as a controlling element allows a cyclic variation in macromolecular shape. That is, when GTP is bound, the macromolecule has an active conformation and when the GTP is hydrolyzed or removed, the molecule resumes its inactive form.

Posttranslational Modification of Proteins

The protein molecule ultimately needed by a cell often differs from the polypeptide chain that is synthesized. There are several ways in which the modification of the synthesized chain occurs:

1. In prokaryotes, fMet is never retained as the NH_2-terminal amino acid. In roughly half of all proteins the formyl group is removed by the enzyme **deformylase**, which leaves methionine as the NH_2-terminal amino acid. In both prokaryotes and eukaryotes the fMet or methionine, and possibly a few more amino acids, are often removed; this removal is catalyzed by a hydrolytic enzyme called **aminopeptidase.** This hydrolysis may sometimes occur while the chain is being synthesized and sometimes after the chain is released from the ribosome. The choice of deformylation versus removal of fMet usually depends on the identity of the adjacent amino acids. That is, deformylation predominates if the second amino acid is arginine, asparagine, aspartic acid, glutamic acid, isoleucine, or lysine, whereas fMet is usually removed if the adjacent amino acid is alanine, glycine, proline, threonine, or valine.

2. Newly created NH_2-terminal amino acids are sometimes acetylated.

3. Some amino-acid side chains may also be modified (Figure 13-16). For example, in collagen a large fraction of the proline and lysine are

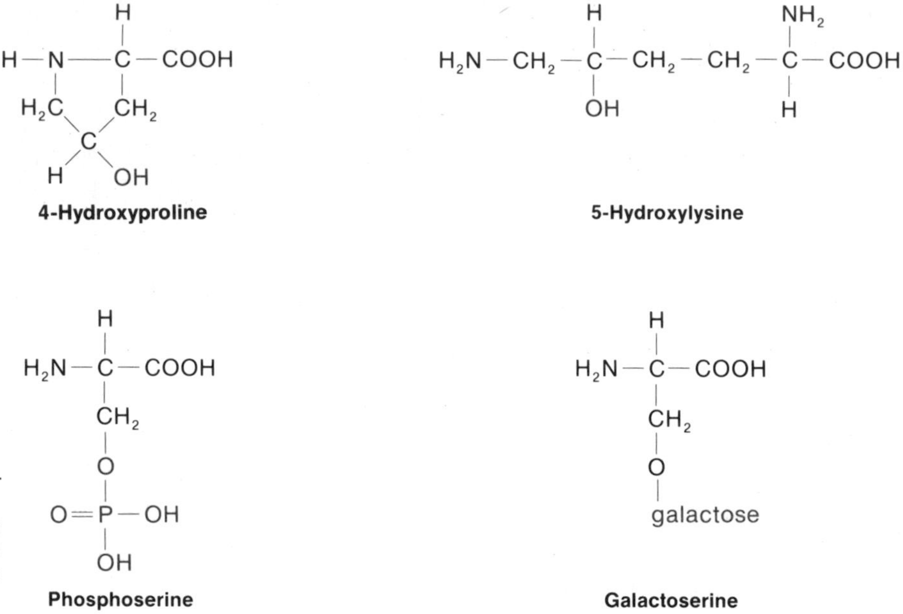

4-Hydroxyproline
5-Hydroxylysine
Phosphoserine
galactose
Galactoserine

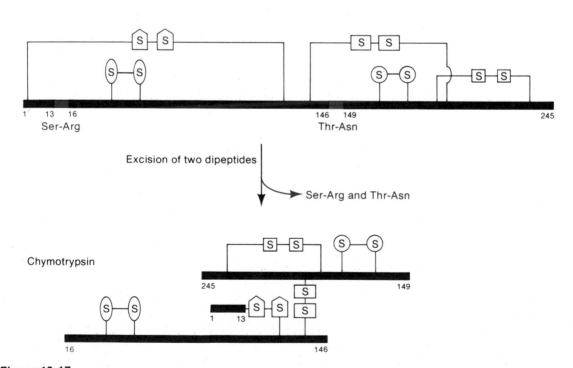

Figure 13-16
Some examples of modified amino acids. The added group is shown in red. Modification occurs after protein synthesis is complete.

1
13
16
Ser-Arg
146
149
Thr-Asn
245
Excision of two dipeptides
Ser-Arg and Thr-Asn
Chymotrypsin
245
149
1
13
16
146

Figure 13-17
Conversion of chymotrypsinogen to chymotrypsin by removal of two pairs of amino acids (shown in red). The various disulfide bonds are given different shapes for reference only. The numbers refer to amino acid positions.

hydroxylated. Phosphorylation of serine, tyrosine, and tryptophan occurs in many organisms. Also, various sugars may be attached to the free hydroxyl group of serine or threonine to form glycoproteins. Finally a variety of prosthetic groups such as heme and biotin are covalently attached to some enzymes.

4. Two distant sulfhydryl groups in two cysteines may be oxidized to form a disulfide bond. This is very common.

5. Polypeptide chains may be cleaved at specific sites. For instance, chymotrypsinogen is converted to the digestive enzyme chymotrypsin by removal of four amino acids from two different sites (Figure 13-17). In some cases the uncleaved chain represents a storage form of the protein that can be cleaved to generate the active protein when needed. This is true of many mammalian digestive enzymes—for example, pepsin is formed by cleavage of pepsinogen. An interesting precursor is a huge protein synthesized in cells infected with poliovirus; this molecule is cleaved at several sites to yield different active proteins and hence is called a **polyprotein.** We will see many examples of polyproteins when animal viruses are described in Chapter 21.

COMPLEX TRANSLATION UNITS

The unit of translation is almost never simply a ribosome traversing a mRNA molecule, but is a more complex structure, of which there are several forms. Some of these structures are described in this section.

Polysomes

After about 25 amino acids have been joined together in a polypeptide chain, the AUG initiation site of the encoding mRNA molecule is completely free of the ribosome. A second initiation complex then forms. The overall configuration is of two 70S ribosomes moving along the mRNA at the same speed. When the second ribosome has moved along a distance similar to that traversed by the first, a third ribosome is able to attach. This process—movement and reinitiation—continues until the mRNA is covered with ribosomes at a density of about one 70S particle per 80 nucleotides. This large translation unit is called a **polyribosome** or simply a **polysome.** This is the usual form of the translation unit in all cells.

An electron micrograph of a polysome and an interpretive drawing are shown in Figure 13-18. The following features of a polysome should be noted:

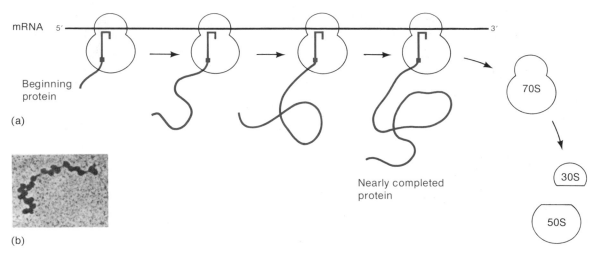

Figure 13-18
Polysomes. (a) Diagram showing relative movement of the 70S ribosome and the mRNA, and growth of the protein chain. (b) Electron micrograph of an *E. coli* polysome. (Courtesy of Barbara Hamkalo.)

1. The ribosomes move along the mRNA in the 5′→3′ direction.

2. There is a gradient in the length of the nascent polypeptide chain increasing in the 5′→3′ direction. This is because a ribosome nearer the 3′ end has been synthesizing protein for a greater time than one nearer the 5′ end.

3. In prokaryotes, the spacing between adjacent ribosomes is approximately constant. The codons being translated are about 80 nucleotides apart. This is greater than the length (30 nucleotides) of the region of the mRNA that is protected from RNase digestion. This discrepancy is not understood.

4. The total number of ribosomes per polysome is proportional to the length of the polypeptide chain and inversely proportional to the spacing between adjacent ribosomes. A rough value is about 1 ribosome per 25 codons. Thus, since typical proteins have a molecular weight ranging from 3 to 5×10^4, the segment of mRNA encoding them would contain about 11–18 ribosomes. In prokaryotes most mRNA molecules are polycistronic, so that polysomes containing 40–50 ribosomes would be expected. Polysomes of this size have been detected, though not often because they are very fragile and easily broken.

The use of polysomes has a particular advantage to a cell—namely, the overall rate of protein synthesis is increased compared to the rate that would occur if there were no polysomes. Clearly, ten ribosomes transversing a single segment of mRNA can make ten times as many protein molecules per unit of time as a single ribosome can.

Coupled Transcription and Translation

An mRNA molecule being synthesized has a free 5′ terminus so that, since translation occurs in the 5′→3′ direction, each cistron contained in the mRNA is synthesized in a direction appropriate for immediate translation. That is, the ribosome binding site is transcribed first, followed in order by the AUG codon, the region encoding the amino acid sequence, and finally the stop codon. Thus, in bacteria, in which no nuclear membrane separates the DNA and the ribosome, there is no obvious reason why the 70S initiation complex should not form before the mRNA is released from the DNA. With prokaryotes this does indeed occur; this process is called **coupled transcription-translation.** This coupled activity does not occur in eukaryotes, because the mRNA is synthesized and processed in the nucleus and later transported through the nuclear membrane to the cytoplasm where the ribosomes are located. In prokaryotes, coupling of transcription and translation is the rule, though a few phage mRNA molecules are first processed after release from the DNA and then translated.

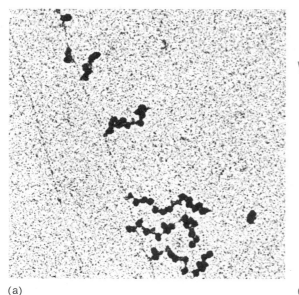

(a)

Initiation site on DNA

5′ mRNA 1

1

mRNA 2

2

mRNA 3

3

mRNA 4

4

(b)

Figure 13-19

(a) Transcription of a section of the DNA of *E. coli* and translation of the nascent mRNA. Only part of the chromosome is being transcribed. The dark spots are ribosomes, which coat the mRNA. (From O. L. Miller, Barbara A. Hamkalo, and C. A. Thomas. 1977. *Science,* 169, 392.) (b) An interpretation of the electron micrograph of part (a). The mRNA is in red and is coated with black ribosomes. The large red spots are the RNA polymerase molecules; they are actually too small to be seen in the photo. The dashed arrows show the distances of each RNA polymerase from the transcription initiation site. Arrows 1, 2, and 3 have the same length as mRNA 1, 2, 3; mRNA 4 is shorter than arrow 4, presumably because its 5′ end has been partially digested by an RNase.

Coupled transcription-translation speeds up protein synthesis in the sense that translation does not have to await release of the mRNA from the DNA. Translation can also be started before the mRNA is degraded by nucleases.

Coupled transcription-translation was first detected by sedimentation analysis, in which it was found that DNA, RNA polymerase, mRNA, ribosomes, and nascent protein chains sedimented as a unit. Definitive evidence came from the remarkable electron micrograph shown in Figure 13-19(a). This micrograph shows a DNA molecule to which is attached a number of mRNA molecules, each associated with ribosomes. The micrograph is interpreted in Figure 13-19(b). Transcription of DNA begins in the upper left part of the micrograph. The lengths of the polysomes increase with distance from the transcription initiation site, because the mRNA is farther from that site and hence longer. Note that the longest polysome is shorter than the total distance from the initiation site; this is probably because some degradation of mRNA has occurred from its 5' end.

SOME NUMERICAL PARAMETERS OF PROTEIN SYNTHESIS

In the analysis of growth rates and of metabolic regulation the numerical values of various rates of synthesis are needed. These values for growth at 37°C are independent of the doubling time of bacteria, as influenced by nutrition, and are listed in Table 13-3.

A typical protein molecule contains about 500 amino acids, so a protein molecule is synthesized in about one-half minute. Note that the polypeptide elongation rate is one-third that of the mRNA elongation rate, which is consistent with coupled transcription-translation and a coding ratio of three nucleotides per amino acid. From the half-life of mRNA and the peptide elongation rate, the average number of amino acids polymerized during the half-life of a mRNA molecule is known to be 1836. If a typical polycistronic mRNA molecule contains information

Table 13-3
Values of some parameters of protein and RNA synthesis in *E. coli*

Parameter	Value at 37°C*
Polypeptide elongation rate	17 amino acids/sec
mRNA elongation rate	55 nucleotides/sec
mRNA half-life	1.5–1.8 min
Amount of mRNA bound per ribosome	80 nucleotides

*37°C is the temperature for optimal growth of *E. coli*. At 30°C, a temperature commonly used in experiments, these values should be multiplied by 0.66.

for synthesizing three proteins, then roughly one copy of each protein molecule is made per half-life. This clearly indicates the need for continual synthesis of mRNA if a bacterium is to make hundreds of copies of a particular protein molecule in the course of one generation.

INHIBITORS AND MODIFIERS OF PROTEIN SYNTHESIS

Many inhibitors of protein synthesis are known. They have been extremely useful in the study of protein synthesis because a particular substance typically blocks a single step in the complex protein synthetic process. The use of an inhibitor as an experimental reagent yields three types of information. First, if a particular step is blocked and other steps are not, then the uninhibited steps must precede the blocked step. By using this technique, the sequence of some of the stages in protein synthesis have been worked out. Second, blocking a step sometimes causes the accumulation of a previously unrecognized molecule. Such a molecule is likely to be a short-lived intermediate obtained from the preceding step. Third, if a step is blocked, the ribosomal protein responsible for that step can often be identified; this can be done either by showing that a particular ribosomal protein binds the antibiotic or, in the case of a microorganism, by first isolating a mutant that is resistant to the antibiotic and then determining which protein is altered.

Antibiotics That Affect Protein Synthesis

Many antibacterial agents **(antibiotics)** have been isolated from fungi. They have been widely used both clinically and as reagents for unraveling the details of protein synthesis and RNA and DNA synthesis. Table 13-4 lists a few of the hundreds of antibiotics that inhibit protein synthesis in bacteria. Some of them have only limited clinical usefulness because they also inhibit the growth of animal cells and hence are toxic to both bacterium and host.

A particularly well-studied antibiotic is **puromycin** (Figure 13-20). Puromycin is chemically similar to the aminoacyl part of a charged tRNA molecule, as shown in the figure, and it competes effectively with charged tRNA molecules for the A site. In addition, puromycin has an α-amino group that can form a peptide bond with the carboxyl group of a growing peptide chain in a reaction catalyzed by peptidyl transferase. When this bond forms, the polypeptide is cleaved from the tRNA in the P site and the puromycin leaves the A site. Puromycin does not bind to the P site, so the polypeptide, which has a terminal puromycin, falls away from the ribosome. Thus, in the presence of puromycin, polypeptide chain elongation is blocked. It is of interest that studies with

Table 13-4
Antibiotic inhibitors of protein synthesis in prokaryotes

Antibiotic	Action
Streptomycin	Binds to the $S12$ protein of the 30S ribosomal sub-unit and thereby inhibits binding of tRNA^{fMet} to the P site. Also causes misreading in a system that is in the act of synthesis
Neomycin, kanamycin	Same as streptomycin
*Chloramphenicol (also called chloromycetin)	Inhibits peptidyl transferase of 70S ribosome
Tetracycline	Inhibits binding of charged tRNA to 30S particle
Erythromycin	Binds to free 50S particle and prevents formation of the 70S ribosome. Has no effect on an active 70S ribosome
*Puromycin	Causes premature chain termination by acting as an analogue of charged tRNA
*Fusidic acid	Inhibits binding of charged tRNA to A site
Kasugamycin	Inhibits binding of tRNA^{fMet}
Lincomycin	Inhibits peptidyl transferase complex
Kirromycin	Binds to EF-Tu; stimulates formation of (EF-Tu)-GTP and binding of ternary complex to ribosome; and inhibits release of EF-Tu
Thiostrepton	Prevents translocation by inhibiting EF-G

*Also active in eukaryotes. See Table 13-5.

puromycin provided the first evidence for the A and P sites. The observation made was that if translocation were prevented in an *in vitro* system by withholding EF-G (and thereby preventing translocation), puromycin could not bind to the ribosome. After translocation, puromycin could bind. Thus, it was concluded that there are two binding sites for peptidyl-tRNA and that puromycin can bind to only one of these.

Another well-studied antibiotic is streptomycin. Its mode of action is described in a later section.

Inhibitors of Protein Synthesis in Eukaryotes

The nontoxic antibiotics that have no effect on eukaryotes are nontoxic to them either because they fail to penetrate the eukaryotic cell membrane (which is quite common) or because they do not bind to

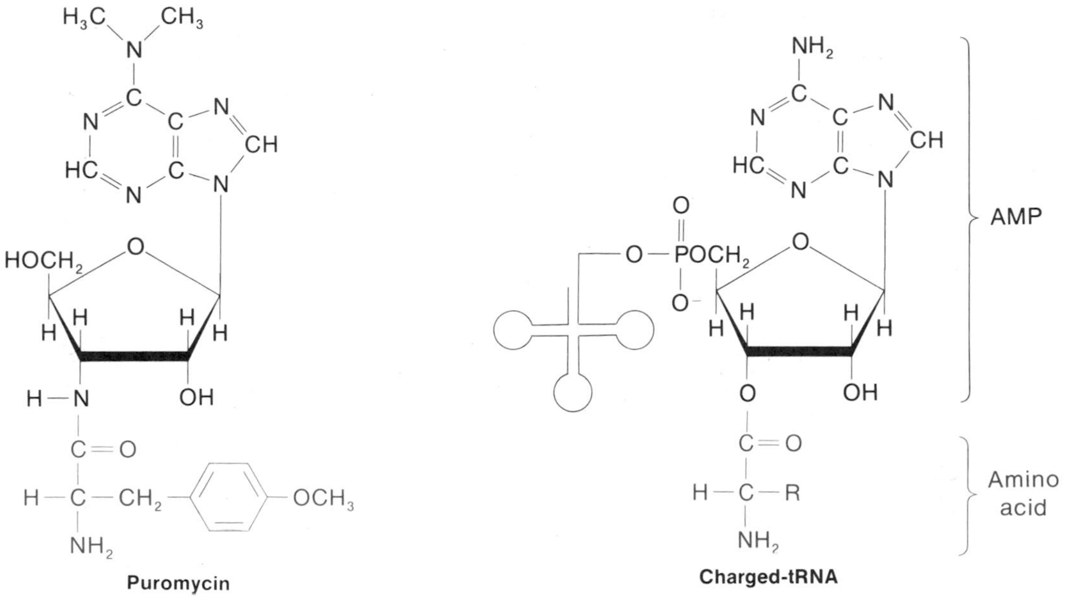

Figure 13-20
Puromycin resembles the aminoacyl terminus of an aminoacyl-tRNA. The cloverleaf represents all of the tRNA except the terminal AMP, whose structure is drawn out. The amino acid in each structure is shown in red.

eukaryotic ribosomes. The former type are effective inhibitors of protein synthesis *in vitro*, in systems obtained either from prokaryotes or eukaryotes. The differences in effectiveness of antibiotics on the two classes of cells *in vivo* is the basis for the use of these drugs. That is, they kill bacterial cells but not animal cells.

Some antibiotics are active against both bacterial and mammalian cells. One example is **chloramphenicol** (also called chloromycetin), which inhibits peptidyl transferase in both bacterial and mitochondrial ribosomes, though normal cytoplasmic ribosomes are unaffected. Such a drug can still be clinically useful if there is a concentration range in which the antibacterial effect is substantial and the toxic effect to the host is weak. However, because of their potential for host toxicity, these antibiotics tend to be used only in serious infections when all else fails.

There are also many drugs that act either mainly or significantly on eukaryotes. Some of these are toxins of bacterial origin while others are synthetic. An example of the former is the toxin produced by *Corynebacterium diphtheriae*, the bacterium that causes diphtheria. This toxin is an enzyme that causes covalent modification of the eukaryotic elongation factor needed for translocation and thereby inhibits that step. **Cycloheximide** is a chemical inhibitor of the peptidyl transferase complex of the 60S subunit and hence inhibits formation of the peptide

Table 13-5
Inhibitors of protein synthesis in eukaryotes

Inhibitor	Action
Abrin, ricin	Inhibits binding of aminoacyl tRNA
Diphtheria toxin	Enzymatic catalysis of a reaction between NAD^+ and eEF2 to yield an inactive factor. Inhibits translocation
*Chloramphenicol (also called chloromycetin)	Inhibits peptidyl transferase of mitochondrial ribosomes. Inactive against cytoplasmic ribosomes
*Puromycin	Causes premature chain termination by acting as an analogue of charged tRNA
*Fusidic acid	Inhibits translocation by altering an elongation factor
Anisomycin	Inhibits peptidyl transferase
Cycloheximide (also called actidione)	Inhibits peptidyl transferase
Pactamycin	Inhibits positioning of $tRNA_f^{Met}$ on the 40S ribosome
Showdomycin	Inhibits formation of the $eIF2$-$tRNA_f^{Met}$-GTP complex
Sparsomycin	Inhibits translocation

*Also active on prokaryotic ribosomes. See Table 13-4.

bond. Substances like cycloheximide are commonly used in cancer chemotherapy.

Table 13-5 lists a few of the better known inhibitors of protein synthesis in eukaryotic cells.

Streptomycin-Induced Misreading and Streptomycin Dependence

Most bacteria cannot grow in the presence of **streptomycin** (Figure 13-21) inasmuch as it interferes in some way with the binding of tRNA^{fMet} to the P site and thereby inhibits the initiation of synthesis of a polypeptide chain. Streptomycin also has another effect that is evident when its concentration is sufficiently low that initiation can still occur. In this case extensive misincorporation of amino acids is observed. This happens because streptomycin alters codon-anticodon recognition—that is, it induces misreading of the code. For example, in an *in vitro* system using poly(U) as mRNA it is found that in addition to phen-

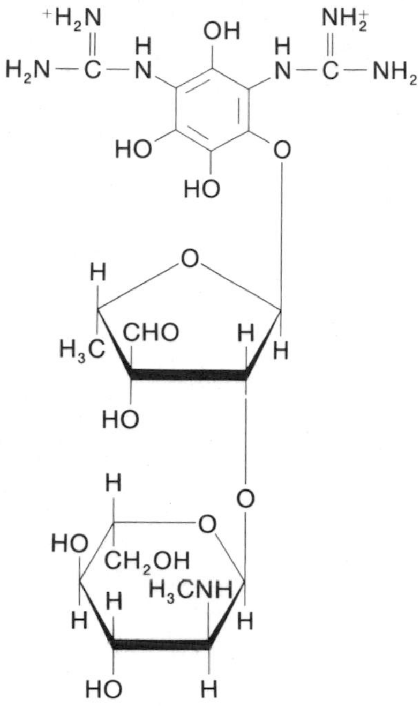

Figure 13-21
Structure of
streptomycin.

ylalanine (UUU codon), four other amino acids are sometimes incorporated—isoleucine (AUU), leucine (CUU), tyrosine (UAU), and serine (UCU). Isoleucine and leucine are substituted much more often than tyrosine and serine, which suggests that the first base of a codon is misread more often than the second base.

Streptomycin-induced misreading also occurs *in vivo*. In Str-s (wild-type) *E. coli,* addition of streptomycin causes both inhibition of initiation of new protein chains (by preventing binding of tRNA^{fMet}) and massive misincorporation of amino acids into proteins that were elongating at the time of addition of the drug. Mutants have been isolated that are resistant to the drug (they are Str-r). These mutants have an altered S12 ribosomal protein that enables them to initiate polypeptide synthesis in the presence of streptomycin. Misreading is also reduced, but still occurs occasionally and causes an effect known as **streptomycin suppression.** For example, suppose a cell contains a protein that is mutant by virtue of a substitution of isoleucine by phenylalanine. If the mutant is grown in a medium containing streptomycin, most of the defective proteins remain mutant, but, because of the induced misreading, a few copies contain isoleucine instead of phenylalanine and will be functional; hence, this strain will express a wild-type phenotype. In this way, streptomycin is said to *suppress* the mutant phenotype.

Streptomycin-dependent *E. coli* mutants have also been isolated. These mutants cannot grow unless a small amount of streptomycin is an ingredient of the growth medium. They are killed, however, by a higher concentration of the drug. These mutations do not map at the *str* locus, but may be located anywhere in the genetic map. The explanation for this odd phenomenon is that a streptomycin-dependent bacterium has a mutation in an unidentified but essential gene and requires streptomycin-induced misreading in order to synthesize a wild-type protein from the mutant gene.

Str-r mutants have other properties that are evident even when streptomycin is not present. Some amber (chain termination) mutations are slightly leaky because occasionally a normal (nonsuppressor) tRNA can read a UAG codon, which may give rise to synthesis of a protein whose activity is 1 to 2 percent of the wild-type value. This leakiness is often experimentally inconvenient when the mutation is to be used to block some biochemical step. If the bacterium is made Str-r, this leakiness is often reduced considerably, even when streptomycin is absent; apparently a mutation in the S12 protein decreases the frequency of misreading of the UAG codon.

PROTEIN SYNTHESIS IN EUKARYOTES

The basic pattern of protein synthesis in eukaryotes is similar to that in bacteria. (This discussion does not apply to mitochondria and chloroplasts, which will be discussed later.) That is, a ribosomal subunit known as **43S$_n$**, which already contains some initiation factors,* first forms a complex with the initiator tRNA and with several factors to form a **43S$_n$ preinitiation complex.** (This is followed sequentially by binding of mRNA and the 60S ribosomal subunit to form an active 80S initiation complex.) Chain elongation then occurs until a termination codon is reached. Each step requires several protein factors. To distinguish these factors from those used by prokaryotes, a lower case "e" precedes the usual symbol. Thus, the first initiation, elongation, and release factors are written eIF1, eEF1, and eR1, respectively.

The process of protein synthesis in eukaryotes is not known in as much detail as in prokaryotes because there are more factors and not all of the factors have yet been purified. There are many differences in detail between protein synthesis in prokaryotes and eukaryotes. Rather than describe the complex (and incompletely understood) sequence of events in eukaryotes, it is more informative to examine the differences between the systems.

---

*The 43S$_n$ subunit can be obtained by dissociation of polysomes and is fairly stable. If purified ribosomes are washed with solutions having a high salt concentration, the factors are removed. Dissociation of these washed ribosomes by removal of Mg^{2+} ions yields the 40S subunit, usually thought of as the small ribosomal subunit, and the large 60S subunit.

Differences Between Protein Synthesis in Eukaryotic and Prokaryotic Cells

The following differences have been noted between protein synthesis in eukaryotic and prokaryotic cells:

1. In eukaryotes the initiating amino acid is methionine and not fMet. The initiating tRNA, which responds only to AUG, is designated $tRNA_{init}^{Met}$ or $tRNA_f^{Met}$ to distinguish it from the $tRNA^{Met}$ used in translating internal AUG codons. This has already been described in Chapter 12.

2. At least five initiation factors plus GTP are required for binding of $tRNA_f^{Met}$ to the preinitiation complex. Two of these are similar to the factors in E. coli that prevent binding of the large particle to the small particle. One of these factors, eIF3, is a large complex containing nine protein subunits and having a molecular weight of 7×10^5. An enzyme called **initiating tRNA hydrolase,** which removes the initial tRNA molecule after the first peptide bond forms, is also present in the complex. In prokaryotes the corresponding enzyme, **tRNA deacylase,** is a ribosomal component.

3. Binding of $tRNA_f^{Met}$ must occur before mRNA can bind, whereas for prokaryotes the mRNA can bind either before or after binding of the initiator tRNA. For binding of mRNA two other initiation factors are needed and ATP must be cleaved to form ADP and P_i. The reason for the cleavage of ATP is unknown. Binding occurs initially at or near the 5′ cap and is mediated by the cap-binding factor (which is unnecessary for uncapped viral mRNA); the mRNA moves with respect to the preinitiation complex until the anticodon of the prior bound $tRNA_f^{Met}$ pairs with the first AUG encountered. No Shine-Dalgarno sequence is needed. The fact that the AUG codon nearest the 5′ terminus is always the initiating codon is a significant difference between eukaryotes and prokaryotes and plays an important role in the regulation of protein synthesis in animal viruses, as will be seen in Chapter 21.*

4. More factors are needed for binding of the 60S subunit than for binding of the bacterial 50S subunit.

5. At least four elongation factors are needed by eukaryotes. These factors probably differ in structure and size in different tissues, often forming aggregates containing as many as 50 monomers. The factors, which in terms of function roughly correspond to the E. coli factors, are listed in the following chart:

---

*A few exceptions are known; in this case, the second AUG is the start codon.

Elongation factor	E. coli factor
eEF1$_\alpha$	EF-Tu
eEF1$_\beta$	EF-Ts
eEF1$_\gamma$	None
eEF2	EF-G

The reactions that are catalyzed in eukaryotic protein synthesis do not exactly correspond to those in prokaryotic protein synthesis though.

6. Little is known about termination in eukaryotes, though release factors have been purified from several systems. Surprisingly, release in *in vitro* systems requires the presence of one of four tetranucleotides— UAAA, UAGA, UGAA, or UAGG.

Two other differences between eukaryotes and prokaryotes in the overall process of protein production should be noted. The most striking is that transcription and translation are not coupled in eukaryotes. The mRNA is synthesized in the nucleus, where it is processed, and then transported to the cytoplasm, where the ribosomes are located. The second point concerns the lifetime of mRNA. In prokaryotes degradation of mRNA occurs continuously in the 5′→3′ direction and while translation is in process. Thus the mRNA is continually being shortened and translation can no longer be initiated once the Shine-Dalgarno sequence has been degraded. The half-life of a typical bacterial mRNA is about 1.8 minutes. Eukaryotic mRNA is very stable—possibly because of the 5′ cap; degradation (by unknown enzymes) occurs very slowly and a typical half-life is several hours. Some of the more stable mRNA molecules have a half-life of a few days. Hemoglobin mRNA, whose half-life is several hours, is translated about 17,000 times on the average before it is degraded.

Protein Synthesis in Mitochondria and Chloroplasts

Eukaryotes contain a variety of small morphologically distinct particles called organelles, which are responsible for carrying out specific functions. These organelles contain DNA and ribosomes, among other things. Two organelles, the mitochondrion (which contains components of the energy-generating system) and the chloroplast (which carries out photosynthesis) have well-characterized protein-synthesizing systems. The most striking properties of these systems is that their ribosomes resemble those in prokaryotes both in size (near 70S) and the number and size of their RNA molecules; for example, their ribosomes have three rRNA molecules, as in prokaryotes, rather than four, which is characteristic of eukaryotes, and the sizes are 5S, 16S, and 23S, as in prokaryotes. Furthermore, many of the individual steps in protein synthesis are the same as those of prokaryotes. Evolutionists have hypothesized that these organelles may be the remnants of ancient prokaryotes that were parasitic and became established in eukaryotic cells. However, the genetic code in mitochondria differs from that of both eukaryotes and prokaryotes (Chapter 12), a surprising fact that does not accord with evolution of these organelles from eukaryotes or prokaryotes.

Bound Ribosomes in Eukaryotic Cells: The Endoplasmic Reticulum

The ribosomes in most prokaryotes are distributed throughout the cell without any apparent spatial organization (except for a minor fraction that is bound to the inner membrane just under the cell wall). However, in most types of eukaryotic cells there are two major classes of ribosomes—attached ribosomes and free ribosomes. The attached ribosomes are bound to an extensive internal network of lipoprotein membranes called the **endoplasmic reticulum** (Figure 13-22). The

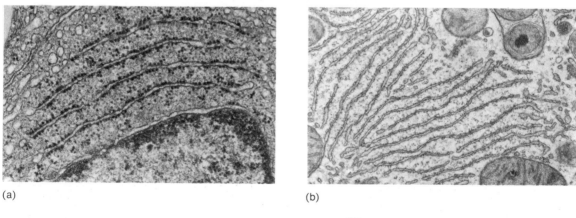

(a)   (b)

Figure 13-22
(a) Electron micrograph of the rough endoplasmic reticulum in the Leidig cell of a guinea pig. The dark circles are ribosomes. (b) The more complex endoplasmic reticulum of the rat liver. (Courtesy of Daniel Friend.) (c) A three-dimensional representation of the rough endoplasmic reticulum. (From S. Wolfe. 1981. *Biology of the Cell*. Wadsworth.)

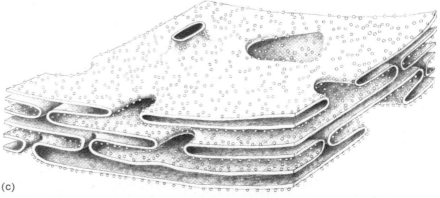

(c)

membranes of the endoplasmic reticulum to which ribosomes are bound are called the rough endoplasmic reticulum; those devoid of ribosomes comprise the smooth endoplasmic reticulum. There is no structural difference between a free and an attached ribosome, and attachment occurs after protein synthesis begins, probably by means of the nascent polypeptide (explained later in the chapter); whether a ribosome does or does not attach apparently depends upon the particular protein that is synthesized.

The endoplasmic reticulum is a highly convoluted membrane system. Sometimes the membranes enclose a discrete region of the cell called a **cisterna.** A large fraction of the system probably forms one or more large, irregularly shaped cisternae. In this sense, the membrane system has an inside and an outside, and ribosomes are only bound to the outside. Proof of this point comes from an experiment explained in Figure 13-23. In this experiment rough endoplasmic reticulum is broken into small fragments. If these fragments are allowed to reunite, small spherical vesicles form and the ribosomes are found only on the outer surface of the vesicles. These coated vesicles can be used as a source of ribosomes in an *in vitro* protein-synthesizing system. When this is done, the newly synthesized protein molecules are found inside the vesicles (Figure 13-24)—this important finding provides a key to understanding the function of the endoplasmic reticulum.

Cells responsible for secreting large amounts of a particular protein (for instance, hormone-secreting cells) have a very extensive endoplasmic reticulum. Furthermore, most (but not all) proteins destined either to be secreted by the cell or to be stored in natural intracellular vesicles such as lysosomes (which contain degradative enzymes) and peroxisomes (which contain enzymes for eliminating hydrogen peroxide) are synthesized by attached ribosomes. These proteins are primarily found in the cisternae of the endoplasmic reticulum. In contrast, most (but not

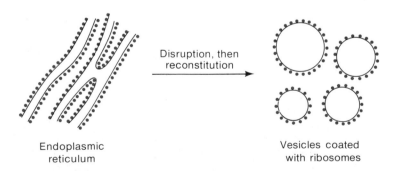

Endoplasmic
reticulum

Disruption, then
reconstitution

Vesicles coated
with ribosomes

Figure 13-23
Evidence that ribosomes are on the "outer" surface of the membranes of the endoplasmic reticulum.

all) proteins destined to be free-floating in the cytoplasm are made on free ribosomes.

The current explanation for the formation of the rough endoplasmic reticulum and the mechanism by which newly synthesized proteins pass through the membrane is called the **signal hypothesis.** There are several variants of this hypothesis, one of which is shown in Figure 13-25. The basic idea of the signal hypothesis is that the signal for attachment of the ribosome to the membrane is a sequence of very hydrophobic amino acids near the amino terminus of the growing polypeptide chain. When protein synthesis begins, the ribosome is free.

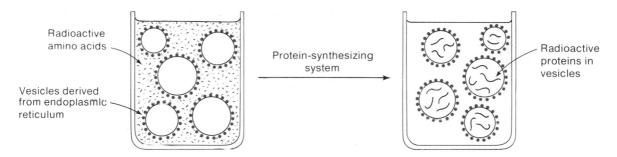

Figure 13-24
An experiment showing that protein synthesized on ribosomes bound to the endoplasmic reticulum passes through the membrane. Vesicles derived from the endoplasmic reticulum and coated with ribosomes are incubated in a solution containing a protein-synthesizing system and radioactive amino acids. Radioactive protein is synthesized; all of the protein is contained in the cavity of the vesicles.

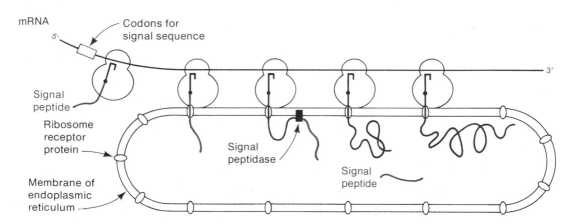

Figure 13-25
Signal hypothesis for the synthesis of secretory and membrane proteins. Shortly after initiation of protein synthesis the amino-terminal sequence of the polypeptide chain binds the translating ribosome to the membrane of the endoplasmic reticulum. The signal sequence is excised by the signal peptidase within the vesicle. When protein synthesis is completed, the protein remains within the vesicle and the ribosome is released.

NH$_2$–Met–Arg–Ser–Leu–Leu–Ile–Leu–Val–Leu–Cys–Phe–Leu–Pro–Leu–Ala– Ala–Leu–Gly ┤Gly–Lys . . .

NH$_2$–Met–Lys–Trp–Val–Thr–Phe–Leu–Leu–Leu–Leu–Phe–Ile–Ser–Gly–Ser–Ala–Phe – Ser┤Arg . . .

Figure 13-26
Signal sequences of two secretory proteins. The hydrophobic amino acids are printed in red. The vertical bars denote the points of cleavage by the signal peptidase.

The hydrophobic amino-terminal sequence interacts with lipophilic membrane components and somehow directs the association of the large ribosomal subunit to a ribosome receptor protein on the membrane surface. As protein synthesis continues, the protein moves through the membrane to the cisternal side of the endoplasmic reticulum. A specific protease termed the **signal peptidase** cleaves off the amino-terminal signal sequence.

Two important experimental results support the signal hypothesis: (1) If an *in vitro* synthesis of several secretory proteins is carried out by using free ribosomes, the resulting proteins contain NH$_2$-terminal sequences about twenty amino acid residues long that are not present in the proteins isolated from intact cells. (2) The amino acids in these NH$_2$-terminal extensions are very rich in hydrophobic amino acids (Figure 13-26).

Many of the structural proteins of the membrane of the endoplasmic reticulum are made by the bound ribosomes. These proteins do not pass through, but remain within, the membrane. Presumably these proteins possess amino acid sequences that prevent transfer through the membrane before the carboxyl terminus is reached. A signal sequence is cleaved from these proteins also.

WHY IS PROTEIN SYNTHESIS SO COMPLICATED?

The reader cannot fail to notice the extraordinary complexity of protein synthesis, and it is natural to ask why it is so complicated. The answer is, surely, at least in part, to guarantee accuracy while allowing protein synthesis to occur rapidly. Most of the ribosomal proteins seem to be designed to form a surface on which two tRNA molecules are precisely positioned both with respect to one another and with respect to the codons. The various protein factors probably also serve this function, though why they are not ribosomal components is not obvious. The need for the dissociation cycle of the 70S ribosome is also unclear. It is difficult to conceive of a system in primordial time being as complicated as one in contemporary life. More likely, a primitive highly error-prone system initially existed and in time was improved as natural selection acted on chance mutation. Whereas the resultant complexity indeed seems horrendous, if a new capability in a system either reduced the fraction of defective molecules that were synthesized and/or increased the rate of polypeptide chain growth, a cell having the new capability would grow slightly faster than other less efficient cells. Thus, in time the new cell type would become the predominant species even if its synthetic mechanisms were very complicated. The critical factors are always the survival value or the growth rate, not simplicity.

Compared to prokaryotes, eukaryotes have 27 more ribosomal proteins and one more RNA molecule, and the ribosomes, as well as all the components of the ribosomes, are larger than those of prokaryotes. Furthermore, eukaryotic protein synthesis requires more accessory factors, as we have seen. Presumably, this larger number of components reflects the greater needs of a cell that is more complex than a bacterium. The greater stability of eukaryotic mRNA compared to prokaryotic mRNA, which is a valuable attribute in slowly growing cells often committed to production of large quantities of a few proteins, is probably in part responsible for the large number of factors in eukaryotic protein synthesis. For example, rapid, short-term changes in the overall rate of protein synthesis necessitated by environmental fluctuations are accomplished in bacteria by changing the rate of production of particular mRNA molecules (see Chapter 14); when synthesis is turned off, the concentration of the mRNA rapidly decreases. This mechanism cannot be operative in eukaryotes, for which environmental responses must be mediated by regulation of translation of stable mRNA molecules. That is, changes in the concentration of a particular protein must often be effected by changing the rate of initiation of protein synthesis. One would expect this to necessitate having a multifactor system that can respond to various regulatory signals. In Chapter 22 some examples of regulated systems in which regulation depends on inactivation of accessory factors will be seen.

POLYPEPTIDE SYNTHESIS WITHOUT RIBOSOMES

Some small polypeptides synthesized in bacteria and in eukaryotes are made without the aid of ribosomes. The simplest one is the tripeptide glutathione (H_2N-Glu-Cys-Gly-COOH). This molecule is synthesized enzymatically in two steps by the reaction sequence

$$\text{Glu} + \text{Cys} + \text{ATP} \xrightarrow{\substack{\text{Glutamyl cysteine} \\ \text{synthetase}}} \text{Glu-Cys} + \text{ADP} + \text{P}_i$$

$$\text{Glu-Cys} + \text{Gly} + \text{ATP} \xrightarrow{\substack{\text{Glutathione} \\ \text{synthetase}}} \text{Glutathione} + \text{ADP} + \text{P}_i$$

Consideration of this reaction shows the advantage of the ribosome-mRNA mechanism. In the synthesis of glutathione the amino acid sequence is determined by enzyme specificity. If this mechanism were the only one available, synthesis of a protein containing 1000 amino acids might require 1000 enzymes. Nonetheless, in bacteria there are a few polypeptides, having as many as 20 amino acids, that are synthesized without ribosomes. The mechanisms used, which are in principle like that for glutathione, probably represent a primordial system that for some reason has not given way to the more efficient and more general ribosome-mRNA mechanism. Polypeptide hormones (having 8 to 12 amino acids) are widespread in animal systems but in these evolutionarily advanced systems they are synthesized on ribosomes.

One of the best understood examples of a polypeptide synthesized without ribosome involvement is the cyclic antibiotic **gramicidin S** (Figure 13-27) made by the bacterium *Bacillus brevis*. This peptide contains two amino acids that are not usually found in proteins—namely, ornithine (Orn) and D-phenylalanine (rather than the usual L isomer). The synthesis of gramicidin S is brought about by two extraordinary enzymes abbreviated as EnzI and EnzII, which are bound together forming a unit called gramicidin synthetase.

Amino acids must be activated in order to form a peptide bond. In normal protein synthesis the activated form is the aminoacyl-tRNA, but in the synthesis of gramicidin S each amino acid is activated by forming a thioester bond with sulfhydryl groups of the enzymes (Enz):

$$\text{Amino acid} + \text{ATP} \rightleftharpoons \text{Aminoacyl-AMP} + 2P_i$$

$$\text{Aminoacyl-AMP} + \text{Enz-SH} \rightleftharpoons \text{Enz-S} \sim \overset{\displaystyle O}{\overset{\displaystyle \|}{C}} - \underset{\underset{\displaystyle R}{|}}{\overset{\overset{\displaystyle H}{|}}{C}} - NH_2 + \text{AMP}$$

Enzyme I has numerous sulfhydryl groups and one molecule each of proline, valine, ornithine, and leucine forms a thioester bond with one of four of the sulfhydryl groups of each enzyme I molecule. The standard amino acid L-phenylalanine forms a thioester with a free enzyme II molecule and is then isomerized to D-phenylalanine. Enzymes I and II then aggregate and the D-phenylalanine on enzyme II is transferred to the proline on enzyme I to form a dipeptide bonded to enzyme I, as shown in the following reaction:

$$\text{EnzI-S} \sim \overset{\displaystyle O}{\overset{\displaystyle \|}{C}} - \text{Pro} - NH_2 + \text{EnzII-S} \sim \overset{\displaystyle O}{\overset{\displaystyle \|}{C}} - \text{Phe} - NH_2 \rightleftharpoons \text{EnzI-S} \sim \overset{\displaystyle O}{\overset{\displaystyle \|}{C}} - \text{Pro} - \text{Phe} - NH_2$$

Dipeptide

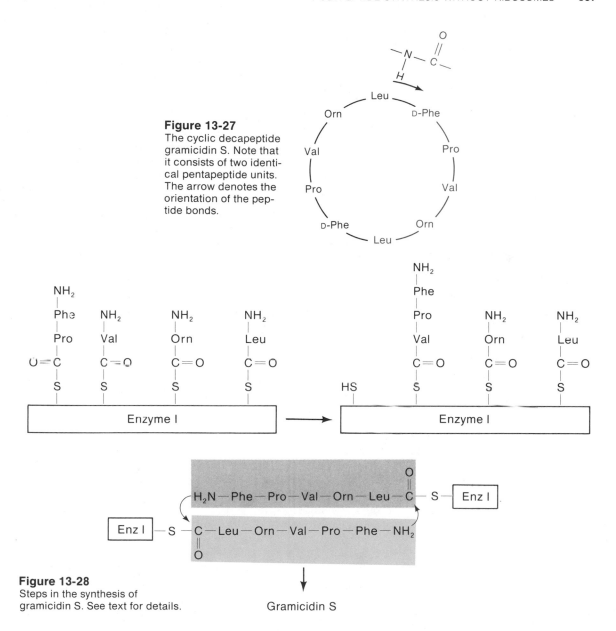

Figure 13-27
The cyclic decapeptide gramicidin S. Note that it consists of two identical pentapeptide units. The arrow denotes the orientation of the peptide bonds.

Figure 13-28
Steps in the synthesis of gramicidin S. See text for details.

Gramicidin S

Enzyme II is no longer needed. Figure 13-28 shows the state of enzyme I after the dipeptide has formed. Peptide synthesis now proceeds by formation of a peptide bond between the activated carbonyl of the dipeptide and the amino group of the adjacent amino acid, as shown in the figure. The process continues with the peptide moving from one sulfhydryl group to the next as each peptide bond is formed. Once the pentapeptide has formed, two EnzI-pentapeptide units interact to form the cyclic decapeptide, gramicidin S.

This remarkable process is very uneconomical, because two enzymes are needed to make a few specific peptide bonds and the information for the amino acid sequence is somehow contained in enzyme I, which specifies the amino acid that can bind to a particular sulfhydryl group. The complexity of such an enzymatic system would have to be very great to synthesize a larger protein. In contrast, in a ribosome-mRNA system, information about sequencing is contained in a single mRNA molecule (one per protein sequence) and the synthetic machinery is the same for all proteins. Thus the ribosome-mRNA system is economical to a cell because of its general applicability.

REFERENCES

Bermek, E. 1978. "Mechanisms in polypeptide chain elongation on ribosomes." *Prog. Nucleic Acid Res. Mol. Biol.,* 2, 63–99.

Brimacombe, R., K. H. Nierhaus, R. A. Garrett, and H. G. Wittman. 1976. "The ribosome of *E. coli.*" *Prog. Nucleic Acid Res. Mol. Biol.,* 18, 1–44.

Caskey, C. T. 1980. "Peptide chain termination." *Trends Biochem. Sci.,* 5, 234–237.

Chambliss, G. (ed.) 1980. *Ribosomes: structure, function, and genetics.* University Park Press.

Clark, B. 1980. "The elongation step of protein biosynthesis." *Trends Biochem. Sci.,* 5, 207–209.

Erdmann, V. A. 1976. "Structure and function of 5S and 5.8S RNA." *Prog. Nucleic Acid Res. Mol. Biol.,* 18, 45–90.

Gold, L., D. Pribnow, T. Schneider, S. Schinedling, B. S. Singer, and G. Storm. 1981. "Translational initiation in prokaryotes." *Ann. Rev. Microbiol.,* 35, 365–404.

Grunberg-Manago, M., and F. Gros. 1977. "Initiation mechanisms of protein synthesis." *Prog. Nucleic Acid Res. Mol. Biol.,* 20, 209–284.

Hunt, T. 1980. "The initiation of protein synthesis." *Trends Biochem. Sci.,* 5, 178–181.

Jagus, R., W. F. Anderson, and B. Safer. 1981. "The regulation of initiation of mammalian protein synthesis." *Prog. Nucleic Acid Res. Mol. Biol.,* 25, 128–186.

Kim, S. 1978. "Three-dimensional structure of transfer RNA and its functional implications." *Adv. Enzymol.,* 46, 279–315.

Kozak, M. 1980. "Evolution of the 'scanning model' for initiation of protein synthesis." *Cell,* 22, 7–8.

Krayefsky, A. A., and M. K. Kukhenova, 1979. "The peptidyl transferase center of ribosomes." *Prog. Nucleic Acid Res. Mol. Biol.,* 23, 2–48.

Kurland, C. G. 1977. "Structure and function of the bacterial ribosome." *Ann. Rev. Biochem.,* 46, 173–200.

Lake, J. 1980. "Ribosome structure and tRNA binding sites." *In* G. Chambliss, ed., *Ribosomes: structure, function, and genetics.* University Park Press.

Montoya, J., D. Ojala, and G. Attardi. 1981. "Distinctive features of the 5′ terminal sequences of the human mitochondrial mRNAs." *Nature,* 290, 465–470.

Nierhaus, K. N. 1982. "Structure, assembly, and function of ribosomes." *Current Topics Micro. Immunol.,* 97, 81–155.

Nomura, M., A. Tissieres, and P. Lengyel (eds.) 1974. *Ribosomes.* Cold Spring Harbor Laboratory.

Ojala, D., J. Montoya, and G. Attardi. 1981. "tRNA punctuation model of RNA processing in human mitochondria." *Nature,* 290, 470–474.

Schimmel, P. R., and D. Soll. 1979. "Aminoacyl tRNA synthetase: general features and recognition of transfer RNA's." *Ann. Rev. Biochem.,* 48, 601–648.

Sherman, F., and J. W. Stewart, 1982. "Mutations altering initiation of translation of yeast iso-1-cytochrome *c*: contrasts between eukaryotic and prokaryotic intiation processes." *In* J. Strathern, E. Jones, and J. Broach, eds., *The Molecular Biology of the Yeast Saccharomyces.* Cold Spring Harbor Laboratory.

Shine, J., and L. Dalgarno. 1974. "The 3′ terminal sequence of *E. coli* 16S ribosomal RNA: complementarity to nonsense triplets and ribosomal binding sites." *Proc. Nat. Acad. Sci.,* 71, 1342–1346.

Steitz, J. A. 1979. "Genetic signals and nucleotide sequences in messenger RNA." *In* R. F. Goldberger, ed., *Biological Regulation and Development.* Vol. 1, pp. 349–399. Plenum.

Steitz, J. A., and K. Jakes. 1975. "How ribosomes select initiation regions in mRNA: base pair formation between the 3′ terminus of 16S rRNA and the mRNA during initiation of protein synthesis in *E. coli. Proc. Nat. Acad. Sci.,* 72, 4734–4738.

Weissbach, H., and S. Pestka, (eds.) 1977. *Molecular Mechanisms in Protein Synthesis.* Academic Press.

Wool, I. 1979. "The structure and function of eukaryotic ribosomes." *Ann. Rev. Biochem.,* 48, 719–754.

14 Regulation of the Activity of Genes and of Gene Products in Prokaryotes

The number of molecules of a protein produced per unit time from a particular gene differs from gene to gene. We have seen in the previous chapters that this can result from varying efficiencies of promoter recognition by RNA polymerase and of initiation of translation. However, this is only one way that the flow of genetic information (**gene expression**) is regulated. For example, the products of many genes are needed only occasionally, and for such a gene the product is usually not present in significant amounts except when circumstances demand it. Thus, not only are there quantitative differences in the amounts of the different gene products that are present but gene activity can apparently be regulated.

In this chapter we consider why gene activity is regulated and the basic mechanisms of metabolic regulation, and we describe numerous examples of well-understood regulated prokaryotic systems.

GENERAL ASPECTS OF REGULATION

In this section some of the general features of regulated systems are discussed.

Why Gene Activity Is Regulated

Natural selection maximizes efficiency. Among the unicellular organisms, any mutation that increases the overall efficiency of cellular metabolism should enable a mutant cell to grow slightly faster than a wild-type organism. Thus, if enough time is allowed, a mutant cell line will outgrow a wild-type one. For example, if in a population of 10^9 bacteria with a 30-minute doubling time one bacterium is altered so that it has a 29.5-minute doubling time, in about 80 days of continued growth, 99.9 percent of the population will have a 29.5-minute doubling time. (In this calculation it is assumed that the growing culture is repeatedly diluted; otherwise, unless the volume were greater than that of the earth, the culture would stop growth in a day or so.) This time may seem very long on the laboratory time scale but it is infinitesimal on the evolutionary time scale. Thus it is reasonable on this basis alone that regulated systems, in which efficiency has been improved, should have evolved.

There are two principal ways for a cell to become more efficient: (1) it can develop a new process that requires less free energy or that simply proceeds at a greater rate, or (2) it can eliminate waste. The second mode is the concern of this chapter.

There are several more-or-less-general rules of intracellular regulation. These are the following:

1. Molecules that are only occasionally needed are synthesized only when the need arises.
2. An enzymatic activity that uselessly consumes energy or consumes a molecule that is the substrate of a second enzyme is usually inhibited.
3. If a cell has a choice of utilizing more than one catabolic pathway for energy utilization, it will choose the pathway that yields the greater amount of energy per unit time.
4. An alteration of a biosynthetic pathway that reduces the production of defective molecules is efficient and therefore valuable.

General Mechanisms of Metabolic Regulation in Bacteria

The most elementary regulatory mechanisms used in bacteria obey the following rule:

A system is turned on when it is needed and off when it is not needed.

This type of on-off activity is accomplished by regulating transcription—that is, mRNA synthesis is either allowed or prevented—or by regulating enzyme activity. Actually, there are no known examples of switching a system completely off. When transcription is in the "off" state, there always remains a basal level of gene expression. This often amounts to only one or two transcription events per cell generation and thus very little protein synthesis. For convenience, when discussing transcription, we will use the term "off," but it should be kept in mind that what is meant is "very low." We will also see examples in this chapter of systems whose activity is switched from fully on to partly on (or even slightly on) rather than to off.

In bacterial systems in which several enzymes act in sequence in a single metabolic pathway, it is often the case that either all of these enzymes are present or all are absent. This phenomenon, which is called **coordinate regulation,** results from control of the synthesis of a single polycistronic mRNA that encodes all of the gene products. There are several mechanisms for this type of regulation, as will be seen in the next major unit of this chapter.

Enzyme activity is commonly regulated by the concentration of the reaction product or, in the case of a biosynthetic pathway, by the product of the reaction sequence. This mode of regulation is called **feedback inhibition** or **end-product inhibition.** Small effector molecules are also frequently used either to activate or inhibit a particular enzyme. Regulation of enzyme activity will be discussed in a later unit of this chapter.

Types of Regulation of Transcription

There are several common patterns of regulation of transcription. These depend on the type of metabolic activity of the system being regulated. For example, in a catabolic (degradative) system the concentration of the substrate of an enzyme early in the pathway (often the enzyme catalyzing the first step) often determines whether the enzymes in the pathway are synthesized. In contrast, in an anabolic (biosynthetic) pathway it is often the case that the final product is the regulatory substance. Even in a system in which a single type of protein molecule is translated from a monocistronic mRNA, the protein may "autoregulate" in the sense that the transcriptional activity of the promoter is determined directly by the concentration of the protein. The molecular mechanisms for each of the regulatory patterns vary quite widely but usually fall in one of two major categories—**negative regulation** and **positive regulation.**

In negative regulation, an inhibitor is present in the cell and the inhibitor keeps transcription turned off. An anti-inhibitor, generally

Table 14-1
Types of regulation

Binding of regulator to DNA	Positive	Negative
Yes	On	Off
No	Off	On

Note: If a system is both positively and negatively regulated, it is "on" when the positive regulator is bound to the DNA and the negative regulator is not bound to the DNA.

called an **inducer,** is needed to turn the system on. In positive regulation, an effector molecule (which may be a protein, a small molecule, or a molecular complex) activates a promoter; an inhibitor is not overridden in positive regulation. Negative and positive regulation are not mutually exclusive, and some systems are both positively and negatively regulated and need two regulators. Properties of negative and positive control are summarized in Table 14-1.

A degradative system may be regulated either positively or negatively. In a biosynthetic pathway, the final product of the pathway often negatively regulates its own synthesis; in the simplest mode, absence of the product encourages synthesis and presence of the product inhibits its synthesis.

Regulation in eukaryotes follows somewhat different rules than does regulation in bacteria—for instance, in eukaryotes all mRNA is monocistronic. The study of regulation in eukaryotes is difficult but has been greatly facilitated by the recombinant DNA technology. Thus, this subject is presented in a later chapter; it is possible to proceed directly to that chapter after completing the present one, if desired.

In the course of this chapter we will examine a variety of regulated bacterial systems. We begin with the best-understood system, the genes responsible for lactose utilization, as study of the lactose system has provided both the language and the principles required to understand regulation.

THE LACTOSE SYSTEM AND THE OPERON MODEL

Metabolic regulation was first studied in detail in the system responsible for degradation of the sugar lactose, and most of the terminology used to describe regulation came from the brilliant genetic analysis of this system by François Jacob and Jacques Monod. We begin with a description of their observations.

Genetic and Biochemical Evidence for the Existence of Two Proteins Required to Degrade Lactose

We now know that in E. coli two proteins are necessary for the metabolism of lactose. These proteins are the enzyme β-galactosidase, which cleaves lactose to yield galactose and glucose (Chapter 2) and a carrier molecule, **lactose permease,** which is required for the entry of lactose into the cell. The existence of two proteins was first shown by a combination of genetic experiments and biochemical analysis.

First, hundreds of Lac$^-$ mutants (unable to use lactose as a carbon source) were isolated. By genetic manipulation some of these mutations were moved from the E. coli chromosome to an F′lac sex plasmid (a plasmid carrying the genes for lactose utilization) and then partial diploids having the genotypes F′lac$^-$/lac$^+$ or F′lac$^+$/lac$^-$ were constructed. (The genotype of the plasmid is given to the left of the diagonal line and that of the chromosome to the right.) It was observed that these diploids always produced a Lac$^+$ phenotype (that is, they made β-galactosidase), which showed that none of the lac$^-$ mutants make an inhibitor that prevents functioning of the lac gene. Partial diploids were also constructed in which both the chromosome and the F′lac plasmid were lac$^-$. Using different pairs of lac$^-$ mutants, it was found that some pairs were phenotypically Lac$^+$ and some were Lac$^-$. This complementation test (Chapter 1) showed that all of the mutants initially isolated fell into one of two groups, which were called lacZ and lacY. (For the sake of simplicity of notation, in the discussion of the lac system, one-letter designations for genetic elements will be used. Thus, the genes lacA, lacI, lacY, lacZ, and the noncoding regions ("sites") lacO and lacP will simply be written a, i, y, z, o, and p respectively.) Mutants in the two groups z and y have the property that the partial diploids F′y$^-$z$^+$/y$^+$z$^-$ and F′y$^+$z$^-$/y$^-$z$^+$ have a Lac$^+$ phenotype and the genotypes F′y$^-$z$^+$/y$^-$z$^+$ and F′y$^+$z$^-$/y$^+$z$^-$ have the Lac$^-$ phenotype. The existence of two complementation groups was good evidence that there were at least two genes in the lac system.

Experiments in which cells were exposed to [^{14}C]lactose showed that radioactive material cannot enter a y$^-$ cell, whereas a z$^-$ mutant readily takes it up. Treatment of a y$^-$ cell with lysozyme, an enzyme that destroys part of the cell wall and makes it very permeable, enables a y$^-$ cell to take up [^{14}C]lactose. This experiment shows that the y gene is probably concerned with lactose transport. Other experiments showed that the z gene is the structural gene for β-galactosidase. For example, immunological analysis of extracts of z$^-$ mutants showed that a protein was present that was structurally similar but not identical to β-galactosidase, though enzymatic activity was lacking. A final important result was obtained by genetic mapping—namely, that the y and z genes are adjacent.

Evidence that the *lac* System Is Regulated: Inducible and Constitutive Synthesis and Repression

The on-off nature of the lactose-utilization system is shown by the following observations:

1. If a culture of E. coli, whose genotype is *lac*+, is growing in a medium lacking lactose or any other β-galactoside, the intracellular concentrations of β-galactosidase and permease are exceedingly low—roughly one or two molecules per bacterium. However, if lactose is present in the growth medium, the concentration of these proteins is about 1 percent of the total cellular protein (about 10^5 molecules per cell).

2. If lactose is added to a *lac*+ culture growing in a lactose-free medium (also lacking glucose, a point that will be discussed shortly), both β-galactosidase and permease are synthesized simultaneously, as shown in Figure 14-1. Furthermore, hybridization of mRNA labeled with ^{32}P at various times after addition of lactose, with DNA obtained from a hybrid phage λ*lac*, which carries the *lac* genes, shows that the addition of lactose triggers synthesis of *lac* mRNA.

These facts led to the view that the lactose system is inducible and that lactose is an inducer. Studies of the mechanism of induction have provided the foundation of our understanding of metabolic regulation.

Figure 14-1
The "on-off" nature of the *lac* system. *Lac* mRNA appears very soon after lactose is added; β-galactosidase and permease appear at the same time but are delayed with respect to mRNA synthesis because of the time required for translation. When lactose is removed, no more *lac* mRNA is made and the amount of *lac* mRNA decreases owing to the usual degradation of mRNA. Both β galactosidase and permease are stable. Their concentration remains constant even though no more can be synthesized. A third protein of the *lac* system, β-galactoside transacetylase, is synthesized coordinately with β-galactosidase and permease. This protein, the product of the *a* gene, will be discussed later.

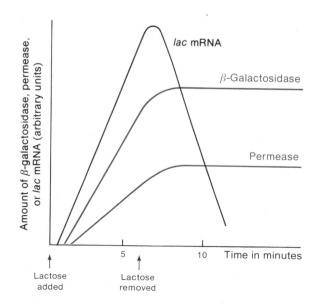

Lactose is rarely used in experiments to study induction because the β-galactosidase that is synthesized catalyzes the cleavage of lactose; thus, the lactose concentration continually decreases and this complicates the analysis of many types of experiments (for example, kinetic experiments). Instead, two sulfur-containing analogues of lactose, in particular isopropylthiogalactoside, IPTG, and thiomethylgalactoside, TMG,

Isopropylthiogalactoside (IPTG)

Thiomethylgalactoside (TMG)

which are effective inducers without being substrates of β-galactosidase, are used. Inducers having this property are called **gratuitous inducers.**

E. coli mutants have been isolated that make *lac* mRNA (and hence β-galactosidase and permease) in the presence as well as absence of an inducer. These unregulated mutants are called **constitutive.** Jacob and Monod, using constitutive mutants, constructed a variety of partial diploid cells and found that these mutants are of two types, which are called *i* and o^c. The characteristics of the mutants are shown in Table 14-2. The i^- mutants behave like ($-$) mutants of most genes and are recessive (entries 3, 4). Since an i^+ cell is turned off for mRNA synthesis

Table 14-2
Characteristics of partial diploids carrying various combinations of *i* and *o* mutants

Genotype	Constitutive(C) or inducible (I)
1. $F'o^c z^+/o^+ z^+$	C
2. $F'o^+ z^+/o^c z^+$	C
3. $F'i^- z^+/i^+ z^+$	I
4. $F'i^+ z^+/i^- z^+$	I
5. $F'o^c z^+/i^- z^+$	C
6. $F'o^c z^-/o^+ z^+$	I
7. $F'o^c z^+/o^+ z^-$	C

and an i^- mutant is on, the i gene is apparently a regulatory gene whose product is an inhibitor that keeps the system turned off. An i^- mutant lacks the inhibitor and thus is constitutive. In an i^+/i^- partial diploid there is one good copy of the i-gene product, so the system is inhibited. Jacob and Monod termed the i-gene product the **lac repressor.** When these observations were first made, it was unclear whether the *lac* repressor was a protein or an RNA molecule that was transcribed from the i gene and does not encode a protein. However, the discovery of amber (chain termination) mutants in the i gene indicated strongly that the repressor is a protein and in 1967 the *lac* repressor was purified. (Properties of this protein will be discussed shortly.) Genetic mapping experiments placed the i gene adjacent to the z gene and established the gene order $i\ z\ y$.

Dominance of o^c Mutants: The Operator

A striking property of the o^c mutants is that in certain cases, they are dominant (entries 1, 2, and 5, Table 14-2). The significance of the dominance of the o^c mutants becomes clear from the properties of the partial diploids, shown in entries 6 and 7. Both combinations are Lac$^+$ as there is a functional z gene. However, entry 6 shows that β-galacto-sidase synthesis is inducible even though an o^o mutation is present. The difference between the two combinations in entries 6 and 7 is that in entry 6 the o^c mutation is carried on a DNA molecule that also has a z^- mutation, whereas in entry 7, o^c and z^+ are carried on the same DNA molecule. Thus, o^c causes constitutive synthesis of β-galactosidase only when o^c and z^+ are on the same DNA molecule. Confirmation of this conclusion comes from an important biochemical observation: an immunological test capable of detecting a mutant β-galactosidase shows that the mutant enzyme is synthesized constitutively in a $o^c z^-/o^+ z^+$ partial diploid (entry 6), whereas the wild-type enzyme is synthesized only if an inducer is added.* A final observation about o^c mutations can be made: genetic mapping experiments show that all o^c mutations map between genes i and z, so that the gene order of the four elements of the *lac* system is $i\ o\ z\ y$. Putting together these observations, Jacob and Monod concluded that o^c mutations define a *site* or a noncoding region of the DNA rather than a gene (because mutations in coding genes should be complementable) and that the o region deter-mines whether synthesis of the product of the adjacent z gene is inducible or constitutive. The o region was named the **operator.**

---

*The test is carried out as follows. Antibody to purified β-galactosidase is prepared. This antibody also reacts with a mutant β-galactosidase as long as the structural differences between wild type and mutant are not too great. This is called **cross-reaction** and the mutant protein is called cross-reacting material, or **CRM.** Thus, the presence of CRM, which can be detected by a variety of standard immunological procedures, is indicative of the presence of a mutant protein.

The Operon Model

The following explanation, called the **operon model** and now known to be correct, was proposed to explain the regulation of the *lac* system (Figure 14-2).

1. The products of the z and y genes are encoded in a single polycistronic mRNA molecule.*
2. The promoter for this mRNA molecule is immediately adjacent to the *o* region. Promoter mutants (p^-) completely incapable of making both β-galactosidase and permease were isolated several years later and shown to be located between *i* and *o*.
3. The operator is a sequence of bases (in the DNA) to which the repressor protein binds.
4. When the repressor protein is bound to the operator, initiation of transcription of *lac* mRNA is prevented.
5. Inducers stimulate mRNA synthesis by binding to the repressor.[†] This binding alters the three-dimensional structure of the repressor so that it cannot bind to the operator. Thus, in the presence of an inducer the operator is unoccupied and the promoter is available for initiation of mRNA synthesis.

This simple model explains many of the features of the *lac* system and of other negatively regulated genetic systems. However, we will see in a later section that this explanation is incomplete as the *lac* operon is also subject to positive regulation.

Background Constitutive Synthesis

As it is stated, the operon model fails to explain two anomalies. First, inducers must penetrate a cell in order to bind to repressor molecules, yet transport of inducers requires permease, and permease synthesis requires induction. Thus, we must explain how the inducer gets into a cell in the first place. There are only two possible explanations—either some inducer can enter a cell without permease or some permease is made without inducer. We will see in a moment that the latter alternative is correct.

A second apparent paradox is that in recent years it has been shown that lactose (galactose-1,4-glucose) does not bind to the repressor and

---

*This mRNA molecule contains a third gene, denoted *a*, that encodes the enzyme **transacetylase**, which catalyzes acetylation of the galactose moiety of lactose. An a^- mutant metabolizes lactose normally; it is Lac$^+$. Transacetylase is not directly involved in lactose metabolism; therefore, it is being ignored at present. Its function will be discussed later.

[†]Because of this observation, with many systems it is common to use the term **derepression** to describe induction.

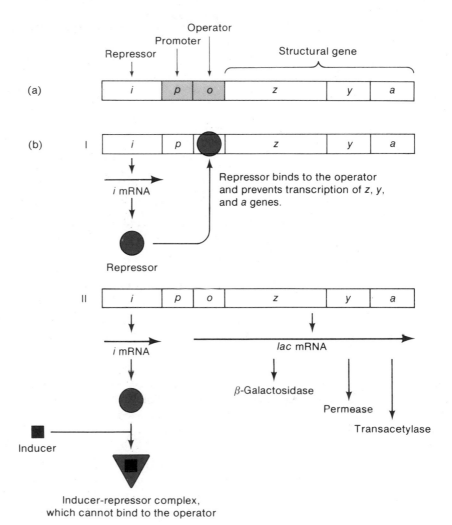

Figure 14-2
(a) Genetic map of the *lac* operon, not drawn to scale: the *p* and *o* sites are actually much smaller than the genes.
(b) Diagram of the *lac* operon in (I) repressed and (II) induced states. The inducer alters the shape of the repressor, so the repressor can no longer bind to the operator.

that the true inducer is a lactose isomer called allo-lactose galactose-1,6-glucose). However, allo-lactose is formed from lactose by the catalytic action of β-galactosidase, so induction of the synthesis of β-galactosidase by lactose requires that β-galactosidase be present. Both of these seeming anomalies have the same explanation—in the uninduced state there is a small amount of *lac* mRNA synthesized (roughly one mRNA molecule per cell per generation). This synthesis, which is called **background constitutive synthesis,** occurs because binding is never infinitely strong. Thus, even though the repressor binds very strongly to the operator, it occasionally comes off and, for the instant that the promoter is free, a RNA polymerase molecule may be able to initiate transcription. This happens about once per cell generation.

Response of a Lac+ Bacterial Culture to Lactose

We are now able to describe in molecular terms the sequence of events following addition of a small amount of lactose to a growing Lac+ culture. Consider bacteria growing in a medium in which the carbon source is glycerol. Each bacterium contains one or two molecules of β-galactosidase and of lactose permease. Lactose is then added. A few lactose molecules are transported into the cell by the single permease molecule and a few molecules of allo-lactose are then made by the single β-galactosidase molecule. An allo-lactose molecule then binds to a repressor molecule that is sitting on the operator and the repressor is inactivated and falls off the operator. Synthesis of mRNA then begins and, from these RNA molecules, hundreds of copies of β-galactosidase and permease are made. This allows lactose molecules to pour into the cell. Most of the lactose molecules are cleaved to yield glucose and galactose, but many molecules are converted to allo-lactose molecules, which bind to and inactivate all of the intracellular repressor molecules. (Repressor is made continuously, though at a very low rate, so that there is usually sufficient allo-lactose to maintain the cell in the derepressed state.) Thus, mRNA is synthesized at a high rate and the concentration of permease and β-galactosidase becomes quite high. The glucose produced by the cleavage reaction is used as a source of carbon and energy. (The galactose formed by the cleavage is converted to glucose by a set of enzymes, the synthesis of which is also inducible. This inducible system, in which galactose is the inducer, is called the *gal* operon. It is discussed in another section of this chapter.)

Ultimately all of the lactose in the growth medium and within the cells is consumed. As a result of repressor synthesis, which has been proceeding unabated, the concentration of active repressor then exceeds the concentration of allo-lactose, and repression is reestablished, thereby eliminating further synthesis of mRNA. In bacteria most mRNA molecules have a half-life of only a few minutes (Chapter 13); in less than one generation there is little remaining *lac* mRNA and synthesis of β-galactosidase and permease ceases. These proteins are quite stable but are gradually diluted out as the cells divide. Note that if lactose were added again to the growth medium one generation after the original lactose had been depleted, cleavage of lactose would begin immediately because the cells would already have adequate permease and β-galactosidase.

Purification of the *lac* Repressor

An important step in proving the principal hypotheses of the operon model was the isolation of the *lac* repressor and the demonstration of its

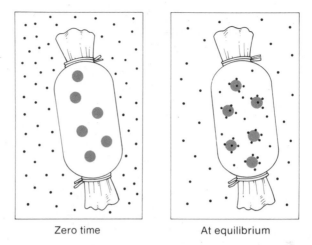

Zero time   At equilibrium

Figure 14-3
Equilibrium dialysis. A dialysis bag filled with a cell extract containing macromolecules (red circles) is placed in a solution containing IPTG (black circles), which can bind to the repressor. At equilibrium, the concentration of free IPTG is the same inside and outside the bag. Because the repressor binds some of the small molecules, the total concentration of small molecules is greater inside the bag than outside.

expected properties. This was accomplished by using the ability to bind radioactive IPTG (one of the gratuitous inducers) as an assay for the repressor. This binding was detected by equilibrium dialysis, as shown in Figure 14-3; by monitoring the IPTG-binding capacity of various protein fractions obtained from a cell extract, the protein was partially purified. The *lac* repressor consists of four identical protein subunits, each containing 347 amino acids and each capable of binding one molecule of IPTG. The crude unfractionated cell extract binds about 20–40 molecules of IPTG per cell, so there are roughly 5–10 repressor molecules per cell. Proof that the IPTG-binding molecule is indeed the repressor came from the observation that the protein is absent in extracts from several i^- mutants.

Since the number of repressor molecules is extremely small, these molecules must be translated from no more than one or two repressor mRNA molecules transcribed per generation time. The number of mRNA molecules is so small that either repressor synthesis itself is regulated or the mRNA is transcribed from a weak promoter. Both mechanisms have been observed for regulation of repressor synthesis in other operons, but for the *lac* repressor the second explanation is correct—that is, repressor mRNA is transcribed constitutively from a weak promoter. The reason for the small number of repressor molecules is made clear from the properties of several mutants in which the weak promoter is converted to a strong promoter. These mutants are uninducible because it is not possible to fill a cell with enough inducer to

Table 14-3
Demonstration of repressor-operator binding by the filter-binding assay

Mixture applied to filter	^{14}C bound to filter
[^{14}C]*lac* DNA	No
[^{14}C]*lac* DNA + repressor	Yes
[^{14}C]*lac* DNA + repressor + IPTG	No
[^{14}C]*lacO^c* DNA + repressor	No

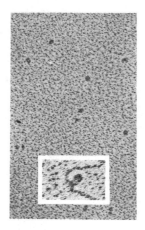

Figure 14-4
Electron micrograph of
E. coli lac repressor
($\times$ 107,000). Inset shows
one repressor molecule
bound to *lac* DNA
($\times$ 215,000). (Courtesy
of Robley Williams.)

overcome repression. These repressor-overproducers have been extremely valuable experimentally because high concentrations of repressor (about 1 percent of the cellular protein) have in turn meant that very large amounts of repressor could be purified—amounts sufficient for physical study and the determination of the amino acid sequence.

With purified repressor the specific binding of repressor to the operator sequence and the inhibition of this binding by an inducer have been demonstrated. The source of *lac* DNA in these studies was a phage variant (a transducing phage, see Chapter 16) that, as a result of genetic manipulations, carried *E. coli lac* DNA. Figure 14-4 shows an electron micrograph of pure *lac* repressor bound to a DNA molecule.

An important procedure for studying repressor-operator binding is the nitrocellulose filter assay. Proteins are retained by these filters but DNA molecules are not; if a mixture of repressor and [^{14}C]DNA is passed through such a filter, ^{14}C will be retained on the filter if the protein and the DNA form a complex. By means of this test the data shown in Table 14-3 have been obtained. These results indicate that the repressor binds to the operator (i.e., it fails to bind to an o^c mutant) and that IPTG prevents this binding, which confirms a major prediction of the operon model.

The Effect of Glucose on the Activity of the *lac* Operon

The function of β-galactosidase in lactose metabolism is to form glucose by cleaving lactose. (The other cleavage product, galactose, is also ultimately converted to glucose by the enzymes of the *gal* operon, as already mentioned.) Thus, if both glucose and lactose are present in the growth medium, then in the interest of efficiency, there is no reason for a cell to induce the *lac* operon. Indeed, cells behave according to this logic; that is, no β-galactosidase is formed until all of the exogenous glucose is consumed (Figure 14-5). The reason for the lack of β-galactosidase synthesis when glucose is present is that no *lac* mRNA is made. There are two ways that this inhibition of *lac* mRNA synthesis might

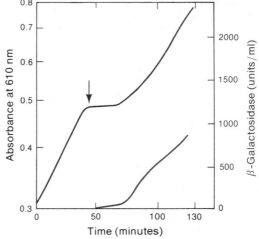

Figure 14-5
An experiment showing the switch from glucose metabolism to lactose metabolism
when a culture containing both glucose and lactose exhausts the supply of glucose. At
the time indicated by the arrow glucose is in such short supply that the cell mass stops
increasing. No β-galactosidase is present because synthesis of *lac* mRNA did not occur
while glucose was present. As the glucose concentration drops, the concentration of
cAMP increases until *lac* mRNA and β-galactosidase synthesis begins (red curve). The
absorbance at 610 nm is a measure of the total mass of the cells; the enzyme activity
is measured in an arbitrarily defined unit.

occur: (1) glucose prevents inactivation of the repressor by the inducer,
or (2) something else is needed for initiating *lac* mRNA synthesis other
than just removing the repressor. We will see in a moment that the
second alternative is the correct one.

The inhibitory effect of glucose on expression of the *lac* operon is
indirect. This was shown by the properties of an *E. coli* mutant unable
to convert glucose 6-phosphate to the next catabolic intermediate in the
glycolysis pathway. This mutant can be induced (e.g., by IPTG) to
synthesize *lac* mRNA even if glucose is present. Thus, rather than
glucose, apparently some glucose catabolite inhibits synthesis of *lac*
mRNA. Because of this result, the effect of glucose has been called
catabolite repression. (This term, as we will see, is a misnomer inasmuch as the process does not utilize a repressor.) In the example in
the previous section, we discussed the induction of the *lac* operon by
the addition of lactose to a minimal medium, in which glycerol is the
carbon source. Glycerol enters the glycolytic pathway at a stage that is
well beyond the reaction of glucose 6-phosphate; hence the catabolite(s)
of interest must precede the glycerol entry step. At the present time,
however, there is no further information about the identity of the actual
catabolite. Nonetheless, the basic mechanism of the glucose effect is
understood, as we shall now see.

Cyclic AMP and the Catabolite Activator Protein

In 1957, Earl Sutherland discovered the substance **cyclic AMP (cAMP)** and found that it is universally distributed in animal tissues (Figure 14-6). Cyclic AMP is synthesized from ATP by the enzyme **adenyl cyclase.** In multicellular eukaryotic organisms cAMP plays an important role in the action of many hormones. It is also present in *E. coli* and its concentration is regulated by glucose metabolism. In a bacterial culture that is starved of an energy source (for example, by incubation in a medium lacking a carbon source), the intracellular concentration of cAMP is very high. If a culture is growing in a medium containing glucose, the cAMP concentration is very low. In a medium containing glycerol or any carbon source that cannot enter the glycolytic pathway, the cAMP concentration is high (Table 14-4). The tentative explanation for these phenomena is that some glucose metabolite that is between glucose 6-phosphate and the glycerol entry point in the glycolytic pathway or a product formed from this metabolite by some other pathway is an inhibitor of adenyl cyclase. The significance of these results is that *cAMP is a mediator of activity of the* lac *operon.*

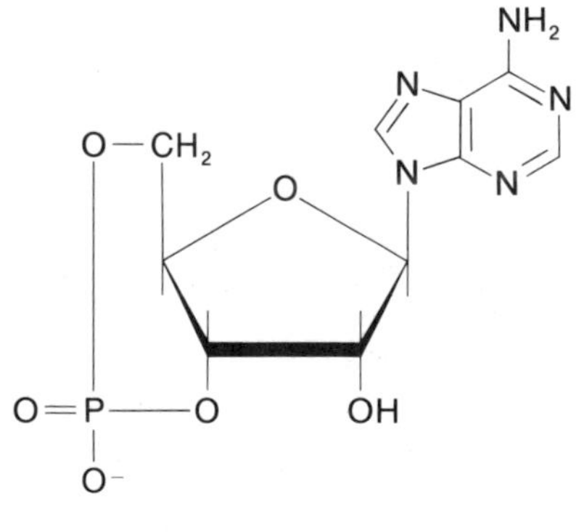

Figure 14-6
Structure of cyclic AMP.

Table 14-4
Concentration of cyclic AMP in cells growing in media having the indicated carbon sources

Carbon source	cAMP concentration
Glucose	Low
Glycerol	High
Lactose	High
Lactose + glucose	Low
Lactose + glycerol	High

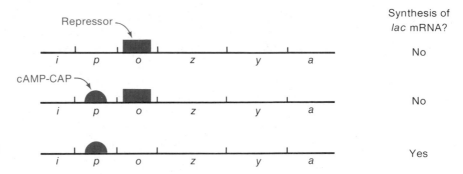

Figure 14-7
Three states of the *lac* operon showing that *lac* mRNA is made only if cAMP-CAP is present and repressor is absent.

E. coli (and presumably other bacteria as well) contain a protein called the **catabolite activator protein (CAP),** which is encoded in a gene called *crp* and which forms a complex with cAMP. Mutants of either *crp* or of adenyl cyclase are unable to synthesize *lac* mRNA, so that both CAP function and cAMP are required for *lac* mRNA synthesis. Further study has shown that CAP and cAMP form a complex, denoted **cAMP-CAP,** and that this is the active complex in the *lac* system. The requirement for the complex is independent of the repression system since *crp* and cyclase mutants are unable to make *lac* mRNA even if there is also an i^- or o^c mutation. We now know that the complex must be bound to a base sequence in the DNA in the promoter region in order for transcription to occur. Thus, *cAMP-CAP is a positive regulator,* in contrast with the repressor, and the *lac* operon is independently regulated both positively and negatively (Figure 14-7).

Other Systems Responding to cAMP

There are many other sugars (e.g., galactose, maltose, arabinose, sorbitol) that are converted to glucose or some intermediate in glycolysis during their degradation. The enzymes responsible for metabolism of each of these sugars are synthesized by inducible operons and, as might be expected, each operon cannot be induced if glucose is present. These are called **catabolite-sensitive operons.** A simple genetic experiment shows that each of these operons is regulated by cAMP-CAP. A single mutant in a sugar operon (e.g., lac^- or mal^-) arises spontaneously at a frequency of roughly 10^{-6}. A double mutant, $lac^- mal^-$, would arise at a frequency of 10^{-12}, which for all practical purposes is unmeasurable. However, double mutants that are phenotypically Lac$^-$Mal$^-$ or Gal$^-$Ara$^-$ do arise at a measurable frequency. These apparent double mutants are not the result of mutations in the two sugar operons but always turn out to be crp^- or are unable to make adenyl cyclase. Furthermore, if a Lac$^-$Mal$^-$ mutant is found, it is always also

Gal⁻Ara⁻, Lac⁻Ara⁻, etc. Biochemical experiments with a few of these catabolite-sensitive operons indicate that binding of cAMP-CAP occurs in the promoter region in each of these systems.

The reason for the requirement for cAMP in the degradation of arabinose (and certain other pentoses) is not obvious since these sugars are not metabolized to glucose. This indicates the more general nature of the role of cAMP. Remember that the concentration of glucose, which is perhaps the optimal carbon source, determines the concentration of cAMP—when the glucose supply is inadequate, cAMP is made in high amounts. Thus, presence of much cAMP is a signal that not enough glucose is available for growth. When the glucose supply is exhausted, the increased cAMP concentration puts a great many sugar operons in a state of readiness so that if any of these sugars is presented to the cell, the response can be rapid. If glucose is present, the signal is absent because such a state of readiness is unnecessary.

Regulatory Region of the DNA of the *lac* Operon

The segment of DNA comprising the *lac* operator has been isolated by the nuclease-protection procedure that we have seen in the study of nucleosomes (Chapter 6) and of RNA polymerase binding sites (Chapter 11). In this procedure purified repressor was adsorbed to DNA that had been isolated from a transducing phage variant (called λ*lac*) that carries the E. coli *lac* operon. Then DNase was added and a short fragment, 24 base pairs long, which survived enzymatic degradation, was isolated. The sequence of bases in this fragment and in a slightly larger fragment isolated in a different way is shown in Figure 14-8. This sequence has a characteristic that will be seen repeatedly in base sequences recognized by regulatory elements—namely, the sequence is an **inverted repeat.**[*] Proof that this palindromic sequence is itself the operator came from studies of operator mutants. A mutant operator sequence could not be isolated by the DNase-protection procedure because the repressor does not bind to a mutant operator. However, by using recombinant DNA techniques (Chapter 20), longer fragments containing the operator were isolated. The operator was then identified in the fragment by locating the operator sequence shown in Figure 14-8; this procedure also defined the base sequence flanking the operator. A mutant operator, which may differ by one or two bases from the wild-type sequence, can be identified by looking for a sequence related to the wild-type sequence between the flanking sequences already identified. This experiment was done by obtaining fragments from several λ*lac* phage particles

---

[*]By this term is meant a sequence such as ABCDEE′D′C′B′A′ in which A and A′ are bases capable of forming the usual base pairs. Such a sequence is also called a **palindrome.** It is thought, and in one case has been proved, that such a sequence may *in vivo* occasionally form hairpin or double stem-and-loop structures of the type shown in Figure 14-9.

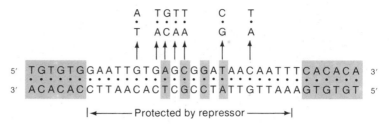

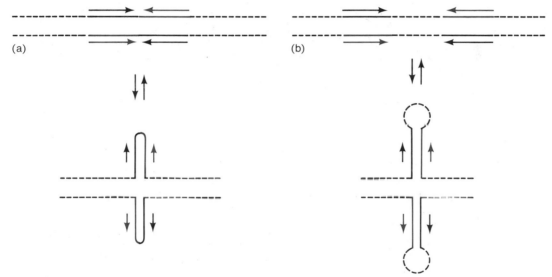

Figure 14-8
Base sequence of the *lac* operator. The symmetrically related regions are shown in red. The arrows point to observed mutational changes that make the operon constitutive.

Figure 14-9
Possible alternative forms of a DNA molecule containing two inverted repeats that are (a) adjacent or (b) separated by a spacer. The horizontal arrows denote orientation of the sequences.

carrying different o^c mutations and observing the base changes shown in Figure 14-8. Note that a single base change either in the inverted repeat sequence or in other base pairs eliminates the ability of the operator to bind the repressor.

The cAMP-CAP binding site has also been isolated by the DNase-protection method. Its sequence is shown below.

Note that this is also an inverted repeat, symmetric about the arrow, except for the two base pairs enclosed in the dotted lines. Two

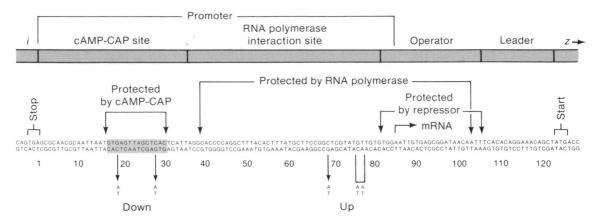

Figure 14-10

DNA base sequence of the regulatory region of the *E. coli lac* operon. The shaded red region indicates the inverted repeat in the cAMP-CAP binding site. Regions protected against DNase digestion by various proteins are shown. Base-pair changes associated with promoter mutations yielding an inactive promoter (down) or an especially active or (cAMP-CAP)-independent promoter (up) are indicated. The start and stop codons in the *z* and *i* gene products respectively are also indicated.

mutations known by genetic and physiological tests to be "promoter-down" mutations (that is, they prevent mRNA synthesis) are base changes in this sequence and these mutations prevent binding of cAMP-CAP; this confirms that binding of cAMP-CAP is a prerequisite for transcription from the *lac* promoter.

A long base sequence containing the cAMP-CAP binding site, the RNA polymerase interaction site and the operator, and flanking sequences has been isolated. This sequence is shown in Figure 14-10. The start point for synthesizing mRNA has been determined by sequencing the 5'-P end of purified *lac* mRNA and finding the complementary sequence in the DNA. The start codon of β-galactosidase and the termination sequences for the *i* gene have also been identified from the known amino acid sequences of β-galactosidase and the repressor. Several features of this entire region should be noted:

1. The initiation site for mRNA synthesis is in the operator and, when the repressor is present, the site is covered.
2. The operator is protected from DNase digestion by bound RNA polymerase.
3 The cAMP-CAP binding site is very near to, but does not overlap, the RNA polymerase interaction site.
4. The regulatory region is 115 base pairs long or roughly 12 turns of the DNA helix.
5. The order of the loci, namely *i p o z*, corresponds to that determined by genetic analysis.

The Mechanisms of Action of the cAMP-CAP Complex and the Repressor

In order for *lac* mRNA synthesis to occur, cAMP-CAP must be bound to the promoter and the operator must not be occupied by a repressor (Figure 14-7). The precise mechanisms by which cAMP-CAP and the repressor respectively stimulate and inhibit transcription are not yet known in detail. However, a reasonable model has been obtained both from a determination of the bases first contacted by RNA polymerase and by the repressor (these complicated experiments will not be described here), as well as the following facts:

1. While RNA polymerase can bind to *lac* DNA *in vitro,* its binding is stimulated by the presence of cAMP-CAP.
2. If RNA polymerase is bound to *lac* DNA, subsequent addition of the repressor cannot displace the polymerase.
3. If the repressor is first bound to the operator, RNA polymerase cannot form a stable complex, but a weak and very transient binding to the RNA polymerase interaction site occurs.

The currently accepted hypothesis for the mechanism of initiation, derived from the information just given, is the following:

1. RNA polymerase forms a loose complex with the *lac* promoter region, like the closed complex described in Chapter 11.
2. Binding of cAMP-CAP probably weakens the double helix in some way and thereby facilitates formation of the stable open promoter complex (Chapter 11).*
3. The repressor is an antimelting protein and thereby prevents formation of an open complex. Alternately, a bound repressor molecule might prevent access of the polymerase to the bases in an open complex.
4. If an open-promoter complex forms, RNA polymerase then initiates synthesis of *lac* mRNA.

Physiological Properties of an Incomplete System

If regulation is valuable to a cell, a mutant in which an operon is unregulated should be disadvantaged in some way. So far the only handicap that has been observed with i^- and o^c mutants is a very slight decrease in growth rate. While this decrease is barely observable in the laboratory, on an evolutionary time scale such a mutation, introducing

---

*The destabilization of a double helix at a site distant from the site of binding of a protein has recently been observed in physical experiments. It is called **telestability**.

regulation to a system and increasing the growth rate slightly, would confer a significant advantage to a cell and thus be selected for.

In certain conditions (which conceivably might arise more frequently in nature), constitutive *lac* mutants grow significantly slower. An analysis of these conditions suggests a possible reason for the presence of the fourth gene of the *lac* operon, the *a* gene, mentioned earlier. This gene encodes the enzyme β-galactoside transacetylase, which can transfer an acetyl group from an acetyl donor to many galactosides, though lactose itself cannot be acetylated; this, we will see, is a significant point.

There exist in nature many substituted galactosides that can be cleaved by β-galactosidase. However, often the substituted galactose moiety produced by the cleavage reaction cannot be metabolized further and accumulates to very high levels. This is often detrimental because several galactose derivatives are inhibitory to normal growth when they are present at high concentrations. An example of such a galactoside (though it is not a natural compound) is the inducer IPTG. It has been found that, when IPTG is present, i^- and o^c mutants that are also a^- grow significantly more slowly than the a^+ counterparts. The difference between the a^- and a^+ cells is that in an a^+ cell, the IPTG is acetylated, which prevents it from being cleaved by β-galactosidase, as shown below:

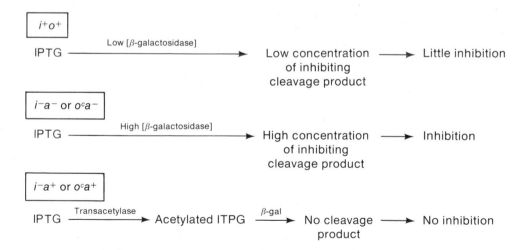

Thus it is likely that the *a* gene plays no role in the breakdown of lactose but instead prevents the accumulation of inhibitory substituted galactose derivatives when β-galactosidase is present in the cell.

It is reasonable to wonder why a cell should expend the energy needed to synthesize transacetylase even when lactose is the only galactoside present, for alternatively this enzyme could be independently regulated. However, independent regulation, uncoupled with β-galactosidase synthesis, is clearly not optimal, since the enzyme is needed only when β-galactosidase is made. It is possible that a secondary regulatory system could be imposed on the *lac* operon but this might be even more costly in energy. It may be that in nature, the substituted galactosides are prevalent in environments in which lactose is the primary carbon source, so that having the *a* gene directly coupled to the *z* gene is reasonable.

Difference in the Amounts of β-Galactosidase, Permease, and Transacetylase Translated from a Single *lac* mRNA Molecule

The ratios of the number of copies of β-galactosidase, permease, and transacetylase in a fully induced cell are $1 : 0.5 : 0.2$. These ratios, which are the results of millions of years of evolution, probably reflect the needs of a cell when exposed to a β-galactoside as the sole carbon source. These differences, which are examples of translational regulation, are achieved in two ways:

1. *Lac* mRNA frequently detaches from its translating ribosome following chain termination. The frequency with which this occurs is a function of the probability of reinitiation at each subsequent AUG codon. Thus, there is a gradient of synthesis from the 5′ terminus to the 3′ terminus of the mRNA. This is true with most polycistronic mRNA molecules.

2. The degradation of *lac* mRNA is initiated more frequently by endonucleolytic scissions in the *a* gene than in the *z* gene. Thus, at any given instant, there are more complete copies of the *z* gene than the *a* gene.

Throughout this chapter, it will be seen that this mode of regulation occurs repeatedly—that is,

The overall expression of activity of an operator is regulated by controlling transcription of a polycistronic mRNA, and the relative concentrations of the proteins encoded in the mRNA are determined by controlling the frequency of translation of each cistron.

Release from Repression in the Absence of an Inducer

If a *lac* operon were placed in a cell lacking repressor, transcription would occur. This situation can be created by transferring the *lac*

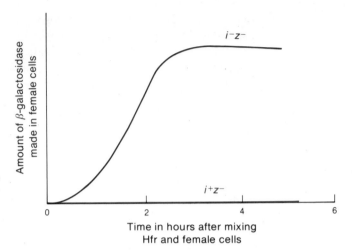

Figure 14-11
Synthesis of β-galactosidase in an i^-z^- or an i^+z^- female accompanying transfer of i^+z^+ DNA from an Hfr cell. No synthesis occurs in the i^+z^- cell, because repressor is always present. Synthesis in the i^-z^- cell terminates since repression is ultimately established.

operon from an Hfr male cell (Chapter 1) having the genotype i^+z^+ to a female cell that is i^-z^- (which makes neither repressor nor β-galactosidase). When the *lac* DNA enters the female cell, which contains no repressor, *lac* mRNA would immediately be made and β-galactosidase would appear in the female (Figure 14-11). Since i^+ mRNA would also be synthesized in the female from the transferred *lac* DNA, repressor would also be synthesized in the female (though at a much lower rate than for β-galactosidase). Thus, the synthesis of β-galactosidase would not continue indefinitely. This phenomenon has been observed for all negatively regulated operons.

Fusions of the *lac* Operon with Other Operons

Gene fusions, which may or may not occur in nature, are of great value in research. One well-studied example is the coupling of the *lac* operon to the *pur* operon, a system reponsible for the synthesis of purines. This operon is regulated in a rather different way from the *lac* operon, though how it is regulated is not relevant to this discussion. The *pur* operon is located "downstream" (in the direction of transcription) from the *lac* operon in the E. *coli* chromosome (Figure 14-12). Between these two operons is a gene *tsx*, which governs sensitivity to the phage T6. Starting with a *tsx-s z*$^+$ cell, a *tsx-r* mutant that was also *z*$^-$ was isolated. This mutant was found to consist of a deletion beginning within the *z* gene, including all of the *tsx* gene, and extending into the *pur* operon past the *pur* operator and promoter. This deletion removes the RNA polymerase termination sequence in the *lac* operon, so that when initiation begins from the *lac* promoter, a mRNA molecule is made that contains the proximal region of the *z* gene and all of the *pur*

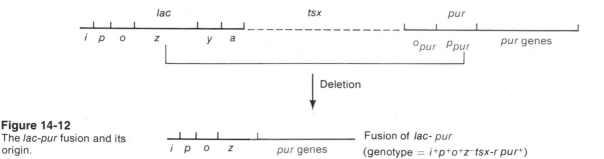

Figure 14-12
The *lac-pur* fusion and its origin.

genes. Thus, in the fused system transcription of the *pur* operon is induced by β-galactosides.

With techniques that allow relocation of the *lac* operon to other parts of the E. *coli* chromosome it has been possible to fuse many genes to the *lac* promoter. This has been a valuable technique for biochemists because the *lac* promoter is a very strong promoter and increased synthesis of proteins transcribed from weak promoters can often be accomplished. In some cases the amount of a particular protein recovered from a cell suspension can be increased a thousandfold.

THE GALACTOSE OPERON

Galactose is another sugar that E. *coli* can utilize as a carbon source. It is metabolized by three enzymes, **galactokinase, galactose transferase,** and **galactose epimerase,** which act in the sequence

$$\text{Galactose} + \text{ATP} \xrightarrow{\text{Galactokinase}} \text{Galactose 1-phosphate} + \text{ADP} + \text{H}^+$$

$$\text{Galactose 1-phosphate} + \text{Uridinediphosphoglucose} \xrightarrow{\text{Galactose transferase}}$$

$$\text{Uridinediphosphogalactose} + \text{Glucose 1-phosphate}$$

$$\text{Uridinediphosphogalactose} \xrightarrow[\text{epimerase}]{\text{Galactose}} \text{Uridinediphosphoglucose}$$

yielding the overall reaction

$$\text{Galactose} + \text{ATP} \longrightarrow \text{Glucose 1-phosphate} + \text{ADP} + \text{H}^+$$

The galactose (*gal*) operon is regulated in principle like the *lac* operon; however, there are several differences in detail that are quite interesting and reflect a dual role of galactose in cellular metabolism.

Genetic Elements of the *gal* Operon

The genes corresponding to the three *gal* enzymes (I, II, III) are *galK, galT,* and *galE,* respectively. Together with an operator *(galO)*, a promoter region *(galP)* and a repressor gene *(galR)*, these genes form an operon.

The most obvious difference between the *gal* and *lac* operons is that *galR,* the structural gene for the repressor, is very far from the cluster consisting of the operator promoter and the *gal* genes (Figure 14-13). Actually, there is no reason for a repressor gene to be adjacent to an operator, since the repressor is a diffusible substance, and in fact, there are as many examples of operons with adjacent repressor and structural genes (e.g., *lac*) as of those with distant repressor genes. The action of the *gal* repressor (which has not yet been purified) is like that of the *lac* repressor in that the $galR^-$ and $galO^c$ mutations confer a constitutive phenotype. The operator and the promoters (more than one, it turns out) are of course adjacent to the structural genes, which is always an optimal arrangement. The inducer is probably galactose rather than a metabolite.

Two *gal* Promoters and the Effect of cAMP-CAP on the Activity of Each One

Galactose is utilized less efficiently than glucose, and one might expect the *gal* operon to be noninducible in the presence of glucose. It is inducible when glucose is present, however, for reasons that will soon be clear. Several promoter mutants have been isolated and, surprisingly, all fall into only two classes. One class of mutant has the property that high-level synthesis of the *gal* enzymes fails to occur only when glucose is present. The other class of mutant is, instead, defective in enzyme synthesis when glucose is absent. Both of these observations are explained by the fact that the *gal* operon has two promoters. The mRNA molecules synthesized from these two promoters differ in length by only five nucleotides, so both transcripts contain the *galK, galT,* and *galE* genes; in fact, the information content of these two mRNA molecules is identical. The properties of these two promoters are best understood by examining the base sequence of the promoter-operator region.

A DNA fragment consisting of 139 base pairs and containing the promoter-operator region has been isolated and sequenced. The sequence of the relevant portion of this fragment is given in Figure 14-14, which shows the start points S1 and S2 for the two *gal* mRNA molecules; these sites have been located by isolating these mRNA

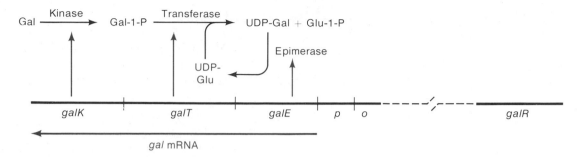

Figure 14-13

The enzymes and elements of the *E. coli gal* operon. Abbreviations: Gal, galactose; Gal-1-P, galactose 1-phosphate; UDP-Gal, uridinediphosphogalactose; Glu-1-P, glucose 1-phosphate; UDP-Glu, uridinediphos- phoglucose. The *galR* repressor gene is very far from the remainder of the operon. Vertical red arrows connect genes and corresponding enzymes.

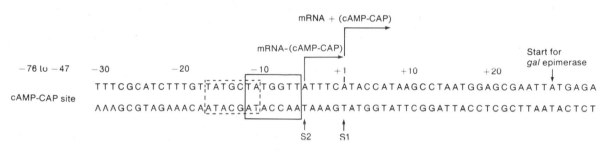

Figure 14-14

Base sequence of the operator-promoter region of the *gal* operon. The +1 refers to the S1 start site for the (cAMP-CAP)-dependent mRNA. The two Pribnow sequences preceding each start site are boxed. S2 is the start site for the (cAMP-CAP)-independent mRNA.

molecules and determining the base sequences of their 5′ termini. S1 and S2 differ functionally in the following way:

1. Transcription from S1 occurs only when glucose is absent and fails to occur in *crp*⁻ and adenyl cyclase mutants. *In vitro* this transcript is made only if cAMP-CAP is present.
2. Transcription from S2 occurs primarily when glucose is present; cAMP-CAP is not required for initiation at S2 and in fact inhibits the transcription from S2.

Apparently, glucose fails to inhibit induction of the *gal* operon because one of the start sites does not require cAMP-CAP and thus remains active even though glucose is present. Hence, the promoter mutation that prevents synthesis of the *gal* enzymes only when glucose is present must be a mutation in S2, and the other mutation, which prevents synthesis when glucose is absent, must be a mutation in S1.

Mechanism of Action of cAMP-CAP and the *gal* Repressor

The site of interaction of cAMP-CAP has been identified by a DNase-protection experiment. The distance of this site from S1 is about the same as the corresponding distance in the *lac* operon. Furthermore, the base sequence shows some degree of symmetry, though not as great as in the *lac* operon. *A priori,* one might expect the cAMP-CAP binding sites in all operons to be identical (since they all bind the same protein) but clearly they are not; in the *lac* and *gal* operons the sequences have some similarities (Figure 14-15) but in the arabinose operon, which will be discussed shortly, it is very different from that in the *lac* operon. The mechanisms by which cAMP-CAP stimulates S1 and inhibits S2 are not clearly known. It is thought that stimulation at S1 occurs in a way that is similar to that in the *lac* operon (namely, facilitation of formation of an open-promoter complex) and that active initiation at S1 may sterically (or otherwise) interfere with initiation at S2.

The mechanism of action of the *gal* repressor is also obscure. All known operator mutations are in the cAMP-CAP binding site, which is so far from the two Pribnow boxes (RNA polymerase binding sites, see Chapter 11) for S1 and S2 that it is very unlikely the *gal* repressor could interfere with RNA polymerase binding as the *lac* repressor does in the *lac* operon. It is thought the *gal* repressor binds to the cAMP-CAP site and that such binding blocks initiation at S1 in two ways—by interfering with the interaction of cAMP-CAP and DNA or by altering the effect of the cAMP-CAP on RNA polymerase binding. Repression at S2 seems to have a totally different mechanism. Apparently, initiation occurs at S2 but, if the *gal* repressor is present, transcription is terminated 10 to 20 bases downstream. How this occurs is unknown.

Possible Value of Two Promoters in the *gal* Operon

Why does the *gal* operon have two initiation sites? The answer is unclear but may have to do with the fact that galactose has two roles in

Figure 14-15
The cAMP-CAP binding sites for the *lac* and *gal* operons. Regions having similar base sequences are shaded in red.

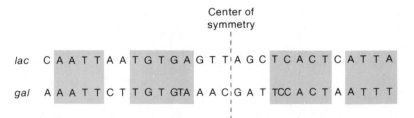

Center of symmetry

lac   C A A T T A A T G T G A G T T A G C T C A C T C A T T A

gal   A A A T T C T T G T GTA A A C G A T TCC A C T A A T T T

cellular metabolism. Galactose not only serves as a carbon source when it is all that is available but the related compound uridinediphospho-galactose (UDPGal) is a precursor in the synthesis of the *E. coli* cell wall. In the absence of exogenous galactose, UDPGal is formed from UDP-glucose in a reaction catalyzed by galactose epimerase, the product of the *galE* gene. Thus, if the cell is to grow, it must be capable of synthesizing the epimerase at all times. This could be accomplished by having a second *galE* gene that is not part of the *gal* operon. Alternatively, since the epimerase is required for cell wall synthesis in very small quantities, the background constitutive synthesis might be sufficient. In fact, the constitutive level of *gal* mRNA synthesis is higher than that of the *lac* operon, and this mRNA is apparently the source of the enzyme when galactose is absent.

Let us suppose that S1 were the only promoter. Since the activity of this promoter depends on cAMP-CAP, the epimerase could not be made constitutively when glucose is present. Suppose instead that S2 were the only promoter. In this case, galactose would fully derepress (induce) the operon even when glucose is present and this would be wasteful. Thus, for the sake of both necessity and economy a cAMP-CAP-independent promoter (S2) is needed for background constitutive synthesis and a cAMP-CAP-dependent promoter (S1) is needed to regulate high-level synthesis; furthermore, the regulation is efficient only if S2 is inhibited by cAMP-CAP.*

The story of the regulation of synthesis of the *gal* enzymes is far from being complete, for there is another regulatory element that has been defined by mutations in a gene called *capR*. (The *capR* gene has nothing to do with the CAP protein.) The phenotype of a *capR⁻* mutant is very complex—namely, elevated production of all of the *gal* enzymes, high sensitivity to ultraviolet light, defective cell division (cells grow to many times the normal length before dividing) and mucoidy (excretion of polysaccharides, which results in a shiny, mucuslike coating on bacterial colonies). The molecular defect in *capR⁻* mutants is not known.

THE ARABINOSE OPERON

Arabinose is another sugar (a pentose) that can serve as a carbon source for metabolism. Three genes *araB, araA,* and *araD,* are needed for arabinose degradation in *E. coli* and form a cluster abbreviated *araBAD*

---

*The jargon term "high-level" is in common usage to mean both high concentration or high rate (the correct meaning is usually obtained from context) in contrast with "low-level," referring to low concentration or low rate.

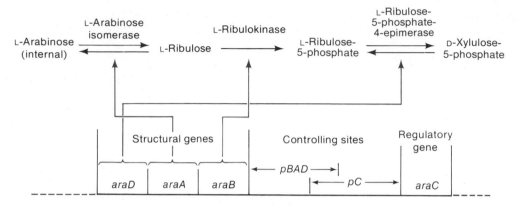

Figure 14-16

A map of the *E. coli ara* operon and the pathway that it regulates. The symbols *pBAD* and *pC* refer to the pro-moters for the *araBAD* and *araC* genes, respectively; the overlap of these promoters is discussed in the text.

(Figure 14-16). These genes encode the enzymes ribulokinase *(araB)*, L-arabinose isomerase *(araA)* and L-ribulose-5-phosphate-4-epimerase *(araD)*. In contrast with the *gal* operon, in which the gene order is identical to the order of the enzymes in the metabolic pathway, the *ara* enzymes are used in the order *araA, araB, araD*. Two other genes, *araE* and *araF*, which are necessary for transport of arabinose across the cell membrane, are quite far from the linked cluster. *AraE* and *araF* code for a membrane protein and for an arabinose-binding protein located between the cell membrane and the cell wall; these two genes are not understood and will not be discussed further.

Adjacent to the *araBAD* cluster is a complex promoter region and a regulatory gene *araC*, whose properties differ markedly from what we have seen in the *lac* and *gal* operons. We will see that the AraC protein functions dually as a positive and negative control element. This contrasts with the *lac* and *gal* repressors, which are solely negative regulators, and the cAMP-CAP protein, which is usually a positive regulator. The *araBAD* genes and the *araC* gene are transcribed in opposite directions with different DNA strands. In the standard orientation of the genetic map the *araBAD* cluster is transcribed leftward from its promoter *pBAD* and the *araC* gene is transcribed rightward from its promoter *pC*.

The *ara* operon is inducible and the inducer is arabinose itself. That is, in a wild-type operon *araBAD* mRNA is made only when arabinose is present. Also, *araBAD* mRNA is not made if glucose is present or by adenyl cyclase-defective and *crp⁻* mutants, which indicates that cAMP-CAP is needed for transcription from *pBAD*.

Positive and Negative Regulation by the AraC Protein

The *araC*-gene product has positive regulatory activity, as shown by the following results:

1. All point and deletion mutations in the *araC* gene, which are denoted *araC⁻*, are unable to synthesize *araBAD* mRNA.

Knowing nothing else, one might also conclude either that an *araC⁻* cell contains a mutant repressor that cannot bind inducer or that the *araC* locus is the promoter. These possibilities are eliminated by the next two results.

2. Partial diploids having the genotype F′*araC⁺*/*araC⁻* are fully inducible.

This means that the *araC⁻* allele is recessive. If *araC⁻* mutants were noninducible repressors, then *araC⁻* should be dominant to *araC⁺*.

3. Partial diploids of the genotype F′*araC⁺araA⁻*/*araC⁻araA⁺* are fully inducible for synthesis of the *araA* product.

If *araC* were a promoter, the *araA⁺* gene, being on the same DNA molecule as the *araC⁻* mutation, could not be transcribed. Since the *araC* product complements in a *trans* configuration, it must be a diffusible molecule.

Other genetic evidence—namely, that there are amber mutants in the *araC* gene—indicates that the *araC* product is a protein. These genetic results have been confirmed by purification of the AraC protein; the amino acid sequence of the protein is also known. Details of the experimental results just given need not be remembered; the important point is that synthesis of *araBAD* mRNA requires a functional AraC protein.

The *araC* product also weakly represses transcription from *pBAD*, a result shown by several complicated genetic experiments that will not be described. In essence, a region of the *ara* operon can be deleted and, if the AraC protein is present (because it is always needed for *araBAD* transcription), some *araBAD* mRNA is synthesized when arabinose is absent. The phenomenon is reminiscent of an operator mutation in the *lac* operon, though there are more differences than similarities. For one thing, the repression is quite minimal compared to that of the *lac* operon in that a *lacOᶜ* mutant gives full expression of the operon, whereas in the *ara* operon, when the repressing activity is eliminated, the amount of *araBAD* mRNA is only 1 percent of the amount found when arabinose is present (that is, in the induced state). The mechanism of this weak repression is not understood. We will see shortly that the AraC protein has still another repressing activity, this time at the *pC* promoter.

The conclusion that both cAMP-CAP and the AraC protein (when acting as a positive regulator) are required to initiate transcription has been confirmed by *in vitro* experiments of two sorts. The first experiments show that no *araBAD* mRNA is synthesized in a reaction mixture containing either (a) only *ara* DNA and RNA polymerase or (b) DNA, RNA polymerase, and only one of the proteins (AraC protein or cAMP-CAP)—that is, both of the proteins must be added. In the second series of experiments, *ara* DNA was examined by electron microscopy (which is capable of visualizing bound RNA polymerase molecules); one result of these experiments was that no RNA polymerase is bound unless the other two proteins are present. How these two proteins act in concert to facilitate *araBAD* transcription will be explained shortly.

Two Forms of the AraC Protein

The dual role of the AraC protein—namely, both as a positive and negative regulator of *pBAD* activity—has been explained by proposing that there are two isomeric forms of the protein. One of these, P_r, is the repressing form and binds to an as-yet-undefined operatorlike site. The other, P_i, is the inducing form; it regulates *araBAD* synthesis by binding to the *pBAD* promoter. P_r and P_i are presumably in equilibrium with one another. In the absence of arabinose, P_r predominates but arabinose, when present, binds to the AraC protein and stabilizes the P_i form. Thus induction by arabinose is pictured as binding of arabinose to P_r, which then leaves its binding site, and production of P_i, which then binds to the promoter.

Many questions about the P_r-to-P_i transition are unanswered. For example, it is not clear how arabinose induces the transition. Physical experiments with purified AraC protein have shown that arabinose does bind to the AraC protein and induces a conformational change, but several other experiments indicate that P_i is not simply an arabinose-AraC protein complex.

Regulation of Synthesis of the AraC Protein

Synthesis of the AraC protein is also regulated. As in the regulation of the *lacI* repressor, the synthesis is self-regulated. That is, the AraC protein itself acts as a negative regulator at *pC* controlling its own synthesis; this is called **autogenous regulation**. Surprisingly, the synthesis of the AraC protein is also regulated by cAMP-CAP. That is, when glucose is present, little AraC protein is made. Thus transcription of the *araBAD* segment and of the *araC* gene are coordinately regulated, which is economical because one protein is not needed without the others.

Table 14-5
Synthesis of β-galactosidase in a *pC-lacZ* fusion strain in various conditions

Genotype	Glucose present	β-Galactosidase made	Conclusions
1. *pC-lacZ*	No	Yes	Transcription of *araC* requires cAMP-CAP
2. *pC-lacZ*	Yes	No	
3. F'*ara*+/*pC-lacZ*	No	Less than in entry 1	AraC protein represses its own synthesis
4. F'*araC*−/*pC-lacZ*	No	Same as in entry 1	

Note: The *lacZ* gene is transcribed from the *araC* promoter *pC*. The *araC* gene is not present in the fusion.

Regulation of the *araC* product has been studied by a novel technique that avoids the necessity of assaying the AraC protein directly, which is quite tedious. By genetic manipulation the *lacZ* gene of the *lac* operon can be moved from its normal location in the *E. coli* chromosome and linked to the promoter for the AraC protein. (This is called a *lacZ-araC* **promoter fusion.**) This rearranged bacterium no longer has an *araC* gene (it has been deleted) but nonetheless the transcriptional activity of the gene can be measured; that is, because the *lacZ* gene is no longer coupled to the *lac* promoter but instead is under control of the *araC* promoter, β-galactosidase is translated only from mRNA initiating at *pC*. Thus, the amount of the enzyme synthesized, which is easily assayed, is a measure of AraC protein synthesis. The amount of β-galactosidase is measured in a culture containing the fusion. If glucose is present in the medium, the amount of enzyme made decreases, which indicates that transcription of the *araC* gene requires cAMP-CAP. If the sex plasmid F'*ara* is added to the strain carrying the fusion, synthesis of β-galactosidase decreases, but if an *araC*− mutation is also carried on the F', there is no decrease in enzyme synthesis. Thus, the AraC protein decreases synthesis of β-galactosidase and hence, by inference, it regulates its own synthesis and must do so by binding to the *araC* promoter. These results are summarized in Table 14-5.

Location of Regulatory Binding Sites in the *ara* Operon

A great deal is known about how the *ara* operon is regulated. The crucial information is the location of the protein binding sites (Figure 14-17). These sites have been located by sequencing experiments. First, the base sequence of the entire *ara* regulatory region extending from within the *araB* gene and through the *araC* gene is determined. Then the DNase-protection method (which we have seen several times in this

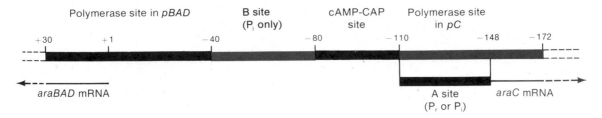

Figure 14-17

The regulatory region of the *ara* operon showing the binding sites for RNA polymerase, cAMP-CAP, and the P_r and P_i forms of the AraC protein. Numbers refer to nucleotides; the start point for *araBAD* mRNA is numbered 1. The A site is contained wholly within the polymerase site in *pC*.

book) is used to locate the regions protected by the cAMP-CAP protein, the AraC protein in the presence and absence of arabinose, and RNA polymerase. The start points for transcription are identified by sequencing mRNA molecules. Other experiments indicate that P_r and P_i compete with RNA polymerase for binding to *pC* and that both cAMP-CAP and P_i are needed for binding of RNA polymerase to *pBAD*. The relevant portions of the regulatory region are shown in Figure 14-17. Five important regions have been detected. These regions and their significance are listed below:

1. The site of binding of RNA polymerase at *pBAD* for *araBAD* transcription.
2. The B site, which is responsible for positive control of *araBAD* transcription by P_i. The important feature of this site is that RNA polymerase fails to bind to *pBAD* unless the B site is occupied by P_i.
3. A single cAMP-CAP binding site common to both *pBAD* and *pC*. This site regulates the expression of the *araBAD* segment at its left and the *araC* gene at its right inasmuch as binding of RNA polymerase is reduced unless the site is occupied by cAMP-CAP.
4. The site of binding of RNA polymerase at *pC* for *araC* transcription.
5. The A site, which is located within site 4 and which can be identified as the operator for the *araC* gene. Both P_r and P_i bind to the A site and in so doing inhibit binding of RNA polymerase to *pC*. Thus, if the AraC protein is in excess, its synthesis is reduced. However, transcription from *pC* is never totally turned off by P_r and P_i, so that if arabinose is present, there is always sufficient P_i to occupy the B site and sustain transcription from *pBAD*.

Note that the *pBAD* promoter includes sites 1, 2, and 3, whereas the *pC* promoter consists of sites 3 and 4.

Response of the *ara* Operon to the Nutritional State of the Cells

Let us now examine the state of the regulatory region in different nutritional conditions (Figure 14-18). In panel (a) of this figure, glucose is present, so there is no cAMP-CAP at its binding site. Arabinose is also absent, so the AraC protein is in the P_r form and occupies the A site. Thus, pC is rarely occupied by RNA polymerase and little AraC protein is made. The system, which is not needed in active form when glucose is present, is quiescent; it is not completely turned off (because it never is) but is as off as it can be. In panel (b) there is no glucose but still no arabinose. Since there is an elevation in the amount of cAMP, the cAMP-CAP site is occupied. P_r still keeps the concentration of the AraC protein low but since it does not compete with RNA polymerase particularly well (not as effectively as P_i), more AraC protein is made. Unless there is another carbon source, the glucose-starved cell is unable to grow.

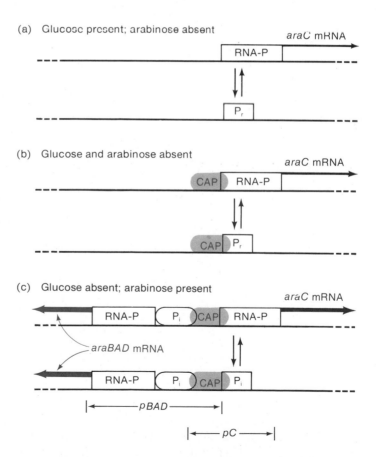

(a) Glucose present; arabinose absent

(b) Glucose and arabinose absent

(c) Glucose absent; arabinose present

Figure 14-18
The state of the regulatory region of the *ara* operon in various nutritional conditions. RNA-P and CAP symbolize RNA polymerase and cAMP-CAP, respectively.

By elevating the concentration of the AraC protein, the cell is in a state of readiness; once arabinose is provided, the P_r will be able to undergo the transition to P_i and the B site can be occupied. In panel (c), arabinose is present and there is no glucose; this is the fully induced situation. The B site is now occupied, as is the cAMP-CAP site, so that RNA polymerase is able to bind to *pBAD*, and *araBAD* mRNA is made. The A site is also occupied by P_i, which competes effectively with RNA polymerase for the *pC* promoter. Wasteful excess synthesis of the AraC protein is thereby avoided but, if the concentration of the protein were to drop to a value at which there is an insufficient amount to maintain transcription from *pBAD*, the A site would be vacant and RNA polymerase would reenter *pC* and make more *araC* mRNA.

If glucose were added to the fully induced system, the *ara* enzymes would no longer be needed. Glucose reduces the concentration of cAMP so that the cAMP-CAP site would be unoccupied and binding of RNA polymerase to *pBAD* would be reduced, as economy would demand. Similarly, if the arabinose were exhausted (and there were no glucose), P_i would revert to P_r, the B site would be vacant, and again synthesis of *araBAD* mRNA, which would be wasteful, would be prevented.

Note that the features of the *ara* operon that make it different from the *lac* and *gal* operons are that it is a system in which two positive regulators (P_i and cAMP-CAP) act in tandem to activate a single promoter *(pBAD)* and in which cAMP-CAP stimulates transcription in both directions (at *pBAD* and *pC*).

THE TRYPTOPHAN OPERON, A BIOSYNTHETIC SYSTEM

The tryptophan *(trp)* operon is responsible for the synthesis of tryptophan. Regulation of this operon is based on the simple principle that when tryptophan is present in the growth medium, there is no need to activate the *trp* operon. Thus, we should expect there to be a regulatory system that turns off *trp* transcription when adequate tryptophan is present and turns it on when tryptophan is absent—in other words, tryptophan, or a related compound, should be active in repression rather than induction. Furthermore, since the *trp* system forms a biosynthetic rather than a degradative pathway, there should be no inhibition by glucose; indeed, there is not, and cAMP-CAP plays no role in the activity of the *trp* operon.

In nature a situation may arise in which there is a small exogenous supply of tryptophan, but not enough to allow normal growth if synthesis of tryptophan were to be totally shut down. There are two ways to avoid tryptophan starvation when there is a suboptimal supply of the amino acid. (1) Repression can be prevented until the external concentration of tryptophan exceeds a critical level. This mechanism is not particularly efficient because more of the biosynthetic enzymes may be made than are needed. (2) A modulating system could be utilized in which the amount of transcription in the derepressed state is determined by the concentration of tryptophan. This mechanism is more efficient than the first mechanism and is the one that is used in the *trp* operon (and in other amino-acid biosynthetic operons as well).

The Elements of the *trp* Operon

Tryptophan is synthesized in five steps, each requiring a particular enzyme. The genes encoding these enzymes in *E. coli* are adjacent to one another in the same order as their use in the biosynthetic pathway; they are translated from a single polycistronic mRNA and are called *trpE, trpD, trpC, trpB,* and *trpA.* The *trpE* gene is the first one translated. Adjacent to the *trpE* gene are the promoter, the operator, and two regions called the **leader** and the **attenuator,** which are designated *trpL* and *trp a* (not *trpA*), respectively (Figure 14-19). The repressor gene *trpR* is located very far from this gene cluster, like the repressor in the *gal* operon. Two other elements also participate in regulation of the *trp* operon, namely, tRNA^{Trp} and tryptophanyl-tRNA synthetase. The genes for these elements are also quite far from the cluster.

The *trp* Repressor-Operator System

Mutations in the *trpR* gene and in the operator cause constitutive initiation of *trp* mRNA synthesis, like what happens in the *lac* operon. The protein product of the *trpR* gene, which is often called the *trp* **aporepressor** does not bind to the operator unless tryptophan is present. The aporepressor protein and the tryptophan molecule join together to form an active repressor that binds to the operator. The reaction scheme is:

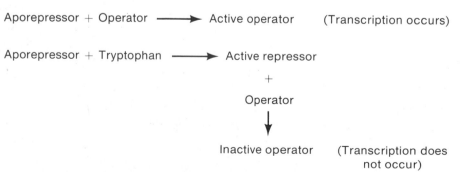

Thus, only when tryptophan is present does an active repressor molecule inhibit transcription. When the external supply of tryptophan

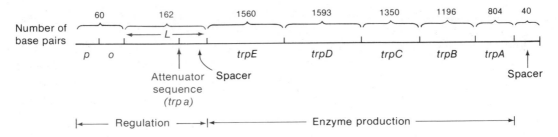

Figure 14-19
The *E. coli trp* operon. For clarity, the regulatory region is enlarged with respect to the coding region. The proper size of each region is indicated by the number of base pairs. *L* is the leader. The regulatory elements are shown in red.

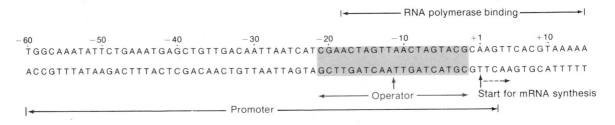

Figure 14-20
The promoter-operator base sequence of the *trp* operon. The operator is shaded in red and the center of symmetry is denoted by a red arrow. The AUG start codon for the first gene in the *trp* operon is at position +162.

is depleted (or reduced substantially), the equilibrium in the equation above shifts to the left, the operator is unoccupied, and transcription begins. This is the basic "on-off" regulatory mechanism.

The promoter and operator regions overlap significantly and binding of the active repressor and RNA polymerase are competitive. *In vitro*, if repressor is added first, RNA polymerase cannot bind, and vice versa. Whether the repressor acts as an antimelting protein, as seems to be the case with the *lac* repressor, is not known.

The base sequence of the promoter-operator region and of the adjacent region through the start point for *trp* mRNA synthesis is shown in Figure 14-20. The operator is an inverted repetitive sequence with 18 of 20 base pairs participating in the symmetry.

The repressor-operator mechanism should be a sufficient on-off switch for the *trp* operon. However, in operons responsible for the biosynthesis of amino acids an additional system allows a finer control in which the enzyme concentration is varied according to the amino acid concentration. This control is effected by (1) premature termination of transcription before the first structural gene is reached and (2) regulation of the frequency of this termination by the concentration of the amino acid. How this is accomplished is described in the following section.

The Attenuator and the Leader Polypeptide

Beginning at the 5′ end of the *trp* mRNA molecule, there are 162 bases before the start codon of the *trpE* gene. This segment of the mRNA is called the leader (a general term used for such regions; see Chapter 11). Within the leader is a sequence of bases (bases 123 through 150) which, if deleted, causes a sixfold increase in the synthesis of the *trp* enzymes in either a derepressed cell or a constitutive mutant cell. Thus, bases 123–150 must have regulatory activity. Furthermore, after initiation of mRNA synthesis, unless there is no tryptophan at all, most of the mRNA molecules are terminated in this region, yielding an RNA molecule consisting of only 140 nucleotides and stopping short of the genes encoding the *trp* enzymes; this explains why deletion of this region results in increased gene expression. This region, in which termination occurs and is regulated, is called the **attenuator.** Careful examination of the base sequence (Figure 14-21) around which termination occurs shows that it contains the usual features of a termination site—namely, a possible stem-and-loop configuration in the mRNA followed by a sequence of eight A · U pairs, as explained in Chapter 11.

A variety of experiments indicate that attenuation requires the presence of charged tRNA^Trp, which suggests that some part of the leader is translated. Examination of the leader sequence shows an AUG codon and a later in-phase UGA stop codon; if translation were to begin at the AUG codon, a polypeptide consisting of fourteen amino acids could be synthesized (Figure 14-22). This hypothetical polypeptide, which has not yet been seen, is called the leader polypeptide. There is some evidence that suggests the peptide may actually be synthesized. For example, there is a ribosome binding site in the mRNA just on the 5′ side of the AUG site. In addition, fusion of the coding region of the leader peptide with the end of the *trpE* gene results in production of a fusion protein having the same NH$_2$-terminal sequence that the leader peptide sequence has.

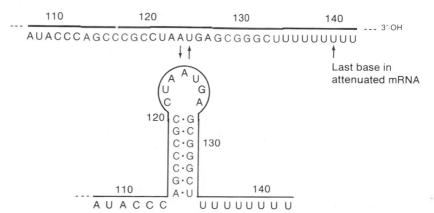

Figure 14-21
The terminal region (right end of *L* in Figure 14-19) of the *trp* leader mRNA . The base sequence given is extended past the termination site at position 140 to show the long stretch of U's. The red bases form an inverted repeat sequence that could lead to the stem-and-loop configuration shown (segment 3–4, Figure 14-24).

	Hypothetical leader polypeptide		TrpE protein

Met Lys Ala Ile Phe Val Leu Lys Gly Trp Trp Arg Thr Ser— Stop   Met Gln Thr Gln
pppAAG…(23)…AUG AAA GCA AUU UUC GUA CUG AAA GGU UGG UGG CGC ACU UCC UGA …(91)… AUG CAA ACA CAA

Figure 14-22
The sequence of the *trp* leader mRNA showing the hypothetical leader polypeptide, the two Trp codons (shaded
red), and the beginning of the TrpE protein. The numbers (23 and 91) refer to the number of bases whose
sequences are omitted for clarity.

(a)   Met Lys His Ile Pro Phe Phe Phe Ala Phe Phe Phe Thr Phe Pro Stop
     5′ AUG AAA CAC AUA CCG UUU UUC UUC GCA UUC UUU UUU ACC UCC CCC UGA 3′

(b)   Met Thr Arg Val Gln Phe Lys His His His His His His His Pro Asp
     5′ AUG ACA CGC GUU CAA UUU AAA CAC CAC CAU CAU CAC CAU CAU CCU GAC 3′

Figure 14-23
Amino acid sequence of the leader peptide and base
sequence of the corresponding portion of mRNA from
(a) the phenylalanine operon and (b) the histidine
operon. The repeating amino acid is shaded in red.

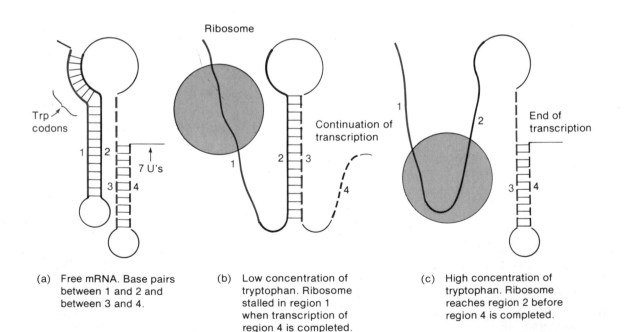

(a) Free mRNA. Base pairs
between 1 and 2 and
between 3 and 4.

(b) Low concentration of
tryptophan. Ribosome
stalled in region 1
when transcription of
region 4 is completed.

(c) High concentration of
tryptophan. Ribosome
reaches region 2 before
region 4 is completed.

Figure 14-24
The accepted model for the mechanism of attenuation in the *E. coli trp* operon.

The leader polypeptide has an interesting feature—namely, at positions 10 and 11 there are two adjacent tryptophan codons. This is significant because in the histidine operon, which also has an attenuator system (i.e., prematurely terminated mRNA), there is a similar base sequence that could encode a leader polypeptide having *seven* adjacent histidine codons (Figure 14-23). We will see the significance of these repeated codons shortly.

Base-pairing in the mRNA Leader

The *trp* leader has been completely sequenced. The most notable features are four segments denoted 1, 2, 3, and 4, which are capable of base-pairing (Figure 14-24) in two different ways—namely, forming either the base-paired regions 1–2 and 3–4 or just the region 2–3. By use of the enzyme RNase T1, which cannot digest base-paired RNA, it has been shown that the two paired regions 1–2 and 3–4 are present in purified *trp* leader mRNA. However, the paired region 3–4, which, by its location, should be in the terminator recognition region, has been shown to be necessary for termination—that is, several mutations in this region both prevent transcription termination *in vivo* and eliminate resistance of that region to RNase T1 *in vitro*. It should be noted that because sequences 2 and 3 are paired in the duplex segments 1–2 and 3–4, then the region 2–3 cannot be present simultaneously with 1–2 and 3–4. This also means—and this is the essential point—that if conditions were somehow right for formation of region 2–3, then neither of the regions 1–2 or 3–4 could be present. Together these facts that have led to a theory of the mechanism of premature termination in the *trp* leader; the theory, which is presented in the following section, is widely accepted though not yet proved rigorously.

The Mechanism of Attenuation

This theory (Figure 14-24) proposes that mRNA termination is mediated through translation of the leader peptide "gene." Because there are two tryptophan codons in this gene, the ability to translate the sequence should be sensitive to the concentration of charged $tRNA^{Trp}$—that is, if the amount of tryptophan is limiting, there will be insufficient charged $tRNA^{Trp}$ and hence translation across the tryptophan codons should be very slow. It is also assumed that transcription and translation are coupled, as is usually true in bacteria, and that all base pairing is eliminated in the segment of the mRNA that is in contact with the ribosome. Figure 14-24 shows that the end of the *trp* leader peptide is in segment 1. Usually a translating ribosome is in contact with about ten bases in the mRNA past the codons being translated, so that when the final codons are being translated, segments 1 and 2 are not paired. In a coupled transcription-translation system, the leading ribosome is not far behind the RNA polymerase. Thus, if the ribosome is in contact with segment 2 when synthesis of segment 4 is being completed, then segments 3 and 4 are free to form the duplex region 3–4 without segment 2 competing for segment 3. The presence of the 3–4 stem-and-loop configuration allows termination to occur when the terminating sequence of seven U's is reached. If there is no added tryptophan, the concentration of charged tRNA^{Trp} becomes limiting and occasionally a translating ribosome is stalled for an instant at the tryptophan codons. These codons are located sixteen bases before the beginning of segment 2. Thus segment 2 is free before segment 4 has been synthesized and the duplex 2–3 region can form. In the absence of the 3–4 stem and loop, termination does not occur and the complete mRNA molecule is made including the coding sequences for the *trp* genes. Thus, if tryptophan is present in excess, termination occurs and little enzyme is synthesized; if tryptophan is absent, there is no termination and the enzymes are made.

Repression versus Attenuation

The *trp* repressor-operator system does not operate as a simple on-off switch but can yield intermediate concentrations of tryptophan. It is known that the synthesis of *trp* mRNA is partially repressed at all times in a cell growing in the absence of added tryptophan because the concentration of the *trp* enzymes is tenfold greater in a cell having a mutant (inactive) repressor than in a wild-type cell. This observation implies that in a wild-type cell, if the endogenous concentration of tryptophan were to fluctuate for any reason, the equilibrium between active and inactive repressor would shift to maintain a usable supply of tryptophan. Thus it is not at all clear why the attenuation system is needed. Possibly the rate of conversion of active repressor to inactive repressor is sufficiently low that the attenuation system allows a more rapid response to a decreasing concentration of tryptophan.

It is also possible that the attenuator is needed when exogenous tryptophan is present, but at a concentration too low to support growth without additional endogenously synthesized tryptophan. For example, an exogenous concentration might be sufficient to increase repression but be insufficient to increase it to the point that endogenous production of tryptophan is triggered, so as to synthesize necessary proteins. In this situation the attenuator system would respond to the lower total endogenous concentration of tryptophan by not terminating ("antiterminating") the mRNA, thereby increasing the synthesis of *trp* enzymes. In this case the attenuator can be viewed as a modulator of environmentally determined fluctuations of internal tryptophan concentrations.

One might also ask why there is a repression system; the attenuator seems to be sufficient to regulate synthesis of the *trp* enzymes without such a system. Possibly the existence of a repressor yields the added economy of preventing unnecessary synthesis of the leader mRNA when a great deal of exogenous tryptophan is present. Indeed, the histidine operon, which also has an attenuator capable of forming three base-paired regions and whose function is virtually identical to that of the *trp* attenuator, has no repressor—that is, the *his* operon is regulated entirely by an attenuator.

At the present time, we do not know why the *trp* operon has a dual regulatory system and the *his* operon does not, and more detailed studies are needed.

Many operons responsible for amino acid biosynthesis are regulated by attenuators equipped with the base-pairing mechanism for competition described for the *trp* operon. So far this has been described for the histidine, threonine, leucine, isoleucine-valine, and phenylalanine operons of the bacteria *E. coli, Salmonella typhimurium,* and *Serratia marcescens,* though in less detail than for the *trp* operon.

THE HISTIDINE-UTILIZATION (*hut*) OPERON

In addition to their role in protein synthesis many amino acids can serve as a source of carbon and nitrogen atoms when a supply of either of these atoms is inadequate. Because of this dual role, the degradation of amino acids must be carefully regulated. The histidine-utilization (*hut*) operon, which is reponsible for the degradation of histidine, is such a regulated system (Figure 14-25). The biochemistry of the degradation is complicated and little information is available about the base sequences of the regulatory regions.* Nonetheless, it is important to be aware of this operon because it is a fine example of a multiply regulated operon, having two promoters, two operators, and two positive regula-

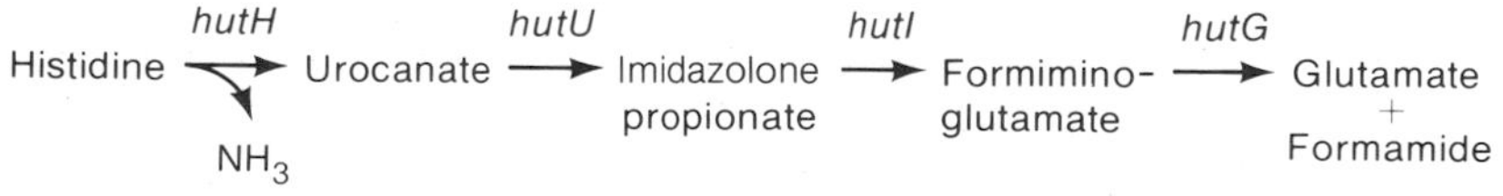

Histidine $\xrightarrow[\text{NH}_3]{hutH}$ Urocanate $\xrightarrow{hutU}$ Imidazolone propionate $\xrightarrow{hutI}$ Formimino-glutamate $\xrightarrow{hutG}$ Glutamate + Formamide

Figure 14-25
The pathway for histidine utilization, showing the genes responsible for each reaction.

*Recently these regions were cloned and isolated, and this represents an enormous step toward an understanding that has been elusive for so many years.

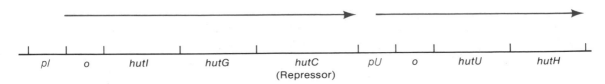

Figure 14-26
Genetic map of the *hut* operon, showing the two mRNA molecules (red).

tors (for each promoter-operator unit). The reason for this complexity will become clear shortly.

The *hut* operon encodes four enzymes, which are specified by the genes *hutG, hutH, hutI,* and *hutU,* and a repressor, which is the product of the *hutC* gene. In *Klebsiella aerogenes* these genes are organized into two transcription units—namely, *hutI hutG hutC* and *hutU hutH* (Figure 14-26).* Both of these units have a promoter and an operator, and transcription is from left to right as drawn in the figure. The operators are presumably identical or nearly so and one *hutC* repressor molecule binds to each operator. Induction is accomplished by addition of histidine, though the true inducer is urocanate, which is formed from histidine by the *hutH*-gene product. Binding of urocanate to the repressor converts the repressor to an inactive form that frees the promoter for each mRNA. However, no *hut* mRNA is made because a positive regulator is also required, as in the *lac* operon. The interesting thing about the *hut* operon is that there are two different positive regulators (as in the *ara* operon), but only one of these regulators is needed for binding of RNA polymerase to the promoters.

The *hut* operon is charged with supplying a cell with carbon and nitrogen when either of these atoms is limiting. If histidine is the sole source of carbon but at the same time the supply of nitrogen is adequate, the operon must remain on; the operon must also be active when there is adequate carbon but nitrogen is limiting. The amount of available carbon is sensed by the cAMP system, as was explained in the discussion of the *lac* operon. That is, if carbon is lacking, cAMP is synthesized and cAMP-CAP is present. There is a binding site for this complex in each promoter of the *hut* operon and occupation of this site is sufficient to allow transcription to occur. The signal for nitrogen limitation is an unknown factor, almost certainly a positive effector, that is active when the nitrogen source is poor. The synthesis of this product is related in a complicated and poorly understood way to the synthesis of glutamine synthetase, the enzyme responsible for building ammonia into nitrogen-containing compounds. Recently several of the components of the nitrogen-regulation system have been isolated and

*The *hut* operon has been studied most carefully in the bacteria K. *aerogenes* and *Salmonella typhimurium* and there is no *hut* operon in E. *coli*.

Table 14-6
Summary of the activity of the *hut* operon in various culture conditions

Glucose present	Histidine present	Nitrogen supply	Carbon supply*	cAMP*	Operon "on" or "off"†
—	No	—	—	—	Off
Yes	Yes	Adequate	Adequate (from glucose)	Low	Off
No	Yes	Adequate	Only from histidine	High	On
Yes	Yes	Only from histidine	Adequate (from glucose)	Low	On
No	Yes	Only from histidine	Adequate (not from glucose)	High	On

Note: A dash means that the entry is irrelevant.

*The relation between the concentrations [ ] of carbon and cAMP is High [C] → Low [cAMP] and Low [C] → High[cAMP].

†Operon is on whenever cAMP is present in high concentration.

an *in vitro* system for transcribing the *hut* operon has been established. Therefore, a clearer understanding of the functioning of the *hut* system should be forthcoming.

An outline of the activity of the *hut* operon is summarized in Table 14-6.

It should be noticed that many textbooks describe a direct role of glutamine synthetase (and its inactivation by the addition of AMP) in the nitrogen-sensing system, but this conclusion has recently been shown to be incorrect.

THE *recA* OPERON

Chapter 9 on DNA repair described the SOS repair system, an inducible system, which is called upon in bacteria when the DNA has suffered extensive damage. A major product of the inducible system is the *recA*-gene product. This protein is active in repair and in genetic recombination (for which it is essential in bacteria) and has several activities: it binds to single- and double-stranded DNA, it has an ATPase activity, and it is a protease. The proteolytic activity is very weak with most proteins, but it is very strong against certain repressors in phage systems and cleaves each repressor at a unique site. We will return to this important point shortly. Following DNA damage of the sort inflicted by ultraviolet irradiation, there is a burst of synthesis of the *recA* mRNA. However, this does not occur if *recA* has mutated to *recA*⁻, which is an observation quite different from what has been seen in the operons described so far. For example, if a cell has a galK$^-$ mutation, addition of galactose still turns on the gal operon even though the galK product that is synthesized is defective and has no enzymatic activity. With the recA operon it appears that the functional recA protein is required to turn the operon on. One possible explanation for this autoregulating activity is that the RecA protein is a positive regulator like the araC product. An alternative, which will be seen to be correct, is that the recA product has inducing activity.

Another gene that is a component of the recA operon is lexA, which probably codes for a simple repressor of the operon. LexA$^-$ mutants have been isolated and are phenotypically RecA$^-$. These mutants are apparently noninducible, for they do not respond to ultraviolet irradiation by increasing the amount of RecA protein synthesis.

A great deal of genetic and in vivo biochemical evidence suggests that the RecA protein has DNA-binding activity in undamaged cells but lacks proteolytic activity. In contrast, purified RecA protein has proteolytic activity. This difference between observations in vivo and in vitro is thought to be due either to activation or loss of an inhibitor during isolation of the protein. A current view is that following DNA damage one of two events occurs: (1) the RecA protein binds to single-stranded regions of the DNA produced in an early step in repair and is thereby converted to an active protease, presumably by a conformational change, or (2) some small effector molecule that is a byproduct of an early stage of repair activates the RecA protein. These alternatives are widely debated at present.

The LexA protein has been isolated. In vitro the purified RecA protein cleaves the LexA protein into two fragments. The product of a mutant, lexA3, has also been isolated. It does not increase the synthesis of the RecA protein as is characteristic of a LexA$^+$ bacterium. In vitro the LexA3 protein is resistant to cleavage by purified RecA protein.

These facts have been combined in the following theory to explain regulation of the recA operon. Normally growing cells contain a constitutive level of RecA protein, which lacks proteolytic activity. Following damage to DNA these few protein molecules are activated by an unknown mechanism to the proteolytic form. The RecA protease then cleaves all of the LexA repressor molecules so that the recA operon becomes derepressed and synthesis of the RecA protein is increased. As long as the activating signal remains, the operon stays on but, when repair is completed, the activating signal disappears and the RecA molecules return to the nonproteolytic form. With only this form present, the LexA protein, which has been continually synthesized, accumulates and reestablishes repression. Derepressed synthesis of RecA protein stops, and as a result of growth and cell division, the RecA protein is gradually diluted back to the constitutive level.

REGULATION OF THE SYNTHESIS OF RIBOSOMES

In the previous sections operons have been described whose control is well understood. Here we consider a complex regulatory system about which there is an enormous amount of information but which, after nearly 30 years, remains a mystery.

The growth rates of all bacterial species vary with the composition of the growth medium. In minimal media having an efficiently utilized carbon source such as glucose, *E. coli* cells divide roughly every 45 minutes at 37°C; with a poorer carbon source such as proline, the doubling time is 500 minutes. In rich media containing glucose, amino acids, nucleic acid bases, vitamins, and fatty acids, a cell does not have to synthesize these substances and hence it can grow very rapidly; typically, the generation time is about 25 minutes in rich media. These different growth rates are achieved by regulating the ability to synthesize protein molecules. This regulation could be accomplished by varying (1) the rate of addition of amino acids to growing polypeptide chains or (2) the rate of initiation of new molecules. The polypeptide chain growth rate is in fact constant, and the overall rate of protein synthesis is instead determined by the number of ribosomes per cell. In rapidly growing cells, septum formation between daughter bacteria lags behind DNA synthesis, so that rapidly growing cells contain more DNA molecules than slowly growing cells. For example, Table 14-7 shows the number of DNA molecules per cell at three different growth rates. The number of ribosomes per cell is not proportional to the number of DNA molecules per cell but also increases with growth rate, as shown in the table. Furthermore, the amounts of tRNA, initiation factors, elongation factors, and aminoacyl-tRNA synthetase are related to the number of ribosomes per DNA molecule. Thus, synthesis of ribosomal proteins, rRNA, and other components of the protein synthesis apparatus must be regulated.

In part at least, synthesis of these substances must be regulated by being organized in operons. Genetic mapping experiments indicate that all of the genes encoding the 53 ribosomal proteins of *E. coli* occupy a

Table 14-7
Some characteristics of *E. coli* growing at different growth rates

Doubling time, minutes	DNA molecules per cell	Ribosomes per DNA molecule
25	4.5	15,500
50	2.4	6,800
100	1.7	4,200
300	1.4	1,450

small region of the chromosome, but it is known that these genes are transcribed as about 20 polycistronic mRNA molecules. The rRNA genes occur in three nearly identical but separated regions near the genes for the ribosomal proteins. These clusters include genes for certain tRNA molecules, as was indicated in Chapter 11; the remaining tRNA genes are found in several other clusters scattered throughout the chromosome. Whether all of these DNA segments are coordinately regulated is not known, but it seems unlikely—it is more reasonable to suppose that the product of one operon would regulate the activity of another operon, because, in that way, a constant ratio of the components produced by the operons might result. Actually, translation of six mRNA molecules encoding ribosomal proteins is autogenously regulated by the concentrations of particular ribosomal proteins (one protein from each polycistronic mRNA), so the rate of ribosome assembly, which is controlled, in part, by the rate of rRNA synthesis (see below), indirectly regulates synthesis of these proteins.

If a bacterial culture is transferred from a growth medium in which growth is rapid to one in which growth is slow (this is called a "downshift"), the ribosome content of each cell decreases from the higher value for the rapid medium to the lower value characteristic of the slow medium. This is reasonable since otherwise each cell would possess more ribosomes than it would need. The decrease is accomplished by allowing DNA synthesis to proceed without synthesis of rRNA. Some signal must be sensed by each slowly growing cell and this signal must inhibit transcription of the rRNA genes. In a downshift, amino acids cannot be synthesized as rapidly in the slow medium as in the rapid medium because of the decreased availability of a supply of carbon atoms. Several lines of evidence implicate a decrease in amino acid concentration as a component of the signal just mentioned. The effect of a downshift can be mimicked by transferring an amino acid-requiring mutant from any growth medium that contains the amino acid to a medium having the same carbon source but lacking the required amino acid. Indeed, following such a transfer, synthesis of rRNA rapidly stops. This inhibition of rRNA synthesis associated with amino acid depletion is called the **stringent response.** The factors responsible for the stringent response are gradually being revealed by studying mutants in which rRNA synthesis continues unabated during amino acid starvation. The phenotype of these mutants is described as **relaxed.** The mutations map in several genes designated *rel*. The only *rel* gene that is understood at all is *relA*.

Accompanying amino acid starvation two unusual nucleotides are produced—guanosine-5'-diphosphate-3'-diphosphate (ppGpp) and guanosine-5'-triphosphate-3'-diphosphate (pppGpp). For historical reasons these compounds are known as **magic spots** 1 and 2, respectively.

The product of the *relA* gene is a protein known as the **stringent factor;** it is an enzyme that is reponsible for the synthesis of ppGpp. The stringent factor is located exclusively in the 50S ribosome; during normal protein synthesis the stringent factor is inactive and virtually no ppGpp is synthesized. Activation of the stringent factor occurs only when two conditions are met: (1) the 50S particle must be in an intact 70S ribosome that is bound to mRNA and engaged in translation, and (2) the A site of the ribosome must be occupied by an uncharged tRNA molecule rather than charged tRNA. In normal growth conditions the concentration of uncharged tRNA is much lower than that of charged tRNA, which means that the A site is rarely occupied by uncharged tRNA. However, in the case of starvation for an amino acid, the tRNA species that normally carry that amino acid will be uncharged. In this way amino acid deprivation leads to production of ppGpp. The ppGpp signals that something has stopped protein synthesis and that additional rRNA should not be made.

A great deal of effort has gone into attempting to understand how ppGpp regulates the synthesis of rRNA; recently some evidence has been obtained for binding of ppGpp to RNA polymerase. It is already known that the promoters for rRNA and tRNA differ from the promoters in other operons such as *lac, gal,* and so forth; specifically, the initiation of transcription of the rRNA genes occurs at four times the rate of a fully induced *lac* operon, making the rRNA promoter the most efficient promoter known. It has been suggested that when RNA polymerase binds ppGpp, it loses the ability to bind to the rRNA promoter, but this attractive hypothesis has not yet been proved.

PHASE VARIATION OF FLAGELLA
IN *SALMONELLA TYPHIMURIUM*

In the systems described so far, a bacterial operon responds to concentrations of external substances such as sugars and amino acids. Most of the biosynthetic and degradative pathways in bacteria use systems that at some level at least resemble the operons already described. In this section we examine a system in which the on-versus-off state is not determined by the availability of the promoter to RNA polymerase but by the physical connection (or lack thereof) of the promoter to the structural genes.

Many bacteria possess flagella (Figure 14-27) — that is, long protein-containing hairlike appendages that propel the bacteria through liquid media or across solid surfaces, usually toward a desired nutrient or away from a deleterious agent. Flagella consist primarily of a large number of subunits of a single protein called **flagellin.**

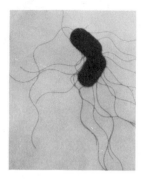

Figure 14-27
Electron micrograph of several cells of *Salmonella typhimurium.* with flagella. (Courtesy of Mel Simon.)

Strains of the bacterium *Salmonella typhimurium* have two genes called *H1* and *H2* that are responsible for the synthesis of two distinct types of flagellin. Both types of flagellin aggregate to form flagella but the flagella are readily distinguishable immunologically. In any particular bacterial cell only one of the two *H* genes is transcribed; thus an individual bacterium makes either H1-type flagella or H2-type flagella. If a bacterium making the H1-type flagella is allowed to grow for many generations and produce a large population of bacteria, it is found that in time bacteria having H2-type flagella arise in the population. The frequency of appearance of the H2-type is about 10^{-5}–10^{-3} switches per cell per generation (the exact frequency depending on the individual strain). If a bacterium with H2-type flagella is selected and allowed to grow, H1 type will appear in time, again at a frequency of 10^{-5}–10^{-3} switches per cell per generation. This phenomenon is called **phase variation.**

The DNA of each gene has been isolated and used as a probe for detecting mRNA in hybridization experiments. The results of these studies indicate that in any given cell only one gene is transcribed. The *H1* and *H2* genes are very far from one another in the *Salmonella* chromosome, which means that some diffusible product must be used as some part of the switching mechanism. This is indeed the case, as we will see.

The *H1* and *H2* genes have been isolated and partially sequenced. There is nothing unusual about the *H1* gene, but in repeated isolates of the *H2* gene it has been found that the gene exists in two forms. At one end of *H2* is a 970-base-pair segment that exists in one orientation in some of the isolates and is inverted in others. Furthermore, only one orientation can be isolated from a population that makes H1-type flagella, whereas the other is isolated exclusively from populations in which H2-type flagellin is made. Thus, it appears that inversion of this 970-base-pair element is related to the activity of the *H1* and *H2* genes. The key to understanding what is happening is that when this element is in one orientation, *H2* is transcribed, and when it is in the other orientation, *H2* is inactive. The explanation for the two states of activity is shown in Figure 14-28. The invertible element of 970 base pairs carries the promoter for *H2*. Thus, in one orientation initiation of mRNA synthesis at this promoter leads to transcription of the *H2* gene, whereas in the off orientation, transcription is away from *H2*. This invertibility accounts for the activation and deactivation of the *H2* gene but not the regulation of *H1*. We suggested earlier that a diffusible gene product plays a role in regulation of this system. When *H2* is transcribed, a polycistronic mRNA is made that encodes not only flagellin but also a protein that inhibits transcription of the *H1* gene. Thus, in the *H2*-on orientation *H2* flagellin is made and *H1* flagellin is not, whereas in the *H2*-off orientation, *H1* flagellin but not *H2* flagellin is made.

(a) *H1* on, *H2* off

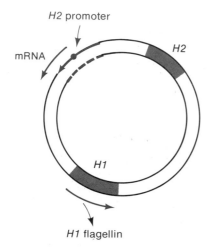

(b) *H1* off, *H2* on

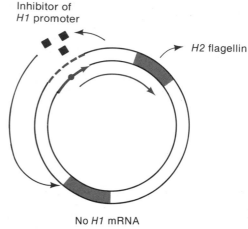

Figure 14-28
Regulation of the two flagellin genes in *Salmonella typhimurium.* The heavy red region is an invertible seg-ment that contains the *H2* promoter. The red arrows represent mRNA molecules.

This system is an example of regulation by means of a **site-specific inversion.** That is, somehow the sequence of 970 base pairs is excised from the chromosome and then reinserted in the opposite orientation. Further analysis of the products encoded in the *H2* polycistronic mRNA indicates that in addition to the *H2* promoter and the inhibitor of *H1* transcription, the invertible element also contains a gene whose product is needed for the inversion.

We have not yet explained what determines whether the inversion does or does not occur. In the systems analyzed earlier in this chapter, gene expression occurs in response to some external influence. It is possible that this is also the case for phase variation, though no environmental factors have been found that affect the switching frequency. It is also possible that inversion is a random event, which in the following situation would have a clear advantage. Consider an animal infected with a small population of bacteria making H1-type flagellin. At the same time that the population was growing, the immune system of the animal would be making antibodies to *H1* flagellin, so in time the population would be destroyed by the antibody. However, if at a significant frequency bacterial variants were to arise that could make H2-type flagellin, these bacteria would survive the antibody.

The advantage of phase variation is different in a significant way from the operons that we have already discussed. In these operons the ability to turn on a needed system or to turn off a wasteful system is advantageous to the individual bacterium. However, phase variation does not enable an individual cell to survive but instead ensures that the *population* is able to make it past a barrage of antibodies. This principle probably underlies many systems in bacteria.

In Chapter 22 it will be seen that site-specific exchanges are used in programming eukaryotic cells—for example, to produce antibodies and to alter the mating type in yeast.

FEEDBACK INHIBITION

In this section a mode of regulation that does not alter transcription is described. If a culture of bacteria in which the *trp* operon has been derepressed is suddenly exposed to tryptophan, repression is rapidly established, and there is little further synthesis of the *trp* enzymes. However, the *trp* enzymes persist and were it not for a second type of regulatory mechanism, wasteful synthesis of tryptophan and needless consumption of precursors and energy would occur. This state of affairs is avoided by **feedback** or **end-product inhibition,** a mechanism by which the activity of the enzymes is turned off by the product of the pathway. In bacteria, feedback inhibition is common in biosynthetic pathways.

In the pathway for tryptophan synthesis the first two reactions are catalyzed by the products of the *trpD* and *trpE* genes (Figure 14-29). These proteins form a tetramolecular aggregate consisting of two subunits of each protein. The *trpE* subunits produce anthranilate (from chlorismate and glutamine) and the *trpD* subunits convert the anthranilate to phosphoribosyl anthranilate (PRA). The *trpE* subunit also contains a binding site for tryptophan that is occupied only when there is an excess of tryptophan. Occupation of this site causes a conformational change in the *trpE* subunit and this change destroys the enzymatic activity (Figure 14-30). The tetramer is an allosteric protein (see Chapter 6) and the conformational change in the *trpE* subunit induces a

Figure 14-29
Sites of action of the *trp* gene products. The products of genes *E* and *D* and of *B* and *A* form an active tetramer (E_2D_2) and a dimer (BA) respectively. The site of feedback inhibition is E_2D_2. The components of the reaction sequences have been abbreviated.

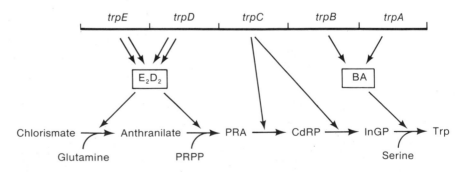

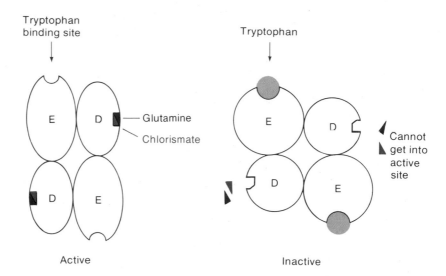

Tryptophan
binding site

E   D   — Glutamine
— Chlorismate

D   E

Tryptophan

E   D   ◀ Cannot
get into
active
site

D   E

Figure 14-30
Inactivation of the an-
thranilate complex E_2D_2
by tryptophan.

Active   Inactive

similar change in the *trpD* subunit, whose activity is thereby inhibited.

The most common type of feedback inhibition is effected by an inhibition of the first enzymatic step of a biosynthetic pathway by the product of the pathway. For example, in the pathway

$$A \xrightarrow{1} B \xrightarrow{2} C \xrightarrow{3} D$$

catalyzed by enzymes 1, 2, and 3, the product D would act on enzyme 1, which is clearly the most economical mode of inhibition. Note that this means that enzyme 1 must have two binding sites—one for the substrate A and one for the product D. Often binding of the product inhibits binding of the substrate. This can be accomplished by an overlap of the binding sites or by a conformational change that weakens or eliminates the substrate-binding site. An example of the second mechanism was described in Chapter 6—namely, the inhibition of aspartyl transcarbamoylase by CTP.

Some biosynthetic pathways are branched—that is, they are responsible for the synthesis of two products from a common precursor. A hypothetical example of such a pathway is the following:

$$A \xrightarrow{1} B \begin{array}{c} \nearrow^{4} E \xrightarrow{5} F \\ \searrow_{2} C \xrightarrow{3} D \end{array}$$

In this type of pathway it would be undesirable for a single product to inhibit enzyme 1 because both pathways would be inhibited. In general, the most economical kind of inhibition prevails—namely, D inhibits enzyme 2 and F inhibits enzyme 4. In this way, neither D nor F inhibits the synthesis of the other. Conversion of A to B is wasteful if both D and F are present; therefore, in many pathways it is found that D and F together inhibit enzyme 1. There are four major ways in which an enzymatic step can be inhibited by two different molecules: (1) enzyme 1 is inhibited by its product B, which accumulates when enzymes 2 and 4 are blocked; (2) enzyme 1 has binding sites for both D and F, and when both sites are filled the enzyme is inactivated; (3) D and F both inhibit enzyme 1 slightly—for example, d-fold and f-fold, respectively, and together they inhibit the enzyme ($d \times f$)-fold; (4) in some pathways, A is converted to B by two different enzymes **(isoenzymes),** one inhibited by D and one by F.

The elegance of feedback inhibition of a branched pathway is shown in Figure 14-31, which shows a well-studied multibranched pathway in which lysine, methionine, and threonine are synthesized from aspartic acid. Note how each amino acid inhibits an enzyme immediately after a branch, namely, enzymes 4, 6, and 8. Three

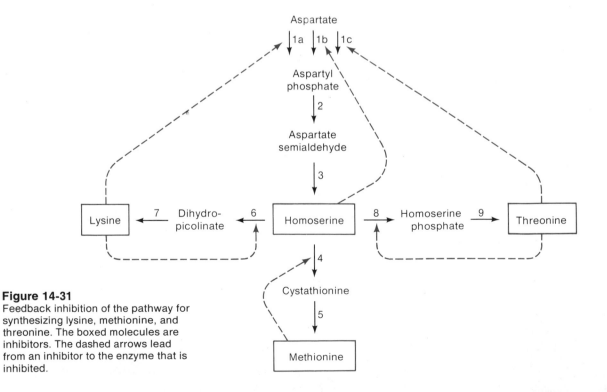

Figure 14-31
Feedback inhibition of the pathway for synthesizing lysine, methionine, and threonine. The boxed molecules are inhibitors. The dashed arrows lead from an inhibitor to the enzyme that is inhibited.

isoenzymes, 1a, 1b, and 1c, are separately inhibited by lysine, homo-serine, and threonine, respectively. The three isoenzymes have different enzymatic activity; each synthesizes an amount of aspartyl phosphate that is needed to form the appropriate amounts of the product that inhibits it. Thus, when an isoenzyme is inhibited, the correct amount of aspartyl phosphate is still made.

AN EFFECT OF A GENE PRODUCT ON TRANSLATION

In various places in this chapter translational regulation has been mentioned. Usually this refers to the fact that the number of copies of each protein translated from a polycistronic mRNA varies from gene to gene. Usually there is a gradient of translation decreasing from the 5′ terminus to the 3′ terminus of the mRNA, and several processes account for this phenomenon—namely, varying efficiencies of initiation of translation, different spacing between chain-termination codons and a subsequent AUG codon allowing the ribosome and mRNA to dissociate, and differential sensitivity of various regions of the mRNA to degradation. A novel mechanism that has only been found in a few bacteriophage species is inhibition of translation of a particular gene by its gene product.

E. coli phage R17 contains only three gene products—two are structural proteins (the A protein and the phage coat protein) and one is an enzyme (replicase). Far more coat protein is needed than replicase,

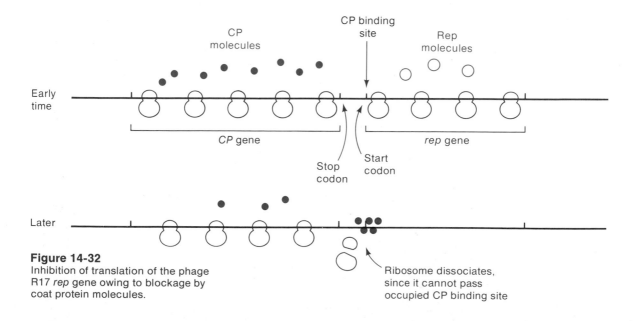

Figure 14-32
Inhibition of translation of the phage
R17 *rep* gene owing to blockage by
coat protein molecules.

which is used only in catalytic amounts. The mRNA molecule has a binding site for the coat protein between the termination codon of the coat protein gene and the AUG codon of the replicase gene. As the coat protein is synthesized, this binding site gradually is filled with protein molecules, blocking the ribosome from translating the replicase region (Figure 14-32).

REFERENCES

Clark, B. F. C., H. Klenow, and J. Zeuthen, (eds.). 1978. *Gene Expression.* Pergamon.

Gilbert, W., and B. Muller-Hill. 1966. "Isolation of the *lac* repressor." *Proc. Nat. Acad. Sci.,* 56, 1891–1898.

Jacob, F., and J. Monod. 1961. "Genetic regulatory mechanisms in the synthesis of proteins." *J. Mol. Biol.,* 3, 318–356.

Johnson, H. M., W. M. Barnes, F. G. Chumley, L. Bossi, and J. R. Roth. 1980. "Model for regulation of the histidine operon of *Salmonella.*" *Proc. Nat. Acad. Sci.,* 77, 508–512.

Lee, N., W. O. Gielow, and R. G. Wallace. 1981. "Mechanism of *araC* autoregulation and the domain of two overlapping promoters, *pC* and *pBAD* in the L-arabinose regulatory region of *E. coli.*" *Proc. Nat. Acad. Sci.,* 78, 752–756.

Lindahl, L., and J. M. Zengel. 1982. "Expression of ribosomal genes in bacteria." *Adv. Genet.,* 21, 53–122.

Little, J., S. H. Edmiston, L. Z. Pacelli, and D. W. Mount. 1980. "Cleavage of the *E. coli lexA* protein by the *recA* protease." *Proc. Nat. Acad. Sci.,* 77, 3225–3229.

Maniatis, T., and M. Ptashne. 1976. "A DNA operator–repressor system." *Scient. Amer.,* January. pp. 64–76.

Miller, J. H. 1980. "Genetic analysis of the *lac* repressor." *Curr. Topics Microbiol. Immunol.,* 90, 1–18.

Miller, J. H., and W. S. Reznikoff (eds.) 1978. *The Operon.* Cold Spring Harbor Laboratory.

Nierlich, D. P. 1978. "Regulation of bacterial growth, RNA, and protein synthesis." *Ann. Rev. Microbiol.,* 32, 393–432.

Oxender, D. L., G. Zurawski, and C. Yanofsky. 1979. "Attenuation in the *E. coli* tryptophan operon: role of RNA secondary structure involving the tryptophan coding region." *Proc. Nat. Acad. Sci.,* 76, 5524–5528.

Pastan, I., and S. Adhya. 1976. "Cyclic adenosine-3'-5'-monophosphate in *E. coli.*" *Bact. Rev.,* 40, 527–551.

Platt, T. 1978. "Regulation of gene expression in the tryptophan operon in *E. coli.*" *In* Miller, J. H. and W. S. Reznikoff (eds.), *The Operon.* pp. 213–302. Cold Spring Harbor Laboratory.

Ptashne, M., and W. Gilbert. 1970. "Genetic repression." *Scient. Amer.,* June, pp. 36–44.

Reznikoff, W., J. Miller, J. Scaife, and J. Beckwith. 1969. "A mechanism for repressor action." *J. Mol. Biol.,* 43, 201–213.

Reznikoff, W., R. Winter, and B. C. Hurley. 1974. "The location of the repressor-binding site in the *lac* operon." *Proc. Acad. Sci.,* 71, 2314–2318.

Safer, B., and W. F. Anderson. 1978. "The molecular mechanism of hemoglobin synthesis and its regulation in the reticulocyte." *CRC Revs. Biochem.,* 6, 261–290.

Ullman, A., and A. Danchin. 1980. "Role of cyclic AMP in regulatory mechanisms of bacteria." *Trends. Biochem. Sci.,* 5, 95–96.

Wallace, R. G., N. Lee, and A. V. Fowler. 1980. "The *araC* gene of *E. coli:* transcription and translation start points and complete nucleotide sequence." *Gene,* 12, 179–190.

Yanofsky, C. 1981. "Attenuation in the control of expression of bacterial operons." *Nature,* 289, 751–758.

15 Bacteriophages

I. LYTIC PHAGES

Bacteriophages, or phages, have played an important role in the development of molecular biology. At the present time a few phages are the most completely understood of any organisms. Because of their lesser complexity than bacteria and higher cells and the availability of an enormous number of mutants, phages have been extraordinarily useful in the study of replication, transcription, and regulation.

GENERAL PROPERTIES OF PHAGES

A bacteriophage is, first of all, a bacterial parasite. By itself, a phage can persist; but a phage can neither grow nor replicate except within a bacterial cell. Most phages possess genes encoding a variety of proteins. However, all known phages use the ribosomes, protein-synthesizing factors, amino acids, and energy-generating systems of the host cell. That the energy-generating system of the host is needed by the phage means that a phage can grow only in a metabolizing bacterium. Thus, when studying the phage life cycle, it is essential to infect a logarithmically growing bacterial culture to insure that all bacteria are metabolically active.

Each phage must perform some minimal functions for continued survival. These are the following:

1. To protect its nucleic acid from environmental chemicals that could alter the molecule (for example, break the molecule or cause a mutation).
2. To deliver its nucleic acid to the inside of a bacterium.
3. To convert an infected bacterium to a phage-producing system which yields a large number of progeny phage.
4. To release phage progeny from an infected bacterium.

These problems are solved in a variety of ways by different phage species, as will be seen in this chapter when the characteristics of several phage species are described in detail. All of these species have certain features in common but differences in detail show the many ways in which specific biological functions can be accomplished.

An important observation to be made in this chapter is of the degree to which an individual phage particle uses parts of the machinery of the cell. Some phage species have fewer than 10 genes and use almost all of the cellular functions, whereas others have 30–50 genes and are less dependent on the host. A few of the largest phage particles have so many of their own genes that, for certain functions such as DNA replication, they need no host genes. Surprisingly, in a few cases phage genes nearly duplicate host genes; this seeming redundancy in function is discussed in a later section concerned with these so-called nonessential genes.

Phage particles also differ in their physical structures from species to species and often certain features of their life cycles are correlated with their structure. These structural differences are described in the following section. However, before beginning that section the reader is urged to review some of the material about phage biology and techniques for detecting and counting phage presented in Chapter 1.

Structures of Phages

There are three basic kinds of phage structures: icosahedral tailless, icosahedral head with tail, and filamentous.* Usually the phage particle consists of a single nucleic acid molecule (which may be single- or double-stranded linear or circular DNA, or single-stranded linear RNA) and one or more proteins. (The single known exception is phage $\phi6$, which contains a so-called segmented genome—that is, each $\phi6$ particle contains three linear double-stranded RNA molecules, whose base sequences differ from one another.) The proteins form a shell, called either the **coat** or the **capsid,** around the nucleic acid; the nucleic acid is thereby protected from nucleases and harmful substances. Figures 15-1

---

*An icosahedron is a quasispherical polyhedron having 20 triangular faces and 12 corners.

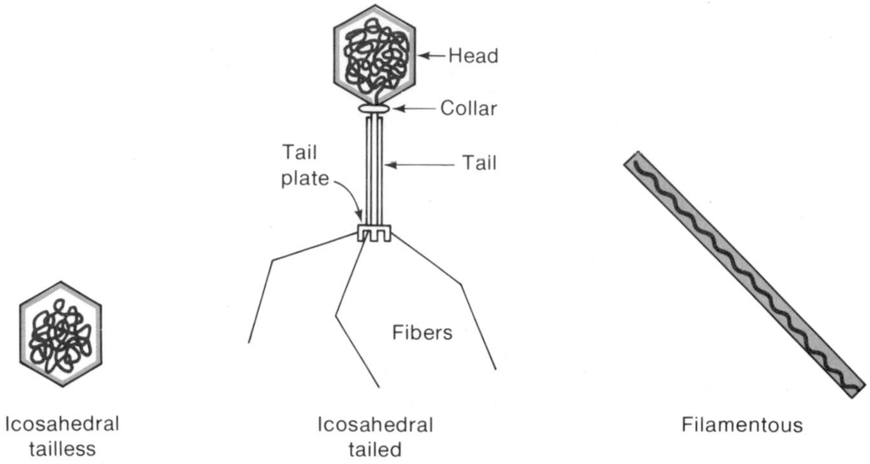

Figure 15-1
Two-dimensional view of three basic phage structures. The tailed phages do not always have a collar and can have from 0–6 tail fibers, the number depending on the phage type.

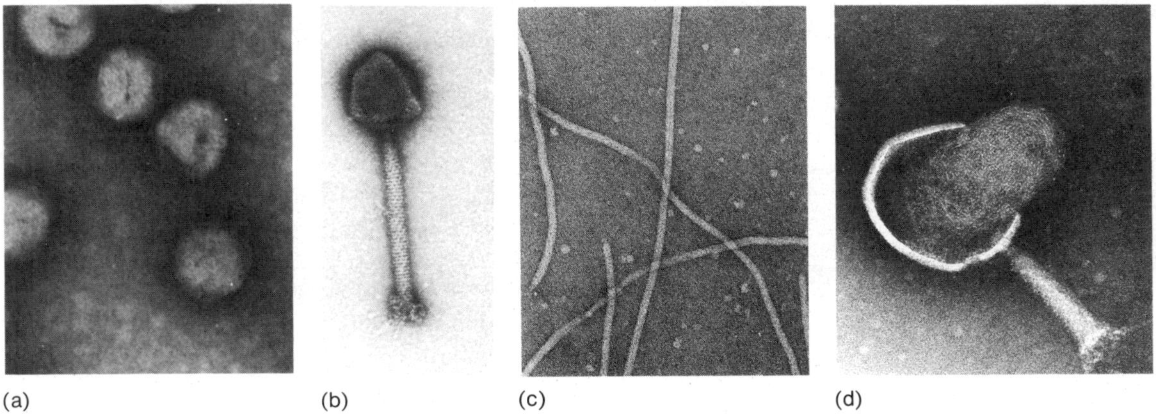

Icosahedral tailless

Icosahedral tailed

Filamentous

(a)   (b)   (c)   (d)

Figure 15-2
The three major morphological classes of phages. (a) Icosahedral, tailless: PM-2; (b) Icosahedral, tailed: SPO1; (c) Filamentous: M13. (d) Phage T4 releasing its DNA from a broken phage head. (Courtesy of (a) C. Brack, (b) B. ten Heggler, and (c,d) Robley Williams.)

and 15-2 show schematic drawings of the three basic structures. The following points should be noted:

1. In both icosahedral tailless and tailed phages the nucleic acid is contained in a hollow region formed by the capsid and is highly compact. In a filamentous phage the nucleic acid is embedded in the capsid and is present in an extended helical form.
2. In a tailed phage there may be a very tiny piece of DNA extending from the capsid into the tail.

3. The tail is a complex structure often terminated by tail fibers.
4. The length of the DNA molecule is very much greater than any dimension of the head.

There are many variations on the basic structure of the tailed phages. For example, the length and width of the head may either be the same or the length may be greater than the width; however, short fat heads are not seen. The tail may be very short (barely visible in electron micrographs) or up to four times the length of the head and it may be flexible or rigid. The baseplate of the tail is often quite complex but it may also be absent; when present, it typically has from one to six tail fibers. Finally, there may or may not be a collar surrounding the head–tail joint.

Stages in the Lytic Life Cycle of a Typical Phage

Phage life cycles fit into two distinct categories—the **lytic** and the **lysogenic** cycles. A phage in the lytic cycle converts an infected cell to a phage factory, and many phage progeny are produced. A phage capable only of lytic growth is called **virulent.** The lysogenic cycle, which has been observed only with phages containing double-stranded DNA, is one in which no progeny particles are produced; the phage DNA usually becomes part of the bacterial chromosome. A phage capable of such a life cycle is called **temperate.*** In this section only the lytic cycle is outlined. The lysogenic cycle will be described in the next chapter, after particular species of virulent phages have been studied in detail.

There are many variations in the details of the life cycles of different virulent phages. There is, however, what may be called a basic lytic cycle, which is the following (Figure 15-3):

1. *Adsorption of the phage to specific receptors on the bacterial surface* (Figure 15-4). These receptors are very varied and serve the bacteria for purposes other than phage adsorption (because this is invariably a lethal event for the bacterium). Some examples are proteins used to transport maltose (*E. coli* phage λ), and proteins engaged in transport of the Fe^{2+} ion (*E. coli* phage T1).

2. *Passage of the DNA from the phage through the bacterial cell wall.* Some types of tailed phages use an injection sequence shown schematically in Figure 15-5. In this process the nucleic acid is transferred into the cell and is never exposed to the medium surrounding the recipient cell. Little is known about the transfer mechanisms for

---

*Most temperate phages also undergo lytic growth under certain circumstances.

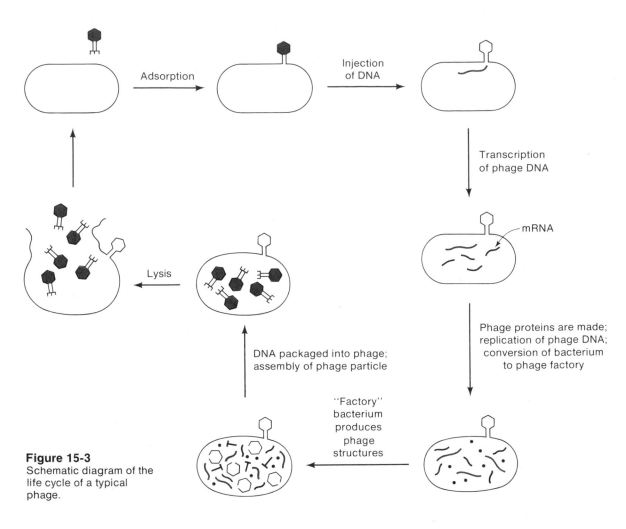

Figure 15-3
Schematic diagram of the life cycle of a typical phage.

Adsorption

Injection of DNA

Transcription of phage DNA

mRNA

Phage proteins are made; replication of phage DNA; conversion of bacterium to phage factory

"Factory" bacterium produces phage structures

DNA packaged into phage; assembly of phage particle

Lysis

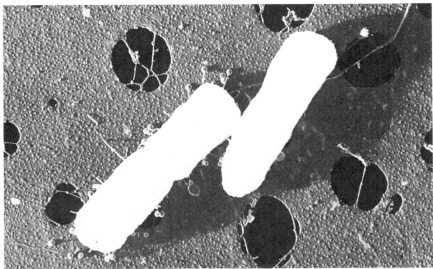

Figure 15-4
An electron micrograph of an *E. coli* cell to which numerous phage particles are adsorbed by their long tails. The phage are λ rather than T4, but the appearance of the two are similar when adsorbed to bacteria. (Courtesy of T. F. Anderson.)

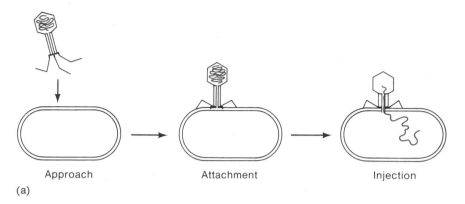

Approach   Attachment   Injection

(a)

Figure 15-5
(a) Injection sequence of a tailed phage. In the injection stage the tail sheath contracts and drives a core protein tube through the cell wall like a hypodermic syringe. (b) Electron micrograph of T4 phage adsorbed to the cell wall of *E. coli,* observed in thin section. The tail sheath is contracted and the core is fixed firmly against the cell wall. The arrow shows a portion of the core projecting through the cell wall. DNA can be seen entering the cell from the two phage at the right. (Courtesy of Lee Simon.)

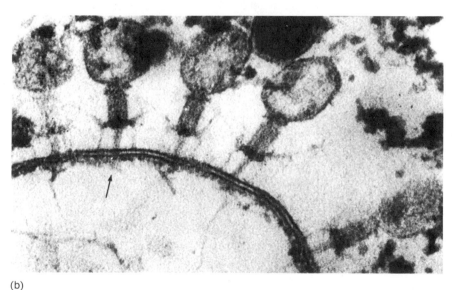

(b)

the other phage types. With tailless phages the nucleic acid is transiently susceptible to nuclease attack, so it is thought that the phage coat may break open and release its nucleic acid onto the cell wall prior to entering the cell.

3. *Conversion of the infected bacterium to a phage-producing cell.* Following infection by most phages a bacterium loses the ability either to replicate or to transcribe its DNA; sometimes it loses both. This shutdown of host DNA or RNA synthesis is accomplished in many different ways depending on the phage species.

4. *Production of phage nucleic acid and proteins.* By several mechanisms the phage directs the synthesis of a replicative system that

specifically makes copies of phage nucleic acid. This programming is accomplished either by synthesis of phage-specific polymerases or by addition of specificity elements to bacterial enzymes. In both cases many bacterial replication proteins are used. RNA-containing phages are quite constrained in this respect—they must encode their own replication enzymes because bacteria do not contain enzymes that replicate RNA. Transcription is almost always initiated by the bacterial RNA polymerase but after the first transcription event either the bacterial polymerase is modified to recognize phage promoters or a phage-specific RNA polymerase is synthesized. Transcription is regulated and phage proteins are synthesized sequentially in time as they are needed. The RNA molecules of phages containing single-stranded RNA serve as their own mRNA, and thus have no need for transcription; these phages are regulated by controlling the efficiency of translation. Several regulatory processes will be described in detail later in this chapter. Usually sufficient nucleic acid and proteins are synthesized for the production of from 50–1000 phage particles, the number depending on the particular phage species.

5. *Assembly of phage particles.* This process is often called **morphogenesis.** Two types of proteins are needed for the assembly process: **structural proteins,** which are present (possibly in modified form) in the phage particle, and **catalytic proteins,** which participate in the assembly process but do not become part of the phage particle. A subset of the latter class consists of the maturation proteins, which convert intracellular phage DNA to a form appropriate for packaging in the phage particle. For the icosahedral phages assembly occurs in several stages: (1) Aggregation of phage structural proteins to form a phage head and, when needed, to form a phage tail; at this point, the tail is not attached to the head. (2) Condensation of the nucleic acid and entry into a preformed head. (3) Attachment of the tail to a filled head. With filamentous phages the nucleic acid and the protein form a phage particle in a single step. The mechanism of condensation is unknown.

6. *Release of newly synthesized phage.* With most phages a phage protein called a **lysozyme** or an **endolysin** is synthesized late in the cycle of infection. This protein causes disruption of the cell wall. Other proteins called **membranases** dissolve the cell membrane and together with the lysozyme cause total destruction of the cell, so that phage are released to the surrounding medium. This disruptive process is called **lysis.** A few filamentous phages release progeny continuously by outfolding of the cell wall; this process is called **budding** and it does not cause major damage to the cell. Cells infected with such filamentous phages can continue to produce virus particles for very long periods of time.

Properties of a Phage-Infected Bacterial Culture

In the preceding section the events following infection of one cell by a phage were described. In the laboratory one usually infects a bacterial culture with a large number of phage particles. In this and the following three sections the parameters needed to describe such an infection are described.

The adsorption of phage to a bacterial culture is a random process and the variation in the distribution of phage among the cells is described by Poisson's Law,

$$P(n) = \frac{m^n e^{-m}}{n!}$$

in which *P(n)* is the fraction of the bacteria to which *n* phage have adsorbed when *m* is the average number of adsorbed phage per bacterium (the multiplicity of infection or MOI). That is, the fraction of bacteria infected with 0, 1, 2, 3, . . . , *i* phage is $P(0)$, $P(1)$, $P(2)$, $P(3)$, . . . , *P(i)*. Thus, if 3×10^8 phage adsorb to 10^8 bacteria ($m = 3$), the values of $P(0)$, $P(1)$, $P(2)$, $P(3)$, . . . , are 0.05, 0.15, 0.22, 0.22, ... Since $P(0) = 0.05$, the sum of $P(1) + P(2) + \cdots (P)i$ must equal $1 - 0.05 = 0.95$. In other words, 95 percent of the bacteria are infected by at least one phage.

Note also that the value of $P(0)$ tells the fraction of the phage particles that have adsorbed. For example, in the infection just described, $P(0)$ should equal 0.05 for $m = 3$, if all of the added phage had adsorbed to the bacteria. However, if, in a particular experiment using 3×10^8 phage and 10^8 bacteria, one observed that 12 percent of the bacteria remained uninfected, then the value of $P(0)$ would be 0.12; using this value one can calculate from the Poisson term $P(0) = e^{-m}$ that $m = 2.12$, which is the true value of the MOI. Thus, $2.12/3 = 0.71$ or 71 percent of the added phage particles actually adsorbed.

It is possible to measure $P(0)$ in a simple way (Figure 15-6). After an adsorption period the bacteria are placed on an agar surface and the fraction of the bacteria able to form a colony is measured. Only uninfected cells can do so. The number of infected cells can be measured in the following way. First, antibodies that inactivate unadsorbed phage are added to the infected culture.* Then, the phage suspension is plated on a lawn of phage-sensitive cells, where each infected cell, providing it is plated before lysis, will produce a *single* plaque.

A cell that can form a plaque in this way is called an **infective center.** The number of unadsorbed phage particles is determined by

---

*Antibodies are obtained from the blood of a rabbit that has been injected with a purified suspension of the phage.

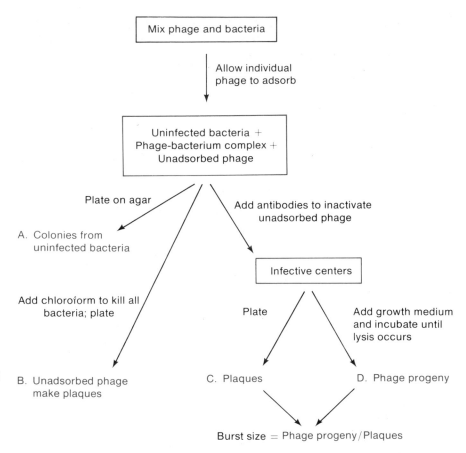

Figure 15-6
Scheme for determining the number of uninfected bacteria (A), unadsorbed phage (B), infective centers (C), the number of progeny phage (D), and the burst size (D/C).

adding chloroform to the culture after the adsorption period; free phage are unaffected by this treatment but chloroform kills infective centers. Thus, the number of plaques formed after chloroform treatment equals the number of unadsorbed phage. In a well-performed experiment, the number of infective centers and free phage and the true MOI should always be measured, because many experimental results depend on the value of the MOI.

The number of phage produced by an infected cell is called the **burst size.** This is an important parameter in many experiments because it is a measure of the efficiency of phage production. It is measured by determining (by plaque counting) the number of phage produced after lysis of the culture and dividing by the number of infective centers (see also Figure 15-6).

Phage multiply much more rapidly than bacteria. That is, bacteria double in one generation time whereas, in one life cycle, the number of phage is increased by a factor equal to the burst size. This is easily seen in the example shown in Table 15-1, in which a single phage whose

Table 15-1
Calculation of the phage and bacterial concentrations after various numbers of bacterial generations*

Number of generations	Concentrations		
		Bacteria/ml	
	Phage /ml	Approximate	Precise
0	1	10^6	10^6
1	10^2	2×10^6	$2 \times (10^6 - 1)$
2	10^4	4×10^6	$(4 \times 10^6) - (2 \times 10^2) - 4$
3	10^6	7.98×10^6	$(8 \times 10^6) - (2 \times 10^4) - (4 \times 10^2) - 8$
4	10^8	1.4×10^7	$(1.6 \times 10^7) - (2 \times 10^6) - (4 \times 10^4) -$ $(8 \times 10^2) - 16$
5	$1.4 \times 10^9 =$ $100 \times (1.4 \times 10^7)$	0	0

*Initially a bacterial culture at a concentration of 10^6 cells/ml is infected with 1 phage. The doubling time of the bacteria and the life cycle of the phage are equal.

average burst size is 100 and whose life cycle lasts 25 minutes infects a one-milliliter bacterial culture with a 25-minute doubling time. In this calculation, it is assumed that adsorption is always instantaneous and complete. Note that, in four generations, the number of bacteria has increased 14-fold, whereas the number of phage has increased 10^8-fold. At this time there are approximately eight times as many phage as bacteria; hence, all bacteria are infected. Thus, after 125 minutes the bacteria are gone and the original phage particle has produced 1.4×10^9 progeny.

Essential and Nonessential Phage Genes

Most of the remainder of this chapter is devoted to a detailed look at particular phages. A distinction will be made between essential and nonessential genes. A gene is usually considered to be nonessential if a mutation in that gene does not prevent plaque formation. If the mutation prevents plaque formation, the gene is essential. These definitions must be interpreted carefully because, whereas a typical burst size for an infected bacterium is 50–100, a burst size of four is usually sufficient for plaque production. Indeed, mutations in many so-called nonessential genes reduce the burst size markedly, though not enough to prevent plaque formation.

A gene may be nonessential for three reasons: (1) An identical gene may be present in the bacterium, or a phage gene may have the same function as a gene present in a bacterium. This duplication may be of some value to the phage, because it might increase the burst size by providing a higher concentration of an essential enzyme; alternatively, in nature there may be hosts that lack the gene and the gene will then be essential for phage growth. (2) The gene does not duplicate a bacterial gene but in some way increases either the rate of phage production or the burst size. In both cases (1) and (2) there is an evolutionary advantage. (3) The gene is not needed in the laboratory but it enables the phage to cope with special situations met in nature. A fourth possibility—that the gene is always unnecessary—is less likely because a truly useless gene would in general lack the survival value needed for the phage to retain it.

We will see that the so-called nonessential genes are a strategy that only the mid-size and large phages can afford.

Specificity in Phage Infection: Adsorption

Many thousands of phages have been isolated. The ability of a particular phage to infect a bacterium is almost always limited to a single bacterial species and often to a few strains of that species.* For example, no phage species that infects the genus *Pseudomonas* can also infect *E. coli*; furthermore, phages that grow in *Ps. fluorescens* generally fail to grow in *Ps. aeruginosa*. Among the *E. coli* phages, which are the most extensively studied phages, extraordinary specificity has been observed; for example, the phage φX174 grows well on *E. coli* strain C but fails to grow on most other laboratory strains. There are exceptions, though; for instance, *E. coli* phage T4 is capable of growing on many strains of *E. coli* and on certain species of the genus *Shigella*.

Several factors contribute to this specificity. One of these is the ability to adsorb. For example, *E. coli* phages will not adsorb to any known species of *Pseudomonas*. This is not unexpected inasmuch as there are specific receptors on the cell wall for phage adsorption. It is of course a distinct disadvantage to a bacterium to be able to adsorb phages, since this invariably kills the bacterial cell; thus, in the course of evolution cell wall components have evolved to be unrecognizable to most phage species. Phages have also evolved in order to be able to recognize some bacterium; if a particular phage had not been able to do so, it would not exist at the present time. An example of this evolutionary process can be seen in a small way in the laboratory.

If 10^8 *E. coli* B cells are infected with 10^{10} T6 phage particles and the infected cells are put on an agar surface, about 100 colonies will form. Cultures prepared from these colonies will have lost the ability to be infected by T6 and phage will no longer adsorb to the mutant cells. (These cells were of course not produced by the infection but existed in

---

*An interesting exception to this rule is the phage PRR1, which can infect *any* Gram-negative bacterium that carries the plasmid RP4.

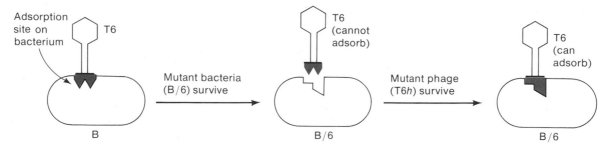

Figure 15-7
The generation of a phage-resistant bacterium, and a phage *h* mutant. Although the tail fiber is modified, for graphic purposes, the tail plate has been modified in the drawing.

the culture and were selected by growth with an excess of phage.) These bacterial mutants are called T6-resistant (Figure 15-7); this is denoted either B/6 (read "B-bar-six") or Tsx-r. If 10^8 Tsx-r cells are used to form a bacterial lawn on agar and 10^8 T6 phage are added, about ten plaques will result. The phage in these plaques carry a mutation in the tail fiber gene and have gained the ability to adsorb to Tsx-r cells. These phage are called *h* mutants (for host range). They usually retain the ability to form plaques on Tsx-s bacteria and hence are said to have an "extended host range."

In a few cases the adsorption problem is the only barrier to cross species infection. For example, phage T4 fails to adsorb to cells of the genus *Aerobacter*. However, by a special technique it is possible to force T4 DNA into spheroplasts of *A. aerogenes,* and phage progeny can thereby be produced.

Specificity in Phage Infection: Host Restriction

In the preceding section the inability of a particular phage to adsorb to most bacterial species was described. With some bacteria there may be a more fundamental barrier—the inability of a "foreign" bacterial RNA polymerase to recognize phage promoters. However, even when adsorption and transcription are possible, with most bacteria there is usually another barrier called **host restriction** and **host modification.** This is a phenomenon in which a bacterium of type X is able to distinguish a phage that has been grown in a type X bacterium from one grown in a different type such as Y and is able to prevent the phage grown in Y from carrying out a successful infection. The notation used to discuss this phenomenon is the following: a phage P grown in a bacterium X is denoted P·X. Host modification and restriction are illustrated by the data in Table 15-2. Note that λ·K, which has been grown in *E. coli* strain K, forms plaques at a low efficiency in strain B.

Table 15-2
The restriction and modification pattern
of *E. coli* phage λ

Bacterial strain	Phage		
	λ·K	λ·B	λ·C
K	1	10^{-4}	10^{-4}
B	10^{-4}	1	10^{-4}
C	1	1	1

Note: Numbers indicate relative plating efficiency.

Thus, λ · K is **restricted** by strain B. The phage population in these rare plaques (λ · B) has been modified by strain B so that the phage grow efficiently in strain B; however, λ · B now fails to grow in strain K—that is, it is restricted. The molecular explanation for this is the following: *E. coli* B contains an enzyme called a **restriction endonuclease** or a restriction enzyme—specifically, it is the EcoB nuclease. This enzyme is a site-specific nuclease in that it cuts DNA strands only within a specific base sequence. Phage λ · K contains this sequence; when its DNA is injected into *E. coli* B, the phage DNA is broken. *E. coli* B also contains this sequence and would destroy its own DNA were the sequence not modified. A site-specific methylating enzyme (EcoB methylase) methylates an adenine in the sequence, thereby rendering the sequence resistant to the EcoB nuclease. When λ · K infects strain B, a few parental phage-DNA molecules in the large population of infected cells are methylated before they are restricted. All progeny DNA molecules are already methylated on one strand and the newly synthesized strands are also methylated rapidly (see Chapter 8); and restriction is avoided. Thus, a small population of phage having the B modification (λ · B) are produced. *E. coli* K also contains a site-specific nuclease called the EcoK enzyme. It attacks a base sequence that is different from the sequence recognized by the EcoB enzyme. An EcoK methylase also protects *E. coli* K from self-destruction, producing the K modification. A λ phage that has always been grown in strain K—namely, λ · K—is methylated in the EcoK-specific sequence and is resistant to the EcoK enzyme. However, λ · B has an unmethylated EcoK sequence, so λ · B DNA is usually broken when a strain K cell is infected. Occasionally, a λ · B DNA molecule escapes restriction and replicates, and its replicas have a methylated K-specific sequence. Thus, the rare progeny phage that result when λ · B successfully infect *E. coli* K are λ · K; they now lack the B modification and are restricted when infecting strain B.

Note in Table 15-2 that λ grown on strain C—i.e., λ · C—also fails to grow well in strains B and K, but neither λ · B nor λ · K is restricted by strain C. The reason for the lack of restriction is that strain C has no restriction nuclease active against any base sequence in λ DNA. The λ • C phage are restricted by both strains B and K because, of course, strain C does not have the B and K methylases.

Host restriction and modification is a widespread process. It probably serves the purpose of destroying foreign DNA. In Chapter 20 we will see that the restriction enzymes are the basis of the recombinant DNA technology.

SPECIFIC PHAGES

On a biological scale of complexity, a phage is a relatively simple form of life. However, phages are sufficiently complex that there is not a single phage type for which all of the molecular details of its life cycle are completely understood. The major progress has been with a small number of phage species that grow in *E. coli, Bacillus subtilis,* and *Salmonella typhimurium.* Special features of each phage have resulted in particular phages being more suited to the study of certain processes. For example, regulation of transcription is best understood in phages λ and T7, the study of morphogenesis has been more successful with phages T4 and λ than with other phages, phages P1 and P22 gave the first clues to understanding transduction, T5 has yielded the best information about DNA injection, T4 has been most profitable in the study of DNA synthesis, and the study of T4 and T7 has shown clearly how a phage takes over a bacterium.

In the following sections several of these phages (and a few others) will be described in detail. We will emphasize those features that are well understood, mentioning other features only briefly. As much as possible, an attempt will be made to give an overview of the life cycle of each phage. Table 15-3 lists some of the important features of several types of phages.

E. COLI PHAGE T4

During the period 1939 to 1941 Milislav Demerec isolated a set of phages, each of which grew in a particular strain of *E. coli* (strain B); these phages were named T1, T2, . . . , and T7. Their study revolutionized genetics because, with these phages, physical and biochemical experiments could be coupled to genetic analysis. The genetic material of these phages was later shown to be double-stranded DNA (Chapter 7). The structures and life cycles of some of these phages differ in significant ways, so the phages are grouped in four classes: T1; T2, T4,

Table 15-3
Properties of the nucleic acid of several phages

Phage	Host	DNA or RNA	Form	Molecular weight, $\times 10^6$	Unusual bases
φX174	E	DNA	ss, circ	1.8	None
M13, fd, f1	E	DNA	ss, lin	2.1	None
PM2	PB	DNA	ds, circ	9	None
186	E	DNA	ds, lin	18	None
B3	PA	DNA	ds, lin	20	None
Mu	E	DNA	ds, lin	25	None
T7	E	DNA	ds, lin	26	None
λ	E	DNA	ds, lin	31	None
N4	E	DNA	ds, lin	40	None
P1	E	DNA	ds, lin	59	None
T5	E	DNA	ds, lin	75	None
SPO1	B	DNA	ds, lin	100	HMU for thymine
T2, T4, T6	E	DNA	ds, lin	108	Glucosylated HMC for cytosine
PBS1	B	DNA	ds, lin	200	Uracil for thymine
MS2, Qβ, f2	E	RNA	ss, lin	1.0	None
φ6	PP	RNA	ds, lin	2.3, 3.1, 5.0*	None

Note: Abbreviations used in this table are: E, *E. coli;* B, *Bacillus subtilis;* PA, *Pseudomonas aeruginosa;* PB, *Ps. aeruginosa BAL-31;* PP, *Ps. phaseolica;* ss, single-stranded; ds, double-stranded; circ, circular; lin, linear; HMU, hydroxymethyluracil; HMC, hydroxymethylcytosine.

*φ6 contains three molecules.

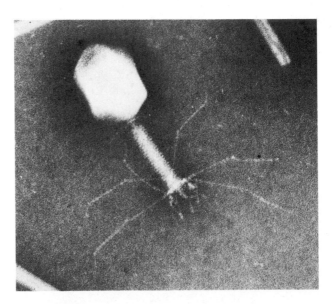

Figure 15-8
An *E. coli* T4 phage. The DNA is contained in the head. Tail fibers come from the pronged plate at the tip of the tail. (Courtesy of Robley Williams.)

T6 (called T-even phages); T3, T7; and T5.* The T-even phages have been especially attractive subjects for study, because they are easily grown in large quantities, many mutants can easily be obtained, and their DNA molecules contain an unusual base that enables the phage DNA to be distinguished from host DNA. In recent years the principal effort with this class has been with phage T4, an electron micrograph of which is shown in Figure 15-8. A great deal has been learned about this phage, though because of its large size and complex life cycle certain features of the life cycle are still poorly understood. Of the other classes there is detailed molecular information only about T5 and T7; the latter will be discussed after the T4 life cycle has been analyzed.

Properties of T4 DNA

The DNA of phage T4 is rather large for a phage (Table 15-3); its molecular weight is about 110 million; it contains about 166,000 base pairs; and it has a length of 55 μm. The phage head is roughly 0.06 by 0.09 μm; therefore, the DNA must be folded tightly when packaged (refer back to Figure 15-2). The DNA contains adenine, thymine, and guanine but no cytosine; instead, there is a modified form of cytosine called **5-hydroxymethylcytosine (HMC),** which base-pairs with guanine (Figure 15-9). This base is further modified by glucosylation—that is, a sugar is coupled to the OH group of HMC. The three T-even phages, which we have said are quite similar, differ chemically in the identity of the sugar (Table 15-4). There is a sufficient amount of the sugar attached to the HMC that T4 DNA is virtually cloaked within these molecules; this has the consequence that purified T4 DNA is somewhat resistant to a variety of DNases.

The DNA of T4 is **terminally redundant** (Figure 15-10(a))—that is, a sequence of bases is repeated at both ends of the molecule. Terminal redundancy, which occurs in many phage species, is demonstrated in the following way (Figure 15-10(b,c)). A DNA sample is treated with a

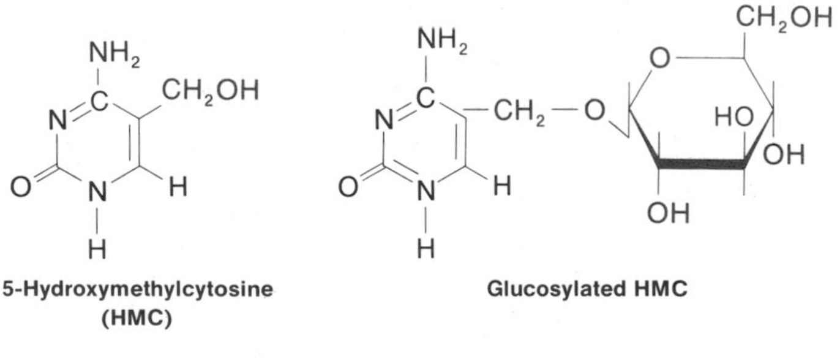

Figure 15-9
Nonglucosylated and glucosylated 5-hydroxymethylcytosine. If the CH₂OH (red) in HMC were replaced by hydrogen, the molecule would be cytosine.

5-Hydroxymethylcytosine
(HMC)

Glucosylated HMC

*T1, T3, T5, and T7 have been called T-odd phages, but because the four phages fall into three classes, the term has no value.

Table 15-4

Glucosylation pattern of 5-hydroxymethylcytosine
(HMC) in T-even phage DNA

Glucosylation	Percentage of total HMC		
	T2	T4	T6
Nonglucosylated	25	0	25
α-Glucosyl	70	70	3
β-Glucosyl	0	30	0
α-Glucosyl-β-glucosyl	5	0	72

(a) Terminally redundant DNA

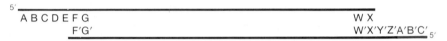

(b) After digestion with a 3′ exonuclease

(c) After circularization of the molecule in (b)

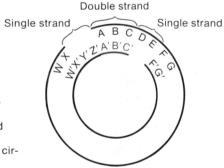

Figure 15-10
A terminally redundant
molecule and its identification by means of exo-
nucleolytic digestion and
circularization. A nonre-
dundant DNA cannot be circularized in this way.

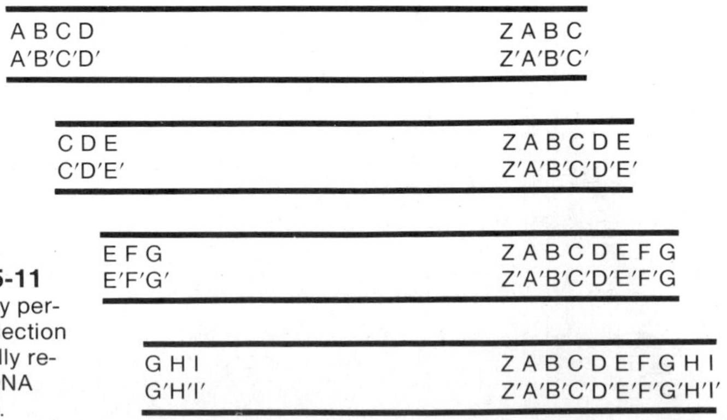

Figure 15-11
A cyclically per-
muted collection
of terminally re-
dundant DNA
molecules.

DNA exonuclease that removes nucleotides one by one only from the 3'-OII terminus of each single strand; the treated DNA is then subjected to a renaturation procedure. Once a number of bases is removed that exceeds the number in the terminally redundant region, a circular structure can form; this structure contains a short double-stranded region, which is the length of the terminal redundancy, flanked by two single strands.* A study of this sort indicated that the terminal redundancy contains about 1600 base pairs.

A remarkable feature of T4 DNA is that, though each phage contains one DNA molecule, the molecules differ from phage to phage, even if the population was produced by replication of a single phage particle and its progeny. A sample of T4 DNA molecules is **cyclically permuted,** the meaning of which is shown in Figure 15-11. The figure indicates schematically that the termini of the DNA molecules in the population can be found at many different bases within an overall sequence—in reality, probably at any point in the base sequence. Note that cyclic permutation is a property of a phage population, whereas terminal redundancy is a property of an individual phage DNA molecule.

Many phage species produce terminally redundant and cyclically permuted populations of DNA molecules. Terminal redundancy can also be present without cyclic permutation but the converse has not been observed. How these properties arise in phage T4 will be seen when packaging of phage DNA is described.

Genetic Organization of the T4 Phage

To date, 135 genes have been identified in the T4 phage. These genes account for about 90 percent of the DNA; thus 15 to 20 genes remain to be found. A genetic map of these genes is shown in Figure 15-12. Genes having related functions are often adjacent and transcribed as part of polycistronic mRNA molecules. This is an efficient arrangement, allowing the synthesis of functionally related proteins to occur at nearly the same time and minimizing the number of regulatory elements required. However, not all functionally related genes are part of single transcription units and some transcription units contain functionally discrete genes. The tendency to cluster related genes is common in many phage systems and occurs almost without exception with *E. coli* phage λ, as will be seen in a later section.

T4 genes can be divided into two classes—82 metabolic genes and 53 particle assembly genes. Of the 82 metabolic genes only 22 genes—

---

*Actually, a circle can form as soon as the number of bases removed exceeds *half* the number in the terminally redundant region (the reader should confirm this). However, these circles are less stable than those described in the text and give no information about the length of the terminal redundancy.

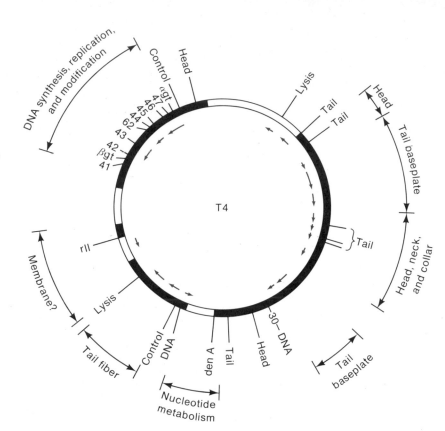

Figure 15-12
Genetic map of phage T4 showing some, but not all, genes. The clustering of genes with related functions should be noted; although the tail baseplate and tail fiber genes form large clusters, other tail genes are distributed throughout the map. The solid and open regions indicate the locations of essential and nonessential genes respectively. Control refers to genes needed to initiate various modes of transcription. The inner red arrows indicate the direction and origins (but not the lengths) of various transcripts.

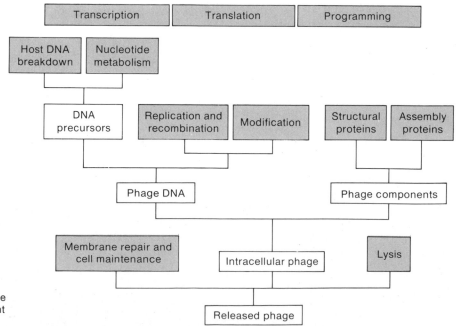

Figure 15-13
Categories of T4 genes (red boxes) classified by function. Products and general functions are in unshaded boxes. The connecting lines show how various functions and components combine for production of different components.

namely, those involved in DNA synthesis, transcription, and lysis—are essential. The remaining 60 metabolic genes duplicate bacterial genes; particles in which these genes are mutated will grow, though occasionally they will have a smaller burst size. Of the 53 assembly genes, 34 code for structural proteins and 19 code for the synthesis of enzymes and protein factors that are required catalytically for assembly. Thus, 17 percent of the DNA of phage T4 encodes essential metabolic functions, 39 percent is necessary for phage assembly, and 44 percent serves nonessential metabolic functions.

A useful way to classify the functions of T4 genes is shown in Figure 15-13. The significance of these functions will become clear as we examine various stages of the life cycle in detail.

The T4 Growth Cycle in Brief

The complete life cycle of T4 is illustrated in Figure 15-14 and outlined below. The times are in minutes at 37°C.

$t = 0$	Phage adsorbs to bacterial cell wall. Injection of phage DNA probably occurs within seconds of adsorption.
$t = 1$	Synthesis of host DNA, RNA, and protein is totally turned off.
$t = 2$	Synthesis of first mRNA begins.
$t = 3$	Degradation of bacterial DNA begins.
$t = 5$	Phage DNA synthesis is initiated.
$t = 9$	Synthesis of "late" mRNA begins.
$t = 12$	Completed heads and tails appear.
$t = 15$	First complete phage particle appears.
$t = 22$	Lysis of bacteria; release of about 300 progeny phage.

The main feature of this list to be noticed at this point is the orderly sequence of events. This sequence is also illustrated in part in the electron micrographs shown in Figure 15-15. In the sections that follow a few of these stages are described in detail.

Turning Off Bacterial Macromolecular Synthesis

Shortly after infection, several phage proteins are synthesized that are capable of efficiently turning off transcription and replication of host DNA. For example, the host RNA polymerase is modified in such a way that host promoters are recognized poorly. Furthermore, the first phage mRNA encodes DNases that rapidly degrade host DNA. Thus, in time the bacterial DNA is almost totally degraded to nucleotides and is not available as a template for replication or transcription. However,

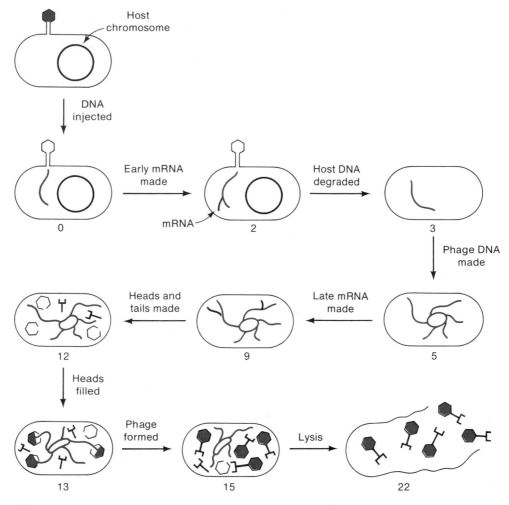

Figure 15-14
A schematic diagram of the life cycle of phage T4. The numbers represent time after injection in minutes at 37°C. For clarity, mRNA is drawn only at the time at which its synthesis begins.

turning off the synthesis of host DNA, RNA, and protein precedes synthesis of these nucleases and also occurs when *E. coli* is infected by phage mutants unable to synthesize these nucleases. Thus another explanation is needed for inhibition of host macromolecular synthesis. By genetic and biochemical manipulation it is possible to prepare phage particles that lack DNA—these particles are called **ghosts.** Adsorption of ghosts to *E. coli* also turns off host macromolecular synthesis, so this inhibition does not require any product synthesized immediately after the infection. There is some evidence that some components of both the cell wall and the plasma membrane are solubilized and released to the

Figure 15-15
Electron microscopic observation of development of bacteriophage T4. The times are indicated in minutes after infection at 30°C. (0 min) Immediately after infection the cell has the morphology of an uninfected cell. The DNA-containing nucleoid is the light region. (5 min) This micrograph shows the nuclear disruption induced by this phage. At this stage the host DNA and its breakdown products are found mainly near the cell wall. The nucleotides of the host DNA will be used for synthesizing phage DNA. (15 min) At about eight minutes after infection the structural proteins used for assembling and maturing the phage start to be made. A few minutes later, the first phage particles appear. From then on, phage appear at a rate of about five particles per minute. It takes about seven minutes to assemble a particle. A prohead (insert) containing little or no DNA is made first. The fine fibrillar material (light region) resembling the bacterial nucleus is the large mass of phage DNA. (30 min) Many finished heads are present. Since this micrograph is of a thin section of the cell the actual number of particles per cell is about 20 times the number seen. (Courtesy of B. Menge, J. v.d. Broek, H. Wunderli, K. Lickfield, M. Wurtz, and Edward Kellenberger.)

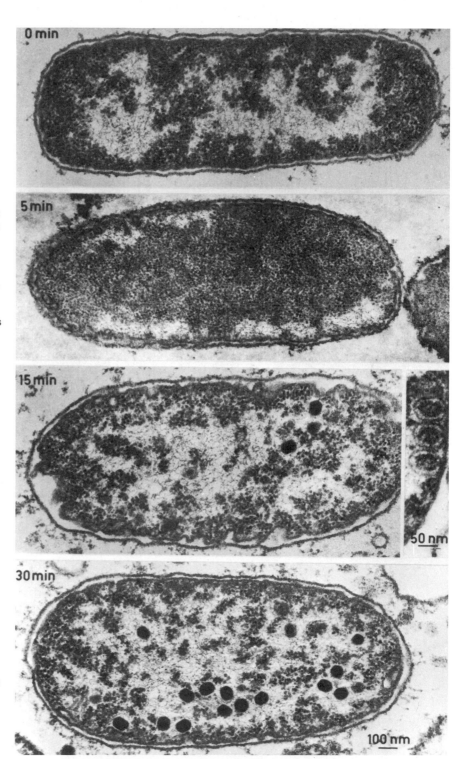

surrounding medium following adsorption of either an intact phage or a ghost. Thus, it is thought but has never been proved that adsorption of a T4 tail alters the *E. coli* plasma membrane and, thereby, secondarily influences host macromolecular synthesis. This might occur either by altering the permeability of the membrane or by affecting the binding of the host chromosome to the membrane; the latter is hypothesized to be essential for host replication and possibly for host transcription.

Transcription of Phage T4 DNA

It is essential that the timing of synthesis of phage proteins be regulated. For example, if T4 lysozyme were made in large amounts early in the life cycle, lysis might occur before progeny phage form. Also, if heads were made before the onset of DNA replication, the injected DNA molecule might be packaged and there would be no increase in the number of phage. The major mechanism for regulating the timing of protein synthesis in phage-infected cells is by regulating mRNA synthesis; in phage T4 this is accomplished by varying the ability of *E. coli* RNA polymerase to recognize particular promoters and to bypass terminators.

Many classes of mRNA molecules are transcribed from T4 DNA. Each class is initiated at a different time and synthesized for a definite time interval. It is convenient to consider two main classes of T4 mRNA—**early mRNA** (transcribed before DNA synthesis occurs) and **late mRNA** (postreplicative):

1. The early mRNA molecules encode enzymes and protein factors needed for DNA synthesis, regulation of mRNA synthesis, genetic recombination, and for a variety of nonessential proteins.
2. The late mRNA molecule contains information for synthesis of the structural proteins of the phage, assembly of the particle, and lysis.

The complete pattern of early transcription of T4 DNA includes the synthesis of more than ten species of early mRNA, many of which have overlapping sequences, and there are several attenuators (similar to those in the *E. coli trp* and *his* operons). The basic pattern that is true of several phages is the following:

Transcription of early mRNA starts at a single class of promoter by means of *E. coli* RNA polymerase (because that is the only polymerase in the cell); then the polymerase is modified by the activity or addition of phage-specified proteins, so that it no longer recognizes host promoters but goes on instead to initiate transcription of later species of mRNA at phage promoters. Successive modifications then provide one means of temporal control of the synthesis of many species of T4 early mRNA.

Figure 15-16
Alteration and modification of *E. coli* RNA polymerase by successive addition of two ADPR molecules (red).

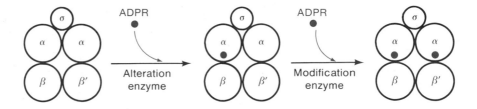

We will see later that other phages ultimately inactivate the *E. coli* RNA polymerase and that a phage-encoded polymerase is made.

The major steps in the sequence of successive modification of *E. coli* RNA polymerase activity by T4-encoded proteins are the following:

1. *Transcription by unmodified* E. coli *RNA polymerase.* Shortly after injection, the first T4 mRNA molecule is synthesized by means of host RNA polymerase.

2. *Alteration.* The T4 phage head contains several small proteins, known as internal proteins, which are injected into a bacterium along with the DNA.* One of these proteins, the **alteration enzyme,** modifies host RNA polymerase by linking a small molecule (**ADP-ribose or ADPR**) to one of the α subunits of the polymerase (Figure 15-16).

3. *Modification.* Shortly after alteration occurs, the RNA polymerase undergoes another change, called modification, in which one more ADPR is added. It is thought that alteration and modification reduce the ability of RNA polymerase to bind the σ factor and thus to initiate transcription at host promoters. However, these are probably nonessential functions.

4. *Specificity change.* A phage-encoded protein, **Mot** (possibly a DNA-binding protein), reduces the ability of RNA polymerase to recognize early promoters and enables it to bind to the next class of promoters.

The net result is an orderly synthesis in time of the early species of mRNA. The changing specificity of RNA polymerase means that certain mRNA molecules are no longer made at the time when continued synthesis of their gene products is not required and that new mRNA molecules are made when needed.

Shortly after termination of synthesis of a particular mRNA species, the proteins encoded in the molecules are no longer synthesized because of the rapid degradation of mRNA. Continuous synthesis of a protein needed for a long period of time can be accomplished either by continued synthesis or by resistance of the encoding mRNA to degradation. The former is a common mechanism.

*These proteins were discovered by Alfred Hershey shortly after the Hershey-Chase experiment was performed. Fortunately for the development of molecular biology, internal proteins contain little sulfur and no injected protein-containing ^{35}S was detected in bacterial cells in this experiment.

Other modifications occur that enable more types of early mRNA molecules to be synthesized. By the time late mRNA synthesis begins, RNA polymerase has become an extraordinarily complicated enzyme—the two α subunits carry two ADPR molecules and four modifying proteins. This is sometimes called "hypermodified" RNA polymerase. However, this enzyme is unable by itself to initiate late transcription, for *late mRNA is synthesized only after DNA replication begins;* in mutants unable to synthesize DNA, late transcription never occurs. This type of programming is, of course, quite efficient since the proteins encoded in the late mRNA molecule—namely, heads, tails, assembly proteins, and lysozyme—are not needed until replication has progressed for some time. The requirement of DNA replication for late mRNA synthesis is not understood but is also observed *in vitro.* For example, neither purified normal RNA polymerase nor the hypermodified polymerase can synthesize late mRNA *in vitro* with DNA isolated from purified phage. These and several other results have indicated that the DNA itself must have some particular conformation, which is normally brought about by replication, before it can serve as a template for transcription. Note that this is a type of regulation that has not been encountered before in this book. In 1983, a purified *in vitro* system was developed in which late transcription occurs. Examination of this system will surely lead to an understanding of this perplexing phenomenon.

Replication of T4 DNA

Five aspects of T4 DNA replication are especially interesting: (1) the source of nucleotides; (2) the synthesis of 5-hydroxymethylcytosine (HMC), which substitutes for cytosine (C); (3) the prevention of incorporation of cytosine; (4) glucosylation of T4 DNA; and (5) the enzymology of replication. Each is discussed below.

1. *Source of T4 DNA nucleotides—degradation of host DNA.* An early event in the T4 life cycle is the degradation of host DNA to deoxynucleoside monophosphates (dNMP). The degradation is initiated by two endonucleases, which are the products of the genes *denA* and *denB*. These two enzymes, which are active only on cytosine-containing DNA, cleave the host DNA to double-stranded fragments, which are then degraded to dNMP by an exonuclease controlled by genes *46* and *47*. These mononucleotides are then built up to dATP, dTTP, dGTP, and dCTP by the usual *E. coli* enzymes and this provides sufficient dNTP to synthesize 30 T4 DNA molecules. DNA precursors are also formed by *de novo* synthesis but, to ensure an abundant supply of dNTP, five phage-encoded enzymes, which are virtually identical in

activity to the *E. coli* enzymes, are also synthesized; these enzymes, which are dispensable, are thymidylate synthetase, nucleoside diphosphate reductase, dCMP deaminase, dihydrofolate reductase, and deoxynucleotide kinase.

2. *Synthesis of 5-hydroxymethylcytosine (HMC).* T4 DNA contains HMC instead of cytosine. However, *E. coli* does not possess enzymes for forming HMC; therefore, this is accomplished by two phage enzymes, which convert dCMP to dHDP:

$$dCMP \xrightarrow{\text{T4 hydroxymethylase}} dHMP$$

$$dHMP \xrightarrow{\text{HM kinase}} dHDP$$

The *E. coli* enzyme, nucleoside phosphate kinase, which forms all triphosphates in *E. coli,* then converts dHDP to dHTP, the immediate precursor of the HMC in the DNA. These pathways are shown in Figure 15-17.

3. *Prevention of incorporation of cytosine into T4 DNA.* In Chapter 8 we saw that the *E. coli* DNA polymerases cannot readily distinguish dUTP from dTTP because the two molecules have the same hydrogen-bonding properties and that an additional enzymatic mechanism is

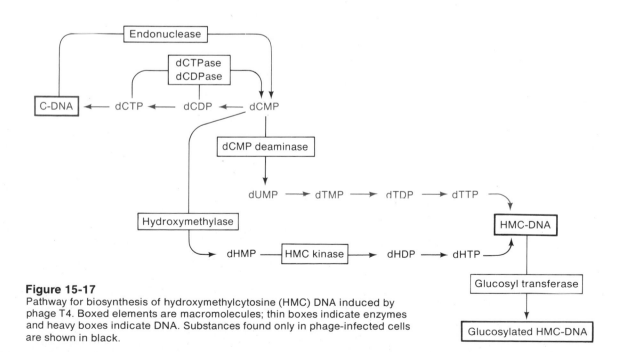

Figure 15-17
Pathway for biosynthesis of hydroxymethylcytosine (HMC) DNA induced by phage T4. Boxed elements are macromolecules; thin boxes indicate enzymes and heavy boxes indicate DNA. Substances found only in phage-infected cells are shown in black.

required to prevent incorporation of uracil into DNA. A similar problem arises with T4 DNA polymerase, for it cannot distinguish dCTP from dHTP, both of which can hydrogen-bond to guanine. It is essential that no cytosine be incorporated into daughter T4 DNA strands, because such cytosine-containing DNA would be a substrate for the T4 endonucleases that degrade host DNA. Also, for unknown reasons, cytosine-containing DNA is not a template for late transcription. *E. coli,* as host, has no use for enzymes that prevent incorporation of cytosine into DNA; thus, the phage must encode these enzymes. For cytosine to become part of daughter DNA molecules, dCMP must be converted to dCDP and then to dCTP. A phage enzyme, dCDP-dCTP pyrophosphorylase, usually called dCTPase, degrades both dCDP and dCTP to dCMP. This process seems somewhat wasteful in that ATP is consumed in forming dCTP; however, the more economical process of preventing formation of dCTP would be difficult to carry out in *E. coli,* since a single enzyme (nucleoside phosphate kinase) is responsible for the production of all triphosphates. Inhibiting the dCMP→dCDP reaction would also be ineffective because in *E. coli* most of the deoxynucleoside diphosphates are formed by enzymatic reduction of ribonucleoside diphosphates by a single enzyme (Chapter 8).

Another phage enzyme, dCMP deaminase, converts dCMP to dUMP (in preparation for synthesis of dTMP) (Figure 15-17). This enzyme duplicates the activity of a similar *E. coli* enzyme (and hence is a product of one of the nonessential genes) but has an interesting economic function. The base compositions of *E. coli* DNA and T4 DNA are 50 percent (A + T) and 66 percent (A + T), respectively. In *E. coli,* the ratio of dTTP and dCTP is about 1 : 1 in keeping with the T : C ratio in DNA. The bacterial and phage dCMP deaminases, acting together, increase the amount of dTMP with respect to dCMP, so the ratio of dTTP to dHTP is 2 : 1, as is the T : HMC ratio in T4 DNA.

Occasionally some cytosine appears in progeny phage DNA. Presumably, the T4 endonucleases degrade this DNA shortly after synthesis; since the cytosine is in only one strand of each daughter double helix, the DNA can be repaired by normal repair synthesis. Also, as is true of *E. coli* DNA synthesis, occasionally uracil is inserted into progeny DNA (by incorporation of dUTP). The uracil residues are removed by the usual uracil-N-glycosylase (see Chapter 8).

4. *Glucosylation of T4 DNA.* The presence of HMC in T4 DNA creates another problem for the phage because *E. coli* possesses an endonuclease that attacks certain sequences of nucleotides containing HMC. To avoid this damage the HMC residues in T4 DNA are glucosylated. This is accomplished by two phage enzymes, α-glycosyl transferase (αgt) and β-glucosyl transferase (βgt), each of which transfers a glucose from uridine diphosphoglucose (UDPG) to HMC that is already in DNA. Thus glucosylation is a postreplicative modification.

The *E. coli* endonuclease is inactive against glucosylated DNA; therefore, glucosylation is a protective device. A simple genetic experiment shows that this is the only essential function of glucosylation. A T4 *αgt⁻* mutant cannot carry out glucosylation, so its newly synthesized DNA is destroyed by the *E. coli* HMC endonuclease. However, if an *E. coli* mutant *(rglB⁻)*, which lacks this nuclease, is used as a host bacterium, both T4 *αgt⁻* and T4 *βgt⁻* mutants grow normally even though nonglucosylated DNA is produced.

5. *The enzymology of T4 DNA replication and the form of replicating DNA.* Many phage genes participate in the replication process and some of these duplicate the functions of *E. coli* genes. However, an *in vitro* system has been reconstituted that contains the four dNTP's and the products of only seven T4 genes, numbered *32, 41, 43, 44, 45, 61,* and *62,* whose properties are listed in Table 15-5. In this system biologically active T4 DNA is synthesized with purified phage DNA as a template; the rate of synthesis approaches that of replication *in vivo* and many of the normal intracellular structures (see below) are formed; it is believed that the system (nearly) reproduces the system used *in vivo*. Thus, T4 DNA replication has no need for any of the *E. coli* replication genes. The seven proteins form a macromolecular aggregate. For instance, the gene-43 product (T4 DNA polymerase), the gene-32 helix-destabilizing protein, and T4 DNA form a stable complex, as do the products of genes 44 and 62. The gene-45 protein joins with the 44/62 pair to form a DNA-dependent ATPase that can probably be compared either to *E. coli* helicases or the DnaB protein. The gene-41 product has primase activity but is also needed for synthesis of the leading strand; this protein also pairs with the gene-61 product. The enzymatic mechanism of T4 DNA replication is not yet known but presumably this will soon be understood as a result of careful study of the *in vitro* system.

The T4 DNA replicating structure is quite complex. One reason is that a mechanism has evolved that enables the rate of synthesis of progeny DNA to be very high (and thereby generates an enormous

Table 15-5
T4 genes essential for DNA replication

T4 gene	Function of gene product
32	Helix-destabilizing, single-strand-binding protein
[41,61]*	Priming
43	DNA polymerase
[44,62]*	Binds to polymerase
45	Binds to polymerase

*The gene products of each pair in brackets are isolated as a tightly bound complex.

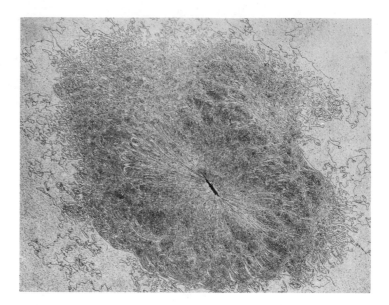

Figure 15-18
An electron micrograph of the replicating complex of T4 DNA. Note how it
resembles the structure of *E. coli* DNA
shown in Figure 6-11. (Courtesy of
Joel Huberman.)

amount of DNA in a short time span). The T4 DNA polymerase adds
about 800 nucleotides per second at a growing fork (roughly comparable
to *E. coli* polymerase III *in vivo*), whereas the apparent growth rate of
T4 DNA *in vivo* is about 5×10^4 nucleotides/second per cell. This rapid
growth is accomplished by a combination of bidirectional replication
from two (or possibly more) replication origins to form replication
bubbles and reinitiation at these origins even before a complete
replication cycle has been concluded. A second reason for the complexity is that T4 DNA frequently undergoes genetic recombination and
in each molecule there occur many events in which the double helix is
broken and rejoined. Linear concatemers of up to six times the length of
a parental molecule can be isolated from an infected cell; these
concatemers are not a result of rolling circle replication but are formed
by genetic recombination between redundant ends. Furthermore, it
seems to be the case that in both singly and multiply infected cells,
recombination between replicating daughter molecules occurs so frequently that all of the unpackaged phage DNA in a single cell is part of
one gigantic interconnected unit containing several hundred replication forks (Figure 15-18). The complexity of the structure has greatly
hampered its study.

Production of T4 Phage Particles

Production of complete phage particles can be separated into two
parts—assembly of heads, tails, and other structures, and packaging of
DNA of a sufficient length to provide a little more than one set of genes
in the phage head.

Assembly of T4 (and many other phages) has been studied by two techniques, both of which require a large collection of phage mutants unable to produce finished particles. In one method, different cultures of cells, each infected with a particular mutant, are lysed and examined by electron microscopy. This procedure shows that heads are made in the absence of tail synthesis and that tails are made by a mutant unable to synthesize heads; head and tail assembly are independent processes. The second technique is a complementation assay of a type widely used to purify proteins. If two extracts of infected cells, one lacking heads and the other lacking tails, are mixed, phage particles form *in vitro* (Figure 15-19). The "headless" extract can also be fractionated and a component can be isolated that allows a "tailless" extract to make tails. In this way, a protein in the tail assembly pathway can be isolated and identified.

These studies have shown that there are two types of components—**structural proteins** and **morphogenetic enzymes.** Some of the structural components assemble spontaneously to form phage structures, whereas others need the help of enzymes. A few host-encoded factors are also needed for head assembly. For T4 (and also for phage λ) the assembly pathway has been almost completely worked out; the known steps for T4 are shown in Figure 15-20.

The mechanism of packaging of T4 DNA in a phage head is not yet elucidated, though the process can be carried out *in vitro*. The basic problem is how the long strand of DNA is tightly folded so that it will fit in the phage head. It is thought that packaging begins by the attachment of one end of a DNA molecule to a protein contained in the phage head. Then either a condensation protein or a small, very basic molecule (such as a polyamine) induces folding (Figure 15-21). The best-understood feature of the process is that the molecule is cut from long concatemers. The cuts are not made in unique base sequences in the DNA because, if they were, T4 DNA could not be cyclically permuted. Instead, the cuts are made at positions that are determined by the amount of DNA that can fit in a head. Presumably a free end of the DNA molecule enters the head and this continues until there is no more room; then the concatemer is cut. This is known as the **headful mechanism** and it explains how both terminal redundancy and cyclic permutation arise (Figure 15-22). The essential point is that the DNA content of a T4 particle is greater than the length of DNA required to encode the T4 proteins. Thus, when cutting a headful from a concatemeric molecule, the final segment of DNA that is packaged is a duplicate of the DNA that is packaged first—that is, the packaged DNA is terminally redundant. The first segment of the second DNA molecule that is packaged is not the same as the first segment of the first phage. Furthermore, since the second phage must also be terminally redundant, a third phage-DNA molecule must begin with still another segment. Thus, the collection of

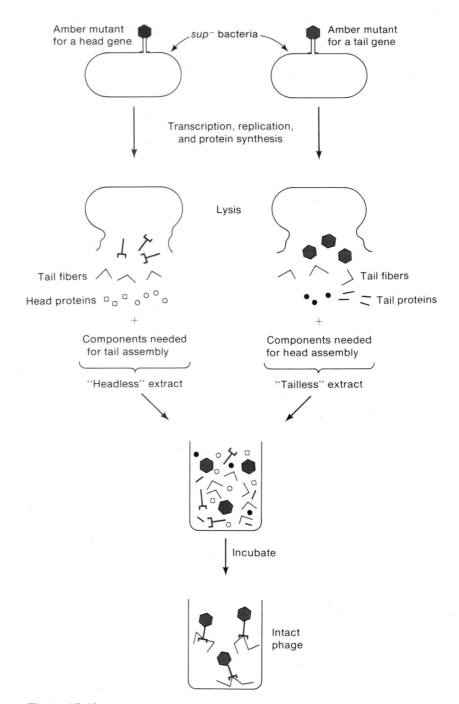

Figure 15-19
Production of intact T4 phage by *in vitro* complementation.

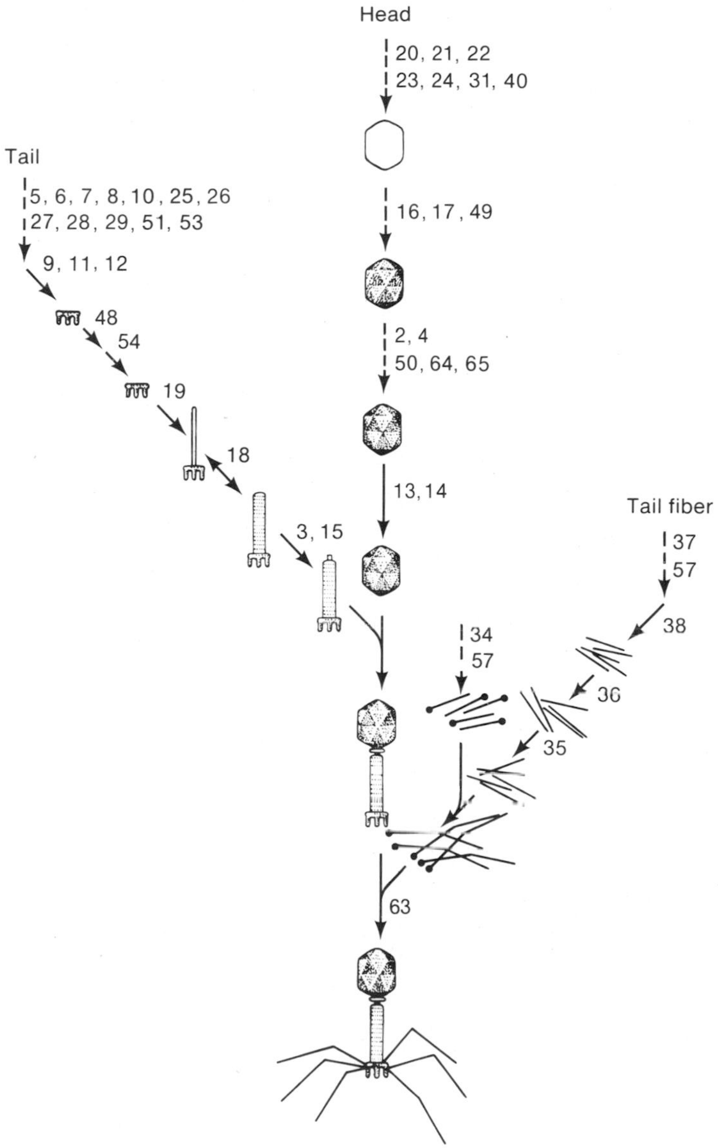

Figure 15-20
Morphogenetic pathway of
T4 phage. (Courtesy of
William Wood.)

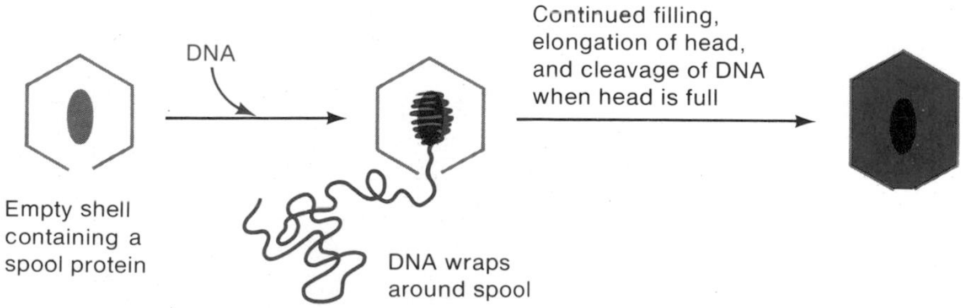

Head

20, 21, 22
23, 24, 31, 40

16, 17, 49

2, 4
50, 64, 65

13, 14

Tail

5, 6, 7, 8, 10, 25, 26
27, 28, 29, 51, 53

9, 11, 12

48

54

19

18

3, 15

Tail fiber

37
57

38

36

35

34
57

63

Figure 15-21
Proposed model for filling
a T4 head. Cleavage and
rearrangement of head
proteins is known to oc-
cur at several stages of
the process.

DNA

Continued filling,
elongation of head,
and cleavage of DNA
when head is full

Empty shell
containing a
spool protein

DNA wraps
around spool

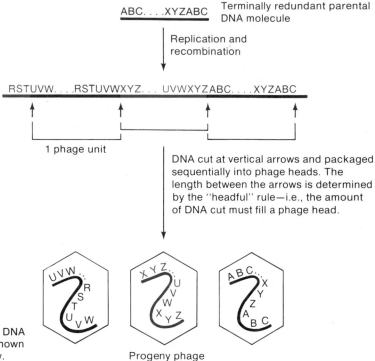

ABC. . . .XYZABC   Terminally redundant parental DNA molecule

Replication and recombination

RSTUVW. . . .RSTUVWXYZ. . . UVWXYZABC. . . .XYZABC

1 phage unit

DNA cut at vertical arrows and packaged sequentially into phage heads. The length between the arrows is determined by the "headful" rule—i.e., the amount of DNA cut must fill a phage head.

Progeny phage

Figure 15-22
Origin of cyclically permuted T4 DNA molecules. Alternate units are shown in different colors for clarity only.

DNA molecules in the phage produced by a single infected bacterium is a cyclically permuted set. There are two important kinds of evidence for the headful mechanism: (1) if a mutant phage contains a deletion for nonessential genes, the terminally redundant region increases in length by the amount of DNA that is deleted; and (2) mutant phage that have heads 0.65, 0.8, 1.2, and 5 times normal volume have 0.65, 0.8, 1.2, and 5 times the normal length of DNA.

E. COLI PHAGE T7

Phage T7 is a midsized phage (Table 15-3). Its DNA has a molecular weight of about 26×10^6 and contains equal amounts of adenine, thymine, guanine, and cytosine, and has no unusual bases. It has a terminal redundancy of 160 base pairs but it is not cyclically permuted. The DNA molecule is encased in a head that is attached to a very short tail (Figure 15-23).

T7 has been of interest for many years because it has a somewhat small number (49–50) of genes, which simplifies analysis of its life cycle. Furthermore, most of the gene products have been identified by gel

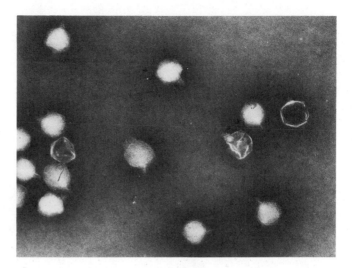

Figure 15-23
Electron micrograph of phage T7.
Note the very short tails. Three of the
particles are empty. (Courtesy of
Robley Williams.)

electrophoresis, 34 have been purified (by early 1982), and many have
been sequenced by chemical means. Of great importance is the recent
elucidation of the complete sequence of 39,930 base pairs of T7 DNA;
this information allows one to locate, by direct inspection, all of the
promoters, termination sites, regulatory sites, spacers, leaders, initiation
codons, termination codons, and genes. In fact, several previously
unknown genes were detected simply by noting the existence of an
AUG start codon and a stop codon in the same reading frame. Furthermore, from the base sequence of each gene the amino acid sequence of
each protein can be deduced.

Organization of the T7 Genes

A nearly complete genetic map of T7 is shown in Figure 15-24. (For
clarity a few genes of unknown function have been omitted.) Of the 41
genes indicated, 34 gene products have been identified and the functions of 26 genes are known. As will be seen in a later section the gene
products are synthesized in numerical order starting with gene *0.3*.

The DNA content of T7 is less than one-fourth that of T4, and it
might be expected that there would be less duplication of host genes. Of
the 82 known metabolic genes of T4, 60 are nonessential; if all known
genes are counted, 39 percent of the T4 genes are nonessential and most
of these duplicate host function. In contrast, for T7, only 2 of the 26
genes (8 percent) of known function are nonessential. Other signs of
economy are made evident by examination of the base sequence. For
example, leaders and spacers between genes are generally only a few

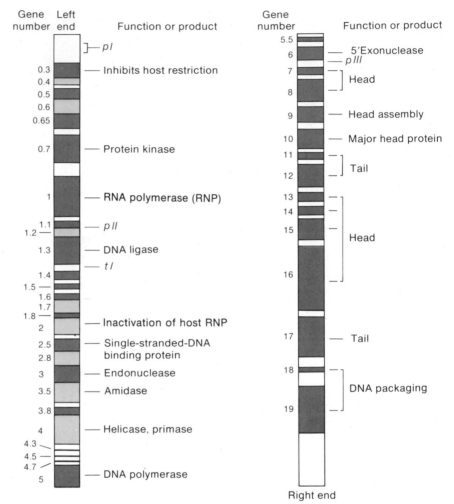

Figure 15-24
Genetic and physical map of phage T7 (drawn vertically instead of the standard left-to-right orientation to save space). Genes are numbered. Genes whose functions are not listed are either nonessential or have an unknown function. The sizes of the blocks are approximately proportional to protein size. Intensity of shading is used only to distinguish adjacent genes and has no other significance. The promoters, *pI, pII,* and *pIII,* and the terminator *tI* are in red.

nucleotides long, when present at all, whereas in bacterial genetic systems hundreds of nucleotides are used for these. For roughly half of the genes to the left of gene 4 the Shine-Dalgarno sequence used for binding mRNA to a ribosome (Chapter 12) is in the coding sequence of the preceding gene. Furthermore, the termination codon of one cistron often overlaps the initiation codon of the next cistron—for example, the sequence UGAUG terminates gene *1.7* and initiates gene *1.8* and UAAUG serves a dual function for genes *0.5* and *0.6*. There are also three cases in which coding sequences of adjacent genes overlap; the maximum overlap is 23 bases, so this overlap is much smaller than that present in phage φX174 (explained later in this chapter) and many animal viruses (Chapter 21). Finally, the catalytic proteins made early in the life cycle tend to be much smaller than the average bacterial protein, ranging from 29 to 883 amino acids.

The genes of T7 are clustered according to function and are arranged to provide a continuous order according to the time of their function. The first gene (reading the genetic map from top to bottom in Figure 15-24, or left to right in the standard orientation of the map) is essential for survival of the phage DNA after infection; *E. coli* strain B, a common host for T7, contains a restriction endonuclease that makes cuts in specific base sequences present in T7 DNA, and the product of the first gene inhibits the activity of this nuclease by directly binding to it. The next few genes control transcription; these are followed by genes needed for DNA replication and then by genes encoding the structural proteins of the phage particle. The final genes are used for packaging of newly synthesized DNA in the phage head.

The T7 Growth Cycle

The life cycle of T7 is outlined below (times are in minutes at 30°C):

$t = 0$	Adsorption of the phage.
$t = 0$–1	Initiation of slow injection of phage DNA.
$t = 2$	Initiation of synthesis of phage mRNA.
$t = 4$	Turning off of host transcription begins.
$t = 8$–9	Initiation of phage DNA synthesis.
$t = 8$–10	Initiation of synthesis of structural proteins; completion of injection of phage DNA.
$t = 15$	First phage appears.
$t = 25$	Lysis and release of progeny phage.

Most of these stages will be discussed in the following sections.

Injection of T7 DNA into a Host Cell

With T4 and for many other phages, the phage DNA is injected into a host bacterium very soon after adsorption, and injection is complete in less than one minute. In contrast, with T7 it takes about ten minutes for injection of all of the DNA. Why T7 injects slowly is unknown (but no more so than the mechanism of injection of any other phage).

Slow injection is a significant feature of the T7 life cycle because a gene cannot be transcribed until it has been injected. In fact, the timing of transcription, which we discuss in the next section, depends in part on the kinetics of injection.

Regulation of Transcription of T7 DNA

The life cycle of T7 is normally divided into two transcriptional stages called early and late. The early stage uses *E. coli* RNA polymerase to transcribe genes *0.3* through *1.3*; the late stage uses a newly synthesized T7 RNA polymerase. Transcription of T7 DNA is temporally regulated. There are three major classes of transcripts, labeled I, II, and III in Figure 15-25. Each of these is transcribed from the same DNA strand, which is termed the *r* strand (because transcription occurs in the rightward direction when the DNA molecule is drawn in a standard orientation). The class I transcripts are synthesized by *E. coli* RNA polymerase and comprise the early mRNA. There are three distinct promoters for the class I transcripts; they are very near one another, so each mRNA includes the same genes. It is convenient when discussing regulation to consider each of these mRNA molecules as a single molecule called transcript I and to refer to a single promoter termed *pI*. The existence of three promoters probably serves only to increase the net rate of mRNA synthesis threefold.

Synthesis of transcript I begins 2 minutes after infection, the delay probably resulting from the slow injection of the DNA. Transcription is mediated by the bacterial RNA polymerase and stops at a termination site just after gene *1.3*. Transcript I is cleaved by *E. coli* RNase III into five mRNA molecules; however, in laboratory conditions the processing is not essential, since if an *E. coli* mutant lacking RNase III is used, the transcript remains intact and phage production seems to be normal. Possibly in nature or in other bacterial hosts processing is required.

Several proteins are translated from transcript I. One is a protein kinase (gene *0.7*) which phosphorylates *E. coli* RNA polymerase,

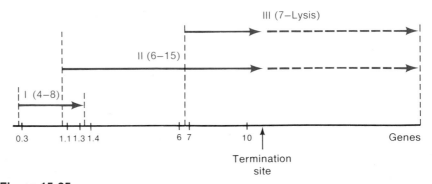

Figure 15-25
A simplified transcription map of T7. The DNA and gene numbers are in black; the three classes of transcripts are in red. The red numbers indicate the time interval (in minutes at 37°C) during which the transcript is made, the termination site indicated by the black arrow is about 90 percent efficient; thus 10 percent of the class II and III molecules are extended further, as indicated by the dashed lines.

inactivating the enzyme. The inactivation reduces transcription of *E. coli* DNA and is the first step in takeover of the bacterium by the phage. This kinase is not necessary in infections by T7 phage of most laboratory strains of *E. coli*; however, T7 gene-*0.7* mutants yield a small burst size compared to wild-type phage, probably because the cell continues to synthesize *E. coli* RNA molecules, which are of no value to T7 and thus a drain of resources. Also, as we will see shortly, another inhibitor of *E. coli* RNA polymerase is made later. A second protein translated from transcript I is T7 RNA polymerase (gene *1*), which is essential because this enzyme is responsible for all further transcription. In contrast with *E. coli* RNA polymerase, which is a complex, multisubunit protein, T7 RNA polymerase is a single polypeptide chain. Presumably, its simplicity is a reflection of the simple demands for initiation and termination placed upon this enzyme compared to the more complicated requirements of the *E. coli* enzyme.

Transcripts II and III also represent classes of mRNA that are initiated at several promoters. These promoters are not recognized by *E. coli* RNA polymerase but require the translation of T7 RNA polymerase from transcript I. However, transcription from these promoters does not begin as soon as T7 RNA polymerase is made. Instead, class II transcripts, whose promoters are much weaker than class III ones, are made first. This delay and the difference in the time of initiation of the two classes both have the same explanation—namely, that T7 DNA is injected slowly and that the promoters for class II enter the cell several minutes before entry of the later promoter.

Transcripts II are initiated upstream of the termination site of class I; therefore, gene *1.3* is contained in both transcripts I and II. The reason for this is unclear, especially since gene *1.3* mutants, which lack T7 DNA ligase, grow fairly well when supplied with *E. coli* DNA ligase—at least in the laboratory. Possibly in nature T7 frequently encounters host strains that cannot supply enough ligase for rapid phage growth and has evolved a way to grow in such hosts, or the T7 enzyme has sufficient advantages that its synthesis for a prolonged time is warranted. The T7 RNA polymerase ignores the site that causes termination of transcript I and each transcript runs rightward to a second termination site (Figure 15-25). This termination site is 90 percent efficient, meaning that 10 percent of the transcripts proceed on to the end of the DNA molecule. Since the termination site is just after gene *10*, which encodes the major head protein, it is believed that its location is a means of increasing the concentration of this gene product with respect to the tail proteins and minor head proteins, whose genes follow the termination site and which are needed in lesser amounts.

A class II transcript gene (gene *2*) makes a second inhibitor of *E. coli* RNA polymerase. Combined with the earlier reduction of activity of the polymerase by the gene-*0.7* protein kinase, the *E. coli* enzyme is completely inhibited. Thus, by eight minutes after infection there is no longer any synthesis of *E. coli* RNA and takeover of the bacterium is completed. Transcript I, which is made by the *E. coli* RNA polymerase, is also no longer made; this is economical because at this point sufficient T7 RNA polymerase has been produced to complete the life cycle and no other products of this transcript are needed.

Transcript II contains genes for DNA replication and also for synthesis of structural proteins. In fact, the structural genes are transcribed very shortly after the replication genes. At first glance this seems inefficient because premature assembly of phage particles and packaging of phage DNA might result in termination of replication or at least a vast reduction of the number of templates for replication. This arrangement contrasts with that found in T4, in which synthesis of the structural proteins is delayed with respect to the replication enzymes. However, the T7 DNA molecule is small and can be replicated rapidly (in nearly 20 seconds), while the maturation proteins (genes 18 and 19) are translated late and head assembly is slow; there is in fact no packaging of DNA before replication forms many DNA molecules.

Once the DNA replication enzymes have been synthesized, there should be no further need for transcription of this region since the proteins are, for the most part, catalytic. However, termination of synthesis of transcript II would also stop transcription of structural genes, whose products are needed in stoichiometric amounts. Efficiency as well as continued synthesis of structural proteins are both attained with two late mRNA molecules (II and III) and by turning off the synthesis of transcript II. Admittedly, the same result could be achieved if transcript II terminated before encountering the promoter for transcript III, but evolution has taken a different course.

Synthesis of transcript I stops 15 minutes after infection. This is a result of a general shutting down of transcription and the relative strengths of the promoters *pII* and *pIII* for transcripts II and III, respectively. That is, overall transcription is reduced and only the stronger promoter *pIII* remains active. The mechanism of the overall reduction of transcription is unknown but requires the action of the gene-3.5 protein (*N*-acetylmuramyl-L-alanyl amidase). This protein is a lysozymelike enzyme and is also responsible for cell lysis.

T7 DNA Replication

Proteins utilized specifically during T7 DNA replication are translated from transcript II. Genes 3 and 6 encode two nucleases responsible for breakdown of host DNA, which furnishes most of the nucleotides for phage DNA synthesis. A T7 DNA polymerase (gene 5) is also synthesized. This enzyme lacks $5' \rightarrow 3'$ exonuclease activity, which is instead provided by the gene-6 protein. The product of gene 4 serves as both a primase and a helicase. A single-stranded DNA-binding protein is also made by gene 2.5.

The mode of replication of T7 DNA is described in detail in Chapter 8 (see Figures 8-46 through 8-48). It is recommended that the reader review that section, but the essential point is that the original parent molecule generates a long, linear concatemer. Terminally redundant unit-molecules are cut from this concatemer by the method described in the following section.

Maturation of T7 Phage

Packaging of T7 DNA proceeds by a mechanism quite different from that in T4 DNA. For example, (1) the DNA is terminally redundant but not cyclically permuted, and (2) a mutant phage containing a genetic deletion contains less DNA than a wild-type phage. The mechanism of maturation of T7 is not known but a proposal that is likely to be valid is shown in Figure 15-26. The essential point is that by generating a nick with 3′-OH and 5′-P termini, the concatemeric DNA is made susceptible

Figure 15-26
Proposed scheme for production of mature T7 DNA from a concatemer. Three redundant regions are designated by showing the base pairs. The steps are the following. I. Nicks are made at each end of a redundant region, generating two 3′-OH termini. II. A DNA polymerase adds nucleotides to one of the 3′ termini produced by nicking, displacing the 5′-terminated parental strand. III. When the displacement reaction reaches the second nick, the two double-stranded fragments separate to yield one completed daughter molecule. IV. Extension of the 3′ end of the incomplete daughter molecule by DNA polymerase forms a second completed daughter molecule. (Model proposed by J. D. Watson.)

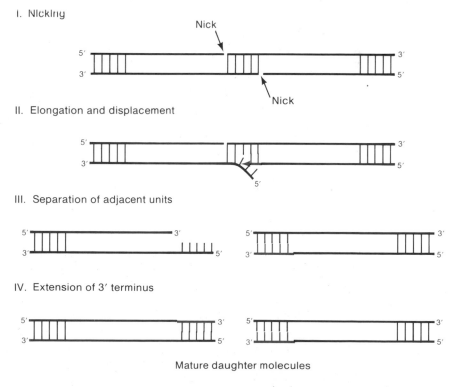

I. Nicking

II. Elongation and displacement

III. Separation of adjacent units

IV. Extension of 3′ terminus

Mature daughter molecules

to the action of enzymes that can form the blunt-ended terminally redundant DNA molecule found in T7 phage particles.

A great deal is known about the formation of T7 heads and tails and *in vitro* assembly. This information can be found in the references listed at the end of the chapter.

TWO PHAGES OF *BACILLUS SUBTILIS*

The two *Bacillus subtilis* phages, PBS-2 and SPO1, have interesting regulatory features relevant to the discussion of phages T4 and T7. These features are described briefly in the following sections.

Phage PBS-2: Lack of a Need for Host Polymerase

The phages discussed so far utilize host RNA polymerase for the initial transcription event since no other RNA polymerase is present at the time of infection. It is essential that a phage set up a program of transcription so that its DNA is transcribed preferentially to the host DNA, and this is usually accomplished by synthesis of a new polymerase (T7) and/or by modification of the host polymerase (T4 and T7). However, *Bacillus subtilis* phage PBS-2 uses a different mechanism. This phage synthesizes a phage-specific RNA polymerase, but the phage RNA polymerase is packaged within the phage head and is injected along with the DNA. This phenomenon is quite rare for phages but, as will be seen in Chapter 21, is not unusual for animal viruses.

Pseudomonas phage ϕ6 contains double-stranded RNA rather than DNA and injects an RNA replicase. This phage is described later in this chapter.

Phage SPO1

Phage SPO1 of *Bacillus subtilis* utilizes a regulatory scheme that combines some of the features of T4 and T7. There are three classes of SPO1 transcripts initiated at three groups of promoters called *PI, PII,* and *PIII. B. subtilis* RNA polymerase, a multisubunit protein similar to the *E. coli* enzyme, initiates transcription at *PI* but not from *PII* or *PIII.* The product of gene *28* is translated from transcript I. Protein 28 is a sigma factor that displaces the host sigma factor. The modified RNA polymerase is no longer able to act at host promoters and at *PI.* However, it initiates transcription at *PII.* Two other proteins, the products of genes *33* and *34,* are made from transcript II. Together they displace protein 28 and form a second modified polymerase that can initiate transcription only at *PIII*. Thus, by this cascading synthesis of RNA polymerase modifiers (that is, sigma factors) the phage achieves an orderly program of synthesis of early, middle, and late mRNA.

E. COLI PHAGE λ

E. coli phage λ (Figure 15-27) has two alternate life cycles—the lytic and the lysogenic. In this chapter we consider only the former; the lysogenic cycle is described in detail in Chapter 16.

At present, phage λ is probably the best understood of the double-stranded DNA phages, especially with respect to the elegant way in which transcription is regulated. The study of λ has made many important contributions to molecular biology, such as the discovery of bidirectional replication, transcription termination and the Rho protein, antitermination proteins, DNA ligase, DNA gyrase, site-specific recombination, circular DNA molecules, break-and-rejoin genetic recombination, and SOS repair, to name just a few. Furthermore, λ has been the system of choice for understanding many other biological processes.

Phage λ DNA and Its Conversion Following Injection

Phage λ is a midsized phage (Table 15-3). It contains a linear double-stranded DNA molecule whose molecular weight is 31×10^6 (50,000 base pairs). The DNA molecule has an unusual structure—at each end of the molecule the 5′ terminus extends 12 bases beyond the 3′-terminal

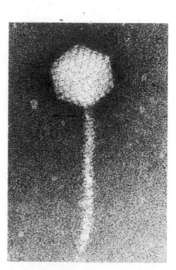

Figure 15-27
Electron micrograph of phage λ. (Courtesy of Robley Williams.)

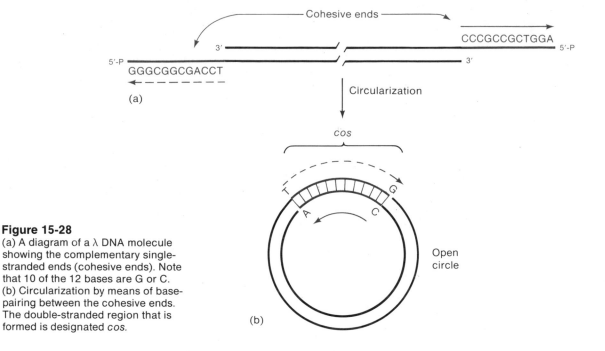

Figure 15-28
(a) A diagram of a λ DNA molecule showing the complementary single-stranded ends (cohesive ends). Note that 10 of the 12 bases are G or C. (b) Circularization by means of base-pairing between the cohesive ends. The double-stranded region that is formed is designated *cos*.

nucleotide. The base sequences of these single-stranded terminal regions, which are known as **cohesive ends,** are complementary one to the other (Figure 15-28). Thus, by forming base pairs between the single-stranded ends, the linear DNA molecule can circularize, yielding an open circle containing two single-strand breaks. No bases are missing in the newly formed double-stranded region of this circle; DNA ligase can convert the molecule to a covalent circle. This circularization is easily performed in the laboratory but also occurs in an infected cell according to the following schedule (time in minutes at 37°C):

$t = 0$	Adsorption of a phage and injection of a linear DNA molecule.
$t = 1-2$	Formation of an open circle.
$t = 3-6$	Sealing by *E. coli* DNA ligase and supercoiling by *E. coli* DNA gyrase.

Organization of λ Genes and Regulatory Sites

Forty-six λ genes have been identified; of these, fourteen are nonessential to the lytic cycle but only seven are nonessential to both the lytic

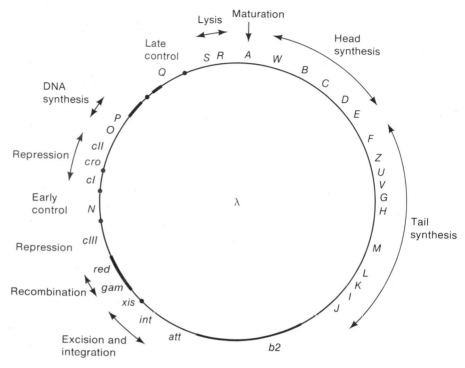

Figure 15-29
Genetic map of phage λ. Regulatory genes and functions are given in red. All genes are not shown. Major regulatory sites are indicated by black solid circles. Regions non-essential for both the lytic and lysogenic cycles are denoted by a heavy line.

and lysogenic cycles. Most λ proteins have been either separated by gel electrophoresis or purified. Many regions of the λ DNA have been sequenced and the base sequence of most of the regulatory sites, promoters, and termination sites are known. From the sizes of the proteins and the DNA molecule one would estimate that 90 percent of the genes have been identified.

The genetic map of λ is shown in Figure 15-29. For reasons that will become clear when lysogeny is discussed, the map is usually drawn as a circle. As with phage T7, genes of the phage λ are clustered according to function. For example, the head, tail, replication, and recombination genes form four distinct clusters. Many λ proteins—for example, regulatory proteins and those responsible for DNA synthesis—act at particular sites in the DNA. In general these proteins are situated adjacent to their sites of action (when there is a single site)—for instance, the origin of DNA replication lies within the coding sequence for gene *O*, which encodes a DNA replication-initiation protein.

Timing in the Lytic Cycle

The schedule of the lytic cycle of λ is fairly complex, probably because certain genes are used in both the lytic and lysogenic cycles. The life cycle in outline is the following (time in minutes at 37°C):

$t = 0$	Phage adsorbs and DNA is injected.
$t = 3$	First ("pre-early") mRNA is synthesized.
$t = 5$	Two classes of early mRNA are synthesized.
$t = 6$	DNA replication begins.
$t = 9$	Synthesis of late mRNA begins.
$t = 10$	Structural proteins begin to be made.
$t = 22$	First phage particle is completed.
$t = 45$	Lysis and release of progeny phage.

Note that the cycle is 45 minutes long rather than the 22–25 minutes we have seen for phages T4 and T7. This longer period is primarily a reflection of the complex regulatory pathways needed by a phage having alternative life cycles.

Requirement for Maintenance of the Integrity of the Host System

An essential feature of the life cycles of phages T4 and T7 is rapid killing of the host and solubilization of the host DNA. Phage λ differs in this respect because in the lysogenic cycle its host must survive. There are λ gene products that slowly turn off host macromolecular synthesis and that can kill the host cell, but these products are nonessential; that is, if these λ genes are genetically deleted, λ develops normally and the burst size is not reduced significantly. (How these gene products work and their roles in the life cycle of λ are not known and they will not be discussed further.) Ultimately, in the lytic phase the infected cell dies as a result of damage to its DNA replication mechanism and lysis.

Later in this chapter we will examine a phage (M13) that is produced continuously by an infected cell and yet never kills the cell; thus, a cell can even serve as a phage-producing factory without losing its viability or ability to replicate.

The Transcription Sequence of λ

In the phages discussed earlier, timing of synthesis of the various mRNA molecules is accomplished primarily by mechanisms that determine the availability of promoters—namely, the synthesis of a new RNA polymerase, the modification of the host polymerase, or the time of entry of the promoter in the host cell. In λ the host RNA polymerase is also

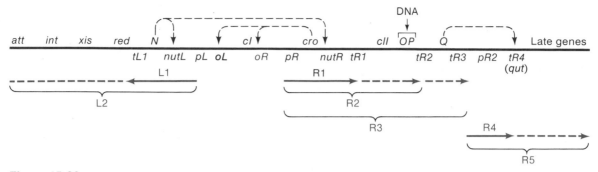

Figure 15-30
A genetic map of the regulatory genes of phage λ. Genes are listed above the line; sites are below the line. The mRNA molecules are red. The dashed black arrows indicate the sites of action of the N, Cro, and Q proteins.

modified but not for the purpose of recognition of phage promoters. Instead, the modification enables the polymerase to ignore termination sites. In fact, the main cause of the delay in synthesis of late mRNA molecules is the time required to transcribe and translate genes that encode antitermination proteins. Figure 15-30 shows a genetic map of λ, which includes the three regulatory genes *cro*, *N*, and *Q*, three promoters *pL*, *pR*, and *pR2* and five termination sites *tL1*, *tR1*, *tR2*, *tR3*, and *tR4*. Seven mRNA molecules are also shown; the L and R series are transcribed leftward and rightward respectively from complementary DNA strands. It is recommended that the reader refer to Figure 15-30 continually during the following discussion.

Since inactivation of host functions is not required (see preceding section), the transcription pattern that might be expected is the following. First, the *O* and *P* genes, whose products are necessary for DNA synthesis, would be transcribed. This would be followed by transcription of the genes encoding the structural proteins, and finally the packaging system and the lytic proteins would be made. This is basically what occurs, with the modification that other small transcripts are formed that encode the regulatory proteins responsible for turning transcription on and off at the appropriate times.

The sequence of transcriptional events is described in the list that follows. The genes, sites, and mRNA molecules referred to are depicted in Figure 15-30 and listed in Table 15-6. Two events having the same number, e.g. 1A and 1B, begin at the same or nearly the same time.

1A. *Synthesis of transcript L1*. This begins at the promoter *pL* and terminates at the site *tL1*. This transcript encodes only the gene-N protein, which is a positive regulatory element needed to allow certain regions of the DNA to be transcribed.

1B. *Synthesis of transcript R1*. This begins at the promoter *pR* and

Table 15-6
Some sites and gene products in phage λ

Site or gene product	Description
Site	
oL, oR	Left and right operators
pL, pR	Left and right promoters
tL (1,2)	Termination sites for leftward transcription
tR (1,2,3,4,5)	Termination sites for rightward transcription
Gene product	
cro	Protein inhibitor of transcription from *pL* and *pR*
cII	Protein delaying late transcription (plays other roles in the lysogenic pathway, as described in Chapter 16)
N	Antitermination protein acting at *tL1*, *tR1*, and *tR2*
O, P	Proteins required for DNA replication
Q	Antitermination protein acting at *tR4*
L	Messenger RNA synthesized in rightward direction
R	Messenger RNA synthesized in leftward direction
R4	Constitutively synthesized mRNA

terminates at the site *tR1*. This transcript encodes the gene-*cro* protein, which is a negative regulatory element serving ultimately to turn off transcription from the promoters *pL* and *pR*.

2. *Synthesis of transcript R2.* The terminator *tR1* is a weak one and about half of the R1 mRNA molecules are extended to *tR2*, a strong terminator. The *cro*-gene protein, whose function is discussed in item 4, is translated from both of these mRNA molecules. The portion of R2 not present in R1 contains the genes O and P, whose products are needed for DNA replication; the O and P products are made at this time from R2 but their concentrations are sufficient for only one round of replication.

3A. *Antitermination at tL1 and synthesis of transcript L2.* Once synthesized, the gene-N protein plus a host protein (called NusA) modify the *E. coli* RNA polymerase; it is then able to ignore the terminator *tL1*. In this way the transcript L1 is no longer made but is extended to form L2. The L2 mRNA is translated to yield several products that are not essential but make phage production more efficient; the function of these proteins is not known. The mechanism of N-mediated antitermination remains a subject of active study. A site in the λ DNA called *nutL* (for N-utilization), which is located "downstream" from *pL*, is apparently required. It has been suggested, though there is little evidence at this time, that the gene-N protein may bind to a *nut* site and be picked up by an RNA polymerase molecule moving along the DNA molecule while transcribing L1 or R1.

3B. *Antitermination at tR1 and tR2 and synthesis of transcript R3 and production of DNA replication proteins.* Once the gene-N protein has been synthesized, the modified RNA polymerase is able to ignore *tR1*. R1 is no longer made; *tR2* is also bypassed, thereby extending R2 to form R3. This molecule terminates at *tR3*, which is a termination site for RNA polymerase even when it is modified. That is, *tR3* is not sensitive to the antitermination effect of the *N* product. The synthesis of R3 provides enough mRNA that the concentrations of the *O* and *P* proteins reach values sufficient for DNA replication. The gene-*cII* protein is also made from R2 and R3; a function of this protein, which is discussed shortly, is to delay late mRNA synthesis.

4. *Turning off of synthesis of transcript L2.* While the preceding events have been occurring, the concentration of the gene-*cro* protein has been increasing. This protein is a repressor that binds to the left operator oL and blocks initiation of transcription at *pL*. Thus, L1 and L2, which at this point are not needed because sufficient N protein has been made, are no longer synthesized.

5A. *Turning off of synthesis of transcript R3.* Somewhat later than the time when the cro protein inhibits transcription from *pL*, the concentration of the gene-*cro* protein increases to the point that the operator oR is occupied and this binding blocks synthesis of transcript R3 from *pR*. The turning off is efficient since there is now an adequate supply of the proteins encoded in R3.

5B. *Translation of the gene-Q protein from transcript R3 and synthesis of late mRNA.* Since the time of infection, a tiny transcript R4 has been synthesized continually from *pR2*, terminating at *tR4*. This transcript does not encode any known genes but is a leader for the late mRNA. Once R3 has been made (item 3), the product of gene Q is made. This protein is responsible for turning on synthesis of the late mRNA molecule that encodes the structural and assembly proteins, the maturation system, and the lysis enzymes. The Q product is also an antiterminator and acts at the site *qut* to enable *E. coli* RNA polymerase to ignore *tR4*. Thus R4 is then extended to form transcript R5, the late mRNA.

Timing of Late Transcription

Since the products of the O, P, and Q genes are translated from the same transcript (R3), it is essential that the late proteins, such as those comprising the DNA packaging system, are not made before a large number of DNA replicas have been produced. Four features of the regulatory scheme separate the time of DNA replication from the time of synthesis of the active elements translated from late mRNA.

1. Some O and P products are translated from R2, thereby giving DNA replication a small head start.
2. Some time is required for the Q product to be made and to reach a concentration sufficient for antitermination to occur; replication is proceeding during this time.
3. Synthesis of R5 is delayed (in an unknown way) by the gene-cII protein.
4. The late transcript is very long and contains many cistrons; therefore, it takes a few minutes to complete its synthesis. Translation of downstream proteins begins several minutes after translation of upstream proteins starts. Finally, the assembly of heads and the establishment of the system that cuts λ DNA in preparation for packaging is complex and time consuming; also, the lytic enzyme is fairly inefficient and a large concentration is needed before lysis can occur.

Ultimately, a large amount of the structural proteins encoded in the late mRNA is needed. Thus the inhibition of Q-mediated antitermination by the gene-cII protein must be relieved if late mRNA synthesis is to proceed at a high rate. This is accomplished by the action of the cro product, which turns off synthesis of R3, as described in item 5A of the preceding section. R3 encodes the gene-cII protein. Since mRNA is unstable, in a matter of minutes no template remains for synthesis of the cII product. This protein is short-lived and, once synthesis of the R3 transcript stops, the concentration of the gene-cII protein quickly drops, Q-mediated antitermination is no longer inhibited, and synthesis of late mRNA begins. (Actually, the delay by the cII product is not essential in that cII⁻ mutants form plaques. However, these mutants lyse sooner than the wild-type phage and only about half the usual number of phage are made.)

Production of a large number of late mRNA molecules requires DNA replication, but this is only a "gene-dosage" effect in contrast to the situation with phage T4. Once the O and P products have been made, DNA replication begins. Replication serves not only to make progeny DNA molecules for the new phage but also to increase the number of copies of the late genes that can serve as a template for RNA polymerase. Thus, while heads, tails, and other particle elements are being synthesized, the rate of synthesis of late mRNA (measured as molecules per minute) increases owing to the increased number of DNA templates. This transcription increases the rate of synthesis of the structural components of the phage and accelerates phage assembly.

Review of Transcription of λ in the Lytic Cycle

Let us now review the essential features of this elegant regulatory system. An important feature is that no λ-specific RNA polymerase is used; the *E. coli* RNA polymerase is used throughout the life cycle (as is true of T4) and is modified by accessory proteins to alter its specificity toward various DNA base sequences. However, at no time is its activity with respect to promoters modified; instead, its ability to terminate transcription at certain termination sites is impaired. Inhibition of transcription occurs, but the inhibition is a consequence of the repressor activity of the *cro* product; however, the gene-*cro* protein eliminates the availability of the promoters *pL* and *pR* by the standard mechanism of repression and has no direct activity on the RNA polymerase. The timing of initiation of transcription of selected regions of λ DNA is thus determined primarily by the time required (1) to transcribe and translate the antiterminating products of genes N and Q, (2) for the *cro* product to block synthesis of the *cII* product, and (3) for the residual *cII* protein to be degraded. Wasted synthesis is avoided by the repressing activity of the *cro* product. Finally, by having all structural components encoded in a single giant mRNA molecule, which is translated sequentially, synthesis of the complete set of components takes many minutes, thereby delaying synthesis of intact heads and of a functional phage maturation system until the DNA replication system has provided many copies of λ DNA. The result of all of these delays is that there are about 30 copies of λ DNA formed before the maturation system is established and about 100 completed phage particles are formed before the onset of lysis.

DNA Replication and Maturation: Coupled Processes

Following synthesis of the O and P products, replication of circular λ DNA begins. There are two modes of λ DNA replication—θ and rolling circle replication (Figure 15-31; also Chapter 8). The θ mode increases the number of templates for transcription and further replication; the rolling circle mode provides the DNA for phage progeny. As the life cycle proceeds, θ replication stops and the rolling circle becomes the predominant form. A fair amount is known about the θ mode and, in fact, an analysis of θ molecules provided the first evidence in any organism for bidirectional movement of replication forks (Chapter 8). The origin of replication has been sequenced. It has a very high A+T content and may have an unusual structure. An early event, which is somehow connected with initiation, is the production of a single-strand break at the origin.

Much less is known about rolling circle replication and how it is initiated. However, for our purposes the most important feature for the present discussion is the relation between replication and maturation of the DNA, since the DNA-cutting mechanism of λ differs from those used by the phages that have already been discussed.

The DNA found in a λ phage particle is linear and has single-stranded termini. These ends are joined when a circle forms, and the double-stranded region so formed is called a *cos* site (for *co*hesive site). Thus, every monomeric λ circle contains one *cos* site; however, a multimeric branch of a rolling circle contains many *cos* sites.

Since the ends of the DNA molecule in the phage particle are always the single-stranded termini, there must be a mechanism for cleaving a *cos* site to generate these termini. Two modes for this cleavage are possible—a site-specific mode and a length-measuring system. The latter should immediately be recognized as unlikely, because the requirement for accuracy in cutting (measuring the length of a DNA molecule containing 50,000 base pairs accurately to one base) is too stringent. A length-measuring system is also ruled out by the observation that the DNA of mutants having only 73 percent or as much as 107 percent of the normal DNA content are packaged efficiently and have single-stranded ends with the same base sequence as the ends of DNA of normal size. Thus, a site-specific cutting system seems likelier—it is called the **terminase** or **Ter system** and the DNA-cutting is often called Ter-cutting.

Figure 15-31 shows the three major species of intracellular λ DNA—circles, θ molecules, and rolling circles. By the time heads and tails have been synthesized and the Ter system is active, rolling circles predominate. Note that since a rolling circle has two classes of *cos* sites—the one in the circle and those in the linear branch, some

Figure 15-31
Three species of λ DNA present at the time maturation begins. The region containing the joined complementary single-stranded termini (of the linear DNA molecule present in the phage head) is called *cos* (for *cohesive* site.

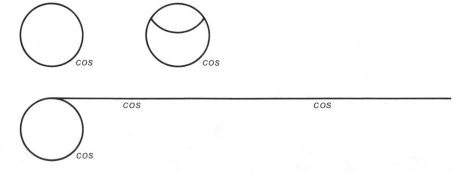

mechanism must exist for preventing cleavage of the one in the circle because, if it were broken, replication would cease. This difficulty is avoided by a site requirement in the Ter system—namely,

> Efficient cleavage of a single *cos* site does not occur; if there are two *cos* sites and both are present on a single segment of DNA, cutting can occur.

A λ DNA molecule can be cut from a linear branch by cleavage of two neighboring *cos* sites, but a pair of cuts in which one *cos* site is in the branch and the other is in the circle cannot be made. The DNA molecule need not be linear for a cut to be made though, inasmuch as a single λ unit can be cut from a dimeric circle that has been formed by genetic recombination. A single cut in a monomeric circle does not occur. The "two *cos* sites" rule explains how the first λ DNA unit is cut from a concatemeric branch of a rolling circle. However, this rule would not allow excision of the second (adjacent) λ unit, because this unit would be flanked by only one *cos* site and a free cohesive end. Hence, only half of the DNA would be usable because only alternate segments of DNA would be packageable (Figure 15-32). The solution to this lack of economy is that a free cohesive end and an adjacent *cos* site are also sufficient for DNA cutting to occur, and allow sequential packaging. Thus, the precise Ter-cutting rule may be stated

> Ter-cutting requires two *cos* sites or one *cos* site and a free cohesive end on a single DNA molecule.

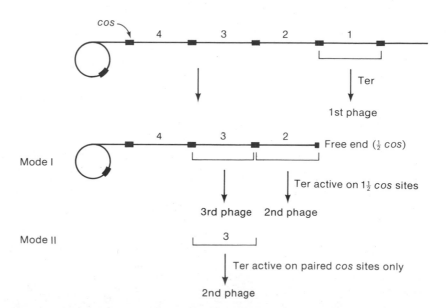

Figure 15-32
Two rules of packaging. In one mode (black) each λ unit is packaged. In the more limited mode (red) alternate units are packaged. The former (black), which is the more economical, is used by λ.

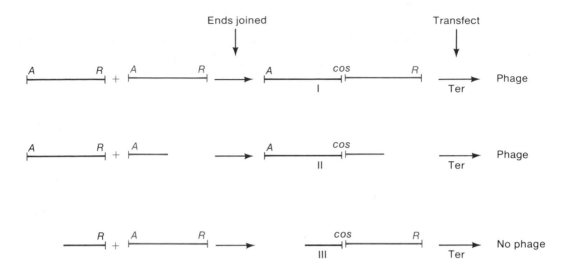

Figure 15-33
This experiment shows that one *cos* site and a free A-gene cohesive end are sufficient for activity of the Ter system. The cohesive ends are depicted by a vertical line. All molecules contained a mutation in the λ *P* gene to prevent DNA replication. Transfection refers to infection of a bacterial spheroplast with phage λ DNA.

An elegant experiment depicted in Figure 15-33 shows that the free end must be the end near the *A* gene (Figure 15-29).* In this experiment three types of λ DNA molecules were prepared by *in vitro* cohesive-end joining of either two intact DNA molecules (I) or one intact molecule plus one fragment containing either the gene-*A* cohesive end (II) or the gene-*R* cohesive end (III). (The fragments were prepared by breaking intact molecules near the center and then separating the fragments bearing a particular end from one another.) The molecules also contained a mutation in the *P* gene, so DNA replication was not possible. Three cultures of *E. coli* were separately transfected (Chapter 1) with these fragments. There was no DNA replication in the infection but transcription occurred that resulted in synthesis of heads, tails, and the elements of the Ter system. Phage would be produced if the Ter system could cut λ units from these hybrid molecules. As shown in the figure, phage were obtained when cells were transfected with the dimer or with the molecule containing a free *A* end, but no phage were obtained when the fragment containing only a free *R* end was used. This and other experiments show that the packaging of λ DNA from the concatemeric branch of the rolling circle is polarized and proceeds from the *A* end to the *R* end.

*The *A* and *R* genes are not significant. We have used these genes to name the two ends of λ DNA to avoid possible ambiguity in the terms left and right.

Cutting at the *cos* sites and packaging of λ DNA are somehow coupled. In fact, the Ter system is virtually inactive unless the Ter proteins are components of an empty λ head. Thus, when a bacterium is infected with a λ mutant unable to cause formation of an intact phage head (for example, an E^- mutant, which fails to make the major head protein), the linear branch of a rolling circle of DNA is not cleaved. With such a mutant, dimers, which contain two *cos* sites, are also not cleaved.

E. COLI PHAGE φX174

Study of *E. coli* phage φX174 has been important for many reasons: it was the first organism to be discovered containing single-stranded DNA as its genetic material and having DNA in a circular form. Its DNA was also the first to be replicated *in vitro* to yield biologically active DNA, and study of its mode of replication led to the hypothesis of rolling-circle replication. This phage is very small, contains only a few genes and is fairly simple for a phage (Table 15-3). Study of the base sequence of its DNA and the amino acid sequence of its proteins first showed that genes could be translated in overlapping reading frames.

E. coli phage φX174 contains one circular, single-stranded DNA molecule (Figure 15-34) consisting of 5386 nucleotides, whose base sequence is known. The DNA is enclosed in an icosahedral head composed of three coat proteins (Figure 15-35) and one internal protein.

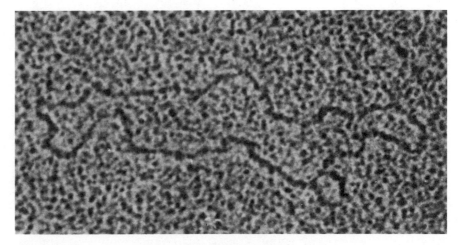

Figure 15-34
A circular single-stranded φX174 DNA molecule. This photo shows the first single-stranded DNA molecules ever seen by electron microscopy. The circumference of the circle is 1.7 μm. (Courtesy of Albrecht Kleinschmidt.)

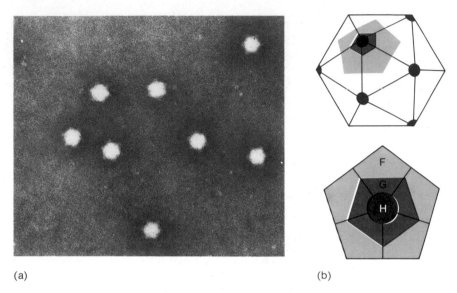

Figure 15-35
Phage φX174. (a) Electron micrograph. (Courtesy of Robley Williams.) (b) Schematic representation of the φX174 subunits in the phage particle (top) and an enlargement of a spike region (bottom), suggesting orientations of gene *H*, gene *G*, and the major coat (gene *F*) proteins.

(a)   (b)

Overlapping Genes

Phage φX174 contains the eleven genes listed in Table 15-7 and each gene product has been isolated. The number of amino acids contained in these proteins exceeds the coding capacity of this very small DNA molecule. However, φX174 has efficiently evolved to have overlapping genes that are translated in different reading frames. This is shown in the genetic map drawn in Figure 15-36; for example, genes *B* and *E* are contained within genes *A* and *D,* respectively. How translation in overlapping reading frames is accomplished is explained in Chapter 12.

Adsorption and Penetration

Phage φX174 adsorbs to *E. coli* by means of a spike protein (Figure 15-35). In contrast with the T phages, which do not inject proteins, the terminal spike protein (of gene *H*) enters the bacterium with the DNA. The *H* product is required in an unknown way for converting the bacterium to a phage factory. Similar proteins, which are termed **pilot proteins,** have been observed with many of the very small phages. The small phages have no mechanism for turning off host macromolecular synthesis, so it has been hypothesized that these proteins somehow guide the bacterial RNA polymerase to the phage DNA and direct the phage DNA to the replication apparatus. There is no evidence for this function but it is believed necessary to account for the speed and

Table 15-7
Phage φX174 genes and functions

Gene	Function
A	RF replication; viral strand synthesis
A*	Turning off of host DNA synthesis
B	Formation of capsid
C	Formation of phage-unit-sized-DNA
D	Formation of capsid
E	Lysis of bacterium
F	Major coat protein
G	Major spike protein
H	Minor spike protein; adsorption to host cell
J	Core protein; entry of progeny DNA into phage particle
K	Unknown

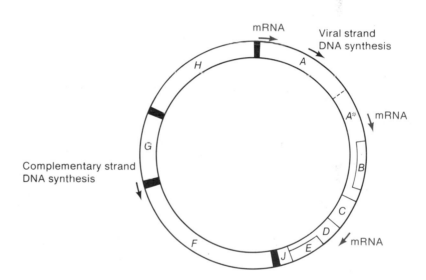

Figure 15-36
Genetic map of φX174 showing the phage genes and the replication (black) and transcription (red) start sites. Spacer regions are blackened.

efficiency with which a small DNA molecule is able to compete with the enormous excess of bacterial DNA.

The DNA of φX174 is not injected into the bacterium. Instead, release of the phage DNA occurs on the cell surface; this is known because, shortly after adsorption of a phage particle to a bacterium, the phage DNA can be digested if DNase is added to the suspending medium. (Such DNase sensitivity is not true of any of the tailed phages.) Removal of the coat protein is coupled to replication of the parental DNA. The mechanisms of uncoating the DNA and transfer of the molecule to the cell interior, and the role of replication in the transfer process are unknown.

Replication of φX174 DNA

In order to discuss replication of single-stranded DNA molecules, the following terminology is used. The strand contained in the virus particle or any strand that has the same base sequence is called a $(+)$ strand; a strand having the complementary base sequence is called a $(-)$ strand. RNA is transcribed from the $(-)$ strand.

Replication of φX174 DNA occurs in several steps (Table 15-8):

1. *Conversion of the parental single-stranded DNA molecule (the viral or $(+)$ strand) to a covalently closed, double-stranded molecule called replication form I* (**RFI**). This occurs before transcription begins and hence depends on host enzymes exclusively. The newly synthesized strand (the $(-)$ strand) is the coding strand—the only strand that is transcribed. The enzymatic mechanism for converting the $(+)$ strand to RFI is described in detail in Chapter 8.

2. *Synthesis of many copies of RFI.* The $(-)$ strand is transcribed and a multifunctional protein, the gene-*A* protein, is made. The *A* product makes a single-strand break in the $(+)$ strand between bases 4297 and 4298 (this is called the replication origin, *ori*) and remains covalently linked to the 5′-P terminus, as shown in Figure 15-37. This nicked molecule is called RFII. *E. coli* proteins (synthesized prior to the infection) then cause the parental $(+)$ strand to be displaced from RFII by the looped rolling-circle replication mode discussed in greater detail in Chapter 8. When one round of replication is completed, the displaced $(+)$ strand is cleaved from the looped rolling circle, recircularized, and used as a template for the synthesis of another $(-)$ strand; the result is a new RFI.

Table 15-8
Replication cycle of φX174

Stage	Time, minutes at 33°C	Event
ss → RF	0–1	Adsorption and penetration; viral single strand converted to parental RF; RF transcribed
RF → RF	1–20	Parental RF replicates and 60 progeny RF are formed
	25	Replication of RF molecules and host DNA stops
RF → ss	20–30	About 35 rolling circles form from progeny RF; from these, about 500 ss molecules are produced; DNA is inserted into phage particles
	40	Lysis occurs

Note: Abbreviations used are: ss, single-stranded phage DNA; RF, circular replication form.

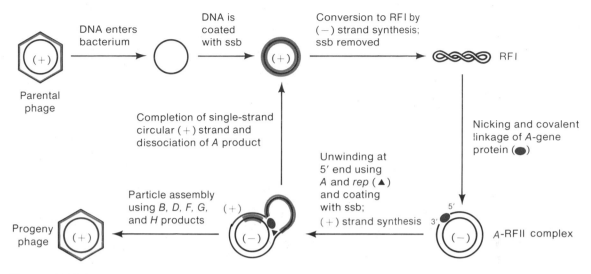

Figure 15-37
The two stages of replication of φX174 DNA. Parental phage DNA is first converted to the double-stranded supercoiled RFI. Then, the gene-*A* protein makes a nick and the looped rolling circle mechanism begins, causing displacement of the progeny (+) strand made during the formation of RFI. In the presence of phage-encoded maturation and coat proteins, progeny (+)-strand circles are encapsidated rather than being converted to RFI. With the exception of the phage-gene-*A* protein, all proteins used in replication are bacterial proteins.

3. *Synthesis of (+) strands for encapsidation.* There is no switch from synthesis of RFI to synthesis of free (+) strands. Instead, the packaging system captures progeny (+) strands before they can serve as templates for further synthesis of (−) strands. The capture is delayed until nearly completed phage heads have been synthesized and begins when there are about 35 copies of RFI. At this point, most of the DNA is engaged in looped rolling-circle replication, producing new RFI and some (+) strands for packaging. Host DNA synthesis finally stops at this time, partly as a result of an unknown function of the *A* protein and partly because all of the host DNA polymerase is engaged in synthesis of φX174 DNA. As more heads are made, all of the displaced (+) strands are packaged and synthesis of RFI stops for lack of a template. The mechanism of packaging is described in a later section.

Transcription of φX174 DNA

Phage φX174 has three promoters, all of which are activated simultaneously. Except for lysozyme, no temporal regulation is required. All transcription is from the (−) strand, so transcription cannot occur until the first (−) strand is made; from then on, RFI synthesis continues unabated until a sufficient number of heads are formed to start packaging. Because a φX174 particle is very small, many rounds of replication (approximately two seconds per round) are completed before the head proteins are synthesized and assembled. (The first completed particle appears 24 minutes after infection.) Lysozyme activity is delayed because translation is slow and the enzyme is not particularly active; lysis does not occur until 40 minutes after infection, by which time about 500 phage particles have been synthesized.

Packaging of φX174 DNA

Packaging of φX174 DNA requires seven phage proteins, four of which are present in the finished particle. The steps of assembly, which are shown in Figure 15-38, are the following:

1. The head protein (gene *F*) and the spike protein (gene *G*) form pentamers that, under the influence of the gene-*B* protein, then form an aggregate called a 12S particle.
2. The gene-*H* proteins are added to complete the spike.
3. The gene-*D* protein forms a frame on which a prohead is built from the 12S units.

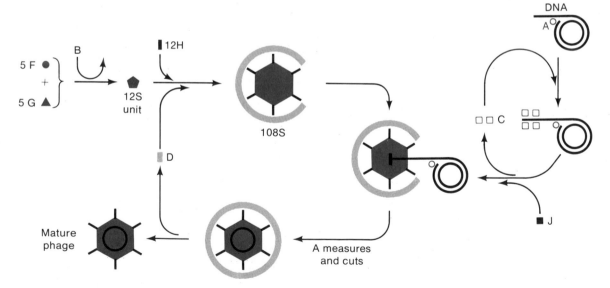

Figure 15-38
Model for assembly of φX174 phage. Letters represent φX174 proteins. The B protein catalyzes aggregation of five molecules of F protein (the major head protein) with five molecules of the G protein (spike protein) to form a 12S unit. The D protein may provide a scaffolding func- tion to form the 108S particle. The C protein facilitates encapsidation of the DNA. The A protein measures a unit length of DNA and forms the circle. The J protein is least well understood.

4. The gene-C protein shown schematically in the figure as binding to the 5′ end of the displaced (+) strand somehow directs the 5′ end into the prohead. The gene-J protein, which probably initially binds to the DNA outside the prohead, in an unknown way organizes the DNA inside the head and consolidates the structure.

5. The gene-A protein cuts the displaced (+) strand from the looped rolling circle and circularizes the displaced strand.

6. The gene-D protein is then removed (probably spontaneously following completion of and closing of the head) leading to formation of the finished phage particle).

At the present time it is possible to synthesize *in vitro* both the first RFI and the daughter RFI molecules and to carry out rolling circle replication. Also, if the seven morphogenetic proteins are added, the packaging reaction can be coupled to rolling circle replication. Since *in vitro* coupled transcription-translation systems are also available, it should be possible to synthesize all of the phage proteins from RFI. When these systems are combined, φX174 will probably be the first complete biological organism synthesized totally from molecularly defined components.

FILAMENTOUS DNA PHAGES

Three filamentous *E. coli* phages, M13, fd, and f1 (Figure 15-39), each of which contains a circular single-stranded DNA molecule, have been carefully studied. They are very similar and a picture of the life cycle of each has been developed by combining information obtained from each

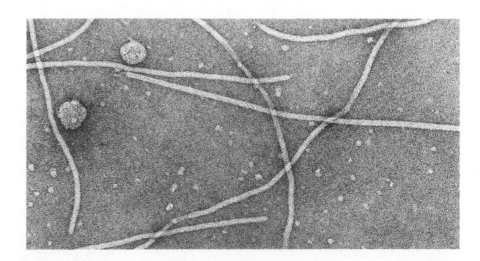

Figure 15-39
Electron micrograph of the filamentous *E. coli* phage M13. The large and small spherical particles are cell debris. (Courtesy of Robley Williams.)

of them. The life cycle, which shows many similarities to that of ϕX174, will not be given in any detail; only the mode of penetration of the phage DNA and release of completed particles will be described.

Of all known virulent phages, only the filamentous *E. coli* phages neither kill the host cell nor cause lysis. The filamentous phages adsorb to the tip of the F pilus. In an unknown way (for which there are many theories), the particle is brought to the bacterial cell wall at the base of the pilus; one idea is that the phage moves down a groove on one side of the pilus and another is that the pilus retracts by adsorption of pilus proteins into the bacterium (retraction is an observed phenomenon with pili). What is certain is that the *entire* particle, in intact form, penetrates the cell wall and comes to rest on the inner cell membrane. There, the particle is uncoated and simultaneously the single-stranded DNA is converted to an RFI; these processes are coupled as with ϕX174. The coat is degraded and the protein subunits are re-used in progeny particles. The entry of coat proteins into the cell is unique to these phages.

The details of the replication cycle of filamentous phages are not known. There is a stage in which RFI is formed by a rolling circle mechanism as with ϕX174. The switch to (+) strand synthesis requires the gene-5 protein, which is a single-stranded DNA-binding protein that coats the tail of the rolling circle. This switch is not coupled to encapsidation as with ϕX174; instead, linear (+) strands coated with gene-5 protein are continually cut from the rolling circle and then circularized in an unknown way. Thus, a large pool of complexes of (+) strand and gene-5 protein are present in the cell.

The major coat protein (gene-8 protein) is synthesized in large quantities and immediately deposited in the inner cell membrane; no gene-8 protein is found free in the cytoplasm. This protein is an α helix having three distinct regions—a basic end, a hydrophobic center, and an acidic end (Figure 15-40). It is hypothesized (on the basis of very little data) that the hydrophobic center enables the proteins to be situated in the hydrophobic membrane, the acidic end faces the cell wall, and the basic end faces the cell interior, perhaps displacing the gene-5 protein and binding to the (+) strand. In a completely obscure way the phage particle, which also contains two other proteins, is assembled in the inner membrane. For each completed particle a bud forms on the cell surface and a phage is extruded (Figure 15-41). This process can continue indefinitely without damage to the cell.

For a typical phage, plaques are the result of lysis of bacteria. However, the filamentous phages form plaques despite the lack of lysis of the host cell. The reason is that infected cells, which continue to grow and divide, do so much more slowly than uninfected cells, presumably because the replication machinery of the host is engaged in making phage DNA for a significant fraction of the host life cycle. This slow

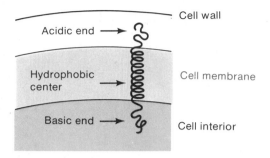

Figure 15-40
Schematic diagram of orientation of the M13
coat protein in the cell membrane.

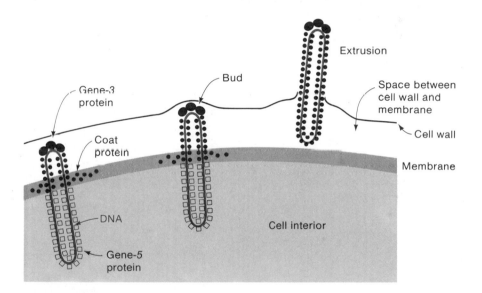

Figure 15-41
Assembly and extrusion of M13 phage. Gene-*5* protein covers the DNA and somehow
enables the DNA to penetrate the cell membrane. Coat protein molecules in the membrane bind to the DNA, displacing the gene-*5* protein molecules. The gene-*3* protein
is attached to one end of the particle and is needed for both adsorption and extrusion.

growth is sufficient to produce a less turbid region in the bacterial lawn,
which is the highly turbid plaque of the filamentous phages. These
plaques must be observed at the right time, for when the bacteria in the
lawn stop growing owing to exhaustion of nutrients, the infected cells in
the plaque continue growth for roughly one generation more, increasing
the turbidity of the plaque to the point that the plaque disappears.

SINGLE-STRANDED RNA PHAGES

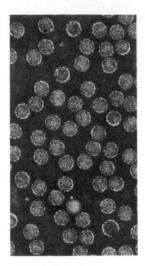

Figure 15-42
Electron micrograph of phage MS-2. Note that a few of the phage particles are empty. (Courtesy of Robley Williams.)

Several single-stranded RNA phages have been isolated; the best known are the *E. coli* phages f2, R17, MS-2, and Qβ. Each of these is a tailless icosahedron like ϕX174 (Figure 15-42). The genetic map of MS-2 is shown in Figure 15-43. Their life cycles, which are very simple, have been elucidated by comparing different features discovered from each phage. The RNA is a single-stranded linear molecule having a great deal of intramolecular hydrogen-bonding, and it consists of about 3600 nucleotides (the sequence is known for R17, MS-2, and Qβ) and contains three genes encoding a coat protein (CP), an attachment protein (A), and an RNA replicase (Rep). The RNA molecule is both a replication template and mRNA, so that (1) neither DNA nor RNA polymerase is needed in the life cycle and (2) regulation must occur at the translational level. A typical burst size is 5000 to 10,000, which is very large compared to the typical burst of one to a few hundred for DNA phages. The particles form huge crystalline arrays (Figure 15-44) within each bacterium; the crystals somehow damage the cell membranes, causing cell lysis in the absence of a lytic enzyme. Thus, the timing of lysis is not precise, as with other plaques; instead, lysis of a bacterial population occurs gradually from 30 to 60 minutes after infection.

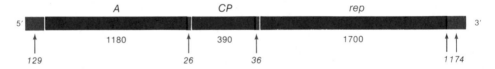

Figure 15-43
The genetic map of phage MS-2 showing the positions and number of bases in each gene. The italicized numbers are the lengths of leaders and spacers.

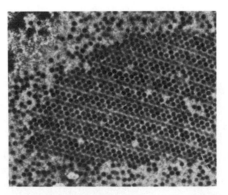

Figure 15-44
A crystalline array of mature MS-2 particles in *E. coli*.

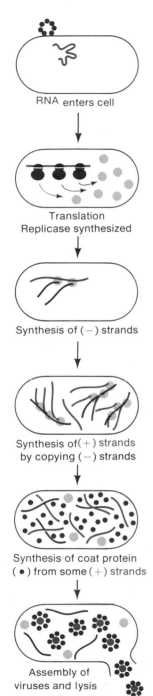

RNA enters cell

Translation
Replicase synthesized

Synthesis of (−) strands

Synthesis of(+) strands
by copying (−) strands

Synthesis of coat protein
(•) from some (+) strands

Assembly of
viruses and lysis

Figure 15-45
Schematic diagram of the
life cycle of an RNA
phage.

A schematic diagram for the life cycle of a RNA phage is shown in Figure 15-45. The individual stages are the following:

1. *Entry to the cell and binding to ribosomes.* Immediately after entry (the mechanism is unknown), a ribosome attaches to a binding site at the beginning of the *CP* gene, which is the centermost gene in the RNA molecule.

2. *Translation and regulation of the amount of each of the three proteins.* As explained in Chapter 13, the ribosomal binding sites for the A and Rep proteins of R17 are blocked by the secondary structure of the RNA. However, translation of the *CP* gene opens up the binding site for the Rep protein. Thus, both proteins are made initially but increasing amounts of the CP protein bind to the *rep* site and block translation of the *rep* gene (see Chapter 14). This inhibition is efficient: about 2×10^6 copies of the CP protein are needed as structural components for 10,000 phage, whereas the replicase is needed only in catalytic amounts. The synthesis of the A protein will be described shortly.

3. *Replication of the phage RNA* (Figure 15-46). Only the Qβ replicase has been studied in detail. It is a tetramer consisting of one Rep molecule and three host proteins—the EF-Ts and EF-Tu translation factors needed for placement of charged tRNA molecules on the ribosome during protein synthesis, and the ribosomal protein S1, whose function is unknown. The replicase copies the viral (+) strand to generate a (−) strand. While synthesis is proceeding, the (−) strand is in contact with the (+) strand only at the polymerization site; for the most part the replicative form is therefore single-stranded. Initiation of several (−) strands occurs before the first (−) strand is complete, and the replication form is branched. The (−) strands are released and immediately used by the replicase to form (+) strands. Some of the (+) strands return to the ribosomes for synthesis of more CP proteins and others are packaged. All progeny contain (+) strands exclusively.

4. *Synthesis of the A protein.* The ribosomal binding site for the A protein is never available on a free (+) strand because of base pairing. However, replication of a (−) strand begins at the 5′ terminus adjacent to the A gene. Just after (+) strand synthesis begins there is a brief period during which the ribosomal binding site for the A protein is free, the complementary segment of the (+) strand having not yet been replicated (Figure 15-47); during this time the *A* gene is translated. Only one or two passages are possible before the binding site becomes closed; therefore, the number of A proteins is maintained roughly equal to the number of (+) strands; this arrangement is economical since each virus particle contains one A protein molecule. Each A protein later becomes

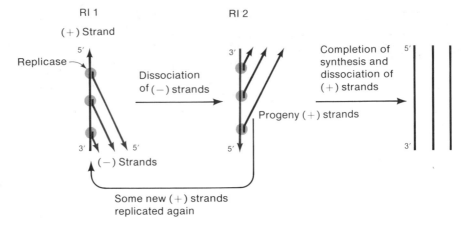

Figure 15-46
Replication of RNA by Qβ replicase. RI stands for replication intermediate.

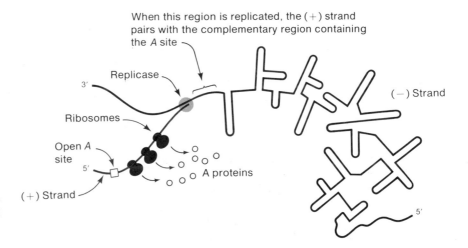

Figure 15-47
Synthesis of gene-*A* protein is permitted by opening the ribosomal binding site on the nascent (+) strand.

bound to one RNA molecule at an unidentified binding site. It is thought that the A protein facilitates the interaction of the RNA with the CP molecules. The A protein remains bound to the RNA and enters the cell in a subsequent infection. Possibly its function is similar to that of the spike protein of the phage φX174.

5. *Particle assembly.* Molecules of the CP protein spontaneously aggregate around the newly synthesized (+) strand and form an icosahedral shell.

6. *Cell lysis.* Lysis occurs after about 10,000 phage particles have formed. However, no phage-encoded lytic enzyme is produced and the lytic mechanism is not known.

A DOUBLE-STRANDED RNA PHAGE: φ6

Very few double-stranded RNA phages are known. The most thoroughly studied one is phage φ6, whose host is the bacterium *Pseudomonas phaseolica.* This phage differs structurally from most bacteriophages in that the protein coat is enclosed in an envelope containing both protein and lipid (Figure 15-48). However, the most striking characteristic of φ6 is that it contains three molecules of double-stranded RNA, having molecular weights of 2.3, 3.1, and 5.0×10^6. Each RNA molecule carries distinct genes and the phage is said to have a **segmented genome.** (In Chapter 21 examples of viruses with segmented genomes are given.) The three RNA molecules collectively encode at least twelve proteins.

Transcription and replication pose several interesting problems and only these features of the φ6 life cycle will be discussed. In appropriate conditions bacterial RNA polymerases can use single-stranded RNA as a template *in vitro,* yielding double-stranded RNA as the product. The reaction cannot go further because double-stranded RNA is a poor template for RNA synthesis. Thus, replication requires an enzyme other than the bacterial polymerase. Double-stranded RNA also fails to serve as a template for translation; if double-stranded RNA is injected into a bacterium, there is no mechanism making use of bacterial enzymes for translation and, hence, no way to produce phage proteins. The single-stranded RNA phages described in the previous section avoid this difficulty by translating the injected single strands first and among the translation products is a phage-encoded RNA replicase that is used to produce progeny RNA strands.

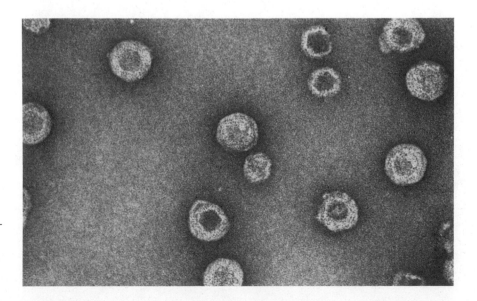

Figure 15-48
Electron micrograph of phage φ6. The larger particles are the complete enveloped particle. The smaller particles lack the lipid envelope. (Courtesy of Robert Haselkorn.)

A double-stranded RNA phage has two alternatives: (1) it must either unwind to produce a coding strand, a process that is energetically quite unlikely, or (2) it can carry into the bacterium an enzyme capable of replicating double-stranded RNA. Phage $\phi 6$ utilizes the second of these mechanisms—that is, the phage particle contains a phage-encoded RNA-polymerizing activity that enters the bacterium with the phage RNA. This activity, which is needed for production of the phage protein coat, has not yet been purified, and it is thought to arise from a conjunction of regions of distinct proteins in the phage head in much the same way that the peptidyl transferase activity of ribosomes results from several ribosomal proteins (Chapter 13).

Phage $\phi 6$ RNA serves as a substrate for the RNA-polymerizing activity of this phage and mRNA molecules are produced by a strand displacement reaction (Figure 15-49). The RNA-polymerizing activity of the phage initiates synthesis at one end of the double-stranded RNA and copies the noncoding strand, displacing the coding strand, which is then used as a mRNA. A second displacement reaction is usually initiated before the first displacement is complete, as shown in the figure. An electron micrograph of a molecule engaged in such a displacement reaction and having one growing single strand is shown in Figure 15-50.

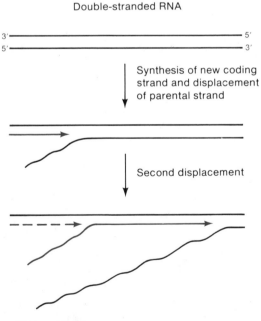

Figure 15-49
Synthesis of mRNA in $\phi 6$ by displacement of a parental RNA strand.

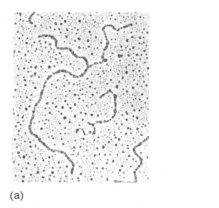 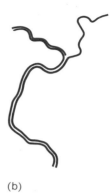

(a)   (b)

Figure 15-50
(a) Electron micrograph showing transcription of φ6 RNA. (b) An interpretive drawing. The red strand is the growing strand. (Courtesy of Robert Haselkorn and Stephen Usala.)

Each of the three phage RNA molecules serves as a template for transcription by the displacement reaction and at least 14 proteins are translated from the mRNA molecules.

The process of replication presents a special problem in that the displacement reaction generates daughter single strands without increasing the number of double-stranded RNA molecules. The possibility that the displaced single strands are then converted to double-stranded progeny RNA has been ruled out experimentally. At present, the mode of replication is not known, though it seems clear that both replication and transcription use the same enzymatic activity.

Production of progeny phage particles requires that one of each of the three double-stranded fragments is packaged in a single particle. How this occurs is unknown. The animal virus reovirus also has a segmented genome; the mode of packaging one of each of its segments is discussed in Chapter 21.

REFERENCES

Adhya, S., S. Garges, and D. F. Ward. 1981. "Regulatory circuits of bacteriophage lambda." *Prog. Nucleic Acid Res. and Mol. Biol.,* 26, 103–118.

Arber, W., and S. Linn. 1967. "DNA modification and restriction." *Ann. Rev. Biochem.,* 38, 467–500.

Casjens, S., and J. King. 1975. "Virus assembly." *Ann. Rev. Biochem.,* 44, 555–611.

Dubow, M. S. (ed). 1981. *Bacteriophage Assembly.* Alan Liss.

Dunn, J. J., and F. W. Studier. 1981. "Nucleotide sequence for the genetic left end of bacteriophage T7 DNA to the beginning of gene 4." *J. Mol. Biol.,* 148, 303–330.

Earnshaw, W. C., and S. R. Casjens. 1980. "DNA packaging by the double-stranded DNA bacteriophages." *Cell,* 21, 319–331.

Echols, H. 1980. "Bacteriophage development." In *Molecular Genetics of Development.* T. Leighton and W. F. Loomis, eds. Academic Press.

Furth, M. E., J. L. Yates, and W. F. Dove. 1979. "Positive and negative control of bacteriophage lambda DNA replication." *Cold Spring Harb. Symp. Quant. Biol.,* 43, 147–153.

Gottesman, M. E., S. Adhya, and A. Dar. 1980. "Transcription antitermination by bacteriophage lambda *N* gene product." *J. Mol. Biol.,* 140, 57–75.

Hershey, A. D. (ed.). 1971. *The Bacteriophage Lambda.* Cold Spring Harbor Laboratory.

Kaiser, A. D., M. Syvanen, and T. Masuda. 1975. "DNA packaging steps in bacteriophage lambda head assembly." *J. Mol. Biol.,* 91, 175–186.

Lewin, B. 1977. *Gene Expression. III. Plasmids and Phages.* Wiley-Interscience.

Luria, S. E., J. E. Darnell, D. Baltimore, and A. Campbell. 1978. *General Virology.* Wiley.

Simon, A., and R. Haselkorn. 1978. "Transcription and replication of bacteriophage $\phi6$ RNA." *Virology,* 89, 206–217.

Rabussay, D., and E. P. Geiduschek. 1977. "Regulation of gene action in the development of lytic bacteriophage." *Comprehensive Virology,* 8, 1–196.

Rosenberg, M., D. Court, H. Shimatake, C. Broad, and D. Wulff. 1980. "Structure and function of an intermediate regulatory region of bacteriophage lambda." *In* J. Miller, ed. *Operons.* Cold Spring Harbor Laboratory.

Stent, G., and R. Calendar. 1978. *Molecular Genetics.* W. H. Freeman and Co.

Studier, F. W. 1973. "Analysis of bacteriophage T7 early RNA and proteins on slab gels." *J. Mol. Biol.,* 79, 237–248.

Szybalski, W. 1977. "Initiation and regulation of transcription and DNA replication in coliphage lambda." *In* J. C. Copeland and G. A. Marzluf, eds. *Regulatory Biology.* Ohio State University Press.

Zinder, N. (ed.) 1975. *RNA Phages.* Cold Spring Harbor Laboratory.

16 Bacteriophages

II. THE LYSOGENIC LIFE CYCLE

Each of the phages discussed in the preceding chapter has a life cycle during which the host cell lyses and releases many progeny phage (the most common outcome) or during which phage are continually released without killing the host (this is true with the filamentous phages). In this chapter an alternate life cycle, the lysogenic cycle, is described. In this cycle phage are not produced and the host survives and divides indefinitely. However, many bacterial generations later, if environmental conditions are right, the lysogenic cycle can be terminated and a lytic cycle started anew. When this occurs, the host cell is killed and progeny phage are released as in a lytic cycle that has been initiated by infection with a phage particle.

One of the fascinating aspects of phage species that can enter a lysogenic cycle is that they can also initiate a lytic cycle of the type described in the preceding chapter. The growth conditions of the bactcrium prior to infection and the multiplicity of infection determine the pathway entered by the phage. This possibility results from elegant regulatory systems that differ from those we have seen in the preceding chapter in that they not only have temporal regulation but are sensitive to environmental conditions, such as the supply of nutrients.

THE LYSOGENIC CYCLE

We begin by describing the general properties of lysogens; then we will examine E. coli phage λ, the best-understood temperate phage.

Outline of the Lysogenic Life Cycle

There are two types of lysogenic cycles. The most common one, for which *E. coli* phage λ is the prototype, follows (Figure 16-1):

1. A DNA molecule is injected into a bacterium.
2. After a brief period of transcription, needed to synthesize an integration enzyme, transcription is turned off by a repressor.
3. A phage DNA molecule is inserted into the DNA of the bacterium.
4. The bacterium continues to grow and multiply and the phage genes replicate as part of the bacterial chromosome.

The second and less common type, for which *E. coli* phage P1 is the prototype, differs from the preceding one in that there is no integration system and the phage DNA becomes a plasmid (an independently replicating circular DNA molecule) rather than a segment of the host chromosome. In this chapter we will primarily consider the first type of life cycle.

Some Useful Terms

The following terms describe various aspects of the lysogenic life cycle.

1. A phage capable of entering either a lytic or a lysogenic life cycle is a **temperate** phage.
2. A bacterium containing a complete set of phage genes is called a **lysogen.** If some phage genes are absent, so that the phage is unable to complete a lytic cycle, the cell is a **defective lysogen.**
3. The process of forming a lysogen by infecting a bacterial culture with temperate phage is **lysogenization.**
4. If the phage DNA is contained with the bacterial DNA, the phage DNA is said to be **integrated.** The process by which this occurs is called **integration** or **insertion.** Phage DNA in plasmid form is nonintegrated. The integrated or plasmid DNA within a lysogen is called a **prophage.** If some phage genes are absent, the prophage is said to be defective.

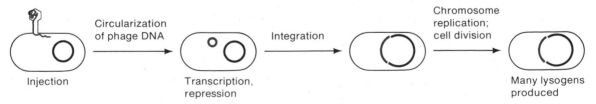

Figure 16-1
The general mode of lysogenization by insertion of phage DNA into a bacterial chromosome.

General Properties of Lysogens and Lysogenization

Two important properties of lysogens are the following:

1. A lysogen cannot be reinfected by a phage of the type that first lysogenized the cell; this resistance to "superinfection" is called **immunity.**
2. Even after many cell generations, a lysogen can initiate a lytic cycle; in this process, which is called **induction,** the phage genes are excised as a single segment of DNA (Figure 16-2).

The molecular mechanism for immunity and the circumstances that give rise to induction will be described shortly.

More than 90 percent of the thousands of known phages are temperate. These phages often are unable to produce bursts as large as the highly virulent phages such as T4 and T7, but compensate by their ability to multiply in environmental conditions that are not suitable for rapid production of progeny. The meaning of this statement will become clear when we examine the possible outcomes for a single phage encountering a population of bacteria in nature.

Let us first consider a bacterial population that is actively dividing. If a phage can infect one cell and multiply (in a lytic cycle), the number of progeny phage will increase rapidly. However, if the bacteria were growing very slowly owing to exhaustion of nutrients in the surrounding medium, the phage infection might abort if the infected cell were to stop growing during the phage life cycle. (Remember that a phage can only grow in a bacterium that is actively metabolizing.)

When bacteria are starved of nutrients, they degrade their own mRNA and protein before they become dormant. Restoration of nutrients enables the bacteria to grow again. This is not true of a phage-infected cell in which the phage life cycle has been interrupted; generally the ability to produce phage is permanently lost, probably because the delicate balance of phage functions is destroyed by the protein and mRNA degradation. Furthermore, the cell dies. This deleterious outcome can be avoided if a lysogenic cycle is possible— that is, if the phage DNA can become dormant. When growth resumes, the genes replicate as part of the chromosome. Even though phage

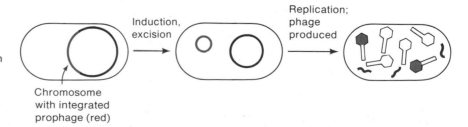

Figure 16-2
An outline of the events in prophage induction. The prophage DNA is in red. The bacterial DNA is omitted from the third panel for clarity.

Induction, excision

Replication; phage produced

Chromosome with integrated prophage (red)

production has been suspended for the time being, because of the induction phenomenon, the potential for phage production remains.

Now let us return to an infection of an actively growing bacterial population in which phage are multiplying rapidly. When the number of phage exceeds the number of bacteria, the phage enter their final cycle of multiplication; after lysis occurs, no further multiplication is possible because there are no more sensitive bacteria. It is possible that years could pass before these phage particles might encounter another sensitive host bacterium and during this time various deleterious agents (e.g., heat, acid, proteases, and protozoa) might damage the particles. In fact, the adsorption apparatus of most phage species is especially susceptible to chemical injury. Until a host cell appears, the phage particles have no chance to increase in number. However, if lysogenization could occur at a high multiplicity of infection (MOI), the phage genes could be maintained indefinitely, since the lysogen would grow whenever nutrients are available.

Indeed, the two conditions that stimulate the lysogenic cycle of a temperate phage are depletion of bacterial nutrients and a high multiplicity of infection.

E. COLI PHAGE λ

The best-understood temperate phage is E. coli phage λ, and its lysogenic cycle is the main topic of this chapter. Before proceeding, the reader is advised to review the description of its lytic cycle presented in the preceding chapter.

Immunity to Infection

When a virulent phage forms a plaque on a lawn of growing bacteria, the plaque is clear because all bacteria in the center of the plaque are killed and lysed. However, phage λ forms a plaque with a turbid center (as do other temperate phages) (Figure 16-3). The turbidity is caused by the growth of phage-immune lysogens in the plaque. Phage and bacteria are usually placed on an agar surface in a ratio of about 1 phage per 10^6 bacteria. The bacteria grow rapidly and the MOI is low, so the lytic cycle ensues. After several lytic cycles, the MOI becomes high and a few cells are lysogenized; since availability of nutrients does not yet limit development of phage or cells, most cells are lysed. When the nutrients in the agar are depleted, the uninfected cells stop growing, and the plaque stops increasing in size. However, since there has been less bacterial growth within the plaque, nutrients are still present here.

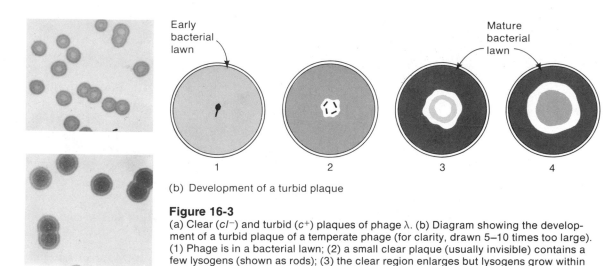

(b) Development of a turbid plaque

Figure 16-3
(a) Clear (*cI⁻*) and turbid (*c⁺*) plaques of phage λ. (b) Diagram showing the development of a turbid plaque of a temperate phage (for clarity, drawn 5–10 times too large). (1) Phage is in a bacterial lawn; (2) a small clear plaque (usually invisible) contains a few lysogens (shown as rods); (3) the clear region enlarges but lysogens grow within the plaque; (4) clear region reaches maximum size and lysogens stop growing as nutrient is exhausted. (Courtesy of A. D. Kaiser.)

(a)

Figure 16-4
The repressor-operator system of λ, showing the two early mRNA molecules. Symbols: *cI*, repressor gene; *p*, promoter; *o*, operator; *L*, left; *R*, right. See also Figure 16-14.

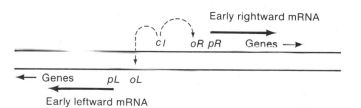

Therefore, the lysogenic cells, which are immune to subsequent infection by λ, continue to grow, forming a turbid center in the plaque.

The resistance of a λ lysogen to infection by λ is called **immunity.** The cause of the phenomenon is the following. Phage λ contains a repressor-operator system (Figure 16-4). The repressor gene is called *cI*; the repressor protein binds to two operators, *oL* and *oR,* which are adjacent to two promoters *pL* and *pR.* The letters L and R mean leftward and rightward and refer to the direction of synthesis of two early mRNA molecules, when the genetic map is drawn in a standard orientation (see Chapter 15, Figure 15-29).

In a lysogen the cI repressor is synthesized continuously and in slight excess with respect to the operators. Both *oL* and *oR* sites contain bound repressor molecules so that *pL* and *pR* are unavailable to RNA polymerase. Thus, in a lysogen, transcription from the two early promoters is prevented. It will be seen later that this is sufficient to keep the prophage in an "off" state; thus the lysogen grows indefinitely. If a normal cell is infected by λ, the two operators of the incoming λ DNA molecule will be unoccupied, since phage repressor has not yet been made and transcription will occur. However, if a phage tries to infect a lysogen, the excess repressor molecules already present in the lysogen bind to the two operators on the infecting DNA molecule before an RNA polymerase can bind to the pL and pR sites. This operator-binding prevents the phage from proceeding into lytic development. This inhibition is referred to as **resistance to homoimmune superinfection.**

What happens to superinfecting DNA? It is able to form a super-coil—phage gene products are not needed for supercoiling—but it cannot replicate. The bacterium is unaffected by the presence of this DNA molecule and grows and divides normally, so that the superinfecting DNA is progressively diluted out.

Mutants of the immunity system have been isolated. The two most important types are cI^- and *vir* **(virulent)** mutants:

1. A cI^- mutant does not make a functional repressor and hence can engage only in a lytic cycle. Thus, a cI^- mutant makes a clear plaque (Figure 16-3(a)).
2. A *vir* mutant carries mutations in both oL and oR and also makes a clear plaque because it cannot establish repression; furthermore, it can superinfect and grow in a lysogen because it is insensitive to the repressor already present in the cell.

There are many temperate phages of *E. coli* other than λ. Two of these related to λ are phages 21 and 434. Each of these phages has its own immune system—that is, its own repressor and repressor-specific operators. Thus, the 434 repressor cannot bind to a λ operator and a λ repressor cannot bind to a 434 operator. Such a pair of phages is said to be **heteroimmune** with respect to one another. A temperate phage can form a plaque on a heteroimmune lysogen because the repressor made in the lysogen does not bind to the operator of the superinfecting phage. This is summarized in Table 16-1. The immunity region of the DNA, which includes the *cI* gene, operators, and promoters (and other

Table 16-1
Ability of different phages to form plaques on homoimmune and heteroimmune lysogens

Superinfecting phage	Lysogen		
	B(λ)	B(*434*)	B(*21*)
λ	−	+	+
434	+	−	+
21	+	+	−

Note: The notation B(P) denotes a bacterium B lysogenic for phage P. + = forms a plaque; − = does not form a plaque.

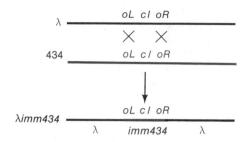

Figure 16-5
Formation of λ*imm434*. The ×'s indicate the
sites of genetic exchange.

elements to be described shortly) is given the genotypic symbol
imm—specifically, *imm*λ, *imm21* and *imm434*. Interesting hybrid
phages, which have been very useful in the laboratory, have been
created by crossing two heteroimmune phages and selecting a recombinant containing the immunity region of one phage and the remaining
genes of the other heteroimmune phage. A prominent example is a λ-434
hybrid, which is genotypically designated λ*imm434* (also, occasionally,
λ*434hy*), shown in Figure 16-5.

Prophage Integration

In the lysogenic cycle of λ and other similar phages, a phage DNA
molecule is inserted (or integrated, to use an equivalent term) into the
bacterial chromosome to form a prophage. In order to form a lysogen,
(1) repression must be established and (2) a prophage must be formed by
insertion of a λ DNA molecule.

When λ DNA integrates, it is inserted at a preferred position in the
E. coli chromosome; this site is between the *gal* operon and the *bio*
(biotin) operon and is called the λ attachment site or, genotypically, *att*.
For most temperate phages integration occurs at one preferred site,
though there are a few phages that either have several preferred sites or
appear to be able to insert their DNA anywhere in the *E. coli*
chromosome. An important feature of the insertion mechanism was
derived from the genetic observation that the gene order in the λ
prophage is a permutation of the gene order of the DNA in the phage
(see Figure 16-6). This permutation was explained by a model (due to
Allen Campbell and widely known as the **Campbell model**) for the
mechanism of integration (Figure 16-7). In this correct model, λ DNA is
circularized first; and then prophage integration occurs as a physical
breakage and rejoining of phage and host DNA—precisely, between the
bacterial DNA attachment site and another attachment site in the phage

Figure 16-6
The order of the genes on the DNA molecule in the phage head (vegetative order) and in the prophage (prophage order). The genes have been selected arbitrarily to provide reference points.

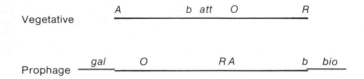

DNA that is located near the center of the phage DNA molecule. A phage protein, **integrase** (its gene designation is *int*), recognizes the phage DNA and bacterial DNA attachment sites and catalyzes the physical exchange. This results in integration of the λ DNA molecule into the bacterial DNA. As a consequence of the circularization and integration, the linear order of the genes in the infecting λ DNA molecule is permuted, as shown in the figure. A question that arose after this model was proposed was: why doesn't the integrase excise the prophage shortly after integration occurs, since the prophage is flanked by two attachment sites? Furthermore, excision would seem to be the preferred reaction since, kinetically, two attachment sites in the same DNA molecule (that is, the host chromosome) ought to interact with each other more rapidly than two attachment sites in different DNA molecules (phage and bacterial DNA). Why this does not happen became clear when it was discovered that, inasmuch as the DNA attachment sites in the phage and the bacterium are not the same, they recombine to form prophage attachment sites that, as a consequence, also differ from the sites joining the bacterial and phage DNA.

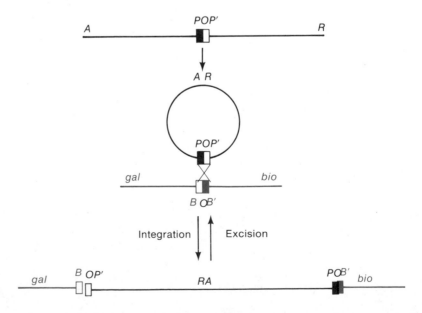

Figure 16-7
The Campbell model for the mechanism of prophage integration and excision of phage λ. The phage attachment site has been denoted *POP'* in accord with subsequent findings. The bacterial attachment site is *BOB'*. The prophage is flanked by two new attachment sites denoted *BOP'* and *POB'*.

All of the λ attachment sites have three different components. One of these is common to all sites and is denoted by the letter *O*.* The phage attachment site is written *POP'* (P for phage) and the bacterial attachment site is written *BOB'* (B for bacteria). Thus, in the integration reaction two new attachment sites, *BOP'* and *POB'*, are generated (Figure 16-7). These are often written *attL* and *attR* to designate the left and right prophage-attachment sites, respectively. Integrase cannot catalyze a reaction between *BOP'* and *POB'*, so that the reaction

$$BOB' + POP' \xrightarrow{\text{Integrase}} BOP' + POB'$$

Bacterium Phage   **Prophage**

is irreversible if integrase is the only enzyme present.

The result of the integration reaction is that the λ DNA is linearly inserted between the *gal* and *bio* loci and henceforth is replicated simply as a segment of *E. coli* DNA. The first evidence for insertion, gained from genetic experiments, showed (1) a permuted gene order in the prophage, (2) genetic linkage between *gal* and the genes to the right of *POP'*, (3) genetic linkage between *bio* and the genes to the left of *POP'*, and (4) a greater distance between the *gal* and *bio* genes in a lysogen than in a nonlysogen. Final proof for insertion came from the following physical experiment. A prophage was formed in a circular plasmid whose molecular weight was 105×10^6 and which contained *BOB'*. The plasmid remained circular but its molecular weight increased by the amount of one λ molecule (31×10^6) to 136×10^6.

Integrase has been purified, the base sequences of the *BOB'* and *POP'* sites are known, and the integration reaction can be carried out *in vitro* (Figure 16-8). Integrase has DNA-binding activity, binding strongly to *POP'*, and is a type I topoisomerase (Chapter 8); that is, it can cause a strand break in one strand of a double helix, rotate one branch of the broken strand about the continuous strand, and then rejoin the ends. In addition to integrase, the reaction requires a host protein called IHF (integrase host factor); this protein also binds to the attachment site, but its biochemical function is not yet known. IHF is a dimer or a tetramer and one of its subunits is the product of a gene called *himA*; the *himA* product is also involved in some way in the choice between lysogenic and lytic pathways. Sequencing data show that the sequences of B, B', P, and P' are quite different and the sequence of the O (as in *BOB'*), often called the **core sequence,** is very rich in A · T pairs (Figure 16-9).

The mechanics of the integration reaction are not yet known in detail. However, several points are clear. First, two single-strand breaks are made in each attachment region in the complementary strands

---

*Another common notation is a raised dot, · , so that the bacterial attachment site, which we write *BOB'*, may be written B · B'. In this book we prefer the O.

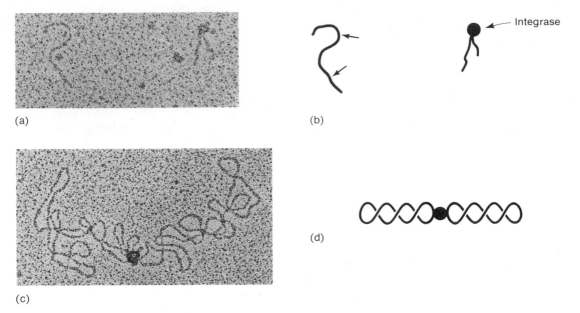

(a)

(b)

(c)

(d) Integrase

Figure 16-8

(a) Electron micrograph showing the integrase reaction carried out *in vitro*. At the left is a fragment of λ DNA that carries *POP′*. At the right is another fragment to which purified integrase is bound. (b) An interpretive drawing of the two molecules in (a). The two arrows indicate the locations of two binding sites for integrase.

The molecule at the right is shorter by the distance between these sites (250 base pairs), indicating that integrase brings the sites together. (c) Two supercoiled λ DNA molecules each carrying *POP′* joined by integrase. (d) Interpretive drawing of the paired molecules in panel (c). (Courtesy of Marc Better.)

```
--- ------------- ---
    GCTTTTTTATACTAA
    CGAAAAAATATGATT
--- ------------- ---
```

Figure 16-9

Base sequence of the core of the λ *att* region.

Figure 16-10

The exchange that occurs in the integrase reaction showing the approximate positions of the strand breaks (arrows). The source of the circled bases in *attL* or *attR* is uncertain; each may have come from the attachment region opposite the one designated by the color. The lower case letters are bases in the flanking sequences *B, B′, P,* or *P′*; the capital letters are in the *O* region.

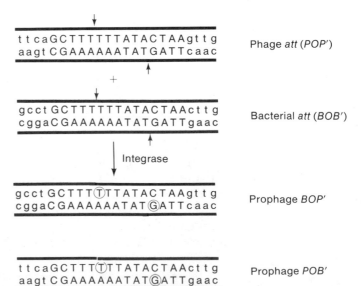

```
      ↓
ttcaGCTTTTTTATACTAAgttg          Phage att (POP′)
aagtCGAAAAAATATGATTcaac
      ↑
      +
      ↓
gcctGCTTTTTTATACTAActtg          Bacterial att (BOB′)
cggaCGAAAAAATATGATTgaac
      ↑
      │ Integrase
      ↓
gcctGCTTT(T)TTATACTAAgttg        Prophage BOP′
cggaCGAAAAAATAT(G)ATTcaac

ttcaGCTTT(T)TTATACTAActtg        Prophage POB′
aagtCGAAAAAATAT(G)ATTgaac
```

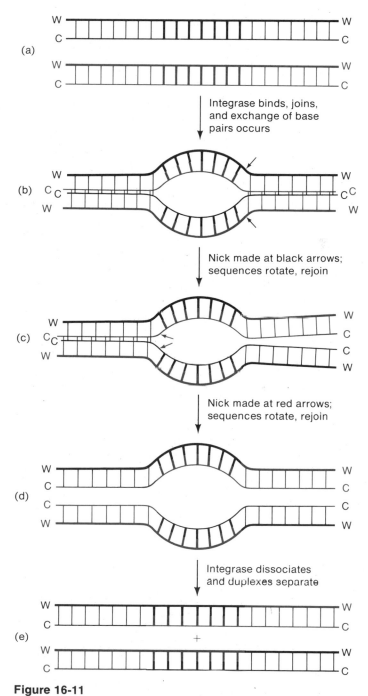

Figure 16-11
A model for the integrase reaction. See text for details. The heavy
bars represent the only base pairs that are broken. (Modified from
a drawing provided by Howard Nash.)

seven base pairs apart, as shown in Figure 16-10. Second, an exchange occurs in which two overlapping joints containing strands from two core sequences are formed, as also shown in the figure. Recently, a model for the integrase reaction has been proposed (Figure 16-11) in which the easily melted, high-A+T core sequence in each attachment site opens and then complementary strands from different attachment sites pair to form a four-stranded segment (panel (b)). Pairs of single-strand breaks in homologous regions are made (panel (c)). Then the topoisomerase activity of integrase causes each double-stranded region to undergo a 270° rotation (panel (c)) after which corresponding strands from the two core sequences are joined (panel(c)). This process (nick–rotate–join) is repeated on the other side of the base-paired region (panel (d)), and the two newly formed core sequences with overlapping joints separate (panel (e)). Further experimentation is needed to determine if this model, which is probably correct in outline, is also correct in detail.

The synthesis of integrase is coupled to the synthesis of the cI repressor. This is efficient because integrase and the repressor are both needed in the lysogenic cycle and neither is needed in the lytic cycle. An outline of the regulatory pathway responsible for this coupling is shown in Figure 16-12. In the absence of a positive regulatory element (the product of the λ gene *cII*) the promoters for both the *int* and *cI* genes are unavailable to RNA polymerase. Shortly after infection, the cII protein is synthesized (see Chapter 15). If the concentration of the protein is high enough, the *cII* product binds to sites near the promoters for the *cI* and *int* genes (designated *pre* and *pI*, respectively) and thereby renders them accessible to RNA polymerase. (In this respect the cII protein is similar to the cAMP-CAP complex in the *lac* operon.) RNA polymerase then transcribes both genes and the gene products are made. We will see later that the immediate cause that determines whether a lysogenic or lytic cycle occurs in an infection is mediated through the concentration of the cII protein.

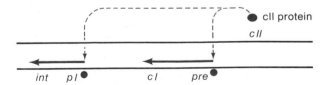

Figure 16-12
The role of the cII protein in stimulating synthesis of *cI* and *int* mRNA from *pre* and *pI* respectively. The mRNA molecules are drawn as red arrows which indicate the direction of RNA synthesis. The arrows are located nearer the DNA strand that is copied. Dashed arrows point to the sites of binding of the cII protein (red dots).

Induction and Prophage Excision

A lysogen is generally quite stable and can replicate nearly indefinitely without release of phage. However, if a lysogenic bacterium were to become damaged, it would be to the advantage of the phage to become derepressed and initiate a lytic cycle. This does occur and the signal for derepression (**prophage induction**) is damage to the DNA. The mechanism of induction is not known but there are two current theories (Figure 16-13). So far, all inducing agents, of which ultraviolet radiation has been studied most extensively, cause DNA damage that results both in activation of the SOS repair pathway (Chapter 9) and the generation of single-stranded regions in DNA. Following ultraviolet irradiation an unknown chain of events occurs that leads to cleavage of the cI

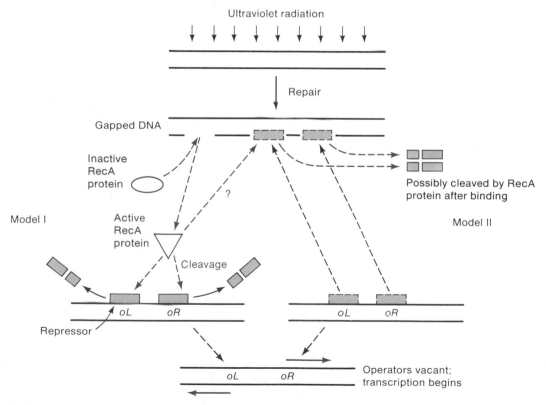

Figure 16-13
Schematic diagram of two possible models for prophage induction. Repressor molecules are shaded. In Model I, an early step after ultraviolet irradiation is the activation of the RecA protein, possibly by binding to ultraviolet repair-induced gaps. The active protein cleaves the repressor, yielding unoccupied operators. In Model II, the repressor leaves the operator because of tighter binding to gapped DNA.

repressor by a protease activity of the *recA* product (see also Chapter 14), a major element in SOS repair and in genetic recombination. Since induction cannot occur in a *recA⁻* mutant, this destruction of the repressor is, according to one theory, the cause of induction. It is proposed that an activator is needed *in vivo* to turn on the protease activity of the *recA* product and that the activator forms as a response to the DNA damage. The activator has been sought for years, but, if it exists, it eludes discovery. According to the second theory, single-stranded DNA, which is known to bind the repressor, simply removes all of the available repressor so that none is available to bind to the operator. This theory is also unproved at present. It is possible that transient binding of repressor to single-stranded DNA renders it susceptible to the protease, which would then not need an activator— that is, the substrate is activated instead. A third possibility, which combines features of the two theories, is that the protease activity of the RecA protein is activated by single-stranded DNA. At any rate, dere-pression does occur and this is the ultimate signal to initiate the lytic cycle inasmuch as the early promoters *pL* and *pR* (Chapter 15, Figure 15-29) are available to RNA polymerase.

A product of transcription from *pL* is a protein called **excisionase** (designated Xis), encoded by the gene *xis*. Excisionase has recently been purified, but only a few of its properties are known. Genetic evidence and binding studies indicate that it forms a complex with integrase, allowing the latter to recognize the prophage attachment sites *BOP′* and *POB′*; once bound to these sites, integrase can make cuts in the core sequence and re-form the *BOB′* and *POP′* sites. The physical role of Xis has recently been elucidated by electron microscopic studies demon-strating a tight complex in which *POB′* becomes wrapped around Xis; this complex is thought to contain Int also. Formation of the complex is required before *POB′* and *BOP′* can interact. Thus, the Xis-Int complex reverses the integration reaction, causing excision of the prophage (Figure 16-7). Note that the reactions between all attachment sites can now be written

$$
BOB' + POP' \underset{\text{Int,Xis}}{\overset{\text{Int}}{\rightleftharpoons}} BOP' + POB'
$$

The Xis-Int complex fails to catalyze the rightward reaction, so that when excess excisionase is present, excision is an irreversible reaction.

Note that the product of excision is an intact *E. coli* chromosome and a circular λ molecule, which is also the arrangement present in an infected cell immediately after infection.

Two Promoters for the Synthesis of the λ cI Repressor

In a lysogenic cycle it is essential that repressor molecules be synthesized more rapidly than DNA replication increases the number of operators. If there were too many operators for the number of repressor molecules, transcription leading to phage development and cell death would occur. Thus the repression system is designed to have an initial burst of repressor synthesis, just after infection, when the lysogenic pathway is to be followed. In contrast, a lysogen contains only a single λ DNA molecule (the prophage), so that less (but some) cI repressor is needed to maintain repression. Clearly, transcription of the *cI* gene must occur in a lysogen (this modifies an earlier statement that in a lysogen the prophage is in an "off" state), but transcription does not need to be very strong.

In λ, high-level and low-level synthesis of the repressor is achieved by the existence of two promoters (Figure 16-14)—(1) the establishment promoter, *pre*, which functions in an infected cell and which requires activation by the cII protein, and (2) the maintenance promoter, *prm*, which functions in a lysogen and is regulated by the repressor itself. Thus, when lysogenization occurs (that is, when the phage has made enough cII protein to activate *pre*), there is a burst of synthesis of the repressor from the mRNA initiated at *pre*, and this protein activates the promoter *prm*. As the amount of repressor increases, transcription of the *cII* gene is repressed and no more cII protein is made. The cII protein is unstable, so that soon its concentration becomes too low to maintain *pre* in an active state and synthesis of *pre* mRNA terminates. Synthesis

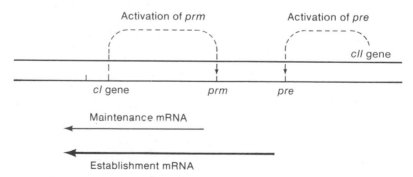

Figure 16-14
The *cI* gene of phage λ, and adjacent regions, showing the two promoters for synthesis of *cI*-encoded mRNA. The heavy black lines are the two DNA strands. The thin and heavy red arrows represent two mRNA molecules, both of which are transcribed from the lower DNA strand. The *cII*-gene product is translated from a mRNA molecule (not shown) transcribed rightward from the upper strand; this protein activates *pre*, yielding the establishment mRNA (heavy red arrow). The cI protein activates *prm*, yielding the maintenance mRNA.

Table 16-2
An outline of the sequence of events in cI repressor synthesis in the lysogenic pathway

---

1. Infection of cell
2. Transcription of rightward mRNA from *pR*
3. Activation of *pre*
4. Transcription and translation of the cI repressor from *pre* (high-level synthesis)
5. Activation of *prm* by the cI repressor
6. Transcription and translation of the cI repressor from *prm*
7. Turning off of *pR* by the cI protein
8. Degradation of the cII protein, so that *pre* becomes inactive
9. Continued synthesis of the cI protein from *prm* (low-level synthesis)

---

of *prm* mRNA continues and throughout successive generations translation of this mRNA maintains a concentration of the repressor sufficient to repress the prophage. As we have already stated, less repressor is needed in a lysogen; for the sake of efficiency, less should be made when transcription is occurring from *prm*. Actually, in the fully "on" state the two promoters are equally active in binding RNA polymerase; however, less repressor is made from the transcript initiated at *prm* because the *pre* transcript has a strong ribosome binding site and the *prm* transcript binds only weakly to the ribosome. Thus, the amount of repressor made from the two mRNA molecules is determined by the efficiency of translation.

A summary of the events in the sequence of repressor synthesis is given in Table 16-2.

We have just said that the repressor is needed to turn on transcription from *prm*. However, when the repressor concentration is very high, transcription from *prm* does not occur because the repressor also negatively regulates its own transcription. This regulatory mechanism enables λ to maintain a fairly constant repressor concentration, which is advantageous for two reasons: (1) In nature bacteria have continually varying growth rates owing to fluctuations in the supply of nutrients. Regulation ensures that the repressor concentration never diminishes so much that induction occurs spontaneously. (2) If unregulated, the repressor concentration might become so great that stimulated induction could not occur when needed.

One might ask why *prm* should be stimulated by the repressor. Since less repressor is made from *prm* than from *pre*, then if λ only had a lysogenic pathway, synthesis of *prm* mRNA could be constitutive. However, if a lytic cycle is also to be possible, then constitutive synthesis of *prm* mRNA should be avoided—such synthesis might allow premature repression to occur and this would prevent initiation of the lytic cycle. With a negatively regulated system, only the burst of cI repressor synthesis from *pre* needs to be regulated because, without that burst, no repressor is made from *prm*.

In order to understand the mechanisms of these regulatory circuits, knowledge of the structure of the operators is required; this is presented in the following section.

Structure of the Operator and Binding of the Repressor and the Cro Product

The operators *oL* and *oR* can be subdivided into six regions: *oL1, oL2, oL3, oR1, oR2,* and *oR3* (Figure 16-15). The base sequences of these regions are similar and each region is capable of binding cI repressor. The affinities of each region for the repressor are not the same and have the order

$$oL1 > oL2 > oL3$$

and

$$oR1 > oR2 > oR3.$$

Binding to *oL1* and *oL2* and to *oR1* and *oR2* is sequential and cooperative. Binding to *oR3* and *oL3* is not cooperative.

At low concentrations of repressor, *oL1* and *oR1* are occupied. Since the promoters *pL* and *pR* are adjacent to *oL1* and *oR1* respectively, when repressor is bound to these sites, RNA polymerase has reduced access to both promoters. The block is even more complete

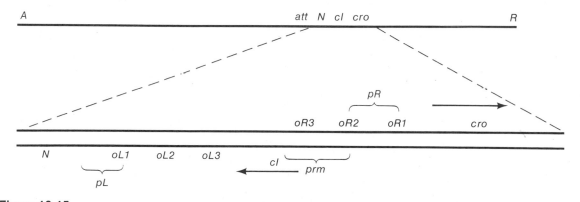

Figure 16-15
A close look at the operator-promoter region of λ. The arrows denote the direction of transcription of the *cro* and *cI* genes. See also Figure 16-4.

when *oL2* and *oR2* are also filled, which is the state of the operators in a lysogen. The promoter *prm* is between *oR2* and *oR3*, so that if the repressor concentration is very high, *prm* is blocked; this is the means by which the repressor negatively regulates its own synthesis. Furthermore, *prm* is not accessible to RNA polymerase unless *oR1* is occupied by repressor. The mechanism of *prm* activation is not known but binding of repressor to *oR1* and *oR2* is the means of positive self-regulation of repressor synthesis. Possibly, binding of the repressor weakens the base-pairing in *prm* and makes it easier for RNA polymerase to form an open-promoter complex, as is the case with positive regulation of the *lac* operon by the cAMP-CAP complex (Chapter 14).

Thus, if a cell infected by a λ particle is destined for lysogeny, the following events occur:

1. Transcription from *pL* and *pR* begins.
2. The cII protein is translated from the *pR* transcript.
3. The cII protein enables RNA polymerase to transcribe from the *pre* promoter so that repressor is synthesized.
4. Repressor binds to *oL1* and *oR1,* turning off transcription and hence synthesis of the cII protein. That protein is unstable and soon there is no transcription from *pre.*
5. Occupation of *oR1* (item 4) causes *prm* to be activated, so the repressor continues to be synthesized from this promoter, albeit at a lower rate.
6. The cI repressor continues to accumulate, so that *oL2* and *oR2* become occupied and repression of transcription from *pL* and *pR* becomes complete.
7. Transcription from *prm* continues unless the repressor concentration becomes so high that the protein also binds to *oR3.* Henceforth, the activity of *prm* is turned on and off to accommodate mild fluctuations in repressor concentration.

As we have described the system, there is no possibility for a lytic cycle. Entry into the lytic cycle requires the *cro*-gene product, which is necessary for several features of the lytic cycle and to prevent synthesis of the cI repressor. The Cro protein is also a repressor that binds to the operators *oL* and *oR*. Its mode of action is based on its affinity for the subsites in the operators; its affinity is opposite to that of the cI repressor—namely,

$$oR3 > oR2 \cong oR1$$

and

$$oL3 > oL2 > oL1.$$

The sequence is the following:

1. The *cro* gene is transcribed from *pR* (Figure 16-15).
2. After transcription of the *cro* gene, the *cII* gene, whose product is required to activate the *pre* promoter, is transcribed.
3. Once the Cro protein is synthesized, it binds to *oR3*; hence, activation of *prm* is prevented.
4. The concentration of the Cro protein increases and *oR2* and *oR1* gradually become occupied. When these sites are filled, RNA polymerase loses access to *pR*, so synthesis of the Cro protein eventually stops.

Items 3 and 4 indicate that the Cro protein represses its own synthesis as well as repressing synthesis of the cII protein.

Most of the genes required for lytic growth are to the right of the *cro* gene. Thus if the Cro protein were an efficient repressor, lysogeny could never be established (because no cI repressor would be made) and a lytic cycle would be impossible (because of lack of transcription from *pR*). However, the Cro protein binds very weakly to *oR* and by the time there is sufficient Cro protein to turn off rightward transcription and to repress *prm*, there is also a good deal of cII protein present in the cell, stimulating transcription from *pre* and repressing synthesis of the Cro protein. If sufficient cI repressor were made early, then synthesis of the Cro protein would be aborted and repression could be established. In fact, whether or not a phage enters a lytic or lysogenic cycle depends on the outcome of competition between synthesis of the Cro protein and the cI repressor; the critical element moderating this competition is the cII protein, as will be seen in the following section.

Decision Between the Lytic and Lysogenic Cycles

The mechanism that determines the choice between the lytic and lysogenic cycles has not yet been fully worked out.* However, it is clear that the cII protein plays a central role. This protein has three important functions—(1) it activates the *pre* promoter, (2) it activates *pI*, the *int* promoter, and (3) it delays synthesis of late mRNA (which encodes phage heads, phage tails, and the lytic system). Several lines of evidence indicate that the amount of active cII protein limits activation of the *pre* promoter when the MOI is low; the reason for this is not known but nonetheless it explains the need for high MOI to obtain efficient lysogenization. It is possible that for the cII protein to activate *pI* and

*This topic, which is a subject of active research, is not completely understood and some of the statements in this section are speculative.

pre it must be a multisubunit protein. Thus, the active form would only exist when a high total concentration of the protein is maintained and this might only be the case when the MOI is high.

Two important elements governing the choice of entry into a lytic or lysogenic life cycle are the *E. coli* genes *hfl* and *himA*. The *hfl* gene apparently encodes (or controls) one of two proteases that destroy the cII protein. Its activity, however, is antagonized by the λ gene-*cIII* protein, which is encoded in the mRNA transcribed from *pL*. The HimA protein plays two roles in lysogenization—(1) it is required in the integrase reaction as a subunit of the integrase host-factor protein, as described earlier, and (2) neither integrase nor the cI repressor is synthesized when a *himA⁻* mutant is infected. The *himA* gene is also involved in a large number of intracellular processes—for example, SOS repair and genetic recombination—and the amount of HimA protein varies with the growth state of the cell. It may be the case that it is the variation of the concentration of the HimA protein and/or of the Hfl protein with the growth state of a cell that imposes the requirement for lysogenization mentioned earlier—that the cells must be slightly depleted of nutrients. How the HimA protein affects the choice between lysis and lysogeny is not known, but it makes good sense for the phage to use the *himA* protein as an environment-sensing element, as this product is also required for the integrase reaction. Thus, if the amount of the HimA protein is too low for integration to occur, the phage senses the futility of establishing repression and synthesizing integrase.

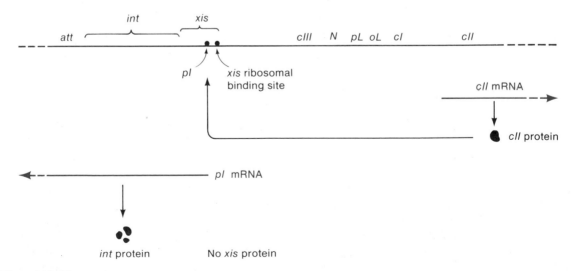

Figure 16-16
Prevention of synthesis of the *xis* protein during lysogenization. Black lines are DNA molecules; red lines are mRNA molecules. The *cII* protein activates *pI*, the *int*-gene promoter, which is downstream from the ribosomal binding site of the *xis* protein.

Excision and Its Avoidance during Lysogenization

In the lysogenic pathway, before cI repressor is made, transcription from *pL* occurs, so that both the *int* and *xis* genes are transcribed. If both integrase and excisionase were present, a prophage that had just been inserted into the chromosome would immediately be excised, which clearly must be avoided. A simple mechanism prevents this excision (Figure 16-16).

As part of the progress along the lysogenic pathway, the activity of the cII protein turns on synthesis of the cI repressor from the *pre* promoter and of integrase from the *pI* promoter. The cI repressor then acts on *pL* and turns off transcription of the *int* and *xis* genes from *pL*. The promoter *pI* is located within the *xis* gene downstream from the ribosomal binding site for synthesis of the *xis* product. Thus, this mRNA molecule provides integrase but not excisionase.

Use of λ to Screen for Carcinogens

Apart from their intrinsic interest, phages have served molecular biology principally as a means of understanding fundamental molecular processes. Recently, λ phage has also proved to be a valuable tool for screening for carcinogens. In Chapter 9 a test (the Ames test) was described in which carcinogens are detected by their ability to induce reversions of certain His⁻ mutations in the bacterium *Salmonella typhimurium*. A similar test (known as the **Devoret test**) utilizes *E. coli* strains lysogenic for phage λ. Lysogens are spontaneously induced at a fairly low frequency and in certain bacterial mutants the frequency is lower. However, many carcinogens cause DNA damage and are inducing agents. The induction can be detected quite simply. If one lysogen is mixed with 10^8 bacteria and put on agar (exactly as might be done in detecting phage by plaque counting), the lysogen, which is not induced by this process, forms a microcolony (of at most 1000 cells) in the bacterial lawn. Occasionally, late in the development of the colony, a cell will be induced, but this does not usually produce a plaque, because depletion of nutrients in the agar limits cell growth, so the released phage cannot multiply (remember: phage multiply only in growing cells). However, if an inducing agent (in this case, a carcinogen) is in the agar, the original lysogenic cell will be induced to make phage at the time all of the bacteria in the agar begin to grow, so that a plaque will form. If 10^5 lysogens are put in the agar, the effectiveness of the inducing agent can be measured by the fraction of cells yielding plaques. This test, which, like the Ames test, uses liver extracts in the agar to activate carcinogens, successfully detects those carcinogens already observed by the Ames test. The Devoret test is more sensitive, though, requiring lower concentrations of carcinogens; thus, it is capable of indicating weak carcinogens whose detection is beyond the sensitivity of the Ames test. Together these tests are being used to examine a very large number of industrial chemicals and food additives.

OTHER MODES OF LYSOGENY

Most temperate phages form lysogens in the way described for λ—namely, a prophage is inserted at a unique site in the host chromosome. Phages have been observed for which there are several chromosomal *att* sites but this is rare. We mentioned at the beginning of the chapter that *E. coli* phage P1 is markedly different in that its prophage is not inserted into the chromosome. Following infection, P1 DNA circularizes and is repressed. In the lysogenic mode it remains as a free supercoiled DNA molecule, roughly one or two per cell. Once per bacterial life cycle the P1 DNA replicates and somehow this replication is coupled to chromosomal replication. When the bacterium divides, each daughter cell receives a P1 circle; how this orderly assortment is accomplished is unknown. The mechanism of prophage maintenance is not as foolproof in phage P1 as in temperate phages that insert their phage DNA into a chromosome; for example, in each round of cell division about 1 cell per 1000 fails to receive a P1 circle. It is not known whether this is due to occasional failure in replication or to imperfect segregation of plasmids into the daughter cells.

SOME PROPERTIES OF LYSOGENS

Lysogenization has been presented as a means of propagating a phage when a supply of sensitive bacteria has been exhausted. The ability to lysogenize also has value to a bacterium, for the bacterium survives the infection and is immune to homoimmune superinfection.

Lysogenization also has other effects on bacteria—that is, often a lysogen obtains properties not present in the original bacterium. This is called **lysogenic conversion.** One example is the resistance to phage T4 gene-*rII* mutants conferred on *E. coli* by a λ prophage. A λ gene, *rex*, also transcribed from the *prm* promoter, blocks an early stage of infection by these mutants. Another example is found in *Corynebacterium diphtheriae*, the causative organism of diphtheria, which can cause the disease only when the bacterium is lysogenic for the phage *β*. A site of insertion may also be within a bacterial gene so that prophage formation inactivates the gene. An example of a phage having this property is Mu, which can form a prophage in any gene and therefore is a powerful mutagenic agent; this phage is discussed in the following section.

MU—AN UNUSUAL PHAGE

E. coli phage Mu is a temperate phage capable of both lytic and lysogenic growth. Its name, Mu, comes from *mutator* inasmuch as Mu lysogens are very frequently mutants. The reason for this is that in the lysogenic pathway, Mu inserts its DNA at randomly distributed sites in the *E. coli* chromosome and often these sites are within *E. coli* genes or regulatory elements, which are inactivated when a Mu prophage divides the gene or element. Another peculiarity of Mu is that insertion of DNA into the host chromosome is an obligatory stage in the lytic life cycle, a feature that is also found in the avian retroviruses (Chapter 21), RNA viruses that cause tumors in birds.

Mu DNA

Mu DNA is linear, neither terminally redundant nor cyclically permuted, and consists of about 38,000 base pairs. However, in a population of Mu phage particles each DNA molecule is different! This is made evident when a sample of Mu DNA is denatured and then renatured. Electron microscopy of the renatured double helices, which invariably contain two single strands derived from different double-stranded molecules, yields the two structures shown in Figure 16-17. All single-stranded regions consist of sequences that are noncomplementary. Further experiments have indicated the following:

1. The noncomplementary sequences at the termini are bacterial DNA segments, which differ from one DNA molecule to the next.
2. If a segment of Mu DNA, e.g., in the α region, is increased in length (by an insertion of foreign DNA), the nonhomologous

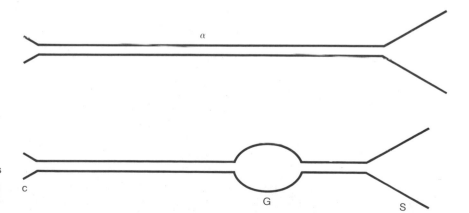

Figure 16-17
The major structures of Mu DNA seen after denaturation and renaturation. The symbols α, c, G, and S refer to distinguishable regions of the heteroduplexes.

Figure 16-18
(a) The structure of a G loop (Figure 16-17) of Mu DNA showing inversion of the G sequence denoted by ABCD. (b) The result of annealing of a single strand of Mu DNA, showing the inverted repeats of the G regions. Only part of the strand is shown.

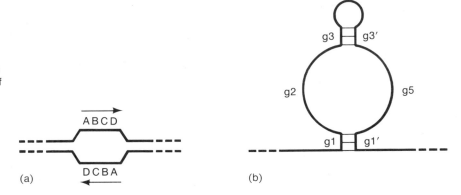

(a)   (b)

region at the left (c) end retains the wild-type length whereas the segment at the right (S) end becomes shorter.

3. Renatured molecules may have an unpaired region called the **G loop.** The two strands of the G loop, which are phage DNA, are identical but one is reversed in direction with respect to the other. That is, a G loop has the structure shown in Figure 16-18(a). If single-stranded DNA is allowed to self-anneal at a low concentration, which avoids formation of interstrand double helices, the structure shown in Figure 16-18(b) is seen. The g1 and g1′ segments are self-complementary. Occasionally in the phage life cycle the G segment becomes inverted by genetic recombination. The normal and inverted configurations are called *flip* and *flop* and annealing of a strand containing *flip* with a complementary strand containing *flop* yields a G loop. A phage having *flop* has a different relation between the DNA and a nearby phage promoter and as a result *flip*-DNA and *flop*-DNA molecules make different tail fibers, which extends the range of bacteria to which Mu can adsorb.

Replication and Maturation of Mu DNA

Points 1 and 2 in the preceding section are explained by the odd mechanism of DNA replication used by Mu.

In the course of DNA replication in a normal infection most, if not all, progeny Mu-DNA molecules are inserted at numerous points into the chromosome. This phenomenon, which is not understood, explains the association of Mu DNA with various pieces of bacterial DNA described in point 1.

Packaging of the DNA probably occurs in the following way (Figure 16-19): The maturation system recognizes the c end of the Mu DNA and

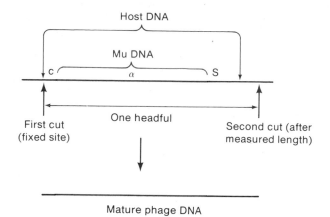

Figure 16-19
Diagram of inserted Mu DNA (red) showing how the cutting process generates host sequences at the termini.

makes a cut in the host DNA about 100 bases to the left of the phage DNA. Packaging begins at this point and the head is filled. The second cut, at the S end, is determined by the headful mechanism (used by T4) and one headful exceeds the length of the phage DNA sequences; this explains point 2. Therefore, note that nothing is particularly unusual about the maturation system; rather, peculiarities in the mode of DNA replication (that is, the insertion of progeny DNA into the chromosome) give rise to the unusual structure of the DNA in the phage head.

The Mu phage will be discussed further in Chapter 19 when transposable elements are described.

TRANSDUCING PHAGES

In the phage systems discussed so far, two rules for packaging DNA into a phage head have been encountered. These are the *headful rule* used by phage T4 (Chapter 15), in which a fixed amount of DNA is packaged, and the *defined termini rule,* in which two cuts are made in a concatemeric DNA molecule at fixed sites, as in the case of phage λ (Chapter 15). In the case of the headful mechanism nothing has been described that would prevent packaging of host DNA if a free end could be provided to initiate the process. However, this is not a common event since phage-encoded nucleases often degrade host DNA very rapidly; little or no host DNA is left by the time packaging occurs. In the case of phage λ, packaging of host DNA should also be rare because it is not likely that the host DNA would contain appropriately spaced base sequences equivalent to *cos* sites. We will see shortly that sometimes a λ phage particle contains host DNA linked to phage DNA.

Several phages are known for which host DNA can be packaged. Such phages are called **transducing phages** and a phage particle containing host DNA is called a **transducing particle.** There are two types of these phages: **generalized transducing phages,** which can produce particles containing only bacterial DNA, and **specialized transducing phages,** which occasionally produce particles containing both phage and bacterial DNA sequences. Both types of transducing particles can inject their DNA into a host bacterium and thereby transfer host DNA from one bacterium to another; this transfer process is called **transduction.** Formation of specialized, but not generalized, transducing particles is a result of lysogenization, but it is convenient to discuss both types together.

Properties of Generalized Transducing Particles

An excellent example of a generalized transducing particle is that produced when phage P1 infects E. coli. Phage P1 has both a lysogenic and a lytic cycle but only the lytic cycle is relevant to a discussion of transduction. The life cycle of P1 is not very different from those of the lytic phages discussed in Chapter 15. The features pertinent to this discussion are: (1) the burst size (which is about 100), (2) the molecular weight of the phage DNA (66×10^6), and (3) the fact that P1 encodes a nuclease that causes degradation of E. coli DNA. This nuclease action occurs very slowly compared to that of T4 so that, when packaging begins, the host DNA consists of fragments whose molecular weights range, for the most part, from 10^7 to 10^8. Packaging of one of these fragments into a phage head occurs in about 1 percent of the infected cells. The P1 packaging system is not very fastidious, so the size of the DNA molecule does not have to be exactly the same in all particles. Furthermore, the head assembly process allows heads of two different sizes to be synthesized. The result is that in the population of particles produced when infected bacteria lyse, there are generalized transducing particles, the major class of which contains a single fragment of DNA having a molecular weight of 20 to 25 million. Fragmentation of the host DNA is a random process and the transducing particles contain fragments derived from all regions of the host DNA. Thus a sufficiently large population of P1-phage progeny will contain at least one particle possessing each host gene. On the average, for any particular gene, there is roughly one transducing particle per 10^6 viable phage. Generalized transducing particles do not produce P1-phage progeny because they contain no P1 DNA; however, the bacterial DNA is injected into the host cell and, if a promoter is present, gene expression can occur.

Let us now examine the events that ensue when a transducing particle infects a bacterium. Consider a transducing particle that has emerged from an infected wild-type E. coli and that contains the gene for leucine *(leu)* synthesis. Such a particle is termed a *leu*+ particle.

Consider also that this *leu*+ particle adsorbs to a bacterium whose genotype is *leu*− and injects its DNA. The bacterium survives because no phage genes are injected and its exonucleases may indeed degrade the injected linear DNA fragment. Another possibility is that the *leu*+ segment can be incorporated into the host DNA by genetic recombination, resulting in replacement of the *leu*− allele by a *leu*+ allele. In this way, the genotype of the host cell would be converted from *leu*− to *leu*+. This is how transducing particles are detected (and how they were discovered). That is, a particle infects a culture of X+ bacteria and the resulting lysate is then used to infect an X− culture. The infected X− bacteria are placed in agar lacking X; if colonies grow, they must be X+, and transduction must have occurred. This entire process is depicted in Figure 16-20.

Another well-studied phage that produces generalized transducing particles is phage P22, which infects the bacterium *Salmonella typhimurium*. Study of this phage provided the first experimental proof that the transducing particles contain only bacterial DNA. The experiment used the density labeling technique of Matthew Meselson and Franklin Stahl in the following way (Figure 16-21). Host bacteria were grown for many generations in growth medium containing ^{15}N and then infected with ^{14}N-labeled P22 in medium containing ^{14}N. The phage progeny were then centrifuged to equilibrium in concentrated CsCl, which separates the particles by density. The centrifuge tube was fractionated by puncturing the tube bottom and collecting successive drops, and each fraction was tested for the presence of both viable phage and

Production of transducing particles

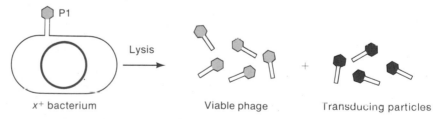

Transduction of *x*− bacterium by P1*x*+

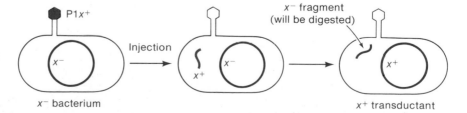

Figure 16-20
Conversion of an *x*− bacterium to an *x*+ genotype by P1 transduction. In the upper drawing there should be about one transducing phage per 10^6 viable phage.

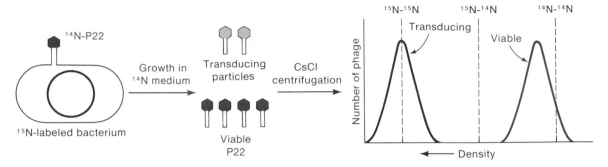

Figure 16-21
Demonstration that P22 transducing particles contain only bacterial DNA. Phage and bacterial DNA are shown in red and black, respectively.

transducing particles. The viable phage had the density of particles containing ^{14}N-labeled DNA (there was a small shift to higher density resulting from incorporation of some ^{14}N-labeled nucleotides obtained from degraded bacterial DNA) but all transducing particles were found at the density of particles having ^{15}N-labeled DNA, indicating that they contained little, if any, phage DNA.

Generalized transducing particles are also found in lysates of E. coli phage λ, but these particles are tailless. They result from occasional packaging of bacterial DNA that has been partially degraded, though it is possible that at low frequency the λ Ter system, which is responsible for cutting concatemers and forming the usual cohesive ends, recognizes base sequences in bacterial DNA. The production of phage particles by all known tailed species proceeds by filling the head with DNA and attaching the tail to the filled head. For λ the tail does not attach to the filled head unless a few nucleotides of the left cohesive end of the phage DNA project from the phage head. Presumably the tail must interact with a particular base sequence in order to join to the head. The bacterial DNA fragments in the transducing particle heads apparently lack the correct base sequence and therefore a tail is not attached. By use of appropriate experimental conditions λ tails can be attached to the filled heads *in vitro*; in this way generalized transducing particles of λ containing bacterial DNA can be prepared.

Use of Generalized Transducing Particles

Generalized transducing particles are valuable experimental tools because they can be used to generate new bacterial strains and to map genes. The simplest use of this procedure is the introduction of a positive allele into a strain that has a negative allele. For example, consider a strain that has the phenotype Leu⁻Lac⁻ and that is to be converted to Leu⁻Lac⁺. One begins by growing P1 on any *lac*⁺ strain. The lysate, which contains a few P1*lac*⁺ particles, is then used to transduce the *leu*⁻*lac*⁻ strain to *leu*⁻*lac*⁺. The latter genotype can be selected by plating infected cells on agar containing lactose as the sole carbon source (and, of course, leucine). The resulting colonies are Lac⁺ transductants having the phenotype Leu⁻Lac⁺.

Transduction can also be used to introduce a negative allele into a strain carrying the positive allele. The selection must be indirect, though, because a mutant carrying such an allele would be unable to form a colony on a selective agar. The method that is used is based on the linkage procedure used in genetic mapping, as described in Chapter 1. The following paragraphs show how linkage is detected and used with a transducing system.

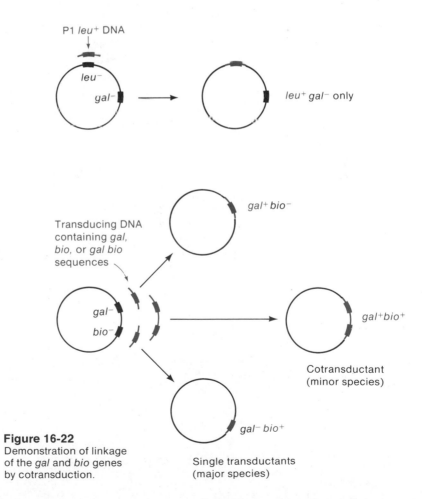

Figure 16-22
Demonstration of linkage of the *gal* and *bio* genes by cotransduction.

Consider a P1 lysate prepared by infecting a $leu^+gal^+bio^+$ bacterium. This lysate contains particles capable of transducing any of these alleles. That is, a leu^- bacterium can be converted to leu^+ or a bio^- strain to bio^+. Also, if a leu^-bio^- strain is transduced, both leu^+bio^- and leu^-bio^+ bacteria can be produced. There would not be any leu^+bio^+ strains, though, because it is exceedingly unlikely (owing to the low frequency of generalized transduction) that a single bacterium would be infected simultaneously with a leu^+ and a bio^+ transducing particle. It is a different situation with a gal^-bio^- bacterium; that is, not only are gal^+bio^- and gal^-bio^+ transductants formed but also gal^+bio^+ bacteria are produced at nearly the same frequency as transduction for a single allele. This happens because the gal and bio genes are separated by only about 15×10^6 molecular weight units of DNA. Since the molecular weight of the transducing DNA is 20–25 million, it is possible to package a single DNA fragment containing both the gal and bio genes. When this occurs, it is said that the gal and bio genes are *linked* (Figure 16-22). Of course, all gal^+ transducing particles will not be bio^+, because the cuts producing the bacterial DNA fragments that are packaged will sometimes be made between the gal and bio genes. Clearly the probability of linkage depends on how near the genes are—the closer they are, the greater is the frequency of linkage. For the gal-bio pair it is found that roughly half of the gal^+ transductants are gal^+bio^+ and half are gal^+bio^-; similarly, with bio^+ transductants there are both bio^+gal^+ and bio^+gal^- strains.

Let us now see how the linkage procedure can be used to introduce a negative allele into a strain. Consider an x^- mutant that has been very difficult to isolate; we assume that it is linked to gene b but not to gene a and that we want a strain whose genotype is a^-x^-. If the mutation b^- can be isolated with ease, the strain a^-b^- is first made. Then P1 is grown on the strain that is b^+x^- and used to transduce a^-b^- to a^-b^+. Some of these a^-b^+ strains will also be a^-x^-, since genes b and x are linked.

Properties of Specialized Transducing Particles

In the generalized transducing systems discussed so far, transducing particles form during a lytic cycle when host DNA fragments are packaged. Since fragmentation is random, all possible host sequences are represented in a heterogenous population of generalized transducing particles, if the population is large enough.

Transducing particles of a different type can also be produced during excision of an integrated prophage. These particles differ from generalized transducing particles in three ways: (1) the transducing particles contain both host and phage DNA linked in one continuous molecule; (2) only one or at most two regions of the host DNA, specifically those regions that flank the prophage, are found in these particles; and (3) a single transducing particle can serve as a template for production of a homogeneous population of identical transducing particles. These particles are called **specialized transducing particles.**

Formation of Specialized Transducing Particles from a λ Lysogen

The mechanism by which specialized transducing particles form is shown in Figure 16-23, which depicts the formation of the galactose- and biotin-transducing forms of phage λ—namely, λ*gal* and λ*bio*.

When λ prophage is induced, an orderly sequence of events ensues in which the prophage DNA is precisely excised from the host DNA. In the case of phage λ this is accomplished by the combined efforts of the *int* and *xis* genes acting on the left and right prophage attachment sites. At a very low frequency—namely, in about one cell per 10^6–10^7 cells—an excision error is made; two incorrect cuts are made—one within the prophage and the other cut in the bacterial DNA. The pair of abnormal cuts will not always yield a length of DNA that can fit in a λ phage head—it may be too large or too small. However, if the spacing between the cuts produces a molecule between 79 percent and 106 percent of the length of a normal λ phage-DNA molecule, packaging can occur. Since the prophage is located between the *E. coli gal* and *bio* genes and because the cut in the host DNA can be either to the right or the left of the prophage, transducing particles can arise carrying the *bio*

Figure 16-23
Aberrant excision leading to production of λ*gal* and λ*bio* phages.

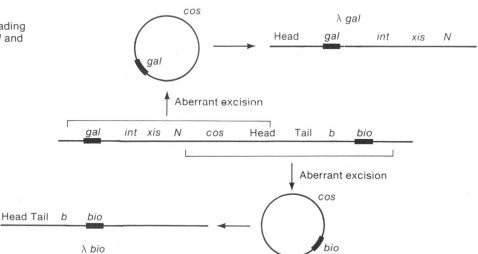

genes (cut to the right) or the *gal* genes (cut to the left). Formation of the λ*gal*- and λ*bio*-transducing particles entails loss of λ genes. The λ*gal* particle lacks the tail genes, which are located at the right end of the prophage; the λ*bio* particle lacks the *int, xis,* . . . genes from the left end of the prophage. The number of missing phage genes of course depends on the position of the cuts that generated the particle and thus correlates with the amount of bacterial DNA in the particle. The missing phage genes come from the prophage ends, but because of the permutation of the gene order in the prophage and the phage particle, the deleted phage genes are always from the central region of the phage DNA, as shown in the figure. The term deletion is usually used to describe a DNA molecule in which wild-type base sequences are absent; however, in a specialized transducing particle the missing phage genes are replaced by bacterial genes. Thus, these are called **deletion-substitution particles.**

Properties of λ*gal* and λ*bio* Particles

Since specialized transducing particles lack phage genes, they might be expected to be nonviable. This is indeed true of the λ*gal* type, which lack genes essential for synthesis of the phage tail and, in the case of the larger deletion-substitutions, the genes for the phage head also. Thus, these particles are incapable of producing progeny; they are defective and this is denoted by the symbol d—thus, a *gal*-transducing particle is written λd*gal* (and sometimes λdg). The λd*gal* particle contains the cohesive ends and all of the information for DNA replication and transcription; it therefore goes through a normal life cycle, including bacterial lysis. In fact, if the head genes are not deleted, the concatemeric branch produced by rolling circle replication is cleaved by the Ter system. However, tails are not added to the filled head, no viable particles are produced in the lysate, and plaques are not formed on a host lawn. Note that there is no discrepancy between the formation of a λd*gal* transducing particle and its ability to reproduce itself. A λd*gal* particle fails to reproduce only because it lacks the tail genes, but such a particle arises from a normal prophage having a full set of genes.

The situation is quite different with λ*bio* particles for these usually lack only nonessential genes—*int, xis,* etc. These genes are needed for the lysogenic cycle but not for the lytic cycle, so these particles are able to replicate and to form plaques. To denote this, the letter p, for plaque-forming, is added; that is, the particle is called λp*bio*.

It is possible for the *bio* substitution to extend through the essential λ gene *N;* such particles are defective and hence are now denoted λd*bio*. By genetic manipulations in which host genes between the *gal* operon and the prophage have been deleted, the *gal* genes have been moved very near to the prophage. From such a lysogen λ*gal* particles have been isolated that lack only the nonessential *b* genes located at the extreme right end of the prophage (Figure 16-23). Such λ*gal* particles are plaque-forming and are denoted λp*gal*.

Specialized Transduction of a Nonlysogen

There are several mechanisms by which a specialized transducing particle can transduce a mutant bacterium. The mechanisms are the same for the λ*gal* and λ*bio* particles, so that we will use only λ*gal* as an example. Consider a Gal⁻ cell that is infected with a lysate resulting from induction of a Gal⁺ culture lysogenic for λ. This lysate will consist overwhelmingly of the normal λ phage but will contain a tiny fraction of the λd*gal⁺* phage. Conditions are used that lead to the establishment of immunity repression; thus the cell is not killed by a lytic response. If these infected cells are plated on agar that distinguishes Gal⁺ from Gal⁻ colonies by their color (e.g., purple for Gal⁺, white for Gal⁻), purple colonies are found at a very low frequency (about 0.001 percent of the infected cells). These purple colonies are of two types (Figure 16-24).

Type I consists of nonlysogenic cells, all of which are Gal⁺. If the colony is dispersed and individual bacteria are allowed to form colonies, all colonies are purple. These contain stable *gal⁺* bacteria and have arisen as a result of two genetic exchanges, as shown in the figure.

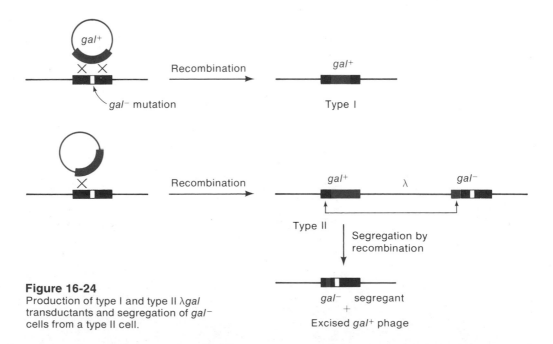

Figure 16-24
Production of type I and type II λ*gal* transductants and segregation of *gal⁻* cells from a type II cell.

Type II cells contain a prophage. If a type II colony is dispersed and individual cells are plated, about 1 percent of the colonies are white (and hence Gal$^-$) and also no longer have a prophage. Type II cells have arisen by a single exchange within the *gal* genes. These cells now contain two copies of the *gal* genes, one *gal*$^+$ and one *gal*$^-$. These transductants are called **heterogenotes.** The type I cells, which contain only one *gal*$^+$ gene, are haploids.

When there are two copies of the *gal* genes in a cell, intramolecular genetic recombination can occur with moderate efficiency. This accounts for the production of Gal$^-$ nonlysogenic cells during continued growth of a type-II *gal*$^+$-transductant, as shown in the figure.

Type II cells are not capable of being induced to produce progeny λd*gal* particles, because the prophage lacks the tail genes. However, transcription and synthesis of many phage-specific proteins do occur.

Specialized Transduction of a Lysogen

A λ lysogen can also be transduced to produce a heterogenote. This occurs at a high frequency in two ways. The first mechanism is the same as that which produces type II cells—namely, a single genetic exchange—except that the exchange can occur either in the *gal* operon or the prophage. Note also that conditions leading to the lysogenic response are unnecessary if the prophage has the same immunity repressor as the λd*gal*, because no phage development can occur. These transductants are heterogenotes both for the *gal* operon and the prophage. They are more unstable than the type II transductants described earlier because the probability of an intramolecular ex-

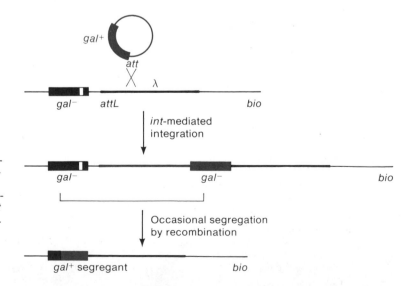

Figure 16-25

Transduction by prophage integration. Segregation by recombination in the *gal* genes occasionally occurs; this segregation can also occur at the *att* sites if some integrase is made. The infecting phage and the prophage must be heteroimmune for the insertion to occur.

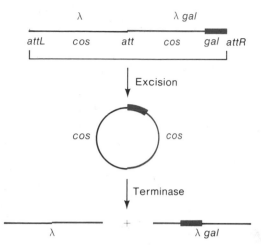

Figure 16-26
Production of λ and λ*gal* from a tandem (λ, λ*gal*) dilysogen.

change leading to production of a *gal*⁻ segregant is greater in that it can occur in both the *gal* and prophage segments of the DNA.

The second type of transduction occurs when the prophage and the λd*gal* are heteroimmune (have different repressors). In this case the λd*gal* can express the *int* gene and a single genetic exchange can occur between the attachment sites of the prophage and the infecting phage (Figure 16-25). This also produces an unstable heterogenote, as shown in the figure. (Note that this Int-mediated mechanism cannot occur with a λp*bio*, because the *bio* substitution replaces the *int* gene.)

High-Frequency-Transducing Lysates

The heteroimmune and homoimmune transductants just described are very useful because these dilysogens contain a λd*gal* prophage and a normal prophage that has functioning head and tail genes and, when induced, these genes can provide the structural components necessary for packaging λd*gal*. Thus, when such a dilysogen is induced, both normal and λd*gal* particles are produced by the mechanism shown in Figure 16-26. Roughly equal quantities of the two particles are produced. Since the DNA content of λd*gal* is generally different from that of normal λ DNA and the protein content is the same, the two particles can be separated by equilibrium centrifugation in a CsCl density gradient.*

The dilysogens just described yield lysates half of whose phage are transducing particles. In laboratory jargon such a lysate is called a high-frequency-transducing, or HFT, lysate. A lysate formed from a single lysogen in which aberrant excision occurs only infrequently is called a low-frequency-transducing, or LFT, lysate.

*The densities of DNA and protein molecules are 1.7 and 1.3 g/cm³ respectively. The density of the phage is determined by the proportion of DNA to protein.

REFERENCES

Campbell, A. 1976. "How viruses insert their DNA into the DNA of a host cell." *Scient. Amer.,* December, pp. 102–113.

Campbell, A. 1977. "Defective bacteriophages and incomplete prophages." *Comprehensive Virology,* 8, 259–328.

Devoret, R. 1979. "Bacterial test for potential carcinogens." *Scient. Amer.,* August, pp. 40–49.

Ebel-Tsipis, J., D. Botstein, and M. Fox. 1972. "Generalized transduction by phage P22 in *Salmonella typhimurium.* I. Molecular origin of transducing DNA." *J. Mol. Biol.,* 71, 433–438.

Hershey, A. D. (ed.) 1971. *The Bacteriophage Lambda.* Cold Spring Harbor Laboratory.

Herskowitz, I., and D. Hagen. 1980. "The lysis-lysogeny decision of phage λ: explicit programming and responsiveness." *Ann. Rev. Genetics,* 14, 399–446.

Howe, M. M. 1980. "The invertible G segment of phage Mu." *Cell,* 21, 605–606.

Landy, A., and W. Ross. 1977. "Viral integration and excision: structure of the lambda *att* site." *Science,* 197, 1147–1160.

Luria, S. E., J. E. Darnell, D. Baltimore, and A. Campbell. 1978. *General Virology.* Wiley.

Miller, H. I., J. Abraham, M. Benedeck, A. Campbell, D. Court, H. Echols, R. Fischer, J. M. Galinda, G. Guarneros, T. Hernandez, D. Mascarenhas, C. Montaney, D. Schindler, U. Schmiessner, and L. Sosa. 1980. "Regulation of the integration-excision reaction by bacteriophage λ." *Cold Spring Harbor Symp. Quant. Biol.,* 45, 439–446.

Miller, H. I., A. Kikuchi, H. A. Nash, and R. A. Weissberg. 1979. "Site-specific recombination of bacteriophage λ: the role of host gene products." *Cold Spring Harbor Symp. Quant. Biol.,* 43, 1121–1126.

Mizuuchi, M., and K. Mizuuchi. 1980. "Integrative recombination of bacteriophage λ: extent of the DNA sequence involved in the attachment site function." *Proc. Nat. Acad. Sci.,* 77, 3220–3224.

Nash, H. A. 1978. "Integration and excision of bacteriophage λ." *Current Topics Microbiol. Immun.,* 78, 171–199.

Nash, H. A. 1981. "Integration and excision of bacteriophage λ: the mechanism of conservative site-specific recombination." *Ann. Rev. Genetics,* 15, 143–168.

Nash, H. A., K. Mizuuchi, L. W. Enquist, and R. A. Weissberg. 1980. "Strand exchange in λ integrative recombination: genetics, biochemistry, and models." *Cold Spring Harbor Symp. Quant. Biol.,* 45, 417–428

Oppenheim, A., and A. B. Oppenheim. 1978. "Regulation of the *int* gene of bacteriophage λ: activation by the *cII* and *cIII* gene products and the roles of the *pI* and *pL* promoters." *Molec. Gen. Genetics.* 165, 39–46.

Ptashne, M., A. Jeffrey, A. D. Johnson, R. Maurer, B. T. Meyer, C. O. Pabo, T. M. Roberts, and R. T. Sauer. 1980. "How the λ repressor and Cro work." *Cell,* 19, 1–11.

Weissberg, R. A., S. Gottesman, and M. E. Gottesman. 1977. "Bacteriophage λ: the lysogenic pathway." *Comprehensive Virology,* 8, 197–258.

17 Mechanisms for Rearrangement and Exchange of Genetic Material

I. PLASMIDS

Plasmids are extrachromosomal circular DNA molecules found in most bacterial species and in some species of eukaryotes. Under normal circumstances a particular plasmid is dispensable to its host cell; for example, sometimes at the time of cell division a plasmid-free daughter cell is formed and such a cell is almost always viable. However, many plasmids contain plasmid genes that may be essential in certain environments. For example, the R plasmids carry genes that confer resistance to numerous antibiotics so that in nature a cell containing such a plasmid can survive in the presence of an antibiotic, whether humanly administered or produced by a fungus. Some plasmids, such as the F′ variants of the F plasmid, carry one or more genes originally acquired from a host chromosome, e.g., the *leu* gene; such a plasmid would be essential in a *leu⁻* cell growing in medium lacking leucine. Plasmids of this and many other types are often detected by the genes that they carry; however, sometimes a plasmid is discovered incidentally as a small circular DNA molecule present in a DNA sample isolated from a cell extract.

Plasmids have many interesting biological properties, which will be described in this chapter. However, they are also extraordinarily useful tools for the molecular biologist, as was indicated in the discussion of the *lac* operon in Chapter 14, and for the field of genetic engineering (Chapter 20).

GENERAL PROPERTIES AND TYPES OF PLASMIDS

Plasmids, like phages, are heavily dependent on the metabolic functions of the host cell for their reproduction, though each plasmid has genes and regulatory sites that distinguish one type of plasmid from another. For example, plasmids normally use most of the replication machinery of the host. However, each type of plasmid has its own genes for regulating the time of synthesis and the number of plasmid copies per cell (which may range from 1 or 2 for the **stringent** or **low-copy-number** plasmids to 10–100 for the **relaxed** or **high-copy-number** plasmids). Furthermore, the segregation of plasmid replicas into daughter bacterial cells during cell division is carefully regulated. That is, if a cell that is ready to divide contains two plasmid DNA molecules, each daughter cell receives one molecule; the failure of this process, which is indicated by the appearance of a plasmid-free cell, occurs at a frequency of only about 1 per 10^4 cells per generation. This stability is conferred by some type of plasmid-chromosome association; this is known from an experiment shown in Figure 17-1, though the nature of the association is unknown.

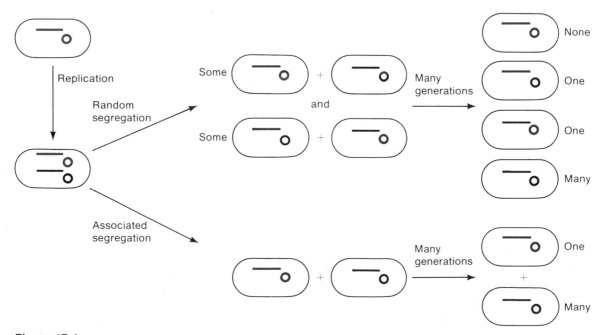

Figure 17-1

An experiment in which a bacterium containing a radioactive chromosome (red) and a plasmid is allowed to grow for many generations in nonradioactive growth medium to yield black molecules. In associated segregation all radioactivity remains within cells containing both a radioactive chromosome and plasmid. Association is observed. "None," "one," and "many" refer to the number of cells of the type shown in that generation.

In this chapter the discussion will be confined to plasmids of E. *coli* except when otherwise noted. Many types of plasmids are found in a variety of E. *coli* strains but three main types—the F, R, and Col plasmids—have been studied. These plasmids share some properties but for the most part are quite different. The presence of an F, R, or Col plasmid in a cell is indicated by the following characteristics acquired by the cell:

1. *F, the sex plasmid.* Ability to transfer chromosomal genes (that is, genes not carried on the plasmid) and the ability to transfer F itself to a cell lacking the plasmid.
2. *R, the drug-resistance plasmid.* Resistance to one or more antibiotics and usually the ability to transfer the resistance to cells lacking R.
3. *Col, the colicinogenic factor.* Ability to synthesize **colicins**—that is, proteins capable of killing closely related bacterial strains that lack the Col plasmid.

Further discussion of each of these plasmid types will be presented throughout the chapter.

Plasmids are known that carry other genes. One clinically important plasmid carries genes for synthesizing an intestinal irritant called an enterotoxin. These plasmids, called Ent, are responsible for travelers' diarrhea and some kinds of dysentery. Another plasmid confers on *Bacillus thuringiensis* the ability to synthesize a product that is toxic to gypsy moths and tent worms. Plasmids in organisms other than bacteria are described in a later section.

PLASMID DNA: PROPERTIES AND MEANS OF ISOLATION

With only a single exception (the killer-plasmid of yeast, which is an RNA molecule) all known plasmids are supercoiled circular DNA molecules (Figure 17-2). The molecular weights of the DNA range from about 10^6 for the smallest plasmid to slightly more than 10^8 for the largest one. Table 17-1 lists the molecular weights for several plasmids that are actively being studied.

Dimers and other polymers of plasmids are frequently found in plasmid-containing bacteria; these forms, which are more prevalent with the smaller plasmids, are generated by genetic recombination.

Plasmid DNA can usually be isolated from bacteria in a simple way. A culture of plasmid-containing bacteria is lysed by adding a detergent and then the lysate is centrifuged. The bacterial chromosome complex, which contains protein and RNA, is very large and compact and moves rapidly to the bottom of the centrifuge tube; the smaller plasmid DNA remains in the supernatant, which is called a **cleared lysate.** CsCl is then

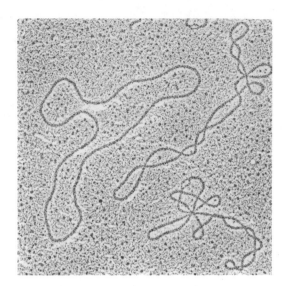

Figure 17-2
One open circular and two supercoiled forms of a plasmid DNA. The open circular form was generated by introducing a nick into a supercoil.

Table 17-1
Several plasmids and selected properties

Plasmid	Mass, $\times 10^6$	No. copies/ chromosome	Self-transmissible	Phenotypic features
Col plasmids				
ColE1	4.2	10–15	No	Colicin E1 (membrane changes)
ColE2 (*Shigella*)	5.0	10–15	No	Colicin E2 (DNase)
ColE3	5.0	10–15	No	Colicin E3 (ribosomal RNase)
Sex plasmids				
F	62	1–2	Yes	F pilus
F'*lac*	95	1–2	Yes	F pilus; *lac* operon
R plasmids				
R100	70	1–2	Yes	Cam+ Str+ Sul+ Tet+
R64	78	(limited)	Yes	Tet+ Str+
R6K	25	12	Yes	Amp+ Str+
pSC101	5.8	1–2	No	Tet+
Phage plasmid				
λdv	4.2	≈50	No	λ genes *cro, cI, O, P*
Recombinant plasmids				
pDM500	9.8	≈20	No	*Drosophila melanogaster* histone genes
pBR322	2.9	≈20	No	High-copy-number
pBR345	0.7	≈20	No	ColE1-type replication
Yeast plasmid				
2μm	4.0	≈60	No	No known genes

added to the lysate and the lysate is centrifuged to equilibrium. Some chromosomal DNA is usually present in the lysate but if the (G+C) content of the plasmid and the chromosomal DNA differ, the two types of molecules form distinct bands that can be separately removed. In the more common case there is no difference in (G+C) content and one must make use of the fact that the plasmid DNA is supercoiled. Then, the CsCl centrifugation step is performed with ethidium bromide in the lysate in order to introduce a difference in the densities of the supercoil and the linear chromosomal fragments (Chapter 4). The supercoiled DNA has a higher density. The only disadvantage of this technique, which is the one most commonly used, is that both nicked circles, which probably form accidentally during isolation, and nonsupercoiled replicating molecules, are discarded with the chromosomal DNA.

The presence of plasmid DNA in a single bacterial colony can be detected in an elegant way by electrophoresis (Figure 17-3). A single

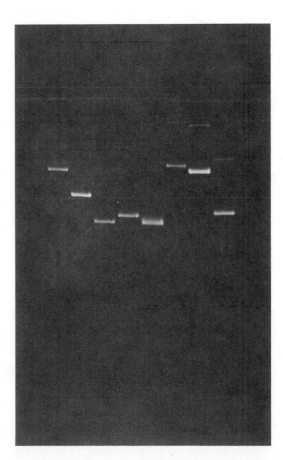

Figure 17-3
A gel electrophoregram showing the migration of eight different plasmids. Movement is from top to bottom. Each vertical column ("lane") represents a single plasmid. After migration the gel is soaked in a solution of ethidium bromide, washed, and then illuminated with near-ultraviolet light. The ethidium bromide, which is bound to the DNA, fluoresces. The single intense band in each lane contains supercoiled DNA molecules; the more slowly moving faint band contains either nicked circles or supercoiled dimers. The apparatus used for gel electrophoresis is shown in Figure 3-14. (Courtesy of Elaine Cocuzzo and Pieter Wensink.)

colony is taken and lysed and then subjected to gel electrophoresis. The bacterial chromosome is very large and cannot penetrate the gel (though some fragments do) but the plasmid DNA can do so. The rate of electrophoretic movement of DNA molecules through a gel increases with decreasing molecular weight so that a plasmid DNA, if present, will form a narrow band at a position in the gel characteristic of its molecular weight. The band is visualized by staining the gel with ethidium bromide, which binds tightly to the DNA and fluoresces upon irradiation with ultraviolet light. From the distance moved in a particular time interval relative to that for plasmids of known molecular weight, the molecular weight of the plasmid DNA can be calculated, as shown in the figure. This screening technique is very useful in recombinant DNA technology, as will be seen in Chapter 20. Plasmid DNA can also be detected in a cleared lysate (see above) by this electrophoretic technique.

The isolation of DNA usually entails a deproteinization step. When the DNA molecules of some E. coli plasmids are isolated without such a step, about half of the supercoiled DNA molecules contain three tightly bound protein molecules. This DNA-protein complex is called a **relaxation complex.** If this complex is heated or treated with alkali, or with proteolytic enzymes, or with detergents, one of these proteins, which is a nuclease, nicks one DNA strand, thereby "relaxing" the supercoil to the nicked circular form (Figure 17-4). This nick occurs in only one strand and at a unique site. During relaxation the two smallest proteins are released but the largest protein becomes covalently linked to the 5'-P terminus of the nick. If, prior to relaxation, the supercoiled DNA is nicked by any of a number of laboratory techniques, the relaxation nuclease is unable to make its site-specific nick, indicating that the nuclease is active only on supercoiled DNA. This nicking is thought to play a role in conjugational transfer of the plasmid to another bacterium (see next section).

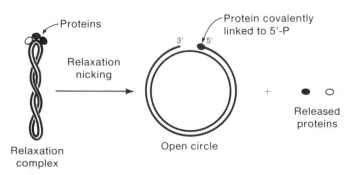

Figure 17-4
Nicking of one strand of a supercoiled DNA relaxation complex.

TRANSFER OF PLASMID DNA

Many plasmids can transfer a replica of the plasmid from a donor cell to a recipient cell. For example, if a single *E. coli* cell containing the F plasmid is added to a culture of growing F^- cells, after several generations of cell growth, a large fraction of the cells contain F. This is a result of transfer of a replica of F from an F^+ cell to an F^- cell without loss of F by the F^+ cell. After transfer, the F replica remaining in the original cell can replicate again, and another replica can be transferred to a second F^- cell. Every recipient acquires F and therefore becomes a donor from which F can be transferred to other F^- cells. Transfer can occur once or twice per cell generation, so F quickly spreads throughout a bacterial population. The ability to conjugate is an example of a successful evolutionary strategy for the plasmid. During bacterial conjugation plasmid genes are replicated while host genes are not; this replication provides a clear advantage for dissemination of the plasmid, which is increased by the ability of plasmids to transfer and be stably inherited by many host species.

An Outline of the Transfer Process and Some Definitions

The plasmid transfer process can be divided into four stages:

1. Formation of specific donor-recipient pairs (**effective contact**).
2. Preparation for DNA transfer (**mobilization**).
3. DNA transfer.
4. Formation of a replicative functional plasmid in the recipient (**repliconation**).

Many plasmids fail to carry out all of these processes, so the following terminology has been developed:

A **conjugative plasmid** is defined as a plasmid carrying genes that determine the effective contact function.
Mobilizable plasmids are those that can prepare their DNA for transfer.
A **self-transmissible plasmid,** such as F, is both conjugative and mobilizable.

Other plasmids, such as ColE1, are mobilizable but nonconjugative. No naturally occurring plasmid is conjugative and nonmobilizable but mutant plasmids may have this property.

A cell may contain two different plasmids—for example, F and ColE1. Since F is both conjugative and mobilizable (it is self-transmissible), F can provide the missing conjugative function to ColE1, which is mobilizable, and ColE1 can be transferred. This process—namely, one in which a nonconjugative plasmid is transferred via the effective contact provided by a conjugative plasmid—is called **donation;** its hallmark is efficient transfer of the nonconjugative plasmid. A conjugative plasmid can also help a nonmobilizable plasmid to be transferred. This process, which occurs at low frequency, occurs via genetic recombination between the two plasmids to form a single transferable DNA molecule; it is called **conduction,** in contrast with donation. The frequency (high versus low) of helped transfer of a non-self-transmissible plasmid is a useful criterion for determining whether a plasmid is mobilizable by donation or conduction.

F, whose molecular weight is about 63×10^6, contains at least 19 genes needed for transfer; these are called *tra* genes. A mutation in any one of these genes eliminates self-transmissibility so presumably this is the minimal number needed for any self-transmissible plasmid. Since this number of genes requires about 15×10^6 molecular weight units of DNA (assuming an average molecular weight of the gene products of 40,000), the small plasmids are not self-transmissible. The smallest known self-transmissible plasmid has a molecular weight of 17×10^6. F will serve as a model for the study of the transfer process.

A Detailed Look at Transfer of F

The first step in effective contact is pair formation between a donor and a recipient cell. This requires a hairlike protein appendage, called a **sex pilus,** on the donor cell (Figure 17-5). The pili on F- and R-containing cells are called **F pili** and **R pili,** respectively. The pilus is a hollow tube (made of a single protein called **pilin**), the tip of which comes in contact with the recipient cell. Twelve genes, of which only one encodes pilin, are needed for synthesis of a pilus. Transfer of DNA through the pilus has been proposed but there is no good evidence to support this notion. Pili retract into the donor cell after pairing, so the pilus probably serves first to bring the pair into initial contact and then to draw the cells together into close contact.

Mobilization begins when a plasmid-encoded protein, which is probably the nicking protein of the relaxation complex, makes a single-strand break in a unique base sequence called the **transfer origin** or *oriT*. (Relaxation complexes have not been detected for all plasmids but presumably there is a protein serving the function just described.) This nick initiates rolling circle replication and the linear branch of the rolling circle is transferred. It is thought that the nicking protein remains bound to the 5′ terminus and that the replication mode is like the looped rolling-circle mechanism used by phage ϕX174 (Chapter 8, Figure 8-40). The sequence of events during transfer is shown schematically in Figure 17-6.

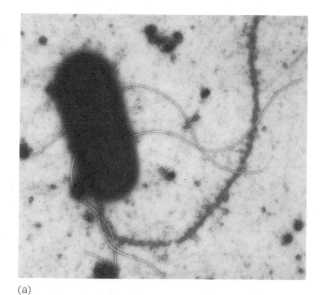

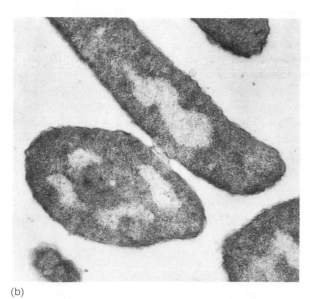

(a)

(b)

Figure 17-5

(a) An *E. coli* cell showing a single sex pilus, which is coated with the F-specific phage R17 to make the very thin pilus visible as a rough dark appendage. The five heavy bright fibers are flagellae. The very faint thin hairs are called fimbrae. (Courtesy of Barry Eisenstein.)

(b) Electron micrograph of two *E. coli* cells during conjugation. The small cell is an *F⁻* cell; the larger cell contains F'*lac.* (Courtesy of Lucien Caro.)

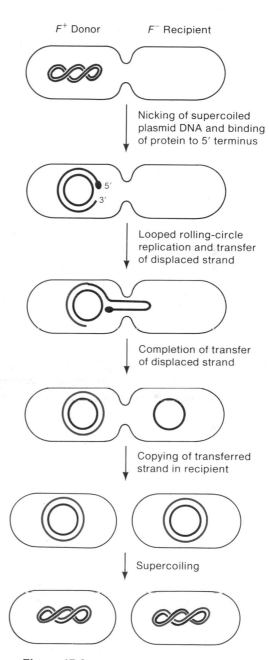

Figure 17-6

A model for transfer of F plasmid DNA from an *F⁺* cell by a looped rolling-circle mechanism. The displaced single strand is transferred to the *F⁻* recipient cell, where it is converted to double-stranded DNA. Chromosomal DNA is omitted for clarity.

DNA synthesis occurs both in donor and recipient cells. The synthesis in the donor, called **donor conjugal DNA synthesis,** serves to replace the single strand that is transferred. Synthesis in the recipient cell **(recipient conjugal DNA synthesis)** converts the transferred single strand to double-stranded DNA. Since the DNA is transferred as a linear molecule and the established plasmid is a circle, some mechanism is needed for recircularization; however, this mechanism is not known.

In the usual situation during transfer the transferred strand is simultaneously replaced by donor conjugal synthesis and converted to double-stranded DNA in the recipient. This would indicate that DNA synthesis and transfer are coupled were it not for the following observations: (1) Transfer occurs even if donor conjugal synthesis is inhibited by appropriate host mutations; (2) donor conjugal synthesis occurs even if transfer is prevented by appropriate plasmid mutations; (3) transfer occurs even though recipient conjugal synthesis is inhibited by mutations in the recipient cell. These observations raise the question of the identity of the motive force for transfer since clearly it is not DNA replication. The question of the driving force for transfer is actually quite subtle because it involves the driving force for unwinding a double helix. In Chapter 8 it was emphasized that DNA polymerase III cannot unwind a helix; consequently, without help, polymerase III alone cannot replicate a double-stranded DNA molecule. Unwinding requires an accessory protein, a helicase. Thus, when the transfer process is set in motion, a helicase must be activated in order to unwind the donor DNA molecule and allow a strand to be transferred. Replication cannot occur without helicase activity but a helicase might be able to act without replication. Therefore, it has been proposed that initiation of transfer requires two events—a nick at the transfer origin and helicase activation. The helicase would then unwind the donor DNA molecule and a single strand would be transferred. Unwinding by a helicase would cause a change in the degree of supercoiling of the plasmid. In Chapter 8, it was pointed out that this geometric constraint is relieved by the action of DNA gyrase, so it might be expected that transfer would be prevented if either helicase or gyrase activity were inhibited. The drug nalidixic acid is a known inhibitor of *E. coli* DNA gyrase and does in fact prevent DNA transfer when it is added to the bacterial growth medium. This supports (but by no means proves) the hypothesis just presented. Unfortunately, no inhibitor of helicase activity is known.

Transformation by Plasmid DNA

In the preceding section the means of natural transfer of plasmid DNA was described—namely, cell-to-cell contact. Plasmid DNA can also be transferred in the laboratory by mixing free plasmid DNA with appropriately treated recipient cells—that is, by transformation. If bacteria are placed in a solution of cold $CaCl_2$, they acquire the ability to take in DNA molecules. This makes it possible to transfer any plasmid that can be isolated to almost any *E. coli* strain. The two requirements are (1) that the recipient bacterium possess replication enzymes that are active on the plasmid, which is invariably the case if donor and recipients are of the same species, and (2) that the host modification-restriction system (Chapter 15) of both the donor and recipient cells should not differ. Currently the $CaCl_2$-transformation procedure is the most common and most important experimental technique for transferring a nontransmissible plasmid from one strain to another.

PLASMID REPLICATION

A plasmid can only replicate within a host cell; one might therefore expect that all plasmids native to the same host species would have the same mode of replication. However, just as with phages, which also replicate only within a host cell, there is enormous variation in both the enzymology and mechanics of plasmid DNA replication. This is explained in the following two sections.

Variations in the Use of Host Proteins

Plasmids rely heavily on the host replication apparatus for their replication. In fact, some plasmids seem to use host gene products exclusively. For example, the small plasmid, ColE1, can be replicated *in vitro* by adding purified ColE1 DNA to a cell extract prepared from cells that do not contain ColE1 or any other plasmid and hence do not contain any plasmid-encoded gene product. (This result shows that ColE1 DNA can replicate without plasmid proteins but of course does not eliminate the possibility that plasmid proteins are used *in vivo*.) Other plasmids require some plasmid gene products. For instance, temperature-sensitive mutants of F are known that fail to replicate at 42°C; however, since bacterial DNA is synthesized normally at this temperature, cells containing the temperature-sensitive plasmid produce F^- daughter cells after several generations of growth at 42°C. Many plasmids do not require all of the host gene products. This may result either from a similar product being encoded in a plasmid gene or possibly from there being a distinct mode of replication. For example, F is able to replicate at high temperatures in an *E. coli dnaA*(Ts) mutant that fails to replicate chromosomal DNA at 42°C; therefore, replication of F does not need the *dnaA* gene.

In addition, some plasmids need host products that the host also needs but in a less stringent way. For example, bacterial mutants in which the activity of polymerase I has been reduced grow normally but are unable to support the replication of the ColE1 plasmid.

Variations in the Mechanics of Replication

All plasmids examined to date replicate semiconservatively and maintain circularity throughout the replication cycle. However, there are significant differences in the replication pattern from one plasmid to the next, as has been seen in phage systems.

Some of the variants are listed below:

1. *Directionality.* Both purely unidirectional and purely bidirectional replication have been observed. In addition, there are plasmids in which both modes are present. For example, the plasmid RK6 replicates first in one direction and then later in the opposite direction from the same origin.
2. *Termination.* In unidirectionally replicating plasmids termination necessarily occurs at the origin after one cycle of replication. Bidirectionally replicating molecules are of two types though, both of which have been observed. In one type, termi-

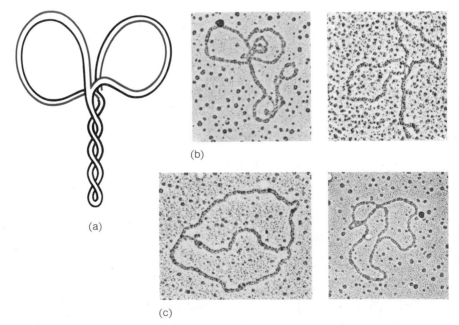

Figure 17-7
Replication of ColE1 DNA.
(a) Diagram of a butterfly molecule. The newly synthesized strands are red.
(b) Electron micrographs of butterfly molecules.
(c) Nicked butterfly molecules, showing that a nick converts a butterfly molecule to a θ molecule. (Courtesy of Donald Helinski.)

(a)

(b)

(c)

nation occurs when the growing forks reach the same region, as with phage λ (Chapter 8). Others have a fixed termination site that is sometimes reached by one growing fork before the other fork reaches it. The signal for termination is unknown.

3. *Replicating form.* In the most carefully studied plasmids, replication occurs by the so-called *butterfly* mode, first observed for animal viruses (Figure 17-7). In a partially replicated molecule the replicated portions are untwisted, as is usually the case in θ replication, but the unreplicated portion is supercoiled. When the replication cycle is completed, one of the circles must be cleaved (possibly by DNA gyrase). The result after one round of replication is one nicked molecule and one supercoiled molecule. The nicked molecule is then sealed and, somewhat later, supercoiled. Whether this is a general mechanism for plasmid replication is not known.

Control of Copy Number

Some plasmids are present in cells in low-copy number—one or a few per cell—whereas others exist in large numbers—from 10 to 100 per cell. The following simple experiment indicates that the copy number is established and regulated by controlling the rate of initiation of DNA synthesis.

If a cell is transformed with a single copy of a low-copy-number plasmid, the plasmid DNA replicates only once or twice before cell division. However, if a cell is transformed with a single DNA molecule of a high-copy-number plasmid, the plasmid DNA replicates repeatedly until the proper copy number is reached.

The generally accepted explanation for the regulation of copy number is that there is a plasmid-encoded repressor that binds to a plasmid operator and thereby inhibits replication. Let us first see how this idea explains maintenance of copy number from generation to generation. In the current theory it is assumed that the repressor is active only as a multisubunit protein and that the monomer-multimer equilibrium is very dependent on concentration. Thus, as a cell grows (enlarges), the repressor concentration drops and inactive monomers form; replication is thereby derepressed, and the number of plasmid DNA molecules doubles. At this point there will exist twice the initial number of repressor genes; therefore, by protein synthesis the repressor concentration also doubles. This results in the formation of active repressor and replication stops. A similar sequence of events would occur if there were initially only one copy of a high-copy-number plasmid—that is, replication would continue until there is sufficient repressor to turn off synthesis. How can this explanation account for differences in copy number? The most likely possibility is that for the high-copy-number plasmids the association of monomers to form a multimeric active repressor requires a higher concentration of monomers than is the case for the low-copy-number plasmids. Thus, only when the number of plasmids per cell is high is the "gene dosage" high enough to produce active repressor. The following experiment supports this view by providing evidence that each plasmid controls its own copy number.

A hybrid plasmid pSC134 was constructed (by recombinant DNA techniques; see Chapter 20) that consists of a complete copy of each of two plasmids, ColE1 and pSC101. The copy numbers for these plasmids are 18 and 6, respectively. The following three facts about pSC134 were obtained (see Figure 17-8):

1. Plasmid pSC134 replicates from the ColE1 origin and has a copy number of 16—roughly equal to that for ColE1.
2. If pSC134 is put into a *polA*⁻ cell, in which ColE1 cannot replicate, the pSC101 origin is used and the copy number becomes 6, the value for pSC101. These two results show that the copy number correlates with the replication origin that is being used.
3. If pSC101 DNA is taken into a bacterium containing 16 copies of pSC134, the pSC101 cannot replicate. This lack of replication shows that the pSC101 repressor is being made by pSC134.

The interpretation of these three results is the following. If there were a copy number less than 6, both pSC101 and ColE1 origins would be active and replication from both origins would increase the number. If the number were greater than 6, the pSC101 origin would be inactive because the pSC101 repressor concentration would be above the inhibitory level. Synthesis of pSC101 would continue, though if it were autogenously regulated, the concentration would not exceed that produced by a copy number of 6. The ColE1 origin would remain active, for it would only be shut off if the copy number were to exceed 16. In a *polA*⁻ cell ColE1 cannot replicate; consequently, the copy number is totally controlled by the concentration of pSC101 repressor, and thus it will not exceed its normal value. This interpretation is completely consistent with the facts, as are several other types of experiments that also support the repressor model of copy number regulation.

The following important point must be understood about the repressor model. Since all of the plasmids in a particular cell are identical, a repressor molecule cannot distinguish one molecule from another. Thus, when the concentration of active repressor is low enough that all DNA molecules are not repressed, the few that are repressed are chosen randomly from the population. This means that if one plasmid replicates to form two daughter plasmids, an individual daughter plasmid has the same probability of replicating a second time as any other molecule has of replicating (a first time). That is, at any

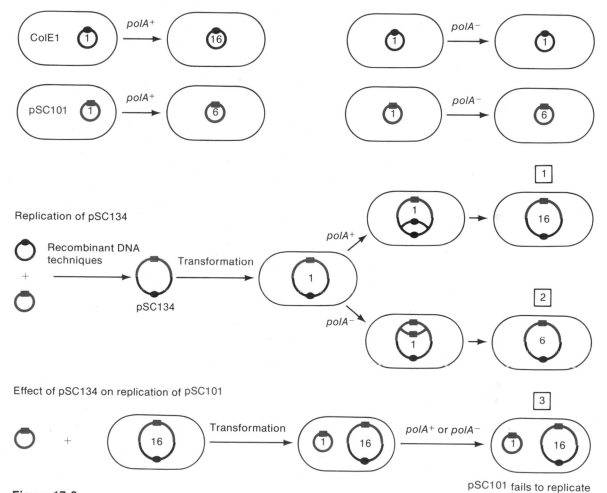

Figure 17-8
Diagram depicting the replication of ColE1, pSC101, and the hybrid plasmid pSC134, starting with one copy per cell. The black solid circle and the red solid square designate the replication origins of ColE1 and pSC101, respectively. The numbers (1, 6, and 16) in the DNA molecules indicate the number of copies in the cell. The boxed numbers (1, 2, and 3) refer to similarly numbered items in the text.

instant, a molecule is chosen for replication by random selection from the entire population of plasmids. This phenomenon has been demonstrated experimentally in a study of the replication of bacteria containing ^{14}N-labeled plasmids in ^{15}N-containing growth medium. If such a population of plasmids is allowed to replicate for several cycles, some $^{15}N^{15}N$-labeled plasmids (which have gone through two cycles of replication) are observed before all $^{14}N^{14}N$-labeled plasmids have been converted to the hybrid $^{14}N^{15}N$ form, which has had only a single round of replication.

Incompatibility

Pairs of closely related plasmids usually cannot be stably maintained in the progeny of a single cell; such plasmids are said to be **incompatible.** The repressor model for initiation of plasmid replication explains this phenomenon.

Let us first consider a cell that contains two plasmids, say, F and ColE1, having different repressors. Replication of each type of plasmid will proceed independently of one another since the repressor of one type (e.g., F) does not regulate the replication of the other type (e.g., ColE1). Thus, F and ColE1 are compatible. Alternatively, one says that they belong to different **incompatibility groups.**

The situation is quite different with two plasmids A and B whose repressors are either identical or are similar enough that the repressor of A can regulate replication of B and vice versa. Let us consider a cell

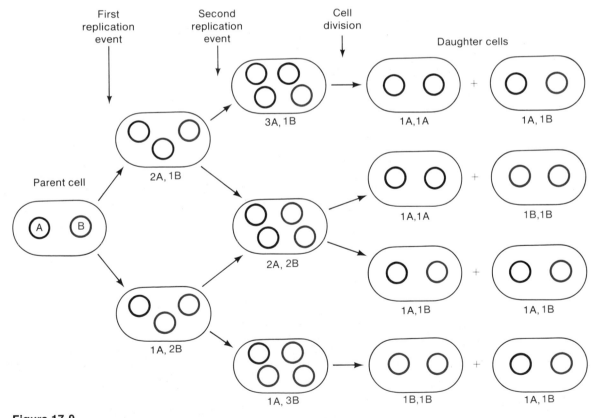

Figure 17-9
Four possible pairs of daughter cells that can arise from division of a cell containing one copy each of the incompatible plasmids A and B. See text for details.

having one copy of A and one of B and which has enlarged sufficiently that derepression is occurring (Figure 17-9). Since the two plasmid copies are selected at random for replication, the result of the first replication event is a cell having either one copy of A and two of B (1A,2B) or one of B and two of A (2A,1B). When the second replication event occurs, each cell will have four plasmids, but depending on the plasmid that is replicated, the plasmid composition may be (1A,3B), (2A,2B), or (3A,1B), as shown in the figure. At this point the cell, which has twice the initial number of plasmids, can divide. The plasmid composition of the two daughter cells will be one of the following:

> (1A,3B) becomes (1A,1B) + (1B,1B).
> (2A,2B) becomes (1A,1A) + (1B,1B) or (1A,1B) + (1A,1B).
> (3A,1B) becomes (1A,1A) + (1A,1B).

Note that two possible types of cells, namely (1A,1A) and (1B,1B), contain only one of the two kinds of plasmids; daughter cells obtained from these cells will of course continue to have only one kind of plasmid. Each cell still containing one A and one B will, in subsequent cell divisions, produce daughter cells lacking one of the plasmids, with 50 percent probability. Thus, as a single cell initially containing two incompatible plasmids divides, the percentage of the progeny population containing only one plasmid type will increase with each generation.

Thus, incompatibility is a result of (1) two plasmids having a similar repressor and (2) the random selection of plasmids for DNA replication.

Replication Inhibition

Plasmid replication is inhibited by various intercalating agents, particularly acridines (e.g., proflavin, acridine orange) without inhibiting chromosomal DNA replication. Such inhibition can lead to loss of the plasmid (acridine curing). The phenomenon, which is detected most easily when the plasmid contains host genes, is illustrated in Figure 17-10. If a culture of a bacterium whose genotype is F'lac^+/lac^- (that is, the chromosome is lac^- while F carries the lac^+ allele) is grown in acridine orange-containing medium, cells continue to grow and divide. However, after one generation the number of Lac$^+$ cells remains constant and the number of Lac$^-$ cells increases. The explanation is the following. At the time of adding acridine orange, most cells contain two copies of F'lac. After one generation each cell contains only one F'lac because plasmid replication is inhibited. Thus, in the next cell division only one plasmid is available for the two daughter cells; as a consequence, plasmid-free bacteria are produced.

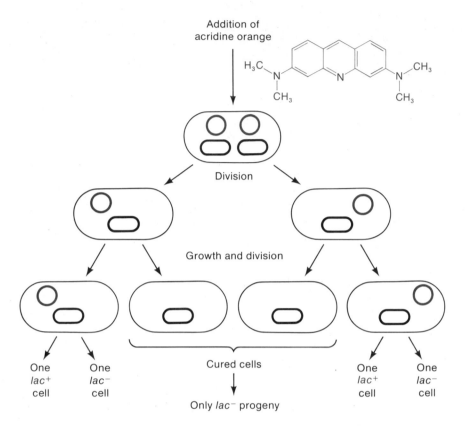

Figure 17-10
Curing of a cell containing F'*lac*+ (red circle) by growth in a medium containing acridine orange.

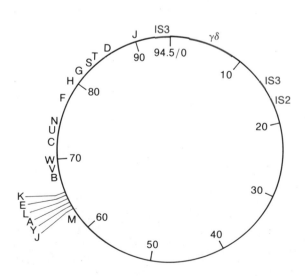

Figure 17-11
A map of the sex plasmid F. Points are given in kilobase pairs. The single capital letters refer to the midpoints of the locations of the corresponding *tra* genes. The sequences $\gamma\delta$, IS2, and IS3 are shown in red.

PROPERTIES OF PARTICULAR BACTERIAL PLASMIDS

In this section we examine the properties of the F, R, and Col plasmids in detail and survey a few other plasmids.

The Sex Plasmid F

F is a circular DNA molecule having a molecular weight of 62.5×10^6 (94.5 kilobases). It has been extensively mapped genetically and physically. Of particular interest are the sequences IS2, IS3, and $\gamma\delta$ shown in Figure 17-11. These regions are called transposons and will be discussed later in this chapter and in Chapter 19.

An important property of F is its ability to integrate into the bacterial chromosome to generate an Hfr cell (Chapter 1). Integration is a reciprocal exchange much like that occurring when phage λ lysogenizes a bacterium (see Figure 16-7). Integration of F into the E. *coli* chromosome has the following characteristics that differ from λ prophage formation: (1) The exchange site in F is not unique; whereas exchange occurs predominantly in the IS3 element located near 94.5 on the physical map, exchange also occurs in the other IS3 element and in $\gamma\delta$, as shown in Figure 17-12. (2) There are many sites in the chromosome at which integration can occur and more than 20 major sites are known (Figure 17-13). The affinity of F for each site is not the same in that the frequency of formation of a particular Hfr strain varies from one site to the next. Each of these sites is believed to be a copy of IS2, IS3, or $\gamma\delta$ in the chromosome, and the affinity represents the probability of exchange with these elements. (3) F can integrate in both clockwise and counterclockwise orientations (Figure 17-13). Since transfer of F to a female cell occurs from a single origin and in a single direction, there are Hfr strains that transfer the chromosome in either direction. In fact, there are a few instances in which F can integrate in either orientation at a single site.

Integration of F into certain bacterial *dna*⁻ mutants gives rise to a phenomenon called **integrative suppression.** As we have mentioned earlier, F does not need the E. *coli* *dnaA* gene product, which is necessary to initiate chromosomal DNA replication, to initiate its own replication. That is, if F, as a free plasmid, is contained in a *dnaA*(Ts) mutant and the temperature is raised to 42°C, initiation of chromosomal replication is no longer possible but F can still replicate. In contrast, in an Hfr strain in which F is integrated, which is also a *dnaA*(Ts) mutant, chromosomal replication can occur indefinitely at the high temperature. However, replication is not initiated at the replication origin of the chromosome but instead is initiated at the *oriT* site for F replication. Thus, integration of F suppresses the DnaA(Ts) phenotype. (Interest-

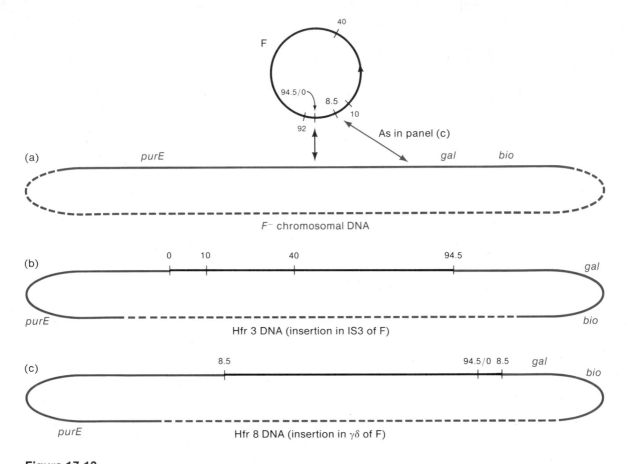

Figure 17-12

Formation of two different Hfr strains. (a) The *F⁻* chromosome and F. The exchange in both molecules occurs at the points indicated by the black double-headed arrow. (b) The result of the exchange shown in panel (a). (c) The result of the exchange indicated in panel (a) by the double-headed red arrow.

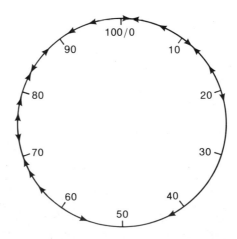

Figure 17-13

E. coli genetic map showing twenty different origins for Hfr transfer. The arrows show the direction of transfer.

ingly, at low temperature initiation occurs primarily at the normal chromosomal origin; why the F origin is virtually inactive is not known.)

Excision of F also occurs, though this is quite rare. Often excision is imperfect and one cut is made at one of the two termini within the integrated F and a second cut is in the adjacent chromosomal DNA (Figure 17-14). This process is akin to the production of specialized transducing particles from λ lysogens (Chapter 16). The F plasmids containing chromosomal genes are called **F′ plasmids.** The terminology used to describe F′ plasmids and their genetic properties are described in Chapters 1 and 14.

F′ plasmids have been very valuable tools in molecular biological research in the following ways: (1) They are used in dominance studies

Figure 17-14
Formation of various F′ plasmids by aberrant excision from a particular Hfr strain. Plasmids I and IV have lost F genes and hence are defective. If the plasmids are replication-defective, they cannot be maintained and hence will not be detected. The usual means of detection of F′ plasmids is by the presence of genes, normally transferred late by an Hfr cell, at a time sufficiently early that the genes could not have been transferred by the Hfr cell. Thus, I and IV are normally not detected because the defects in these plasmids are defects in transfer; similarly, a type II plasmid will not be found because it only contains early genes.

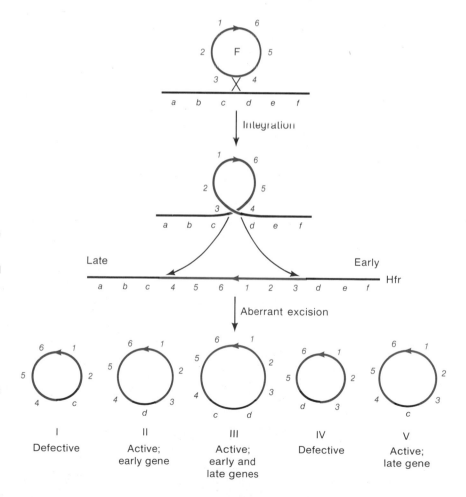

done with partial diploids, as was seen in the analysis of the *lac* operon (Chapter 14). (2) Before the advent of the recombinant DNA technology (Chapter 20), F′ plasmids provided a source of DNA of a particular chromosomal gene; it was even possible by special techniques to isolate the particular gene. (3) They were used to study prophage integration. For instance, the first physical evidence for linear integration of λ DNA was an experiment in which a strain containing F′*att*λ was lysogenized; it was shown that the molecular weight of F′*att*λ increased by the weight of one λ DNA molecule and that the DNA of the prophage-containing F′*att*λ was still a continuous circle; in this experiment, evidence was also presented for the existence of dilysogens containing two adjacent prophages.

The Drug-Resistance (R) Plasmids

The drug-resistance, or R, plasmids, were originally isolated from the bacterium *Shigella dysenteriae* during an outbreak of dysentery in Japan and have since been found in *E. coli* and various species of *Salmonella, Vibrio, Bacillus, Pseudomonas,* and *Staphylococcus.* Their defining characteristics are that they confer resistance on their host cell to a variety of fungal antibiotics and are usually self-transmissible. Most R plasmids consist of two contiguous segments of DNA (Figure 17-15). One of these segments is called **RTF (resistance transfer factor);** it carries genes regulating DNA replication and copy number, the transfer genes, and sometimes the gene for tetracycline resistance *(tet),* and has a molecular weight of 11×10^6. The other segment, sometimes called the **r determinant,** is variable in size (from a few million to more than 100×10^6 molecular weight units) and carries other genes for antibiotic resistance. Resistance to the drugs penicillin (Pen), ampicillin (Amp), chloramphenicol (Cam), streptomycin (Str), kanamycin (Kan), and

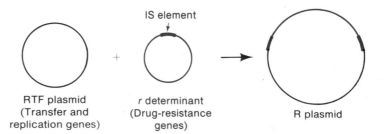

Figure 17-15
The components of an infectious R plasmid. In Chapter 18 we will describe the mechanism for formation of two IS elements in the R plasmid from the single one present in the *r*-determinant.

sulfonamide (Sul), in combinations of one or more, appears commonly. Small drug-resistance plasmids, lacking the ability to transfer but still containing the *tet* gene, are also known; one of these, pSC101, whose molecular weight is 5.8×10^6, is commonly used in the recombinant DNA technology (Chapter 20). The two-component R plasmids are reminiscent of F′ plasmids, but the drug-resistance genes are not acquired by integration of RTF followed by aberrant excision; instead, they result from acquisition of transposons, as will be seen in Chapter 19.

Small drug-resistance plasmids, not usually called R plasmids though they are quite similar, are also found in Gram-positive bacteria. One of these, the penicillinase plasmid of *Staphylococcus aureus,* has been studied in detail and is discussed in several references given at the end of this chapter.

The Colicinogenic or Col Plasmids

Col plasmids are *E. coli* plasmids able to produce colicins, proteins that are capable of preventing growth of a bacterial strain that does not contain a Col plasmid.* There are many types of colicins, each designated by a letter (e.g., colicin B) and each having a particular mode of inhibition of sensitive cells (Table 17-2). Colicin production is detected by an assay similar to that for detecting phage. A colicin-producing cell is placed on a lawn of sensitive cells; the colicin inhibits growth of nearby bacteria, producing a clear area, known as a **lacuna,** in the turbid layer of bacteria.

Table 17-2
Properties of several *E. coli* colicins

Colicin	Action of colicin
Colicin B Colicin Ib	Damages cytoplasmic membrane
Colicin E1 Colicin K	Uncouples energy-dependent processes by an unknown effect on the cell membrane
Colicin E2	Degrades DNA
Colicin E3	Cleaves 16S rRNA

*The Col plasmids are one class of a general type of plasmid called a bacteriocinogenic plasmid, which produce bacteriocins in many bacterial species. Bacteriocins, of which colicins are one example, are proteins that bind to the cell wall of a sensitive bacterium and inhibit one or more essential processes such as replication, transcription, translation, or energy metabolism.

Colicin production is inducible and the inducing agents are those that induce phage production from a lysogen—for example, ultraviolet light. As in prophage induction, the colicin producer is killed but the mechanism of this killing is unknown. A small fraction of every population of cells containing a Col plasmid also produces colicin constitutively. Thus, the presence of the Col plasmid has survival value for the population, which is thereby able in nature to compete with colicin-sensitive cells more effectively for a common food supply. There is, of course, no selective value for the individual cells that produce the colicin, as they die in the process.

Colicins are probably of two types—true colicins and defective phage particles. The latter class is inferred from studies of many purified colicins. Only a few colicins are simple monomeric proteins; others look like phage tails when examined by electron microscopy and are thought to be gene products transcribed from remnants of ancient prophages. The hypothesis is that repeated mutation has resulted in loss of the genes for replication, head production, and lysis, but genes encoding a repressor system and the tail proteins have survived intact. Presumably these phages shared with T4 the property that adsorption without DNA injection causes an inhibition of macromolecular synthesis.

The Col plasmids range in size from a molecular weight of a few million, for those that are not self-transmissible, to more than 60×10^6, for the self-transmissible plasmids.* The best-studied Col plasmid is ColE1, whose molecular weight is 4.2×10^6. It is used extensively in recombinant DNA research (Chapter 20) and in an *in vitro* DNA replication system.

ColE1 is a mobilizable but nonconjugative plasmid and has been used in genetic studies to demonstrate that transfer from a ColE1 donor cell to a recipient cell requires a plasmid-encoded nuclease, encoded by the *mob* gene, and a specific base sequence called *bom* (basis of mobility). Evidence for this conclusion comes from an experiment shown in Figure 17-16, in which the compatible plasmids F and ColE1 cohabit a single bacterium and F donates the conjugal functions that ColE1 lacks, thereby enabling ColE1 to be transferred. Panel (a) shows the state of ColE1 in an F^- cell. The *mob* gene is transcribed, the *mob* product nicks the *bom* site, and the ColE1 supercoil is converted to a nicked circle. Transfer cannot occur because ColE1 lacks the ability to form pili and hence to form conjugate pairs. In the absence of transfer, the process may be reversed so that there is an equilibrium between the supercoiled and nicked form. In panel (b) the cell contains both F and ColE1. F causes synthesis of the pilus and the transfer apparatus; and

*Some of the larger plasmids, such as ColVK, are actually F plasmids that carry genes for colicin production.

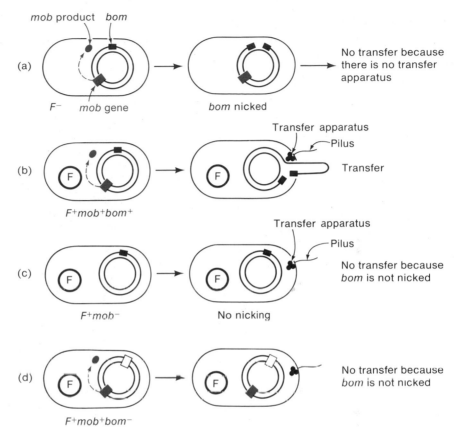

Figure 17-16
Conditions in which F can mobilize a ColE1 plasmid. The *mob* gene and the *mob* product are drawn in red. The *bom* site in the DNA is drawn as a solid box, when functional, and as an open box, when mutant or deleted. Transfer only occurs when ColE1 makes an active *mob* product that acts on a functional *bom* site and F provides the transfer apparatus.

ColE1 is transferred. Mutants (*mob⁻*) of ColE1 that fail to be transferred when F donates the transfer apparatus have been isolated. Genetic analysis shows that the *mob⁻* mutation is recessive, which indicates that the *mob* gene encodes a protein. Furthermore, the mutant plasmid, when isolated, is not in the form of a relaxation complex. The failure of F to help the *mob⁻* mutant to transfer is shown in panel (c), in which the F pilus and transfer apparatus are formed but nicking of the ColE1 DNA does not occur. Another ColE1 mutant, having the Mob⁻ phenotype, but which has a *cis*-dominant mutation, is a deletion that removes the *bom* site. The *mob* protein is made but nicking fails to occur and there is no transfer, as shown in panel (d).*

---

*At the time of this writing it has not yet been proved that the *bom* site is the origin of transfer; the *bom* mutant does fail to form a nicked relaxation complex. Nor is there proof that transfer occurs by means of rolling circle replication, as has been shown in the figure. The possibility has been raised (though not based on solid evidence) that recircularization of the single strand could occur in the donor and that this molecule could be transferred, as in the extrusion of M13 phage DNA. This possibility must be considered until proved to be incorrect.

Other Bacterial Plasmids

Two other classes of plasmids deserve special mention. Various bacterial strains of the genus *Pseudomonas* are able to utilize an extraordinarily large number of organic compounds as energy sources. The number is so great that the number of enzymes required by the cell would exceed the coding capacity of the chromosomal DNA, if all genes were carried in the chromosomes. However, many of these genes are carried on large plasmids; thus, a bacterium able to utilize camphor as a carbon source carries a plasmid encoding the necessary enzymes; a bacterium unable to do so is not a mutant carrying a defective gene but simply one not having the plasmid. Hence, no single *Pseudomonas* bacterium is able to use all of the carbon sources; specific ability is distributed over different strains and the metabolic genes are distributed among a variety of plasmids.

A crown gall tumor found in many dicotyledonous plants is caused by the bacterium *Agrobacterium tumefaciens*. The tumor-causing ability resides in a plasmid called Ti. In an infected plant some of the bacteria enter and grow within the plant cells and lyse there, releasing their DNA in the cell, and from this point on, the bacteria are no longer necessary for tumor formation. By an unknown mechanism a small fragment of the Ti plasmid, containing the genes for replication, becomes integrated into the plant cell chromosomes. The integrated fragment breaks down the hormonally regulated system that controls cell division and the cell is thereby converted to a tumor cell. This plasmid has recently become very important in plant breeding because specific genes can be inserted into the Ti plasmid by recombinant DNA techniques; sometimes these genes can become integrated into the plant chromosome, thereby permanently changing the genotype and phenotype of the plant. It is believed that new plant varieties having desirable and economically valuable characteristics derived from unrelated species can be developed in this way.

PLASMIDS IN EUKARYOTES

Circular DNA molecules have been observed in animal cells that have been maintained in culture in the laboratory. These may be plasmids or they may be contaminating viruses. There is little information about how widespread such elements are. It might be expected that if plasmids do exist in higher cells, they may be much less common there than in bacteria; with the complex sexual systems in higher cells there may be little opportunity for the transfer of plasmid DNA.

Plasmids are known in yeast, a unicellular eukaryote. One of the more intriguing ones is the **killer particle,** which is a double-stranded

RNA molecule of molecular weight about 15×10^6. It is the only known plasmid that does not contain DNA. This particle contains ten genes for replication and several others for synthesis of the killer substance, a colicinlike material.

The best-studied yeast plasmid is the so-called 2μm plasmid, a DNA molecule having a length of 2μm and a molecular weight of 4×10^6. There are about sixty copies of this plasmid per cell and all are found in the yeast nucleus. This location is consistent with the fact that for replication the 2μm plasmid uses the same enzymes needed for replication of the yeast chromosomes. Furthermore, the plasmid DNA is coated with histones, as is all nuclear DNA. The plasmid DNA apparently does not integrate into the host DNA.

The base sequence of the 2μm plasmid includes two inverted repeats, each containing 599 base pairs; these are separated by two unique sequences, containing 2346 and 2774 base pairs, as shown in Figure 17-17. There are two classes of the plasmid in a cell, which differ by the orientation of one of the unique sequences. These classes, which are equal in number, are interchangeable by frequent recombinational events between the inverted repeats, as shown in the figure. The interconversion is catalyzed by a plasmid gene called *flp*. The function of the interconversion is unknown and in the laboratory flp^- plasmids replicate within the host yeast cells, apparently without defect. It seems likely, though, that in nature interconversion is of value since 40 percent of the coding capacity of the plasmid is devoted to the *flp* gene.

Replication of the 2μm plasmid has one characteristic that is quite different from other plasmids. It is a high-copy-number plasmid and presumably the copy number is regulated by a repression system related to the one described earlier for *E. coli* plasmids pSC101 and ColE1. As explained on page 714, when this mechanism prevails, plasmids are selected at random for replication, but for the 2μm plasmid this is not the case. Instead, each DNA molecule is replicated once, and

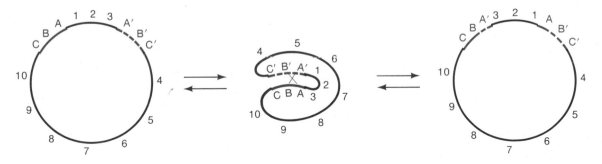

Figure 17-17
Mechanism of interchange of the two forms of the yeast 2μm plasmid. The red segments are the inverted repeats. The four segments are not drawn to scale.

no more, in each division cycle of the host cell. How this is regulated is unknown.

The 2μm plasmid constitutes a useful system for studying DNA replication in yeast. However, at present its greatest value is as a vector for cloning foreign genes in yeast by the recombinant DNA techniques described in the following section and in Chapter 20.

USE OF PLASMIDS IN GENETIC ENGINEERING

Chapter 20 is concerned with recombinant DNA technology and its role in genetic engineering. Here a brief preview is given, in which the functions of plasmids are emphasized.

Enzymes exist that can produce a single cut in a nonessential gene of a plasmid. A second linear piece of DNA, which may even come from a different organism, can be attached to the two free ends of the cleaved plasmid DNA to re-form a circle, and the circle can then be sealed with DNA ligase. Let us assume that (1) the added DNA contains the gene for β-galactosidase, (2) the genotype of the plasmid is tet^+ (resistant to tetracycline) and (3) the plasmid has a copy number of 50. This DNA can be used to transform a cell that is Lac$^-$Tet$^-$, whence a Lac$^+$Tet$^+$ cell can be selected. If the plasmid is a high-copy-number element, the cell will now contain fifty copies of the *lac* gene and thus should be able to make an enormous amount of β-galactosidase. This exemplifies one important use of plasmids in genetic engineering—namely, to increase the yield of a desired protein.

REFERENCES

Broach, J. R. 1982. "The yeast plasmid 2μm circle." *In* J. Strathern, E. Jones, and J. Broach, eds. *The Molecular Biology of the Yeast Saccharomyces.* Cold Spring Harbor Laboratory.

Broda, P. 1979. *Plasmids.* W. H. Freeman and Co.

Bukhari, A. I., J. A. Shapiro, and S. L. Adhya. 1978. *DNA Insertion Elements and Plasmids.* Cold Spring Harbor Laboratory.

Clark, A. J., and G. J. Warren. 1979. "Conjugal transmission of plasmids." *Ann. Rev. Genetics,* 13, 99–125.

Clewell, D. B. 1970. "Properties of a supercoiled DNA-protein relaxation complex and strand specificity of the relaxation event." *Biochemistry,* 9, 4428–4440.

Clowes, R. C. 1972. "Molecular structure of bacterial plasmids." *Bact. Rev.,* 36, 361–404.

Clowes, R. C. 1973. "The molecule of infectious drug resistance." *Scient. Amer.,* April, pp. 18–27.

Deonier, R. C., and N. G. Davidson. 1976. "The sequence organization of the integrated F plasmid in two different Hfr strains of *E. coli.*" *J. Mol. Biol.,* 107, 207–222.

Gurney, D. G., and D. Helinski. 1975. "Relaxation complexes of plasmid DNA and protein. III. Association of protein with the 5′ terminus of the broken DNA strand in the relaxed complex of plasmid ColE1." *J. Biol. Chem.,* 8796–8803.

Hayes, W. 1968. *The Genetics of Bacteria and Their Viruses.* Blackwell.

Holloway, B. W. 1979. "Plasmids that mobilize bacterial chromosomes." *Plasmid,* 2, 1–19.

Kingsman, A., and N. Willets. 1978. "The requirement for conjugal DNA synthesis in the donor strand during F′Lac transfer." *J. Mol. Biol.,* 122, 287–300.

Kornberg, A. 1980. *DNA Synthesis.* W. H. Freeman and Co.

Lewin, B. 1977. *Gene Expression. 3. Plasmids and Phage.* Wiley-Interscience.

Low, K. B., and R. D. Porter. 1978. "Modes of gene transfer and recombination in bacteria." *Ann. Rev. Genetics.,* 12, 249–287.

Meynell, G. 1973. *Drug-Resistance Factors and Other Bacterial Plasmids.* MIT Press.

Mitsuhashi, S. (ed.). 1977. *R Factor: Drug Resistance Plasmid.* University Park Press.

Novick, R. P. 1969. "Extrachromosomal inheritance in bacteria." *Bact. Rev.,* 33, 210–235.

Novick, R. P. 1980. "Plasmids." *Scient. Amer.,* December, pp. 102–129.

Willets, N., and R. Skurray. 1980. "The conjugation system of F-like plasmids." *Ann. Rev. Genetics,* 14, 41–76.

18 Mechanisms for Rearrangement and Exchange of Genetic Material

II. RECOMBINATION BETWEEN HOMOLOGOUS DNA SEQUENCES

Evolution depends on the generation of genetic variation. The first step in creating genetic variation is mutation. The mutations from different lineages are combined in mating and emerge recombined in the progeny. It seems that processes have evolved to enrich the variation by enabling novel combinations to be produced by shuffling large pools of mutants that have accumulated in a particular species in the course of many generations. In diploid organisms this rearranging is done by sexual reproduction. Chromosomes from each parent segregate in forming gametes (egg and sperm cells) and then reassociate in forming progeny.

A second kind of shuffling also occurs in the formation of the gametes—namely, the breakage and rejoining of two members of a pair of chromosomes, a process originally called **crossing over** and now called **recombination.** The mechanisms by which recombination occurs comprise the subjects of this and the following chapter.

The detailed mechanisms we shall discuss have been derived from studies with microorganisms—principally bacteria, phages, and yeast, and, in one case, fungi. The general principles uncovered in these studies probably apply to the higher organisms as well, though there may be differences in detail, as we have seen with many biochemical mechanisms. The reader interested in the genetics of the yeasts, fungi, and higher organisms should consult the references listed at the end of the chapter.

TYPES OF RECOMBINATION IN MICROORGANISMS

Genetic variation is important in the evolution of microorganisms. However, these organisms are haploid; consequently, different alleles of a particular gene are not usually present in the same cell. Thus, mechanisms have arisen that enable microorganisms to acquire genetic information—namely, DNA—from other cells and to incorporate that DNA, or a fragment of the DNA molecule, into the cellular DNA. One example of this process is **transformation** (Chapter 7); here, DNA released to the environment by death of a donor bacterium enters a recipient cell. A second example is **transduction** (Chapter 16), in which a phage carrying cellular DNA picked up by growth in a donor bacterium injects this DNA into a recipient bacterium. Finally there is **conjugation** (Chapter 17), in which a plasmid or DNA from an Hfr cell (Chapter 1) is transferred from one cell to another by cell-cell contact. Except for plasmid transfer, the genetic information introduced into a recipient cell must be incorporated into the hereditary apparatus of the new cell if this DNA is to be maintained throughout successive cycles of cell division. In Chapter 17 it was seen that plasmid maintenance is simply a result of replication of the plasmid DNA, which, owing to its circularity and its replication origin, is a complete replicating unit. However, in each of the other cases, the transferred DNA is not capable of continual replication and its maintenance requires that the DNA become physically inserted into the host DNA, usually by a substitution mechanism. This process of insertion, which invariably requires some type of breakage of the DNA of the recipient and rejoining of the donor DNA to the recipient DNA, is called **genetic recombination.**

In our study of prophage integration (Chapter 16), one type of genetic recombination was examined—namely, the site-specific recombination mediated by the phage λ *int* gene. This process is characterized by an exchange between two short identical DNA sequences (15 base pairs) that must be flanked by four nonidentical sites (B, B', P, and P'), which contribute in an essential way to the specificity of the exchange. However, the processes just described (transformation, transduction, and integration of DNA transferred by conjugation) require extensive regions (probably hundreds of base pairs) whose sequences are identical or nearly identical. That is, the two interacting sequences must be **homologous.** Thus, these processes are called **homologous recombination** (Figure 18-1). Another essential feature of this type of recombination is the necessity for a host gene, which in E. *coli* is called *recA*. The *recA* product, which is discussed later in the chapter, is needed for an early step in DNA-pairing. Similar genes have been found in other bacteria, so this type of recombination has been termed rec dent.

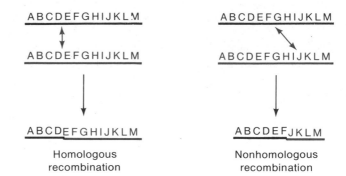

Figure 18-1
The distinction between homologous and nonhomologous recombination. The horizontal lines represent segments of double-stranded DNA; the letters represent base pairs. The red arrows indicate the points of genetic exchange.

Table 18-1
Characteristics of four types of genetic recombination

Type	Homology required?	RecA product required?	Sequence-specific enzyme required?
Site-specific	Yes	No	Yes
Homologous	Yes	Yes	No
Transposition	No	No	Yes
Illegitimate	No	No	?

A third type of recombination, in which particular segments of DNA, known as **transposable elements,** seem to move from one part of a DNA molecule to another or from one chromosome to another in a diploid cell, has been documented. This phenomenon, called **transposition,** does not require homology nor is it dependent on the *recA* gene. Hence, transposition is a kind of *recA*-independent recombination. A fourth type of recombination, which requires neither the *recA* product nor transposable elements, occurs between nonhomologous sequences; it is called "illegitimate" recombination, a name which is simply a reflection of our total lack of knowledge of the process. Properties of these four types are summarized in Table 18-1.

In this chapter we will be concerned with homologous (*recA*-dependent) recombination. The third and fourth types are discussed in Chapter 19.

BASIC FEATURES OF HOMOLOGOUS RECOMBINATION

In all types of recombination two DNA molecules or two segments of the same DNA molecule interact; this interaction is called **pairing.** The essential feature of homologous recombination of the type discussed in this chapter is the following:

The proteins responsible for the pairing do not require specific base sequences, though they do need two segments having identical or nearly identical sequences to provide the means of recognition between two DNA molecules by base-pairing.

With this in mind, an elementary model of genetic recombination, such as that shown in Figure 18-2, might be proposed. In this model base pairing between complementary strands generated by staggered single-strand breakage in different strands enables fragments to join together. Unfortunately, simple models such as this one do not explain how genetic recombination occurs; even the more complicated models that have been proposed have failed to explain the characteristics of many common recombinational events. However, we will examine some of these models because they illustrate how one goes about thinking about recombination.

Many physical experiments have shown four facts quite clearly:

1. Parental DNA molecules are broken and rejoined.
2. The joining requires homologous base-pairing.
3. Overlap regions (also called lap joints) are present at several stages of recombination.
4. DNA replication is not necessary for the recombinational event, though in certain phage systems postrecombinational replication is needed to generate a molecule that can be packaged and a repairlike replication may be needed to complete recombination in higher organisms.

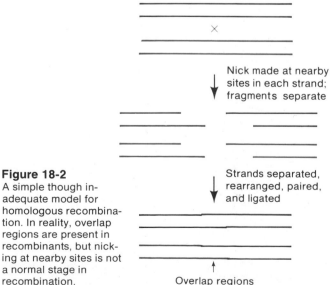

Figure 18-2
A simple though inadequate model for homologous recombination. In reality, overlap regions are present in recombinants, but nicking at nearby sites is not a normal stage in recombination.

Nick made at nearby sites in each strand; fragments separate

Strands separated, rearranged, paired, and ligated

Overlap regions

In recent years, *in vitro* recombination systems have been developed and these have helped to elucidate some of the steps in genetic recombination, such as **branch migration, single-strand invasion,** and **trans-cutting.** Furthermore, many recombination-deficient mutants have become available, making it possible to block recombination *in vivo* at certain stages. All of these results have led to the development of ingenious and testable models for recombination, which we will discuss shortly.

SPECIFIC FEATURES OF HOMOLOGOUS RECOMBINATION

In this section we look closely at several features of homologous recombination. In a later section, after several genetic phenomena have been described, these features will be integrated in the form of several popular theories of genetic recombination.

Breakage-Rejoining and Heterozygosis

In homologous recombination between two DNA segments both segments are broken and the fragments are rejoined via complementary base-pairing. A grossly oversimplified (but instructive) example of breakage and rejoining was shown in Figure 18-2. It should be noted that the overlap region—that is, the region in which pairing has generated the recombinant molecule—need not have the identical base sequence in each strand because, if the overlap is large enough, a few mismatched bases would not weaken the structure and hence could be tolerated and repaired. Indeed, this does occur, for the overlap region generally includes several hundred to several thousand base pairs. Furthermore, the single-strand breaks that generate the two components of the overlap region need not be precisely in the same positions in both double-stranded molecules, as shown in Figure 18-3. For example, they might be situated as in panel (a), in which case gaps remain; assuming that there is a free 3′-OH and a 5′-P group, these gaps can be filled directly by DNA polymerase I and sealed by DNA ligase. Alternatively, the breaks could be situated as shown in panel (b), in which case there is excess DNA, which could be trimmed away by an exonuclease.

By this oversimplified mechanism, recombination between two marked sites would occur as shown in Figure 18-4(a). If the cross involved three genetic markers a,* b,* and c,* a phenomenon that is shown in Figure 18-4(b) can occur. In this example, the breaks are situated in such a way that the overlap region has a mismatched pair of bases caused by pairing of b and b.* That is, the overlap region is **heterozygous,** specifically, a b/b^* heterozygote. (Note that whenever there is a b/b^* heterozygote, the "outside" markers are in a recombinant arrangement.)

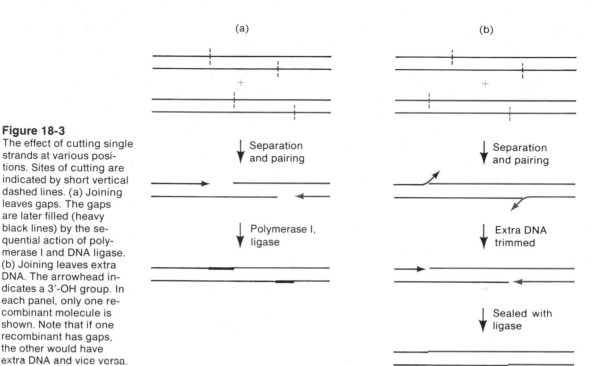

Figure 18-3
The effect of cutting single strands at various positions. Sites of cutting are indicated by short vertical dashed lines. (a) Joining leaves gaps. The gaps are later filled (heavy black lines) by the sequential action of polymerase I and DNA ligase. (b) Joining leaves extra DNA. The arrowhead indicates a 3'-OH group. In each panel, only one recombinant molecule is shown. Note that if one recombinant has gaps, the other would have extra DNA and vice versa.

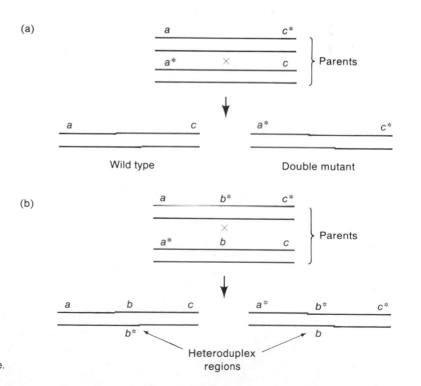

Figure 18-4
An example of recombination between genetically marked parents. (a) Exchanges between the two markers. (b) An exchange straddles a marker and hence produces heteroduplex molecules. A superscript * indicates a mutant allele.

Figure 18-5
Two ways to correct a
heteroduplex. (a) By replication. (b) By mismatch
repair.

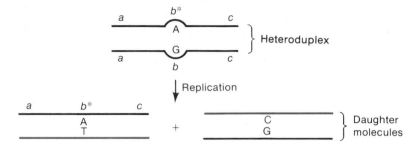

(a) Resolution of a heteroduplex by replication

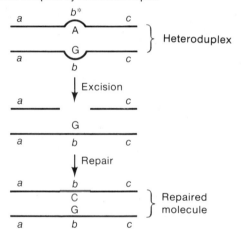

(b) Resolution of a heteroduplex by mismatch repair

A heteroduplex in a DNA molecule is an unstable state. When a
heteroduplex whose genotype is ab/b^*c replicates, the b and b^* alleles
segregate into two daughter molecules (Figure 18-5(a)); these have the
genotypes abc and ab^*c. The state of heteroduplicity can also be resolved by **mismatch repair;** that is, a tract of bases including one base of
the b-b^* mismatch can be excised and then replaced according to the
standard base-pairing rule to yield either of the base pairs $b \cdot b'$ or
$b^* \cdot b^{*\prime}$, in which the prime denotes complementarity (Figure 18-5(b)). If
this occurs, the ab/b^*c heteroduplex would yield the genotype abc or
ab^*c, depending on the strand that is repaired, but not both genotypes.
Repair of such mismatches is fairly common; it is detected genetically
by an apparent loss of a genetic marker among the progeny of a single
genetic event—for example, in the cross just considered, the progeny
would either have the genotype abc, in which case the marker b^* would
be lost, or ab^*c, in which case b would be lost.

Branch Migration

In Chapter 4 **breathing** of DNA was described. Breathing refers to the phenomenon of random and transient breaking and reforming of hydrogen bonds.

Branch migration, which refers to the displacement of one base-paired strand in a double helical region by another strand also able to base-pair, is a phenomenon based on breathing. It can be understood by examining the synthetic DNA polymer shown in Figure 18-6. The W strand of this molecule is continuous. Two single strands complementary to W, I and II, whose combined length exceeds that of the W strand, are both hydrogen-bonded to the W strand, as shown. The heavy black region and the red segment of the C strands have the same base sequence and are complementary to the dashed region of the W strand. When breathing temporarily opens the right terminus of strand I, one of two events can occur—strand I can snap back or strand II can increase the length of its hydrogen-bonded region, as shown in panel (b) of the figure. Breathing can occur again, resulting in restoration of the state of the molecule in panel (a) or in further change, as in panel (c). Breathing never truly stops, so the state of the molecule shifts continually. In a model of recombination to be described later, branch migration will play an important role.

Single-Strand Invasion

Single-strand invasion is also dependent on breathing. Consider a circular DNA molecule and a single-stranded fragment that is homologous to a base sequence in the circle (Figure 18-7). If breathing occurs in the homologous sequence, it is possible for the single strand to form base pairs with the circle, thereby creating a triple-stranded region. By branch migration the entire single strand can be assimilated, as shown in the figure. Single-strand invasion does not occur spontaneously *in vitro*. However, if the double-stranded circle is supercoiled, the under-

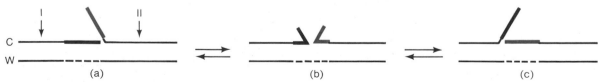

Figure 18-6

Branch migration. The heavy regions (both black and red) are both complementary to the dashed segment of the W strand. In panel (a) the heavy black region is hydrogen-bonded to the dashed segment. The free end of the black region occasionally becomes unbonded (owing to breathing), allowing the red region to pair with the dashed segment (panel (b)). By continued breathing more of the red region can become base-paired (panel (c)) or the molecule can revert to the state in panel (a). Of course, many intermediate states of the type in panel (b) are possible.

Figure 18-7
Single-strand assimilation. Invasion of a circular DNA molecule by a complementary single-stranded fragment to form a displacement loop (D loop).

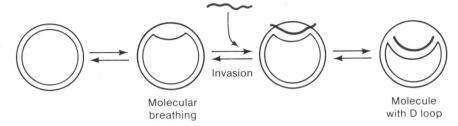

Molecular breathing

Invasion

Molecule with D loop

winding associated with supercoiling increases the size and lifetime of the single-stranded bubbles (see Chapter 4), and if the temperature is raised to between 60 and 70°C, single-strand assimilation occurs efficiently and spontaneously. Of course, such a high temperature would not be utilized in an intracellular process, but in the presence of the product of the E. coli recA gene, assimilation into a supercoil occurs at room temperature. We will see shortly that single-strand invasion may be one of the earliest steps in the pairing of two DNA molecules, and catalyzing this interaction may be one of the principal roles of the RecA protein in homologous recombination.

Reciprocal and Nonreciprocal Exchanges

Consider a cross between two phages whose genotypes are a^+b^- and a^-b^+. If an exchange occurs and both recombinants a^+b^+ and a^-b^- are recovered, the exchange is said to be **reciprocal** because the reciprocal pair of recombinants is recovered. If, however, only one of the reciprocal pair is found, the exchange is nonreciprocal. There is an enormous difference between being genetically nonreciprocal and being physically nonreciprocal. This distinction is exemplified in Figure 18-8. In panel (a), an exchange occurs between markers a and b. No DNA is lost; the exchange is physically reciprocal and both recombinant types are found, and thus, the exchange is also genetically reciprocal. In panel (b), breaks occur on both sides of marker a; the products in the first row are physically reciprocal. However, a^+/a^- heterozygotes are generated in the overlap region (second row). This is significant because physical experiments indicate that overlaps are very large. If there were no mismatch repair, each DNA molecule would replicate and two parental genotypes and the two recombinant genotypes would result; the exchange would appear to be reciprocal. If mismatch repair were to occur prior to replication and the negative allele were converted to the positive allele in both products of the first row of (b), then the products would be one parental molecule and one recombinant molecule, as shown in the third row; genetically, the

(a) Genetically and physically reciprocal exchange

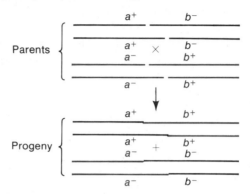

(b) Physically reciprocal but genetically nonreciprocal exchange

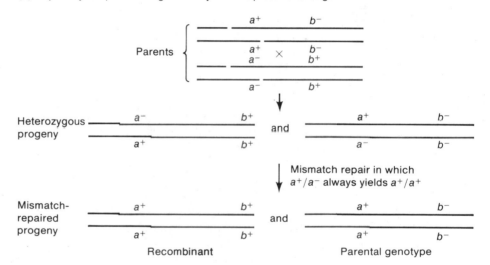

Figure 18-8
The meaning of reciprocal and nonreciprocal exchanges.

exchange would be nonreciprocal. Actually, there are many ways by which a physically reciprocal exchange can show genetic nonreciprocity; however, there are no simple models that can generate genetic reciprocity from an exchange that is physically nonreciprocal. The purpose of determining whether an exchange is genetically reciprocal is to put constraints on hypotheses about the physical event.

Crosses are usually performed with large populations of organisms and this procedure often masks the fact that individual events are nonreciprocal. This can be seen by examining panel (b) of Figure 18-8. In the last row, mismatch repair generates genetic nonreciprocity. In this figure it is assumed that mismatch repair always corrects a^+ to a^-. If, instead, a^- were always converted to a^+, nonreciprocity would also be seen; but if the direction of correction were random, four outcomes would be possible—$a^+b^+ + a^+b^-$ (nonreciprocal), $a^-b^- + a^-b^+$ (nonreciprocal), $a^+b^+ + a^-b^-$ (reciprocal), and $a^+b^- + a^-b^+$ (parental); if the cross were done with a population of organisms and if the entire progeny population were examined, the number of a^+b^+ recombinants and the number of a^-b^- recombinants would be identical and the exchange would appear reciprocal. In fact, when a cross is done *en masse*, exchanges almost always appear to be reciprocal. Nonreciprocity is a property not of a population but of individual recombination events; its detection usually requires that one isolate individual events for study. Two ways to do this are described in the following section.

EXAMINING THE PROGENY OF A SINGLE RECOMBINATIONAL EVENT

Yeasts and other fungi provide a simple way to examine the progeny molecules of certain types of recombinational events. Cells of these organisms exist in both a haploid and a diploid phase. When a diploid cell undergoes meiosis to yield haploid cells (four for yeast and most fungi and eight for a few species of fungi), recombination occurs between homologous chromosomes. Among the *Ascomycetes*, which are the organisms most commonly studied, the four haploid cells **(ascospores)** derived from a single diploid parent cell form in a sac called an **ascus**; thus the products of a single recombinational event are delivered in a single package that can be analyzed.

With yeast the genetic procedure is the following. Cells from a diploid culture whose genotype is known are spread on an agar surface on which they are starved for nitrogen compounds, a procedure that induces meiosis. The asci (plural of ascus) can be seen with a light microscope. Asci are selected and transferred by micromanipulation to a second agar surface. Each ascus is broken and the four ascospores are physically separated and allowed to grow into colonies. The genotype of each colony is determined by a variety of genetic techniques. By this tedious procedure, called **tetrad analysis,** all of the products of an individual recombinational event are obtained. This is certainly the most powerful, strictly genetic technique currently available for studying recombination.

In phage systems a different technique known as the **single-burst experiment** is done. In such an experiment a population of bacteria is infected with two different genetically marked phage. After the phage are adsorbed, the infected cells are diluted in growth medium to a concentration of about 0.05 infected cells per milliliter, and one-milliliter aliquots are dispensed into hundreds of test tubes. According to the

Poisson distribution, at this concentration 95.1 percent of the tubes will not contain infected cells, 4.8 percent will contain one infected cell, and 0.1 percent will contain more than one infected cell. Thus, 4.8/4.9, or 98 percent, of the tubes containing infected cells will contain only one infected cell. The cell in each tube is allowed to lyse and the contents of each tube are placed in a single petri dish on agar with indicator bacteria, so that plaques will form. The genotypes of the plaques are determined and if recombinants are present (they usually are if the infecting phage is λ or T4), their genotypes give information about the number and location of the genetic exchanges that have occurred during the infection. Typically 100 infected cells are studied, which means that one must use about 2000 petri dishes of which roughly 1900 will contain no plaques. Besides its wastefulness, the other disadvantage of the single-burst technique is that only about half of the progeny phage DNA in an infected cell is packaged and some genotypes may be lost. Thus, in a cell in which a reciprocal event has occurred that generated only one copy of each reciprocal recombinant, there is a 50 percent chance that one recombinant will not appear in the burst; the events would then be judged nonreciprocal. To reduce the effect of this problem, the contents of many infected cells must be examined and the data must be subjected to statistical analysis.

Both yeast and the phages T4 and λ have been studied extensively by the techniques just described. For all three organisms nonreciprocity has been inferred from statistical analyses of the data. Physical studies of the mechanics of recombination have suggested, though not proved, that the actual physical exchange of DNA may frequently be reciprocal and that the nonreciprocity results from later events such as mismatch repair. We will see in a later section that this information has been incorporated into several models for recombination.

RECOMBINATION BETWEEN A DNA FRAGMENT AND A CHROMOSOME

The two DNA molecules engaging in genetic recombination may be intact chromosomes (both circular and linear), double-stranded fragments, or single-stranded fragments. In this section we examine some of the features of recombination between a fragment and an intact molecule.

Recombination in Bacterial Transformation

In bacterial transformation, exogenously added DNA recombines with chromosomal DNA. In transformation with *Pneumococcus* (the best-

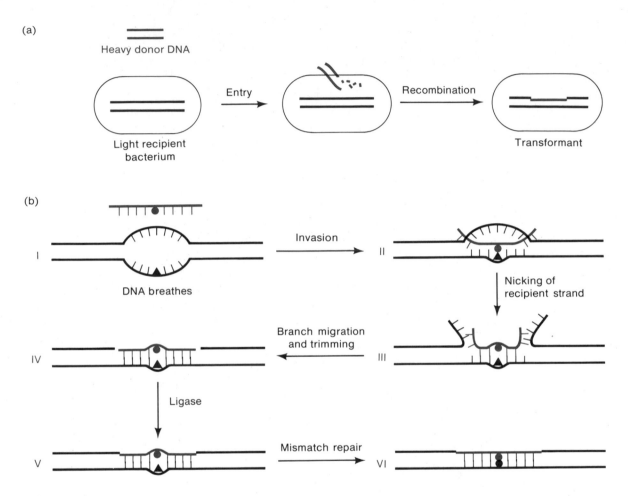

Figure 18-9
Bacterial transformation. (a) Conversion of double-stranded donor DNA to a single strand, which is then inserted. (b) A proposed mechanism for integration of single-stranded donor DNA.

understood case) uptake of the transforming DNA into the cell is accompanied by digestion of one of the input strands; consequently, only a single strand is available for recombination with the recipient DNA. The single strand is, in an unknown way, protected from nuclease attack—possibly it is coated with single-stranded DNA-binding protein. Physical experiments in which density-labeled (heavy) DNA is used to transform a bacterium whose DNA is light show that a single strand of the donor DNA, or a portion thereof, is linearly inserted into the recipient DNA (Figure 18-9(a)). The currently accepted model for this process is shown in Figure 18-9(b). Aided by a bacterial protein, probably equivalent to the *E. coli* RecA protein, the incoming single-stranded fragment invades the DNA of the recipient. By an unknown mechanism (see the following section on trans-cutting), the displaced single strand is cut. This process is very efficient. Branch migration then occurs, which gradually increases the fraction of the invading strand that is base-paired to the recipient strand. Trimming enzymes remove the free ends, which may be on donor or recipient DNA, and ultimately ligase seals the nicks. The result is a heterozygous region containing a mismatched base pair. The outcome of the process—that is, whether the donor marker is or is not recovered in the progeny cells—depends on whether mismatch repair occurs and, if so, whether the donor or recipient base is removed. For some markers, known as high-efficiency markers, either the mismatch is always corrected to match the donor genotype or there is little or no repair. In the latter case, following cell division one cell has the donor genotype and one has the recipient genotype. Because the plating conditions used to detect recombinants are usually chosen to allow growth only of recombinants (for example, an antibiotic is added or a necessary amino acid is deleted), the colony that forms consists exclusively of the recombinant bacteria. For the low-efficiency markers, mismatch repair usually removes the mismatched base from the donor and the cell retains the recipient genotype. Proof that this is the correct explanation for the two types of markers comes from the recent isolation of mutants presumably defective in mismatch repair; when these mutants are transformed, all markers appear to be transformed as high-efficiency markers because one of the two daughter cells following cell division always has the donor genotype.

Recombination in Bacterial Conjugation

In matings between *E. coli* Hfr and F^- cells single-stranded DNA is also incorporated into the F^- chromosome. The production of a recombinant requires both the RecA protein and a dimeric enzyme, which is the product of the genes *recB* and *recC* and is called the RecBC protein. This protein has several enzymatic activities including a powerful nuclease activity. An elegant experiment shown in Figure 18-10 proves that the action of the RecA protein precedes that of the RecBC protein and that the latter is not needed for strand invasion. If a culture of Hfr cells having the genotype *lacZ*$^-$ is mated with F^- cells carrying a different *lacZ*$^-$ mutation, β-galactosidase is made shortly after mating begins. Synthesis of β-galactosidase also occurs with *recBC*$^-$ F^- cells, in which no *lac*$^+$ recombinants will form. Thus, the donor and recipient DNA must be in a double-stranded complementary array from which *lac*$^+$ mRNA can be transcribed before the *recBC* product has acted. No β-galactosidase is made if the F^- cell is *recA*$^-$. Thus, strand invasion must occur in a *recBC*$^-$ cell but not a *recA*$^-$ cell and the RecBC enzyme

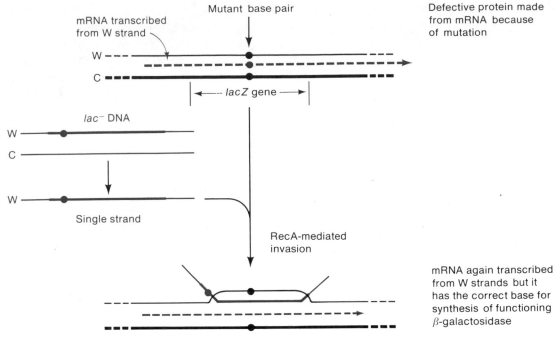

Figure 18-10
An experiment that suggested that the RecA protein but not the RecBC protein is needed for strand invasion. The dashed red line denotes mRNA. The heavy red line represents the coding segment of the W strand of the *lacZ* gene.

must be a finishing enzyme. At present the function of the RecBC enzyme remains a mystery.

Numerous properties of Hfr $\times$ F$^-$ crosses suggest that the mechanism of recombination is not as simple as that shown in Figure 18-9 for transformation.

Recombination in Transduction

Recombination between a double-stranded fragment and the chromosome is exemplified by both specialized and generalized transduction in which a portion of the linear double-stranded transducing DNA is inserted into the bacterial chromosome (Figure 18-11). Note that transduction differs from bacterial transformation in that injection by the phage delivers double-stranded DNA to the cell, whereas in transformation the process of DNA uptake converts the external double-stranded DNA to internal single-stranded DNA. In *E. coli* both the RecA and the RecBC proteins are required for transduction. This process differs from those that have been discussed so far in that not only are two exchanges required but since both strands of the transducing DNA

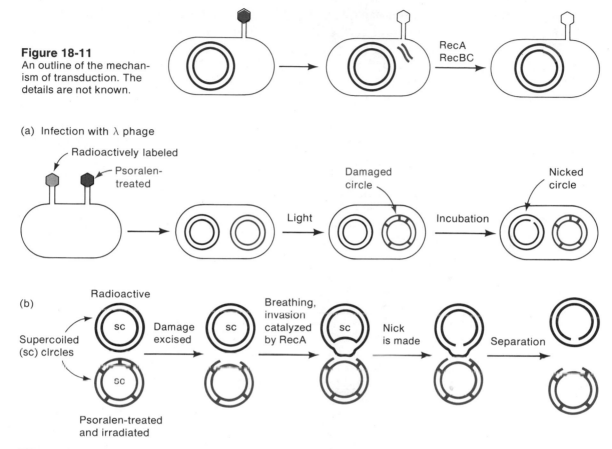

Figure 18-11
An outline of the mechanism of transduction. The details are not known.

Figure 18-12
Trans-cutting. (a) An experiment that demonstrates trans-cutting. (b) An interpretation of the phenomenon. The black molecule is invaded by the single strand of the red DNA molecule. This interpretation has not yet been proved. Supercoiled DNA molecules are drawn as circles marked with sc for the sake of clarity.

are integrated, two double-strand breaks are needed in the chromosome because the DNA is linear. The mechanism of recombination in transduction is very poorly understood and will not be discussed further.

Trans-Cutting

In the transformation mechanism suggested in Figure 18-9, a nick in the displaced recipient strand is needed for final replacement of this DNA by donor DNA. We will see in more general models of recombination that such a nicking event must always be invoked at some stage. Recently, nicking of an invaded DNA molecule has been observed—it is termed **trans-cutting**. The experiment is the following (Figure 18-12(a)).

A lysogenic cell (phage λ) is infected with two identical λ phage that are homoimmune to the prophage. Since replication is prevented by the immunity system of the prophage, the incoming phage DNA molecules simply become supercoiled. One of the λ DNA molecules is radioactive and the other, which is nonradioactive, has previously been exposed to a reagent called psoralen. After infection with the two phage, the cell is exposed to near-ultraviolet light, which causes damage only in the psoralen-treated DNA. Following a brief incubation of the infected cells in growth medium, it is found that the radioactive DNA molecule becomes nicked—if the cell is both RecA+ and carries the excision-repair genes (Chapter 10). One interpretation of this experiment is shown in Figure 18-12(b). Damage is excised from the irradiated, psoralen-treated DNA, which leaves a gap. During breathing of the radioactive DNA (which has large single-stranded regions because it is supercoiled, as explained in Chapter 4), the RecA protein mediates invasion of the radioactive double-stranded DNA by the single-stranded DNA in the gap of the psoralen-treated DNA. This pairing stabilizes the single-stranded region of the invaded radioactive DNA molecule enough for a nuclease to nick one of the radioactive strands. At present, there is good evidence to support this interpretation, though the nuclease has not yet been isolated.

RECOMBINATION BETWEEN INTACT DNA MOLECULES

In the examples discussed so far, one of the two DNA molecules participating in a recombinational event has been a DNA fragment. This is a less common situation than the topic of this section—namely, recombination between intact molecules. Since, with only a few exceptions, intracellular intact DNA molecules are circular, a problem that we have not yet encountered is recombination of molecules that have no free ends. This type of recombination has been studied in a great many laboratories with the result that a plethora of models has been advanced, only a few of which will be reviewed. There are two distinct types of models—(1) those in which a cut generates a free end and then the free end invades an uninterrupted molecule, and (2) a recent novel idea in which homologous pairing occurs prior to nicking. All of the models attempt to take into account the following facts:

1. One stage in the recombination mechanism includes overlapping joints.
2. The overlap is very long.
3. Recombination can be either reciprocal or nonreciprocal.
4. Mismatch repair is a common terminal event.

It is likely that apart from the different models, there are at least two different types of homologous recombination. One of these is found in phage, bacterial, and fungal systems; it does not require extensive DNA synthesis (except of the gap-filling type), it utilizes the RecA protein (in *E. coli*) or a protein having similar properties, and it is probably the most common type of recombination. The other class is *recA*-independent (utilizing phage-encoded genes) and is replication-dependent; so far it has been studied carefully with *E. coli* phages λ (the Red system), T7, and T4, but it is not understood. The former, *recA*-dependent class, is the only one discussed in this section.

Holliday Structures

An early model of recombination formulated by Robin Holliday predicted the existence of a unit consisting of two molecules connected by crossed single strands, as shown in Figure 18-13. This structure is called the **Holliday intermediate.** (At this point we will not discuss how

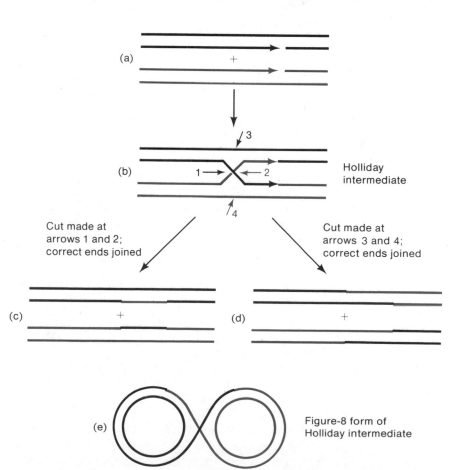

Figure 18-13
Formation of the Holliday intermediate (b) from two nicked DNA molecules (a) and the structure of two sets of recombinants (c and d). (e) is the same as (b) except that the components are circular; note that the figure-8 consists of two monomeric single-stranded DNA circles and one dimeric single-stranded DNA circle.

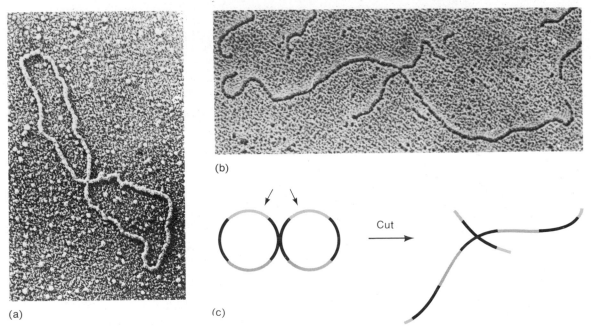

(a)

(b)

(c)

Figure 18-14

(a) Electron micrograph of a figure-8 molecule formed by recombining two φX174 monomeric circles. (b) An X form obtained by cutting a figure-8 molecule with a restriction endonuclease that makes a single cut at a unique position. (c) An interpretation of the molecules in panels (a) and (b). (Courtesy of Robert Warner.)

it arises.) If two cuts are made at corresponding positions in the crossed region (the **Holliday junction**), the pair of recombinant molecules shown in part (c) of the figure result. If, on the other hand, two cuts are made at corresponding positions in the unscathed strands, the molecules shown in part (d) will result. Type (c) and type (d) molecules both have overlaps that might be heterozygous if a genetic marker were in the region. For simplicity the molecules have been drawn as linear molecules in parts (a)–(d) of the figure. Part (e) shows for a pair of circular molecules the configuration of the Holliday structure (part (b)). Thus, the Holliday intermediate for a pair of circles is a figure-8. Such molecules have been observed (Figure 18-14(a)) by electron microscopy of DNA isolated from cells containing many copies of a plasmid and from RecA⁺ cells infected with phage φX174. Proof that these figure-8 molecules do indeed have the predicted structure comes from treatment of the figure-8 with a restriction endonuclease that makes a single cut at a unique position in each circle. If the two circles of a figure-8 are joined in homologous regions, the molecule should be cleaved into an X-shape having two pairs of arms all of equal length (Figure 18-14(b,c)).

Further proof that the Holliday structure is a true recombination intermediate comes from a more detailed look at the figure-8 junction.

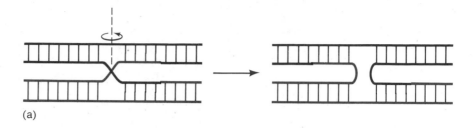

(a)

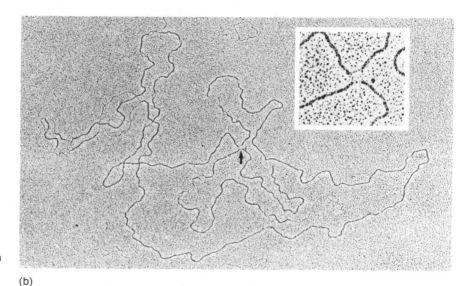

(b)

Figure 18-15

(a) Rotation of the upper half of a Holliday structure about the axis shown (holding the bottom half in place) to generate an open-box structure. (b) Electron micrograph of a recombination intermediate between a circular λ phage DNA molecule and a linear DNA fragment. The Holliday junction is indicated by an arrow and is enlarged in the inset. The molecules have been partially denatured in order to identify corresponding regions of the circular and linear molecules. (c) An interpretation of the region surrounding the Holliday structure. The drawing at the left shows the open-box configuration seen in (b); the drawing at the right shows a possible crossed structure. (Courtesy of Manuel Valenzuela.)

(c)

One circular component of the structure shown in panel (b) of Figure 18-13 can be rotated through 180°, as shown in Figure 18-15(a). If all single-stranded regions of the structure are extended, a close look at the junction would have the rectangular appearance shown in the figure. If an X form, such as that shown in Figure 18-13(b), is examined by electron microscopy in the presence of formamide (which causes single-strand extension), a molecule of the type shown in Figure 18-15(b) is found. These electron micrographs plus other evidence confirm the notion that the Holliday structure is an intermediate in *recA*-mediated recombination.

In the next section we examine one way in which a Holliday structure might arise.

The Strand Transfer Model

In the original version of the Holliday model it was assumed that strand exchange results from a break-switch-join event in which both strands were initially broken at the same position. This is a reciprocal event. Through the years evidence has accumulated that the exchange is not always reciprocal—the most notable observation is that often the filled-in region or the overlap region, shown in panels (c) and (d) respectively of Figure 18-13 as coming from both parents, is usually derived from only one of the interacting DNA molecules. Thus, numerous attempts have been made to generate a model that includes this observation yet still has a Holliday structure as an intermediate. One such model is the strand transfer model, which we now describe.

Before explaining the strand transfer model, it is necessary to understand the phenomenon of **DNA isomerization** shown in Figure 18-16. Isomerization is the mechanism that enables the Holliday structure I to become structure V. In the unstacking process each double helix bends near the Holliday junction so that the strands in the junction become straight. The right half (as drawn) is then rotated through 180° (top to bottom) to generate structure III. The upper portion (branches A and b) is now rotated through 180° (right to left) to generate structure IV. Restacking of the bent helices yields structure V, and the isomerization is then complete. Note what has happened in the isomerization process—the crossover strands in the Holliday structure have become the backbone strands and the backbone strands have become the crossed strands. Neither extensive unstacking nor the rotational diffusion just described have been directly demonstrated in physical experiments; however, by use of molecular models it has been shown that the bends and rotations depicted in Figure 18-16 are accomplished simply and without any steric problems or strain.

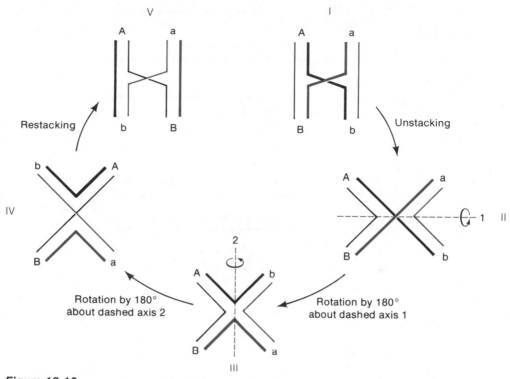

Figure 18-16
The mechanism of isomerization of a Holliday intermediate; structures I and V are interconvertible by isomerization. (After a model by M. Meselson and C. Radding. 1975. *Proc. Nat. Acad. Sci.*, 72, 358.)

The strand transfer model is shown in Figure 18-17. The steps are the following:

1. *Nicking.* By an unknown mechanism a single-strand break having a 3′-OH group and a 5′-P group is formed on one strand of one of the double helices. It is possible (and even likely) that this is a random event in the sense that all intracellular DNA molecules may be continually subjected to nicking (by nucleases) and sealing (by ligase). Alternatively, there may be a mechanism for bringing together homologous molecules, and nicking is a response to this association. Note that in this model nicking occurs in only one strand, in contrast with the two-strand nicking of the Holliday model.

2. *Displacement.* E. coli DNA polymerase I is able to act at a nick at which there is a free 3′-OH group, add nucleotides to the 3′ terminus, and displace the strand having the 5′ terminus (see Chapter 8). This

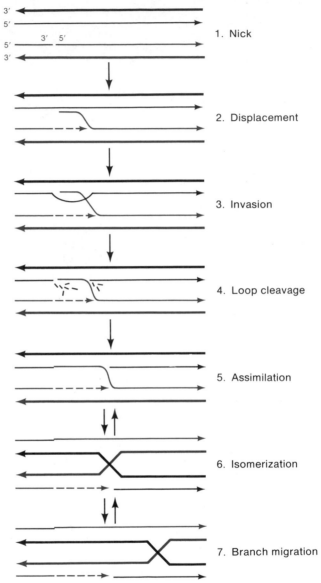

Figure 18-17
The strand transfer model proposed by M. Meselson and C. Radding. (1975. *Proc. Nat. Acad. Sci.*, 72, 358.) The dashed lines represent newly synthesized DNA.

displacement reaction can proceed indefinitely and thereby could produce a long 5′-P-terminated single-stranded appendage.*

*At this point the reader might be disturbed by the fact that when we began to discuss models of recombination, the topic was to be replication-independent processes. The strand transfer model is considered to be of that type for three reasons: (1) The DNA synthesis that occurs is fairly small—in fact, an extension of about 50 bases would be quite sufficient for the invasion process described in the following step, though it is likely that several hundred bases are usually involved; (2) the synthesis does not require DNA polymerase III, the replication enzyme, and probably very little of the standard replication apparatus; (3) the replication is repair replication.

3. *Invasion*. The single-stranded appendage can invade a double helix at a bubble (produced by breathing or possibly formed by a DNA-binding protein). This process apparently requires the RecA protein, whose mode of action will be discussed in a later section.

4. *Loop cleavage*. In response to invasion the displaced strand is removed. This reaction requires at least one endonucleolytic cut and, if there is only one cut, probably 3'-OH- and 5'-P-specific exonucleases as well. It is possible that additional nucleotides other than those in the displaced strand are removed exonucleolytically. This process may be an example of trans-nicking in the initial step.

5. *Strand assimilation*. The 3'-OH group of the gap formed in step 4 can be extended by DNA polymerase I and joined to the invading strand by DNA ligase. Furthermore, the 5'-P end of the gap can be enlarged to allow more of the invading strand to base-pair with the continuous strand of the invaded molecule. Note that a heteroduplex region (that is, a double-stranded segment containing one black strand and one red strand) is present in only one of the two double helices.

6. *Isomerization*. Isomerization of the type shown in Figure 18-16 produces the Holliday structure and also puts the arms flanking the strand crossover in a recombinant configuration.

7. *Branch migration*. Branch migration can occur toward the right or toward the left. In either case, heteroduplex regions can be formed in both double helices.

At this point the Holliday intermediate shown in Figure 18-13(b) has been reached and recombinants can form as shown in later panels of that figure. An important point is that the strand-crossover point in the Holliday intermediate can be very far from the site at which strand invasion occurs, so there is no requirement for a nick occurring after stage 7 of Figure 18-17 to be in the same position as that in stage 1, as was proposed in the original Holliday model (Figure 18-13). This means that

> In a genetic cross in which recombination ultimately occurs between two genetic markers, the initial nick-and-invasion event need not be between those markers.

The strand transfer model is supported by many experimental results. However, certain phenomena are at odds with it and the model certainly will have to be modified.

In the next section a very different way to create the Holliday intermediate will be described.

A "Pair, Then Cut" Model of Recombination

In the strand transfer model and in most other models it is assumed that strand breakage *precedes* pairing. However, a model has been proposed in which prior strand interruption is not needed. In this model (Figure 18-18) two homologous regions unwind (either through breathing or with assistance of accessory proteins) and form heteroduplex pairs. That is, if $A_1 \cdot T_1$ is a pair in molecule 1 and $A_2 \cdot T_2$ is a pair in molecule 2, the pairs $A_1 \cdot T_2$ and $A_2 \cdot T_1$ form. The spatial arrangement of the structure (III in the figure) is difficult to visualize in a planar diagram, but model-building shows that intercoiled heteroduplexes are struc-

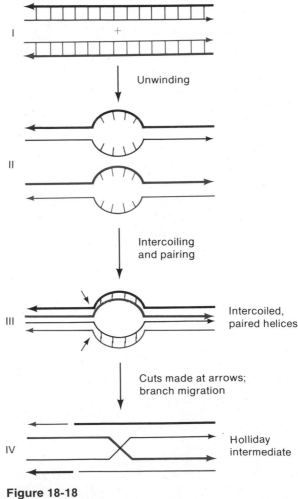

Figure 18-18
A "pair, then cut" model.

turally feasible and are stabilized both by hydrogen bonds and base-stacking. Two nicks in the intercoiled structure at the sites indicated by the arrows generate a Holliday intermediate. These nicks have to be at corresponding points in the two helices, which is a disadvantage of the model, but recent studies with several topoisomerases, e.g., DNA gyrase and the ω protein of *E. coli,* both of which have nicking-closing activity, suggest that these enzymes might be able to coordinate these events. There is little evidence to support this model and a great many experiments demonstrate that DNA that is nicked (either by an external agent or in the process of repair of base damage) and DNA that possesses gaps or free ends is more prone to engage in homologous recombination. These results support models of the strand-transfer type and are not simply incorporated into a "pair, then cut" model. However, until there is evidence to the contrary, "pair, then cut" models must be considered carefully.

Resolution of Figure-8 Molecules

In Figure 18-13 it was shown that a Holliday intermediate formed from two linear molecules can be resolved in two different ways. If the cuts that separate the duplexes are in the crossed-strand region, the result is equivalent to a single-strand insertion (panel (c)), whereas, if the cuts are elsewhere, two molecules with overlap joints result (panel (d)). If the recombining molecules are circles instead of linear molecules, there are also two possible outcomes, but this time they are structurally distinct.

Figure 18-19 shows three ways that a figure-8 molecule can be cleaved. For clarity, we have rotated the lower loop of the figure-8 through 360° to form the open structure II. We assume first that the structure is susceptible to strand breakage and that breaks occur only in single-stranded regions. This means that the breaks must be in one of the four segments indicated by the arrows. Second, we assume that the breaks always occur in pairs because one break alone cannot resolve the figure-8. There are three possible pairs: (1) opposed in the dimer strand, (2) opposed in the monomer circles, and (3) adjacent (one in a monomer circle and the other in the dimer strand). The consequences of these three pairs are two monomers, a dimer, and a rolling circle, respectively, as shown in the figure. These three forms may play very different roles in different systems. For example, if the original participants are chromosomes of either bacteria or eukaryotes, which only have monomer length, either formation of the dimer and of a rolling circle must be prevented or some further processing is required. On the other hand, in a phage system the formation of a monomer may be a dead end; for example, with phage λ, since two *cos* sites are needed for packaging (Chapter 15), only a dimer or a rolling circle (which continues

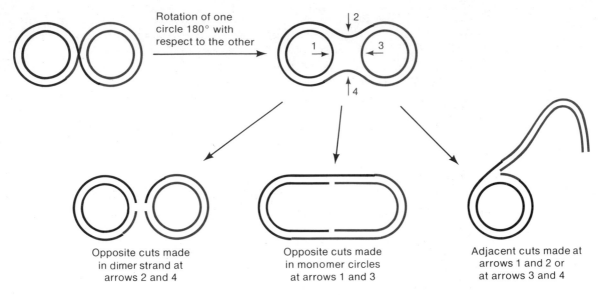

Figure 18-19
Resolution of a figure-8 molecule.

to replicate) could lead to recombinant phage. At present there is no evidence whatsoever for recombination-mediated rolling circle formation in any system. However, dimerization of small plasmids frequently occurs in a RecA$^+$ bacterium. There is little doubt that figure-8 molecules are at least one intermediate in genetic recombination, but neither how they are formed nor how they are converted to recombinant molecules is known with certainty.

PROPERTIES OF THE RecA AND RecBC PROTEINS

In *E. coli* there are several pathways for homologous recombination. The major one uses the RecA and RecBC proteins and possibly other proteins, such as the single-stranded DNA-binding protein, which participate in other reactions as well. Both the RecA and RecBC proteins have been purified and an enormous effort has gone into a study of their DNA-binding and enzymatic properties. These studies have yielded hints of the biological roles of these proteins but have not yet provided clear information.

The RecA Protein

The RecA protein has two principal chemical activities: (1) It is an ATP-dependent, single-stranded DNA-binding protein, and (2) it is a

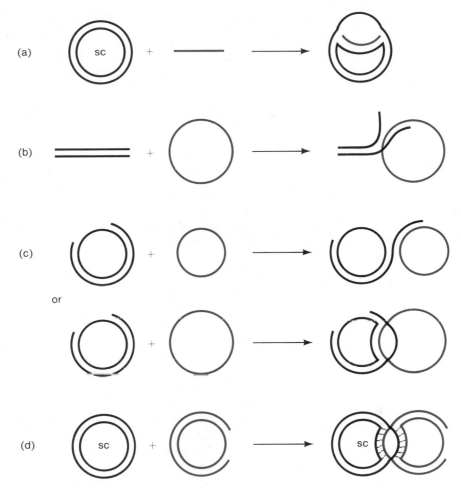

Figure 18-20
Four interactions mediated by the RecA protein; sc indicates that the circle is supercoiled.

protease. The protease activity is thought to be required in the regulation of its synthesis by virtue of its ability to cleave the *lexA* repressor, as described in Chapter 14. Its DNA-binding activity is the feature that is relevant to recombination.

The principal function of the RecA protein is to catalyze strand invasion. Figure 18-20 shows several *recA*-mediated DNA-DNA interactions that have been carried out *in vitro*. The structures shown are very stable and are held together by the normal base pairs between homologous base sequences; the evidence for this conclusion is that the structures can be dissociated only by heating to the normal melting temperature of the DNA. Each of the interactions shown requires the two types of DNA molecules, the RecA protein (and no other proteins) and ATP. The ATP is cleaved during the interaction, implying that strand invasion may be an energy-requiring process but the precise role of the ATP is unknown. There are three requirements for the four interactions shown in the figure:

1. One molecule must be single-stranded or have a reasonably long single-stranded region. This molecule is shown in red in the figure.
2. The other molecule (shown in black) must have either a free end or a single-stranded region. In panels (a) and (d), the single-stranded regions of the covalent circle are provided by the supercoiling; if the molecules are not supercoiled, the interaction does not occur.
3. One of the molecules must have a free end.

Details of the mechanism of interaction are not known but a simple way to think about this interaction would be to say that the RecA protein enables a free end to invade the duplex molecule at a homologous single-stranded region. Since the RecA protein does not itself have unwinding activity, it cannot form the single-stranded region in the invaded molecule, which would explain requirement 2. However, the elegant experiment that follows (Figure 18-21) shows that whereas a free end is needed to form the stable structures of Figure 18-20, it is not needed for the primary interaction between the two DNA molecules.

The experiment utilized the following two DNA molecules: (I) the double-stranded replicating form of the small DNA phage G4 (containing about 6000 base pairs), which had been cleaved by a restriction enzyme to generate a linear, double-stranded DNA molecule, and (II) a single-stranded hybrid molecule consisting of several thousand bases of M13 phage DNA in which 274 bases of G4 DNA have been inserted (Figure 18-21). These two molecules have only the 274-base G4 sequence in common, so no interaction between an end of the G4 double-stranded molecule and any region of the M13-G4 hybrid is possible. Nonetheless, the RecA protein causes an interaction between these molecules, indicating that homologous free ends are not necessary for the two molecules to be joined. This conclusion was confirmed by a second experiment in which uncleaved supercoiled G4 DNA was used; again the molecules held together. The complexes between M13-G4 and G4 differ in an important respect from those shown in Figure 18-20; namely, they are not as stable and can be dissociated at a temperature about 20°C below the normal melting temperature of the DNA. This result suggests that the interaction is of a different type from that shown in Figure 18-20. Clearly, the interaction between M13-G4 and G4 cannot include an extensive segment of paired bases since the double-stranded DNA would have to be unwound. (Remember: only ten base pairs are broken per turn of the helix.) However, recognition certainly involves base-pairing. The low thermal stability of the product suggests that the base-paired region is very short, perhaps less than ten base pairs (the

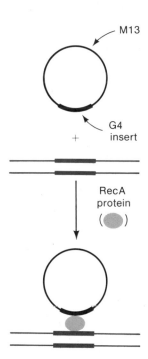

Figure 18-21
RecA-mediated association of a linear form of G4 DNA with single-stranded M13 DNA containing a 274-base insert of G4 DNA. Only the regions shown as the heaviest lines are capable of base-pairing, for all other regions are non-homologous.

codon-anticodon interaction requires only three). Possibly this three-stranded interaction is of the type shown in panel III of the "pair, then cut" model of Figure 18-18. Further study is needed before the nature of this interaction is understood.

The importance of the experiment just described is that it makes a clear statement that the RecA protein can mediate a weak interaction not needing free DNA ends. Thus, strand-strand recognition does not require a free end. Presumably the interacting region can move along the helix (by some type of branch migration) until a free end is encountered; then, unwinding from the free end can occur and a proper Watson-Crick double helix having many base pairs, such as is implied in Figure 18-20, can be formed between one strand of each participating molecule. One may speculate further that the stimulatory role of artificially created single-strand breaks, which has been documented in many biological systems, may be to decrease the time required before a free end is met and thereby allow stable helix formation to occur before the weakly interacting DNA molecules dissociate.

So far, we have said nothing about how the RecA protein accomplishes all of this. The reason is simply that we do not know, though one recent observation gives a hint. It has been found that under certain conditions a very weak complex forms between single-stranded DNA, double-stranded DNA, and the RecA protein, even when there is no homology between the two DNA molecules. It has been suggested that when there is homology, the RecA protein causes an interaction between any two regions of the two DNA molecules and that it allows the molecules to drift back and forth with respect to one another until homologous sequences come into register.

The RecBC Protein

The RecBC protein is a multifunctional protein. It has an endonuclease activity, and a powerful exonuclease activity against both single- and double-stranded DNA. The exonuclease activity was detected first, and so the protein is also called exonuclease V. Under certain experimental conditions the nuclease activities can be inhibited and the enzyme instead exhibits the ability to unwind double-stranded DNA. ATP is hydrolyzed when unwinding occurs. The unwinding that occurs when the nuclease activity is inhibited is unusual, as shown in Figure 18-22. The enzyme binds to one end of the DNA molecule and unwinds the DNA. It moves along one strand but remains in contact with two regions of that strand, allowing the end of the strand to pass through the enzyme, thus forming the single-stranded loop shown in panel (b). As the length of the single-stranded tail increases, homologous base-pairing occurs with the other strand to form the doubly-looped structure

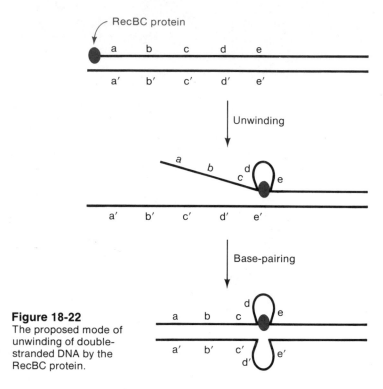

Figure 18-22
The proposed mode of unwinding of double-stranded DNA by the RecBC protein.

shown in panel (c). Possibly this is one mechanism for producing single-stranded regions *in vivo*; however, it is unlikely that this is the mechanism for providing single-stranded regions for *recA*-mediated pairing in view of the experiment described earlier that implied that RecA protein activity precedes RecBC protein activity. If the nuclease activity is not inhibited, nicking occurs in the single strands several thousand nucleotides part. The RecBC protein also forms a complex with DNA polymerase I but the properties of this complex are not known. The enzymatic activities of the RecBC enzyme have been incorporated into many models for homologous recombination but its precise role has not been demonstrated experimentally for any particular stage of recombination.

PROPERTIES OF SYSTEMS LACKING RECOMBINATION PROTEINS

Mutant cells lacking the ability to recombine have an additional phenotype—namely, they are exceedingly sensitive to damage to their DNA. For example, both *recA*⁻ and *recBC*⁻ bacteria are killed by much smaller doses of ultraviolet light than are rec^+ cells; the data for a $recA^-$ mutant can be seen in Figure 9-10 in Chapter 9. The main source of sensitivity is that in both $recA^-$ and $recBC^-$ cells a major repair pathway—recombinational repair (Chapter 9)—is absent.

The DNA of $recA^-$ cells is unstable—that is, it is continually degraded and resynthesized. The degradation is most striking following irradiation with ultraviolet light; then more than half of the bacterial DNA is enzymatically degraded by the $recBC$ enzyme to short nucleotides. This rampant degradation of DNA is called the **reckless** phenotype.

Another feature of $recA^-$ cells is their slow growth compared to rec^+ strains; this is indicative of other defects, most of which are unknown. Rec^+ *E. coli* divide roughly every 25 minutes at 37°C in rich growth media. In contrast, a typical $recA^-$ mutant may have a doubling time of 40–60 minutes. Part of the slowness of growth is probably due to reckless DNA degradation. The growth defect is more extreme for $recBC^-$ cells, which typically divide every 100–120 minutes. These growth defects are undoubtedly the result of many processes that are either inoperative or functioning inefficiently in rec^- cells.

In phage systems the ability to recombine is sometimes related to DNA replication. For example, the *Salmonella* phage P22 contains a linear DNA molecule that must be circularized before DNA replication can begin. The P22 DNA molecule does not have complementary single-stranded ends but it is terminally redundant. Circularization is accomplished by a recombinational event between the terminally redundant sequences. This event is catalyzed by the P22 *erf* gene, and erf^- amber mutants are unable to grow in a $recA^-$ host that lacks an amber suppressor.

E. coli phage λ is an example of a system in which recombination genes play a vital role, though it is not at all obvious that the main function of these genes is recombination. For instance, λred^- phage, which cannot produce recombinant progeny phage in a $recA^-$ host, are nevertheless able to grow, but the number of phage produced by an infected cell that is either $recA^+$ or $recA^-$ is about one-half that produced in an infection with a red^+ phage. Thus, recombination, or at least the function of the *red* gene of λ, is required for production of about 50 percent of the progeny phage. Another λ gene called *gam* serves primarily to inactivate the host RecBC protein early in infection. A red^-gam^- phage grows even more poorly than a red^+gam^- phage and if the host is $recA^-$, there is virtually no phage production. The ability to produce phage is restored to a $recA^-$ host if it is also $recBC^-$, indicating that active RecBC protein is detrimental to development of λ. Thus, λ grows better if there is a functional *red* gene product and a functionless RecBC protein. We will not attempt to unravel this very complex phenomenon (for which there are many models, all of which are inadequate) but merely point out that each of these mutations *(red⁻, gam⁻, recA⁻)* contributes in some way to aberrant DNA synthesis. For example, in an infection with a λ*red⁻* mutant, λ DNA replication is turned off before the infectious cycle is completed. When λ*red⁻gam⁻* infects a *recA⁻* host, rolling circle replication is seriously impaired; also, aberrant products of the replication are found. The details of the structure of these products and of those molecules that are absent are not known, but it is clear that the recombination genes in λ have a function beyond that of being able to recombine genetic markers.

OTHER *E. COLI* RECOMBINATION SYSTEMS

As we have mentioned, there are several pathways in *E. coli* for homologous recombination. Each of these requires the RecA protein and the one we have been discussing also requires the RecBC protein. The other recombination pathways are made evident in an Hfr × F⁻ cross in which the F⁻ cell is *recBC⁻*. That is, if the F⁻ cell is *recA⁻*, the recombination frequency is reduced at least 10^6-fold but if the F⁻ cell is *recA⁺recBC⁻*, the recombination frequency is reduced only 100-fold. The residual recombination with the *recBC⁻* cell depends on several genes; the best-studied and main ones are *recE* and *recF*. (The discussion that follows is complex, and it is recommended that the reader refer frequently to Table 18-2 and Figure 18-23. Beginners may choose to pass on to the next chapter.) In normal cells the RecF pathway is very inefficient because it is inhibited by exonuclease I, whose synthesis is determined by the gene *sbcB*. The explanation for these phenomena, depicted in Figure 18-23, is thought to be the following. When recombinants form by the RecBC pathway, an essential intermediate called Y is converted by the RecBC protein to a recombinant; this conversion may occur in several stages. Y can be formed in an unspecified way or by conversion from X to Y by the *sbcB* product. Thus, a *recBC⁻sbcB⁺* mutant has 1 percent of the wild-type activity because the RecBC pathway is not functional. The RecF pathway requires an intermediate Z, which is formed in an unspecified way from X. As long as exonuclease I is active, X is converted to Y, leaving little Z for the RecF pathway. The low frequency (1 percent) of recombination represents the activity of the RecF protein on the small amount of Z that is formed from X. A *recBC⁻sbcB⁻* mutant lacks exonuclease I; hence X is not converted to Y; all X is instead converted to Z. Thus, a *recBC⁻sbcB⁻* mutant shows normal recombination by the RecF pathway. The intermediate Z can also be formed by a process that utilizes exonuclease VIII. This enzyme is the product of the gene *recE* whose activity is repressed by the *sbcA* gene. Thus, a *sbcA⁻sbcB⁺recBC⁻* mutant also

Table 18-2
Recombination proficiency of several *E. coli* mutants

Genotype	Is recombination proficient?	Active system
recA⁻	No	None
recBC⁻	No	None
sbcB⁻	Yes	RecBC
recBC⁻sbcB⁻	Yes	RecF
recBC⁻recE⁻	No	None
recBC⁻sbcA⁻	Yes	RecE, RecF
recBC⁻sbcA⁻recE⁻	No	None

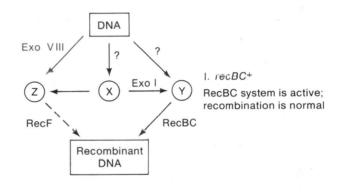

I. *recBC⁺*
RecBC system is active;
recombination is normal

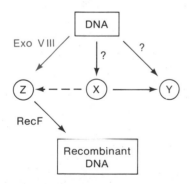

II. *recBC⁻*
Only the RecF system
is active, but weak

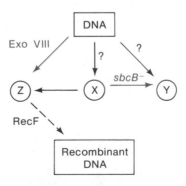

III. *recBC⁻sbcB⁻*(Exo I⁻)
RecF activity is strong;
recombination normal

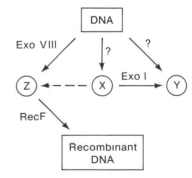

IV. *recBC⁻sbcA⁻*(Exo VIII⁺)
RecF activity is strong;
recombination normal

Figure 18-23
Pathways for recombination with different combinations of mutations. A red arrow indicates a reaction that does not occur. A dashed arrow denotes a weak reaction. The question marks indicate that the enzymes responsible for the step are not known.

shows normal recombination frequencies because Z is made directly by exonuclease VIII rather than from X; the RecF pathway is fully active. Of course, an $sbcA^- recE^- sbcB^+ recBC^-$ mutant is recombination-deficient (again 1 percent recombination) because Z can only be made from the little bit of X that survives exonuclease I. If $recF$ activity is also removed, there is no pathway to recombination and a $recBC^- recF^-$ mutant is fully as deficient as a $recA^-$ mutant. The systems and interactions depicted in Figure 18-23 are complex. However, there are even more rec genes ($recJ$ through $recM$) not shown here. The relations between the products of genes $recJ$ through $recM$ and the gene products mentioned earlier are not yet known.

REFERENCES

Alberts, B., and C. F. Fox (eds.) 1980. *Mechanistic Studies of DNA Replication and Recombination. Vol. 19. ICN-UCLA Symposium on Molecular and Cellular Biology.* Academic Press.

Cunningham, R. P., T. Shibata, C. Das Gupta, and C. M. Radding. (1979). "Single strands induce RecA protein to unwind duplex DNA for homologous pairing." *Nature,* 281, 191–195.

Eisenstark, A., 1977. "Genetic recombination in bacteria." *Ann. Rev. Genetics,* 11, 369–396.

Fox, M. S. 1978. "Some features of genetic recombination in procaryotes." *Ann. Rev. Genetics,* 12, 47–68.

Holliday, R. 1964. "A mechanism for gene conversion in fungi." *Genet. Res.,* 5, 282–304.

Hotchkiss, R. D. 1974. "Models of general recombination." *Ann. Rev. Microbiol.,* 28, 445–468.

Lacks, S. 1962. "Molecular fate of DNA in genetic transformation in *Pneumococcus.*" *J. Mol Biol.,* 5, 119–131.

McEntee, K., G. M. Weinstock, and I. R. Lehman. 1979. "Initiation of genetic recombination catalyzed *in vitro* by the RecA protein of *E. coli.*" *Proc. Nat. Acad. Sci.,* 76, 2615–2619.

Meselson, M., and C. M. Radding. 1975. "A general model for genetic recombination." *Proc. Nat. Acad. Sci.,* 72, 358–361.

Potter, H., and D. Dressler, 1976. "On the mechanism of genetic recombination: electron-microscopic observation of recombination intermediates." *Proc. Nat. Acad. Sci.,* 73, 3000–3004.

Radding, C. M. 1978. "Genetic recombination: strand transfer and mismatch repair." *Ann. Rev. Biochem.,* 47, 847–880.

Sigal, N., and B. Alberts, 1972. "Genetic recombination: the nature of a crossed-strand exchange between two homologous DNA molecules." *J. Mol. Biol.,* 71, 789–793.

Smith, H. O., D. B. Danner, and R. A. Deich. 1981. "Genetic transformation." *Ann. Rev. Biochem.,* 50, 41–68.

Stahl, F. W. 1979. *Genetic Recombination.* W. H. Freeman and Co.

Stahl, F. W. 1979. "Specific sites in generalized recombination." *Ann. Rev. Genetics,* 13, 7–24.

Valenzuela, M., and R. B. Inman. 1975. "Visualization of a novel junction in bacteriophage lambda DNA." *Proc. Nat. Acad. Sci.,* 72, 3024–3028.

Weissberg, R. A., and S. Adhya. 1977. "Illegitimate recombination in bacteria and bacteriophage." *Ann. Rev. Genetics,* 11, 451–473.

West, S. C., E. Cassuto, and P. Howard-Flanders. 1981. "Homologous pairing can occur before DNA strand separation in generalized genetic recombination." *Nature,* 290, 29–33.

Wilson, J. H. 1979. "Nick-free formation of reciprocal heteroduplexes: a simple solution to the topological problem." *Proc. Nat. Acad. Sci.,* 76, 3641–3645.

Mechanisms for Rearrangement and Exchange of Genetic Material

III. TRANSPOSABLE ELEMENTS

In the previous chapters we examined properties of recombination systems in which exchanges occur between homologous DNA sequences. In bacteria, exchanges of this type depend on the *recA*-gene product or its equivalent. Exchanges between short homologous regions flanked by extensive nonhomologous recognition sequences were described in Chapter 16; these exchanges utilize site-specific recombination enzymes such as the integrase of phage λ. In this chapter we consider another type of recombination, called **transposition.** This process is usually detected as the appearance of a linear segment of DNA, called a **transposable element,** already contained in a cell, at a second site in the DNA molecule of the same cell. The new site of insertion is a base sequence with which the element has no homology, and insertion is not dependent on the bacterial *recA* gene. The process is called transposition because in early studies the element appeared to move or transpose from one location to another. The term is a misnomer because, when transposition occurs, one copy of the transposable element *remains* at the original site and a second copy appears at the new site. Thus DNA replication is obligatory in the transposition process, in contrast with homologous recombination. Furthermore, movement of the transposable element is only one possible outcome of this type of recombination.

Transposition usually occurs at low frequency compared to homologous recombination, but it is of interest because it accounts for many deletions and gene inversions found in bacteria and also is a valuable procedure for generating mutants of use to the researcher.

TERMINOLOGY

Transposable elements are usually called **transposons.** Some transposons contain easily recognizable bacterial genes, particularly those for antibiotic resistance. For historical reasons a member of this class is designated by the abbreviation Tn followed by a number (e.g., Tn5), which distinguishes different transposons. By international agreement all newly discovered bacterial transposons for which names must be assigned are designated in this way even if no recognizable bacterial gene is present. When it is necessary to refer to the genes carried on a transposon, these are indicated by standard genotypic designations, e.g., Tn1(amp^+), in which amp^+ indicates that the transposon carries the genetic locus for resistance to ampicillin. The transposons first discovered did not contain any known host genes; and for historical reasons, they were called **insertion sequences** or **IS elements** and designated IS1, IS2, and so forth. Transposons have occasionally been designated in nonstandard ways—e.g., $\gamma\delta$, the element contained in the F plasmid. There is not yet a consistent system for naming transposons in eukaryotes; for instance, one transposon in *Drosophila* is called copia, another 497, etc.

A transposon is frequently located within a particular gene and this creates a mutation in that gene, which in accord with standard rules for naming mutants is given a number. The following notation is used to designate mutation 135 in the *galT* gene produced by transposon number 4:

$$galT135::Tn4$$

Note the use of the double colon.

INSERTION SEQUENCES

The first transposable elements that were discovered, in 1967, were the IS elements, and how they were discovered is of some interest. It is now known that they are just one type of transposon but nonetheless some of the essential characteristics of transposons and the various means of detecting transposons can be easily seen by examining this special class.

Discovery of IS Elements

IS elements were first detected by the isolation of a class of highly **polar mutations** in the E. coli gal and lac operons—each mutation mapped in the first gene of the operon but production of proteins by the downstream genes was totally eliminated. This polarity was eliminated if the bacteria carried a rho(Ts) mutation (temperature-sensitive for growth); in each case the polarity apparently resulted from the presence of a Rho-dependent termination site. Another important property of these polar mutations was that they could not be reverted by base-analogue or frameshift mutagens, so the mutations could not be base substitutions.* Physical studies of plasmids containing such polar mutations showed that in each case a segment of DNA had been inserted in the operon. In several phage systems (e.g., T4 and λ) apparent insertions arise by a replication error, which generates a tandem duplication of a base sequence. However, several lines of evidence indicated that the inserted segments did not simply consist of a duplication of an adjacent sequence. Thus, it was not at all obvious where the sequences had come from, and it was hypothesized that they were mobile segments that moved from one region of a chromosome to another. A direct demonstration of their mobility will be described shortly.

Size and Properties of IS DNA

Specialized transducing particles carrying the polar mutations just described were studied by electron microscopy. Hybridization of the DNA from a transducing particle bearing the mutation with the DNA of a transducing particle lacking a polar mutation showed a loop (Figure 19-1), which indicated that the mutation is an insertion whose length equals the length of the loop. Comparison of different transducing

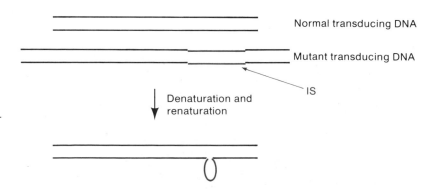

Figure 19-1
A hybridization test showing that an IS element is a segment of inserted DNA. After renaturation half of the molecules have re-formed the normal and mutant transducing molecules (only the heteroduplex has been drawn).

*Less-polar mutations had been known since 1960, but as these were chain termination mutations, they were usually revertable by base analogues.

particles containing polar insertions showed that the loop of the DNA of one particle could hybridize with the loop of another, indicating that the same sequence is present in each of the mutated sites.

Studies of a large number of transducing particles containing genes from various regions of the *E. coli* chromosome into which IS elements were inserted and of plasmids containing insertion sequences showed that there are several different IS elements. Some of these elements have been isolated from transducing-particle DNA and their base sequences have been determined. Those elements producing polar mutations contain Rho-dependent transcription stop signals (except for IS1) and also chain-termination mutations in all possible reading frames, which accounts for the polar effects. An important observation about the base sequence of IS elements is the presence of inverted repeat sequences at the termini of the element (Figure 19-2). These sequences contain 16–41 base pairs, depending on the transposon. At many sites of insertion (for example, in the *E. coli gal* operon, the sequences can be inserted in either the left-to-right or right-to-left orientation. One might expect this to be the case for all sites because of the symmetry of the inverted repeat sequences, but in fact there are also many insertion sites at which only one orientation has been observed. Some of the more important properties of a few of the IS elements are listed in Table 19-1.

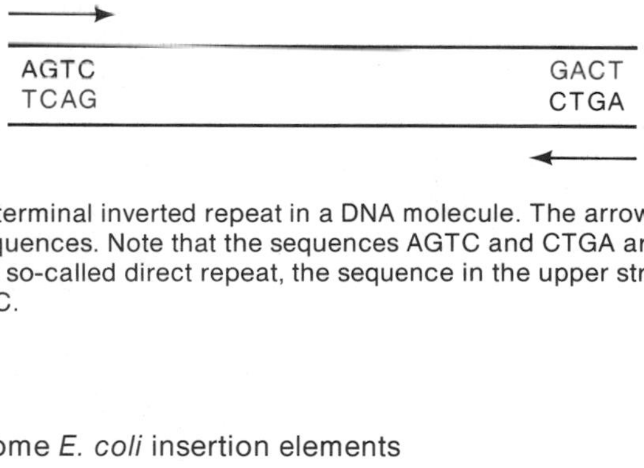

```
                    ───────►
        ┌─────────────────────────────────────────┐
        │  AGTC                          GACT       │
        │  TCAG                          CTGA       │
        └─────────────────────────────────────────┘
                                     ◄───────
```

Figure 19-2
An example of a terminal inverted repeat in a DNA molecule. The arrows indicate the inverted base sequences. Note that the sequences AGTC and CTGA are *not* in the same strand. In a so-called direct repeat, the sequence in the upper strand would be AGTC . . . AGTC.

Table 19-1
Properties of some *E. coli* insertion elements

Element	Number of copies and location	Size, base pairs	Sequence data
IS1	5–8 in chromosome	768	Complete
IS2	5 in chromosome, 1 in F	1327	Complete
IS3	5 in chromosome, 2 in F	Approximately 1400	Nearly complete
IS4	1 or 2 in chromosome	Approximately 1400	Termini only
IS5	Unknown	1250	Termini only
$\gamma\delta$	1 or more in chromosome, 1 in F	5700	Termini only

Location of IS Elements in *E. coli* DNA and in the DNA of Some Phages

In the preceding section we implied that the IS elements causing polar insertions come from within the bacterium. In fact, there are many copies of the IS elements in *E. coli* DNA, as indicated in Table 19-1. The number of copies of each IS element has been evaluated by hybridization studies and their locations in the *E. coli* chromosome have been determined by electron microscopy of self-annealed or heteroduplex DNA obtained from plasmids or specialized transducing particles carrying particular regions of the *E. coli* chromosome. For instance, two IS1 elements flank one of the *arg* loci. Electron microscopic studies have also shown that some plasmids lacking chromosomal genes contain IS elements. For example, IS2, IS3, and γδ are present in F and these elements mark sites in F at which F integrates into the *E. coli* chromosome to form Hfr strains (see Figures 17-11 and 17-12 in Chapter 17). Moreover, in some R plasmids IS elements are located at the boundary between the region determining antibiotic resistance and the genes for DNA replication and conjugal transfer (Figure 17-15). The significance of these sites in plasmids will be discussed later.

IS elements or segments thereof have also been found in several *E. coli* phages. For example, IS1 is a normal component of phage P1 and various IS elements have been located in particular mutants of λ. Some tantalizing matching of sequences in different organisms has also been detected. For example, a decanucleotide near the end of IS1 is identical to a decanucleotide in the prophage attachment site of phage λ, and IS1 carries sequences near its termini that are similar to sequences found in the leftward promoter of λ (*pL*) and in the promoter for synthesis of tRNA^{Tyr}. The significance of these observations is unknown.

TRANSPOSONS

The IS elements just described are a special class of transposon. In this section we examine the more general features of transposons and how they are detected.

Detection of Transposition

In the preceding section it was explained that IS elements are normally detected biologically by virtue of polarity effects on the gene in which insertion has occurred. This is not the case for a typical bacterial transposon, for these elements frequently contain recognizable bacterial genes—for instance, genes for antibiotic resistance; it is via these genes that transposition is usually detected. We will consider two examples of transposition, shown in Figures 19-3 and 19-4.

Figure 19-3 shows transposition detected with a temperate phage that has acquired an amp^+ (ampicillin-resistance) gene from a $recA^-$ cell containing an amp^+ R plasmid in which the amp^+ gene is carried on a transposon. The $recA^-$ cell is chosen for the experiment to eliminate all possibility of homologous recombination. The cell is infected with the phage and in a small number of progeny phage (perhaps 1 in 10^7) the transposon carrying the amp^+ gene has been picked up. The resulting phage population is used to infect an amp^- (ampicillin-sensitive) culture of bacteria at a ratio of about 1 phage per 10 bacteria in a growth medium containing ampicillin. Uninfected bacteria are killed. More important, cells infected with a phage lacking the amp^+ gene will not yield progeny phage because of premature lysis

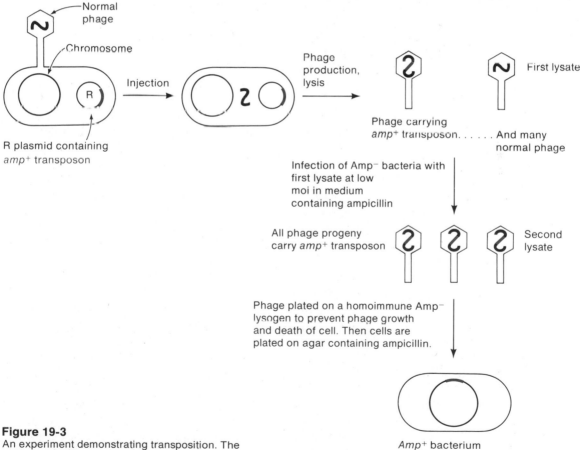

Figure 19-3
An experiment demonstrating transposition. The transposon is shown in red. See text for details.

induced by the drug; however, if the phage carries the amp^+ gene, the bacteria survive and progeny amp^+ phage are produced. Thus, in the presence of ampicillin, only the λ carrying the amp^+ transposon can grow; the lysate therefore consists of a homogeneous population of λamp^+ phage.

To demonstrate that the amp^+ gene is part of a transposon, the λamp^+ phage are used to infect an amp^-recA^-(λ) lysogen. Since the lysogen is homoimmune, the phage are unable to develop and the bacteria survive. The infected cells are then plated on agar containing ampicillin. The amp^+ phage enables the cells to grow, but, since the phage cannot replicate (because of the presence of the phage λ repressor), in each division cycle one daughter cell does not inherit a λamp^+ DNA molecule, and a bacterial colony cannot form. However, in a few bacteria, transposition does occur, the bacterial chromosome acquires the amp^+ transposon, and all daughter cells do inherit an amp^+ gene. Thus, formation of a colony indicates that transposition has occurred. Note that the resulting amp^+ bacteria could be reinfected again by a normal phage to yield a second population of phage, a few of which would contain the transposon.

This procedure can be used to generate mutants of particular genes. For example, a transposon might insert in the *lac* gene. Thus, if the amp^- lysogen used in the last stage of the figure were lac^+ and a lac^- mutant were desired, one would look among hundreds of Amp$^+$ colonies for one that is Amp$^+$Lac$^-$. This could easily be done by plating the infected cells on color-indicator agar containing lactose and ampicillin; the Lac$^+$ and Lac$^-$ colonies can be distinguished by their different colors. This technique is a useful one for preparing defective mutants since the fraction of drug-resistant colonies that are mutated in a particular gene is generally higher than the fraction of the mutant cells induced by most mutagens. The mutants obtained are of course absolute defectives, rather than conditional mutants, because the coding sequence is interrupted by the transposon.

In the second experiment, shown in Figure 19-4, an R plasmid having an amp^+ transposon is transferred by conjugation to a $recA^-$ recipient bacterium that carries a transposon having the kan^+ (kanamycin-resistance) gene. On continued growth in the presence of both antibiotics, transposition occasionally occurs and yields an R plasmid carrying both drug-resistance markers and hence both transposons. The presence of this two-transposon plasmid is detected by transferring the plasmid by conjugation to a Kan$^-$Amp$^-$ bacterium and plating on agar containing both antibiotics. Transposition can be observed again if desired because in the course of many generations of growth one of the transposons (for example, the amp^+ transposon) might transpose to the chromosome. If this occurred and the plasmid was removed from the cell by genetic means (e.g., acridine curing, Chapter 17), the cell would

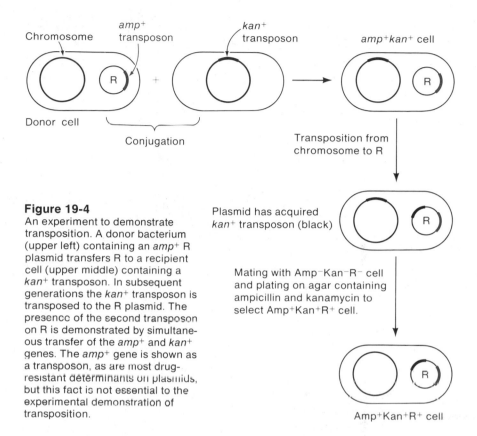

Figure 19-4
An experiment to demonstrate transposition. A donor bacterium (upper left) containing an *amp*+ R plasmid transfers R to a recipient cell (upper middle) containing a *kan*+ transposon. In subsequent generations the *kan*+ transposon is transposed to the R plasmid. The presence of the second transposon on R is demonstrated by simultaneous transfer of the *amp*+ and *kan*+ genes. The *amp*+ gene is shown as a transposon, as are most drug-resistant determinants on plasmids, but this fact is not essential to the experimental demonstration of transposition.

be Amp+. This transposition between a chromosome and an R factor and also between two drug-resistance plasmids (which is commonly observed) is a major route to the formation of the multiple drug-resistance plasmids that have been responsible for drug-resistant infections in hospitals.

In both experiments shown in Figures 19-3 and 19-4 the antibiotic-resistant phage and the plasmid containing the two markers could be hybridized to the original phage or plasmid molecule and examined by electron microscopy; loops such as those in Figure 19-1 would be seen, indicating the presence of an inserted DNA sequence.

Location of Sites of Insertions of Transposons

When transposition occurs, a particular transposon can usually be inserted at one of a very large number of positions. For example, if a wild-type *recA*− *E. coli* culture is infected with a phage carrying a transposable antibiotic-resistance marker, a transposition event can be detected by the production of antibiotic-resistant bacteria that carry a mutation in another gene, as shown in the lower part of Figure 19-3. For example, an infected $lac^+ leu^+ amp^-$ culture can yield both $lac^- amp^+$ and $leu^- amp^+$ mutants. If many hundreds of Amp$^+$ bacterial colonies are examined and tested for both nutritional requirements and ability to utilize particular sugars as a carbon source, it is observed with most of the large transposons that a colony can be found bearing a mutation in almost any gene that is examined. This indicates that insertion sites for transposition are scattered throughout the E. coli chromosome. However, some of the smaller transposons (e.g., the IS elements) seem to have a limited number of insertion sites.

A similar experiment shows that insertion can occur in many sites even in a small DNA segment (molecular weight of about 10×10^6). For example, in an experiment using phage P22, such as that shown in the upper part of Figure 19-3, many different P22 mutants carrying an antibiotic-resistance marker were isolated. To locate the sites of insertion, the DNA from these phage mutants was isolated, hybridized separately with wild-type P22, and examined by electron microscopy; the position of the loop (Figure 19-1) indicates the insertion site. Loops were found in many sites, as shown in Figure 19-5.

Although there are certainly many sites of insertion in E. coli, P22, and λ, it is clear that the locations are not randomly distributed; certain positions seem to be excluded and others (called "hotspots") are used repeatedly. This can be seen by examining the insertion sites in very small regions whose base sequences are known. Such studies show that there is a rather broad range of insertion specificities from one transposon to the next. At one extreme is IS4, for which 20 independent insertion events have been observed in the galT gene, all of which are probably at the same nucleotide site. At the other extreme is the phage

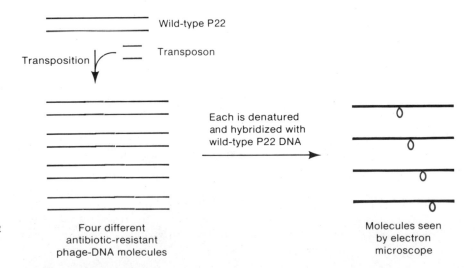

Figure 19-5
An experiment showing that a particular transposon can be inserted at several different sites in phage P22 DNA.

Wild-type P22

Transposition

Transposon

Each is denatured and hybridized with wild-type P22 DNA

Four different antibiotic-resistant phage-DNA molecules

Molecules seen by electron microscope

Mu, which will be discussed later and which probably can insert almost anywhere. Most transposons exhibit insertion specificities between these extremes. For example, sequences of the insertion sites of transposon Tn9 in the final 160 base pairs of the *lacZ* gene have been determined; of 28 independent insertions, 16 different positions have been noted but 5 of these are represented by multiple occurrences of insertion. The transposon Tn10 is even more specific; for example, 22 different sites have been noted in the *his* operon of *Salmonella typhimurium* but 40 percent of all known transposition events have occurred at a single nucleotide position. Little is known about the cause of hot spots for insertion.

Types of Transposons

Transposons have been divided into three classes: (1) the IS elements and composite structures containing them, (2) the Tn3 class, and (3) the phage Mu class. These classes have the following properties:

1. *IS elements and composite transposons.* The IS elements already described are small segments of DNA 750–1500 base pairs long (Table 19-1). Genetic analysis and the identification of start and stop codons in the base sequence indicate that each IS element encodes at least one protein, which is necessary for transposition. Some elements contain a second protein, which probably plays some regulatory role. A variety of transposons carrying antibiotic-resistance genes, which themselves are components of other transposons, consist of the transposable element flanked by two identical or nearly identical copies of an IS element. These transposons are called composite type I transposons. The IS elements in these composite units can be in an inverted or direct repeat configuration (Figure 19-6). Since the two ends of an IS element are themselves inverted repeats, the relative orientation (inverted or direct)

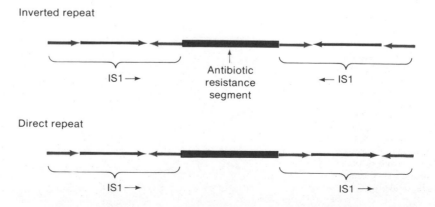

Inverted repeat

IS1 →   Antibiotic resistance segment   ← IS1

Direct repeat

IS1 →   IS1 →

Figure 19-6
Two composite type-I transposons flanked by IS1 in either inverted or direct repeat.

Table 19-2
Properties of selected composite type I transposons of *E. coli*

Element	Genes carried*	Size in base pairs	Terminal IS element and size in base pairs	Relative directions of terminal IS element
Tn5	kan	5400	IS50 (1500)	Inverted
Tn9	cam	2638	IS1 (768)	Direct
Tn10	tet	9300	IS10 (1400)	Inverted
Tn204	cam, fus	2457	IS1 (768)	Direct
Tn903	kan	3100	IS903 (1050)	Inverted
Tn1681	ent	2088	IS1 (768)	Inverted

*cam, choramphenicol; *ent*, enterotoxin; *fus*, fusidic acid; *kan*, kanamycin; *tet*, tetracycline.

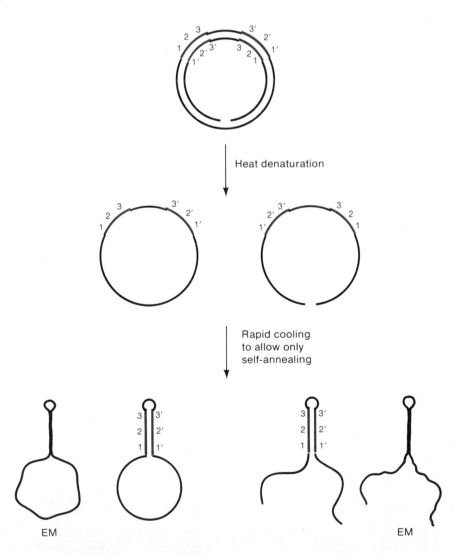

Figure 19-7
Formation of a stem-and-loop structure by denaturation and intramolecular annealing of a plasmid carrying an inverted repeat. The configurations labeled EM indicate the appearance of such a molecule in an electron microscope; light and heavy lines represent single- and double-stranded regions, respectively.

of the flanking IS elements of a composite transposon does not alter its terminal sequences—that is, they remain the same in the inverted repeat array. In most cases the ability to transpose is determined or regulated by the terminal IS elements. Several well-studied composite transposons are listed in Table 19-2.

Most of the composite type I transposons were first detected as genetic elements, which could be transposed from one plasmid to another or to a phage. When the DNA of some of these plasmids was denatured and renatured, single-stranded circles containing a stem-and-loop structure were observed by electron microscopy. Such a structure is indicative of an inverted repeat, as shown in Figure 19-7. Hybridization analysis between these plasmids and other molecules carrying known IS elements (for example, IS1) showed that some of the plasmids contained the IS elements. In this way it was shown that Tn9 (Table 19-2) contains two IS1 sequences in a direct repeat array. Other plasmids did not have regions that hybridized with known IS elements but, since the stems of the stem-and-loop structures that were seen had roughly the same size, it was assumed that these transposons are also flanked by uncharacterized IS elements. Double circles were seen also after annealing. These contained two stem-and-loop structures coming from a single site in the DNA, which also is consistent with an inverted repeat array. Some plasmids showed in hybridization experiments that a known IS element is present yet no stem-and-loop structure was seen. These proved to have the IS element in direct repeat.

2. *The Tn3 transposon family.* The Tn3 family of transposons consists of quite large elements (about 5000 base pairs). Each transposon carries three genes, one encoding β-lactamase (which confers resistance to ampicillin) and two others needed for transposition (which will be discussed later). All Tn3-like transposons contain short (38-base-pair) inverted repeats and none are flanked by IS-like elements. Recently it has been suggested that some transposons, which have been considered to be in the Tn3 family (e.g., Tn501, Tn551, Tn1721, and $\gamma\delta$) may comprise a fourth family.

3. *Transposable phages.* Two related phages, Mu and D108, use transposition as part of their normal mode of production of phage progeny. Early in infection these phages integrate their DNA into the chromosome, at any one of an enormous number of sites, creating a typical transposon sequence—namely, phage DNA flanked by a duplicated target sequence. Using a mechanism that is probably common to transposition, multiple copies of the phage DNA are transposed to many sites in the chromosome. Packaging of phage DNA ultimately occurs from these chromosomally located units. As explained in Chapter 16, a headful is packaged and this amount is greater than the length of the coding sequence of the phage; thus, the termini of the DNA within the phage particles always contains bacterial DNA.

TRANSPOSITION

The end result of the transposition process is the insertion of a transposon between two base pairs in a recipient DNA molecule. Base sequence analysis of many transposons and their insertion sites reveals that there is no sequence homology, which is consistent with the lack of a requirement for the *E. coli recA* system. However, the sequences of the regions in which the inserted element joins the recipient DNA has yielded several surprises, which we document in this section.

Duplication of a Target Sequence at an Insertion Site

A general characteristic of the insertion process is that insertion of a transposon always involves the duplication of a short base sequence (3–12 base pairs long) in the recipient DNA molecule, called the **target sequence,** and the inserted transposon is sandwiched between the repeated bases. This arrangement is shown in Figure 19-8. We repeat, emphatically, that *only one copy of the duplicated target sequence is present in the recipient DNA prior to insertion of the transposon and it is not present in the transposon itself.*

The length of the target sequence varies from one transposon to the next but *is the same for all insertions of a particular transposon.* For example, an inserted IS1 is always flanked by a nine-base-pair sequence, whereas Tn3 is always flanked by a duplication of five base pairs. Note that *for a particular transposon, the target sequence is different for each insertion site; only the length of the duplicated sequence is constant.*

The following basic mechanism, whose details are unknown and which is shown only in outline (Figure 19-9), is believed to be responsible for the duplication of the target sequence.

An enzyme, possibly encoded in the transposon, selects a target sequence (in an unknown way) in the recipient DNA and makes two single-strand breaks, one in each strand, at the ends of the target sequence. The transposon DNA then is attached to the free ends generated by the nicks, so that one strand of the transposon is joined to one strand of the target sequence and the other strand of the transposon is joined to the other strand at the opposite end of the target sequence. This joining leaves two gaps, one across from each strand of the target sequence. The gap is then filled in (presumably by DNA polymerase I) and the nicks are sealed (presumably by DNA ligase); thus, the two strands of the recipient DNA are again continuous and there are two copies of the target sequence flanking the transposon. Note that no homology is utilized in joining the terminal sequence of the transposon to the recipient DNA.

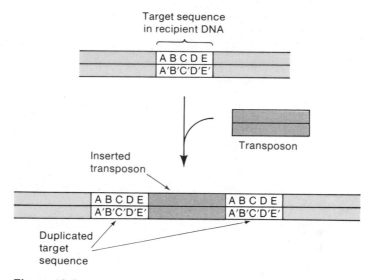

Figure 19-8
Insertion of a transposon generates a duplication of the target
sequence.

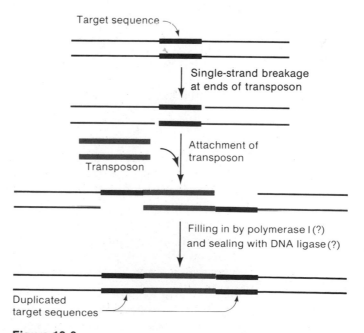

Figure 19-9
A schematic diagram indicating how target sequences might be
duplicated.

Later in this chapter more detailed proposals for generating the duplicated target sequence will be described.

Since every transposon that is integrated is flanked by a duplicated target sequence and since this is true even of those transposons maintained at fixed sites (for example, those in the F plasmid), it is reasonable to inquire whether this duplication is also necessary for transposition from a donor site. This is definitely not the case since DNA sequences have been constructed (using recombinant DNA techniques) in which a transposon is not flanked by a directly repeating sequence, and these transposons are still able to transpose.

The Structure of Transposons and the Nature of the Transposon–Target-Sequence Joint

Segments of DNA of known base sequence containing a transposon have been isolated by cleavage with restriction nucleases (Chapter 20) and the base sequence of the target sequence and the adjacent transposon have been determined. Two important facts have emerged from these analyses:

1. The transposon sequences adjacent to the two target sequences are identical for every insertion of a particular transposon (Figure 19-10). Each kind of transposon has a specific sequence, though. This means that a transposon has well-defined ends and the transposition process consists of joining the ends of the transposon to the ends produced by nicking at opposite sides of the target sequence.

2. With the exception of the transposable phage Mu, which will be discussed shortly, the two termini of the transposon, defined as above, consist of a single base sequence in an inverted repeat array. This is also shown schematically in Figure 19-10. Note that for transposon I the sequence at the left end of the upper strand and the right end of the lower strand is 1234, reading each time in the 5′→3′ direction. Depending on the transposon, the number of bases in the inverted repeat is 8–38; a few of these sequences are shown in Figure 19-11. The match at the two ends is not always perfect, though, as indicated by the vertical arrows. For Mu the termini are not inverted repeats, though there is partial symmetry of a much more complex sort.

A great deal of genetic evidence indicates that the termini of a transposon are essential for transposition. For example, single base-pair changes, small deletions, and small insertions in the inverted repeats have been isolated and these alterations invariably prevent transposition.

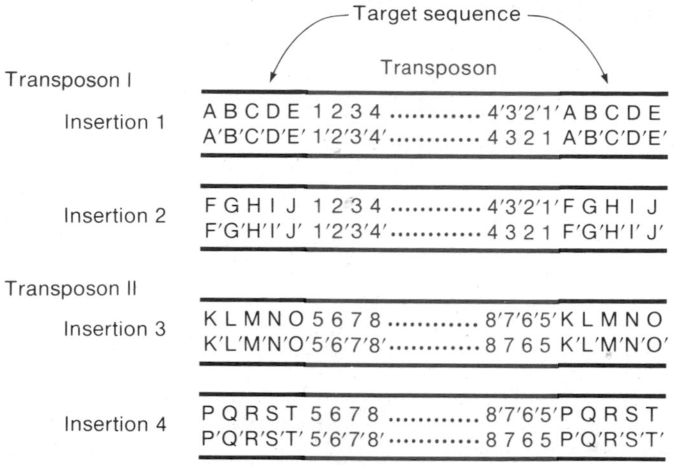

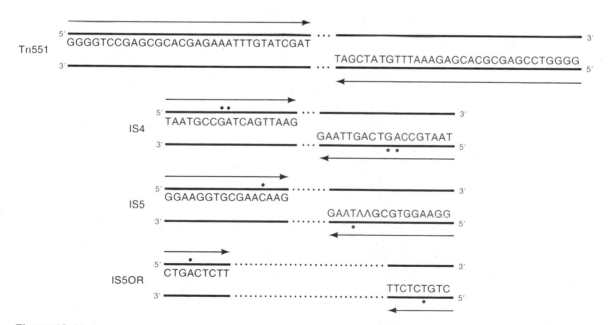

Figure 19-10
Diagram showing that two insertions by the same transposon have the same sequences at the junction of the transposon and the target sequence. That is, 1234 is at the left terminus of all insertions by transposon I. For transposon II the sequence is always 5678. Each number and each letter represents a base; the prime denotes a complementary base. The identifications are merely symbolic; 5678 designates a sequence different from 1234 and *not* one that follows 1234 in numerical order.

Figure 19-11
The terminal sequences of four transposons showing the inverted repeats. The sequence in red, if read in the direction of the arrow, is the same as the sequence in black in the direction of the arrow, except for the few imperfections denoted by the dots.

THE MECHANISM OF TRANSPOSITION

So far nothing has been said about the molecular mechanism of transposition, mainly because its mechanism is not known. In the following sections a transposon-mediated phenomenon called **cointegration** is described, which provides major clues for understanding transposition. We will then examine the presumed role of cointegration in the transposition process and discuss a few hypotheses for its mechanism.

Cointegration

Let us consider a cell possessing two plasmids, one of which contains a transposon. At a frequency of about 10^{-7} events per cell generation, the two plasmids fuse to form a single plasmid called a cointegrate. (This process is sometimes also called **replicon fusion.**) This hybrid plasmid is not simply a fusion of the two plasmids, because it contains two (not just one) copies of the transposon. Both copies are in the same orientation (that is, in direct repeat) and precisely at the junction between the donor and recipient plasmid sequences (Figure 19-12). As far as is known, all transposons are capable of mediating cointegration, though the cointegrates are not always stable.

The formation of cointegrates is significant for two reasons. First, cointegrates are thought to be an essential intermediate in the transposition process and, second, the fact that the cointegrate has two copies of the transposon, whereas only one copy was present before cointegration occurred, clearly indicates that *DNA replication of the transposon has occurred.* These two points will be examined further in the next two sections.

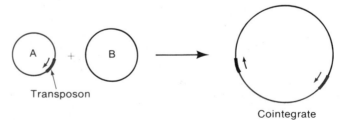

Figure 19-12
Formation of a cointegrate by transposon-mediated fusion of two plasmids A and B. Two copies of the transposon are generated in the process. The arrows denote an arbitrary direction to show that both copies of the transposon in the cointegrate are in direct repeat.

Replication of a Transposon in the Course of Transposition Between Two Plasmids

In the previous section it was shown that a plasmid containing a transposon can fuse with a second plasmid to form a cointegrate—that is, a structure containing two copies of the transposon in direct repeat. A more common observation is actual transposition, which is shown in Figure 19-13. That is, after many generations of growth of a bacterium containing both plasmid A, which has a transposon, and plasmid B, which lacks a transposon, cells are produced in which plasmid B (now called B′) also possesses the transposon. An important finding is that if this cell has succeeded in retaining both plasmids, *a copy of the transposon will be contained in both plasmids*. Thus the original transposon sequence is duplicated in the transposition process, indicating that *transposition is a replicative process*, as stated in the preceding section.

Another informative observation is that if a cell containing a cointegrate (which of course already possesses two copies of the transposon) is grown for many generations, cells arise that lack the cointegrate and contain both plasmids, each of which contains a copy of the transposon. This again suggests that cointegration may be the first step in transposition; this view is examined in the next section.

The Cointegrate as an Intermediate in Transposition of Tn3

Certain experimental results with transposon Tn3 suggest quite strongly that formation of a cointegrate is an intermediate step in the transposition process.

The structure of Tn3 is shown in Figure 19-14. Tn3 has a pair of inverted repeats, like all transposons, and these flank three genes. The three gene products have been isolated and from their size and the base sequence of Tn3, we know that the genes utilize all of the bases between the inverted repeats. The leftmost gene encodes a gigantic protein (1015 amino acids), denoted TnpA, which is a transposase and which is

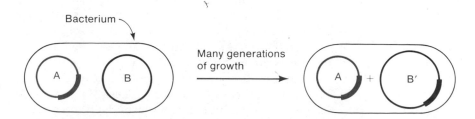

Figure 19-13
Transposition of a transposon from plasmid A to a second plasmid B.

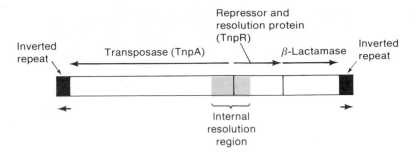

Figure 19-14
The physical map of transposon Tn3. There is a total of 4957 base pairs; each inverted repeat contains 38 base pairs. The three genes are indicated above the upper arrows, which show the direction of transcription. The internal resolution region is discussed in the text.

responsible for formation of cointegrates. The rightmost gene encodes β-lactamase, an enzyme that inactivates ampicillin. The central gene encodes a small protein (185 amino acids) called TnpR, which has two functions—negative regulation or repression of the synthesis of TnpA, and promotion of a site-specific exchange that resolves cointegration in a second step of the transposition process. Near the boundary of the genes encoding TnpA and TnpR is a sequence of DNA called the **internal resolution site** and it includes the base sequence at which TnpR acts.

Genetic experiments have yielded the following information. All $tnpA^-$ mutants are unable either to transpose or to form cointegrates. Also, $tnpR^-$ mutants and deletions of the internal resolution site form cointegrates yet are unable to transpose. That is, in a cell containing a plasmid with a $tnpR^-$ copy of Tn3 and a second plasmid lacking Tn3, the process shown in Figure 19-12 can occur, but the process shown in Figure 19-13 does not occur. These observations suggest that Tn3-mediated transposition occurs by a two-step process—first, TnpA induces formation of a cointegrate and then TnpR promotes a site-specific exchange at the internal resolution site. A schematic diagram of this sequence of events is shown in Figure 19-15.

Some transposons do not carry a TnpR-like site-specific recombination system but are still able to transpose. These systems presumably use a homology-dependent exchange system since, once a cointegrate has formed, there are two identical sequences—namely, copies of the transposon—so that any homology-dependent process could carry out the exchange shown in Figure 19-15. This view is supported by the fact that a mutant of Tn3 lacking TnpR or the internal resolution site can carry out the transposition process of Figure 19-13 (that is, reduction of a cointegrate to individual plasmids) if a $recA$ gene is present in the host cell; in fact, if the $recA^+$ allele is introduced by a transducing particle

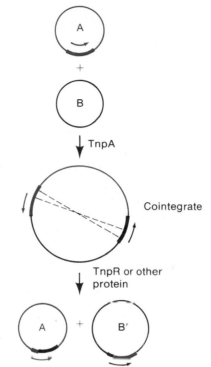

Figure 19-15
A model for transposition utilizing a cointegrate intermediate.

into a *recA*⁻ cell in which a cointegrate has already formed, transposition can occur.

Two Models for Transposition

The mechanism of transposition is not known, but current proposals abound. The facts that must be accounted for are the following: (1) the retention of the transposon by the donor at the initial site of the transposon; that is, that the process is replicative, (2) the existence of cointegrates, (3) the generation of a short repeated sequence of target DNA on each side of the newly integrated element, and (4) the requirement of some transposons for an internal resolution site or, in others, for a homology-dependent exchange.

Many models for transposition have been proposed; some are simply subtle variations of one another and others are quite different. Two features are common to all models, namely:

1. Two single strands are joined together (ligated) without base pairing to indicate the position of joining.
2. A replication fork is created by the ligation event.

The reader should note these features in the two models to be described.

One widely favored model is shown schematically in Figure 19-16. In this model the cointegrate is taken as an intermediate. After describing this model we shall consider a second model in which formation of a cointegrate is not an intermediate but an alternative to simple insertion of the transposon. The second model probably does not explain the transposition of Tn3 but may apply to other transposons for which there is no evidence of an internal resolution site.

We consider a donor plasmid containing a transposon and a recipient plasmid. (Neither the donor nor the recipient need to be plasmids or even circular. However, the figure is drawn that way because it is the arrangement most often studied.) In the recipient there is a short base sequence, the target sequence, at which integration of the transposon is to occur. No assumption is made concerning how the transposon and the target sequence are to be linked. The following steps, numbered as in Figure 19-16, have been proposed:

I. A pair of single-strand breaks is made (by transposase?) in the target sequence. These breaks are staggered, so that if the target sequence is broken, two complementary single strands result.

II. Two single-strand breaks are made in different strands at opposite ends of the transposon; this generates two free ends. Each free end is attached by a single strand to a protruding strand of the staggered cut in the target site; this generates two replication forks.

III. Replication begins by synthesis of a strand complementary to the protruding end of the target sequence. When replication is completed, a cointegrate is formed containing two copies of the transposon.

IV. Finally double-stranded exchange occurs between the two copies of the transposon. This exchange might be in a terminal repeat, though in Tn3, at least, it is probably in the internal resolution site. The result of the exchange is to separate the cointegrate into the donor and recipient units, each of which now has a copy of the transposon. In the recipient the transposon is flanked by a direct repeat of the target site; the length of this repeat is the number of base pairs between the staggered cuts.

We next consider another popular type of model called an **asymmetric model.** The one that will be explained is a synthesis of several that have been proposed. The asymmetric models differ from the model just described in two major respects: (1) the cuts at the end of the transposon are not made simultaneously but at two different stages of the insertion process, and (2) a cointegrate is not an intermediate but one of two alternative products (the other being a simple insertion) that could

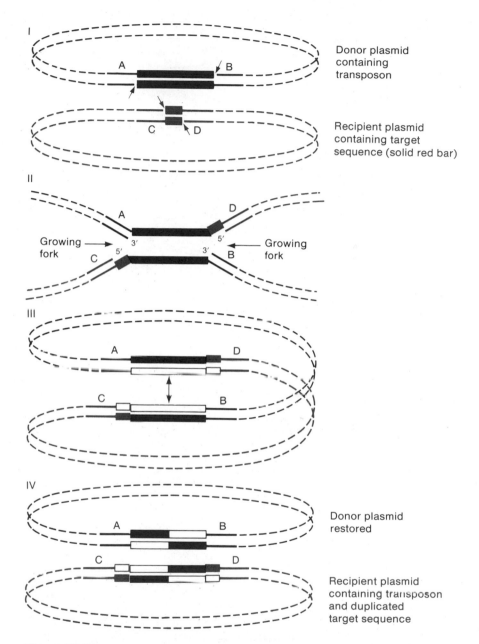

Figure 19-16
A model for transposition, proposed by J. A. Shapiro (see, in References, S. N. Cohen and J. A. Shapiro, 1980). (I) Nicks are made at sites in the DNA molecule indicated by arrows. (II) Strand separation between nicks and joining of nonhomologous strands. Two replication forks result. (III) DNA synthesis (hollow lines) forms this structure. A site-specific strand exchange occurs between the internal resolution sites (arrows), or possibly a homology-dependent strand exchange occurs anywhere in the transposon. (IV) Plasmids separate. Capital letters denote base sequences.

Figure 19-17

An asymmetric model for transposition that can be terminated by simple insertion or formation of a cointegrate (row IV, left and right panels respectively). The two structures in row III are identical; two copies are drawn for clarity in showing the cuts and joints needed to achieve the alternate structures of row IV.

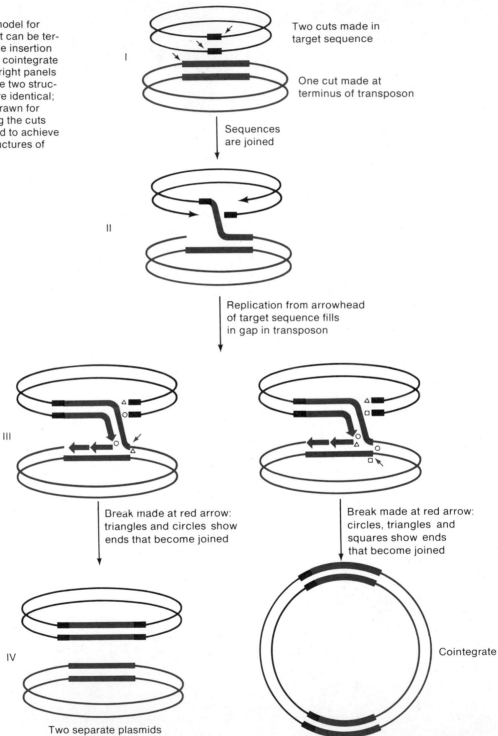

occur at the end of the process. This model is shown in Figure 19-17. The following steps, numbered as in the figure, occur:

I. The target sequence is cleaved in two places but only one cut is made at the left end of the transposon.

II. Joining of the free end of the transposon to one end of the target sequence occurs.

III. Replication then occurs to form the intermediate shown. Note that replication terminates at the end of the transposon; presumably some base sequence would signal this termination.

IV. After replication has ceased, two kinds of cuts and joints can occur, giving rise either to simple insertion with the usual duplication of the transposon (left side of row IV) or a cointegrate (right side of row IV).

There is little substantial evidence at present to support this model, which should simply be taken as a prototype for asymmetric models.

A great deal of research effort is currently being expended to determine the validity of the many models (of both main types) that have been proposed.

OTHER GENETIC PHENOMENA MEDIATED BY TRANSPOSONS

Transposons mediate a variety of genetic phenomena such as gene rearrangement and plasmid-chromosome integration, both of which will be described in this section.

Genetic rearrangement has been very important throughout evolution and may have been mediated by transposons. For example, possibly transposons were responsible for bringing together genes that were originally distant in the chromosome and are now organized in coordinately regulated operons. It is also possible that new proteins were formed in a single giant step forward when two separate base sequences became linked.

Genetic rearrangement can also be important within a single cell generation. For example, transposable elements can serve as switches, turning genes on and off as a consequence of insertion at a particular location or by inverting sequences containing promoters.

An important point to be noticed in the discussions that follow is that each event is a transposon-mediated process that occurs without the final resolution step of Figure 19-16 (panel IV). However, these events should not be thought of as aberrant, for they occur with all transposons that lack a resolution system (for example, the IS elements) and with transposons capable of resolution but in which resolution is infrequent.

Replicon Fusion

Examination of Figure 19-16 shows that, if the resolution step (panel IV) does not occur, the two circular molecules shown in panel I fuse by transposon-mediated recombination to form a single circular molecule (panel III). This new DNA molecule contains all of the genes of each of the original molecules plus two copies of the transposon. Note, however, that in this process each copy of the transposon is not flanked by a duplicated target sequence; instead, one copy of each target sequence is adjacent to each transposon. This type of recombination has been observed between two plasmids, one of which carries a transposon; this is called **plasmid fusion** and it is a special case of a more general phenomenon called **replicon fusion**—that is, the joining of two complete replicating systems mediated by a transposon.

Deletions and Inversions Caused by Transposons

When a transposon is present in a chromosome or plasmid, the frequency of nearby deletions is about 100–1000 times greater than the value observed if the transposon were not present. When such a deletion occurs, it is often found that (1) the transposon is still present, (2) the deletion begins at the end of the element, and (3) the transposon is bounded by only one copy of the target sequence (that is, there is no sequence in direct repetition surrounding the transposon). These features are shown in Figure 19-18. This phenomenon is easily explained in Figure 19-19, which is based on the model of Figure 19-16.

In this figure the array shown in Figure 19-16 is repeated—that is, segments A and B are separated by a transposon and segments C and D surround a target sequence, as shown in panels I and 1 of the two figures. Panel 3 shows the array surrounding the two copies of the transposon after steps I through III of Figure 19-16 are completed. If each of the segments in panel 1 were circles, the two segments in panel 3 would be fused into one circle, as explained in the preceding section.

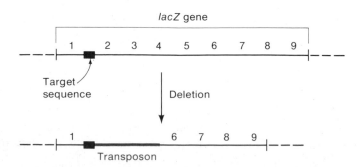

Figure 19-18
Characteristics of a deletion mediated by a transposon. Segments 2, 3, 4, and 5 are deleted and replaced by a transposon.

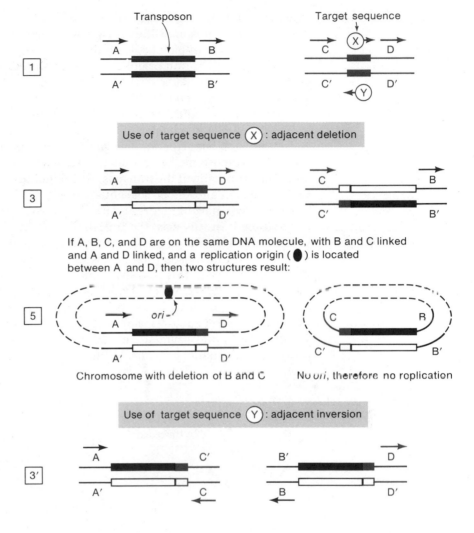

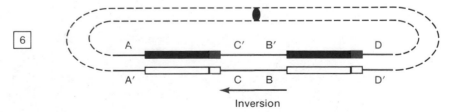

Figure 19-19
Schematic diagram showing how genetic deletions and inversions can form in accordance with the Shapiro model of Figure 19-16. Panel 1 has the same array as I in Figure 19-16; 3 and 3′ correspond to III in Figure 19-16, except that two different orientations of a target sequence are used. Solid arrows indicate the direction of transcription. Other symbols are those used in Figure 19-16.

However, if both segments are part of a single DNA molecule (for example, a chromosome), the transposition process would yield the two molecules shown in panel 5. Note that only one molecule in this panel contains the replication origin and that this molecule has also lost the segment containing B and C; that is, this molecule is deleted for B and C. The smaller circle containing B and C does not have a copy of the replication origin and therefore cannot replicate. Thus, if this entire process occurred in a bacterium, the smaller circle would not multiply as cell division occurred and progeny bacteria would only contain the deleted chromosome. Note that the segment deleted is immediately adjacent to one side of the transposon and includes all material between the transposon and the target sequence.

The lower part of Figure 19-19 shows how genetic inversions could be produced. In this case the initial gene array is exactly that which yielded the deletion but the target sequence has an orientation opposite to that of the transposon (that is, the strands ligated in panel I of Figure 19-16 are reversed); this is equivalent to insertion of the transposon in the opposite orientation. The transposition process then gives rise to the arrays shown in panel 3′. Note that the orientations of A and C are inverted with respect to the starting array, as are the sequences C and D. Thus, if the initial segments come from a single DNA molecule, the array shown in panel 6 results. In this case, no DNA is lost (that is, there

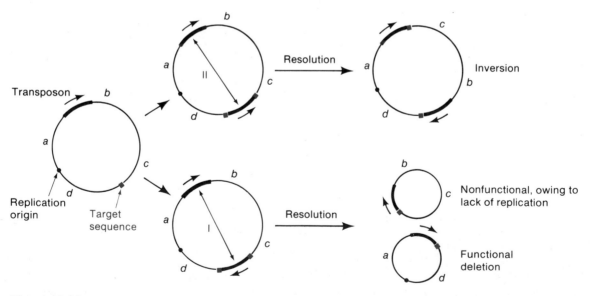

Figure 19-20
Model for production of genetic deletions and inversions. The transposon DNA is inserted into the target sequence in orientation I or II. The circle could be a plasmid or the chromosome. Resolution by a site-specific strand ex- change at sites indicated by double-headed arrows or by exchange in homologous sequences yields a deletion from I and an inversion from II.

is no deletion), but the segment between one side of the transposon and the target sequence is inverted with respect to the other segment—that is, there is a **genetic inversion.**

Although most of the transposon-mediated inversions are formed in this way—that is, without the resolution step—it is possible that others are formed by a mechanism using resolution or perhaps homologous *recA*-mediated recombination. A simple mechanism by which this might occur is shown in Figure 19-20. Here transposition occurs at a target sequence (indicated by a square in the left circle of the figure) in the molecule containing the transposon. Insertion can occur in either a direct repeat orientation (I in the figure) or an inverted repeat (II in the figure). With a transposon capable of resolution a site-specific exchange could occur within the two transposons; alternatively, a homology-dependent exchange mediated either by a transposon gene or by the RecA protein, if the cell is RecA$^+$, may occur. If the transposons are in a direct repeat array, the exchange would generate two circles. As in the model shown in Figure 19-19, since the chromosome or a plasmid generally contains a single replication origin, only the element possessing the origin can persist; this element has lost a DNA segment and thus is a deletion mutant. If the transposons are in the inverted repeat array, the result is instead an inverted sequence, as shown in the right portion of the figure.

Note that the end results of the schemes shown in Figures 19-19 and 19-20 are the same.

PROCESSES UTILIZING HOMOLOGY BETWEEN TWO COPIES OF A SINGLE TRANSPOSON

Several specific processes are known to be the result of *recA*-mediated homologous recombination acting on multiple copies of transposons. Two of these, formation of Hfr cells and gene amplification, are described.

Role of IS Elements in Hfr Formation

Hfr cells are formed by integration of F. Several lines of evidence (principally the determination of the structure of F′ plasmids) indicate that an integrated F sequence is always flanked by two copies (in direct repetition) of one of the IS elements found in the F plasmid. This and the fact that the IS elements in F are also found in the chromosome (Table 19-1) indicates that Hfr formation involves transposons.

There are two ways by which Hfr cells might arise in a cell containing F. One way is by homologous recombination between two identical IS elements, one in the chromosome and one in F. If the recombinational event is physically reciprocal, this mechanism would yield flanking IS elements and might utilize the bacterial Rec system, a transposon gene product, or conceivably some product of a gene in F. Alternatively, an Hfr cell could arise by formation of a cointegrate mediated by an IS element in F, and by duplication of a target sequence in the chromosome (i.e., replicon fusion). It is important to note that the consequences of these two mechanisms are the same—namely, F flanked by two copies of the IS element in direct repeat. It is likely that Hfr strains arise by both mechanisms for the reasons that follow.

There are many sites of integration of F. However, at some sites, called **hot spots,** integration occurs more frequently than at other sites (called minor sites). The base sequences of several regions of the chromosome that include hot spots have been determined and in each case an IS element has been found. Furthermore, most Hfr cell lines have been rec^+F^+ cells, so that it seems likely that these Hfr strains were produced by $recA$-mediated homologous recombination. Integration at the minor sites occurs at a much lower frequency and often in $recA^-$ cells; the resultant Hfr strains probably arise by the cointegrate pathway.

Gene Amplification in Bacteria

Another phenomenon for which $recA$-mediated recombination between multiple copies of transposons has been invoked as an explanation is gene amplification in bacteria.

Some, but not all, R plasmids have the property that if a bacterium containing the plasmid is exposed to an antibiotic concentration considerably lower than the maximum concentration tolerated by the cell, over a period of many generations the cell becomes resistant to ever-increasing concentrations of the antibiotic. This results from a gradually increasing number of antibiotic-resistance segments in the R plasmid. That is, the R plasmid increases in size owing to repeated tandem duplications of the antibiotic-resistance genes. This process requires an active bacterial Rec system, showing that duplication is not a transposition process. Based on the fact that the antibiotic-resistance segment of R plasmids are flanked by transposons, two models for gene amplification shown in Figure 19-21 have been developed. (There are other models but these two are especially simple.) Note that the two basic requirements of the models are that the genes to be amplified are flanked by identical sequences and that the Rec system acts effectively on pairs of homologous regions. There is no proof for these models at present.

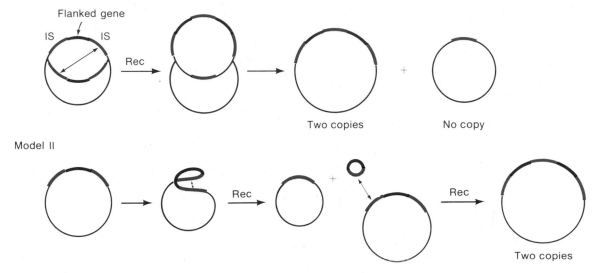

Figure 19-21
Two possible models for gene amplification. Rec-mediated recombination occurs between IS elements as shown by the two-headed arrow. In Model I recombination occurs during replication. In Model II there are two Rec-mediated events. The second event is between an excised minicircle and another copy of the original plasmid.

TRANSPOSONS IN EUKARYOTES

The first evidence for the existence of transposable elements came not from bacteria but from ingenious genetic studies of maize by Barbara McClintock. Her observations indicated that the activities of certain genetic loci in corn are disturbed by a controlling element that inhibits the activity only when the element is adjacent to the locus; change in level of gene activity seemed to occur by movement of the element to and from its inhibiting position. Recently, physical data have shown that the controlling element is a transposon.

Transposons have been found in the yeast *Saccharomyces cerevisiae* and the fruit fly *Drosophila melanogaster*. One of the yeast elements, δ, has been isolated. It has 338 base pairs and there are about 100 copies per haploid cell. A transposonlike element, Ty1, has also been isolated in yeast. It consists of approximately 5100 base pairs, flanked by two copies of δ. There are about 35 copies of Ty1 per haploid cell. Furthermore, polar mutations are associated with insertion of Ty1 and a transcriptional start site is near the end of δ. In the few cases in which there is base sequence data, insertion of Ty1 is accompanied by a flanking five-base-pair sequence—the signature of a transposon in E. coli. In *Drosophila* about 3 percent of the DNA consists of about 5000 different inverted repeats. Most of these sequences contain about 50–200 base pairs. A transposon called **copia** contains about 4400 base pairs and is flanked by two copies of a 300-base-pair element in direct repeat. Transposition of copia affects the activity of certain genes, presumably by movement of regulatory sites, such as promoters and other protein-binding sites, within the transposon. Copia is also flanked by a duplication of five base pairs in direct repetition.

A number of genetic results with *Drosophila* can be explained by the existence of transposons. For example, there are mutations that inhibit nearby recombination events, are polar, and have been shown to be insertions.

An important possibility is that some (or perhaps all) of the RNA tumor viruses are themselves transposons; however, the only evidence for this view at present is that several integrated tumor viruses are flanked by 600-base-pair sequences in direct repetition and, in addition, a duplication of a short target sequence; this is discussed further in Chapter 21.

REFERENCES

Bukhari, A. I. 1981. "Models of DNA transposition." *Trends Biochem. Sci.,* 6, 56–60.

Bukhari, A. I., J. A. Shapiro, and S. L. Adhya. 1977. *DNA Insertion Elements and Plasmids.* Cold Spring Harbor Laboratory.

Calos, M. P., and J. H. Miller. 1980. "Transposable elements." *Cell,* 20, 579–595.

Campbell, A. 1981. "Evolutionary significance of accessory DNA elements in bacteria." *Ann. Rev. Microbiol.,* 35, 55–84.

Campbell, A., M. Benedeck, and L. Hefferman. 1979. "Viruses and inserting elements in chromosomal evolution." *In* A.S. Dion, ed. *Concepts of the Structure and Function of DNA, Chromatin, and Chromosomes.* Symposia Specialists.

Cohen, S. N. 1976. "Transposable genetic elements and plasmid evolution." *Nature,* 263, 731–738.

Cohen, S. N., and J. A. Shapiro. 1980. "Transposable genetic elements." *Scient. Amer.,* February. pp. 40–49.

Green, M. M. 1980."Transposable elements in Drosophila and other Diptera." *Ann. Rev. Genetics,* 14, 109–120.

Kleckner, N. 1977. "Translocatable elements in procaryotes." *Cell,* 11, 11–23.

Kleckner, N. 1981. "Transposable genetic elements." *Ann. Rev. Genetics,* 15, 341–404.

McClintock, B. 1965. "The control of gene action in maize." *Brookhaven Symp. Biol.,* 18, 162–184.

Nevers, P., and H. Saedler, 1977. "Transposable genetic elements as agents of gene instability and chromosomal rearrangements." *Nature,* 268, 109–115.

Starlinger, P. 1977. "DNA rearrangements in procaryotes." *Ann. Rev. Genetics,* 11, 103–126.

Starlinger, P., and H. Saedler. 1976. "IS elements in microorganisms." *Current Topics Microbiol. Immunol.,* 75, 111–152.

20 Recombinant DNA and Genetic Engineering

In many instances in the past forty years some technical development has led to a giant step forward in understanding molecular biology. For instance, x-ray diffraction enabled researchers to have a detailed view of the structure of macromolecules, and the technique of CsCl density-gradient centrifugation opened the door to understanding the mechanics of DNA replication. The most recent development has been a technique for joining DNA molecules together *in vitro,* namely, recombinant DNA technology. This basic technique and its many variants have revolutionized the study of eukaryotes and have provided sources of particular proteins in quantities hitherto considered to be nearly impossible to obtain.

Up to now, this book has been primarily concerned with observations, concepts, and biological mechanisms, and little has been said about techniques. However, the recombinant DNA technology, or **genetic engineering,** as it is also called, is currently such a major research effort that it deserves a detailed look.

THE JOINING OF DNA MOLECULES

The basic procedure of recombinant DNA technique consists of two stages—(1) joining a DNA segment (which is of interest for some reason) to a DNA molecule that is able to replicate, and (2) providing a milieu that allows propagation of the joined unit (Figure 20-1). When this is

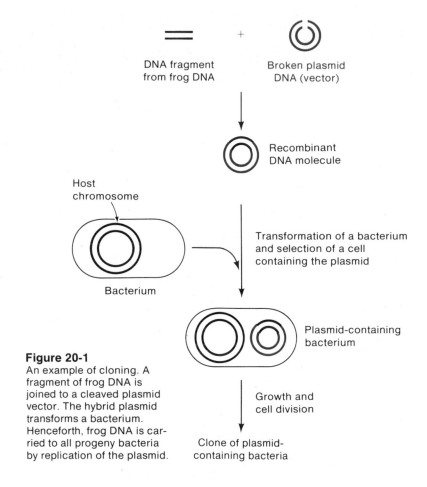

Figure 20-1
An example of cloning. A fragment of frog DNA is joined to a cleaved plasmid vector. The hybrid plasmid transforms a bacterium. Henceforth, frog DNA is carried to all progeny bacteria by replication of the plasmid.

done, the genes on the donor segment are said to be **cloned** and the carrier molecule is the **vector** or **cloning vehicle.** In this section both the procedures used for joining and the many types of vectors in use are described.

Vectors

To be useful, a vector must have three properties:

1. It must be able to replicate.
2. There must be some means of detecting its presence.
3. There must be some way to introduce vector DNA into a cell.

The three most common types of vectors in use are plasmids, *E. coli* phage λ, and viruses, because the DNA of each of these vectors has the third of the aforementioned properties. For example, DNA molecules can enter the cells of many bacterial species if the cells are suspended in a cold $CaCl_2$ solution, and if the molecule is plasmid DNA, the plasmid will sometimes become permanently established in the bacterium. In most experimental protocols about 30 DNA molecules enter each cell and in about 0.1 percent of the cells the plasmid becomes stably established. The plasmid-containing cells are easily isolated by procedures to be described later in the chapter. When phage λ DNA is used, about one infected cell in 10^4 produces progeny phage. A transformation procedure has also been developed for the yeast, *Saccharomyces*, and its plasmids. Animal viral DNA also readily penetrates animal cells; no special treatment is required, though both the amount of DNA entering each cell and the rate of penetration are enhanced in dilute solutions of $Ca_3(PO_4)_2$. A small fraction of animal cells infected in this way produces progeny viruses. As a result of either of these processes (which are called **transformation** for plasmid DNA and **transfection** for phage or viral DNA), a recombinant DNA molecule can be maintained and propagated, either by growth of a cell containing a plasmid or by infection of cells with phages or viruses.

Only a small number of vectors are in common use; some of these vectors and their properties are listed in Table 20-1.

Table 20-1
Some cloning vehicles in use at present

Designation	Genotype or Characteristics
Plasmid	
pSC101	*tet+*
ColE1	*immE1*
pMB9	*tet+immE1*
pBR322	*tet+amp+*
pBR325	*tet+amp+cam+*
Yeast 2μm	None
Phage	
λgt4·λB	Can lysogenize; is thermally inducible.
λ−Charon	*lacZ+*
λΔz1	Insertion occurs in *lacZ* gene.
M13mp2	Contains *lacP*, *lacO*, and *lacZ*; single-stranded DNA; useful for sequencing.
Virus	
SV40	Virus infects animal cells. Maximum size of added DNA is no more than 3×10^6 molecular weight units.

Abbreviations: *tet*, tetracycline; *imm*, immunity; *amp*, ampicillin; *cam*, chloramphenicol; *lacP*, *lac* promoter; *lacO*, *lac* operator; *lacZ*, gene for β-galactosidase. For antibiotic genes, + indicates resistance.

Restriction Endonucleases

A restriction endonuclease is an enzyme that recognizes a specific base sequence in a DNA molecule and makes two cuts, one in each strand, generating 3'-OH and 5'-P termini. Roughly 400 different restriction enzymes have been purified from about 250 different microorganisms. All but a few of these enzymes recognize sequences having rotational (dyad) symmetry. That is, the recognition site generally has a sequence whose form is

$$
\begin{array}{c}
\text{A B C} \mid \text{C}' \text{B}' \text{A}' \\
\text{A}' \text{B}' \text{C}' \mid \text{C B A}
\end{array}
\quad \text{or} \quad
\begin{array}{c}
\text{A B X B}' \text{A}' \\
\text{A}' \text{B}' X \text{B A}
\end{array}
\quad \text{or} \quad
\begin{array}{c}
\text{A B} \mid \text{B}' \text{A}' \\
\text{A}' \text{B}' \mid \text{B A}
\end{array}
$$

in which the capital letters represent bases, a prime indicates a complementary base, X is any base, and the dashed line is the axis of symmetry. Sequences having more than six bases are known but none have been observed containing fewer than four bases.

There are two different types of restriction enzymes: **type I enzymes,** which recognize a specific sequence but make cuts elsewhere; and **type II enzymes,** which make cuts within the recognition site.* We will only be concerned with type II enzymes in this chapter. All restriction enzymes make two single-strand breaks, one break in each strand. There are two distinct arrangements of these breaks: (1) both breaks at the center of symmetry (generating **flush** or **blunt ends**), or (2) breaks that are symmetrically placed around the line of symmetry (generating **cohesive ends**). These arrangements and their consequences are shown in Figure 20-2. Table 20-2 lists the sequences and

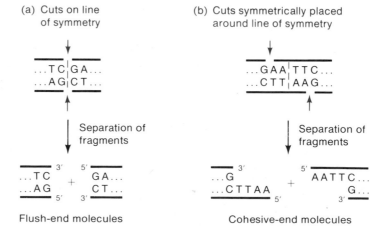

Figure 20-2
Two types of cuts made by restriction enzymes. The arrows indicate the cleavage sites. The dashed line is the center of symmetry of the sequence.

*Recently, type III enzymes have been recognized. These are like type I enzymes in their activity but consist of multiple subunits encoded *in different genes.*

Table 20-2
Some restriction endonucleases and their cleavage sites

Microorganism	Name of enzyme	Target sequence and cleavage sites
Generates cohesive ends		
E. coli	EcoRI	G A A T T C C T T A A G
Bacillus amyloliquefaciens H	BamHI	G G A T C C C C T A G G
B. globigii	BglII	A G A T C T T C T A G A
Haemophilus aegyptius	HaeII	Pu G C G C Py Py C G C G Pu
Haemophilus influenza	HindIII	A A G C T T T T C G A A
Providencia stuartii	PstI	C T G C A G G A C G T C
Streptococcus albus G	SalI	G T C G A C C A G C T G
Xanthomonas badrii	XbaI	T C T A G A A G A T C T
Thermus aquaticus	TaqI	T C G A A G C T
Generates flush ends		
Brevibacterium albidum	BalI	T G G C C A A C C G G T
Haemophilus aegyptius	HaeI	(A) G G C C (T) (T) C C G G (A)
Serratia marcescens	SmaI	C C C G G G G G G C C C

Note: The vertical dashed line indicates the axis of dyad symmetry in each sequence. Arrows indicate the sites of cutting. The enzyme TaqI yields cohesive ends consisting of two nucleotides, whereas the cohesive ends produced by the other enzymes contain four nucleotides. The enzyme HaeI recognizes the sequence GGCC whether the adjacent base pair is A·T or T·A, as long as dyad symmetry is retained. Pu and Py refer to any purine and pyrimidine, respectively.

cleavage sites for twelve useful restriction enzymes, nine of which generate cohesive sites and three of which yield flush ends.

An important point about these restriction enzymes is that since an enzyme recognizes a unique sequence, *the number of cuts made in the DNA from an organism is generally small.* For example, a typical bacterial DNA molecule, which contains roughly 3×10^6 base pairs, is cut into a few hundred fragments. Smaller DNA molecules such as phage or plasmid DNA molecules may have fewer than ten sites of cutting (frequently, one or two, and often none). We will see shortly that plasmids having a single recognition site for a particular enzyme are especially valuable as vectors.

Because of the specificity just mentioned,

> A particular restriction enzyme generates a unique family of fragments for a particular DNA molecule.

Another enzyme will generate a different family of fragments from the same DNA molecule. (Some enzymes—isoschizomers—have the same specificity and generate identical families.) Figure 20-3(a) shows the sites of cutting of *E. coli* phage λ DNA by the enzymes EcoRI and BamI. The family of fragments generated by a single enzyme is usually detected by agarose gel electrophoresis of the digested DNA (Figure 20-3(b)). The fragments migrate at a rate that is a function of molecular weight and their molecular weights can be determined by reference to fragments of known molecular weight run concurrently.

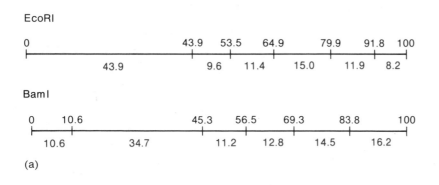

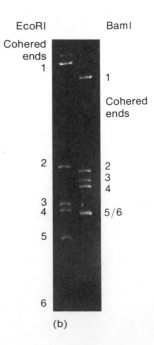

Figure 20-3

(a) Restriction maps of λ DNA for EcoRI and BamI nucleases. The vertical bars indicate the sites of cutting. The black numbers indicate the percentage of the total length of λ DNA measured from the gene-*A* end of the molecule. The red numbers are the lengths of each fragment, again expressed as percentage of total length. (b) A gel electrophoregram of EcoRI and BamI restriction-enzyme digests of λ DNA. The bands labeled cohered ends contain molecules consisting of the two terminal fragments joined by the normal cohesive ends of λ DNA. Numbers indicate fragments in order from largest (1) to smallest (6). Bands 5 and 6 of the BamI digest are not resolved. (Courtesy of Dennis Anderson and Lynn Enquist.)

Determining the Sites of Cuts; Restriction Mapping

The positions of the cuts in λ DNA — that is, the **restriction map** — may be determined in the following way (Figure 20-4).

1. The electrophoretic patterns of enzymatic digests of the linear and circular forms of the DNA are compared. The pattern for the circles lacks the two bands corresponding to the terminal fragments and contains a new band that is formed by the joined ends. Thus, the terminal fragments are identified as those that are missing from the circle digest. In a typical digest one or two cuts will not have been made in every molecule; some faint bands formed by the uncut fragments will be seen. The intensity of these bands can be increased by decreasing the reaction time, generating what is called a **partial digest.** In such a digest other bands are weakened compared to the normal digest. The molecular weight of each enhanced band is invariably the sum of the molecular weights of two fragments contained in the weakened bands, which indicates that the two fragments are adjacent in the original uncleaved molecule. If the number of cuts is small (for example, six), this procedure may be sufficient for unambiguous ordering of the fragments.

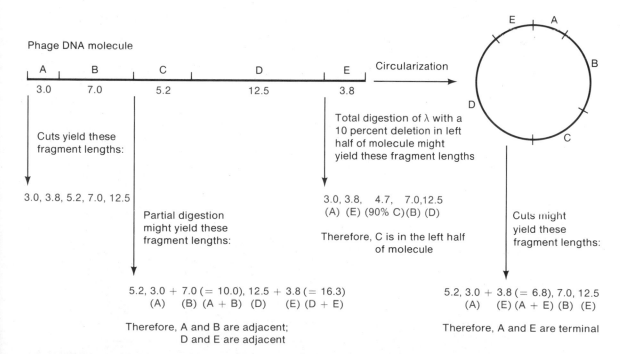

Figure 20-4
A possible scheme for determining the positions of restriction cuts in a phage DNA molecule.

2. The position of a fragment in the complete DNA molecule can also be identified by examining genetic deletion mutants. If the deletion does not eliminate a restriction site, a new fragment (some of whose DNA is deleted) appears and this is smaller, by the size of the deleted DNA, than the corresponding fragment from the wild-type undeleted molecule. If, instead, the deletion spans a restriction site, two bands are absent and a new band appears; the molecular weight of the new band equals the sum of the molecular weights of the missing fragments minus that of the deletion. Experiments of the types just described yield the pattern of cuts (the restriction map), as shown in the figure.

3. Although the method just described usually works, the most common method for deriving restriction maps is to use two different restriction enzymes. The procedure is to take three samples of a particular DNA species, treat each of two of these with a separate enzyme and one sample with both enzymes (a **double digest**), and then compare the three sets of fragments. An example of the sets of fragments that might be generated is shown in Figure 20-5. The value of this procedure is that it yields restriction maps for both enzymes simultaneously. Let us assume, as would usually be the case, that the terminal fragments in each digest are identified either by the circularization procedure used in Figure 20-4 or by labeling the 5′ termini with ^{32}P (with polynucleotide kinase) prior to enzymatic digestion and identifying the fragments that are radioactive. With this information one can then identify the fragments that are adjacent to the terminal fragments. Let us first determine which fragment in digest I is adjacent to fragment A. In the double digest the cut forming fragment F (of digest II) must divide the fragment of digest I adjacent to A, thereby generating

Digest I

Figure 20-5
Three restriction digests useful in determining restriction maps. Enzyme I yields digest I, enzyme II yields digest II, and enzymes I and II together yield the double digest. To simplify the discussion, the fragments are given in order, though the order would be unknown until the data were analyzed.

Digest II

Double digest

a small fragment whose length is 2.3 − 1 = 1.3 and another fragment of length x; thus, 1.3 + x must equal the length of fragment B, C, or D. Excluding the terminal fragments in the double digest, namely those of length 1 and 1.7, the only possible value of x is 0.7; thus, the fragment adjacent to A must have a length of 1.3 + 0.7 = 2, which is the length of B. Hence, B is adjacent to A. Now we identify the fragment adjacent to F in digest II. The right end of B must divide the fragment adjacent to F. Since A + B = 3 and F = 2.3, then 3 − 2.3 + a fragment of length p must equal the length of a nonterminal fragment in digest II. The only possible value of p is 2.5, so the fragment adjacent to F has a length of 3.2 and must be G. We now turn our attention to the right end of the molecule. E divides the fragment adjacent to J. Thus, for some fragment of length q in the double digest, 5 − 1.7 + q must equal the length of a fragment in digest II. The only possible value of q is 5.4 and the length of the fragment adjacent to J is 8.7—that is, fragment I. This process can be continued to generate the complete restriction maps for both enzymes, as the reader should verify. The process is sometimes more complicated if fragments having the same size are present.

Complementary Single-Stranded Termini on Fragments Generated by Many Restriction Enzymes

One of the most exciting events in the study of restriction enzymes was the discovery in 1972 by Janet Mertz and Ronald Davis at Stanford University that fragments produced by many restriction enzymes spontaneously circularize if they are stored in a buffer of high ionic strength and under conditions suitable for renaturation. This observation was the first evidence that some restriction enzymes generate DNA fragments with cohesive ends. This experiment is shown in Figure 20-6.

As we have already said, some restriction enzymes do not produce single-stranded cohesive ends. Such blunt-ended fragments, of course, cannot circularize.

Joining of Restriction Fragments Having Cohesive Ends

A particular restriction enzyme recognizes only a single base sequence. Thus, the fragments generated by a particular enzyme acting on one DNA molecule have the same cohesive ends as the fragments produced by the same enzyme acting on a different DNA molecule (as long as both DNA molecules have sequences recognized by the enzyme). Just as a fragment can be circularized because its two cohesive termini have complementary base sequences, two different fragments can pair by virtue of complementary cohesive ends. Therefore, fragments from the

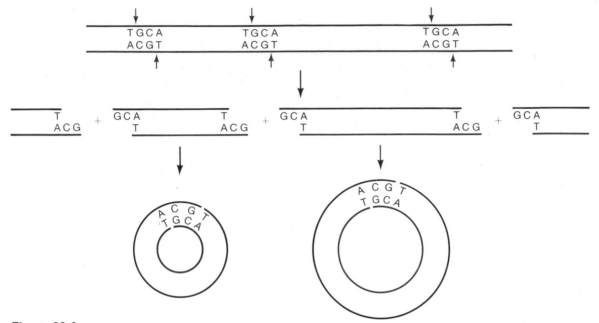

Figure 20-6
An experiment showing that restriction fragments can circularize. Arrows indicate cleavage sites.

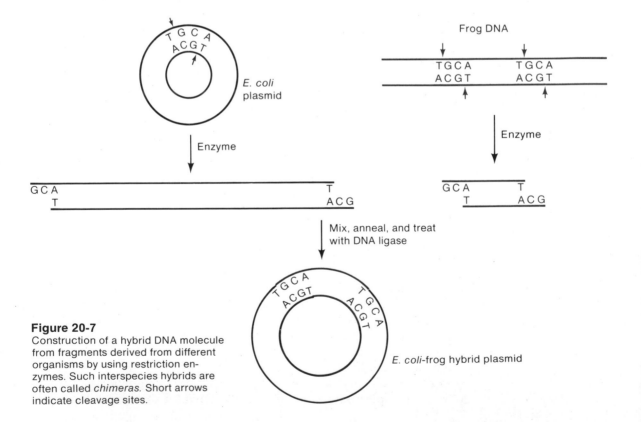

Figure 20-7
Construction of a hybrid DNA molecule from fragments derived from different organisms by using restriction enzymes. Such interspecies hybrids are often called *chimeras.* Short arrows indicate cleavage sites.

DNA molecules of two different organisms (for example, a bacterium and a frog) can be joined, as shown in Figure 20-7. Furthermore, if the joint is ligated after base-pairing, the fragments are joined permanently.

Joining does not always produce a functional DNA sequence. For example, consider a linear DNA molecule that is cleaved into four fragments A, B, C, and D. Reassembly could of course yield the original molecule, whose sequence we will assume to be ABCD, but since B and C have the same pair of cohesive ends, a molecule with the sequence ACBD could form with the same probability as ABCD. Other molecules having multiple copies of one of the fragments—for example, ABBD or ACCD—are also possible. Thus, only one-fourth of the resulting sequences will be functional. Similarly, if there were five fragments A, B, C, D, and E, the molecules A(BCD)E, A(BDC)E, A(CBD)E, A(CDB)E, A(DBC)E, and A(DCB)E, and those with duplicated and triplicated fragments, are all equally likely. The probability of obtaining a functional sequence when reannealing fragments obtained from two different DNA molecules is of course lower. Thus, if plasmid DNA is cleaved into four fragments and frog DNA is cleaved into 1000 fragments, most assembled plasmids will be scrambled and nonfunctional. On the other hand, if a functional plasmid is selected by genetic means and if an insertion is possible, say, between B and C, the probability of having a plasmid containing any frog DNA fragment will be quite high; if a *particular* frog DNA fragment is desired, the probability will be about 1000 times lower.

Joining of DNA Fragments by Addition of Homopolymers

The field of recombinant DNA research began in 1972, just before the properties of restriction enzymes were understood, when Peter Lobban and Dale Kaiser of Stanford University developed a general method for joining any two DNA molecules. Their method used the enzyme **terminal nucleotidyl transferase,** which is an unusual DNA polymerase obtained from animal tissue. This enzyme adds nucleotides (by means of triphosphate precursors) to the 3′-OH group of an extended single-stranded segment of a DNA chain. The process is remarkable in that *it does not need a template strand.* In order to generate the extended single strand one need only treat the DNA molecule with a 5′-specific exonuclease to remove a few terminal nucleotides. In a reaction mixture consisting of exonuclease-treated DNA, dATP, and the enzyme, poly(dA) tails will form at both 3′-OH termini (Figure 20-8). If, instead, dTTP were provided, the DNA molecule would have poly(dT) tails. Thus, two molecules can be joined if poly(dA) tails are put on one DNA molecule and poly(dT) tails on the second molecule and the poly(dA) is allowed to anneal to the poly(dT), as shown in the figure.

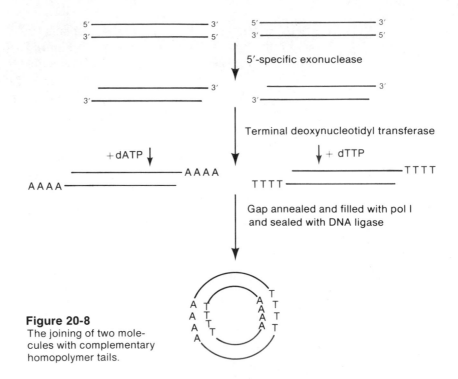

Figure 20-8
The joining of two molecules with complementary homopolymer tails.

Completion of the joined molecule is accomplished by gap-filling with DNA polymerase I and sealing with DNA ligase.

This method, which is called **homopolymer tail-joining**, is useful when DNA molecules lacking complementary ends are to be joined. Such molecules may be the result of digestion with restriction enzymes that yield blunt ends or may be prepared by mechanical breakage of large DNA molecules; c-DNA (complementary DNA), which is a DNA molecule prepared in the laboratory by copying an RNA template and which is extremely important in the recombinant DNA technology (to be described shortly), also has blunt ends.

Blunt-End Ligation

E. coli phage T4 encodes a DNA ligase, which is produced in large quantities in an infected cell. This enzyme is a typical ligase in that it seals nicks in double-stranded DNA having 3'-OH and 5'-P termini, but has the additional property of joining two DNA molecules having completely base-paired ends (Figure 20-9). How this reaction occurs is unknown, but it is a useful alternative to homopolymer joining.

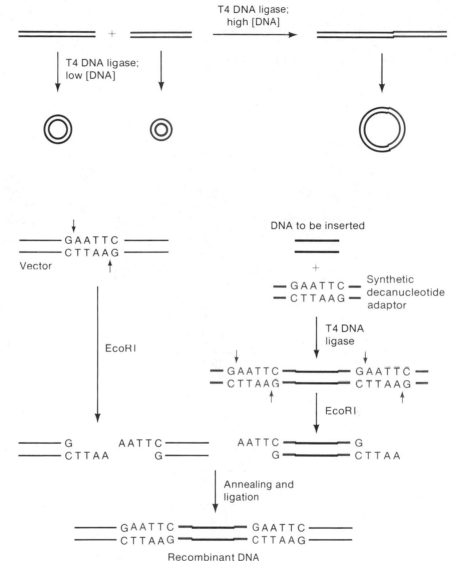

Figure 20-9
Two blunt-end joining reactions. At low concentrations of DNA intramolecular circularization is favored.

Figure 20-10
Formation of recombinant DNA through use of an adaptor. The short arrows indicate the sites of cutting by the EcoRI enzyme.

Joining with Adaptors

In some cases it is useful to be able to join one molecule with blunt ends to a second molecule produced by a restriction endonuclease that generates single-stranded termini. This is possible if a short DNA segment (an **adaptor**) containing a restriction site is coupled to both ends of the blunt-ended molecule. How this is done is shown in Figure 20-10, in which a blunt-ended molecule is inserted at the EcoRI site of a vector. This procedure is useful because the sequence contained in the blunt-ended fragment can at a later time be recovered from the vector by treatment with a restriction enzyme that cuts the site in the adaptor. Such recovery is not possible if joining is done via homopolymer tails.

Generation of Fragments by Hydrodynamic Shear

Consider a gene G that is to be cloned in a plasmid. This is accomplished by inserting a DNA fragment containing G into the plasmid DNA. The simplest method is to treat both the plasmid DNA and the DNA containing G with the same restriction enzyme and anneal the resulting fragments. However, it may be that every known restriction enzyme that cuts the plasmid also makes a cut within gene G and thereby inactivates the gene. Clearly some means of cleavage other than the use of restriction enzymes is required; it has been found that the gene inactivation caused by site-specific cleavage can be eliminated if the donor DNA containing G is fragmented not enzymatically but mechanically, instead, by hydrodynamic shear forces.

If a DNA solution is passed very rapidly through a hypodermic needle or a small orifice or is stirred vigorously, the DNA molecules are broken into fragments ranging in molecular weight from 2×10^6 to 20×10^6. In some DNA molecules, gene G, whose molecular weight might typically be about 10^6, will be cleaved but, on the average, the gene will remain intact.

The fragments resulting from shearing usually have blunt ends. In order to be joined efficiently to a plasmid broken by a restriction enzyme, an appropriate adaptor may be added, as was shown in Figure 20-10, though blunt-end joining can be carried out.

INSERTION OF A PARTICULAR
DNA MOLECULE INTO A VECTOR

As we have described the procedures so far, a collection of fragments obtained in various ways (for example, by restriction enzyme digestion or shearing) is allowed to anneal with a vector, yielding a large number of hybrid vectors. If a particular gene is to be cloned, the vector possessing that gene must be isolated from the class of all vectors possessing foreign DNA. Although for many genes this procedure is adequate, there are certainly many cases in which the clone of interest is either so rare or so difficult to detect that it would be preferable if, prior to hybridization, the fragment containing the gene of interest could be purified. In this section three procedures are described in which a particular DNA molecule can be cloned.

c-DNA and the Use of Reverse Transcriptase

Proof of correct insertion of a particular gene into a vector is most easily detected by observing expression of the gene. For example, the *lac* operon can be detected in a plasmid by virtue of its ability to synthesize β-galactosidase. However, knowing whether a particular eukaryotic gene has been inserted into a bacterial DNA vector can be a problem because, in general, a eukaryotic gene cannot be transcribed and translated into a functional protein. By the technique described in this section the insertion of some eukaryotic genes can become a quite straightforward procedure.

Let us assume that we wish to clone the rabbit gene for β-globin (a subunit of hemoglobin) in a bacterial plasmid.* The recombinant plasmid could be formed by treating both rabbit cellular DNA and the *E. coli* plasmid DNA with the same restriction enzyme, mixing the fragments, annealing, ligating, and finally infecting *E. coli* with the DNA mixture containing the hybrid plasmid. The main problem is to identify the bacterial colony containing this plasmid. If the gene to be cloned is a bacterial *gal* gene, one can use a Gal⁻ host and select a Gal⁺ colony on an appropriate agar. However, it is not possible for a bacterium containing the globin gene to synthesize globin because the globin gene contains introns (Chapter 11) and all known bacteria lack the enzymes needed to process eukaryotic mRNA molecules. Other tests might be performed to find the desired cell, such as hybridization of bacterial DNA with purified globin mRNA, but since only about one plasmid-containing cell in 10^5 will contain the globin gene, this method will be very tedious. Therefore, a more convenient procedure is desirable. One useful procedure, which is applicable to certain classes of genes, is described in the remainder of this section.

Some types of animal cells, of which one example is the reticulo-cyte, which produces globin, make only one or a very small number of proteins. In these cells the specific mRNA molecules, which are already processed, constitute a large fraction of the total mRNA synthesized in the cell and, for this reason, mRNA samples can usually be obtained that consist almost exclusively of a single mRNA species—in this case, globin mRNA. If genes of this type—that is, those whose gene products are the major cellular proteins—are to be cloned, the purified mRNA of each type of cell can serve as a starting point for creating a recombinant plasmid containing only the gene of interest.

The animal viral enzyme **reverse transcriptase** is able to use an RNA molecule as a template and synthesize a double-stranded DNA copy (for the mechanism, see Chapter 21). A double-stranded DNA

---

*The phrase "to clone a particular gene" is common usage to mean "to form a clone containing a particular gene."

molecule prepared in this way is called **complementary DNA** or **c-DNA.** If the RNA molecule that forms the template has been processed (that is, the introns have been removed) before it is isolated, the corresponding c-DNA will contain an uninterrupted coding sequence. Thus, if the purpose of forming the recombinant DNA molecule is to synthesize a eukaryotic gene product in a bacterial cell and if processed mRNA can be isolated, then c-DNA formed from processed mRNA is the material of choice to be inserted. The c-DNA generally has blunt ends and is usually joined by the adaptor method shown in Figure 20-10.

Insertion of Completely Synthetic Genes

Higher organisms (for example, man) produce many polypeptide hormones, which are of great medical value and which typically contain fewer than 20 amino acids. In most cases, neither genes nor mRNA molecules for polypeptide hormones have been isolated. However, it is feasible to synthesize some genes by using ingenious techniques for synthesizing short DNA molecules of defined sequence. The amino acid sequences of most of the peptide hormones are known, so that a particular base sequence can be deduced from the genetic code. The gene that is synthesized need not have precisely the same base sequence as the natural gene because (1) the redundancy of the code allows substantial substitution of codons without altering the information content, and (2) the natural gene may contain introns. To make a functioning synthetic gene, the base sequence must of course be initiated by a start codon and terminated by a stop codon, both of which can be included in the DNA molecule. Once the gene is synthesized, it is joined to a vector by any of the methods described so far for incorporation of blunt-end fragments. At the present time, the largest DNA molecule synthesized in this way—that encoding a human interferon gene—has 514 base pairs, which is sufficient for synthesizing a protein having 170 amino acids.

Shortly we will describe the requirements for expression of a synthetic gene, when we examine the cloning of the somatostatin gene.

Insertion of a Particular DNA Fragment Obtained from a Large Donor Molecule by Restriction Cutting

The study of the regulation of gene expression in phages has been helped considerably by the cloning of particular phage genes in a plasmid. Cloning of a phage gene is facilitated if one can first purify the fragment containing the gene. For instance, Figure 20-4 showed a restriction map for a particular phage; suppose that one wants to clone a gene that is known to be in fragment C. One way to accomplish this is to

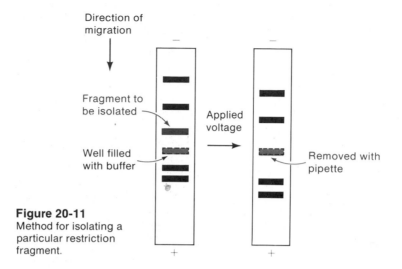

Figure 20-11
Method for isolating a
particular restriction
fragment.

add the complete enzymatic digest to a cleaved vector, allow the DNA to anneal, and seek a cell containing a plasmid that has the gene of interest. Since there are five fragments, at most one-fifth of the plasmids containing phage DNA will have the gene. In some cases, it is valuable to increase this fraction. This can be done by purifying fragment C and using only the purified fragment in the annealing mixture. The most direct way to isolate a fragment is to extract it from a gel after electrophoretic separation. This can be done in the following way (Figure 20-11). The restriction digest is electrophoresed for a period of time, the electric field is turned off, and the DNA bands are located by soaking the gel in a buffer containing ethidium bromide (see also Figure 20-3(b)). The fluorescence of ethidium bromide is markedly enhanced when the molecule binds to DNA, so the DNA can be located by shining near-ultraviolet light on the gel and noting the positions of intense fluorescence. A small well is cut in the agar ahead of the band of interest and filled with buffer. The electric field is applied again and the motion of the fragments is followed visually. When the fragment of interest migrates into the buffer-containing well, the electric field can be turned off again and the buffer, now containing the DNA fragment, can be removed.

DETECTION OF RECOMBINANT MOLECULES

The joining of DNA molecules is theoretically a straightforward process, as we have seen in the preceding sections. However, if a vector is cleaved by a restriction enzyme and annealed with an unfractionated collection of restriction fragments, many types of molecules result—examples are a self-annealed vector that has not acquired any fragments; a vector with one or more fragments; and a molecule consisting only of many joined fragments. Molecules of the third class cannot stably transform $CaCl_2$-treated cells because, in general, such molecules lack a replication origin and replication genes; thus, molecules of this class are not a problem, other than that they consume DNA fragments uselessly. However, some means is needed to ensure that a plasmid (or a phage) detected after $CaCl_2$ transformation does possess an inserted DNA fragment. Furthermore, even if this can be demonstrated, there is no guarantee that a plasmid containing donor DNA has the DNA segment of interest. Thus, some test is required for detecting the desired plasmid. Similar problems exist for phage and viral vectors. In this section several procedures that provide solutions to these problems are described.

Methods That Insure That a Plasmid Vector Will Contain Inserted DNA

The plasmid pBR322 is the most widely used plasmid for cloning DNA. It is small, having a molecular weight of 2.9×10^6, and has two different antibiotic-resistance markers—namely, resistance to tetracycline (tet^+) and to ampicillin (amp^+). Thus, plasmid-containing transformants are easily detected by growth of a DNA-transformed culture on agar containing one of these antibiotics. Also, pBR322 has seven different types of restriction-enzyme cleavage sites at which foreign DNA can be inserted.

A common procedure for detecting insertion is **insertional inactivation,** which is carried out as follows (Figure 20-12). The BamHI and SalI sites are within the *tet* gene. Thus, insertion at either of these sites yields a plasmid that is $amp^+ tet^-$, because the *tet* gene is inactivated. If wild-type (Amp$^-$Tet$^-$) cells are transformed with a DNA solution in which the cleaved pBR322 has been annealed with restriction fragments and the cells are plated on agar containing ampicillin, surviving colonies, which have to be Amp$^+$, must possess the plasmid. These colonies are then further tested for sensitivity to tetracycline. Because pBR322 carries the *tet$^+$* allele, an Amp$^+$ colony will also be Tet$^+$ unless the *tet$^+$* allele has been inactivated by insertion of foreign DNA. Thus, an Amp$^+$Tet$^-$ cell must contain not only pBR322 DNA but donor DNA as well.

This procedure can be simplified by growing the transformed cells, prior to plating, in a medium containing cycloserine and tetracycline. Cycloserine kills growing cells, but Tet$^-$ cells are merely inhibited, not killed, by tetracycline. Thus, in this growth medium Tet$^+$ cells (which

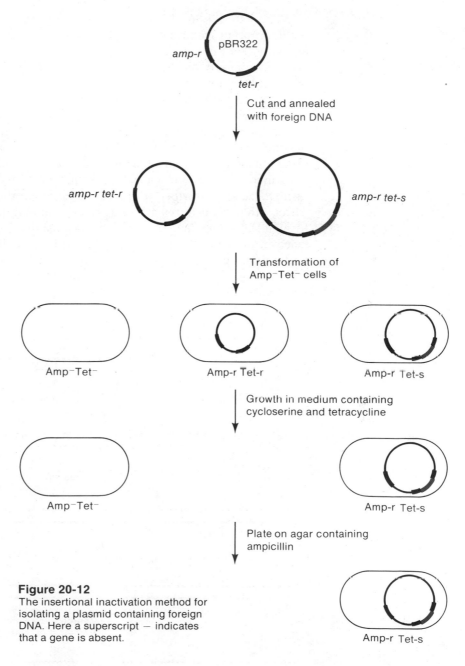

Figure 20-12
The insertional inactivation method for isolating a plasmid containing foreign DNA. Here a superscript − indicates that a gene is absent.

grow) are killed and Tet⁻ cells (which are inhibited) survive. Plating of cells treated in this way on agar containing ampicillin yields Amp⁺Tet⁻ colonies; these all possess pBR322 containing a donor DNA fragment.

The PstI site in pBR322 is in the *amp* gene. Thus, the insertional inactivation procedure can also be used with insertion at this site, and an Amp⁻Tet⁺ colony is thereby selected.

A commonly used restriction enzyme is the EcoRI enzyme. Since pBR322 does not contain an EcoRI site in either the *amp* or *tet* gene, a new plasmid pBR325 was constructed (by recombinant DNA techniques), which possesses a gene for chloramphenicol resistance in which there is an EcoRI site. Thus, insertion at the EcoRI site of this plasmid is detected by the loss of resistance to chloramphenicol. Many other plasmids, several of which have been designed like those just described, are commonly used with the insertional inactivation procedure; these are described in references given at the end of the chapter. Three other procedures select against reconstituted plasmids lacking foreign DNA. One is to treat a cleaved plasmid with **alkaline phosphatase,** an enzyme that removes 5′-P termini and leaves 5′-OH groups. Following this treatment a functional plasmid cannot be reconstituted unless foreign DNA is inserted, as shown in Figure 20-13. A second procedure, the use of homopolymers for joining, automatically selects

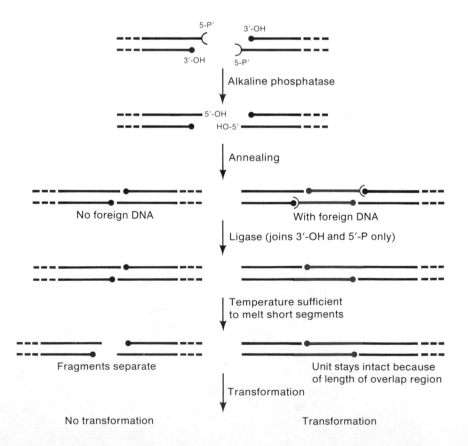

Figure 20-13
Method to prevent reconstitution of a cleaved plasmid by treatment of the plasmid DNA with alkaline phosphatase.

against re-formation of the plasmid because the two 3′-OH termini of the cleaved plasmid both have homopolymers containing the same base. A third procedure utilizes a special type of plasmid called a cosmid; this method is the following.

Cosmids (cos-site-carrying plasmids) are novel vectors that combine the features of a plasmid and phage λ to increase the probability of selecting a recombinant plasmid carrying foreign DNA. A typical cosmid is a circular ColE1 plasmid (Chapter 17) carrying both the gene for resistance to the drug rifampicin (*rif-r*) and the *cos* site of phage λ. There are two cleavage sites for the HindIII restriction enzyme, which separate the *rif* gene from the *cos* site and the ColE1 region (Figure 20-14).

In Chapter 15 the *cos*-site-cutting (Ter) system responsible for packaging phage λ DNA into the phage head was described. The Ter system can act on a DNA molecule only if the following two conditions are satisfied: (1) the DNA molecule must contain two *cos* sites, and (2) the *cos* sites must be separated by no less than 38,000 base pairs and no more than 54,000 base pairs. The use of cosmids as vectors is based on a technique for packaging λ DNA *in vitro* and the requirements just stated.

The cosmid shown in Figure 20-14 cannot serve as a substrate for *in vitro* packaging because the DNA has only one *cos* site. If the cosmid is treated with the HindIII restriction enzyme, the resulting linear fragments still cannot be packaged. Panel (a) of the figure shows that if the molecules are annealed with one another, among the many resulting molecules, one is a dimer (or higher multimer) containing two *cos* sites. However, the Ter system still cannot act on this molecule because the *cos* sites are separated by only a few thousand base pairs. Panel (b) shows that if a mixture of cleaved cosmid DNA and HindIII restriction fragments obtained from the DNA of another organisms (e.g., camel DNA) is allowed to reanneal, of the many molecules that arise, one is a linear dimer having two *cos* sites separated by a sufficient amount of foreign DNA that their distance is suitable for the λ packaging system. Thus, *in vitro* packaging yields a collection of transducing particles containing a linear fragment of recombinant cosmid DNA, terminated at each end by the normal cohesive ends of λ DNA. Note that packaging only occurs if the cosmid contains foreign DNA (though it would be possible to package a multimer containing many copies of the *rif* gene in tandem). Panel (c) shows the result of infecting Rif-s *E. coli* with the cosmid-carrying transducing particles. (There is no phage development because the particles contain no phage genes.) The cosmid DNA is injected and circularizes via the λ cohesive termini. Once the cosmid DNA has circularized and has been ligated, it can replicate, using the ColE1 replication system. Growth of the infected cells on agar containing rifampicin selects cells containing the *rif* gene, the ColE1 region, and the foreign DNA.

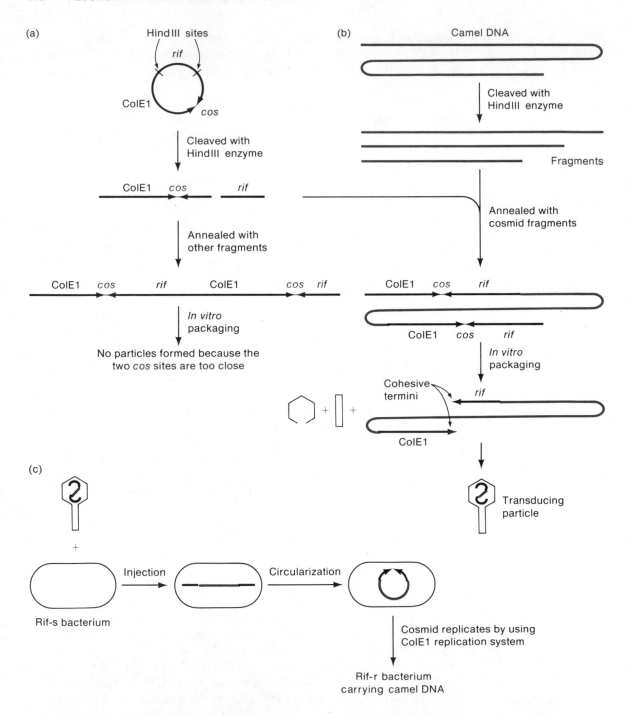

Figure 20-14
Use of a cosmid as a vector. (a) Cleavage of cosmid. (b) Formation of transducing particle. (c) Transduction.

Methods That Insure That a Phage Vector Will Contain Foreign DNA

There are two procedures in use for cloning genes in *E. coli* phage λ. Both are based on the fact that λ has several centrally located genes (namely, the *b2, int, xis, red, gam,* and *cIII* genes) that are not needed for lytic growth. These genes are in the so-called nonessential region (see Chapter 15).

The λ variant λgt4 · λβ has two EcoRI restriction sites near the termini of the nonessential region (wild-type λ has five EcoRI sites but, in forming λgt4 · λβ, three of these sites were removed by genetic manipulation). To use this vector the DNA is cleaved with the EcoRI enzyme and the three fragments are separated by gel electrophoresis. The terminal fragments are isolated and the central fragment is discarded (Figure 20-15). The terminal fragments can be joined via the complementary single-stranded termini but the resulting DNA molecule, whose length is 72 percent that of wild-type λ DNA, will be noninfective, because the minimum length of DNA that can be packaged in a λ phage head is 77 percent of the wild-type value. However, if foreign DNA is inserted, the resulting hybrid DNA becomes infective (that is, produces progeny phage, because the DNA can be packaged). Thus, if an *E. coli* culture is infected with DNA annealed from a mixture of the two terminal fragments and foreign DNA, any phage that is produced will contain foreign DNA.

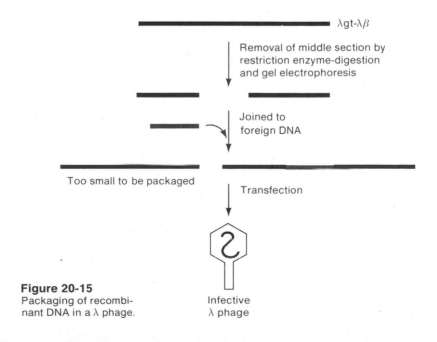

Figure 20-15
Packaging of recombinant DNA in a λ phage.

Another λ variant contains a section of the *lac* operon that includes the *lacZ* gene, the *lacP* promoter, and the *lacO* operator. There is a restriction site in the *lac* operator, so that insertion of foreign DNA into the phage at that site yields a phage with constitutive synthesis of β-galactosidase. There are several substrates of β-galactosidase that are noninducing and produce colored compounds if cleaved by β-galactosidase. If such a substance is incorporated into agar, and a mixture of *lacO*$^+$ and *lacO*$^-$ (constitutive) cells are plated on the agar, the *lacO*$^-$ colonies will be colored because only these cells synthesize β-galactosidase. Similarly, if *lacO*$^+$ and *lacO*$^-$ phages are plated on a lawn of *lacO*$^+$ cells on agar containing such a substance, *lacO*$^+$ phage will produce colorless plaques but *lacO*$^-$ phage (which contain inserted foreign DNA) will yield colored plaques.

Physical Methods of Identifying a Plasmid Containing Foreign DNA

For some experiments a particular fragment of foreign DNA is not required but instead any piece will do. For example, in some studies of the genetic organization of eukaryotic DNA, cloned segments of independent regions of DNA are needed but the regions need not carry particular genes. Such a plasmid might be detected by its having a higher molecular weight than the wild-type plasmid. A particular increase in molecular weight is also a useful criterion when a specific fragment of c-DNA is to be inserted, if the c-DNA is unique, having been prepared from a purified mRNA.

A straightforward method for showing an increase in size is to isolate the plasmid DNA and measure its length. A direct electron-microscopic measurement would, in general, be tedious for screening purposes and instead a simple electrophoretic technique, already described in Chapter 17 and depicted in Figure 17-4, is used. Lysates of single colonies of plasmid-containing cells are electrophoresed (usually sixteen different colonies at a time); there is enough plasmid DNA in a single colony to be visible as a single band moving far ahead of the chromosomal DNA. The plasmid DNA moves a distance related to its molecular weight—the larger the DNA, the smaller the distance moved in a given time interval, so that the larger plasmid DNA molecules, which contain donor DNA, move more slowly than those lacking donor DNA; the colonies containing donor DNA are then easily identified.

If the plasmid is thought to contain a particular segment of foreign DNA, this can be confirmed by two different hybridization procedures, if mRNA is available. If radioactive mRNA is used and it is complementary only to the desired foreign DNA, the total DNA content from a bacterium containing the plasmid is denatured and fixed to a filter, and the radioactive mRNA is added. After renaturation the filter is washed

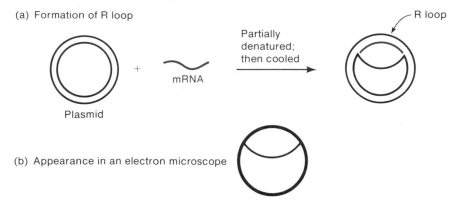

(a) Formation of R loop

Plasmid   +   mRNA   → Partially denatured; then cooled →   R loop

(b) Appearance in an electron microscope

Figure 20-16
Formation of an R loop. Since circular DNA molecules are invariably supercoiled, a single-strand break (not shown) is usually introduced prior to partial denaturation so the molecule will be more extended and more easily visualized.

and the presence of bound radioactivity indicates that the plasmid contains integrated foreign DNA complementary to the tester mRNA. If the mRNA is complementary to host DNA (for example, if an *E. coli* gene is cloned in *E. coli*), the plasmid DNA must be purified prior to hybridization.

When purified plasmid DNA is available, an elegant electron-microscopic procedure, called **R-looping,** can be used (Figure 20-16). Plasmid DNA is partially denatured in a buffer containing the mRNA, in which a DNA-RNA hybrid is more stable than double stranded DNA. Thus, if complementary RNA is present, this RNA will invade the helix, producing a bubble containing a single-stranded (DNA) branch and a double-stranded (DNA-RNA) branch.

Mass Screening for Plasmids Containing a Particular DNA Segment

The simplest procedure for detecting a particular foreign gene is complementation. In the first such experiment reported, a particular gene in the histidine biosynthetic pathway of yeast was detected by recombining a plasmid with fragments of the yeast chromosome and transforming a particular His⁻ *E. coli* mutant with a collection of recombinant plasmid DNA molecules, some of which contained the yeast gene, and then selecting for His⁺ colonies by growth on agar lacking histidine. This method works only for genes that are able to complement and is not generally successful when animal DNA is cloned in bacteria, because usually either the animal DNA is not expressed or there is no corresponding bacterial gene. The two procedures described in this section are more generally applicable.

The **colony** or *in situ* **hybridization assay** allows detection of the presence of a particular gene (Figure 20-17). Colonies to be tested are transferred from an agar surface to a filter paper. A portion of each

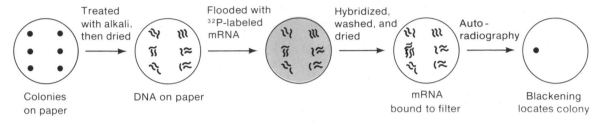

Figure 20-17
Colony hybridization. The reference plate, from which the colonies on paper were obtained, is not shown.

colony remains on the agar, which constitutes the reference plate. The paper is treated with NaOH, which lyses the cells and releases denatured DNA, and then the paper is dried, which fixes the denatured DNA to the paper. The dry paper is flooded with ^{32}P-labeled mRNA, which is complementary to the gene being sought, and is subjected to conditions that favor renaturation. The paper is then washed to remove unbound [^{32}P]mRNA; bound radioactivity remains on the paper only if [^{32}P]mRNA has hybridized. The paper is dried and placed on autoradiographic film, and blackening of the film locates the position of the colony of interest, which can then be selected from the reference plate.

A similar assay is done with phage vectors. Phage are plated on a lawn of bacteria and after plaques develop, paper is placed on the agar. Some phage from each plaque stick to the paper. The paper is then treated with alkali as above in order to fix and denature the DNA, and the hybridization procedure just described is used to detect the plaque whose phage contain the gene of interest. The desired plaque is then removed from the reference plate.

If the protein product of the gene of interest is synthesized, two immunological techniques allow the protein-producing colony to be identified. If the protein is excreted, the **radioactive antibody test** (Figure 20-18) is used. In this test a plastic disk to which antibodies to the gene product are bound is pressed onto the colonies. If the protein is present, it will bind to the antibody. The disk is then placed in a solution containing radioactive antibody, which sticks to the bound protein on the disk. The disk is then washed and autoradiographed. The location of the radioactivity shows which colony synthesized the gene product.

In some cases, insertion of a gene displaces the stop codon for a protein already encoded in the plasmid and couples the gene of interest to the plasmid gene. The result is a hybrid protein molecule called a **fused protein,** which begins at the amino terminus with a portion of the plasmid protein sequence and terminates with the protein of interest. The immunological procedure just described also usually works with such fused proteins.

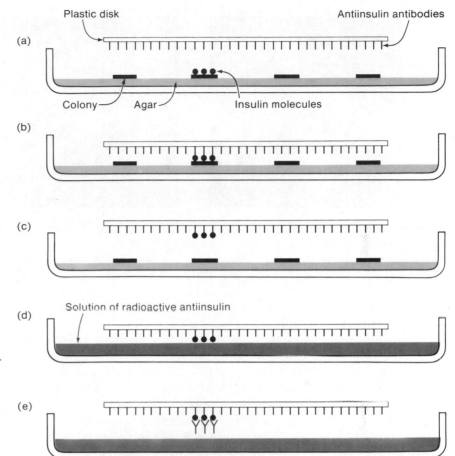

Figure 20-18
Radioactive antibody test. (a, b) A plastic disk coated with antiinsulin antibody is touched to the surface of colonies on a petri dish. (c) Insulin molecules from an insulin-producing colony bind to antibodies. (d) The disk is then dipped into a solution of radioactive antiinsulin. (e) Radioactivity adheres to the disk at positions of insulin-producing colonies.

Fusion is sometimes of particular value if the protein of interest is not normally excreted by the bacterium. In this case, it is useful to use the plasmid pBR322 and to insert the desired gene into the plasmid *amp* gene, which encodes the enzyme β-lactamase. Tet$^+$ cells can then be selected by their resistance to tetracycline (these cells will contain pBR322) and insertion of foreign DNA can be detected by insertional inactivation—that is, by selecting for the Amp$^-$ character among the Tet$^+$ cells. Once the gene of interest has been inserted, its protein product will fuse to β-lactamase, which is excreted; thus, the β-lactamase carries the protein of interest through the cell wall. The protein can then be detected by the radioactive antibody test just described or the test that follows.

An **immunoprecipitation test** can also be used to identify a protein-producing colony by adding an antibody to the protein to the agar. If the protein is excreted, an antibody-antigen precipitate called

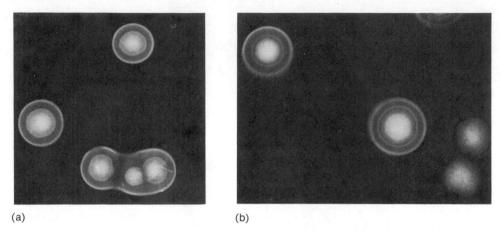

(a)   (b)

Figure 20-19

The immunoprecipitation test. Colonies making β-galactosidase are formed on agar containing antibody to the enzyme. (a) Precipitin bands form around each colony. (b) A mixture of Lac$^+$ colonies (with bands) and Lac$^-$ colonies (no bands).

precipitin forms a visible white ring around the colony producing the protein (Figure 20-19). Two slight modifications of this procedure allow one to detect a protein that is not excreted. In one modification a host cell is used that is lysogenic for $\lambda cI857$, a λ mutant containing a heat-sensitive repressor that is inactivated by heating to 42°C. A master plate containing colonies to be tested is replica-plated onto agar containing antibody. When the colonies on the replica plate become visible, the temperature is raised to 42°C. This induces lysis of many of the cells about one hour later, thereby releasing the cell contents. The protein of interest then reacts with the antibody and forms a precipitin ring at the site of the colony. In the second modification, agar containing both antibody and the enzyme lysozyme is poured on the colonies and allowed to harden. The lysozyme lyses the cells on the surface of the colony, releasing the intracellular proteins; again, the one of interest forms a precipitin ring.

GENE LIBRARIES

Many laboratories utilize recombinant DNA techniques repeatedly as a means of isolating particular genes or DNA segments from a single organism. For example, the spatial organization of genes in the fruit fly *Drosophila* has been studied by analyzing a collection of cloned DNA fragments (this will be discussed in Chapter 22). It is quite time-consuming to go through the complete cloning procedure each time a new

DNA segment is needed. Thus, collections have been established of hybrid plasmid-containing bacteria that are of sufficient size that each segment of the cellular DNA is represented once (or, on occasion, twice) in the collection. Such collections are called **colony banks** or **gene libraries.** In this section we describe how a gene library is established and how to determine how many clones are needed in order that each sequence will be represented.

A library of yeast DNA sequences has been established at the University of California at Santa Barbara. The vector is the ColE1 plasmid. To form the library, purified ColE1 DNA was cleaved with the EcoRI nuclease, which makes a single cut in this DNA molecule. The yeast DNA was not cleaved with a restriction enzyme; doing that would have prevented many genes from being in the library as intact units inasmuch as some genes would be cleaved by the enzyme. Instead, yeast DNA was broken at random by hydrodynamic shear forces (by rapidly stirring a solution of yeast DNA). In order to link the ColE1 and yeast DNA molecules, poly(dT) tails were added to the 3'-OH termini of the cleaved ColE1 DNA and poly(dA) tails were added to the yeast DNA fragments. The molecules were thus joined, and bacteria transformed. Cells containing the ColE1 factor are resistant to colicin E1 added to the agar, so plasmid-containing cells were readily isolated. The colonies were then transferred to a set of plastic blocks each of which contained 96 wells filled with a culture-storage medium. The blocks were stored at $-80°C$, a temperature at which the cells remain viable indefinitely in the particular storage medium. The complete set of blocks comprise the library. It has been observed that at least 70 percent of the colonies contain yeast DNA. The following simple calculation enables one to determine how many colonies are required to make a complete library.

Consider a DNA sample containing fragments of such a size that each fragment represents a fraction f of the DNA of the organism. The probability P that a particular sequence is present in a collection of N colonies is

$$P = 1 - (1 - f)^N$$

or

$$N = \ln(1 - P)/\ln(1 - f).$$

Let us assume that we want a probability of 0.99 that every sequence is in the library—that is, $P = 0.99$. If the donor DNA is haploid yeast (whose molecular weight is about 10^{10}) and the average molecular weight of the sheared DNA is 8×10^6, then $f = 8 \times 10^6/10^{10} = 0.0008$ and

$$N = [\ln(1 - 0.99)]/[\ln(1 - 0.0008)] = 5754.$$

Thus, if the library contains about 5800 colonies, there is a 99 percent probability that any yeast gene will be present in at least one colony. Furthermore, if a few colonies are selected at random for further study, it is exceedingly likely that their yeast DNA sequences do not overlap. For cells with a large DNA content (such as *Drosophila*) the number of colonies is about 300,000; this can be reduced roughly *n*-fold by increasing the size of the constituent fragments by a factor of *n*.

If a clone containing a particular DNA segment is needed, it can in principle be found by any of the screening procedures described earlier in this chapter.

Gene libraries have also been established with λ phage as a cloning vehicle, by using suspensions of each recombinant phage maintained in multiwell blocks. Phage libraries are limited by the maximum size of the DNA that can be inserted in a single phage particle. The minimum number of phage required to establish a complete library for an organism is determined by the maximum size of the foreign DNA that can be inserted in a single phage particle. For the λ cloning vehicles available, no more than 8×10^6 molecular-weight-units of DNA can be added, so the minumum number of phage required to maintain a library for *E. coli*, yeast, and *Drosophila* would be 2900, 5800, and 300,000, respectively.

PRODUCTION OF PROTEINS FROM CLONED GENES

One of the goals of genetic engineering is the production of a large quantity of a particular protein that is otherwise difficult to obtain (for example, proteins for which there are only a few molecules per cell or proteins that are produced by bacteria that, owing to their pathogenicity, are hazardous to handle). The method is simple in principle. The gene is coupled to a vector containing a promoter and tests are performed to ensure that the gene is in the correct orientation so that the coding strand will be transcribed. The vector that is chosen is usually a high-copy-number plasmid or an actively replicating phage (such as λ), so that a large number of copies of the gene will be present in the cell. This may allow synthesis of a gene product to reach a concentration of about 1 to 5 percent of the cellular protein, or 10 to 50 mg protein per liter of culture, if the protein is made at all.

In practice, the goal of producing large quantities of a desired protein is not always met, because of a variety of theoretical and technical problems. Some of these problems and how one seeks solutions are explained in the following sections. Some successful procedures will also be described.

Problems in the Production of Eukaryotic Proteins

If a protein is to be derived from a bacterial or phage gene, the procedures for protein production are straightforward. However, if the gene is from a eukaryote, there are special problems—namely,

1. The eukaryotic promoter may not be recognized by a bacterial RNA polymerase.
2. The mRNA transcribed from eukaryotic genes lacks the Shine-Dalgarno sequence needed for binding to bacterial ribosomes.
3. The mRNA may contain introns that must be excised.
4. The gene product often must be processed.
5. Eukaryotic proteins are often recognized by bacterial proteases as foreign and are cleaved.

Problems 1 and 2 can usually be eliminated by coupling the gene to the *lac* or λ promoter, as described in the following section. The intron problem is avoidable if fully processed mRNA can be isolated from the eukaryotic cell because, in that case, c-DNA can be prepared and joined to a vector. Alternatively, if the amino acid sequence is known and the protein is not too large, the gene can be synthesized directly. However, at the present time, in general, neither the mRNA can be isolated nor can the gene be synthesized. The fact that many finished proteins are modified forms of the initial translation product limits the usefulness of bacteria for the production of eukaryotic proteins, though in some cases (soon to be described) the processing can be carried out *in vitro* after the protein is isolated. Bacterial proteases that destroy eukaryotic proteins can presumably be eliminated (partially, at least) by mutation. Some examples of animal proteins synthesized in E. *coli* will be described shortly.

Recently, the 2μm plasmid of yeast (Chapter 17) has been used as a cloning vehicle for eukaryotic genes. Studies done so far suggest that most of the problems just described may be either absent or solved for this system. Yeast RNA polymerase is able to recognize many animal promoters and processing systems in yeast are apparently able to remove introns from animal genes and to process some animal proteins. Furthermore, there is much less degradation of foreign eukaryotic proteins, which is interesting because it suggests that many eukaryotic and prokaryotic proteins possess distinctive features that have hitherto not been recognized. Apart from the enormous potential utility of the yeast system, which has not yet (1982) been adequately exploited, these results suggest that the evolutionary transition between prokaryotes and eukaryotes was a biochemical event of major significance in which a certain uniformity was imposed on cells with and without an enclosed nucleus.

Expression of a Cloned Gene

Expression of a cloned gene requires that the gene be both transcribed and translated. Let us see how one designs a system to be sure that this occurs.

Often a restriction-enzyme cleavage site separates the coding portion of a gene from its promoter, and unless another promoter is provided, the gene cannot be transcribed after it is inserted into a vector. If the vector contains a promoter near the insertion site, this promoter can be used. (Ordinarily this promoter would be the RNA-polymerase-binding site for an adjacent gene in the vector.) For many vectors such a promoter is not available; then a promoter must also be inserted into the cloning vehicle. When doing so, it is clearly advantageous to insert a strong promoter in order to maximize transcription. A common procedure is to redesign a plasmid—by recombinant DNA techniques—so that it contains either the promoter-operator region of the *E. coli lac* operon (Chapter 14) or the *cI* repressor-*oL*-*pL* region of *E. coli* phage λ (Chapter 15). Both of these systems allow transcription of the cloned gene to be controlled. For instance, in the first case, transcription can be initiated by the addition of an inducer of the *lac* operon (such as lactose or IPTG) and can be terminated by the addition of glucose (to depress the concentration of cyclic AMP). With the λ system, if a temperature-sensitive repressor (called *cI857*) is used, transcription begins when the temperature is raised above 40°C, which inactivates the repressor, and is terminated by lowering the temperature to 34°C (at which point the repressor renatures).

In the course of restriction-cutting, the coding sequence of the gene of interest may also be separated from the ribosomal binding site, a not uncommon event. Thus, this site must also be provided. The inserted *lac* and λ systems just described are well designed in this respect—also in that they contain strong ribosomal binding sites.

Coupling a restriction fragment containing a gene of interest to a functioning promoter and a ribosomal binding site does not guarantee that the gene will be correctly expressed (Figure 20-20). A problem is that both ends of a restriction fragment have the same single-stranded termini, so that insertion into a vector can occur in two possible orientations. Since only one DNA strand of any gene is a coding strand, only one orientation will yield useful mRNA. Insertion in both orientations occurs with equal probability; therefore, one must select a recombinant vector that has the gene in the correct orientation. If synthesis of the desired gene product occurs, the correct orientation is detected easily. However, if there is no synthesis, methods described in earlier sections must be used.

In the following sections two successful attempts to synthesize eukaryotic proteins are described. The student should note how

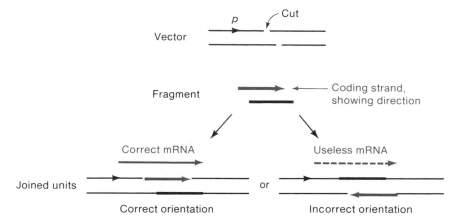

Figure 20-20
Insertion of a cloned gene in two orientations with respect to a promoter p in the vector.

detailed knowledge of gene expression gained from fundamental research on bacterial regulation and protein synthesis contributed to the successes.

Synthesis of Somatostatin

Somatostatin is a 14-residue polypeptide hormone synthesized in the hypothalamus. It has been made in *E. coli* by a procedure that is applicable to most short polypeptides. Through chemical techniques, a double-stranded DNA molecule is synthesized that contains 51 base pairs; the base sequence of the coding strand is

TAC-(42 bases encoding somatostatin)-ACTATC;

the base sequence of the corresponding mRNA is

AUG-(42 bases encoding somatostatin)-UGAUAG.

Thus, this mRNA has a methionine codon (which will not be used for initiation), a coding sequence for somatostatin, and two stop codons. The vector is the plasmid pBR322 modified to contain the *lac* promoter-operon and a portion of the *lacZ* gene encoding the NH$_2$-terminal segment of β-galactosidase. The vector is cleaved at a site in the *lacZ* segment (Figure 20-21) by a restriction enzyme that leaves blunt ends; the synthetic DNA molecule is then blunt-end-ligated to the cleaved plasmid. When the *lac* operon is induced, a protein is made which consists of the NH$_2$-terminal segment of β-galactosidase *coupled by methionine* to somatostatin, and which is terminated by the repeated

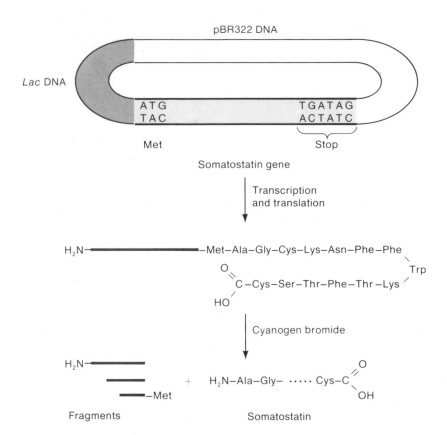

Figure 20-21
Synthesis of somatostatin from a chemically synthesized gene joined to the plasmid pBR322 *lac*.

stop codons of the synthetic DNA. This protein is purified and treated with cyanogen bromide, a reagent that cleaves proteins only at the carboxyl side of methionine. Thus, the methionine linker remains attached to a β-galactosidase fragment and somatostatin is released. Following the methionine-coupling by cyanogen bromide cleavage is a useful trick for separating any polypeptide from a bacterial protein to which it is fused, as long as the polypeptide itself does not contain methionine.

Synthesis of Proinsulin and Insulin

Insulin is a hormone that regulates sugar metabolism. It is clinically important in the treatment of diabetes and is normally obtained from pig or cow pancreas. Insulin itself is not the gene product but is formed

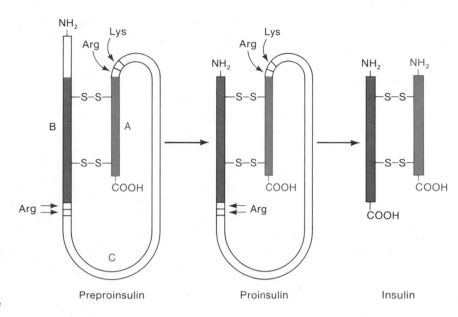

Figure 20-22
Conversion *in vivo* of pre-proinsulin to proinsulin and then to insulin. The positions of four amino acids pertinent to the discussion in the text are noted. A third disulfide bond joining two parts of the A segment is not shown.

in several stages (Figure 20-22). The gene product is preproinsulin, which is first cleaved to yield proinsulin. Then, insulin is formed by first making two disulfide links between terminal segments (the A and B segments) of proinsulin and afterwards removing the internal (C) segment. The result is a molecule consisting of two chains having 21 and 30 amino acids respectively, joined by two disulfide bonds.

Insulin has been synthesized in *E. coli* in two different ways. One of these procedures is similar to that for somatostatin but requires additional steps. Two DNA segments, one encoding the A chain and the other encoding the B chain, are first synthesized, and each is then separately cloned in two different bacteria by fusion to the β-galactosidase gene. The two fused proteins, β-gal–A and β-gal–B, are isolated next, and from these proteins the A and B chains are purified. The A and B chains are then mixed and disulfide bonds are allowed to form. The disulfide bonds form randomly between the six cysteines present in the A and B chains, so the yield of active insulin is low.

In the second procedure, DNA encoding proinsulin is cloned in the pBR322 plasmid. This procedure starts with isolation of mRNA from pancreatic cells of a pig. Although the mixture contains all species of cellular mRNA, this is enriched with preproinsulin mRNA, since this is a major product of pancreatic cells. From this RNA, c-DNA can then be made by reverse transcriptase. The DNA is next fragmented by the restriction enzyme PstI and inserted into a PstI restriction site within the β-lactamase gene of the pBR322 plasmid (Figure 20-23), whose genotype is *tet*+*amp*+. After transformation of a Tet⁻Amp⁻ *E. coli* culture, cells are isolated that are Tet⁺Amp⁻, to ensure that some DNA has been inserted into the plasmid gene. The Tet⁺Amp⁻ colonies are then screened for the production of insulinlike material by testing for reaction with radioactive antiinsulin antibody, and productive colonies are found. Base-sequencing of PstI fragments (see Figure 20-23) obtained from the plasmid has shown that a portion of the preproinsulin gene is lost in the PstI cleavage of the DNA, though the proinsulin sequence is present. The end product of *in vivo* transcription and translation is a hybrid (H₂N-β-lactamase)-(proinsulin-COOH) molecule. The presence of the C segment of the proinsulin causes correct folding of the proinsulin segment, so the proper disulfides form. *E. coli* does not process the molecule further. On the other hand, the β-lactamase fragment is so large that bacterial proteases do not recognize the insulin as foreign and the proinsulin segment survives.

Insulin can be cleaved from the hybrid protein *in vitro*. Note in Figure 20-22 that there are three arginines in or near the ends of the C segment. These are useful because trypsin cleaves peptide bonds only

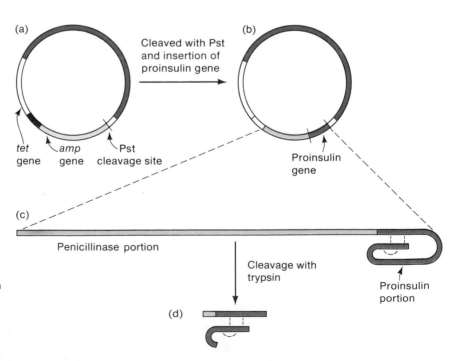

Figure 20-23
Production of pig insulin in *E. coli*. (a) The vector plasmid pBR322. (b) pBR322 containing the proinsulin sequence. (c) The hybrid gene product showing the proinsulin portion (red) and the disulfide bonds (dashed). (d) The result of trypsin digestion of the hybrid gene product.

on the carboxyl side of arginine and lysine. A very brief treatment of the β-lactamase–proinsulin hybrid (brief so that the protein is not completely fragmented by cleavage at every trypsin-sensitive site) occasionally makes one cut at the correct place between the A and C segments and another cut that leaves one or two extra arginines at the carboxyl terminus of the B chain. A cut is also needed to separate the B chain from the β-lactamase segment. There is neither a lysine nor an arginine at the required position, so that a few amino acids from the β-lactamase remain appended to the B chain and the insulin is not perfect; nonetheless, this modified molecule has biological function. It is not clear at present whether this modified form will be clinically useful.

Synthesis of Other Eukaryotic Proteins from Cloned Genes

Several other eukaryotic proteins have been synthesized in bacteria. A yeast enzyme active in histidine biosynthesis is expressed efficiently in *E. coli* from a λ phage vector. Chicken ovalbumin has also been made in *E. coli*. Of particular interest is human interferon, an antiviral and possibly anticancer substance (see Chapter 21). Interferon has been made in *E. coli*, though the yield at present is a small number of interferon molecules per bacterium; in contrast, about 100,000 copies of β-galactosidase are present in a fully induced *lac*+ cell. The rabbit β-globin gene has been cloned in SV40 virus; when this hybrid virus infects monkey kidney cells, rabbit β-globin is synthesized.

In principle, any eukaryotic protein can be synthesized in a properly constructed bacterium; a great many investigations are in progress. The potential for the synthesis of medically important substances warrants the enormous effort.

APPLICATIONS OF THE RECOMBINANT DNA TECHNOLOGY

So far we have discussed the utility of the recombinant DNA technology mostly in terms of the production of useful proteins, and this is certainly of great importance. The technology is also of use in two other ways—(1) as a means of altering the genotype of an organism, and (2) as a research tool.

Bacteria can carry out an enormous number of chemical reactions but no one species can do everything. However, by genetic engineering, organisms can be created that combine the features of several other bacteria. For example, several genes from different bacteria have been inserted into a single plasmid that has then been placed in a marine bacterium, yielding an organism capable of metabolizing petroleum; this organism has been used to clean up oil spills in the oceans. Furthermore, many biotechnology companies are at work designing bacteria that can synthesize industrially important chemicals. Organisms capable of making large quantities of expensive antibiotics are used to manufacture these drugs more cheaply, possibly resulting in lower consumer prices. Bacteria have been designed that are able to compost waste more efficiently and to fix nitrogen (to improve the fertility of soil) and an enormous effort is currently being expended to create organisms that can convert biological waste to alcohol. This type of genetic engineering will surely have a great impact on world economy, environmental quality, and the quality of life.

Altering the genotypes of plants is also an important application of recombinant DNA technology. In Chapter 17 we discussed the bacterium *Agrobacterium tumefaciens* and its plasmid Ti, which produces crown gall tumors in plants. These tumors result from integration of the plasmid DNA into the plant chromosome. It is possible by genetic engineering to introduce genes from one plant into this plasmid and then, via infection of a second plant with the bacterium, transfer the genes of the first plant to the second plant. Attempts are being made to perform plant breeding in this way. An example is the attempted alteration of the surface structure of the roots of grains such as wheat, by introducing certain genes from legumes (peas, beans), in order to give grains the ability of the legumes to establish root nodules of nitrogen-fixing bacteria. If successful, this would eliminate the need for the addition of nitrogenous fertilizers to grain-growing soils.

Recombinant DNA technology has also had a tremendous impact on basic research. Throughout this book, examples have been given of mutant bacteria that have simplified systems and made them more amenable to study (for example, the numerous mutants in the *lac* operon). Such mutants have been made by standard genetic techniques (e.g., mutagenesis and genetic recombination); for simple mutants this is straightforward but for mutants required to have many genetic markers (which may be very closely linked) the frequency of mutants produced can be so low that their isolation is a very tedious job. Recombinant DNA techniques can simplify mutant construction, since fragments containing desired genetic markers can be purified and combined in a test tube. This saves time, labor, and often enables mutants to be constructed that could not in practice be formed in any other way.

The greatest impact of the new technology on basic research is in the study of eukaryotes, in particular, eukaryotic regulation. Experiments of the type seen in Chapter 14, to study the regulation of operons in bacteria, are made possible by the use of mutations in promoters, operators, and structural genes. This approach is not feasible with eukaryotes, because eukaryotes are diploid and, hence, mutants are very difficult to isolate. Furthermore, except for the unicellular eukaryotes such as yeast, there is no simple way to do multiple genetic manipulations with eukaryotic cells.

Gene activity in bacteria also has been monitored by studying the synthesis of mRNA in a variety of conditions; the only convenient assay is DNA-RNA hybridization, which requires a source of DNA enriched for the gene being studied. Transducing particles (formed either by standard genetic methods or by recombinant DNA techniques) have been the source of the DNA in such experiments—for example, *lac* mRNA is assayed by hybridization with λ*lac* DNA. However, transducing particles do not exist for eukaryotic systems. By recombinant DNA techniques, a gene whose regulation is to be studied is first cloned in a phage or a plasmid vector and then the phage or a cell containing the plasmid is allowed to grow. This provides the researcher with a large supply of the DNA of that gene. The vector containing that DNA sequence is said to be a **mRNA probe** because its DNA, in denatured form, can be added to a cell containing radioactive mRNA to probe for a particular mRNA; the mRNA is assayed by its specific ability to hybridize with the DNA. The usual technique would be to apply the denatured probe-DNA to a nitrocellulose filter and use a filter-binding assay for the specific mRNA—that is, mRNA will not bind to a nitrocellulose filter unless it hybridizes to the denatured DNA. This simple technique has revolutionized the study of eukaryotic gene regulation and is the source of a great deal of the information presented in Chapter 22.

THE RECOMBINANT DNA DEBATE

In the late 1970's a considerable number of scientists, politicians, and interested lay persons argued about the safety of recombinant DNA research, a question first raised by the scientists themselves. It was immediately recognized that it would be a trivial task to create obviously harmful organisms. For example, the β-lactamase gene could be introduced into *Streptococcus pyogenes*, the causative organism for "strep" throat and rheumatic fever; if such a hybrid bacterium escaped from the laboratory into nature, penicillin, the drug of choice for the treatment of streptococcal infections, would be rendered useless. More alarming is the possibility of introducing the botulin gene from *Clostridium botulinis* into *E. coli*; such a strain could induce the highly lethal state of botulism (death within a day of infection) if the bacterium became established in the human large intestine (the natural residence of *E. coli*). By international agreement the construction of these obviously dangerous bacteria was banned. The debate continued, however, centered on the possible hazards of constructing bacterial strains even with foreign genes that are believed *not* to cause obvious harm when released in nature. It has been argued that scientists have no way of predicting how a eukaryotic gene will be expressed in *E. coli*. Various scenarios began to be advanced, of which the following is an example. A piece of animal DNA containing a gene of interest might also contain a gene that is normally unexpressed in the animal cell but that is expressed in a bacterium. A human could be infected with the recombinant bacterium and if this gene product were excreted, it might cause disease. Because of lack of information one could not prove whether such a scenario would be likely or unlikely and two camps arose—those who wished to see all recombinant DNA research banned, and those who felt that safety could be assured if appropriate measures were taken. Neither group denied the great benefits of recombinant DNA research to science and to humanity, but the former group contended that the benefits did not justify the risk.

The result of the controversy was a set of U.S. government regulations—nearly a thousand pages in length—called the *National Institutes of Health Guidelines for Research Involving Recombinant DNA Molecules,* and commonly known as the NIH Guidelines. These guidelines define various means of preventing escape of laboratory organisms into the environment **(physical containment)** and of preventing growth of an organism if it were to escape **(biological containment).** The former consists of various means of isolating work areas and of waste disposal. Biological containment refers to the creation of biological defects in the host cell that will introduce growth requirements not found in nature. An important genetically designed bacterial host of this type is *E. coli* χ1776 (created in the U.S. bicentennial year). This bacterium has several nutritional requirements for substances not commonly found in sewage, soil, or the human gut; it is killed by ionic detergents usually found in sewage and by the bile salts present in the human digestive tract; and it cannot grow at or above 37°C (the body temperature of humans). Thus, without the occurrence of several genetic reversions, it is exceedingly unlikely that χ1776 could successfully grow in humans. Other defective hosts are also in use, as well as defective phage vectors. Furthermore, all plasmid vectors in use are not self-transmissible, so it is unlikely that they would be transferred to a nondefective bacterium. Most of the precautions have been widely accepted (but not altogether happily by some scientific groups).

As time has passed and more knowledge has been gained, there has been a steady lessening of the concerns and a gradual relaxation by the National Institutes of Health of some of the restrictions. It seems now that the concerns about hypothetical health hazards have been unwarranted, though these concerns were socially and politically valuable. To date, no adverse effect of recombinant DNA research, or even hint thereof, has been reported and the great potential benefit of the new techniques is recognized by layman and researcher alike. Many new questions have come forth, though, as genetic engineering has become industrially important and a device for economic profit. Some of the questions are these: Should corporations have the right to patent new life forms created in their laboratories? (The U.S. Supreme Court has affirmed their right). To what extent should the profit motive shape the direction of future research? (Will the government support only "pure" research, leaving projects with economically profitable goals to the corporations?) Will industrial competition diminish free exchange of scientific information and ideas? How will nonprofit research institutions and universities be able to compete for the best young researchers, given the financial difficulties common to such institutions and the high salaries offered by the richer corporations? Questions such as these are not simple to answer and are widely debated.

REFERENCES

Abelson, J., and E. Butz. 1980. "Recombinant DNA." *Science*, 209, 1317–1438.

Anderson, W. F., and E. G. Diacumakos. 1981. "Genetic engineering in mammalian cells." *Scient. Amer.*, July, pp. 106–121.

Beers, R. F., and E. G. Bassett (eds.) 1977. *Recombinant Molecules: Impact on Science and Society.* Raven.

Brown, D. D. 1973. "The isolation of genes." *Scient. Amer.*, August, pp. 20–29.

Cohen, S. N. 1975. "The manipulation of genes." *Scient. Amer.*, July, pp. 24–33.

Curtiss, R. 1976. "Genetic manipulation of microorganisms: potential benefits and hazards." *Ann. Rev. Microbiol.*, 30, 507–533.

Gilbert, W., and L. Villa-Komaroff. 1980. "Useful proteins from recombinant bacteria." *Scient. Amer.*, April, pp. 74–94.

Hofschneider, P. H., and W. Goebel. 1982. "Gene cloning in organisms other than *E. coli*." In Arber, W., et al. (eds.) *Current Topics in Microbiology and Immunology*, vol. 96. Springer-Verlag.

Lobban, P., and A. D. Kaiser. 1973. "Enzymatic end-to-end joining of DNA molecules." *J. Mol Biol.*, 78, 453–471.

Maniatis, T., R. C. Hardison, E. Lacy, J. Lauer, C. O'Connell, D. Quon, G. K. Sim, and A. Efstratiadis. 1978. "The isolation of structural genes from libraries of eucaryotic DNA." *Cell*, 15, 687–701.

Mertz, J., and R. Davis. 1972. "Cleavage of DNA: RI restriction enzyme generates cohesive ends." *Proc. Nat. Acad. Sci.*, 69, 3370–3374.

Public Health Service, U. S. Department of Health, Education, and Welfare. 1976. *National Institutes of Health Guidelines for Research Involving Recombinant DNA Molecules.* U. S. Government Printing Office.

Scott, W. A., and R. Werner. 1977. *Molecular Cloning of Recombinant DNA.* Academic Press.

Seeberg, P. H., J. Shine, J. A. Martial, R. D. Ivarie, A. Morris, A. Ullrich, J. D. Baxter, and H. M. Goodman. 1978. "Synthesis of growth hormone by bacteria." *Nature*, 276, 795–798.

Setlow, J., and A. Hollaender (eds.) 1979–1981. *Genetic Engineering: Principles and Methods.* Vols. 1–3. Plenum.

Sinsheimer, R. L. 1977. "Recombinant DNA." *Ann. Rev. Biochem.,* 46, 415–438.

Smith, D. H. 1979. "Nucleotide sequence specificity of restriction endonucleases." *Science,* 205, 455–462.

Wu, R. (ed.) 1979. *Recombinant DNA. Methods Enzymol.,* vol. 68. Academic Press.

21 Eukaryotic Viruses

Hundreds of viruses are known, each of which infects a particular species of animal or plant cells. Many of these are responsible for serious diseases but some are merely rather benign intracellular parasites. Originally, viruses were studied as a means of understanding disease, but it quickly became clear that they possessed a rich variety of interesting biochemical features. Just as a byproduct of the study of phages was the elucidation of many fundamental processes in bacteria, so virus research has provided a window into eukaryotic cells. Some phenomena common to eukaryotes—for example, gene splicing and polyproteins—were first observed while studying viruses. In addition, unique biochemical processes (such as reverse transcription) were discovered.

The student may be struck by an apparently greater variety of reproductive strategies used by viruses than by phages. However, this difference may be illusory, because all parasites must be adapted to the biochemistry of their hosts; the viruses that have been studied reproduce in hundreds of different hosts, whereas study of phages has been confined almost exclusively to three species of bacteria.

BASIC STRUCTURE OF EUKARYOTIC VIRUSES

There are four morphological types of viruses—naked icosahedral, naked helical, enveloped icosahedral, and enveloped helical (Figure

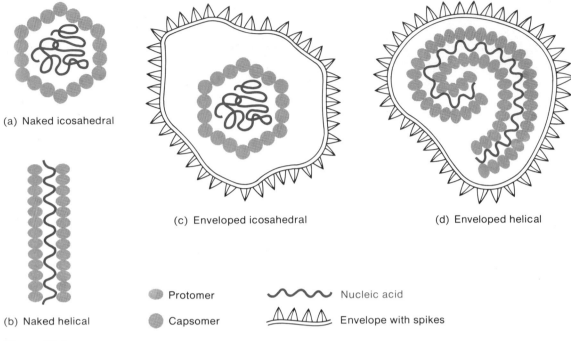

(a) Naked icosahedral

(b) Naked helical

(c) Enveloped icosahedral

(d) Enveloped helical

Protomer

Capsomer

Nucleic acid

Envelope with spikes

Figure 21-1
Four basic structures for viruses.

Table 21-1
Nucleic acids of several viruses and a viroid

Type of nucleic acid	Representative virus	Molecular weight $\times 10^6$	Approximate number of genes
Single-stranded DNA	Adeno-associated	1.5	4–5
	Parvovirus	1.5	4–5
Double-stranded DNA	Polyoma, SV40	3	4–5
	Papilloma	6	9
	Adenovirus 2	23	30
	Herpes simplex	100	150
	Vaccinia	160	240
Single-stranded RNA 1–2	Potato spindle viroid*	0.12	none (?)
	Satellite tobacco necrosis	0.4	1–2
	Tobacco mosaic	2	6
	Polio	2.5	7–8
	Influenza	4†	12
Double-stranded RNA	Reovirus	15†	22
	Rice dwarf	15	22

*Viroids are infectious unencapsidated RNA molecules; they are discussed later in this chapter.

† These virions contain several RNA fragments. The total molecular weight is given. The range for the individual fragments is 0.03×10^6 to 3.4×10^6.

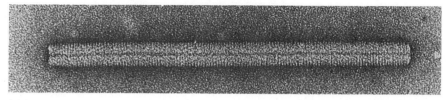

(a)

Figure 21-2
Electron micrographs of tobacco mosaic virus (TMV) (a) An intact particle showing the disclike arrangement of the proteins and the hollow core. (b) Virus particles have been broken, and fragments have been deposited "end-on" on the support film, clearly showing the hollow core. The magnification in (b) is about 1/3 that of (a). (Courtesy of Robley Williams.)

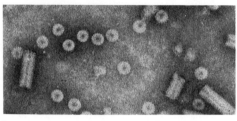

(b)

21-1). Each consists minimally of a coat covering a nucleic acid. The complete virus particle is called a **virion.** A particular species of virus usually contains only a single type of nucleic acid, which may be single- or double-stranded RNA or single- or double-stranded DNA (Table 21-1). Usually the virion contains a single nucleic acid molecule, but some virions contain several molecules, each containing one or occasionally two genes.

The coding capacity of the nucleic acid molecule of the naked viruses is small; therefore, the coat of a naked virion consists primarily of one (plant virus) or two (animal virus) structural proteins. This building block is called a **protomer.** (On occasion, individual enzyme molecules are also contained in the coat.)

In a naked helical virus the protomers are arranged in a helical array and surround one molecule of single-stranded RNA. An electron micrograph of the most thoroughly studied naked helical virus, tobacco mosaic virus, is shown in Figure 21-2.

Because of their importance as agents of human and plant disease, naked icosahedral viruses have likewise been widely studied. In these virions the protomers are organized aggregates called **capsomers.** The capsomers contain either five protomers—**pentons**—or six protomers—**hexons.** In animal icosahedral viruses two different polypeptide chains are used to form pentons and hexons but in plant viruses (and RNA phages) both pentons and hexons use the same protein. The capsomers are organized in several ways to form an icosahedral (20-faced) capsid. The capsid encloses the nucleic acid, and the capsid-nucleic acid unit is called a **nucleocapsid.** If no envelope is present, the nucleocapsid itself is the virion.

Figure 21-3
Several icosahedral viruses. (a) Human wart virus. (b) Wound-tumor virus. (c) Tomato bushy stunt virus. (Courtesy of H. F. Howatson (a) and Robley Williams (b) and (c).)

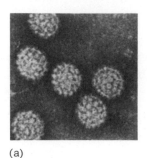 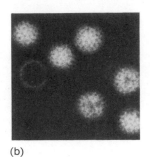
(a)

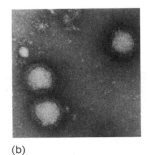 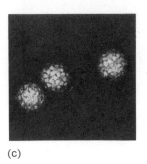
(b)

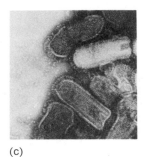 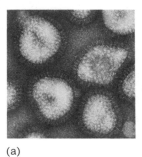
(c)

Figure 21-4
Several enveloped viruses. (a) Influenza virus. (b) Sindbis virus. (c) Vesicular stomatitis virus. (Courtesy of Robley Williams.)

(a)

(b)

(c)

Table 21-2
Some enzymes contained in virions

Enzyme	Virus	Function
Reverse transcriptase	Retrovirus	Makes DNA from single-stranded RNA; converts single-stranded DNA to double-stranded DNA; RNase H
Double-stranded RNA transcriptase	Reovirus	Makes single-stranded RNA from double-stranded RNA
DNA transcriptase	Poxvirus	Makes mRNA from DNA
Single-stranded RNA transcriptase	Myxovirus, rhabdovirus, paramyxovirus	Makes single-stranded RNA from single-stranded RNA
DNA exo- and endonuclease	Poxvirus, retrovirus	Degrades DNA to oligo- and mononucleotides
Nucleoside triphosphate phosphotransferase	Many enveloped viruses	Phosphate exchange
Adeno DNA endonuclease	Adenovirus	Nicks double-stranded DNA
Neuraminidase	Myxovirus, paramyxovirus	Cleaves cell surface polysaccharides
Protein kinase	Retrovirus, myxovirus, paramyxovirus, herpesvirus	Phosphorylates proteins
tRNA aminoacylase	Many retroviruses	Charges tRNA
DNA ligase	Many retroviruses	Seals single-strand breaks in DNA

Electron micrographs of several different icosahedral virions are shown in Figure 21-3.

Empty capsids are often found in partially purified virus samples, indicating that the nucleic acid is not essential for assembly of a capsid. However, the structure of an empty capsid is slightly different from that of a nucleocapsid.

The enveloped virions do not have well-characterized structures. In these, the nucleocapsid, which may be either helical or icosahedral, is encased in a loose membranous envelope, which is roughly spherical. The shape of the virion is variable because the envelope is not rigid. The envelope is derived from the cell membrane. In an enveloped helical virion the nucleocapsid is coiled within the envelope. Figure 21-4 shows several examples of enveloped viruses. Many virions contain virus-encoded enzymes not present in the host cell (Table 21-2). Sometimes these enzymes are nucleic acid replicases or are factors needed for adsorption of the virion to the host cell; however, in many cases, the precise role of the enzyme is unknown.

VIRAL NUCLEIC ACIDS

Viral nucleic acids have an enormous range in molecular weight, as shown in Table 21-1. Furthermore, viral DNA and RNA have many forms.

Viral DNA

Single-stranded linear DNA is found in one small family of viruses. Double-stranded DNA molecules—linear, circular-nonsupercoiled, and circular-supercoiled—are all common, just as they are in phages. However, there are also unusual forms that have not been observed in other organisms. For example, the DNA of vaccinia virus is double-stranded with closed ends (Figure 21-5).

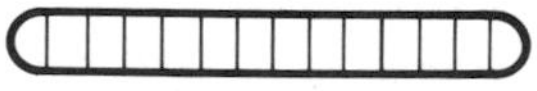

Inverted-repetition base sequences are an essential feature of the linear DNA molecules (Figure 21-6). For example, all linear single-stranded DNA molecules as well as the double-stranded DNA of adenovirus have an inverted repetition at the termini; these repeated sequences are essential for replication, as we will see later. The inverted terminal sequences in herpesviruses are unusual in that the DNA molecules in any population of herpesvirus are always present in four forms; herpesvirus DNA consists of two segments, each of which can be in two orientations, as shown in the figure.

Figure 21-5
The closed hairpin structure of vaccinia virus DNA. The short vertical lines represent base pairs.

(a) Inverted repetitions at termini

a b c   c′ b′ a′
Parvoviruses

a b c   c′ b′ a′
Adenoviruses

a′ b′ c′   c b a

(b) Internal inversions of herpesvirus

a′ b′ d′   e b a a′c′f′   g c a
               La   Sh
a b d   e′b′a′ a c f   g′c′a′

a′ b′ e′   d b a a′c′f′   g c a
               La   Sh
a b e   d′b′a′ a c f   g′c′a′

a′ b′ d′   e b a a′c′g′   f c a
               La   Sh
a b d   e′b′a′a c g   f′ c a′

a′ b′ e′   d b a a′c′g′   f c a
               La   Sh
a b e   d′b′a′ a c g   f′ c′a′

Figure 21-6
The internal structure of base sequences in viral DNA. In part (b) the abbreviations La and Sh represent large and short, respectively. The lower case letters refer to sequences of bases; a prime indicates a complementary sequence.

(a) ————————————————
Single-stranded RNA

(b)
Inverted terminal repetition

(c) —— —— —— —— ——
Segmented single strands

(d)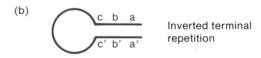
Diploid single strands with associated tRNA

(e)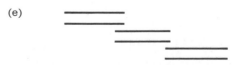
Segmented double strands

Figure 21-7
Forms of viral RNA. In panel (b), the loop is drawn very much larger in relation to the stem than it actually would be. Note the unusual parallel hydrogen-bonded structure in panel (d). In panel (e), in some instances the segments have common sequences.

Viral RNA

Several forms of viral RNA are known, as shown in Figure 21-7. The most common form is single-stranded RNA; in a few viruses—as, for example, in Sindbis virus—an inverted terminal repetition can form a double-stranded region and this region plays a role in replication, as is the case with single-stranded DNA viruses. Many mammalian retroviruses are diploid and have two RNA molecules that are identical except for their 5′ termini. The two strands are held together near the 5′ termini by hydrogen bonds. Retroviruses having diploid RNA also contain two host tRNA molecules, which are hydrogen-bonded to the viral RNA and function as primers in replication. Some single-stranded RNA viruses contain several nonidentical single strands each carrying separate genetic information (influenza virus is an example); such viruses are said to possess a segmented genome. Alfalfa mosaic virus is an especially interesting example of a virus with a segmented genome: its four segments of RNA are packaged into four distinct virions; each virion has a defined RNA content (Figure 21-8). The RNA species in the alfalfa mosaic virus, which are called B, M, Tb, and Ta RNA, have molecular weights of 1.1, 0.8, 0.7, and 0.3 $\times$ 10^6, respectively. Packaging of these fragments occurs as shown in the figure. Successful infection requires that one RNA of each type enter the cell. Viruses with segmented genomes, the components of which are packaged in distinct virions, are called **heterocapsidic** viruses; if all fragments are present in a single virion, the virus is called **isocapsidic**. About 20 different heterocapsidic viruses are known at present.

Plant viruses containing segmented genomes are called **multiple-component viruses** or **coviruses.** The term is not used for the segmented animal viruses.

The single-stranded RNA molecules in viruses are of two types: $(+)$ strand and $(-)$ strand.

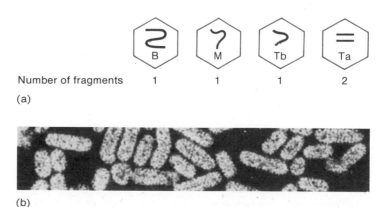

(a)

Figure 21-8
(a) The four virions of alfalfa mosaic virus, a multiple component virus (covirus). (b) Electron micrograph showing that the particles having different contents of nucleic acid have the same width but different lengths. (Courtesy of J. Hull.)

(b)

A viral RNA molecule is said to consist of a (+) strand if it has the same polarity as viral mRNA and contains codon sequences that can be translated into viral protein.

A (−)-strand RNA molecule is a noncoding strand and thus it must be copied by an RNA-dependent RNA polymerase to yield a translatable mRNA.

Viruses having each of these types of RNA necessarily have different life cycles.

Usually the 5′ end of each virion (+) strand or viral mRNA molecule has a cap similar to that found in mRNA molecules in eukaryotic cells. (Capping was in fact first discovered in viral RNA.) The single known exception is poliovirus RNA, which is uncapped. In viruses having segmented genomes each fragment is capped. There is no cap on the 5′ ends of double-stranded viral RNA molecules, though mRNA molecules copied from these duplex molecules are capped. Neither the viral RNA nor the mRNA copied from the double-stranded RNA viruses (e.g., reovirus) has a poly(A) tail.

The 3′ terminus of the single-stranded animal virus RNA molecule, has a poly(A) tail, which is usually shorter than the tail found in animal cellular mRNA. The 3′ termini of many single-stranded plant virus RNA molecules (e.g., turnip yellow mosaic virus, bromegrass mosaic virus) are very unusual. The 3′-terminal base sequence resembles tRNA and in fact *in vitro* these molecules can accept an amino acid from tRNA aminoacyl synthetases. However, aminoacylation has not been observed *in vivo*. The properties of the termini of viral RNA molecules are summarized in Table 21-3.

Table 21-3
Properties of the termini of viral RNA molecules

Type of RNA	5′ Cap	3′-Poly(A)	tRNA-like 3′ end
Animal viruses			
Single-stranded	Usually	Yes	No
Segmented, single-stranded	Yes, for each segment	Yes, for each segment	No
Double-stranded	No	No	No
mRNA made from double-stranded RNA	Yes	Yes	No
Plant viruses			
Single-stranded	No	No	Many viruses

Viruses having double-stranded RNA always have segmented genomes and are isocapsidic. The best-studied viruses of this type are the wound-tumor plant viruses, which have twelve segments, and reovirus, an animal virus having ten segments. The life cycle of reovirus will be discussed later.

This section has showed that there is great variation in the kinds of nucleic acids present in the RNA viruses. We will see that the various structures require different strategies for protein and mRNA production and their regulation, primarily because all eukaryotic mRNA molecules are monocistronic; this property of the mRNA also introduces different requirements for nucleic acid replication and protein processing.

THE BASIC LIFE CYCLE OF VIRULENT VIRUSES

The life cycle of all virulent viruses consists of the following stages:

1. Adsorption.
2. Entry of the nucleic acid into the cell.
3. Transcription, translation, and replication.
4. Maturation of particles.
5. Release of particles.

The life cycle typically takes 6–48 hours (in contrast with 20–60 minutes for phages), and 10^3–10^5 virions are released per cell per generation. In a later section tumor viruses, which have a different life cycle, will be described.

Viral life cycles are so varied that no single mechanism can account for any one of the stages just listed. Nonetheless, in the following sections some of the major characteristics of these stages and a few common mechanisms are described briefly. The mechanisms of transcription, translation, and replication will be discussed in detail elsewhere in this chapter.

Adsorption of a Virus to a Host Cell

Adsorption occurs in a great many ways and will not be discussed in this book. Its study has been hampered by the difficulty of isolating the receptor unit on the cell surface. However, it is likely that the receptor for all animal viruses is a surface glycoprotein. The adsorption sites on animal viruses are short spikes, made of a single protein for the enveloped viruses, and either spikes or flexible hairs for the naked viruses. Very little is known about adsorption of plant viruses.

Entry of Viral Nucleic Acid into a Host Cell

The number of mechanisms for transport of the viral nucleic acid into a cell may be as great as the number of viruses. For some viruses only the nucleic acid enters the cell; others carry their own polymerase, which is essential, so that both nucleic acid and protein molecules (or possibly nucleoproteins) are transported across the cell membrane. With a few enveloped animal viruses there is some evidence that an early stage of penetration is fusion of the membranous envelope of the virus with the cell membrane; for a while the nucleocapsid is surrounded by cell membrane components. In both enveloped and naked viruses, proteolytic digestion of the capsid, called **uncoating,** is a step following the entry stage. Earlier textbooks described several detailed models for the mechanism of uncoating but recently each of these models has been shown to be incorrect. At present we have no real understanding of the full process of entry but it is likely that the process is always mediated by a specific interaction between a virus-encoded protein in the capsid or envelope and a particular membrane protein.

Maturation and Release of Progeny Virus Particles

The life cycles of virulent animal and plant viruses do not always terminate with programmed cell death and lysis, in contrast with the life cycles of most virulent bacteriophages. Upon infection by some viruses an infected cell will become generally unhealthy, grow very slowly, and perhaps ultimately die, but it is equally common for cells infected with certain viruses to grow indefinitely, releasing progeny viruses continually. (Often such an infected cell synthesizes and releases specific proteins—for example, the adenovirus penton—that may be responsible for pathogenic effects in an infected organism.)

Virus particles are assembled from viral nucleic acids and proteins and are produced in very large numbers. The mode of assembly of tobacco mosaic virus has already been described in detail in Chapter 6. In this section we will see that for some viruses maturation and release are coupled processes.

There are three basic mechanisms of release of viruses:

1. Release from vacuoles formed under the cell membrane.
2. Lysis.
3. Budding.

The first two are characteristic of the naked viruses, whereas budding is the rule for the enveloped viruses. Each of these mechanisms exhibits many variations. In part the differences are imposed by the location of viral assembly—some viruses replicate and are assembled in the

Figure 21-9
Release of virions by formation of a vacuole. An aggregate of virions—a viral inclusion—forms under the cell membrane in a vacuole. The membrane ruptures and virions are released. The cell may or may not remain intact.

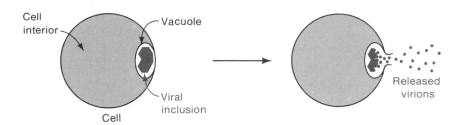

(a)

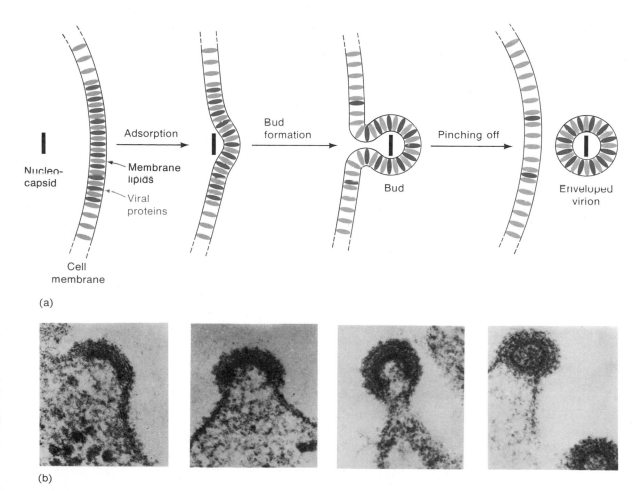

(b)

Figure 21-10
(a) Formation of an enveloped helical nucleocapsid by budding. In reality, the membrane is much thinner than the width of the nucleocapsid, and the lipids and viral proteins are not in an array as ordered as that shown.
(b) Electron micrograph of the budding process. (Courtesy of W. Bernhard.)

cytoplasm, whereas others are nuclear viruses. Cytoplasmic particles often accumulate in vacuoles located just under the cell membrane and are released by rupture of these vacuoles (Figure 21-9) but without disruption of the entire cell. This usually occurs 1–2 days after infection but some viruses, known as rapid-release viruses, are released 6–8 hours after infection. The nuclear viruses generally do not cause vacuolization. Instead, the viruses form large quasicrystalline arrays in the nucleus and sometimes also in the cytoplasm. About 24–48 hours after infection the cell starts to disintegrate and the virions are released. Some viruses that multiply in the cytoplasm also lyse cells by this disintegration mode. The enveloped viruses are released by **budding** (Figure 21-10). In this mode of release virus-encoded proteins, which are synthesized in the cytoplasm, are incorporated into the cell membrane, causing a reorganization of the membrane structure. The nucleocapsid is thought to adhere closely to these viral proteins. This causes the membrane to form a protruding sphere around the nucleocapsid. In an unknown way (possibly by steps used in penetration), the cell membrane re-forms under the bud, releasing the enveloped nucleocapsid. Thus, the envelope is a portion of the cell membrane that contains viral proteins and no cellular proteins.

Takeover of a Host Cell by a Virus

In our discussion of the phage life cycle a stage was defined in which the phage "takes over" the bacterial host, converting the bacterium exclusively to a phage-making machine. Part of this takeover is a cessation or reduction of synthesis of host DNA, RNA, and protein early in the life cycle of the phage. Such a stage is notably absent in the outline of the life cycle of a typical animal virus. Events of this type are evident in the life cycle of some viruses, but this is certainly not generally true. In fact, often the virus and host coexist for some time and only in the late stages of infection are host functions impaired.

In the life cycles of some viruses there are definite periods of time in which the host is severely altered, but it is not clear that this is necessary for optimal production of viruses. Three types of alterations have been observed in a few systems:

1. Inhibition of host RNA and DNA synthesis sometimes occurs very late in the life cycle, well after viral nucleic acid replication has begun.
2. Impairment of the synthesis of ribosomes sometimes occurs; transcription of rRNA genes occurs normally and processing of the rRNA precursor is inhibited.
3. Inhibition of cellular protein synthesis is a common event, though the extent and timing of the inhibition varies widely with

different viruses; its mechanism is known for poliovirus and adenovirus. In a poliovirus infection the initiation of host polypeptide chains is inhibited. The defect is in the formation of the host mRNA-ribosome-tRNA^{Met} initiation complex. As mentioned in the section on viral RNA structure, polio virion RNA, which is a (+) strand, lacks a cap. Apparently a viral protein, which has not yet been isolated, interferes with the formation of an initiation complex with capped RNA. There is no interference with uncapped RNA. For adenovirus, cellular mRNA is made in the nucleus but this RNA is not processed, which prevents transport to the cytoplasm. Thus this mRNA is never translated. Translation of long-lived host mRNA existing in the cytoplasm prior to infection also does not occur.

ASSAYING VIRUS PARTICLES

In performing any experiment it is desirable to know the ratio of viruses and cells—the multiplicity of infection. Viruses can be counted by electron microscopy, though this has two disadvantages: (1) for technical reasons the measured value can be as much as three times too great or too small, and (2) there is no way of knowing what fraction of the particles are infective. For this reason assays similar to plaque formation by phages have been developed. Some of these assays are described in this section.

Biological Assays for Viruses

Many animal cells can be grown as a monolayer in a petri dish. When growth is complete, a confluent layer of cells results which can be stained with various dyes to yield a colored layer. Shortly after the cells are added to a petri dish, viruses can be added and some of these will infect cells in the growing layer. If the virus is a type that either causes cell death (for example, poliovirus), or at least significant inhibition of cell growth, there will be a region in the layer in which there are few or no cells when the monolayer has stopped growing. Thus there will be colorless circular regions—**plaques**—in the stained background (Figure 21-11). In contrast with phages for which plaque formation takes 3–12 hours, formation of a viral plaque requires 3–21 days depending on the particular virus, because of the slow growth of animal cells.

Some viruses fail to form plaques because they do not kill cells fast enough. However, if a virus (for example, influenza virus) stimulates hemagglutinin synthesis, infectious particles can be assayed. **Hemagglutinins** are glycoproteins formed on the surface of infected cells that

(a)

(b)

Figure 21-11
Three biological methods for detecting viruses. (a) Plaques of adenovirus in a confluent background of stained animal cells. (Courtesy of Alvin Winter.) (b) A focus of Rous sarcoma virus. The dark region is a cluster of cells—each cell infected with virus—growing in numerous layers on a monolayer of uninfected chick embryo cells. (Courtesy of Howard Temin.) The plaques in (a) and the focus in (b) range in size from 1–3 mm. (c) Tobacco leaf infected with tobacco mosaic virus, showing lesions of dried tissue. (Courtesy of Heinz Fraenkel-Conrat.)

(c)

cause an infected cell to bind a red blood cell. Thus, if a confluent layer of cells containing foci of infection, which are not visible as plaques, is overlaid with a suspension of red cells, and then washed, the foci can be identified as red spots on the surface of the monolayer.

Before the plaque assay was developed, certain viruses (examples are vaccinia virus and herpesviruses) were assayed by the **pock method.** If the chorioallantoic membrane of a 10-day chick embryo is infected with viruses and then cultured for 36–72 hours, opaque foci called pocks appear on the transparent membrane. The pocks are quite visible and easily counted.

Tumor viruses (for example, Rous sarcoma virus) can also be titrated on confluent cell layers. These viruses cause infected cells to continue to grow for several days after the confluent layer has ceased growth. The regions of growth, called **foci,** appear as raised clusters on the monolayer (Figure 21-11).

Plant viruses (for example, tobacco mosaic virus) are titrated by abrading the surface of a leaf and then spreading a droplet of a virus suspension on the abraded surface. Dead spots on the leaf (called **lesions**) are formed; each lesion is the result of infection by a single virus particle (Figure 21-11).

Efficiency of Detecting a Virus

The efficiency of detection of viruses by each of the bioassays just described is low and varies with the individual virus (Table 21-4). When compared with the total number of particles seen by electron microscopy, 0.01–10 percent of animal virus particles produce a visible infection. (Semliki forest virus and herpes simplex virus are notable exceptions; the best preparations give almost 100 percent efficiency of infection when assayed under optimal conditions.) The plant viruses are even less efficient, as the table indicates. There are three possible causes of the low efficiency of infection:

1. Most of the particles may be noninfective owing to defective particle assembly or damage of some sort after assembly.
2. Conditions used to detect the virus might, for unknown reasons, limit the ability to form a plaque, pock, focus, or other end point.
3. Successful infection requires the cooperation of many functional particles. It is very important to determine whether alternative (3) is correct if one is to understand the molecular basis of the life cycle. In fact, apart from the multicomponent plant viruses in which different particles carry distinct RNA molecules (Figure 21-8), it is almost always true that successful infection by an animal virus requires only one functional virus particle, as is also true of bacteriophage infection. This is proved by the fact that the number of plaques (or other foci of virus growth) is directly proportional to the number of particles added. The argument for this conclusion can be seen most easily by considering a hypothetical case in which two particles are needed for successful infection.

Table 21–4
Ratio of virus particles to infectious units, as assayed by various methods

Virus	Ratio*
Animal viruses	
Semliki forest virus	1
Poxvirus	1–100
Influenza	7–10
Reovirus	10
Adenovirus	10–20
Poliovirus	30–1000
SV40	100–200
Plant virus	
Tobacco mosaic virus	$5 \times 10^4 - 10^6$

*The wide range for a particular virus indicates different efficiencies of various assay procedures.

The Poisson distribution states that if the ratio of the total number of virus particles to cells is m (which in ordinary assay conditions on a petri dish is very small), then the proportion $P(k)$ of cells infected by k particles is

$$P(k) = e^{-m}m^k/k!$$

Thus, if two virions are needed for successful infection, there will be two types of uninfected cells—those infected with no viruses, for which the proportion is e^{-m}, and those infected with one virus, for which the proportion is me^{-m}. Thus, the proportion $P(0)$ of cells that do not produce plaques is $e^{-m} + me^{-m}$ or $e^{-m}(1 + m)$. If m is small, this expression can be approximated as $P(0) = 1 - (1/2)m^2$. Since the proportion of infected cells P_i is $1 - P(0)$, then $P_i = (1/2)m^2$, which means that the number of plaques increases as the square of the number of added virions. It can easily be shown that if more than two virions are required for a successful infection, the number of plaques will increase at a power of m greater than two. On the other hand, if only one virion is needed, the fraction of uninfected cells will be e^{-m} and the fraction of infected cells will be $1 - e^{-m}$. If the number of added virions is increased q-fold, the fraction of infected cells will be $1 - e^{-qm}$. The ratio $(1 - e^{-qm})/(1 - e^{-m})$ is the factor by which the number of plaques increases when q-fold more viruses are added. For small values of m this expression equals q. Thus, when the number of added virions increases q-fold, the number of plaques increases q-fold. Since for small m this conclusion is independent of the value of m—that is, independent of the number of added particles—a linear curve such as that in Figure 21-12 is proof that one virion can initiate an infection.

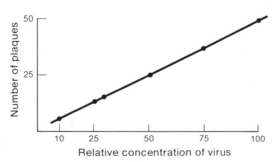

Figure 21-12
Data showing that the number of plaques is proportional to the number of added particles. A constant volume of a virus suspension at the concentration indicated is added to a single petri dish inoculated with sufficient cells to form a confluent layer.

CLASSIFICATION OF ANIMAL VIRUSES

Animal viruses will be the major topic of this chapter. So many animal viruses are used in molecular biological research and their names are so complicated that the list in Table 21-5 is provided for further reference. This table lists the major families of animal viruses, the particular viruses encountered most frequently in research, and the derivation of their names.

Table 21-5
Viruses used in molecular biological research

Virus family	Derivation of name	Type of nucleic acid	Representative virus	Tumor-forming?
Adenovirus	*Adenoid* tissue	ds DNA	Adenovirus	Some
Herpesvirus	*Herpein,* Gr., "to creep"	ds DNA	Herpes zoster (chicken pox) Herpes simplex I (cold sores)	Some
Poxvirus	Pock-forming	ds DNA	Vaccinia, fibroma, myxoma	Some
Picornavirus	*Pico-* ("little") + "RNA"	ss RNA	Polio, ECHO, mengovirus	No
Myxovirus	*Myxo,* Lat., "mucus"	ss RNA	Influenza	No
Rhabdovirus	*Rhabdos,* Gr., "rod"	ss RNA	Vesicular stomatitis	No
Togavirus, or arbovirus	*Toga,* Lat., "covering" or "coat"; arthropod-*borne*	ss RNA	Yellow fever, Sindbis, Semliki forest, Eastern equine encephalitis	No
Reovirus	*R*espiratory *e*nteric *o*rphan	ds RNA	Reovirus	No
Polyoma*	Lat., *Poly-* ("many") + *-oma* ("tumor")	ds DNA	Polyoma, SV40	Yes
Papilloma*	*Papilla,* Lat., "nipple" or "wart" (wart-forming)	ds DNA	Shope papilloma	Yes
Retrovirus	*Retro,* Lat., "reverse"	ss RNA	Avian leukosis, murine leukemia, Friend leukemia, Moloney, Rous sarcoma, mouse mammary tumor	Yes

Note: ss, single-stranded; ds, double-stranded.

*Papilloma, *polyoma,* and the *vacuolating* viruses together form the papovavirus family.

ANIMAL RNA VIRUSES

Several points must be considered when studying animal RNA viruses:

1. The mechanism of production of mRNA is the principal feature by which the RNA viruses can be distinguished, in the sense that the particular mechanism determines the basic pattern of replication and translation.

2. All eukaryotic mRNA molecules are monocistronic and eukaryotic ribosomes are rarely able to translate more than one polypeptide chain from a single mRNA molecule; however, all known RNA viruses direct the synthesis of several proteins.

3. The (−)-strand RNA viruses need to synthesize a (+) strand before translation can occur, yet host enzymes that can copy an RNA molecule do not exist.

4. A single-stranded virion RNA molecule cannot serve as a template for direct synthesis of virion RNA; thus, a replication intermediate containing or consisting of a complementary strand must be made.

Replication and the production of viral mRNA are interdependent processes for many RNA animal viruses and are best discussed together. There are four main mechanisms for producing mRNA, and many variations. These four mechanisms are depicted in Figure 21-13. It is convenient to select a particular virus of each type and examine the mode of replication, production of mRNA, and translation for each. The viruses that have been selected are also indicated in the figure.

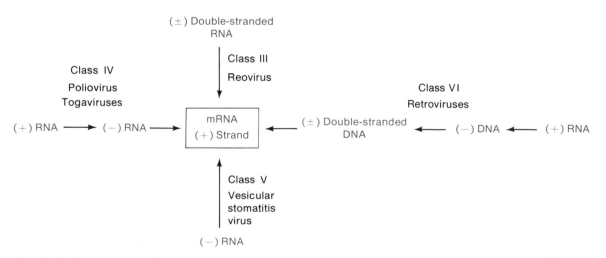

Figure 21-13
Four mechanisms for producing mRNA. The taxonomic classes corresponding to these mechanisms are indicated, as are the viruses discussed in the text. Classes I and II are DNA viruses and are discussed elsewhere.

Viral Polymerases

Before we begin our discussion of individual viruses, a few comments about polymerases are needed. Many virions carry their own polymerases. The (−)-strand RNA viruses could not otherwise function because animal cells do not contain RNA-dependent RNA polymerases and the infecting (−) strand is by definition not translatable. A virion polymerase is also essential for the double-stranded RNA viruses because the (+) strand in the duplex is unavailable for ribosome binding and translation, and no known cellular enzyme uses double-stranded RNA as a template for RNA synthesis. There is no obvious need for a virion-bound polymerase for (+)-strand RNA viruses. However, the retroviruses, which have a DNA intermediate (because of a unique feature of their life cycles), are in need of a virion-bound enzyme because no known animal cell synthesizes DNA on an RNA template.

Vesicular Stomatitis Virus (VSV)

Vesicular stomatitis virus, or VSV (Figure 21-14), is a pathogen of cattle. It contains a single (−)-strand RNA molecule having a molecular weight of 4×10^6 and lacking both a 5′ cap and a 3′-poly(A) tail. Five proteins are encoded in this molecule. These proteins are the following:

N	The major capsid protein.
L, NS	Found in the capsid and needed for RNA synthesis.
M	The envelope protein.
G	The spike protein.

For synthesis of these five proteins VSV must form (+)-strand mRNA and the problem of the inability of animal cells to translate polycistronic mRNA must be overcome. The solution developed by VSV is to synthesize five different mRNA molecules, as is shown in Figure 21-15. These mRNA molecules are synthesized by the combined action of the L and NS enzymes, which are brought into the cell on

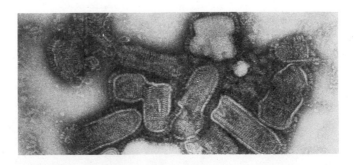

Figure 21-14
Electron micrograph of vesicular stomatitis virus. (Courtesy of Robley Williams.)

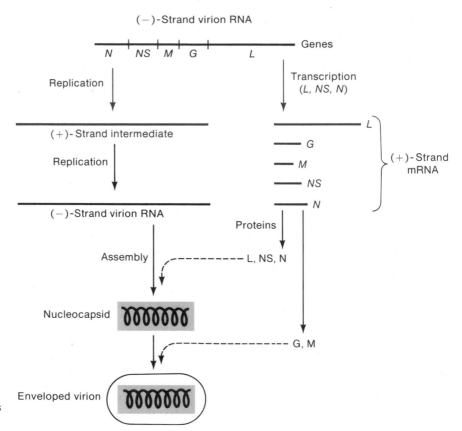

Figure 21-15
Schematic diagram for replication and transcription of vesicular stomatitis virus.

(−)-strand virion RNA that is coated with N protein molecules, derived from the capsid during uncoating. Each of these mRNA molecules is capped, has a 3′-poly(A) terminus, and contains only one cistron. It is not known with certainty whether a single large (+) strand is synthesized and then cleaved into five fragments or if the five molecules are made by successive starting and stopping. Preliminary evidence suggests that distinct (+)-strand molecules are synthesized by a sequence of successive starting and stopping events.

Replication of VSV also proceeds by first copying a (+) strand from the virion mRNA. These full-length strands are also made by the action of the L and NS proteins but they are not cleaved (serving no mRNA function); instead, they are immediately copied, again by the L and NS enzymes, to make (−)-strand virion DNA. At the time (−)-strand RNA is made, a lot of N protein has been translated from the mRNA strands made earlier; the virion RNA and the N protein rapidly associate to form nucleocapsids.

The processes of synthesizing (+)-strand mRNA molecules, which are cleaved, and (+)-strand templates for making (−)-strand virion RNA are somehow separated. This is known because mutants exist that inhibit one process but not the other.

Segmented (−)-Strand Viruses

The mode of replication and transcription of the segmented (−)-strand viruses, of which influenza virus is a member, is not well understood. However, since each segment contains only one cistron, no special mechanism is needed to generate monocistronic mRNA molecules. The problem these viruses face instead is how to obtain a complete set of fragments in each nucleocapsid. There are two possibilities: (1) a special mechanism selects one molecule of each type for each nucleocapsid, or (2) the molecules are packaged at random so that most virions are incomplete and thus defective. In the case of influenza virus it seems likely that packaging is random, though this is not a firmly established fact.

Poliovirus

Poliovirus (Figure 21-16), one of the best-understood viruses, contains a single (+)-strand RNA molecule having a molecular weight of 2.5×10^6. The virus replicates in the cytoplasm of sensitive cells, forming large crystalline inclusions of mature virions (Figure 21-17). The virions are

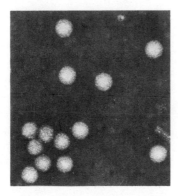

Figure 21-16
Electron micrograph of poliovirus. (Courtesy of Robley Williams.)

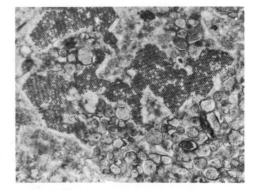

Figure 21-17
Crystalline inclusions of mature polio virions in the cytoplasm of an infected cell. (Courtesy of J. Meyer, M. Weiss, and R. Böni.)

released when the damaged cell lyses. The virion RNA has a 3′-poly(A) tail but the 5′ end is uncapped; instead the 5′ end is covalently linked to a small protein, whose function is unknown. The RNA molecule encodes seven proteins—four coat proteins, an RNA replicase, and two proteins of unknown function. No enzymes are carried in the capsid. The virion RNA has some mRNA activity but does not direct the synthesis of all of the proteins. At no time is the RNA segmented and therefore some mechanism is needed to generate several proteins from one mRNA molecule.

Immediately after entry into the cell, the virion RNA strand is partially translated and an RNA replicase is synthesized. The replicase is processed from a larger polypeptide chain, as we will soon see. This enzyme catalyzes synthesis of (−)-strand RNA (Figure 21-18) whose sole function is apparently to serve as a template for further synthesis of (+)-strand RNA. Copying of the (−) strand begins at the 3′ terminus. Initiation of synthesis of a second (+) strand does not await completion of the first strand and usually five (+) strands are simultaneously copied

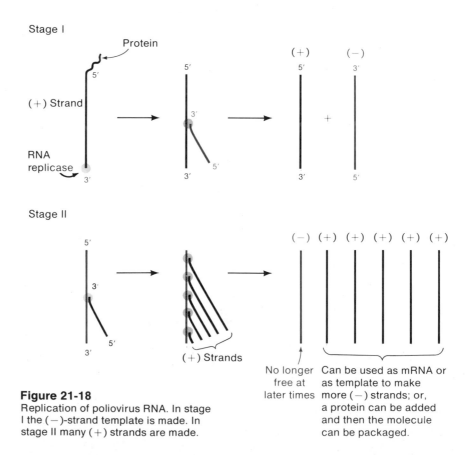

Figure 21-18
Replication of poliovirus RNA. In stage I the (−)-strand template is made. In stage II many (+) strands are made.

from a single (−) strand, to yield a multibranched replication interme-diate. The (+) strands that are formed serve three functions: (1) as mRNA, (2) as a template for further synthesis of (−) strands, and (3) as virion RNA for progeny viruses. These (+) strands lack the 5′-terminal protein, which is added during the packaging process.

Four features of the poliovirus replication process require further comment. These features are listed and discussed below:

1. *The nascent RNA is not totally base-paired to the template.* In the molecule shown in Figure 21-18, consisting of one intact (+) strand and an incomplete **(nascent)** (−) strand, it might be expected that the nascent RNA would base-pair with the complementary segment of the (−) strand to form a double-stranded region. It is believed, but not yet proved, that the nascent RNA is coated with a protein that prevents this pairing.

2. *Double-stranded RNA is found in infected cells.* The existence of double-stranded RNA originally led to the idea that the first stage of replication was a conversion of the (+)-strand virion RNA to double-stranded RNA. This view was dispelled by several lines of evidence and it is now believed that double-stranded RNA forms when a progeny strand, such as in the first molecule of stage II of Figure 21-18, becomes hydrogen-bonded to the template, possibly owing to lack of the RNA-binding protein that normally inhibits base-pairing. It is thought that the double-stranded RNA has no function in the virus life cycle.

3. *A replication intermediate containing a long, nicked double-stranded segment with single-stranded branches can be isolated.* This structure is only found when the RNA is extensively purified and the protein that prevents base-pairing of the nascent RNA is removed. The structure is believed to be an artifact of isolation that arises by the mechanism shown in Figure 21-19.

4. *The amount of poliovirus RNA increases exponentially for several hours and then increases linearly with time.* Figure 21-18 shows the replication of poliovirus RNA occurring in two stages. At the end of stage I, which is responsible for synthesis of (−)-strand RNA, there are two molecules. In stage II, (+) strands are made with the (−)-strand template synthesized in stage I. The parental (+) strand can be used as a template again to synthesize another (−) strand, as can the (+) strands made in stage II. The figure shows that in stage II several (+) strands are being synthesized simultaneously. However, at early times in the infection there is a deficiency of replicase molecules, so the first time stage II is used, only one progeny (+) strand forms. Thus, after the first part of stage-II synthesis is completed, there are four RNA molecules. A

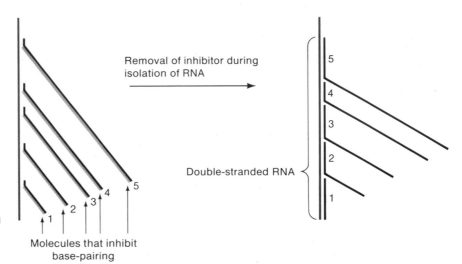

Removal of inhibitor during isolation of RNA

Double-stranded RNA

Molecules that inhibit base-pairing

Figure 21-19
Mechanism for producing an artificial replication intermediate during the isolation process.

repeat of simultaneous rounds of stages I and II yields eight molecules, a third yields sixteen, and so forth. Furthermore, there is a significant time delay between each round of replication because only one minute is required to synthesize a poliovirus RNA molecule, yet the apparent doubling time is fifteen minutes. This doubling, which is the exponential process, continues for about two hours until there are about 600 molecules. During these rounds of doubling, the number of replicase molecules has gradually increased by repeated translation and processing, as has the number of capsid proteins. The multibranched replicative form then predominates, turning out (+) strands linearly with time. The newly made (+) strands are either being translated or being packaged, so (−) strands are only infrequently made. Thus, at late times, the major synthesis is production of (+) strands with linear kinetics.

Poliovirus mRNA encodes seven proteins. Some special mechanism is needed to produce these proteins since, as has been said before, in eukaryotes only a single polypeptide chain can be translated from a mRNA molecule. An indication that there are unusual features of protein synthesis in poliovirus-infected cells came from two observations:

1. Before the number of poliovirus genes was known, it was found by pulse-labeling infected cells with [35S]methionine (which labels proteins) that ten poliovirus proteins were synthesized and that the total molecular weight of these proteins exceeds the coding capacity of the mRNA. The possible explanations for this observation are that genes overlap (which was considered to be a preposterous idea at the time) and that some proteins are

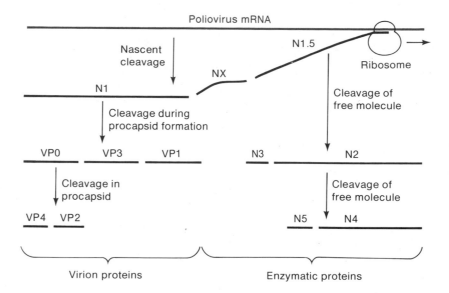

Figure 21-20
Synthesis of poliovirus proteins by successive cleavage of precursors. The labels VP and N stand for *virion p*rotein and *n*onvirion protein, respectively.

cleavage products of others. It turns to be the latter explanation, as reported below.*

2. In a pulse-chase experiment using [^{35}S]methionine some labeled proteins disappeared during the chase and the label simultaneously reappeared in the form of proteins of lower molecular weight than those labeled initially. These data clearly indicated that cleavage of poliovirus proteins occurs.

Several pulse-labeling experiments have led to the following picture of the production of poliovirus proteins (Figure 21-20). A single giant polypeptide, called a **polyprotein,** is translated from the mRNA. Two cleavages, those surrounding protein NX, are made in the polyprotein immediately after the two peptide bonds, which are to be cleaved, have formed. Thus, proteins N1 and NX are released from the growing chain of the polyprotein before protein N1.5 is fully synthesized. These cleavages are called nascent cleavages since they are made in a growing (nascent) polypeptide chain. Once N1.5 is released, a small NH$_2$-terminal fragment (N3) is removed, converting N1.5 to N2, which is again cleaved to form N4 and N5. The proteins prefixed by N have not been well studied but at least some of them are known to be involved in synthesis of RNA. The RNA replicase has not yet been purified and it is not known whether replicase is one of these proteins or is an aggregate consisting of several of these polypeptides.

*At the time these experiments were done, the one gene-one enzyme hypothesis was doctrine. The possibility that genes might overlap was never considered seriously until it was clearly demanded by both genetic and base-sequencing data for *E. coli* phage φX174.

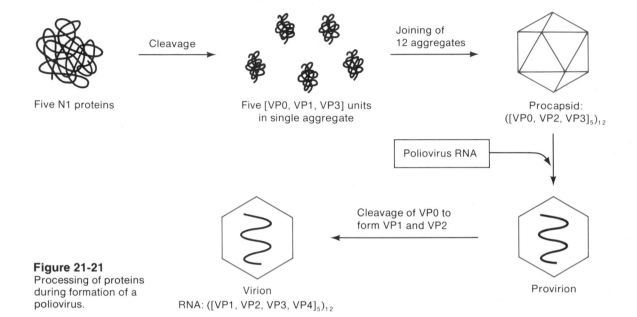

Figure 21-21
Processing of proteins during formation of a poliovirus.

Five N1 proteins

$\xrightarrow{\text{Cleavage}}$

Five [VP0, VP1, VP3] units in single aggregate

$\xrightarrow[\text{12 aggregates}]{\text{Joining of}}$

Procapsid: $([VP0, VP2, VP3]_5)_{12}$

Poliovirus RNA

Provirion

$\xleftarrow[\text{form VP1 and VP2}]{\text{Cleavage of VP0 to}}$

Virion
RNA: $([VP1, VP2, VP3, VP4]_5)_{12}$

Protein N1 is cleaved in a series of steps during morphogenesis of the capsid (Figure 21-21). Five N1 protein molecules aggregate to form a pentamer. Then, cleavage of each N1 occurs to form five units, each of which consists of one VP0, VP1, and VP3 protein and which is called a propenton. Twelve of these propentons aggregate to form the procapsid, an empty particle having the size and shape of the virion. A poliovirus (+) strand enters the procapsid to form a provirion. Finally, VP0 is cleaved to form VP2 and VP4, and a subtle rearrangement of VP1, VP2, VP3, and VP4 occurs to generate the intact virion.

In summary, poliovirus uses a single mRNA molecule and cleaves the single polypeptide translated from it to form the individual viral proteins, in contrast with vesicular stomatitis virus, which synthesizes monocistronic mRNA molecules from (−)-strand virion RNA.

In the following section we examine the togaviruses, which—like poliovirus—have (+)-strand virion mRNA but combine features of poliovirus and vesicular stomatitis virus in their life cycles.

Togaviruses

The togaviruses contain a single (+) strand of RNA that typically encodes about eight proteins; this RNA molecule contains two AUG start signals (Figure 21-22). Without being processed, the viral RNA is used as an mRNA from which a polyprotein (polyprotein I) is translated by initiation of synthesis at AUG_1, the start codon nearest the 5′

terminus. Polyprotein I is then cleaved to yield proteins needed for RNA replication. These replication proteins are then used to synthesize a (−) strand from the viral RNA template. This (−) strand serves two functions: (1) It is a template for synthesis of more (+) strands (RNA I, in the figure), which are ultimately assembled into progeny virions. (2) It is a template for formation of a short mRNA molecule (mRNA II), which contains the AUG_2 start codon but not the AUG_1 codon. It is thought that one of the proteins cleaved from the polyprotein I is responsible for the production of mRNA II. At present it is not known whether mRNA II is cleaved from an intact (+) strand of RNA I or if synthesis of this

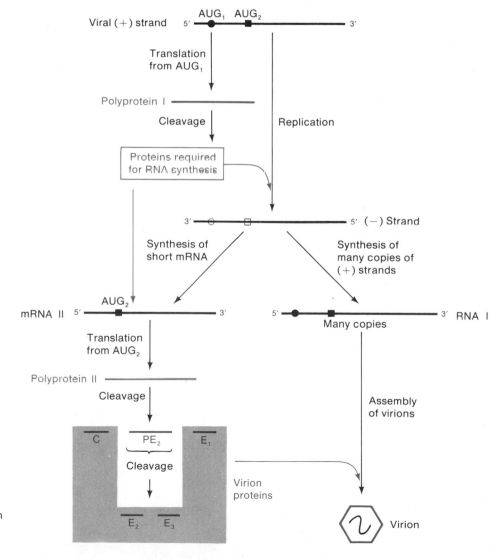

Figure 21-22
Life cycle of a togavirus. The viral strand is translated at early times to yield replication proteins. A (−)-strand replica of the viral strand is copied at a late time to yield a shortened mRNA molecule (lacking AUG), which is translated into virion proteins.

fragment is initiated at a special internal replicase-recognition site. The mRNA II is translated to yield polyprotein II, which is cleaved to yield the proteins C, PE_2, and E_1. Protein PE_2 is cleaved further to generate E_2 and E_3. Proteins C, E_1, E_2, and E_3 are then assembled to form a capsid, into which RNA is packaged. Note that having two temporal classes of mRNA (that is, the "early" viral mRNA and the "late" mRNA II) enables the togaviruses to regulate the amounts of replication enzymes and virion proteins that are synthesized. That is, a single copy of the viral RNA is used initially to make the catalytic proteins needed in small concentrations and synthesis of the structural proteins from mRNA II is delayed until the structural proteins are needed. Then the structural proteins are made in large quantities because there are many mRNA II molecules. This presumably should make togavirus production more efficient than picornavirus (poliovirus) production. Greater efficiency of virus production has not been observed; instead, the togaviruses are able to grow on a greater variety of hosts (both insects and vertebrates) than support picornaviruses (mammals only). In fact, it has been generally observed that separation of replication and transcription (as in vesicular stomatitis virus), and increased regulation of the relative amounts and timing of synthesis of the various proteins, are normally accompanied by an increased host range.

Reovirus

Figure 21-23
Electron micrograph of reovirus. (Courtesy of Robley Williams.)

Reovirus (Figure 21-23) contains ten double-stranded RNA molecules and exemplifies a third type of viral genetic system, somewhat more complex than what we have so far seen. The most striking difference between reovirus and other viruses is that the virion RNA is not released from the virion as free RNA molecules at any time in the course of infection.

The ten virion RNA fragments, each of which encodes a single polypeptide chain, are joined as part of a complex nucleoprotein core that is contained within an icosahedral shell. The shell is removed after the virion enters a cell and an RNA replicase, which is one of the components of the core, is activated (Figure 21-24). The replicase catalyzes synthesis of (+)-strand molecules (that is, mRNA) from each of the double-stranded segments. Capping of each of the (+) strands also occurs, catalyzed by enzymes within the core. A poly(A) tail is not added, making reovirus mRNA the only known viral mRNA that does not have a poly(A) tail. The newly synthesized (+) strands are then extruded from the core (Figure 21-25) but the double-stranded segments remain in the core. Each of the (+) strands is separately translated into a single polypeptide chain. Some of the ten different polypeptides are cleaved to form the finished proteins.

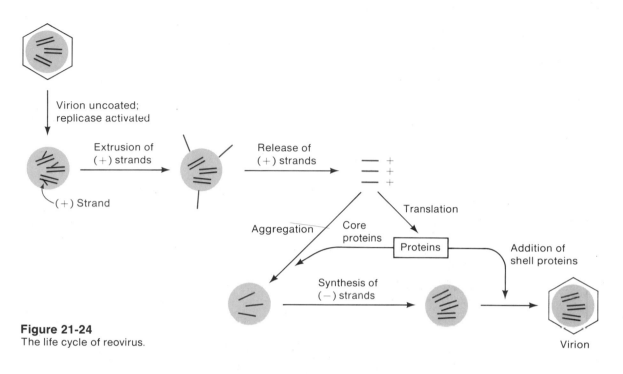

Figure 21-24
The life cycle of reovirus.

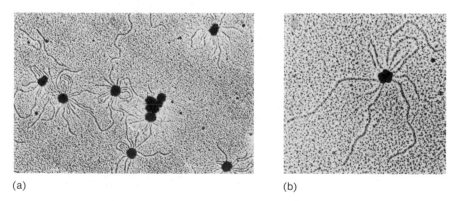

Figure 21-25
(a) Reovirus cores. (b) An enlarged view. The mRNA molecules appear as threads emerging from the dark cores. (From N. M. Bartlett, S. G. Gillies, S. Bullivant, and A. R. Bellamy. 1974. *J. Virol.*, 14: 324.)

(a)   (b)

Replication requires the production of the ten double-stranded segments. This is accomplished by copying (+) strands rather than by semiconservative replication of double-stranded RNA. Formation of new virions requires that one of each of the ten fragments be packaged in the capsid. In reovirus the assortment is regulated and precedes replication. In a way that is not yet understood, one each of the ten (+) strands is joined with core proteins to form a precore and the (+) strands are copied sequentially in a definite order, which is probably determined by the core proteins. The result of this replication is an aggregate containing a complete set of double-stranded segments and the core proteins. (Note that at no time are there free (−) strands.) The finished core then rapidly forms and the icosahedral shell assembles around it.

A striking feature of the reovirus life cycle is that the (+) strand of the virion RNA is never used. The (−) strand is the template for synthesis of all (+) strands and *progeny* (+) strands are templates for the double-stranded virion RNA.

Retroviruses

Retroviruses contain two identical single-stranded RNA molecules linked in a complex. These viruses are unique among all known viruses in that the growth cycles have a stage in which the flow of information is reversed (that is, RNA→DNA) with respect to the usual mechanism (DNA→RNA). Another unusual characteristic of the growth cycles of all retroviruses is that there is an obligatory step in which the DNA-containing intermediate is inserted into the host chromosome, like phage Mu. Most of the retroviruses cause tumors and for this reason have aroused special interest. In this section we will not be concerned with their tumor-producing ability and will consider only the replicative mode in which progeny virions are produced.

The best-understood retrovirus, at this time, is **Rous sarcoma virus, or RSV,** which infects chickens. In 1964 Howard Temin observed that the multiplication of RSV could be prevented by inhibitors of DNA synthesis (amethopterin, cytosine arabinoside) and by actinomycin D, an inhibitor of DNA-dependent RNA synthesis; on the basis of this observation he proposed that the viral growth cycle includes a DNA intermediate, which he called a **provirus.** This hypothesis, which included the novel notion that DNA can be made from RNA, was considered quite doubtful until 1970 when Temin and David Baltimore independently demonstrated the presence in several avian and murine (mouse) retroviruses of the enzyme **reverse transcriptase,** an enzyme that makes double-stranded DNA from a single-stranded RNA template.

Reverse Transcriptase

Reverse transcriptase has three enzymatic activities:

1. It copies an RNA molecule to yield a double-stranded DNA-RNA, using a primer and joining deoxynucleoside 5′-triphosphates in a 3′-5′ linkage.
2. It can copy a primed single strand of DNA to form double-stranded DNA.
3. It degrades RNA in a DNA-RNA hybrid (RNase H activity).

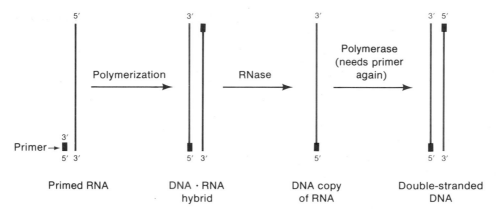

Figure 21-26

A hypothetical sequence of events that could enable reverse transcriptase to convert a single-stranded RNA to double-stranded DNA. The primer, which could be either RNA or DNA, is shown hybridizing to the 3′ termi- nus of the template. Terminal priming is not known to occur in nature, so the actual mechanism must be more complicated than the sequence shown.

In vitro, if an appropriate primer is provided, a single strand of RNA can be converted to a complementary DNA strand; subsequently, this single strand can be converted to double-stranded DNA, as shown in Figure 21-26. For the simple scheme shown in the figure, an oligodeoxynucleotide primer would be needed to avoid the problem of the replication of the termini of linear DNA molecules, as was discussed in Chapter 15 in connection with the replication of phage T7. Such a terminal primer is not available in nature, so much more complicated schemes are needed. The scheme for RSV will be discussed shortly.

The Growth Cycle of Rous Sarcoma Virus: An Outline

The basic growth cycle of a retrovirus is shown in Figure 21-27. The growth cycle is divided into two phases. In phase I reverse transcriptase, which is contained in the virion, catalyzes the formation of a double-stranded DNA provirus from the single-stranded virion RNA. An essential step in the development of the virus is integration of the double-stranded DNA provirus into the host chromosome. It is thought that a circular molecule is integrated but evidence for this idea is meager. The gene order in the inserted DNA is the same as the gene order in the viral RNA, in contrast with phage λ (Chapter 16), so the sites of circularization and integration of the viral DNA must be the same. In the second phase, transcription occurs. The transcripts are processed to yield mRNA molecules, from which two finished proteins and two polyproteins are translated. The two polyproteins are then cleaved, yielding six more polypeptide chains, two of which aggregate to form

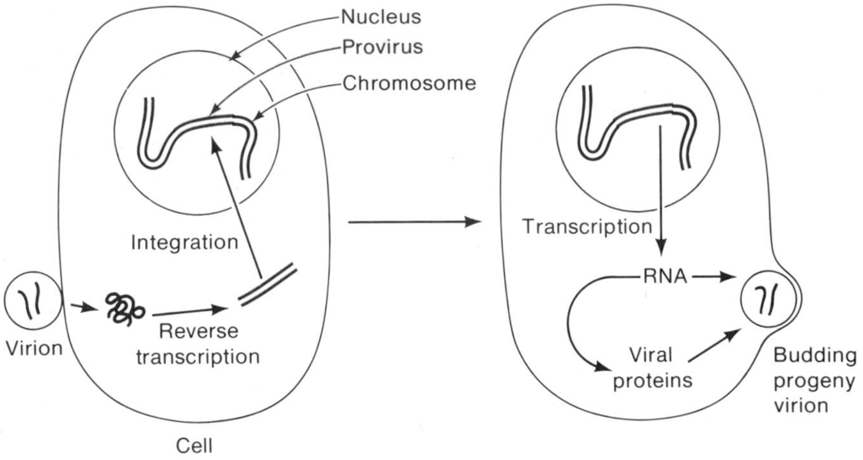

Figure 21-27
Life cycle of a retrovirus. Virion RNA enters the cell and by reverse transcription is converted to double-stranded DNA, which is integrated into the host chromosome, becoming a provirus. Transcription then produces RNA for protein synthesis and packaging into progeny virions.

reverse transcriptase. Four of these proteins form the capsid and two form the envelope. In a way that is not understood, virion RNA is synthesized. Completed virions are extruded continuously from the infected cell by budding without any obvious harm to the cell.

In the following sections various features of the biology of Rous sarcoma virus are examined further.

Properties of RSV Virion RNA

As mentioned earlier, RSV does not contain a single RNA strand but four strands hydrogen-bonded together (Figure 21-28). The larger pair of

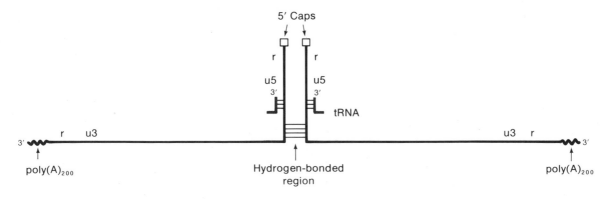

Figure 21-28
Structure of Rous sarcoma virus RNA. Note that the hydrogen-bonding between the two monomers creates parallel rather than antiparallel strands. Detailed understanding of this unusual region is poor. The symbols, r, u3, and u5 represent base sequences described in the text; DNA copies of these sequences are denoted R, U3, and U5, respectively.

strands (called the 38S or monomer strands) carry all of the genetic information and are identical. The two small strands are host tRNA^{Trp} molecules. This species of tRNA is found in all avian retrovirus virions; mouse retroviruses carry host tRNA^{Pro} instead. The location of the tRNA molecules is of particular significance for replication, as we shall see shortly. The 3′-terminal section of each tRNA molecule is hydrogen-bonded to the 38S strand 101 bases from the 5′ terminus of the 38S strand in a region termed pbs.

Another important feature of the Rous sarcoma virus RNA is that each 38S molecule contains two identical, or very similar, noncoding sequences of 21 bases. These sequences are designated r (Figure 21-28) and are located immediately adjacent to the cap at the 5′ termini and to the poly(A)$_{200}$ sequence at the 3′ termini of the 38S strands. Two other noncoding terminal sequences, which will be discussed shortly, are u5, located next to the r sequence at the 5′ terminus, and u3, next to the 3′-terminal sequence.

Conversion of Virion RNA to a Double-Stranded DNA Intermediate

All intermediates in the conversion of virion RNA to a double-stranded DNA molecule contain only one 38S RNA strand. Presumably the units separate shortly after entry into the cytoplasm.

Synthesis of RSV RNA precedes transcription and translation. This is possible because the virus carries its own replicase, reverse transcriptase, in the virion. This enzyme, like many replicating enzymes, needs a primer; the retroviruses have solved the priming problem by including a primer, namely tRNA, in the virion. The primer is at the 5′ end of the 38S RNA. As a consequence, only a small segment of DNA can be polymerized from the 3′-OH group of the tRNA before the growing DNA strand reaches the 5′ terminus of the template RNA molecule.

The exact mechanism of conversion of viral RNA to a double-stranded DNA intermediate is not known. The main feature to be accounted for is the difference among the bases of the viral RNA and the DNA copy, as shown in Figure 21-29; the sequence U3-R-U5, which is a combination of the 5′ r-u5 segment and the 3′ u3-r segment of the RNA, is found at both ends of the double-stranded DNA molecule—*the DNA is terminally redundant.* The sequence U3-R-U5 is commonly denoted **LTR** for *long* terminal repeat. The proposed scheme, which embodies the main features of the conversion reaction, is shown in Figure 21-30. The following steps are shown:

1. Reverse transcriptase extends one of the tRNA molecules from its 3′-OH terminus to the 5′ end of the template, resulting in a DNA copy of r-u5.

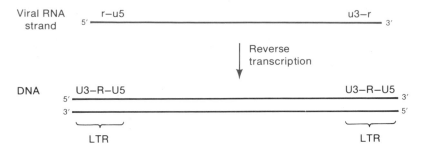

Figure 21-29
The arrangement of terminal sequences (lower case letters) on one 38S viral RNA strand and on the DNA copy (capital letters). The sequence U3-R-U5 is called LTR, or, long terminal repeat.

2. The RNA bases corresponding to the r-u5 sequence are removed (perhaps by another activity of reverse transcriptase), which renders the complementary sequence R'-(U5)' in the newly synthesized DNA available for hydrogen-bonding. The cap and poly(A) tail are also assumed to be removed at this point.

3. The tRNA and its complementary sequence melt, separating the newly synthesized DNA from the RNA. The DNA segment then binds to the other end of the RNA by forming hydrogen bonds between r and R'.

4. A DNA polymerase adds nucleotides to the 3' terminus of the DNA, forming a nearly completely double-stranded molecule.

5. The sequence u3-r at the 3' terminus of the RNA is digested away.

6. A DNA polymerase again adds nucleotides to the 3' terminus of the RNA, completing the duplex and forming the first LTR.

7. All RNA is digested, presumably by an RNase activity of reverse transcriptase.

8. The first LTR is melted and the short fragment forms a (PBS)(PBS)' duplex at the opposite end of the molecule. This pairing generates a template for formation of the second LTR.

9. A DNA polymerase adds nucleotides to both 3' termini, forming a complete duplex terminated at each end by an LTR.

Note that the molecule formed in step 8 has two complementary LTR sequences. As an alternative to step 9, these sequences could hydrogen-bond, forming a duplex LTR. Extension of each 3' terminus by a DNA polymerase would generate a double-stranded circle carrying only one copy of LTR. Such molecules have also been observed in infected cells, though linear molecules greatly outnumber circles.

The final molecule shown in Figure 21-30 is integrated into the chromosome and there it is transcribed. At some point in the life cycle a

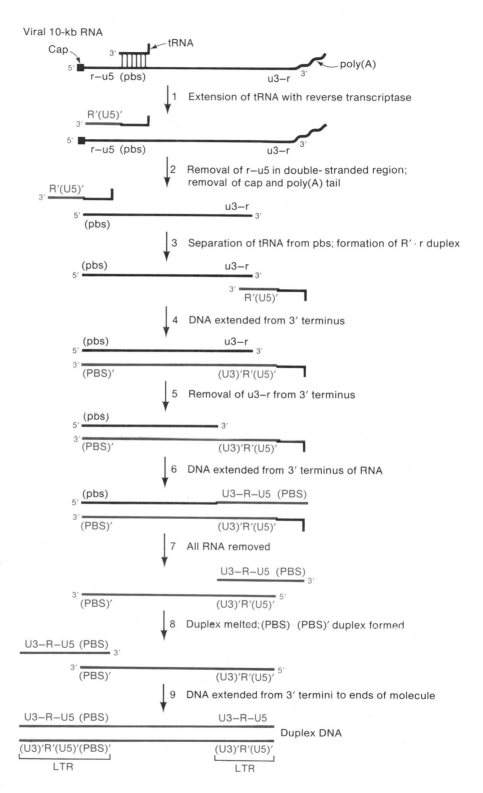

Figure 21-30
Hypothetical scheme for generating a duplex DNA molecule having two terminal LTR sequence from Rous sarcoma viral RNA. The sequence pbs is complementary to the tRNA; lower and upper case letters and numbers refer to RNA and DNA sequences, respectively. A prime indicates a complementary sequence. RNA and DNA are black and red, respectively.

template must be copied to regenerate virion RNA. This apparently occurs as a transcriptional step, whose details are unknown.

It is sometimes difficult to relate the observed intracellular structures to the life cycle of the virus because only about 1 percent of RSV particles are infective and experimentally one must infect cells with more than 100 particles per cell in order to have a sufficient number of infected cells to study. Since 99 percent of the particles are noninfective (defective), then if every RNA molecule undergoes some alteration in an infected cell, 99 percent of the molecules observed might have developed from defective particles that infect cells. Thus, perhaps the less frequently observed circles are the only productive intermediates. Clearly a great deal more work is needed before replication of this virus is understood.

Integration of Viral DNA

The mechanism of formation of the provirus is not known in any detail, but the structure of the provirus suggests a possible and somewhat surprising mechanism. Integration has been studied by cloning proviruses from many infected cells into *E. coli* phage λ by means of recombinant DNA techniques. Proviral DNA was isolated and cleaved by restriction enzymes, and the base sequences in the regions surrounding the two junctions between host and viral DNA were determined. These sequences were then compared with sequences at the termini of unintegrated virion RNA. This experiment has not been carried out with RSV but conducted instead with seven other retroviruses. The results for spleen necrosis virus (SNV) are depicted in Figure 21-31. Panel (a) shows a schematic diagram of the virion RNA without the 5′ cap and the 3′-poly(A) tail. These terminal regions are

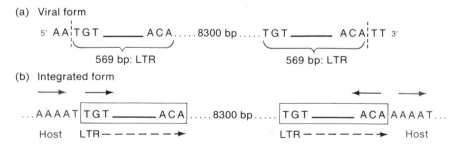

Figure 21-31
Sequences in unintegrated and integrated spleen necrosis virus DNA. Viral and host sequences are black and red, respectively. Pair of arrows of the same kind (black, red, dashed) represent corresponding sequences. If the arrows point in the same direction, the sequence is a direct repeat; if they point in the opposite direction, the sequences are inverted.

omitted because they are not copied in the reverse transcriptase replication scheme. A long 569-base segment (the LTR) terminates each strand, and three terminal bases at each end are arrayed in an inverted repetition. The two vertical dotted lines indicate the limits of the base sequence present in the provirus; that is, the terminal TT and AA bases are not integrated. Panel (b) shows the structure of one strand of the integrated proviral DNA. Three important features are:

1. The 569-base sequence is repeated.
2. The complete inserted region including the repeated 569 bases is terminated by 5'-TG and CA-3'.
3. The host sequences at each joint are sequences of five bases in direct repetition.

Thus the sequence of the integrated DNA may be written as

$$\overbrace{\qquad\qquad}^{\text{LTR}} \qquad\qquad \overbrace{\qquad\qquad}^{\text{LTR}}$$
$$5H{-\!-}3V{-\!-}563V{-\!-}3V{-\!-}8300V{-\!-}3V{-\!-}563V{-\!-}3V{-\!-}5H$$

in which H and V stand for host and virus respectively. The following pattern can be observed: (1) viral DNA terminated by two long sequences (LTR) in direct repeat, each of which contains a short viral sequence (3V) in inverted repeat; and (2) a short host sequence in direct repeat. The pattern has been observed for every species of retrovirus examined; only the length and base sequences of the short repeated sequence vary from one species to the next. Furthermore, the integrated DNA has the termini 5'-TG and CA-3' for all species of retrovirus observed so far, which suggests that there may be a common integration mechanism for all retroviruses. When independent proviruses of a single species have been examined, it has been found that the directly repeated short host sequence differs in each isolate; thus, integration can occur at many sites in the host DNA.

The sequences of the provirus and the adjacent regions are reminiscent of the properties of transposons. In Chapter 19 it was pointed out that movable genetic elements are, after transposition, always bounded by a small direct repeat of cellular DNA (4–12 base pairs) next to an inverted repeat of element DNA (2–1500 base pairs). Furthermore, the yeast element TY1, the *Drosophila* element copia, and the integrated DNA of phage Mu all end with the dinucleotides TG and CA. It is far too improbable that these are chance similarities; it has therefore been hypothesized that the mechanism of integration of retroviruses is similar to the mechanism of transposition. There is no evidence that transposition actually occurs, since the hallmark of transposition is not only a duplication of the target sequence but also a duplication of the transposon itself. In the retroviruses, a chromosomal target sequence is duplicated but one does not know whether the double-stranded DNA molecule synthesized by reverse transcriptase is itself inserted or whether a copy of it is transposed from the nucleoplasm of the cell to a chromosomal site. Nonetheless, on the basis of the sequence data, Howard Temin has suggested that retroviruses are derived from ancient transposable elements that have gained the ability to replicate and be encapsidated.

Once integrated, the proviral DNA is there to stay—it is not excised. The presence of the provirus does not inhibit cell division, so daughter cells inherit the integrated provirus. These daughter cells (and cells in subsequent generations) continue to produce active virions.

Transcription of RSV

Transcription begins the second phase of the RSV life cycle, as shown in Figure 21-27. There is a single transcript (designated 38S RNA) which includes all of the viral genes (Figure 21-32). Some of these transcripts

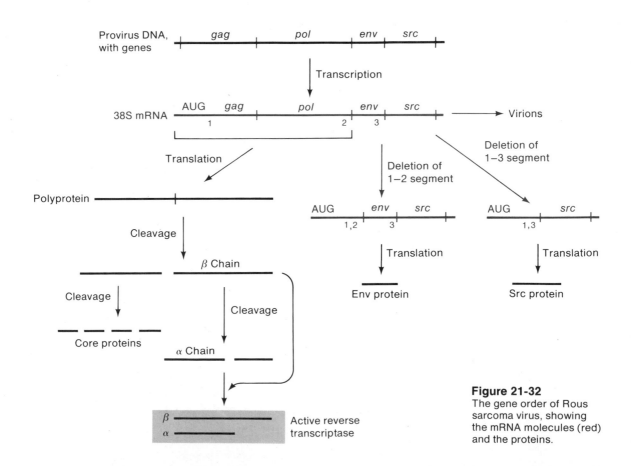

Figure 21-32
The gene order of Rous sarcoma virus, showing the mRNA molecules (red) and the proteins.

remain intact and are used both as a source of viral RNA and as a mRNA for synthesis of a polyprotein from which the Gag protein and the Pol proteins are cleaved. Many of the 38S molecules are cleaved to yield mRNA molecules that encode the products of the *env* and *src* genes. The Gag protein is cleaved into four smaller proteins that are contained in the viral core. The Pol protein, which is the β subunit of reverse transcriptase, is also cleaved to yield two fragments; one of these fragments, called the α chain, joins with the β subunit to form active reverse transcriptase. The *env*-gene product is the viral protein of the envelope of the virion. The Src protein is needed for the tumor-producing response and will be discussed in the section on tumor viruses. Mutants of the *src* gene produce active virus but are not able to induce transformation to tumor cells.

It was mentioned earlier that insertion of viral DNA (actually the activity of reverse transcriptase) is essential for successful completion of the life cycle of the virus. One reason for this requirement has become clear from an analysis of the base sequences of the LTR and the 38S mRNA molecule. The coding sequence for the *gag* gene begins to the right of the U5 sequence shown in Figure 21-29. However, the promoter for the 38S mRNA is in the U3 region. Thus, in viral RNA the promoter is disconnected from the coding region. Construction of the LTR by joining U3 to the R-U5 sequence brings the promoter to a position upstream from the coding sequence, thereby allowing synthesis of 38S mRNA.

ANIMAL DNA VIRUSES

One might expect that DNA viruses would lack some of the problems we have seen with the RNA viruses, since both replication and transcription of DNA viruses should follow the same patterns seen with the DNA bacteriophages. Furthermore, DNA viruses should not require special enzymes, such as reverse transcriptase: theoretically a DNA virus could consist solely of a DNA molecule that specifies one or two capsid proteins. The tiny parvoviruses do follow such a simple pattern, but we will see that the life cycles of most DNA viruses are considerably more complex. Some DNA viruses have evolved with more than 100 genes, which gives them greater versatility than those with only a few genes, and these viruses are enormously complicated. We shall examine only a few of the many types of DNA viruses in this section, selecting only those whose life cycles are reasonable well understood. These viruses are the parvoviruses, the papovaviruses, and adenovirus. For discussion of the well-studied herpesviruses, poxviruses (for example, vaccina), and papilloma viruses, the reader should consult the references at the end of the chapter.

Parvoviruses

Parvoviruses are of two types—autonomous and defective. The autonomous viruses are able to produce progeny by using only the products of their own genes and of the host. The defective parvoviruses need some additional gene product supplied neither by their own nucleic acid nor by the host; they are able to replicate only when the host cell is infected with another DNA-containing virus—for example, an adenovirus—which presumably provides the necessary protein.

Parvoviruses contain one very small piece of single-stranded DNA. The DNA, which typically consists of about **4800** bases, encodes three proteins, whose molecular weights are about 61,000, 64,000, and 83,000. The total of the three molecular weights exceeds the coding capacity (in a single reading frame) of **4800** bases (a maximum protein molecular weight of 176,000) by 32,000; however, three reading frames are used. (At present, the parvoviruses are the only example, other than E. coli phage ϕX174, of viruses having overlapping genes.) The three viral proteins are only used to construct the capsid, for the parvoviruses utilize host enzymes exclusively for transcription and replication; as a result, these viruses are able to replicate only during the stage of the life cycle of the host in which DNA replication occurs, and viral DNA replication takes place only in the nucleus of the host cell. Presumably certain host enzymes that are essential for viral replication are available only at that time and in that location.

Replication of both autonomous and defective viruses are similar, though details differ. The replication of adeno-associated virus (AAV), a defective virus, is best understood; it is this scheme that is described in this section.

Having single-stranded DNA, parvoviruses face the usual problem of providing a primer for initiation of DNA synthesis. However, the arrangement of base sequences in the DNA is such that the DNA primes its own replication. The single strand contains two terminal inverted repetitions that allow the termini to fold and form terminal hairpin loops, as shown in Figure 21-33. One hairpin loop has a 3′-OH-terminal base-paired nucleotide; this gives the exact condition needed for priming. The figure shows the stages in the DNA replication sequence that ensue after priming; this mode of replication is sometimes termed the **hairpin transfer model.** Initially the 3′-OH-terminated nucleotide is extended. When the growing chain reaches the 5′ terminus, a displacement reaction occurs and the 5′-terminated hairpin is opened, allowing completion of synthesis of the growing chain to the 5′ end. At this stage the molecule is a completely double-stranded hairpin (II). A nick is then made at the site indicated in the figure by the arrow and extension again occurs, generating a double-stranded linear molecule. Then terminal

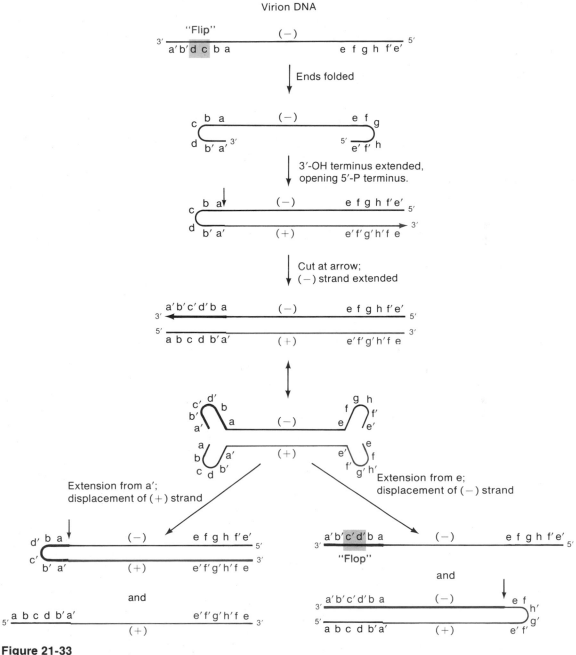

Figure 21-33

A scheme for replicating parvovirus DNA, which generates a sequence inversion (shaded red region) in a progeny (−) strand. The short vertical arrows along the strands indicate sites at which site-specific cleavage occurs, to generate a 3′-OH group from which a new chain can grow. All molecules shown are capable of further growth and extension and will generate other sequence inversions in the terminal segments.

breathing occurs, which allows the terminal inverted repeats to form two short hairpin loops. Once a 3′-terminated hairpin loop has formed, a 3′-OH primer is again available for extension. There are two such preprimed ends and one or both can be used. If both are used simultaneously, very large structures will arise (and have been observed). We will be concerned only with those molecules having only a single active primer. Extension from the 3′ end at the left side of the intermediate shown in the figure yields molecules II′ and V. Extension from the other 3′ end yields molecules I′ and IV. Molecule I′ is a (−) strand like the parental molecule but has a short inverted segment— namely, its 3′ terminal sequence is a′b′c′d′ba, whereas in the parent the sequence is a′b′dcba. Inversions also are present in the other three molecules. These two sequences are called "flip" and "flop" and it is the presence of both sequences among progeny viruses that supports the replication scheme shown in the figure. The replication process probably proceeds by a nicking of II′ and IV at the sites of the arrows, thus generating a molecule of type III, which would be converted to the quadruple-hairpin intermediate. Molecule V can also follow the pathway I-II-III. As the model has been drawn, the starting molecule is a (−) strand. However, to some extent, molecule V, a (+) strand, is packaged and the particles containing these molecules are also infective. Replication of a (+) strand is no different from the scheme shown and both (+) and (−) strands would result. Thus, in a mass sample, both (+) and (−)-strand viruses are found in both the flip and flop configurations. Flip and flop particles are not observed with the autonomous parvoviruses, so another replication scheme is thought to be used by them.

Little is known about the mode of transcription of the parvoviruses.

Papovaviruses: Polyoma and SV40

Polyoma virus and simian vacuolating virus 40, or SV40 (Figure 21-34), are two well-characterized papovaviruses. Both induce tumors very efficiently when infecting certain hosts; in other hosts they are virulent, resulting in a bursting of the infected cell and release of progeny viruses (lytic cycle). In this section we will consider the lytic cycle only.

Properties of Papovavirus Virions and DNA

The virions of all papovaviruses are naked icosahedrons containing a single circular DNA molecule; when purified, the DNA is supercoiled (see below). The capsid consists of a major capsid protein VP1 and two minor proteins VP2 and VP3. In contrast with the viruses discussed so far, the DNA is associated with all of the host histone proteins except for

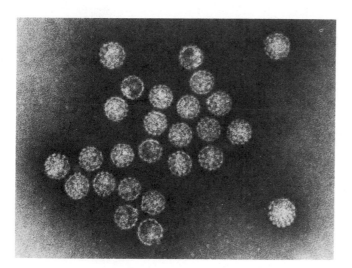

Figure 21-34
Electron micrograph of SV40 virus. (Courtesy of Robley Williams.)

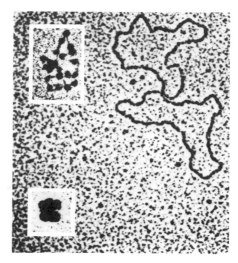

Figure 21-35
SV40 DNA in various degrees of conden-
sation with host histones. (Courtesy of
Edward Kellenberger.)

histone H1. These proteins cause the DNA molecule to contract and
form the beaded nucleosomal structure characteristic of eukaryotic
chromatin (see Figure 21-35 and Chapter 6). These beaded structurs are
called **minichromosomes.** Since histones when bound cause a slight
unwinding of the double helix, and since they are bound throughout the
replication cycle, at the completion of a round of replication daughter
molecules are slightly underwound. This underwinding is not evident
in a minichromosome as it is taken up by the winding of the DNA
around the histones; however, when the DNA is purified and the histones are removed in a deproteinization step, such as phenol extraction,
the DNA assumes the supercoiled configuration.

The DNA molecules of both polyoma and SV40 have been completely sequenced; each contains 5224 base pairs.

In the sections that follow, only SV40 will be discussed. At the end
of the discussions a few comments will be made about the minor
differences between SV40 and polyoma.

Life Cycle of SV40

SV40 grows lytically in monkey cells. As with many viral infections, the
SV40 life cycle can be divided into early and late phases. Five
virus-encoded proteins (see Table 21-6) are synthesized. Following
infection there is a long (12 hour) latent period in which the viral DNA is

Table 21-6
Proteins encoded by SV40 virus

Protein	Timing of synthesis	Function
T antigen	Early	Initiation of DNA replication
t antigen	Early	Unknown
VP1	Late	Major capsid protein
VP2	Late	Minor capsid protein
VP3	Late	Minor capsid protein

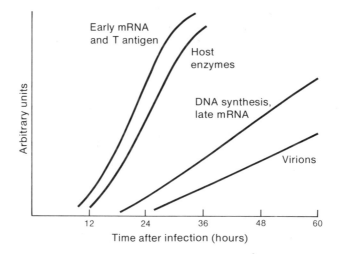

Figure 21-36
The time scale for various processes that occur in the life cycle of SV40.

inactive; during this period the SV40 DNA gradually migrates to the nucleus of the host cell. Neither transcription nor replication can occur outside of the nucleus, presumably because the necessary polymerases are only contained in the nucleus.

In the early phase of infection the major viral proteins synthesized are the T and t antigens (Figure 21-36).* Following synthesis of early mRNA the synthesis of several host enzymes is stimulated—DNA polymerase, thymidine kinase, dTMP kinase, and dCMP deaminase. These host enzymes are all part of the pathway for DNA synthesis; T and t antigens are also necessary for viral DNA synthesis, as will be seen shortly. Several hours later, host DNA synthesis is also stimulated, but it is not clear whether this is essential to the life cycle of the virus or just an inevitable concomitant. Approximately 12 hours after synthesis of viral and host replication enzymes, viral DNA synthesis begins. Shortly

*The names T and t antigen are derived from an observation made by immunofluorescence microscopy that these proteins are located in the nuclei of cells that have been converted to tumor cells by viral infection.

after the onset of viral DNA synthesis, viral late mRNA is synthesized. Viral DNA synthesis is apparently a prerequisite to late mRNA synthesis, as is the case with *E. coli* phage T4. The molecular basis for the requirement for DNA synthesis is unknown.

After 24 hours, late proteins and viruses aggregate and form virions. The assembly process is fairly well understood but will not be described in this book.

After about three days the infected cells lyse, releasing 10^5 virions per cell. Lysis probably results from mechanical damage caused by very large crystalline arrays of virions rather than by a lysing enzyme.

Physical Map of SV40 DNA

Figure 21-37 shows a physical map of SV40, which includes the location of the replication origin and the mRNA molecules. This map has been constructed by cleaving virion DNA with many restriction enzymes, ordering the fragments to form a restriction map, and hybridizing fragments with labeled SV40 mRNA isolated from infected cells. The gene products corresponding to the mRNA molecules have been identified by using purified mRNA in an *in vitro* translation system and identifying the proteins synthesized by a particular mRNA molecule. The enzyme EcoRI makes a single cut in SV40 and the position of this cut has been chosen as the reference or zero point of the map.

Figure 21-37
Physical map of SV40 showing the location of the five genes, T, t, VP1, VP2, and VP3, the replication origin, and the early and late mRNA molecules.

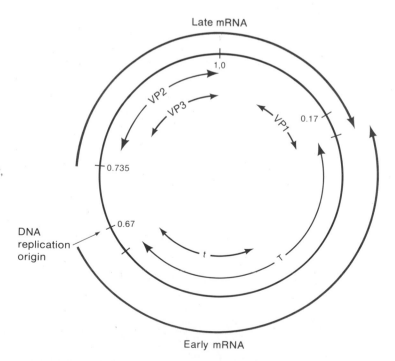

Late mRNA

1,0

VP2

VP3

VP1

0.17

0.735

DNA replication origin

0.67

t

T

Early mRNA

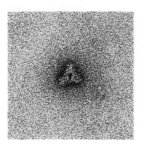

Figure 21-38
T antigen molecule of SV40 viewed by negative-contrast electron microscopy. (Courtesy of Robley Williams.)

SV40 DNA Replication

Replication of SV40 DNA is fairly well understood except for the mechanism of initiation. The replication origin is at position 0.67 of the SV40 map. This is the site of binding of the T antigen (Figure 21-38). This protein is essential for the initiation of SV40 DNA synthesis; this was first demonstrated by the fact that a temperature-sensitive mutant A(Ts) (the gene for T antigen was originally called the A gene) fails to initiate DNA synthesis at a high temperature. T antigen is an ATPase; its exact function in initiation is unknown but other of its functions will be discussed in the section concerned with tumor viruses.

An interesting observation suggested that the replication origin must have an unusual structure. When supercoiled (deproteinized) virion DNA is exposed to the enzyme S1 nuclease, an enzyme that attacks only single-stranded DNA, a single cut is made at the origin.* The enzyme is inactive against the nonsupercoiled form of the DNA. In Chapter 4 it was pointed out that a significant fraction of the base pairs of a supercoiled DNA molecule are broken at any moment and that these melted regions are continually changing. It would be expected that a high-A + T region would remain single-stranded most of the time if the DNA is supercoiled; from the result of S1 cleavage it could be guessed that the replication origin of SV40 would be rich in A · T pairs. This idea was confirmed when the base sequence of SV40 DNA was completely determined. Figure 21-39 shows the base sequence of the replication origin. This region does have an unusual structure, consisting of a 27-base-pair, G+C-rich sequence made up of an inverted repetition of 13 base pairs (pink in the figure) surrounding a single G · C pair (arrow) and flanked by 17 A · T pairs. A T antigen tetramer binds first to the high-G · C palindrome; then two additional T antigen tetramers bind cooperatively to fill the G · C region and then the flanking A · T region. Replication begins at the single G · C pair on the axis of symmetry of the inverted repetition. It is of interest that this is also very near the site of initiation of late mRNA synthesis, as will soon be discussed.

Position
0.67
↓

```
CAGAGGCCGAGGC GGCCTCGGCCTCTG CATAAATAAAAAAAATTA
GTCTCCGGCTCCG CCGGAGCCGGAGAC GTATTTATTTTTTTAAT
```

Figure 21-39
Base sequence at the replication origin of SV40, showing an inverted repetition of 13 base pairs separated by a single G·C pair at position 0.67.

*S1 nuclease is an endonuclease isolated from the fungus *Aspergillus oryzae*. It acts on single-stranded DNA and RNA and on the single-stranded branches of a denaturation bubble. It is a valuable reagent in molecular biological research.

Study of viral DNA replication has been made possible by a DNA fractionation procedure (developed by B. Hirt and called a Hirt extraction) that consists of sequential treatment of infected cells with the detergent sodium dodecyl sulfate (SDS) and a high molarity of NaCl. After an overnight incubation at 0°C, a complex containing SDS, NaCl, cellular protein, and high-molecular-weight DNA (hence, host DNA) precipitates, leaving a supernatant that contains most of the viral DNA and virtually no host DNA. The viral DNA, which consists of both replicating and nonreplicating molecules, can then be examined directly by electron microscopy. The following important features of the replication of SV40 DNA have been observed:

1. Replication is bidirectional and, like *E. coli* phage λ, terminates by collision of the growing forks at a point opposite the origin.
2. Partially replicated DNA molecules contain intact parental strands, and the unreplicated portion remains supercoiled (Figure 21-40). This intermediate, in which the hybrid progeny loops appear open, is sometimes termed a **butterfly molecule.** Some

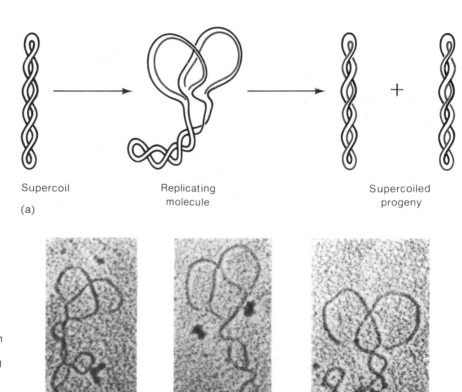

Supercoil

(a)

Replicating
molecule

Supercoiled
progeny

Figure 21-40
Replication of the DNA of a papovavirus. (a) A replicating molecule retains its superhelical form in the unreplicated portion. (b) Three replicating molecules. (From R. Jaenisch, A. Mayer, and A. Levine. 1971. *Nature.* 233: 72–75. Reprinted by permission. Copyright, Macmillan Journals Ltd.)

(b)

unwinding of the parental helix is necessary, and without some relief of the torque developed by the unwinding, the structure would become a tightly tangled mass. This torque is relieved by a topoisomerase but the particular enzyme has not yet been identified.

3. Late in the life cycle, large amounts of long concatemeric SV40 DNA are found. These molecules could be branches of rolling circles, yet rolling circles have rarely, if ever, been isolated.* These concatemers can be used to infect monkey cells; it is found that they are cleaved into monomers and then function to produce progeny virus. The significance of the concatemers is unknown but it is possible that, like *E. coli* phage λ, they are the precursors to virion DNA.

Except for initiation, SV40 DNA replication has been carried out *in vitro* (that is, one must start with a partially replicated molecule). The following facts have emerged from these *in vitro* studies:

1. One daughter strand is made continuously and the other discontinuously.
2. Each precursor fragment on the lagging strand is primed by a decaribonucleotide chain (10 bases) synthesized by a primase like the *E. coli* enzyme. The base sequences of the primers vary widely, indicating that a special sequence is not recognized.
3. Removal of the primer is not (as in *E. coli*) coupled to joining of the precursor fragments but comes at a much earlier time.
4. The polymerizing enzyme is DNA polymerase α, the major host-DNA replication enzyme.

At present purified T antigen is being used in attempts to initiate DNA synthesis *in vitro*.

Transcription in an SV40 Infection

Transcription of SV40 DNA is a complex process involving temporal (early-to-late) regulation and extensive mRNA processing, which selectively removes start and stop codons. The early and late mRNA molecules are transcribed from different strands of the DNA molecule; they are synthesized in opposite directions. Furthermore, they are probably initiated at several nearby sites in regions adjacent to the replication origin (Figure 21–37). Figure 21-41(a) shows that a single early mRNA molecule, which encodes both the T and t antigens, is synthesized from a promoter to the left of position 0.66 in the physical

---

*Circles with short linear branches have on occasion been observed. However, if the length of a branch is less than the circumference of the circle, it is difficult to know whether the molecule is a rolling circle or a theta molecule that is broken in a replication fork.

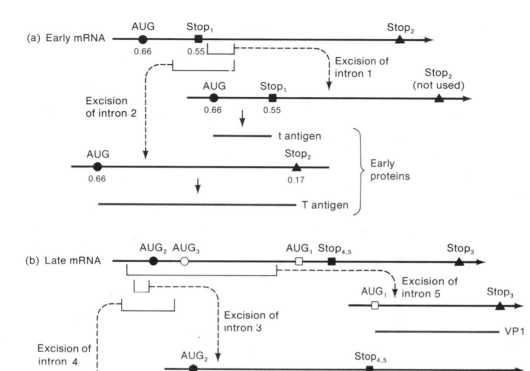

Figure 21-41

Processing of SV40 (a) early and (b) late mRNA. Horizontal brackets indicate excised introns; a dashed arrow from above leads from an unprocessed mRNA molecule to the processed molecule lacking the intron indicated by the arrow. Open and solid squares, circles, and triangles designate identical positions on each mRNA molecule. The mRNA is black; protein is red.

map. A single AUG codon is responsible for initiation of synthesis of both proteins. This mRNA is processed in two different ways to yield monocistronic mRNA molecules, one encoding T antigen and the other encoding t antigen. In one mode of processing, intron 1 is excised, to yield $mRNA_1$, from which the t antigen is translated by starting at the AUG and terminating at $stop_1$. For synthesis of T antigen, intron 2 is removed; it has the same 3' terminus as intron 1 but has a 5' terminus nearer to the 5' end of the mRNA. Excision of this larger intron removes the $stop_1$ codon, so that translation continues to $stop_2$, to yield the T antigen, which is much larger than t antigen. Note that the NH_2-terminal segments of t and T antigens are identical.

Three proteins—VP1, VP2, and VP3—are encoded in the late mRNA. The pattern of splicing determines which protein is translated, and the use of overlapping reading frames enables the virus to store a great deal of information in a rather small segment of DNA. Only one late mRNA is transcribed from an initiation site at position 0.735 in the map. Figure 21-41(b) shows the three modes of splicing that generate the three monocistronic mRNA molecules needed to translate the three proteins. Note that no protein is translated from unspliced RNA. VP2 and VP3 are translated in the same reading frame and both have the same carboxyl termini. Thus, VP3 is actually a fragment of VP2; it is translated from a mRNA fragment from which the start codon for translation of VP2 has been removed by processing enzymes. VP1 is read in a different reading frame from V2 and V3. This reading frame is established by excising a very large segment of RNA that contains the AUG codons for both VP2 and VP3. The only remaining start signal establishes the reading frame for VP1.

Note that no gene products of SV40 are cleaved from polyproteins.

The pattern of replication of polyoma viral DNA is the same as that for SV40 DNA, and transcription is almost identical. The major difference is that polyoma has three (rather than two) modes of processing early mRNA molecules, which enables it to synthesize a third protein called middle-t antigen. The function of this protein is discussed in the section concerned with tumor viruses.

Location of Mature Virions

SV40 is a vacuolating virus; that is, late in infection the infected cell accumulates membranous, liquid-filled sacs called **vacuoles.** Viral DNA is synthesized in the nucleus and transported to the cytoplasm where the proteins are made. Virion assembly occurs between cytoplasmic membranes that delineate the vacuoles (Figure 21-42). Later, the membranes break down as the cell disintegrates and the particles are released from the cell.

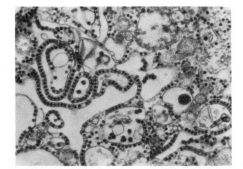

Figure 21-42
Mature SV40 virions in cytoplasmic membranes. (Courtesy of W. Hecker.)

LARGE DNA VIRUSES

So far, viruses having only a few gene products have been described and consequently the number of their mRNA molecules has been small. There are many large viruses and, as expected, these have very complex life cycles—sufficiently complex that in no case can one say that the life cycle is understood. In this section we will only present a few highlights of three viruses—adenovirus, herpes simplex, and vaccinia.

Adenoviruses

The adenoviruses form a large group of well-studied, naked icosohedral viruses (Figure 21-43), each containing one double-stranded DNA molecule. The molecular weights of the DNA molecules of different viruses are $23-30 \times 10^6$. The DNA encodes at least 20 proteins of which 15 proteins are contained within the virion. Four virion proteins are associated with the DNA—one interacts with the phosphates, a second organizes the DNA into a multilobed form and binds the DNA to the twelve pentons of the capsid, a third is covalently linked to the 5'

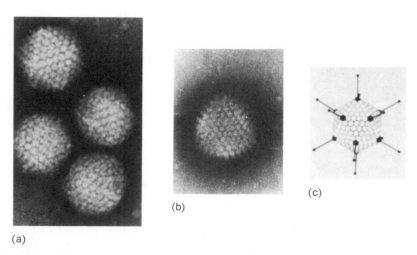

(a)

(b)

(c)

Figure 21-43
Adenovirus particles. (a) Electron micrograph of several virions, taken by the negative-contrast technique. Each virion consists of 252 capsomers (which appear as white circles in the micrograph) arranged to form an icosahedron. Of this number, 240 of the capsomers have six neighbors and are called hexons (represented as white spheres in the diagram in (c)); 12 vertex capsomers have five neighbors and are called pentons; each vertex capsomer consists of a base and a projection, known as a spike. (b) An electron micrograph of a single particle, in which the spikes can be seen. (c) A plastic model of the virion showing hexons (white spheres), pentons (black spheres), and spikes. (Courtesy of M. Wurtz.)

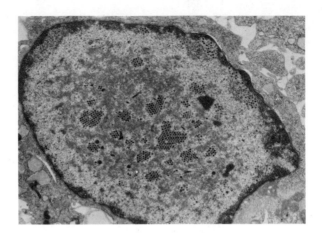

Figure 21-44
Adenovirus crystals in the nucleus of an infected L cell. (Courtesy of Alvin Winter and Keith Brown.)

termini of the DNA molecule, and a fourth has an unknown function. The linked protein will be discussed in detail shortly. The virus forms large crystalline arrays in the nuclei of infected cells (Figure 21-44).

Much of the initial interest in adenovirus arose as a result of observations that most adenoviruses, which cause respiratory infections in humans and simians, are oncogenic (tumor-forming) in newborn rats and mice. Furthermore, rodent cells in culture are converted to transformed cells (see later section) by adenovirus. Thus, the adenoviruses have been studied as models of viral carcinogenesis. Following the discovery of mRNA splicing in adenovirus, it became clear that these viruses are also useful for studying eukaryotic transcription; because they use host replication proteins exclusively, they should yield information about eukaryotic DNA replication.

The Termini of Adenovirus DNA

Adenovirus DNA is a linear molecule with an inverted terminal repetition that extends precisely to the end of the molecule (Figure 21-45). The size of this sequence ranges within the adenovirus family from 103 to 162 base pairs. There is considerable homology from one virus to another within the first 50 bases, which are exceedingly rich in A · T pairs. This homology, evident throughout the adenovirus family in the face of extensive sequence-divergence elsewhere in adenovirus DNA, suggests that the sequence is of importance; indeed it is probably used in the initiation of DNA replication.

Another striking feature of the terminus of adenovirus DNA was discovered when it was noted that DNA isolated from the virion spontaneously forms circles and oligomers. This is reminiscent of phage λ DNA, whose complementary single-stranded termini can produce these structures, but adenovirus DNA does not have single-stranded

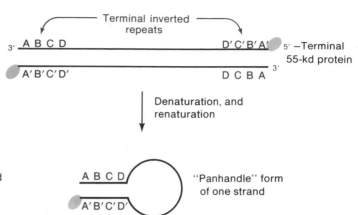

Figure 21-45
The structure of adenovirus DNA and the panhandle form that has been created in the laboratory by successive denaturation and renaturation.

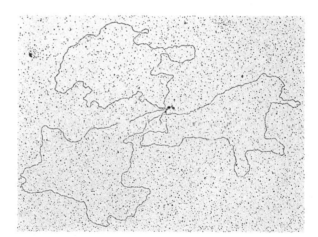

Figure 21-46
Demonstration of the formation of multimers of the adenovirus 55-kd proteins. Three adenovirus DNA molecules each bearing two terminal proteins have circularized by means of dimerization of the proteins and then, again by the protein-protein interaction, the three DNA molecules have joined together in a trifoliate form. (Courtesy of Ellen Daniell.)

termini. Treatment of the DNA with a protease eliminates the ability to form these structures. The explanation for circularization and oligomer formation is that each 5′ terminus of adenovirus DNA is covalently linked to a terminal protein (the 55-kd protein) and these proteins can form dimers and even higher multimers (Figure 21-46).* The protein-DNA linkage is a phosphodiester bond between the OH group of a serine residue in the 55-kd protein and the 5′-phosphoryl group of the terminal deoxycytidine of each DNA strand. The protein plays an important role in DNA replication. (Terminal proteins have been observed in other organisms. One of the best-understood is in *Bacillus subtilis* phage 29; other organisms include parvovirus H1, mouse minute virus, and hepatitis B virus.)

*The symbol kd designates "kilodalton"; such terms are often used in naming a protein whose molecular weight has been evaluated before knowledge of its function has been obtained.

Replication of Adenovirus DNA

Studies of the replication of adenovirus DNA have showed the following:

1. Initiation occurs at either terminus of the DNA molecule. Since the two ends are inverted repetitions, they are indistinguishable and initiation occurs at both ends with equal frequency. Furthermore, some molecules are initiated at both ends. An RNA primer is not used for initiation; how this need is bypassed will be discussed shortly.

2. Growth of the daughter strand continues to the other end of the molecule. In the course of this growth, a parental strand is displaced by the growing daughter strand. In a molecule initiated at both ends, both parental strands are displaced. This means that replication intermediates have extensive linear single-stranded branches, a molecular state that is rather unusual in phage and bacterial systems. Such intermediates comprise the bulk of the intracellular DNA during the replication phase of adenovirus (Figure 21-47).

Figure 21-48 shows the currently accepted model for adenovirus DNA replication. There are two pathways for synthesis. In the major pathway, synthesis and displacement go to completion (via a type I intermediate—see the figure), generating a daughter duplex molecule

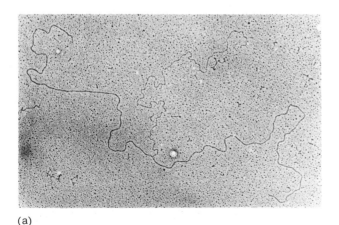

(a)

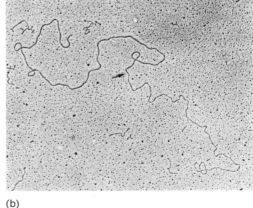

(b)

Figure 21-47
Electron micrographs of replicating adenovirus DNA (a) A type I molecule in which a single-stranded branch is being displaced from a parental duplex. (b) A type II molecule in which a fully displaced single strand is being converted to double-stranded DNA; the arrow shows the position of the growing 3′ terminus. (Courtesy of Thomas Kelley.)

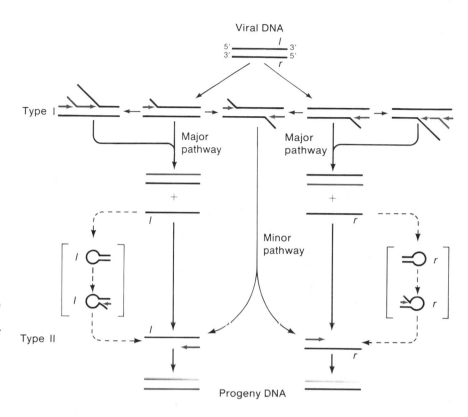

Figure 21-48
Scheme for replication of adenovirus DNA. Heavy arrows indicate the major pathway; light arrows, minor pathway; dashed arrows, hypothetical reactions. See the text for details.

and releasing a parental single strand. The latter then serves as a template for synthesis of a complementary strand, so that two semiconservatively replicated molecules result. Complementary synthesis begins at the 3′ end of the released strand and generates a type II intermediate, which is observed in electron micrographs as a double-stranded segment to which a long single strand is attached. Continued growth of this intermediate yields the complete double-stranded molecule. The minor pathway results because initiation occasionally occurs at both ends of the molecule. In such a doubly initiating molecule, when two oppositely moving replication forks meet, the parental strands can no longer be held together by base-pairing and will separate, forming two type II molecules whose replication is completed by simple extension; this event is termed "fork annihilation." The bracketed forms in Figure 21-48 will be discussed shortly.

The mechanism of chain elongation of adenovirus DNA differs from the other double-stranded phage and viral systems we have considered so far. The *l* and *r* strands of adenovirus DNA are replicated separately, each in the direction appropriate for DNA polymerase activity. Thus, as the model has been conceived, there is no need for discontinuous replication—that is, for precursor fragments. There is in fact considerable experimental support for continuous replication of both strands.

Initiation of replication requires special consideration for two reasons: (1) initiation at termini creates the same types of problems faced by *E. coli* phage T7 (Chapter 15), and (2) an RNA primer is not used. Furthermore, as the model has been presented, there are two classes of initiation events—those occurring at the termini of double-stranded DNA (generating type I molecules) and those at the 3′ terminus of a single strand (generating type II molecules). It has been suggested that the different kinds of initiation may be more apparent than real for, by virtue of the inverted terminal repetition, a single strand should be able to form a terminal double-stranded region, as shown in the brackets in Figure 21-48. These regions are known as "panhandle" forms. The forms will have a terminal double-stranded region identical to that found in linear double-stranded adenovirus DNA; initiation of a panhandle will therefore be identical to initiation of virion DNA.

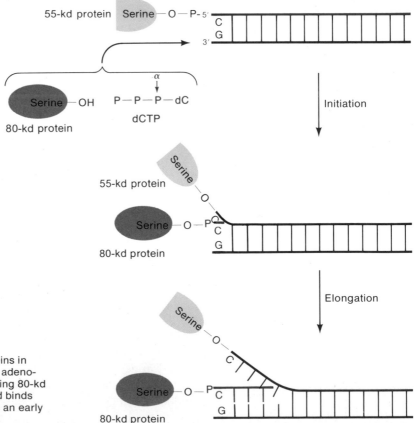

Figure 21-49
The functions of terminal proteins in the initiation and elongation of adenovirus DNA. Possibly, the incoming 80-kd protein to which dCTP is linked binds to the terminal 55-kd protein in an early recognition step.

Although panhandle molecules can be created in the laboratory by successive denaturation and renaturation (Figure 21-45), they have not yet been isolated from infected cells, and it is not known if they exist at all *in vivo*.

Priming is accomplished by a protein molecule from which the terminal 55-kd protein (which is linked to nongrowing viral DNA) is derived. The 55-kd terminal protein is synthesized from a viral gene as a larger 80-kd protein, whose molecular weight is 80,000. The 5′ terminus of a growing single strand is covalently linked to the 80-kd protein. This protein is used in initiation by the mechanism depicted in Figure 21-49. The primary initiation event is the formation of an ester linkage between the α-phosphoryl group (the phosphate attached to deoxyribose) of dCTP and the β-OH group of a serine residue in a 80-kd protein molecule. This linkage is concomitant with association of the 80-kd protein with the terminus of the parental DNA. Such association might be mediated via binding to a particular base sequence in the template strand or binding to the 55-kd protein on the other parental strand. The latter possibility is believed to be correct since DNA freed of the 55-kd protein is a poor template for initiation *in vitro*. Growth of the daughter chain then proceeds from the 3′-OH group of the dCTP primer. At a later time a viral protein cleaves the attached 80-kd protein, converting it to a 55-kd protein.

Transcription of Adenovirus DNA

The transcription pattern of adenovirus DNA is very complex. Not only are there early and late classes of mRNA but, in contrast with the smaller viruses discussed so far, there are at least five early transcription units, each initiating at a unique promoter. Furthermore, these five units are not activated simultaneously, initiation covering a period of about three hours, and several of these units are negatively regulated by their gene products. Finally, each early mRNA is spliced prior to translation. (In fact, mRNA splicing was first discovered in studies of adenovirus.) The pattern of late mRNA production is even more complex. One example of this complexity is that the synthesis of late mRNA requires that viral DNA replication begin. In contrast with the case of phage T4, it is only the onset of DNA replication, rather than replication itself, that is needed, because if replication is inhibited shortly after it starts, late mRNA continues to be made. Actually, late mRNA is initiated continually but prior to the onset of DNA replication it is terminated about 5000 bases after the initiation codon. Late mRNA is also extensively processed to form many different mRNA molecules and this processing, which is very complex, is a splendid example of maximizing the information content of a segment of DNA. There is a single late promoter and a single stop signal. However, there are four different sites at which the mRNA can be cut and a poly(A) tail added.

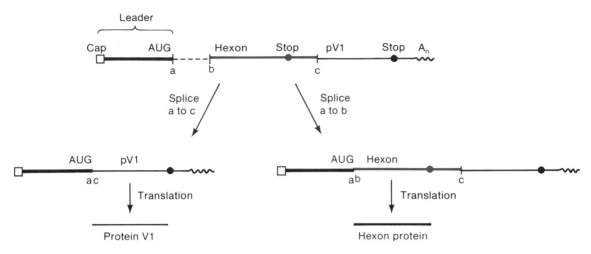

Figure 21-50
An example of one type of splicing used in adenovirus late mRNA.

Each of these five fragments has an AUG codon and hence can be translated. However, each fragment contains several adjacent introns and several pairs of cuts and splices can be made; an example of the type of alternate splicing that can produce several distinct mRNA molecules is shown in Figure 21-50. This multiple pattern allows the synthesis of at least 14 late proteins to be under the control of a single promoter. Note that each time late mRNA is initiated, only one mRNA molecule is made. Without processing, only the protein translated from the AUG codon nearest the 5' terminus is synthesized. Since the synthesis of every other protein requires that one or more splicing events occur, the rate of synthesis or the final concentration of each late protein is determined by the probabilities both of each splicing event and of the addition of a poly(A) tail. Since the probability of each event is not the same, different protein concentrations result. This type of translational control is common in eukaryotic systems but does not occur in prokaryotes; in the latter, the probability of translation of different cistrons in a polycistronic mRNA is determined by the strength of the ribosome binding sites and the proximity of a start codon to a preceding stop codon.

Herpesviruses

Herpesviruses are very large encapsulated viruses and contain more than 100 genes. They differ from all other DNA viruses in two respects: (1) They contain one large double-stranded DNA molecule, which is apparently fragmented during infection. Each fragment is separately replicated and daughter fragments are somehow reassembled into a virion. (2) They have three classes of mRNA, called α, β, and γ, but the timing of the synthesis of each class is independent of DNA replication. Preliminary data suggest that a gene product translated from one of the α-type molecules triggers transcription of the β class and a gene product translated from β mRNA initiates transcription of the γ class. As is true of all viruses, the transcribed mRNA molecules are extensively spliced to yield functional mRNA.

Vaccinia

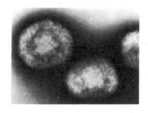

Figure 21-51
Electron micrograph of vaccina virus. (Courtesy of Robley Williams.)

The poxviruses, of which vaccinia is an example, are the largest viruses known, some having more than 200 genes (Figure 21-51). As with very large bacteriophages, the poxviruses synthesize a great many enzymes and factors needed for DNA and protein synthesis; thus they can profit by turning off host macromolecular synthesis. In fact, it is this turnoff that ultimately causes disruption of the host cell and release of progeny virions. Of particular interest is the mechanism of DNA replication of vaccinia virus. The DNA molecule of vaccinia is a continuous circular single strand that is almost completely base-paired to form a doubly-closed hairpin (Figure 21-5). Alternately, it may be thought of as a double-stranded linear molecule with sealed ends. Priming occurs at a replication origin; the helix then opens to allow the leading strand to advance and precursor fragments are synthesized as components of the lagging strand. When the molecule is almost completely replicated, a single-strand break occurs at the origin to yield a double-strand break in the parent-daughter hybrid. On copying the last base pair, another break is made at a point exactly opposite from the origin, yielding two complete double-stranded molecules. How these molecules acquire the doubly-closed hairpin form is unknown.

ASSOCIATED VIRUSES

It has been observed with many DNA and RNA viruses that a sample of progeny viruses obtained from an infected cell culture contains particles of two types having different sizes. These types can be separated either by velocity sedimentation or equilibrium centrifugation; the smaller of the two types is found to be uninfective. Such particles are said to be defective. Since they are not infective, one might wonder how they get into a freshly propagated virus sample. The explanation is that these defective viruses are capable of reproduction if they infect a cell that is also infected with the larger particle. The defective viruses are called **associated viruses;** the best-studied are the RNA-containing Rous-associated virus (RAV) and the DNA-containing adeno-associated virus (AAV). Associated viruses are of two types. Some are simply deletion mutants of the larger viruses; others are totally different viruses. In both cases their defects are complemented by viral genes of the infective particle. This phenomenon has been seen on occasion in bacteriophage systems. For instance, *E. coli* phage P4 lacks many essential genes but grows well if the host cell is also infected with phage P2.

INFECTIOUS DNA

Under certain conditions some bacteria are able to be infected with phage DNA, which can initiate an infectious cycle. The process usually requires that the bacteria be converted to spheroplasts by exposure to lysozyme (see Chapter 1) before addition of the DNA or that the DNA be added to cells suspended in a solution of $CaCl_2$ (see Chapter 20).

Purified nucleic acids of many DNA viruses are also infectious and usually the cells do not require any pretreatment or special solutions to take in the DNA. Negative-strand viral RNA is not infectious by itself because of the lack of essential enzymes not present in the host and usually carried into the cell with the capsid—for example, a lack of replicase or reverse transcriptase. Positive-strand viral RNA is frequently infectious. Plant viral RNA is also infectious.

The ability to initiate a life cycle with purified viral nucleic acid has been quite important in experimental virology. Most animal viruses engage in genetic recombination very infrequently, if at all, so that for many years it was difficult to create the kinds of multiple mutants that have been so valuable in studying phages. However, by using recombinant DNA techniques, multiple mutants of DNA viruses are obtained in a straightforward way. Two mutant molecules are cleaved by the same restriction enzyme, fragments are purified, and then by mixing the appropriate fragments, a DNA containing both mutations can be formed. Infection with purified DNA yields doubly mutant viruses.

INTERFERON

Interferon is a glycoprotein that is synthesized and excreted by all vertebrate cells in response to an infection by an RNA virus. The excreted interferon enables uninfected cells in a population of cells to survive infection temporarily. (Cells that are temporarily invulnerable to viral infection are said to be in an "antiviral state.") Although induction of interferon synthesis occurs only with RNA viruses, all animal viruses, both DNA and RNA, become inhibited when infecting a

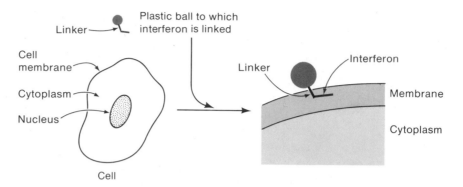

Figure 21-52
A test to show that interferon acts in the cell membrane. An interferon molecule is covalently linked to a microscopic plastic ball by means of a short (10–20 Å) hydrophobic linear molecule called a linker. The linker is shorter than the thickness of the cell membrane. When this unit is added to cells, an antiviral state is achieved, even though the plastic ball cannot enter the membrane.

cell exposed to interferon. Only a very small concentration (10^{-11} M or about 10^{-9} g/cm³) of interferon is needed to produce the antiviral state. Interferon is active even if it is covalently linked to a synthetic particle that cannot penetrate a cell, which suggests that interferon acts on the cell surface (Figure 21-52). Still other studies indicate that interferon binds tightly to the outer membrane of a cell, which suggests that this membrane is the site of its action.

Interferon is thought to be the first line of defense rallied against RNA viruses by most vertebrates since it is usually produced by the infected cells more rapidly than the immune system of an infected animal produces antibodies. This notion is suggested by the result of the following experiment, which has been done with many variations. A mouse is infected with a virus known to grow in mouse cells in culture but which generally causes no obvious symptoms of illness in the animal. If the mouse is also injected with an antibody (independently prepared in a rabbit or a goat) directed against mouse interferon, the virus-infected animal, which presumably made interferon during the infection, becomes very sick and may die because of cellular damage caused by extensive viral growth. Thus, unless the interferon made by the mouse in response to the infecting virus remains active, the virus infection will become well-established and will be too far along to be inhibited by the antivirus antibody made by the mouse several days after infection.

The two questions about interferon that have dominated the field of interferon research are the identity of the component in the viral life cycle that induces an infected cell to synthesize interferon and the mechanism by which interferon inhibits subsequent viral synthesis. A clue to the identity of the inducing substance is that interferon is induced almost exclusively by RNA viruses. Furthermore, induction of interferon synthesis never precedes replication except perhaps in the case of the double-stranded RNA viruses. These two observations have suggested that double-stranded RNA might be the inducer of interferon synthesis, and this notion has been tested by exposing uninfected cells to various synthetic polynucleotides. It has been found that synthetic single-stranded polyribonucleotides such as poly(C) and poly(U) are poor inducers, whereas the double-stranded polyribonucleotides induce interferon synthesis efficiently. The best inducer known to date is poly(I) • poly(C), in which I is inosine. The conclusion that double-stranded RNA is the only inducer is not a rigorous conclusion because single-stranded DNA also induces synthesis, albeit poorly. At present, however, it is generally agreed that the double-stranded RNA present in cells infected with RNA viruses is probably the inducer. (It should be remembered though that double-stranded RNA is not a replicative form and may only be a useless by-product in the life cycle of an RNA virus.)

The antiviral effect of interferon results from inhibition of translation. Three independent pathways of inhibition, which are depicted in Figure 21-53, are listed below.

1. The first pathway, which involves a weak inhibition, is a state of readiness. In a completely obscure way a deficiency of tRNA is created. This may weaken the cell somewhat; certainly it decreases the efficiency with which a viral infection can become established. Weakening of the exposed cells would be detrimental to an animal, for a while at least, if all of the cells were similarly weakened; however, inasmuch as the exposed cells are only in the immediate environment of virus-infected cells, weakening of these cells has survival value for the animal. This first pathway probably accounts for the weak inhibition of DNA viruses.

The second and third pathways directly abort a viral infection. These pathways are inactive in an uninfected, interferon-treated cell; they become activated by the addition of double-stranded RNA.

2. In the second pathway, a ribonuclease is activated that solubilizes viral mRNA. The activation can be demonstrated *in vitro* in cell extracts of interferon-treated cells following addition of a high concentration of double-stranded RNA to the extract. (Neither double-stranded DNA nor a DNA-RNA hybrid will suffice.) A synthetase, which has not yet been purified, links three molecules of adenosine triphosphate to form the oligonucleotide pppA2′p5′A2′p5′A (which is abbreviated 2,5 A) as indicated in Figure 21-53. This oligonucleotide

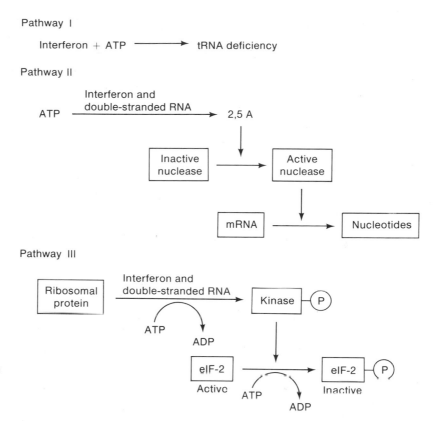

Figure 21-53
Three pathways for inhibition of translation by interferon. Pathways II and III are major ones.

activates a preexisting inactive host RNase, which then degrades mRNA to nucleotides; this rapidly terminates the infectious cycle.

3. The third pathway ensues when the concentration of double-stranded RNA is low. A preexisting ribosomal protein in the host is phosphorylated, which converts this protein to a protein kinase. The protein kinase then phosphorylates the initiation factor eIF-2 and thereby prevents all protein synthesis from occurring.

Research on interferon was for a long time hampered by the inability of laboratory workers to obtain an adequate supply of the purified substance: the number of molecules made by an infected cell is quite small. The difficulty was overcome when several different interferon genes were cloned in *E. coli* by recombinant DNA techniques; soon there will be ample material.

One important benefit of the greater supply of interferon is the possibility of testing it as a clinically useful antiviral agent. It has also been suggested that it might also have antitumor activity in cases in which the tumors are of viral origin.

PLANT VIRUSES

Productive studies of plant viruses, all of which are RNA viruses, have lagged behind research with animal viruses. This has primarily been because of the enormous multiplicity of infection that is needed to initiate viral development, and the requirement of slowly growing plants as hosts. In the past few years methods have been developed for growing plant viruses in plant cells that have been partially stripped of their cell walls (protoplasts) and, as a result, details of the life cycles of some plant viruses are now being gathered. For most plant viruses the molecular weight of the gene products exceeds the nonoverlapping coding capacity of the viral RNA, but this discrepancy can be explained by the existence of overlapping genes. Figure 21-54 shows the arrangement of the protein genes on the RNA of **tobacco mosaic virus (TMV),** a particularly well-studied plant virus. It is thought that the viral proteins VP1 and VP2 are translated from a single mRNA molecule without cleavage, but the lack of existence of a polyprotein has never been proved experimentally. The coat protein cannot be translated *in vitro* from virion RNA and appears late in infection from a mRNA that is identical in length to the virion RNA but different in shape. It is thought that newly synthesized but never packaged TMV RNA has an open ribosomal binding site and that the site is closed in virion RNA. If so, synthesis of the coat protein would be regulated in the same way as synthesis of the coat protein of E. *coli* phage M13 (Chapter 15).

Plant viruses do not induce cell lysis. As a result, enormous numbers of particles (up to 10^7) can form within a plant cell (Figure 21-55). Subsequent infection of sensitive cells usually occurs by direct intercellular transport via pores and connecting tubules. The concentration of particles is sometimes so high that crystals form that have enabled x-ray diffractionists to make detailed studies of virion structure. Because of the ease of purifying TMV, a great deal of work has been devoted to structural studies.* One of the most impressive

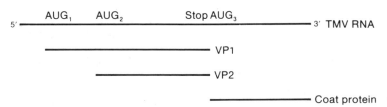

Figure 21-54
The three proteins translated from tobacco mosaic virus (TMV) RNA.

---

*The reader may remember that the first work showing that a single base change causes a single amino acid change was done with TMV coat protein.

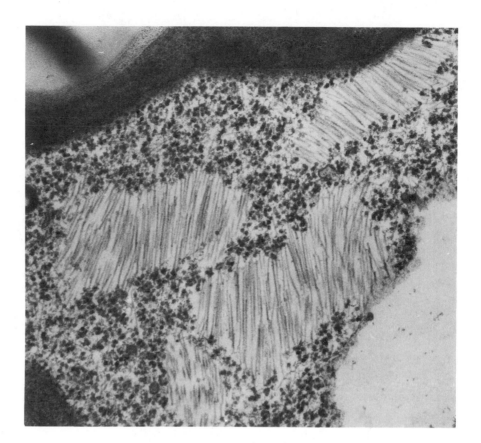

Figure 21-55
Λ cross-section of a to-
bacco leaf, showing
arrays of TMV particles.
(Courtesy of Robley
Williams.)

breakthroughs in the study of virus structure occurred in 1955 when
Heinz Fraenkel-Conrat and Robley Williams showed that infective
TMV particles could be reconstituted from purified RNA and coat
protein molecules. Various intermediate forms were examined by x-ray
diffraction and the sequence of events described in Chapter 6 were
elucidated.

TUMOR VIRUSES

In 1911 Peyton Rous demonstrated that sarcomas (muscle tumors) could
be induced in chickens by a virus now called the Rous sarcoma virus.
Since then it has been found that there are many tumor-inducing
(oncogenic) viruses in animals, though in humans they appear to be
rare. Viruses having this property are found in almost all families of
DNA viruses but among the RNA viruses only the retroviruses produce
tumors.

A common observation when DNA tumor viruses are studied is that a particular virus may cause a productive infection in cells of one species and a tumor in another. For instance, polyoma virus multiplies efficiently in mouse cells, yielding about 10^5 progeny per cell, yet in hamsters and rats tumors are produced. Types of cells in which infection is not productive are called **nonpermissive.** When a culture of nonpermissive cells is infected with a DNA tumor virus, the infection is abortive in most cells and only a small fraction of the population is converted to a tumor cell. On the other hand, for retroviruses most infected permissive cells are tumor cells. We will see that the difference in the oncogenic potential of the DNA viruses and the retroviruses may simply be the relative probability of forming a provirus; that is, in the growth cycle of a retrovirus, integration of DNA is obligatory, whereas, in the case of a DNA virus, integration occurs only sometimes (and for unknown reasons).

The principal breakthrough in the study of tumor induction by viruses was the discovery of the **transformed cell** by Renato Dulbecco and Marguerite Vogt. These are cells that differ in a variety of properties from normal cells and can often form a tumor if injected into an appropriate animal. Transformed cells are described in the following section.

Transformed Cells

Animal cells, other than cells in blood or cells of the immune system, grow on surfaces. If a piece of a typical tissue is placed on a glass or specially treated plastic surface, various types of fibroblast cells grow out from the tissue mass. Alternatively, the tissue can be treated with the protease trypsin, which separates the tissue into individual cells; if the resulting cell suspension is allowed to grow, the fibroblasts in the suspension adsorb to the surface and grow. (Most of the other cell types grow very poorly or not at all.) Such a culture, which is called a **primary culture,** has the property that the cells divide and grow along the surface until a monolayer of cells is formed; then growth stops. This cessation of growth is called **density-dependent growth;** it was formerly termed **contact inhibition** because it was believed that the growth inhibition was caused by the cells coming in contact with one another. When growth stops, it is observed that most of the cells are in an orderly array (Figure 21-56(a)). These cells can be removed from the surface by a brief treatment with trypsin. If the cells are then placed on a new surface and fresh growth medium is provided, again a confluent monolayer forms.

If a primary culture that is nonpermissive for growth of a particular tumor virus is infected with that virus, the result is quite different. In this case, when confluency is reached, some of the cells, which have

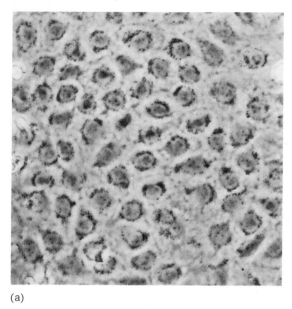

(a)

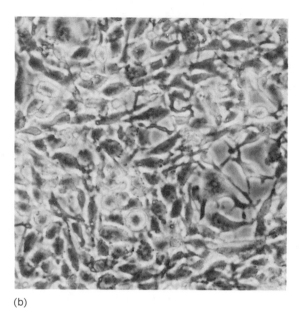

(b)

Figure 21-56

Confluent layers of (a) normal mouse 3T3 cells and (b) 3T3 cells transformed by polyoma virus. The transformed cells not only have a different shape, but also grow in less-organized fashion than normal cells. (Courtesy of Walter Eckhart.)

been growing in somewhat disordered arrays, continue to grow and ultimately form a jumbled, multilayered mass on top of the confluent background. If these cells are removed and dispersed by exposure to trypsin and then allowed to grow on a fresh surface, the resulting culture shows disordered, multilayered growth (Figure 21-56(b)). Cells having this property are said to be **transformed.***

Cells transformed in this way have several other properties that distinguish them from normal cells. One of the most intriguing is that *transformed cells are immortal.* By this is meant the following. If human tissue is cultured to confluency, and then dispersed and recultured several times, it is found that after about fifty generations almost all of the cells lose the ability to grow and gradually disintegrate. Perhaps only one cell (at most) of the 10^{15} progeny of a single cell will have gained the ability to multiply indefinitely. The number of generations of growth before a culture contains no living cells varies with the species, and the few cells that survive are mutants. In contrast, transformed cells obtained by infecting a primary culture are able to multiply indefinitely.

*Many chemical carcinogens cause transformation, but this phenomenon will not be discussed.

Transformed cells also differ from normal cells in that

1. They can grow in suspension (rather than only on a surface).
2. They require a lower concentration of blood serum in the growth medium.
3. They have an altered surface and a more fluid cell membrane.
4. Very often the number of chromosomes exceeds the normal diploid number.

However, the two most outstanding properties of viral transformed cells are the following:

1. Each transformed cell contains one or more integrated proviruses.
2. Transformed cells often form tumors if the cells are injected into an animal of the same species (i.e., transformed mouse cells into a mouse).

Little is known about how a virus induces the transformed state and especially why the transformed cell induces tumors. Furthermore, an unusual phenomenon has been observed in the life cycles of some viruses when the outcomes of normal infection are compared to infection with pure DNA—namely, differences are noticed in requirements for particular genes. For example, certain temperature-sensitive mutants of the T antigen of polyoma virus fail to transform at high temperatures when infection is by a virus yet transform at elevated temperatures when viral DNA is used. At present, we will have to be content with a description of some of the requisite viral genes, and this will be presented in a later section.

Detection of Integrated Viral DNA

In the study of a virus-induced transformation system, one usually wants to ascertain that a transformed cell contains integrated viral DNA and to determine the number and locations of the proviral DNA sequences. The methodology for these determinations is described in this section.

The first indication that virus-transformed cells contains viral DNA came from hybridization experiments. That is, ^{3}H-labeled RNA transcribed *in vitro* from purified polyoma DNA was found to hybridize with DNA isolated from polyoma-transformed cells but not with DNA from untransformed cells. The viral DNA could either be integrated or be in plasmid form. A simple experiment distinguished these alternatives (Figure 21-57). In this experiment cellular DNA was isolated from polyoma-transformed cells and then broken at random positions into fragments with a mean molecular weight of 50×10^6 and a broad range

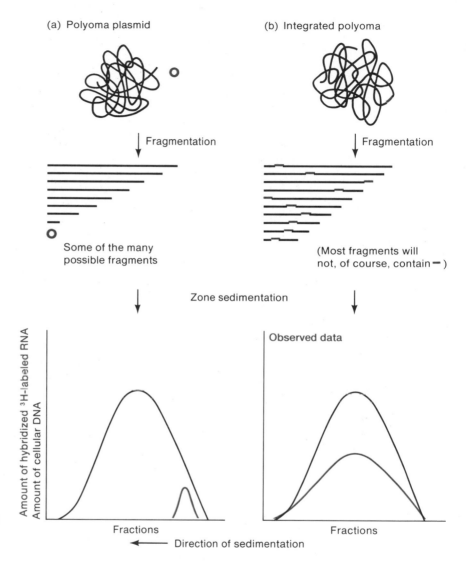

(a) Polyoma plasmid

(b) Integrated polyoma

Fragmentation

Fragmentation

Some of the many
possible fragments

(Most fragments will
not, of course, contain ━)

Zone sedimentation

Observed data

Amount of hybridized ³H-labeled RNA
Amount of cellular DNA

Fractions

Fractions

◀━━━━ Direction of sedimentation

Figure 21-57
Demonstration by zonal
centrifugation that poly-
oma DNA is integrated
into cellular DNA. The
data in the lower section
of panel (b) are observed;
those in panel (a) are
hypothetical.

of total molecular weights, generally larger than polyoma DNA (molecular weight, 4.8×10^6). The fragments were fractionated by zonal
centrifugation and each fraction was tested for the presence of polyoma
DNA by denaturation and hybridization with ³H-labeled polyoma
mRNA. If the viral DNA was a plasmid, all of the polyoma DNA would
sediment as a single component with a molecular weight of 4.8×10^6.
However, if the polyoma DNA was integrated and the cellular DNA was
broken at random, a fragment containing polyoma DNA could be of any
size; it was found that ³H-labeled polyoma mRNA would hybridize to

fragments of any size. The data clearly indicated that polyoma DNA had become integrated, as shown in the figure.

More elegant experiments done with SV40 virus demonstrated conclusively that SV40 DNA is integrated. These experiments also enabled one to count the number of proviruses, and to determine whether integration occurs at a unique position in the chromosome and in the viral DNA. The method in each of these experiments was to fragment DNA from transformed cells with a restriction enzyme (which makes cuts only in a particular nucleotide sequence), separate the fragments by gel electrophoresis, and identify the fragments containing viral DNA by denaturation and hybridization with ^{32}P-labeled denatured viral DNA.

The first experiment (Figure 21-58), which is analogous to the centrifugation experiment, showed that the viral DNA was integrated at

Figure 21-58
An experiment showing that SV40 DNA is integrated in a transformed cell. The black solid lines represent cellular DNA; the dashed lines represent large segments that are irrelevant to the experiment; SV40 DNA is shown in red. The x's are sites of cutting by the restriction enzyme. If the SV40 DNA were a plasmid (0), electrophoresis would show a single band migrating at the same speed as pure SV40 DNA. If a single provirus were present (I), it would be expected that an electrophoretic band would move less far (have higher molecular weight) because the DNA would be linked to cellular DNA. If there were two proviruses (II), there would be two bands, because in general the restriction sites would be spaced differently around the two proviruses. A supercoiled SV40 DNA standard is used because its migration rate is almost identical to that of linear, unit-length SV40 DNA.

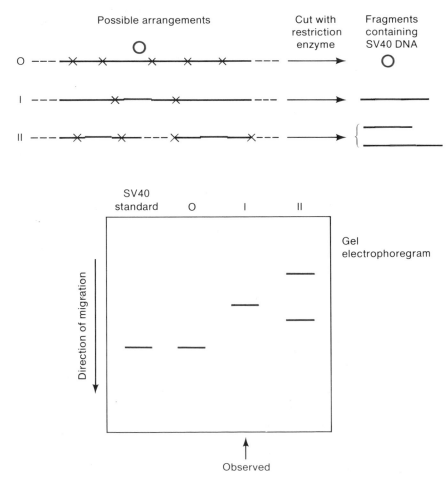

a single site in a transformed cell. Total DNA from a transformed cell was digested with a restriction enzyme that makes no cuts in SV40 DNA but many cuts in cellular DNA, and was then electrophoresed. The figure shows the data expected for an SV40 plasmid and for a provirus located in one or two different positions. The observed data corresponded to integration at a single site. This experiment has been repeated with DNA isolated from many different clones of transformed cells. The electrophoregrams have usually had a single band, though a few of the clones had two bands. Thus, one may conclude that usually a provirus is integrated at only one site but that some transformed cells contain two proviruses at different sites. When many clones transformed by a particular virus were examined, it was found that the band or bands of one clone were very rarely at the position for other clones, as shown in Figure 21-59. This means that there are many potential integration sites; possibly, the sites are randomly disposed. This situation is of course different from that with lysogenic phages.

If a restriction enzyme that makes several cuts in the provirus is used, it is possible to determine whether insertion occurs at a particular point in the viral DNA. How this is done is shown in Figure 21-60. DNA is isolated from many transformed clones and digested. Because the enzyme cuts the viral DNA, fragments smaller than the size of uncut SV40 viral DNA will always be present. These fragments are a subset of the fragments obtained by digesting circular viral DNA with the same enzyme. If the integration event occurs at a unique site, the same fragments will be found for all clones. If the site is variable and the enzyme makes n cuts, $n-1$ fragments will be seen and there will be n subsets. This is the observed result for SV40-transformed cells, which indicates that SV40 does not have a fixed site of integration.

Figure 21-59
Position of bands in a gel for different provirus locations. (a) A portion of the cellular DNA showing four possible positions of a provirus in four different transformed cell lines. (b) Restriction fragments containing a provirus for the four different cell lines. The fragments have different sizes because the proviruses are not always the same distance from the nearest restriction sites. (c) The gel. Each band (I–IV) represents a separate clone of transformed cells.

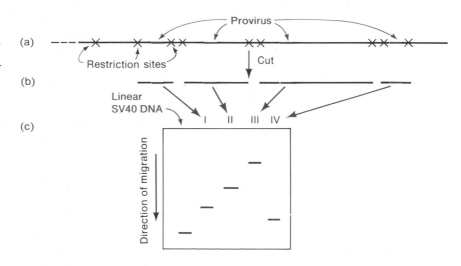

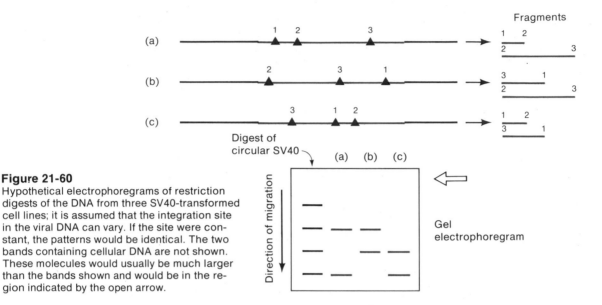

Figure 21-60
Hypothetical electrophoregrams of restriction digests of the DNA from three SV40-transformed cell lines; it is assumed that the integration site in the viral DNA can vary. If the site were constant, the patterns would be identical. The two bands containing cellular DNA are not shown. These molecules would usually be much larger than the bands shown and would be in the region indicated by the open arrow.

Many cells transformed by either SV40 or polyoma virus have more than one copy of the provirus yet the cleavage test shown in Figure 21-58 yields a single band, which indicates that there is only a single integration site. This apparent discrepancy is explained by analyzing the results of cleavage with a restriction enzyme that makes one cut in the viral DNA. Three bands are usually found and the major band is exactly the size obtained by cleaving intact virion DNA. This fragment arises because there are two proviruses integrated in tandem, as shown in Figure 21-61. Tandem (and higher order) integration is fairly common.

It is clear that a provirus can be integrated at many different sites in cellular DNA. An interesting question is whether these sites all reside on the same chromosome. The answer, which is that they do not, was obtained by applying the technique of **cell fusion.** Sendai virus has the property, which is extremely useful in the laboratory, of inducing fusion of Sendai-infected cells.* Fusion can occur between cells of both the same and of different species—for instance, mouse and human cells can fuse. Fused mouse-human cells usually have a complete set of mouse chromosomes but only a few human chromosomes. Normal mouse cells have also been fused with SV40-transformed human cells, and clones resembling mouse cells but with the transformed phenotype

---

*Sendai virus is an enveloped DNA virus that does not kill its host but promotes such a weakening of the host cell membrane that two infected cells frequently merge to form one fused cell or **heterokaryon.** As a precaution against uncontrolled infection in human subjects, nonviable or inactive Sendai virus is used, for it promotes fusion also.

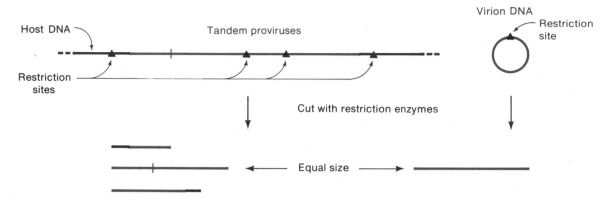

Figure 21-61
Demonstration of tandem proviruses by cutting with a restriction enzyme that makes a single cut in viral DNA.

have been isolated. Mouse and human chromosomes, as well as individual human chromosomes, are easily distinguished by their morphology during metaphase, so that transformed mouse clones can be examined to determine which human chromosomes are present. Analysis of many clones has clearly indicated that there must be at least four chromosomes in which an SV40 provirus can be located. It is possible that if a large enough number of clones is examined, it will be found that any chromosome can carry a provirus.

Structure of a Provirus

When phage DNA integrates into a bacterial chromosome, the resulting prophage is a faithful copy of the phage DNA, though it may be permuted. The main point is that phage DNA sequences are neither lost nor gained during integration. We have seen that when a retrovirus integrates, a few bases at either end of the virion base sequence are lost and a large segment of viral DNA (the LTR) is duplicated. Thus, integration of retroviral DNA is accompanied by subtle rearrangements; however, no information (that is, coding sequences) is lost, which is of course quite essential since, with retroviruses, only integrated DNA can be a source of progeny virus.

The situation is quite different with the DNA-containing tumor viruses. Integration is not an essential part of their life cycles and in fact is found only when an infection is abortive. Thus, when integration occurs, there is no selective pressure to retain all of the coding sequences other than those needed to maintain the integrated state, because the ability to integrate does not have survival value for the virus. In fact it has been found that integration is usually accompanied by loss of genetic information. Several lines of evidence support this idea.

One way to determine whether information is lost is to cause a transformed cell to excise the integrated DNA. If wild-type particles are produced, the DNA must contain all of the genetic information. A special technique is needed to induce excision because transformation only occurs in a nonpermissive cell—that is, in a cell in which mature virions cannot be made. How then can virus development be induced? Usually a cell is nonpermissive because it lacks certain factors for viral DNA replication and transcription. Thus, if these factors could be provided by another cell or virus, then if the virus possesses an excision system, production of virions from a transformed cell could take place. Fusion of transformed nonpermissive cells with permissive cells mediated by Sendai virus (see preceding section) enables such complementation to be carried out. The technique used for induction of virus production from polyoma or SV40-transformed cells is the following:

Transformed cells (i.e., nonpermissive cells derived from a single clone) are mixed with uninfected permissive cells and then the mixture is incubated with inactivated Sendai virus. The cells fuse and components of the permissive cells make up for the deficit in the nonpermissive transformed cells. About 24 hours after fusion with Sendai virus, circular SV40 or polyoma viral DNA molecules are present in the hybrid cells. If different transformed cell lines are fused with a single type of permissive cell, virions are produced by some, but not all, transformed cell lines. Thus, some cells that carry a provirus have lost the ability to form virions. Furthermore, even with virion-producing cell lines, the particles are not always infective. Restriction analysis of the viral DNA circles produced from cell lines that do not yield infective virions indicate that segments of viral DNA are either missing or sometimes inverted. Examination of many SV40 and polyoma-transformed cell lines indicates that the sole requirements for excision are the presence in the integrated DNA of the replication origin and the gene encoding the viral T antigen. Thus, excision of integrated DNA without virion production might reflect a lack of a gene encoding a virion protein. Furthermore, production of a noninfective virion might occur if the gene encoding some other essential protein were absent.

The deficiency in virus production just described might also be owing to errors in excision. To investigate this, various integrated DNA molecules from SV40-transformed cells have been cloned in *E. coli* phage λ and the DNA has been subjected to restriction analysis and sequencing of selected segments. The result of these analyses is that in the integrated DNA there are deletions and inversions, and even gene rearrangements. Thus, integration of SV40 and polyoma is accompanied by extensive scrambling of viral DNA. This situation contrasts strikingly with the retroviruses and phages. The significance of the scrambling is unknown but clearly integration with these viruses is a very complex process.

A more extreme case of loss of genetic information is seen in adenovirus transformation. Most of the adenoviruses will transform some species of host cell and the transformed clones always contain integrated adenovirus DNA. However, it is never possible for a cell transformed by an adenovirus to release virions or even adenovirus DNA, even when the Sendai fusion technique is used, because the provirus is never an intact adenovirus DNA molecule. Analysis of many adenovirus-transformed cells by hybridization with a set of adenovirus mRNA molecules (hybridized one at a time) has shown that many segments of viral DNA are absent. Study of adenovirus-transformed cells has shown that a particular small segment of adenovirus DNA is always present in an adenovirus-transformed cell. This important finding suggests that a particular gene or set of genes is responsible for the transformed phenotype. The general topic of transformation genes is examined in the following sections.

Relation between Integration and Transformation

There are basically two ways that viral integration can lead to transformation: (1) integration can occur within a regulatory host gene, thus inactivating the gene and deregulating the cell, or (2) a viral gene product, expressed in the provirus, can directly or indirectly cause transformation. The multiplicity of integration sites and the requirement for a particular segment of adenovirus DNA to be present in a cell, as well as other findings to be discussed indicate that the second explanation may often be correct.

Tumor viruses differ from lysogenic phages in that tumor viruses do not possess repressors and, in fact, transcription of the integrated DNA occurs continually. A phage repressor is needed in a lysogen in order to prevent activation of functions that kill the host bacterium. This is unnecessary with retroviruses since they do not kill the host cell—that is, cells infected with retroviruses (for example, Rous sarcoma virus) continue normal growth while continually shedding progeny viruses by budding. The oncogenic DNA viruses avoid the need for a repressor in another way. All oncogenic DNA viruses have an early and a late phase of transcription and, for those viruses having functions that are lethal to the host cell, these functions are confined to the late phase. Thus, the transformation process does not require a repressor; instead, *transformation is confined to those cell types in which late transcription or perhaps a late processing mechanism cannot occur— that is, to nonpermissive cells.*

The minimal requirement for transformation is that there be no active killing function. Clearly, failure to kill the cell is not sufficient to explain the complex cellular changes accompanying transformation, so that for many years there has been an active search for tumor-inducing viral genes. Several genes of this sort have been found in retroviruses. These genes are described in the following section.

Protein Kinases and Transformation by Retroviruses

In 1970 temperature-sensitive transformation mutants of Rous sarcoma viruses (RSV) were isolated. Mutant viruses grew productively equally well at both permissive and nonpermissive temperatures yet they transformed only at permissive temperatures. Moreover, if cells transformed at permissive temperatures were shifted to a nonpermissive temperature, the transformed cells reverted back to normal cells after one or two days. Since nontransformed, transformed, and revertant cells all produced the same virus, these experiments clearly dissociated the replicative function from the transforming function of RSV. At the same time, nonconditional deletion mutants of RSV were isolated that had lost the transforming function altogether. The nontransforming mutants replicated and formed infectious particles that were physically like wild-type RSV. The gene responsible for the transforming function was called *src* (for sarcoma-producing). Comparison of the RNA of wild-type RSV and the nontransformable mutants proved the variants to be *src*⁻ deletion mutants that lacked 1.6 kb of RNA. This provided the biochemical basis for the analysis of the *src* gene.

Growth of a *src* deletion mutant shows that *src* is a nonessential gene for virus production. It is surprising that a tiny virus carrying only a few genes could afford to have a nonessential gene. The explanation may be the following. RSV, like all retroviruses, is dependent on the host replication apparatus and can only infect when the host cell is in its DNA-synthetic phase. In an animal, most cells are resting and hence cannot support infection by a retrovirus. However, if the virus has a gene that causes cells to have repeated growth cycles even when they are in close contact (which is the primary phenotype of a transformed cell), such a virus can reproduce more often and has a definite advantage in nature. This advantage would not be detected in laboratory cell cultures because most of the cells in the culture would be growing.

Since *src*-deletion mutants replicate well, one might ask where the *src* gene came from. A reasonable guess is that the "original" RSV lacked the *src* gene and that the gene was acquired from the DNA of an ancestral host cell. By this hypothesis one might expect that the src gene or some closely related base sequence remains to this day in avian cells. Such sequences have been sought by hybridizing ^{32}P-labeled src RNA or DNA with denatured host DNA isolated from normal cells. Sequences identical to that of the src gene have been found in the DNA of all vertebrates examined to date, supporting the idea that the src gene is a normal cellular gene that has been acquired by the virus in the course of evolution.

Similar transformation genes, which are homologous to host sequences, have been found in several other retroviruses (feline sarcoma and chicken sarcoma viruses).

Probably the most unusual feature of the src gene is that a single virus-encoded protein is able to cause the large number of effects associated with the transformed phenotype. The explanation, as will be seen, is that the src-gene product appears to be an enzyme capable of modifying a large number of proteins.

The src-gene product (and possibly also the products of the transformation genes of some, but not all, retroviruses) is a protein kinase—that is, an enzyme that transfers a phosphate from ATP to one or more amino acids in a protein molecule. Protein kinases are common enzymes; they regulate such a wide variety of cellular events that it is not surprising that high concentrations of a protein kinase (such as exist for the Src protein in an infected cell) could cause a major disruption of cell activities and cell shape. The protein kinases associated with viral transformation proteins phosphorylate tyrosine. Numerous studies with temperature-sensitive mutants of the transformation genes of retroviruses show that transformation and tyrosine-phosphorylating activity go hand in hand. At permissive temperatures the kinase is active, the cells are transformed, and some cellular proteins contain phosphorylated tyrosine. At nonpermissive temperatures there is neither kinase activity nor transformation and there are no phosphorylated tyrosines in the cellular target proteins.

In order to understand the significance of protein phosphorylation, the target proteins must be identified. Several proteins containing phosphotyrosine have been isolated but the functions of these proteins have not yet been identified. Of particular interest is a cytoplasmic protein (P36), which constitutes 0.5 percent of the total cellular protein; however, it is not known whether this protein is structural, enzymatic, or even plays a role in cell division.

A protein present in blood serum and called the **epidermal growth factor (EGF)** may be related to the Src phenomenon. EGF is a stimulator of cell division for a wide variety of cells in culture and presumably plays a similar role in animals. EGF binds to the plasma membranes of sensitive cells and, in so doing, markedly stimulates the enzymatic activity of a protein kinase within the cell membrane. This kinase also phosphorylates tyrosine. It is not the same as the src-gene product but the two proteins are antigenically related. (That is, an antibody to the Src protein also inactivates the membrane kinase.) Furthermore, the protein P36 is phosphorylated after EGF binds to cells. However, EGF does not transform cells.

Several proteins whose functions are known and which may be related to transformation are phosphorylated by the Src kinase. One of these is **vinculin.** The shapes of most cells are determined by the arrangement of filaments of actin, a fibrous protein also found in muscle. Transformation usually is accompanied by a rounding of the cell and a disorganization of actin filaments. The membrane protein, vinculin, is responsible for anchorage of actin filaments to the inner side of the cell membrane. Vinculin is located in patches **(adhesion plaques)** on the cell membrane that are thought to be responsible for intercellular adhesion, another property that is markedly weakened in transformed cells. Vinculin is phosphorylated (on its tyrosines) by the Src kinase, which is concentrated in the adhesion plaques. It has not yet been proved but it is possible that phosphorylation of vinculin might both destabilize the actin linkages and break down the adhesion between cells.

Transformed cells also show a striking increase in the rate of glycolysis. These changes are not understood but recent experiments implicate protein kinases. Whether the phenomenon involves the Src kinase is not known.

An important question that remains is why a viral product that closely resembles a cellular enzyme should lead to transformation. Possibly the effect results from increasing gene dosage in that, because of continual transcription of the provirus, a transformed cell makes about 50 times as much viral enzyme as a normal cell would make of the cellular enzyme. Also, it is possible that the specificity of the viral enzyme is somewhat different from that of the cellular enzyme, so that either the viral kinase phosphorylates proteins that are not affected by the cellular kinase or the viral kinase is not subject to regulation by effector molecules that control the specificity and timing of the activity of the host kinase. Clearly, more research on the role of kinase is needed.

Transformation by retroviruses is not exclusively determined by the Src kinase. Retroviruses cause different types of tumors—sarcomas in fibroblast cells, carcinoma in epithelial cells, and leukemia in blood cells. Fibroblasts and epithelial cells are mutually adhering cells and the blood cells are nonadherent. The src gene is only needed for transformation of fibroblasts. That is, src⁻ mutants of several retroviruses are able to cause leukemia, though they are unable to transform fibroblasts. A *leuk* gene has been identified but it has not yet been characterized.

Transformation Proteins in DNA Viruses

Transformation proteins have only very recently been shown to exist in DNA viruses. The best evidence for these proteins is for adenovirus, polyoma, and SV40.

Cells transformed by adenovirus contain only fragments of adenovirus DNA, and a specific region of adenovirus DNA is always present in a transformed cell. A very significant result is that if adenovirus DNA is cleaved by a restriction enzyme and the fragments are purified, a small fragment that is 8 percent of the size of the intact DNA molecule and that contains the specific region can induce transformation by transfection. It is thought that at least one of the two proteins carried on this fragment is a transformation protein. However, these genes encode essential early proteins needed for replication and hence are fundamentally different from the transformation proteins of retroviruses.

Some years ago it was observed that the polyoma gene *a* and the SV40 gene *A*, both of which encode T antigens, are necessary both for replication of the viral DNA in a permissive host and for transformation of a nonpermissive host. However, a relation between DNA replication and transformation proved to be much more complex than first thought for several reasons, one of which is that these genetic loci encode several proteins (as indicated for SV40 in Figure 21-37) whose coding sequences overlap. The SV40 early region encodes two proteins, t antigen and T antigen, and the polyoma locus encodes three — t antigen, middle-t antigen, and T antigen. Study of SV40 mutations of the *A* gene in the t and T regions has shown that in certain cell lines only T antigen is needed for transformation. (In some cell lines t antigen is also needed but the reason is unknown.) Furthermore, T antigen is a multifunctional protein having several activities located in different parts of the molecule. A defective mutant in the T region has been isolated which prevents the viral DNA from replicating at the nonpermissive temperature, yet transformation occurs at that temperature. Thus, for SV40, transformation does not require that the viral DNA replicate.

The situation with polyoma virus is more complex. Several features of cell growth contribute to the transformed phenotype. Three characteristics that are relevant to this discussion are that a transformed cell (1) can grow in agar rather than on a surface, (2) can grow if a low concentration of fetal serum is present — normal cells require a high concentration — and (3) can form piles of jumbled cells on a confluent surface of cells. A point mutation in the middle-t antigen fails to transform by criteria (1) and (3), yet growth in low concentrations of serum is possible. On the other hand, mutation in the T antigen prevents growth in a low serum concentration. In a critical experiment different segments of gene 3 were cloned and the cloned DNA was used to transfect nonpermissive cells. A complete gene 3 yielded transformed cells. When the fragments were used, it was again found that transformation did not need the t antigen, that T antigen was needed to allow growth in a minimal serum concentration, and that middle-t antigen was responsible for most of the features of a transformed cell. Thus, middle-t antigen can be thought of as a transformation protein. Whether t antigen is sufficient to enable a cell to form a tumor in an animal is not yet known but information should be forthcoming.

Virus-induced protein kinases have been sought in cells transformed by DNA viruses, but to date, none have been found with certainty. A recent report suggests that the T antigen of polyoma virus may be a protein kinase.

An important question concerning transformation is that of distinguishing a requirement for integration from a requirement for transformation. This distinction is easily made with retroviruses—that is, integration but not transformation occurs with src^- mutants—but with the DNA viruses it has proved to be difficult. It is probably the case that transformation is not merely the result of integration but that integration is the means of carrying a transformation gene from one generation to the next.

Viruses and Cancer

There is no doubt that many cancers in animals are caused by viruses. In humans the case is less clear cut. The best case can be made for the Epstein-Barr virus, which is found in all patients having Burkitt's lymphoma, a form of cancer occurring mainly in a small region of Africa. (This virus is also the cause of infectious mononucleosis in humans, though acquiring this disease early in life does not predispose one to getting Burkitt's lymphoma at a later time.) It is difficult to prove unequivocally that the association is meaningful because only 95 percent of the population carry the virus and less than 1 percent have the disease and because one cannot do the experiment of infecting humans lacking the viruses with the viruses. Other candidates are nasopharyngeal carcinoma, a tumor restricted to certain Chinese populations, and hepatoma, which appears to be associated with infectious hepatitis B virus.

The question of whether all, or even most, cancers are of viral origin is also a difficult one. In the early 1970's the answer was thought to be yes. However, as more information has been gathered, this conclusion seems less likely (but still debatable). Perhaps the next decade will provide a clear answer.

VIROIDS

Several diseases are known which are clearly infectious but for which neither viral nor microbial agents can be isolated. It is now known that some of these are caused by **viroids,** naked pieces of intracellular RNA. The best-studied viroid is the **potato spindle tuber viroid (PSTV),** which consists of 359 nucleotides (Figure 21-62). These tiny RNA molecules are circular single strands in which 262 bases (73 percent) are base-paired. The RNA is sufficiently large to code for a protein of molecular weight 10^4 but no viroid protein has ever been found in infected cells. Furthermore, purified viroid RNA shows no mRNA activity with *in vitro* protein-synthesizing systems. The mechanism of replication of viroid RNA is unknown. RNA replicases have been isolated from uninfected plant cells and these are able to replicate viroid RNA if a primer is provided. Replication *in vivo* is inhibited by actinomycin D, a

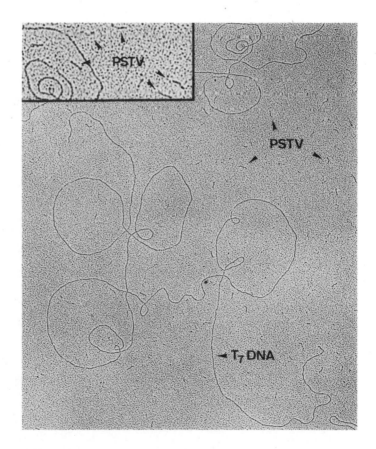

Figure 21-62
Electron micrograph of potato spindle tuber viroid (PSTV) and *E. coli* phage T7 DNA (molecular weight $= 25 \times 10^6$, length $= 12.5 \ \mu$m). The inset shows an enlargement. Note that PSTV and the DNA have the same width. (Courtesy of T. Keller and J. M. Sogo.)

drug whose only known biochemical activity is to prevent transcription from a DNA template, but no DNA intermediate has been found. There is some homology between viroid DNA and host DNA, so that it is currently thought that an RNA primer is transcribed from host DNA and that this transcription is inhibited by actinomycin D. There is no information about the means of causing disease. It has been suggested, however, that the viroid RNA hybridizes to essential segments of host DNA and thereby interferes with transcription of these segments. Since viroids are never encapsidated, the means of transmission is unknown.

To date, all known viroids are of plant origin. However, evidence is accumulating that some animal diseases may be caused by viroids.

REFERENCES

Bishop, J. M. 1978. "Retroviruses." *Ann. Rev. Biochem.*, 47, 35–88.

Bishop, J. M. 1980. "The molecular biology of RNA tumor viruses: a physician's guide." *New England. J. of Med.*, 303, 675–682.

Botchan, M., J. Stringer, T. Michison, and J. Sambrook. 1980. "Integration and excision of SV40 DNA from the chromosome of a transformed cell." *Cell*, 20, 143–152.

Butler, P. J. G., J. T. Finch, and D. Zimmer. 1977. "Configuration of tobacco mosaic virus during virus assembly." *Nature*, 265, 217–219.

Butler, P. J. G., and A. Klug. 1978. "The assembly of a virus." *Scient. Amer.*, November, pp. 62–69.

Caspar, D. L. D., and A. Klug. 1962. "Physical principles in the construction of regular viruses." *Cold Spring Harb. Symp. Quant. Biol.*, 27, 1–24.

Cold Spring Harbor Laboratory. 1979. *Viral Oncogenesis. Cold Spring Harb. Symp. Quant. Biol.*, Vol. 44.

Crawford, L. V. 1980. "The T antigens of simian virus 40 and polyoma viruses." *Trends Biochem. Sci.*, 5, 39–42.

Diener, T. O. 1979. *Viroids and Viroid Diseases.* Wiley.

Diener, T. O. 1981. "Viroids." *Scient. Amer.*, January, pp. 66–73.

Eckhardt, W. 1977. "Genetics of polyoma virus and simian virus 40." *Comprehensive Virol.*, 9, 1–26.

Eckhardt, W. 1981. "Polyoma T antigens." *Adv. Cancer Res.*, 35, 1–26.

Fenner, F., B. R. MacAuslan, C. A. Mims, J. Sambrook, and D. O. White. 1974. *The Biology of Animal Viruses.* Academic Press.

Friedmann, R. M. 1977. "Antiviral activity of interferon." *Bacter. Rev.*, 41, 543–567.

Graessman, A., M. Graessman, and C. Mueller. 1981. "Regulation of SV40 gene expression." *Adv. Cancer. Res.*, 35, 111-151.

Henle, W., G. Henle, and E. T. Lennette. 1979. "The Epstein-Barr virus." *Scient. Amer.*, July, pp. 48–59.

Holmes, K. C. 1980. "Protein-RNA interactions during the assembly of tobacco mosaic virus." *Trends. Biochem. Sci.*, 5, 4–7.

Hynes, R. 1980. "Cellular location of viral transformation proteins." *Cell,* 21, 601–602.

Joklik, W. K. (ed.). *Principles of Animal Virology.* Appleton-Century–Crofts.

Kitamura, N., B. L. Semler, R. G. Rothberg, G. R. Larsen, C. J. Adler, A. J. Dorner, E. A. Emini, R. Hanecak, J. J. Lee, S. van der Werf, C. W. Anderson, and E. Wimmer. 1981. "Primary structure, gene organization, and polypeptide expression of poliovirus RNA." *Nature,* 291, 547–553.

Knight, C. A. 1975. *Chemistry of Viruses.* Springer-Verlag.

Lebeurier, G., A. Nicolaieff, and K. E. Richards. 1977. "Inside-out model for self-assembly of tobacco mosaic virus." *Proc. Nat. Acad. Sci.,* 74, 149–153.

Lebowitz, P., and S. Weismman. 1979. "Organization and transcription of the simian virus 40 genome." *Current Topics in Microbiol. and Immunol.,* 87, 43–172.

Luria, S. E., J. E. Darnell, D. Baltimore, and A. Campbell. 1978. *General Virology.* Wiley.

Mitra, S. 1980. "DNA replication in viruses." *Ann. Rev. Genet.,* 14, 347–398.

Nathans, D. 1979. "Restriction endonucleases, simian virus 40, and the new genetics." *Science,* 903–909.

Shimotuno, K., S. Mizutani, and H. M. Temin. 1980. "Sequence of retro provirus resembles that of bacterial transposable elements." *Nature,* 285, 550–554.

Silverstein, S. C., J. K. Christman, and G. Acs. 1976. "The reovirus replicative cycle." *Ann. Rev. Biochem.,* 45, 375–408.

Temin, H. M. 1972. "RNA-directed DNA synthesis." *Scient. Amer.,* January, pp. 24–33.

Temin, H. M. 1980. "Origin of retroviruses from cellular movable genetic elements. *Cell,* 21, 599-600.

Tooze, J. 1981. *The Molecular Biology of Tumor Viruses.* Cold Spring Harbor Lab.

Weinberg, R. A. 1980. "Integrated genome of avian viruses." *Ann. Rev. Biochem.,* 49, 197–226.

Williams, R. C. and H. W. Fisher. 1974. *An Electron Micrographic Atlas of Viruses.* Thomas.

Ziff, E. B. 1980. "Transcription and RNA processing by the DNA tumor viruses." *Nature,* 287, 491–499.

22 Regulation in Eukaryotes

The regulatory systems of prokaryotes and eukaryotes are quite different from each other. Prokaryotes are generally free-living unicellular organisms that grow and divide indefinitely as long as there are appropriate environmental conditions and an adequate supply of nutrients. Thus, their regulatory systems are geared to provide the maximum growth rate in a particular environment—except when growth would be detrimental. This optimization is accomplished by limiting cell synthesis and enzymatic activity to that necessary to achieve the maximum growth rate—that is, by eliminating waste—and with mechanisms that enable a cell to respond rapidly to changing conditions. Because no membrane encases the DNA of a prokaryote, the DNA is always available to receive signals present in the cytoplasm. Thus, the on-off aspect of protein synthesis is often regulated by controlling initiation of transcription, as was seen in Chapter 14, in which several operons were examined.

The eukaryotes, except for the unicells such as yeast, algae, and protozoa, have different requirements. In a developing organism—for example, in an embryo—a cell must not only grow and produce many progeny cells but also the cells must undergo considerable change in morphology and biochemistry and maintain the changed state—that is, they must differentiate. However, while growing and dividing, these cells have a somewhat easier time than bacteria in that the composition and concentration of their growth media do not change in time—examples of such media are blood, lymph, or other body fluids, or, in the case of marine animals, sea water. Finally, in an adult organism, growth and cell division are stopped in most cell types and each cell needs only to maintain itself. Many other examples could be given; however, the main point is that because the needs of a typical eukaryotic cell differ from those of a bacterium, the regulatory mechanisms of eukaryotes and prokaryotes are basically different. From this reasoning one might expect the regulatory mechanisms of freely growing unicellular eukaryotes to resemble those of bacteria, but this is not the case. Somehow, in ancient times enclosure of the nucleus in a membrane was a great step forward in evolution and was accomplished by, or perhaps preceded, significant changes in gene organization and regulation.

This chapter is concerned primarily with eukaryotic cells that form organized tissue. Most of the information comes from detailed studies of the DNA of mammals, amphibians (toads of the genus *Xenopus*), insects (the fruit fly *Drosophila*), birds (usually the chicken), and echinoderms (the sea urchin). An unusual feature of gene expression in the yeast *Saccharomyces* will also be presented.

The study of regulation in multicellular eukaryotes has been hampered by the inability to obtain regulatory mutants, to isolate and manipulate genes, and to detect specific mRNA molecules. All of this is being changed by the recombinant DNA technology, which enables DNA fragments to be cloned (Chapter 20). The cloned fragments have been used to study genome organization and as probes for detecting RNA. The information gained has produced a dramatic increase in the pace of research. The framework for understanding the new facts is not yet clear; thus, this chapter is presented as a compendium of observations illustrating the great variety of regulatory mechanisms used by different cells.

SOME IMPORTANT DIFFERENCES IN THE GENETIC ORGANIZATION OF PROKARYOTES AND EUKARYOTES

There are numerous differences between prokaryotes and eukaryotes with regard to transcription, translation, and the spatial organization of DNA. Some of these differences were mentioned in earlier chapters and others will be discussed in this chapter. Six of these differences will come up repeatedly in discussing regulation of eukaryotic gene expression; these are the following:

1. In a eukaryote only a single polypeptide chain can be translated from a completed mRNA molecule (Chapter 13); thus, operons of the type seen in prokaryotes are not found in eukaryotes.
2. The DNA of eukaryotes is bound to histones (to form chromatin)

and to numerous nonhistone proteins. Only a small fraction of the DNA is bare.

3. A significant fraction of the DNA of eukaryotes consists of a few base sequences that are repeated hundreds to millions of times. Some sequences are repeated in tandem but most repetitive sequences are not.

4. A large fraction of the base sequences in eukaryotic DNA is untranslated.

5. Eukaryotes possess mechanisms for rearranging DNA segments in a controlled way and for increasing the number of genes when needed.

6. Genes and gene products are usually not collinear in eukaryotes; that is, there are introns in most eukaryotic genes.

We shall see throughout this chapter how these features are incorporated into particular modes of regulation. A point to be noted is how eukaryotic genes are subject to regulation at more points in their biosynthetic pathways than bacterial genes.

ORGANIZATION OF EUKARYOTIC DNA

A large fraction of eukaryotic DNA consists of repeated base sequences that do not encode proteins. This feature is examined in the following sections.

Noncoding Sequences in Eukaryotic DNA

Various genetic techniques allow one to estimate the number of genes in an organism. This number can then be compared to the DNA content of one cell in order to estimate the fraction of the DNA that consists of coding sequences. For instance, it is estimated that E. coli contains about 1500 genes. If a typical protein contains 500 amino acids, 3000 bases would be needed to encode one protein, or 4.5×10^6 bases for 1500 proteins. An E. coli chromosome contains 8×10^6 bases, so roughly half of the sequences would encode proteins, a value that is consistent with the size of the leaders, spacers, and regulatory regions seen in sequenced genes. Genetic data for the fruit fly *Drosophila melanogaster* suggest that it possesses twice as many genes as E. coli. However, a fruit fly cell contains 20 times as much DNA; therefore, only about 5 percent of the DNA of *Drosophila* consists of coding sequences. Similar data for mammalian cells show that a typical cell contains about 600 times as much DNA as found in E. coli but only 20 times as many genes. Thus, only about 2 percent of mammalian DNA consists of coding sequences.

These genetic estimates could be off by a factor of two, but nonetheless it is clear that a huge fraction of the DNA of eukaryotes must be devoted to regulation (or other functions), so we should not be surprised to discover that regulatory mechanisms are very complex in this class of cell.

Satellite DNA

If the DNA of a bacterium is isolated and randomly fragmented into about 500–1000 pieces and then centrifuged in a CsCl density gradient, the DNA forms a single fairly narrow band. This is because the buoyant density of DNA is proportional to the (G+C)-content of the DNA and the average base composition of segments containing several hundred base pairs does not vary much from one section of the DNA to the next. However, if DNA from a crab is analyzed, two bands are found (Figure 22-1). The major band has a base composition of 58 percent A+T, whereas the minor band, which is called **satellite DNA,** is 97 percent A+T, a most unusual base composition. Satellite bands have been observed in the DNA of many organisms and may comprise 1–30 percent of the total DNA, averaging about 15 percent. An important characteristic of all satellite bands is that they contain **repeated base sequences** having various lengths. For example, in the cow a 1400-base-pair sequence is repeated again and again and forms a satellite band, whereas in some monkey species a 172-base-pair sequence is repeated in the satellite band. The crab satellite DNA repeats only two bases—that is, the DNA has the sequence ATATAT . . . , only occasionally interrupted by a G or a C (1 in 30 base pairs). Of particular note is a species of fruit fly, *Drosophila virilis*, which has three satellite DNA

Figure 22-1
The result of equilibrium centrifugation in CsCl of the DNA of the bacterium *E. coli* and the crab *Cancer borealis*. As is common with bacteria, the DNA has a narrow range of densities. The crab DNA consists of two discrete fractions, one of very low density. This minor band is called satellite DNA.

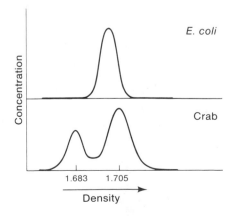

bands, each containing a distinct but very closely related repeating heptanucleotide, namely,

$$5'\text{-A C A A A T T-}3'$$

$$5'\text{-A C A A A C T-}3'$$

$$5'\text{-A T A A A C T-}3'$$

These sequences are repeated 10^7 times in each cell of this species!

The technique of ***in situ* hybridization** has been used to localize the satellite DNA in the chromosome. In this technique cells are fixed to a glass microscope slide and treated with alkali to denature the DNA in the chromosomes. (It is remarkable that the chromosomes remain visually intact.) A droplet containing ^{3}H-labeled RNA transcribed *in vitro* from purified satellite DNA is placed on the cells on the slide and allowed to renature with the chromosomal DNA. After renaturation is complete, the cells are washed to remove satellite RNA that is not chromosome-bound and are then autoradiographed. Grains in the film, indicating the presence of radioactive material, are found mainly over the centromeres of each chromosome—that is, at the site of attachment of the chromosome to the mitotic spindle (Figure 22-2). This location of satellite DNA is of unknown significance but suggests that the DNA may play an important role in mitosis.

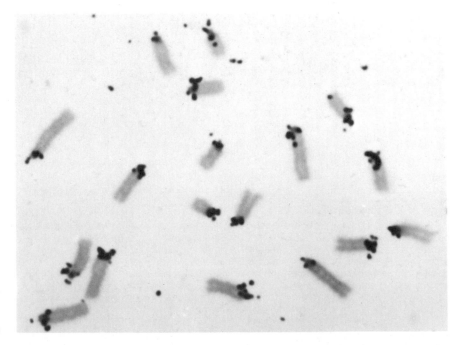

Figure 22-2
Autoradiogram of meta-phase chromosomes of a mouse cell; radioactive RNA transcribed from satellite DNA by RNA polymerase has been hybridized to the chromo-some to show the loca-tion of the satellite DNA. (Courtesy of Joseph Gall.)

Unique and Repetitive Sequences

Repeated sequences are found in all eukaryotes except unicellular organisms; however, these sequences do not always show up as satellite bands because their base compositions may not be very different from the average base composition of the bulk of the DNA. The extent to which repeated sequences exist is easily shown by C_0t analysis (Chapter 4). In this technique, DNA is extracted from an organism (*without* fractionation of the sort used to demonstrate satellite DNA), then denatured and renatured, and the kinetics of renaturation are determined. Since renaturation is a concentration-dependent process, repeated sequences (which are at a higher concentration simply by virtue of being repeated) renature more rapidly. Figure 22-3 shows a C_0t curve for bacterial and mouse DNA. Close examination of C_0t curves from many species shows that there are four classes of sequences—**unique** (single copy), **slightly repetitive** (1 to 10 copies), **middle repetitive** (10 to several hundred copies), and **highly repetitive** (several hundred to several million copies).* The bounds of these classes are arbitrary, so one must be careful when making generalizations about the classes. The unique sequences account for most of the genes of the cell but frequently for only a few percent of the DNA. Some examples of the slightly repetitive class are the genes encoding histones and the tRNA genes. In some cases these sequences are not perfectly repetitive: the gene products may have slight differences in amino acid composition—

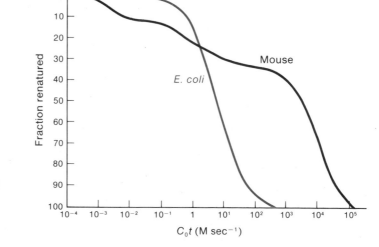

Figure 22-3
C_0t curves for *E. coli* and mouse DNA. The sequences are unique in *E. coli,* so the curve has a single step. In mouse DNA 10 percent consists of about one million copies of a 300-base-pair sequence, 30 percent consists of 10^3–10^4 copies of several sequences, and 60 percent consists of unique sequences.

*Note that since one usually studies a diploid cell, unless the DNA is isolated from eggs, there are actually two copies of a unique sequence.

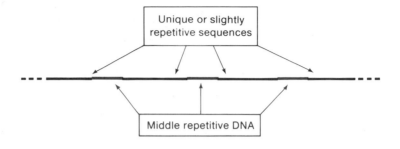

Figure 22-4
Arrangement of middle repetitive DNA in a eukaryotic chromosome.

for example, the genes encoding the embryonic, fetal, and adult forms of the β chain of hemoglobin, as we shall soon see. The middle repetitive sequences are generally *not* coding sequences and are thought to play a role in regulation. The highly repetitive sequences include the short sequences in satellite DNA and sequences of normal length in certain genes, such as ribosomal RNA genes, that exist in very large numbers. There are in fact very few different highly repetitive sequences, but the number of copies is so great that these sequences may account for 20 percent (or more) of the mass of the DNA.

The arrangement of a significant fraction of the middle repetitive DNA is striking. That is, in many organisms a great deal of the middle repetitive DNA alternates in a fairly regular fashion with the unique sequences (Figure 22-4). Moreover, there are generally several thousand middle repetitive sequences having an average length of about 300 base pairs; these occur several hundred times in the genome and comprise about 20 percent of the total DNA. The significance of this arrangement has not yet been established but one interpretation of this finding is that the middle repetitive sequences are regulatory sequences responsible for turning on and off the activity of many genes. This interpretation is expressed in a model for the regulation of eukaryotic genes, which will be described shortly. However, first it is necessary to understand the concept of gene families.

GENE FAMILIES

In prokaryotes closely related genes are often organized in operons and are transcribed as part of a polycistronic mRNA. Thus, the entire system is under control of a single promoter and the system can be turned on and off by controlling the availability of the promoter. This method is of no use in eukaryotes without modification inasmuch as eukaryotic DNA is monocistronic. Many related eukaryotic genes can be functionally grouped as a set of genes called a **gene family.** Activity of members of the set is usually but not always correlated. Recombinant DNA tech-

(a) Simple multigene family

Single-stranded
5S rRNA

(b) Complex multigene families

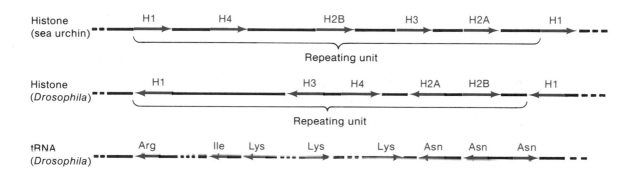

Histone
(sea urchin)

Repeating unit

Histone
(Drosophila)

Repeating unit

tRNA
(Drosophila)

(c) Developmentally controlled complex multigene family

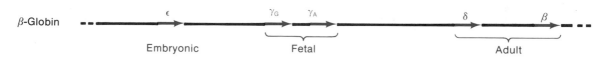

β-Globin

Embryonic   Fetal   Adult

Figure 22-5
Five examples of gene families. The arrows indicate the direction of transcription. Genes are shown in red; spacers are black.

niques have been used to clone gene families and on occasion it has been found that all genes in a family can be cloned on a single plasmid, which clearly indicates that such a family comprises a gene cluster. More often, a cluster of related genes is sufficiently large that cloning of the entire cluster on one plasmid is unlikely and it is instead found that two or three genes of a family are cloned as a unit. This partial recovery is useful in determining the order of the genes. For example, if one clone contains genes *A* and *B*, another *B* and *C*, and a third *C* and *D*, the gene order in the chromosome must be *ABCD*. Cloning experiments have also shown that many genes whose products are related but not coordinately regulated are also clustered; the term gene family has therefore been expanded to include any cluster of functionally related genes (such as the tRNA genes).

Gene families are currently classified as simple multigene families, complex multigene families, and developmentally controlled complex multigene families. An example of each type is shown in Figure 22-5.

Simple Multigene Families

A simple multigene family is one in which one or a few genes are repeated in a tandem array. The simplest example is the set of genes for 5S rRNA, which has been studied most carefully in the toad *Xenopus*. This set of rRNA genes forms a gigantic array in which each 5S rRNA gene sequence is separated from an adjacent gene sequence by a spacer to form a gene cluster. The sizes of the spacers vary from two to six times the length of the 5S rRNA gene and the spacers contain middle repetitive sequences. Each 5S rRNA gene is transcribed as a separate RNA molecule that is processed to generate the finished 5S rRNA molecule. In addition to clustering of the 5S rRNA genes the clusters themselves exist in several copies in *Xenopus*, each copy containing hundreds to thousands of 5S rRNA genes.

The 5.8S, 18S, and 28S rRNA genes of *Xenopus* comprise another simple multigene family, in which a single transcription unit containing all three rRNA sequences is separated by spacers. The 5.8S, 18S, and 28S rRNA molecules are cleaved from the transcript by processing enzymes. The number of copies of this three-gene rRNA sequence ranges from a few hundred to several thousand per diploid cell of various types.

The rRNA genes provide an interesting contrast between prokaryotes, unicellular eukaryotes, and multicellular eukaryotes. For example, in *E. coli*, 5S, 16S, and 23S rRNA molecules comprise a single transcription unit from which each molecule is cleaved. In yeast, on the other hand, the rRNA molecules form a single *repeated* unit separated by spacers; however, the 5S rRNA is transcribed separately from the 5.8S–18S–28S unit. (This grouping probably has evolutionary significance since the 5.8S rRNA of eukaryotes corresponds functionally to the 5S rRNA of prokaryotes.) Finally, in the multicellular eukaryotes the 5S RNA genes comprise a distinct family, as has just been seen.

Complex Multigene Families

The complex multigene families generally consist of a cluster of several related genes, each transcribed independently and separated by a spacer. Three examples are shown in Figure 22-5(b); each example has features that indicate that there are several types of complex multigene families.

First, let us examine the histone-gene family of the sea urchin. In this animal there are five histone proteins synthesized by five genes, contained in a small (7-kd) DNA segment and separated by spacers. This five-gene unit is tandemly repeated about 1000 times. Each of the five genes is separately transcribed as a single monocistronic RNA (it comes as a surprise that this RNA does not have introns). Note that each gene is transcribed in the same direction—that is, from the same DNA strand. In a way that is not known, the rate of transcription and translation of each gene is regulated, first, so that histones are made at the right time for chromatin replication and, second, so that equimolar amounts of histones H2A, H2B, H3, and H4, and half as much H1 (the molar proportion in chromatin), are produced. The purpose of having multiple copies of the five-gene unit is undoubtedly to enable the cell to make large amounts of histone rapidly when DNA replication is occurring. Recently, it has been shown that in a particular cell all of the copies of the five-gene unit are not transcribed; different units are copied in different tissues and at different specific stages of embryonic development, suggesting that there may be subclasses of specific histones formed at different times. (This gene family may prove to be a member of the developmentally regulated class.)

Examination of the histone-gene family of *Drosophila* shows that, in contrast with the sea urchin, the five genes are transcribed in both directions, as shown in the Figure 22-5. In yeast the organization of the histone genes differs even more; there are two quite distant gene clusters, each containing one H2A gene and one H2B gene, whereas the other histone genes are separate and scattered throughout the chromosome.

The tRNA gene cluster of *Drosophila* shown in the figure has two characteristics not seen in the histone clusters—namely, there are several copies of the same gene (for example, three tRNA^{Lys} genes) in the cluster, and not every tRNA gene is in the cluster. Each of the genes in this cluster is separately transcribed and both directions of transcription are used. Other tRNA genes are in clusters found elsewhere in the DNA, even in other chromosomes. The clustering of tRNA genes is not a general rule for the unicellular eukaryotes; for example, 175 of the 360 known yeast tRNA genes have been cloned in plasmids, and in only a few cases are two tRNA genes contained in one recombinant plasmid. Furthermore, the eight tRNA^{Tyr} genes in yeast are located on six different chromosomes.

Developmentally Controlled Complex Gene Families

Hemoglobin is a tetramer containing two α subunits and two β subunits. However, there are several different forms of both α and β subunits

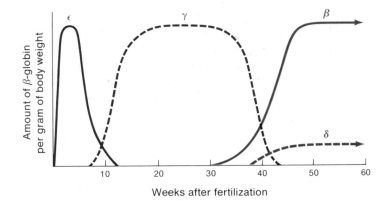

Figure 22-6
The variation of the concentrations of the different β-like globins in the course of development of a human.

differing by only one or a few amino acids, and the forms that are present depend on the stage of development of an organism. For example, the following temporal sequence shows the subunit types present in humans at various times after conception (Figure 22-6):

	Embryonic (<8 weeks)	Fetal (8–41 weeks)	Adult (birth → henceforth)
α-like	$\zeta_2 \rightarrow \zeta_1$	α_1 and α_2	α_1 and α_2
β-like	ϵ	γ_G and γ_A	β and δ

During the embryonic period the ζ_2 α-like chain, which appears first, is gradually replaced by the ζ_1 form. In the fetal and adult stages, the α_1 and α_2 types are both present but α_2 predominates; the two β-like types, γ_G and γ_A (called so because one has a glycine at a site at which the other has an alanine), are roughly equimolar. In the adult stage, there is fifty times as much of the β type as the δ form. The result of these changes is that embryonic hemoglobin contains two ζ_2 chains and two ϵ chains $(2\zeta_2,2\epsilon)$, whereas after birth 98 percent contains two α_2 and two β subunits (customarily called hemoglobin A) and 2 percent contains two α_2 and two δ units (hemoglobin A$_2$).

Both the α- and β-like genes form separate clusters. The β cluster is shown in Figure 22-5(c). A remarkable property of each cluster is that the order of the genes is the order in which they are expressed in development. (This is not true of all developmentally controlled gene families.) Neither the significance of the different forms nor the way in which synthesis is programmed is known.

The human α and β globin clusters have been cloned and analyzed for repetitive sequences. The base sequences of some regions have also been determined. The most important observation that has emerged from this analysis is that *repetitive sequences are interspersed between the individual genes.* This fact is taken to support the notion mentioned earlier that repetitive sequences have some regulatory function. Another observation, which is not yet understood but which is tantalizing, is the following. One repetitive sequence found in both the α and β clusters, and which can hybridize with a similar sequence in the β cluster in the rabbit, is one of a collection of sequences that differ by only one or a few bases (such as the variants of the Pribnow box in E. *coli*) for which there are 3×10^5 copies per human cell. This sequence also closely resembles the replication origins of SV40 and polyoma viruses! When cloned β globin DNA is used as a template for *in vitro* transcription, many of the repetitive sequences are transcribed. Those that are transcribed are homologous with an abundant species of small nuclear RNA molecules found in hemoglobin-producing cells. It has been hypothesized that these nuclear RNA molecules are components of a regulatory circuit.

ABUNDANCE CLASSES OF EUKARYOTIC RNA MOLECULES

We have seen in discussing prokaryotes that the concentrations of individual species of RNA molecules vary from one operon to another. In an "on" state the range of concentration from the most abundant species to the least abundant species is at most a factor of a few hundred. In eukaryotes the variation is much greater. Furthermore, it is likely that the synthesis of the most abundant RNA molecules is regulated by mechanisms that differ from those for the other classes of RNA. Thus, it has been useful to define (arbitrarily) three classes of eukaryotic RNA molecules: **complex-class** RNA molecules, which appear at a concentration of one to several copies per cell; **moderately prevalent** RNA molecules, for which there may be several hundred copies per cell; and **superprevalent** RNA molecules, for which there may be 10^4 copies per cell. Molecules of this latter class are synthesized in developing eggs and in cells committed to making an enormous amount of a single protein such as hemoglobin.

Little is known about the regulation of the complex-class and the moderately prevalent RNA molecules. A widely debated proposal that attempts to develop a general mechanism for such regulation is described briefly in the next section. The modes of regulation of the synthesis of the superprevalent RNA molecules differ from one gene to the next and are of several major types. For some systems the regulatory circuits are reasonably well known, though not in detail. These systems will be described in a later section.

REGULATION OF THE SYNTHESIS OF COMPLEX-CLASS AND MODERATELY PREVALENT RNA

A major feature of the prokaryotic operon model and its many variants is that, adjacent to each promoter, a base sequence is acted upon by some regulatory element, and this action determines whether transcription occurs. The high frequency with which regulatory sites and promoters are adjacent and the simplicity of this regulatory mechanism have made it hard to envision any system in which the regulatory site and the gene promoter are separated. Although we will see that in eukaryotes other mechanisms serve in rather special circumstances (for example, turning a system permanently on in the course of differentiation), a general mechanism of the operon type has been sought. Many models that include *cis*-acting regulatory sequences have been proposed. None of these models has proved to be correct for any particular gene or gene family but it is worth examining the kinds of proposals that have been made in order to have some framework for thinking about regulation of eukaryotic gene activity.

A major problem in understanding regulation in eukaryotes is to explain how sets of genes that are very far apart in the DNA (even in different chromosomes) can be turned on and off simultaneously. One suggestion is that each gene in a set is preceded by the same regulatory sequence, so that a single regulatory element can activate or deactivate all of the genes in the set. A very elementary form of a model for regulation might be the following. Adjacent to each gene of a set of separated but related genes a common regulatory sequence (an operator) might be present. If a system were negatively regulated, a single type of repressor would occupy each operator and the system would be turned on by an effector molecule that inactivates all such repressors. If the system were positively regulated, an effector molecule would activate a single type of regulatory element, which would bind to the sites and activate the promoter. This simple model cannot be correct for eukaryotes because there are many examples in which a gene can be activated by one of *several* distinct signals and other examples in which a gene appears to be a member of two or more regulated units. To accommodate these observations, it has been suggested that there may be intermediate regulatory elements that are themselves turned on by some signal. A grossly oversimplified way by which this might occur is shown in Figure 22-7.

Any model for regulation in eukaryotes must include the fact that it is common in eukaryotes for certain gene products to be synthesized in response to several different signals. For instance, two different hormones might induce synthesis of the same protein. Moreover, there are sets of genes that are activated in different combinations by different

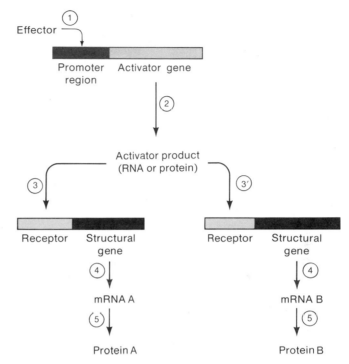

Figure 22-7
A simple way by which a eukaryotic gene might be regulated. Each step is numbered in order of its occurrence. Receptors might be activated either by the activator mRNA in step 3 or an activator protein in step 3′. This overly simple model is unlikely to be generally correct.

signals. For example, one signal might activate genes P and Q, another might activate Q and R, and a third might activate P and R. These phenomena can be accounted for by assuming that some genes are adjacent to two (or more) receptors, only one of which needs be occupied to cause activation (a multicomponent receptor) or by having single receptors and multicomponent signal molecules. How such systems might function is shown in Figure 22-8.

As we have already stated, models of this type are surely oversimplified. (No attempt has been made to explain any regulatory event in molecular terms.) The important point for the reader is to see how distant genes could possibly be coordinately regulated and to recognize the need for several stages of regulation.

In recent years evidence has accumulated that suggests that in eukaryotes it is often not the synthesis of complex-class or moderately prevalent RNA but the availability of constitutively synthesized mRNA

I. Multiple activators

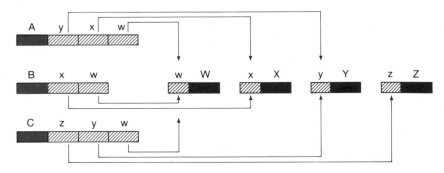

II. Multiple receptors

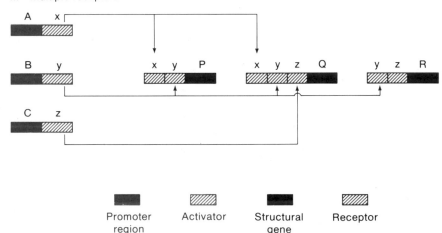

Figure 22-8
A possible organization of multiply regulated gene families. In the upper panel, units having several activator genes can activate several structural genes. In the lower panel, a single activator turns on several structural genes, each having the same receptor. (After a proposal by R. Britten and E. Davidson.)

to the translation system that is controlled. It has been observed that most nuclear RNA, having both unique and repetitive sequences, is degraded very rapidly, and it has been hypothesized (though there is little evidence at present) that inhibition of degradation of particular RNA molecules may be a crucial event in regulation of gene activity.

REGULATION OF SUPERPREVALENT RNA

Superprevalent mRNA molecules are transcribed from only a small fraction of eukaryotic genes and have many special modes of regulation. The production of such RNA molecules is usually associated with a change in the stage of differentiation—for example, an adult erythroblast produces a huge amount of mRNA from which adult globins may be translated, whereas little or no globin is produced by precursor cells that have not yet become erythroblasts. In some cases differentiation only confers on cells the ability to produce a large quantity of a particular protein in response to an external signal—for example, cells of the oviduct synthesize ovalbumin only when stimulated to do so by the estrogenlike hormones.

There are many ways in which cells are programmed to synthesize a large quantity of a particular gene product. These regulatory modes fall into two categories—those in which the genes themselves are altered and those in which the expression of the genes is modulated. Examples of the former category are **loss, amplification, and rearrangement of genes.** The latter category consists of **transcriptional, posttranscriptional, and translational control.** Each of these modes will be discussed in the following sections.

Gene Loss

One way to eliminate the activity of certain genes as a cell differentiates is to eliminate the genes from the cells. Some protozoans, insects, and crustaceans do this but in a gross way—that is, entire chromosomes are lost. In these organisms only those cells destined to produce germ cells maintain the complete DNA complement. Chromosome loss has not been observed in the higher organisms.

Many protozoa have two types of nuclei—a small body called the **micronucleus** and a large body called the **macronucleus.** Usually, the germ cells contain only a micronucleus, but as the cell differentiates (as many protozoans do), the micronucleus undergoes a variety of changes, the result being a cell possessing both a micronucleus and a macronucleus. The micronucleus contains inactive DNA and is simply a storehouse of genetic information for the production of future germ cells. The DNA molecules of the macronucleus are the templates for all transcription.

The formation of the macronucleus of the protozoan *Oxytrichia* has been studied carefully. The DNA of the micronucleus in the germ cell is cleaved during differentiation into linear fragments ranging in size from 0.5 to 20 kb. Most of the DNA of the micronucleus is completely degraded during the fragmentation. In a way that is not understood, these fragments replicate repeatedly until the cell contains roughly 1000 copies of 17,000 different fragments that constitute the macronucleus. At this point, the macronucleus has several hundred times as much DNA as the micronucleus yet more than half of the DNA sequences are absent. The mechanisms for this selective gene loss is not yet known, though recent experimental results seem to be yielding significant facts. A particularly interesting observation is that most of the fragments have the same base sequence at each end. This terminal repetition might represent a site of cleavage by a site-specific enzyme like a restriction endonuclease.

Gene loss has not yet been detected in the higher organisms and it is thought that this mode of regulation may be confined to organisms fairly low in the evolutionary scale. It is of course impossible to determine with certainty that gene loss does not occur at all in the higher organisms since loss of one gene in 50,000 might be difficult to detect. However, the following experiment, performed with frog eggs, suggests that no *essential* gene is lost during development.

In this experiment the nucleus of a fertilized frog egg was removed by microsurgery in such a way that the egg remained intact. A second nucleus was also taken from a cell in the intestine of a tadpole. The intestinal nucleus was injected into the nucleus-free egg. Many of these repaired eggs grew and divided and, remarkably, develop into living frogs. If genes essential for growth and development had been lost during differentiation, the egg would have lacked information to form a living frog. Although some nonessential DNA might be lost, it seems reasonable to conclude from this experiment that the nucleus of the differentiated cell had not lost any genetic material; it is said that the nucleus of the differentiated cell is **totipotent.**

Gene Amplification

Gene amplification is a means by which a cell can produce vast quantities of a specific gene product. The best-understood example of gene amplification occurs in the development of the oocytes (eggs) of the toad *Xenopus laevis,* in which the genes for rRNA increase in number about 4000-fold. The precursor to the oocyte, like all somatic (nongerm) cells of the toad, contains about 500 rRNA-gene (rDNA) units; after amplification there are about 2×10^6 copies of each unit, which amounts to about 75 percent of the nuclear DNA of the oocyte. This enables the oocyte to synthesize 10^{12} ribosomes, which are required for the large amount of protein synthesis that occurs during the various cell cleavage stages following fertilization.

Prior to amplification the rDNA region consists of 500 rDNA units arranged in tandem. Each unit contains one gene each for 18S and 28S RNA and the genes are separated by a spacer (Figure 22-9, upper portion). The size of the spacer is not constant but varies from one unit to the next. The entire rDNA segment is contained in a single **nucleolus** (a nuclear body containing rDNA, found in most animal cells). During amplification, over a three-week period during which the oocyte develops from a precursor cell, the amount of rDNA increases 4000-fold and the number of nucleoli increases to several hundred (Figure 22-10).

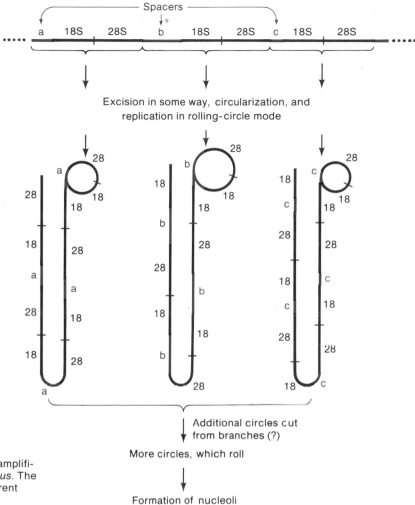

Figure 22-9
A hypothetical mechanism for amplification of rDNA genes in *Xenopus*. The
spacers, a, b, and c, have different
lengths.

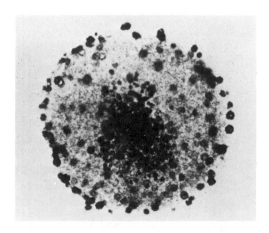

Figure 22-10
Micrograph of an isolated nucleus
from a *Xenopus* oocyte stained to
show the hundreds of nucleoli that
are formed in the amplification of
the ribosomal RNA genes. (From
D. D. Brown and I. B. David. 1968.
Science, 160: 272.)

Both during and after amplification the total amount of rDNA no longer consists of a single contiguous DNA segment but instead is present as a large number of small circles and replicating rolling circles. The precise mechanism of formation of these structures is not known but some information about it is based on the size of the spacer between the 18S and 28S components contained in the nonreplicating and rolling circles. Figure 22-9 shows a schematic diagram of the rDNA contained in three rolling circles. The essential feature of these molecules is that each molecule contains a spacer having only one size. If a large segment of chromosomal rDNA containing adjacent rDNA units were excised and replicated, the rolling circles would have spacers of different sizes. This and several other experimental results suggest that a single rDNA unit is excised from the chromosome, as shown in the figure. The excised molecule circularizes (or perhaps is excised as a circle) and begins to replicate by the rolling circle mode, generating long branches.

Curiously, in most of the rolling circles the circular portion does not contain one rDNA unit but often two to four and as many as sixteen. It is unclear how a monomeric excised circle can be converted to a multimeric circle. This question may be related to the following point. An rDNA unit has a length of 4.3 μm and requires roughly one-half hour to form a unit-sized branch by rolling circle replication. Gene amplification proceeds for about 72 hours, and in this time each monomeric segment can generate 144 copies. Since all monomers are not excised from the chromosome, the amplification must be less than a factor of 144. However, the observed value is roughly a factor of 4000, so at least some of the excised monomers must be replicated to yield a template for further DNA synthesis. There are several possible ways that this might occur. For example, a circle might replicate several times via the θ mode prior to initiation of rolling circle replication; however, θ molecules are rarely seen. Another possibility is that DNA segments, which may often contain several rDNA units, could be excised from the linear branch of a rolling circle and then circularize. This could account for the existence of the circle mentioned above containing tandem repeats of single rDNA units. Clearly, more work is required to elucidate this phenomenon.

Ultimately, each replicating unit forms one of the many nucleoli of the mature oocyte (Figure 22-10). Amplification stops at this stage. It is of interest that not all rDNA genes are amplified in a particular oocyte and that the particular genes that are amplified vary from one oocyte to the next.

Once the oocyte is mature, the excess rDNA serves no purpose and it is slowly degraded. Following fertilization the chromosomal DNA replicates and mitosis ensues. This occurs repeatedly as the embryo develops. During this period the extra chromosomal rDNA does not replicate; degradation continues and by the time several hundred cells have formed, none of this DNA remains.

Amplification of rRNA genes during oogenesis has been observed in a large number of organisms including insects, amphibians, and fish, and in protozoa during formation of the macronucleus. Recently, amplification of a gene that produces protein has also been observed in *Drosophila;* the genes that produce chorion proteins (a component of the sac that encloses an egg) are amplified in ovarian follicle cells just before maturation of an egg. In this case, as with the rRNA genes, amplification enables the cells to produce a large amount of protein in a short time. We will see later that if a large amount of protein is to be synthesized but a long time is available for the synthesis, gene amplification is unnecessary, for this can be accomplished by increasing the lifetime of the RNA.

Gene Rearrangement: I. Joining a Promoter to a Structural Gene

Regulatory gene rearrangements are of two sorts: (1) those in which a structural gene and a promoter are put into contact or separated, and (2) those in which different segments of transcribed DNA are brought together in order to change the base sequence of the coding region. In Chapter 14 an example of the first type was presented. That is, in phase variation of flagella loci in the bacterium *Salmonella* the orientation of a structural gene is inverted, so either the coding or the noncoding strand of the DNA is downstream from the promoter. There are also several examples in corn (*Maize*) and the fruit fly (*Drosophila*) of gene inactivation by insertion of a transposable element between a structural gene and its promoter and transposition of a promoter to a structural gene. Details of these processes are unknown and the movements probably should not be called regulatory events because, as far as is known, they are random and not a response to some signal.

The best-understood system in which a structural gene is activated by movement to its promoter is that which controls the mating type of the yeast *Saccharomyces cerevisiae*. Two mating types are present in haploid yeast, *a* and *α*, controlled by the *HMa* and *HMα* loci respectively. One type *a* cell and one type *α* cell can efficiently merge to form an *aα* diploid cell, whereas two cells having the same mating type mate only very infrequently. The *HMa* and *HMα* loci are nonhomologous gene clusters (two genes are known with certainty in each cluster) located on the same chromosome, though not near one another. Thus, the loci are not allelic. Each haploid cell contains both *HMa* and *HMα* gene clusters but only one set is expressed in a particular cell; if, for example, *HMa* is expressed, the cell is mating type *a*. On occasion, a type *a* cell switches its phenotype to type *α* (and vice versa), a phenomenon called **mating-type interconversion.** This can only occur if the cell contains a gene *HO*. (Cells having the genotype *HO* are called heterothallic, meaning that they can change genotype, whereas mutants or *HO* cells—often written *ho*—are called **homothallic,** meaning that their genotype is stable.)

Several lines of evidence have led to the model for mating-type gene expression shown in Figure 22-11. In this model, called the **cassette model,** it is hypothesized that the *HMa* and *HMα* genes are inactive in the positions shown in the figure. Between these positions is a region called the **mating-type locus,** which is the target site into which either the block of *HMa* or *HMα* genes can insert like a cassette into a tape recorder. This insertion is directed by the *HO* gene in the sense that a switch in mating type cannot occur in a cell lacking a functional *HO* gene, as just mentioned. Adjacent to the mating-type locus is a promoter. When either an *HMa* or *HMα* cassette is inserted at this locus, transcription of the mating-type genes can occur from this promoter. An important fact is that when this insertion occurs, the original silent copy of the locus remains at the inactive site.

Superficially, this process resembles transposition in that one copy of an *HMa* or *HMα* unit moves to the mating-type locus and a master copy remains at the original site (Chapter 19). However, the gene cluster moves to a fixed site on the chromosome and this is not usually the case for transposition. Furthermore, when an *HMa* or *HMα* cluster moves to the mating-type locus, the locus is already occupied by one cluster (that is, the mating-type locus is never vacant), so entry of one unit into the mating-type locus must be accompanied by removal of the resident unit. There is some evidence that the removed unit is discarded. This process resembles a phenomenon called **prophage substitution** in lysogens. For example, if a λ lysogen is superinfected by a heteroimmune phage λ*imm434* under conditions leading to repression of the infecting phage λ*imm434*, a fairly common event is excision of the *imm*λ

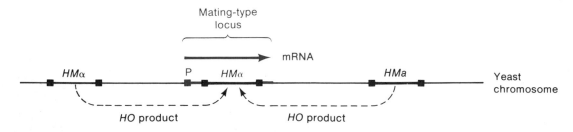

Figure 22-11
The cassette model for regulation of the mating type of yeast. The dashed lines show possible movement of the *HMa* and *HMα* genes to the mating-type locus; the movement requires the *HO*-gene product. A cell containing the yeast chromosome shown would be mating type α since *HMα* is in the mating type locus. (From J. B. Hicks, J. N. Strathern, and I. Herskowitz. 1977. *DNA Insertion Elements, Plasmids, and Episomes.* Cold Spring Harbor Laboratory. With permission of the publisher.)

prophage by the Int and Xis products of the infecting *imm434* phage and insertion of the λ*imm434* DNA. Thus, by analogy it has been proposed that mating-type interconversion is a *site-specific recombinational event* that may be mediated by an enzyme like the λ integrase. Recently, the mating-type locus has been cloned, and an enzyme that *in vitro* produces a double-strand break at that locus has been detected.

Gene Rearrangement: II. The Joining of Coding Sequences in the Immune System

Activation of gene expression by juxtaposition of different segments of structural genes is best exemplified by the production of **immuno-globulins (antibodies);** this is the topic of this section.

Humans are able to synthesize more than 10^6 different antibody molecules—"different" in the sense of being an antibody to a different target molecule or **antigen.** If each antibody molecule were encoded in a distinct gene, a significant fraction of the DNA of a single cell would have to be used for antibody synthesis. The amount of DNA required could be reduced substantially if an antibody contained three subunits A, B, and C, and there were 100 different A genes, 100 different B genes, etc. In this way, $100 \times 100 \times 100 = 10^6$ different proteins would be generated by 300 genes. If each of the genes were active in every cell—that is, if 300 different proteins were synthesized in every antibody-producing cell—then each cell would produce 10^6 different antibodies. However, it is known that a mature antibody-producing cell makes only one type of antibody, so there must be some mechanism that is responsible for programming each cell.

In this section it will be seen that such programming does occur. In outline, the germ-line cells contain the set of all genes whose coding sequences can be combined to generate all possible antibody molecules. In the course of cell division and development of the immune system most of these genes are removed from the daughter cells and discarded. However, the removal is programmed in such a manner that in the adult a cell will exist that is capable of producing each (but only one) type of antibody. This population consists of roughly 10^6 cells, a small fraction of the 10^{12} somatic cells in an adult human. The ability to produce large quantities of a particular antibody comes about by a complex response of the immune system to a particular antigen; in this response, the antigen triggers multiplication of the cell capable of producing the particular antibody that can neutralize that antigen. As a result of this multiplication, an antibody-producing clone forms, in which each cell can synthesize the particular antibody.

The immune system is very complex in that it consists of many cell types and can respond to a large number of signals. In this section we will be concerned primarily with how the base sequence of an antibody-producing gene is formed. This will be explained in five parts by describing (1) the properties of the amino acid sequence of an antibody molecule, (2) the genes whose coding sequences are joined, (3) the base sequence of a spliced gene, (4) possible signals for gene splicing, and (5) the final splicing event. Following this we discuss briefly how a clone of antibody-producing cells may be instructed to produce a specific antibody.

Properties of the Amino Acid Sequence of an Antibody Molecule

Immunoglobulin G (IgG) is one of several classes of immunoglobulins; it is a tetramer containing two **L chains** and two **H chains** (Chapter 6). Determination of the amino acid sequences of many different IgG molecules (different in that they respond to different antigens) has shown that both the L and H chains comprise two distinct segments. One of these, the **constant** or **C region,** has a nearly invariable amino acid sequence (there is one amino acid that can vary) for a particular class of immunoglobulins. The other segment is called the **variable** or **V region** because its amino acid sequence differs between any two IgG molecules that recognize different antigens. The locations and the sizes of the V and C regions of the L and H chains are shown in Figure 22-12. In most classes of immunoglobulins, the C regions are used in the transport of the molecules across the membrane of the antibody-producing cell to the blood stream and the lymphatic system and in the final disposal of the neutralized antigen. The V regions are responsible for recognition and binding of a specific antigen. Note in Figure 22-12 that the V regions of the L and H chains have the same number of amino acids, though their amino acid sequences are quite different.

There are actually two distinct types of L chains. These are called κ and λ and have significantly different amino acid sequences. Functionally these types have many similarities and their differences are unimportant for the present discussion. We will only describe how the amino acid sequence of the κ type is determined; the basic mechanism for the λ type is similar, differing only in detail.

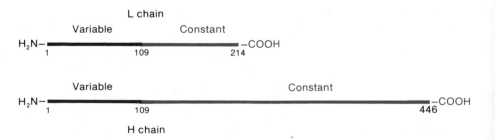

Figure 22-12
The positions and lengths of the variable and constant regions of the IgG L and H chains. The numbers refer to the amino acid position counted from the H_2N terminus.

Gene sequence in an embryonic cell

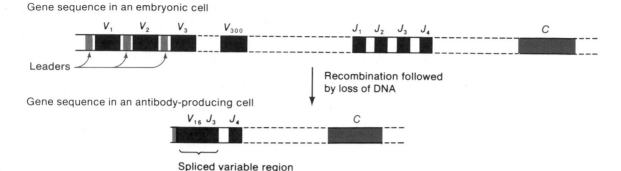

Figure 22-13

DNA splicing in the formation of the variable region of an L chain. The V_{16}–J_3 joint has been removed, presumably by a site-specific recombination event. The V genes to the left of V_{16} have been removed either by homologous recombination or by site-specific recombination within the leader sequences.

Genes Whose Coding Sequences Are Joined

Many genes can be used to form a κ-type L chain. These genes are of three types—V, J, and C. There are roughly 300 different V genes (which are responsible for the synthesis of the first 95 amino acids of the variable region), 4 different J genes (which encode the final 12 amino acids of the variable region and join the V and C regions), and 1 copy of the C gene (which encodes the constant region). In an embryonic cell the V genes form a tight cluster, the J genes form a second tight cluster quite far from the V-gene cluster, and the C gene follows not far after the J-gene cluster. All of the genes are on the same chromosome, as shown in Figure 22-13. Note that each V gene is preceded by leader regions where transcription can be initiated; the J and C genes are not preceded by leaders.

The Base Sequence of a Spliced Gene

Genetic regions encoding particular IgG molecules have been cloned (by recombinant DNA techniques) from various mouse cell lines, each producing a particular IgG molecule.* The V-J-C region has also been cloned from mouse embryo cells, which have not yet been committed to antibody synthesis and thus presumably contain an unaltered master set of genes for antibody synthesis. For each clone obtained from an

---

*These cell lines were derived from cancerous tumors called **myelomas.** A myeloma is a tumor in which a single cell programmed to make one type of antibody has multiplied to yield a large clone of cells, each of which produces only that antibody. A single tumor can be isolated and individual cells can be grown indefinitely in cell culture; the cells secrete a large quantity of a pure antibody that can be recovered from the culture medium.

antibody-producing cell line it has been found that a large segment of the embryonic DNA sequence is not present and that the missing segment is always a sequence between the particular V gene that encodes the first 95 amino acids of the V region and the J gene that encodes the last 12 amino acids of the V region. This is explained by a gene rearrangement in which DNA between the particular V and J genes is deleted. An example of one such rearrangement is shown in Figure 22-13. Many different gene sequences encoding particular IgG proteins have been cloned and it has been found that in each clone a different segment of DNA is not present. For example, in the figure the DNA between V_{16} and J_3 is absent; in another clone there might be a V_{210}-J_1 junction instead. In both cases a V gene and a J gene have become *spliced* together to form a complete gene for the variable region. Note that this is DNA splicing and not the RNA splicing that has been discussed in Chapters 11, 15, and 21.

Studies of the base sequence of cloned IgG DNA sequences show that the junction between a particular V gene and a particular J gene is not always the same. That is, the two terminal triplets of the juxtaposing V and J genes can exchange at any one of four sites that yields a triplet. The meaning of this statement is shown in Figure 22-14. In the example shown there are three possible amino acids at the joint, which adds diversity to the number of possible variable regions.

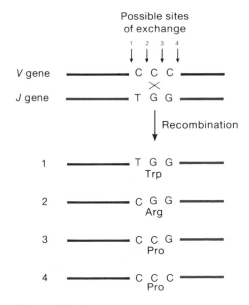

Figure 22-14
Four possible junctions at a *V-J* joint giving rise to three different amino acids.

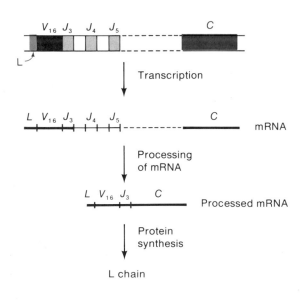

Figure 22-15
Production of the L chain of a particular IgG molecule. Solid regions indicate coding sequences used to generate the final L chain.

Since there are 300 different V genes, 4 different J genes, and (on the average) 2.5 different amino acids at the junction, there are, then, $300 \times 4 \times 2.5 = 3000$ different variable regions. The H chain genes are organized in a similar but not identical way (there are four different types of genes) and there are about 5000 variable regions in the H chains. Thus, since each IgG molecule contains two identical H chains and two identical L chains, there are about $3000 \times 5000 = 1.5 \times 10^7$ different IgG molecules can be formed from the V_L, J_L, V_H, and J_H genes. This is ample to account for the diversity of antibody molecules.

Possible Signals for Gene Splicing

The splicing of the V and J genes just described is thought to occur by means of a site-specific recombination mechanism. The signal for the V-J joining is probably a pair of specific base sequences to the right and left of the V and J genes, respectively. A base sequence

V gene–CACAGTG–11 bases–ACATAAAC

consisting of a 7-base-pair palindrome, an 11-base-pair spacer, and an 8-base-pair $(A+T)$-rich sequence is immediately to the right of each V gene. To the left of each J gene is the sequence

GTTTTTGT–22 bases–CACTGTG–J gene

consisting of (reading right to left) of a 7-base-pair palindrome identical (except for the fourth base) to the one that is adjacent to the V gene, a 22-base-pair spacer, and an 8-base-pair $(A+T)$-rich sequence almost complementary to the one near the V gene. Presumably, those sequences are recognized by the splicing system. Neither the enzymes required for splicing nor the nature of the chemical exchange nor the chemical events that trigger splicing are known.

The Final Splicing Event: RNA Splicing

Figure 22-15 shows that the DNA splicing event does not fully generate an L chain sequence since (1) the spacer between the J and C gene remains and (2) the actual L chain has amino acids derived from only one V gene and one J gene and the spliced DNA usually contains many V and J genes. The correct amino acid sequence is obtained by a final RNA-splicing event, as shown in the figure. Note how the particular V-J joint determines the RNA splicing pattern. For example, the RNA removed is always a segment between the leader and the V gene and between the C gene and the right end of the J gene in the V-J joint.

Diversity of H Chains

The mechanism of generating amino-acid-sequence diversity in the H chains is similar to that for the L chains, though it is not known in as much detail. The H chain system also has many V and J genes but it is more complicated than the L chain system in two ways: (1) there is an additional large set of genes, called D, between the V and J genes, so that many V-D and D-J joints are possible and (2) there are eight different C regions. Choice of the C region determines the particular class of immunoglobulin (for example, IgD, IgG, IgM, and so on). It is interesting that the seven-base palindrome described above, which signals V-J joining in the L chain, is also present in corresponding positions in the C chain system.

Diploidy in Antibody-Producing Cells

An interesting point arises because antibody-producing cells are diploid: since only one type of IgG molecule is produced by a particular cell, then either the same rearrangement occurs in both chromosomes or one of the V-J-C regions is silent. The evidence available at present suggests that during development of an antibody-producing cell from an embryonic cell, gene rearrangement stops as soon as one complete functional gene is assembled. Genes still in the embryonic configuration and incompletely spliced variable region genes are either not expressed or their products cannot function.

Programming

A major question is whether DNA recombination occurs *in response to* a particular antigen. The best evidence suggests that the events occur at random in the course of development and that when differentiation is complete, there are about 10^6 different antibody-producing cell types, each able to make a single type of antibody. How, then, does an organism know when to make a huge amount of a particular antibody in response to a particular antigen? The answer to this question is not known with certainty but the following mechanism, called **clonal selection,** is probably correct—at least in outline. Each antibody-producing cell type makes a small amount of specific antibody. Some of this antibody becomes bound to the cell surface. When such a cell is exposed to the specific antigen that can react with the specific antibody, a complex antigen-antibody network forms, because each antibody molecule can interact with two antigen molecules and each antigen can interact with two antibody molecules (Chapter 6). These interactions, coupled with an energy-requiring movement of the surface antibody molecules, cause the formation of a single antigen-antibody network called a **cell cap** (Figure 22-16). The cell membrane then folds in around the cap and the cap is taken into the cell (this process is called

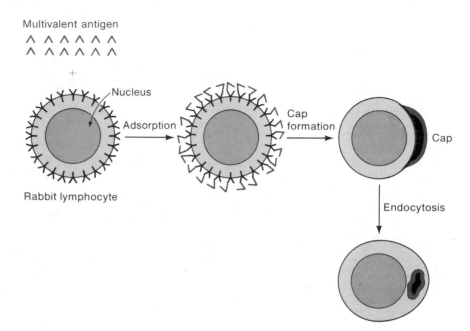

Multivalent antigen

Adsorption

Nucleus

Rabbit lymphocyte

Cap formation

Cap

Endocytosis

Figure 22-16
The capping phenomenon. Rabbit lymphocytes that have made immunoglobulin (IgG)
(spread uniformly over the cell surface) are exposed to an antigen. The antigen and IgG
molecules combine and, by an unknown energy-requiring process, the IgG molecules
move toward one another. If the antigen is multivalent (has multiple binding sites), a
complex called a cap forms. By a process called endocytosis the cap is taken into the
cell for processing, after which the cell becomes programmed to make large quantities
of the antibody. If the antigen is coupled with fluorescein, a fluorescent cap can be
seen with a fluorescence microscope; if the antigen is radioactive, an autoradiogram
can be prepared and the film will be blackened over one region of each cell.

endocytosis). Then, in a way that is totally obscure, cell division is
stimulated and extensive production of antibody ensues. Thus, a clone
of cells is generated which synthesizes a large amount of the specific
antibody.

Transcriptional Control: I. Initiation

The most common means of regulation in prokaryotes is transcriptional
control, as was seen in Chapter 14, in which various operons were
examined. This is also common in eukaryotes, though, because of
experimental difficulties, fewer eukaryotic systems are understood in
detail.

To establish that a set of genes is transcriptionally controlled one
must demonstrate that in the absence of a particular signal, no mRNA
copies of the genes of interest are present in the cell, and that their lack
is not a result of mRNA degradation. A particular mRNA is usually

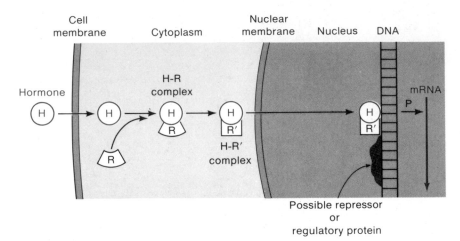

Figure 22-17
Schematic diagram showing how a hormone H reaches the DNA and triggers transcription by binding to a cytoplasmic receptor. A regulating protein may prevent the H-R′ complex from reaching a promoter (P) or may stimulate binding to the promoter.

identified by hybridization with DNA containing the gene. In bacteria the most common source of the gene is a transducing particle, but since there is no equivalent with eukaryotes, such studies were not possible until cloning techniques were developed. The technique (already described in Chapter 20) is the following. Pure mRNA is isolated from a differentiated cell that makes a large quantity of the particular mRNA. Using reverse transcriptase, c-DNA (a DNA copy of an RNA molecule) is made and this DNA is cloned. The cloned DNA can then be used in a hybridization experiment as a probe for detecting complementary radioactive mRNA.

Of the known regulators of eukaryotic gene activity those studied most extensively are the **hormones**—extracellular substances, either small molecules, polypeptides, or small proteins, that are carried from hormone-producing cells to target cells. It has been found that many of the steroid hormones (e.g., estrogen, progesterone, aldosterone, glucocorticoids, and androgens) frequently act by turning on transcription. If a hormone regulates transcription, it must somehow signal the DNA. The penetration of a target cell by a hormone and its transport to the nucleus is a complex process that is understood only in outline, as shown in Figure 22-17. Steroid hormones (H) are hydrophobic molecules and pass freely through the cell membrane. A target cell contains a specific cytoplasmic receptor (R) that forms a complex (H-R) with the hormone. The receptor R usually undergoes some modification (conformational or chemical) after the H-R complex has formed; the modified form of the receptor is denoted R′. The H-R′ complex then passes through the nuclear membrane and enters the nucleus. In the nucleus either the H-R′ complex or possibly the hormone alone binds to chromatin and transcription begins. The binding is very likely to be mediated by specific DNA-binding proteins, though it is possible that in some systems direct binding to the DNA occurs.

A well-studied example of induction of transcription by a hormone is the stimulation by **estrogen** of the synthesis of ovalbumin in the chicken oviduct. When chickens are injected with estrogen, oviduct tissue responds by synthesizing ovalbumin mRNA. This synthesis continues as long as estrogen or an estrogenlike hormone (e.g., diethyl-stilbesterol) is administered. Once the hormone is withdrawn, the rate of synthesis decreases. Both before giving the hormone and sixty hours after withdrawal, no ovalbumin mRNA is detectable. Experiments of this sort must be interpreted with caution to avoid a potential artifact arising from the existence of introns. Specific mRNA, such as oval-bumin mRNA, is detected by hybridization of labeled mRNA with a plasmid containing a copy of c-DNA. Suppose a primary mRNA molecule had a great many large introns. This might interfere with hybridization with c-DNA. If, compared to processed mRNA, the primary transcript binds poorly to c-DNA, the ability to detect mRNA might only indicate that processing rather than *de novo* synthesis has occurred. There are two straightforward ways to demonstrate that a hormone stimulates transcription rather than processing. One way is to show that the newly synthesized RNA contains introns, which can be done if cloned intron DNA of the natural gene is used in the hybridization test. The other method is to measure the size of the RNA that appears. If transcription is being turned on by the hormone, the primary transcript, which is larger than the processed mRNA, should be detected; furthermore, it should appear sometime before the processed mRNA is observed. If only processing were stimulated, there would be no change in the amount of the large mRNA that is detected; only the smaller processed mRNA will appear to increase in concentration. It was just such experiments that indicated that estrogen stimulates transcription of ovalbumin mRNA.

When estrogens are given to chickens, only the oviduct synthesizes mRNA; other tissues fail to be stimulated because they lack the cytoplasmic hormone receptor. This type of deficiency is invariably the cause of insensitivity to a particular hormone. How the receptor is synthesized in some but not all cells is not known; this is another problem of differentiation.

We have said nothing so far about how a hormone that turns on transcription might do so. Without invoking complicated details we may examine at least five possible basic mechanisms:

1. A hormone binds directly to the DNA and facilitates binding of RNA polymerase or of a protein that is needed for the polymer-ase interaction with the DNA.
2. A hormone binds to an effector protein (such as the CAP protein in the *lac* operon) that must bind to a site on the DNA prior to RNA polymerase binding.

3. A protein already bound to the DNA is activated by the hormone and the activated form is needed for RNA polymerase binding.
4. A hormone is an inducer that inactivates a repressor.
5. A hormone causes a structural change in chromatin that makes the DNA available to RNA polymerase.

In the first three mechanisms, transcription is positively regulated; in the fourth and fifth, it is negatively regulated. (See Chapter 14 for a review of positive and negative regulation.) At present only a few hormone-induced systems have been studied and there is no clear picture yet.

The fifth point, that there may be a change in chromatin structure, must be examined more closely, because in eukaryotes RNA is not synthesized from free DNA but from chromatin.

In one case—namely the synthesis of α-amylase in the rat—transcriptional control occurs by selection of particular promoters. The rat salivary gland produces considerably more of the enzyme than the rat liver. In both organs the same coding sequence is transcribed, but transcription begins at two distinct promoters, roughly 2800 base pairs apart. One of these promoters (that used in the salivary gland) is stronger than the other. Of unknown significance is the fact that different splicing events are needed to join the coding regions to the leader.

Transcriptional Control: II. Termination

Control of the termination of transcription is a well-known phenomenon in prokaryotes—examples are the gene-N-regulated antitermination of E. coli phage λ, and the attenuation systems of the tryptophan and histidine operons. Control of termination also occurs in cells infected with adenovirus, which suggests that this mechanism of regulation may be used in uninfected cells also. In an adenovirus infection a particular promoter (P16) is used for both early and late mRNA but the termination sites and the sites of polyadenylation differ.

One example has been observed recently of regulation of termination in eukaryotic cells. Two forms of certain immunoglobulins are known that differ only in their COOH-terminal regions. One form contains several additional amino acids. These two forms are translated from two mRNA molecules that are initiated at the same site in the DNA but terminated at two different sites. The two proteins have different biological functions. It is not known if the choice of the termination site is regulated or if it is random, but, a regulated system is more consistent with what is generally known. Regulation may occur via Rho-like factors, though such factors have not yet been identified in eukaryotes.

Transcriptional Control: III. Active Chromatin

A great deal of evidence suggests that the structure of chromatin in regions in which transcription occurs is different from that of untranscribed chromatin. It is said that these regions contain **active chromatin.** The principal observation is that if purified chromatin is treated with the enzyme DNase I, the transcribed regions are especially sensitive to degradation by this enzyme. For example, the globin region is preferentially digested if the chromatin is obtained from chicken reticulocytes, which synthesize globin, whereas the ovalbumin region is preferentially digested if the chromatin is from chicken oviduct tissue, which synthesizes ovalbumin. The sensitive chromatin structure seems to be fairly universal—integrated animal virus genes and ribosomal DNA are also especially sensitive to DNase I. Sensitivity is not related to the rate of transcription, because the degradation rate of any transcribed region is constant and not dependent on the abundance of the RNA transcribed from that region. Furthermore, the DNase I sensitivity probably reflects a *potential* for transcription rather than that transcription is occurring, because globin DNA remains sensitive in mature chicken erythrocytes in which transcription of globin DNA no longer occurs.

DNase I sensitivity requires the presence of two nuclear proteins called HMG-14 and HMG-17.* It is not clear that these proteins play any role in gene activity—their binding (and hence the DNase I sensitivity) may only be a reflection of an altered nucleosome structure (which is known to exist) in the active region. In fact, these proteins only bind to regions that can be transcribed.

There is little information about the structure of active chromatin but there is some preliminary evidence that the structural change may be a consequence of acetylation of two histones, H2A and H2B; this is currently an active field of research.

Inasmuch as transcribable chromatin has a different (but unknown) structure, it may be that a gene or gene family is turned on (directly or indirectly) by an effector molecule such as a hormone when induced to do so by a change in nucleosome structure in that region, thereby forming active chromatin.

Posttranscriptional Modification

Preliminary evidence indicates that posttranscriptional modification is a major mode of regulation of the complex-class and moderately prevalent RNA molecules. This is probably also true for the superprevalent

*HMG stands for high mobility group, a class of proteins having a high electrophoretic mobility.

mRNA molecules, though no examples are known. In this section we consider how it might occur.

Posttranscriptional modification might control synthesis of particular proteins in several ways. One way is to regulate transport of RNA molecules from the nucleus to the cytoplasm. In one carefully examined system (and presumably in others as well), it has been found that nuclear RNA is not transported across the nuclear membrane to the (cytoplasmically located) ribosomes until all introns are excised.

Furthermore, intron excision does not occur until the nuclear RNA is capped and the poly(A) tail is added. Thus, controlled inhibition of capping, or poly(A) addition, or intron excision could regulate gene expression.

Nuclear proteins might also serve a regulatory function. An interesting, poorly characterized collection of these has been studied in this regard. Nuclear RNA rarely exists intracellularly as free RNA molecules; instead, the RNA molecules, often prior to completion of transcription, form complexes with these nuclear proteins to form ribonucleoprotein (RNP) particles. Neither the structure nor the properties of these particles is well understood but a lot of evidence suggests that processing, transport to the cytoplasm, and attachment to ribosomes occurs while the RNA is in the complex. Possibly, complex formation is a prerequisite for these processes; it has been suggested that these proteins may be the processing enzymes. It is also possible that some of these proteins are receptors for signal molecules such as hormones. One could easily imagine a positive role for these proteins—for example, a particular protein might only be active in processing when it has a bound effector molecule.

Another type of posttranscriptional modification is regulation of alternate modes of RNA splicing. There are examples (especially in viral systems, as was seen in Chapter 21) of multiple modes of splicing of a particular transcript to yield different mRNA molecules. Possibly, synthesis of two particular gene products A and B could be determined by regulating the modes of splicing. It might be the case that both products are always made but two different concentrations are needed. If a particular processing molecule (for example, a nuclease) is needed for each mode, then the ratio of the two types of processing molecules could determine the ratio of the gene products. Alternatively, a cell might need product A in one circumstance and product B in another. Synsthesis of one or the other could be accomplished by an A-stimulating effector that activates processing mode A or a B-type effector for mode B. Examples of these modes of regulation are not yet known, but such modes are certainly consistent with the overall regulatory features that have been observed. So far, only two examples of regulation by differential processing are known, other than in viral systems. One of these is the differential synthesis of the hormone calcitonin in different tissues and the other is the synthesis of two forms of the μ chain of immunoglobulins. In both cases, the differential processing includes distinct patterns of intron excision (i.e., splicing) but these are necessitated by an earlier event in which differential poly(A) sites are selected from the primary transcript. That is, when the poly(A) site nearer the promoter is selected, a splice site used in the larger mRNA is not present, so that a different splice pattern results. Thus, slightly different proteins are translated.

Translational Control

By translational control is meant the regulation of the number of times a finished mRNA molecule is translated. There are three ways in which translation of a particular mRNA may be regulated: (1) by the lifetime of the mRNA; (2) by the probability of initiation of translation; and (3) by regulation of the rate of overall protein synthesis. In this section we will see examples of each of these modes of regulation.

Extension of the Lifetime of mRNA

The silk gland of the silkworm *Bombyx mori* predominantly synthesizes a single type of protein, silk fibroin. Since the worm takes several days to construct its cocoon, it is the *amount* and not the rate of fibroin synthesis that must be great; hence, the silkworm can manage with a fibroin mRNA molecule that is very long-lived .

The fibroin gene is thought to have a strong promoter, and when a signal (of which the exact nature is unknown) is received to initiate transcription, about 10^4 fibroin mRNA molecules are made in a period of several days. An "average" eukaryotic mRNA molecule has a lifetime of about three hours before it is degraded. However, fibroin mRNA survives for several days during which each mRNA molecule is translated repeatedly to yield 10^5 fibroin molecules. Thus, each gene is responsible for the synthesis of 10^9 protein molecules in four days. If a fibroin-producing cell were diploid, 2×10^9 protein molecules would be synthesized per diploid set of chromosomes. The whole silk gland makes 300 μg or 10^{15} molecules of fibroin in four days; therefore, some 5×10^5 diploid cells would be needed. However, the silk gland has an unusual feature in that it develops from a single diploid cell. When the chromosomes of the original cell double and separate by mitosis, the cell does not divide; hence, a tetraploid cell results. Chromosome doubling without cell division continues until the silk gland is mature; at this time the gland consists of a single giant cell with a ploidy of approximately 10^6. Each of the 10^6 sets of chromosomes contains a single active fibroin gene, which is responsible for the synthesis of 10^9

fibroin molecules; in this way, the $10^6 \times 10^9 = 10^{15}$ protein molecules are made by a single exceptional cell. How cell division is inhibited during formation of the silk gland is unknown.

In the synthesis of casein (the major protein of milk) in mammary glands, the lifetime of the mRNA is also regulated. When the hormone prolactin is present, the concentration of casein mRNA becomes elevated. This may be in part a result of increased initiation of transcription (a role for hormones discussed in an earlier section) but is primarily a result of an increased lifetime and continued synthesis of the mRNA. When the hormone is withdrawn, the concentration of casein mRNA decreases because the RNA is degraded much more rapidly. How prolactin alters the lifetime of casein mRNA is unknown.

Production of a large amount of a single type of protein by means of prolonged mRNA lifetime is common in highly differentiated cells. For example, cells of the chicken oviduct, which makes ovalbumin (for egg white), contain only a single copy of the ovalbumin gene per haploid set of chromosomes, but the cellular mRNA is long-lived.

It is instructive to compare this mode of protein production, which is sometimes called **translational amplification,** with the gene amplification utilized to generate the 10^{12} rRNA molecules needed in the *Xenopus* oocyte. The silkworm is able to produce a huge number of protein molecules because there are two stages of amplification—that is, the template for protein synthesis is amplified 10^4-fold by synthesis of 10^4 mRNA molecules and then each template is amplified 10^5-fold by being translated repeatedly. However, in the synthesis of rRNA, transcription is the final step—there is no translation. Thus, in order to get a large overall multiplicative factor an amplification step must occur prior to transcription—that is, the genes themselves must be amplified, as is the case.

In the silkworm, fibroin synthesis need not be rapid, as we have stated. However, in other organisms and in other cellular systems accelerated protein synthesis is needed and, in these cases, translational amplification is inadequate. For example, in the cleavage of fertilized eggs, the histone needed to form new chromatin must be synthesized at a rate of about 3×10^5 molecules per minute to keep pace with DNA synthesis. Using ordinary translation rates, about 1000 molecules can be synthesized per minute. The necessary high rate of protein synthesis is achieved here by having several hundred initial copies of each histone gene. Note that this is not an example of the type of gene amplification discussed earlier in this chapter since this copy number is present in all cells.

Differential half-lives of specific mRNA molecules play a role in the life cycle of adenovirus (Chapter 21). Transcriptional unit 1B produces two mRNA molecules—a 14S species and a 20S species—which differ because of distinct modes of processing of the primary transcript. The ratio of the amount of 14S mRNA to 20S mRNA is considerably greater late in the growth cycle of the virus than at early times. This is because of an increased half-life of the 14S mRNA.

Control of the Initiation of Translation

Another example of translational regulation is that of **masked mRNA.** Unfertilized sea urchin eggs maintain a state of readiness by storing large quantities of mRNA for many months. This RNA is inactive, but within minutes after fertilization, translation of these mRNA molecules begins. The mechanisms for stabilization of the mRNA protection against nucleases and activation are unknown. It is notable that the mRNA is stored in the form of ribonucleoprotein particles. Here the regulation involves the timing of translation.

Since mature unfertilized eggs need to maintain themselves but do not have to grow or change their state, the rate of protein synthesis in eggs is generally low. This is not due to an inadequate supply of mRNA but to a limitation of some element required to initiate translation. This conclusion is derived from an unusual type of experiment, the data for which is shown in Figure 22-18. Toad eggs are sufficiently large that small volumes of liquid can easily be injected into them with a hypodermic syringe. If globin mRNA (isolated from a reticulocyte) plus a radioactive amino acid are injected into a toad egg, radioactive globin is synthesized. Figure 22-18(a) shows the result of injecting increasing amounts of globin mRNA. As more mRNA is added, more radioactive globin is made, but the amount of normal egg proteins synthesized decreases. This result suggests that in the egg the ability to utilize mRNA is saturated—as more exogenous mRNA is added, it competes with

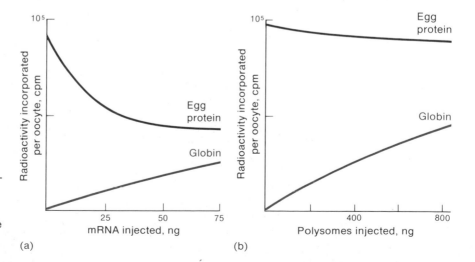

Figure 22-18
Effect of injected reticulocyte globin mRNA (panel (a)) or reticulocyte polysomes (panel (b)) on incorporation of radioactive amino acids into oocyte proteins.

(a)   (b)

endogenous mRNA for the protein synthesis apparatus. However, if reticulocyte polysomes, which contain globin mRNA, are injected instead of purified globin mRNA, synthesis of egg proteins is only slightly reduced, as shown in Figure 22-18(b). Thus, there must be an ample supply of amino acids, elongation factors, and termination factors. Together these experiments show that the limiting factor is either the number of ribosomes or some molecule needed to form a polysome. There are more than enough ribosomes, so that some other element of the system is implicated. The element, which is still unidentified, has been termed the **recruitment factor.** One method that can be used to identify the element is to fractionate toad embryonic tissue (in which protein synthesis is active) and to inject into an egg a portion of the fractions, one by one, with the globin mRNA. Experiments of this sort are in progress.

The synthesis of some proteins is regulated by direct action of the protein on the mRNA. For instance, the concentration of one type of immunoglobulin is kept constant by self-inhibition of translation. This protein, like all immunoglobulins, consists of two H chains and two L chains. The tetramer binds specifically to H-chain mRNA and thereby inhibits initiation of translation. How L-chain synthesis is regulated is unknown.

Regulation of the Rate of Overall Protein Synthesis

Cells that synthesize only a single protein sometimes regulate synthesis of that protein by modulating overall protein synthesis. The production of hemoglobin by reticulocytes is regulated in this way. Reticulocytes are the penultimate cell type in the line of differentiation leading to a red blood cell. A reticulocyte has lost its nucleus and utilizes the globin mRNA molecules transcribed from the nucleus of its precursor cell (the erythroblast) to synthesize the α- and β-globin subunits. Other enzymatic systems are responsible for making hemin, the prosthetic group of hemoglobin. There is one hemin molecule per globin subunit, and for optimal efficiency each type of globin (α and β) should be synthesized at twice the rate of synthesis of hemin. Maintaining the correct ratio of the rates is accomplished by a two-phase regulatory system in which (1) hemin represses its own synthesis, and (2) hemin inactivates an inhibitor of globin synthesis. With this system, if there is more hemin than globin, hemin synthesis is shut off and globin synthesis is stimulated, but, on the other hand, if there is not enough hemin, globin synthesis is repressed and hemin synthesis is turned on. In this way, the amounts of hemin and globin are kept in balance. The mechanism by which the synthesis of hemin is regulated is unknown, but how hemin controls globin synthesis is well understood, as we will see.

Rabbit reticulocytes do not have a nucleus and hence have no DNA. Therefore, globin synthesis in these cells cannot be transcriptionally

Figure 22-19

Regulation by hemin of protein synthesis in reticulocytes. Proteins are boxed; active forms are in red; inactive forms in black. The reaction shaded in red leads to protein synthesis. Red arrows (dephosphorylations) mark the path of promotion of protein synthesis; black arrows trace inhibition of protein synthesis.

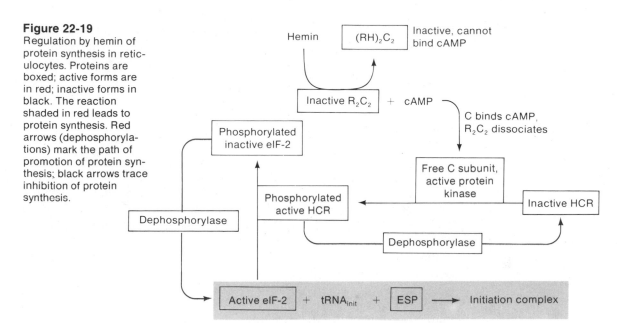

regulated. Furthermore, since all of the mRNA is cytoplasmic, it must have been fully processed, so regulation must be translational. The mode of regulation is depicted schematically in Figure 22-19. Initiation of translation requires the formation of a ternary complex between $tRNA_{init}$, the initiation factor eIF-2, and a third protein called eIF-2-stimulating protein, or ESP. Reticulocytes contain a protein kinase (a protein-phosphorylating enzyme) called the **hemin-controlled repressor,** or HCR. This inhibitor phosphorylates the small subunit of eIF-2 and thereby prevents eIF-2 from complexing with ESP. HCR is also activated by a cAMP-dependent protein kinase; this kinase contains two pairs of identical subunits called R and C and is denoted R_2C_2. The R and C subunits have regulatory and catalytic roles respectively. The two R subunits are each able to bind cAMP. When cAMP is present, R_2C_2 dissociates and the C subunits, which now have the ability to phosphorylate HCR, are released. The active HCR then phosphorylates eIF-2, thereby preventing initiation of protein synthesis. Hemin binds to the R subunit, causing a structural change in R that prevents R from binding cAMP. Thus, if there is excess hemin, HCR is not activated and protein synthesis can begin.

It should be noted that hemin binds to both globin and HCR, though binding is stronger to globin. Thus, when globin is synthesized, it binds hemin, thereby reducing the concentration of free hemin, and this allows HCR to be activated. Globin synthesis is then inhibited until more hemin is made. Without the binding of hemin to R_2C_2, the dephosphorylases deactivate HCR and thereby activate eIF-2, thus allowing globin synthesis to catch up to the hemin. Note that globin synthesis is not specifically turned on and off but that overall protein synthesis is modulated. This is possible only in a cell such as this one in which only a single type of protein is made.

Polyproteins and Posttranslational Modification

It has been pointed out repeatedly that only a single polypeptide chain can be translated from a completed eukaryotic mRNA molecule. However, there are numerous examples of the production of several proteins from one mRNA, as we saw when studying animal viruses in Chapter 21. This is the result of posttranslational cleavage of a polyprotein.

The best-known examples of nonviral polyproteins are the hormone precursors and **vitellogenin,** the egg-yolk-protein precursor (which is discussed in the next section). For example, insulin consists of two polypeptide chains, termed A and B chains, joined by disulfide bonds. The A and B chains are cleaved from a single polyprotein (see Figure 20-22, Chapter 20). As another example of polyprotein modification, when the protein precursor of adrenocorticotropic hormone is cleaved, not only that hormone but other active polypeptide hormones are excised from the polyprotein. With other hormones, polypeptides are stored and, in response to unknown signals, are cleaved when the individual components are needed.

The **zymogens** (Chapter 2) are examples of posttranslational modification. These proteins are precursors to digestive enzymes (for example, pepsinogen for pepsin and chymotrypsinogen for chymotrypsin) and are activated when needed by cleavage in particular peptide sequences.

Other posttranslational modifications are the coupling of sugars (to form glycoproteins) and lipids (to form lipoproteins). Usually the proteins have no biological activity until these molecules are added; thus, it is possible that activity could be regulated by controlling addition of these substances, though no examples of this are known.

REGULATION OF THE SYNTHESIS OF VITELLOGENIN: AN EXAMPLE

Vitellogenin is the protein precursor to the egg yolk proteins **phosvitin** and **lipovitellin,** found in amphibians and birds. Vitellogenin is an especially interesting protein because its synthesis is subject to regulation at several points, and many of the features of regulation that have been discussed can be seen to act in a single pathway. Vitellogenin synthesis has been studied most carefully in the toad *Xenopus laevis.*

There are at least four genes in the vitellogenin gene family. These are called *A1, A2, B1,* and *B2.* The translation products of the class-A genes differ from the products of the class-B genes by about 20 percent of the amino acids. The product of a subgroup-1 gene differs from the product of a subgroup-2 gene by about 5 percent of the amino acids. Synthesis of each of the four gene products is regulated by a common system. However, the cleavage pattern of the polyprotein gene products differ. We will have more to say about these four genes at the end of this section.

Overall Pathway for the Synthesis of Yolk Proteins

The overall pathway for the production of phosvitin and lipovitellin is shown in Figure 22-20. The three main stages are the following:

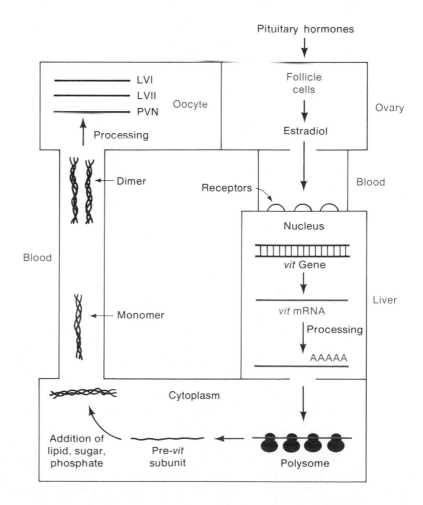

Figure 22-20
Scheme summarizing the stimulation by estradiol of the synthesis of vitellogenin and its processing to form phosvitin (PVN), lipovitellin I and II (LV), and phosvettes I and II (PVT). The results of processing of a type-A vitellogenin in the oocyte are shown. For type B there would be four proteins—LVI, LVII, RVTI, and PVTII.

1. Vitellogenin is synthesized in the liver of a mature female; synthesis is stimulated by female sex hormones.
2. It is then secreted into the bloodstream and thereby transported to the ovaries.
3. In the ovary it is selectively taken up by developing eggs (oocytes) in which the vitellogenin is cleaved into the yolk proteins.

Details of the synthesis of vitellogenin are listed in the following section.

Specific Steps in the Synthesis of Yolk Proteins

The synthesis of the yolk proteins is regulated by hormones acting to control transcription. The transcript is processed, a polyprotein is synthesized and modified, and finally the individual polypeptide chains are cleaved from the modified polyprotein. The individual steps are the following (Figure 22-20):

1. The initial stimulus for vitellogenin production comes from the hypothalamus (a gland at the base of the brain), which triggers the pituitary gland to secret the hormone gonadotrophin.

2. Gonadotrophin is taken up by cells of the ovarian follicle, which encloses the developing egg. The follicle cells respond by synthesizing estrogen, another hormone, which is secreted into the bloodstream.

3. The estrogen binds to receptors in the plasma membrane of liver cells, which respond by synthesizing vitellogenin. This effect of estrogen makes the synthesis of vitellogenin an especially useful experimental system because the liver cells of both hens and roosters can synthesize vitellogenin but only hens synthesize estrogen. Thus, if roosters are used, it is possible to turn the system on at will by injection of estrogen into the bloodstream and this makes it possible to look carefully and in a controlled way at early events.

4. The estrogen is carried to the nucleus of the liver cell where it turns on transcription of the vitellogenin gene. The mode of stimulation of vitellogenin synthesis by estrogen differs slightly from the basic mechanism of hormone stimulation already described. At the time of first exposure to estrogen the nucleus of each liver cell has about 200 estrogen receptors. During the first 12 hours the number of receptors increases to 1000 as a result of *de novo* synthesis; this number is retained by the cells for eight days. This increase is thought to explain the

following phenomenon. If an animal is given a small dose of estrogen on day 1 and then it is given a second injection several days later when vitellogenin synthesis is no longer occurring, it is found that the rate of synthesis of vitellogenin mRNA is much greater after the second dose than after the first one.

5. The transcript is capped, poly(A) is added, introns are removed by processing enzymes, the mRNA is transported to the cytoplasm, and polysomes form.

Interestingly, the lifetime of vitellogenin mRNA is increased by the presence of estrogen. As long as estrogen is present, the half-life of the mRNA is about 24 hours; if, after continuous administration of estrogen for several days, the hormone is withdrawn, the half-life changes to 3 hours. The biochemical role of estrogen in determining the half-life has not been established.

6. Translation of the processed mRNA occurs. Oddly, translation does not begin immediately, though overall protein synthesis continues normally. It is thought that one or more translation factors are needed to initiate translation of vitellogenin mRNA and that these factors are also produced in response to estrogen. The evidence for this is the following: if the hormone is added to a male animal and then, after vitellogenin synthesis has begun, it is withdrawn, synthesis of vitellogenin stops because no more vitellogenin mRNA is being made and the preexisting mRNA has been degraded. However, if estrogen is given again, vitellogenin is synthesized as soon as vitellogenin mRNA is made. That is, there is no delay in translation, because the protein factors (which have not yet been identified) are already present.

7. The capacity for translation increases. Once vitellogenin synthesis begins, the liver cells gain the capacity to synthesize all proteins more rapidly than before the administration of estrogen. This is shown by an experiment in which one compares the ability of two extracts of liver cells, obtained from animals before and after estrogen is administered, to support poly(U)-mediated synthesis of polyphenylalanine. It is found that much more polyphenylalanine is made by an extract of cells from an estrogen-treated male than from an untreated male. Thus, there are two effects on translation that enable the liver cells to produce a huge amount of vitellogenin: (i) overall protein synthesis is accelerated and (ii) certain cytoplasmic factors enable vitellogenin mRNA to compete very effectively with all other mRNA molecules, so that vitellogenin is the major protein made.

8. The final step occurring in liver cells is the modification of the protein by the addition of phosphate groups, lipids, and sugars.

9. The completed vitellogenin monomer is secreted into the bloodstream where the modified protein molecules aggregate to form a dimer, which is the usual molecule thought of as vitellogenin.

10. The dimer is taken from the blood by the oocytes. The dimer is a polyprotein and in the oocyte it is cleaved. The cleavage patterns of class-A and class-B proteins differ. Class-A polyproteins are cleaved to form phosvitin (molecular weight $M = 35,000$), lipovitellin I ($M = 115,000$), and lipovitellin II ($M = 31,000$); class-B polyproteins form lipovitellin I and II, phosvette I ($M = 14,000$), and phosvette II ($M = 19,000$). The phosvettes have been discovered only recently and little is known of their function or of the significance of the multiple forms of vitellogenin. In both cases, however, the proteins are, after cleavage, assembled in the yolk platelet to yield a mature egg.

Coordinated Synthesis of Eggwhite Proteins

The eggs of amphibians and birds contain both yolk and white. It seems reasonable that the synthesis of both yolk proteins and white proteins should be coordinately regulated, since one is not needed without the other. This does in fact occur and the regulator is again estrogen, which not only turns on synthesis of vitellogenin but also turns on the synthesis of ovalbumin (the major white protein, as described earlier in this chapter).

Possible Significance of the Vitellogenin Introns

The vitellogenin system has another informative feature; namely, the base sequence of the $A1$ and $A2$ genes suggests how the two sets of genes might have arisen and also suggests a possible reason for the existence of introns in eukaryotic cells.

Both $A1$ and $A2$ genes contain 33 introns and thus there are 34 exons for each gene. Two facts are quite striking: (1) the sizes of the corresponding $A1$ and $A2$ exons are the same—for example, the third exon of $A1$ has the same size as the third exon of $A2$—and the base sequences of corresponding pairs of exons are homologous (but not strictly identical). This conservation of exon sequence and size suggests that the duplicity of genes $A1$ and $A2$ might have arisen by an ancient duplication (of the type described in Chapter 19) of an ancestral gene. Furthermore, since the positions of corresponding $A1$ and $A2$ introns are the same, the ancestral gene probably contained introns before the duplication occurred. It has been speculated that such duplications might have been responsible for the formation of all gene families.

In contrast with the conservation of exon sequences, the sizes of corresponding *A1* and *A2* introns differ markedly. For example, the total number of bases in all of the *A1* and *A2* introns are 21,000 and 16,000, respectively (the exons contain only 6000 bases). Also, the base sequences of corresponding introns differ; the most common difference involves insertion, deletion, and inversion—possibly resulting from the action of transposable elements. Although the size and sequence of corresponding introns differ, the pattern of intron excision is, as far as one can tell, identical in *A1* and *A2* mRNA; thus, introns can tolerate a great deal of change without affecting the final RNA product, and the fact that the introns are different means that there is no selective pressure to maintain the intron sequence. To account for the properties

(a) Intramolecular recombination between duplicated genes

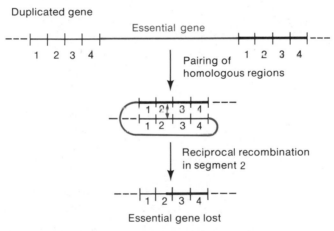

(b) Reduced intramolecular recombination between duplicated genes having different introns

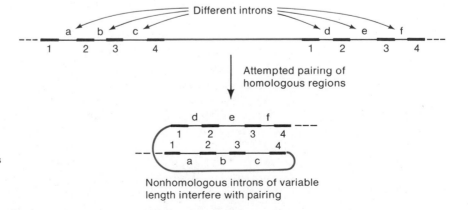

Figure 22-21
Recombination between duplicated genes. (a) An essential gene (red) located between a pair of duplicated genes is lost by intramolecular recombination between homologous segments (arbitrarily numbered 1, 2, 3, and 4. (b) The presence of nonhomologous introns (a–f) of variable site interferes with pairing of exons (numbers).

of the introns in the *A1* and *A2* genes, it has been hypothesized that there may be a definite advantage to having different intron sequences in corresponding genes and, in fact, for having introns.

The fact that gene duplications exist clearly indicates that duplications are important. For example, having several copies of a gene enables a cell to make the gene product at a higher rate, as we have already seen; there are other reasons also, though most of the reasons are probably unknown at present. However, gene duplication may trigger other problems in a cell; intramolecular genetic recombination between nonadjacent duplicated genes can, by a reciprocal exchange between homologous regions of the genes, convert two gene copies to one and, in so doing, discard possibly essential DNA between the two genes, an event which is likely to be lethal (Figure 22-21). The existence of nonidentical introns reduces the homology between the two genes and should reduce the probability of recombination markedly. Thus, introns may serve to prevent gene loss.

The reader should recognize the highly speculative nature of the preceding discussion.

REGULATION OF PROTEIN ACTIVITY BY CYCLIC AMP

Up to this point, we have discussed the regulation of synthesis of eukaryotic proteins. The activities of many proteins and most enzymes are also regulated. It is impossible to review this kind of regulation, which is more properly found in biochemistry texts. However, since we have been discussing hormones and cyclic AMP (cAMP), it seems appropriate to describe, in outline at least, how certain hormones and cAMP work in concert to regulate the activity of many enzymes.

Hormonal Regulation of the Synthesis of cAMP

The function of many hormones is to regulate metabolic processes such as glucose metabolism and calcium utilization. This is done through binding of hormones to specific receptors in the cell membranes of target cells, as we have described earlier. However, many of these hormones are unable to penetrate the target cells and, instead, the binding of a hormone to a membrane receptor induces intracellular synthesis of cAMP; the cAMP rather than the hormone then causes the desired metabolic effect—usually by indirect action on a target molecule.

Figure 22-22 shows the usual way that a hormone stimulates a target cell to produce cAMP. Three components are necessary—the hormone

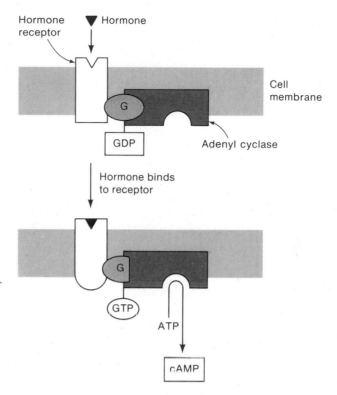

Figure 22-22
Synthesis of cAMP initiated by binding
of a hormone to a hormone receptor
contained in a cell membrane. On binding the hormone, the receptor undergoes a shape change that causes the
G protein to change shape also. The
G protein releases its GDP and binds a
GTP and, by virtue of protein-protein
contact with adenyl cyclase, activates
cyclase; the active cyclase converts
ATP to cAMP.

receptor, the **G protein,** and the enzyme **adenyl cyclase,** which forms
cAMP from ATP. The G protein in its normal (inactive) state contains
bound guanosine diphosphate (GDP). When the specific hormone
receptor binds a hormone molecule, the receptor catalyzes an exchange
of the bound GDP for guanosine triphosphate (GTP). The GTP form of
the G protein then binds to adenyl cyclase and causes it to synthesize
cAMP, which builds up in the cytoplasm of the target cell.

Cyclic AMP rarely acts directly on target enzymes but is an allosteric activator of numerous protein kinases that are found in various
cell types. The active protein kinase then phosphorylates the target
enzyme; the phosphorylation may either activate or inactivate the target
enzyme, depending on the particular enzyme.

The cAMP effect is a more-or-less nonspecific one in that cAMP
activates several protein kinases and these enzymes can probably
phosphorylate many different intracellular proteins. The specificity of
the hormone effect is determined solely by the specific interaction of the
hormone with the hormone receptor in the cell membrane. There is, of
course, a second level of specificity in that all cell types do not contain
the same protein kinases and the same target molecules.

Regulation of Energy Metabolism in Skeletal Muscle by cAMP

One protein kinase, an enzyme found in skeletal muscle, is well understood. This kinase is a tetramer consisting of two pairs of identical subunits (Figure 22-23). Two subunits are regulatory and two are catalytic, and the catalytic subunits are inactive in the tetramer. When cAMP is synthesized, it binds to the regulatory subunits. This binding causes a conformational change, and a dimer consisting of the two cAMP-containing regulatory subunits dissociates from the tetramer. The catalytic subunits then become an active kinase, as was seen earlier in the case of the regulation of globin synthesis by hemin; this tetramer is not the same as R_2C_2, though. Next, the kinase both inactivates the enzyme glycogen synthetase and activates the enzyme **phosphorylase kinase** by phosphorylation. The activated phosphorylase kinase then phosphorylates the enzyme **phosphorylase,** which activates this enzyme. This is summarized in the following reactions in which P_i refers to inorganic phosphate:

$$\text{I. Glycogen synthetase} + P_i \xrightarrow{\text{Protein kinase}} \text{Inactive glycogen synthetase}$$

$$\text{IIA. Inactive phosphorylase kinase} + P_i \xrightarrow{\text{Protein kinase}}$$
$$\text{Active phosphorylase kinase}$$

$$\text{IIB. Inactive phosphorylase} + P_i \xrightarrow{\text{Active phosphorylase kinase}}$$
$$\text{Active phosphorylase}$$

The enzymes glycogen synthetase and phosphorylase are important in the energy metabolism of muscle because they catalyze the following reactions:

$$\text{III. (Glycogen)}_n + \text{UDPglucose} \xrightarrow{\text{Glycogen synthetase}} \text{(Glycogen)}_{n+1}$$

$$\text{IV. Glycogen} + P_i \xrightarrow{\text{Phosphorylase}} \text{(Glycogen)}_{n-1} + \text{Glucose 1-P}$$

Figure 22-23
Activation of skeletal muscle protein kinase by cAMP. Binding of cAMP to the regulatory subunit causes dissociation of the tetramer, yielding two active catalytic subunits.

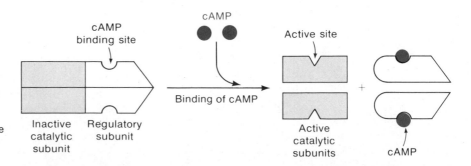

in which (glycogen)$_n$ refers to a polymer containing n glucose monomers.

To understand the effect of cAMP on energy metabolism in muscle, one must remember that the concentration of cAMP and glucose are related in the following way:

$$[\text{High glucose}] \rightarrow [\text{Low cAMP}]$$
$$[\text{Low glucose}] \rightarrow [\text{High cAMP}]$$

Thus, when glucose is limiting, the concentration of cAMP is high and protein kinase is active. Both glycogen and phosphorylase are phosphorylated (reactions I and IIB); glycogen is not made (reaction III is inhibited) but instead is broken down as an energy source (reaction IV is activated). When glucose is in excess, the concentration of cAMP is low and protein kinase is inactive. Thus, glycogen synthetase remains active (reaction I is inhibited) and it converts the excess glucose to glycogen (reaction III); phosphorylase is inactive (reaction IIB is inhibited) and therefore glycogen is not degraded (reaction IV is inhibited). Thus, the concentration of cAMP, via its effect on protein kinase, maintains an adequate supply of glucose for the activity of skeletal muscle when glucose is limiting and converts glucose to the storage form of glycogen when there is more glucose than is needed.

Consequence of Hormone Removal

A mechanism for restoring a hormone-stimulated cell to the unstimulated state must also exist. This occurs in the following way. The hormones continually move on and off the receptor protein. When the hormone is no longer available, the receptor no longer can activate the G protein. The G protein is also a GTPase, continually hydrolyzing GTP to GDP. Thus, unless the hormone is present, the G protein reverts to the inactive form and adenyl cyclase cannot be activated. Without continual production of cAMP, the concentration of cAMP drops owing to continuous hydrolysis of cAMP to AMP by cAMP phosphodiesterase.

REFERENCES

Axel, R., T. Maniatis, and C. F. Fox. 1979. *Eukaryotic Gene Regulation.* Academic Press.

Bostock, C. 1980. "A function for satellite DNA?" *Trends Biochem. Sci.,* 5, 117–119.

Britten, R. J., and D. Kohn. 1968. "Repeated segments of DNA." *Scient. Amer.,* April, pp. 24–31.

Brown, D. 1981. "Gene expression in eukaryotes." *Science,* 211, 667–674.

Davidson, E. H., and R. J. Britten. 1979. "Regulation of gene expression: possible role of repetitive sequences." *Science,* 204, 1052–1059.

Gough, N. 1981. "The rearrangement of immunoglobulin genes." *Trends Biochem. Sci.,* 6, 203–208.

Gurdon, J. B. 1974. *The Control of Gene Expression in Animal Development.* Harvard.

Herskowitz, I., and Y. Oshima. 1982. "Control of cell type in *Saccharomyces cerevisia:* mating type and mating-type conversion." *In* J. Strathern, E. Jones, and J. Broach eds. *The Molecular Biology of the Yeast Saccharomyces.* Cold Spring Harbor Laboratory.

Hood, L. E., I. L. Weissman, and W. B. Wood. 1978. *Immunology.* Benjamin.

Kedes, L. H. 1979. "Histone genes and histone messenger." *Ann. Rev. Biochem.,* 48, 837–870.

Kolata, G. 1981. "Gene regulation through chromosome structure." *Science,* 214, 775–776.

Leighton, T., and W. F. Loomis, (eds.) 1982. *The Molecular Genetics of Development.* Academic Press.

Lewin, B. 1981. *Gene Expression. 2. Eukaryotes.* Wiley.

Long, E. O., and I. B. Dawid. 1980. "Repeated genes in eukaryotes." *Ann. Rev. Biochem.,* 49, 727–766.

Marx, J. L. 1981. "Antibodies: getting their genes together." *Science,* 212, 1015–1017.

Ochoa, S., and C. de Haro. 1979. "Regulation of protein synthesis in eukaryotes." *Ann. Rev. Biochem.,* 48, 549–580.

O'Malley, B. W., and W. T. Schroeder. 1976. "The receptors of steroid hormones." *Scient. Amer.,* February, pp. 32–43.

O'Malley, B. W., H. W. Towle, and R. J. Schwartz. 1977. "Regulation of gene expression in eukaryotes." *Ann. Rev. Biochem.,* 11, 239–275.

Revel, M., and Y. Goner. 1978. "Posttranscriptional and translational controls of gene expression in eukaryotes." *Ann. Rev. Biochem.,* 47, 1079–1126.

Safer, B., and W. F. Anderson. 1978. "The molecular mechanism of hemin synthesis and its regulation in reticulocytes." *CRC Rev. in Biochem.,* 6, 261–290.

Sakano, H., H. Hüppi, G. Heinrich, and S. Tonegawa. 1979. "Sequences in the somatic recombination of immunoglobulin light chain genes." *Nature,* 280, 288–294.

Seidman, J. G., and P. Leder. 1978. "The arrangement and rearrangement of antibody genes." *Nature,* 276, 790–795.

Tata, J. R. 1976. "The expression of the vitellogenin gene." *Cell,* 9, 1–14.

Wahli, W., I. B. Dawid, G. U. Ryffel, and R. Weber. 1981. "Vitellogenesis and the vitellogenin gene family." *Science,* 212, 298–304.

Weissbrod, S., and H. Weintraub. 1979. "Isolation of a subclass of nuclear proteins responsible for conferring a DNase I-sensitive structure in globin chromatin." *Proc. Nat. Acad. Sci.,* 76, 630–634.

Index